The
Botany of Worcestershire

The
Botany of Worcestershire

An account of the

Flowering Plants, Ferns, Mosses, Hepatics, Lichens
Fungi, and Fresh-Water Algae, which grow or have
grown spontaneously in the County of Worcester

With an Introduction and a Map

By

John Amphlett, M.A., S.C.L. (of Clent)

and

Carleton Rea, B.C.L., M.A.

With the assistance of many friends

The Mosses and Hepatics contributed by
J. E. Bagnall, A.L.S.

With later additions

EP Publishing Limited
1978

First published by Cornish Brothers Ltd.,
Birmingham, 1909

Republished 1978 by
EP Publishing Limited
East Ardsley, Wakefield
West Yorkshire, England

ISBN 0 7158 1338 2

British Library Cataloguing in Publication Data
Amphlett, John
   The botany of Worcestershire.
   1. Botany – England – Worcestershire
   I. Title II. Carleton, Rea III. Bagnell, J E
   581.9′424′4   QK306
   ISBN 0–7158–1338–2

Please address all enquiries to EP Publishing Limited
(address as above)

Printed in Great Britain by
The Scolar Press Limited
Ilkley, West Yorkshire

DEDICATED

BY TWO OF HIS SUCCESSORS

TO THE MEMORY OF

WILLIAM MATHEWS

WITHOUT WHOSE WORK ON THE COUNTY BOTANY

OF WORCESTERSHIRE THIS BOOK WOULD

NOT HAVE BEEN WRITTEN

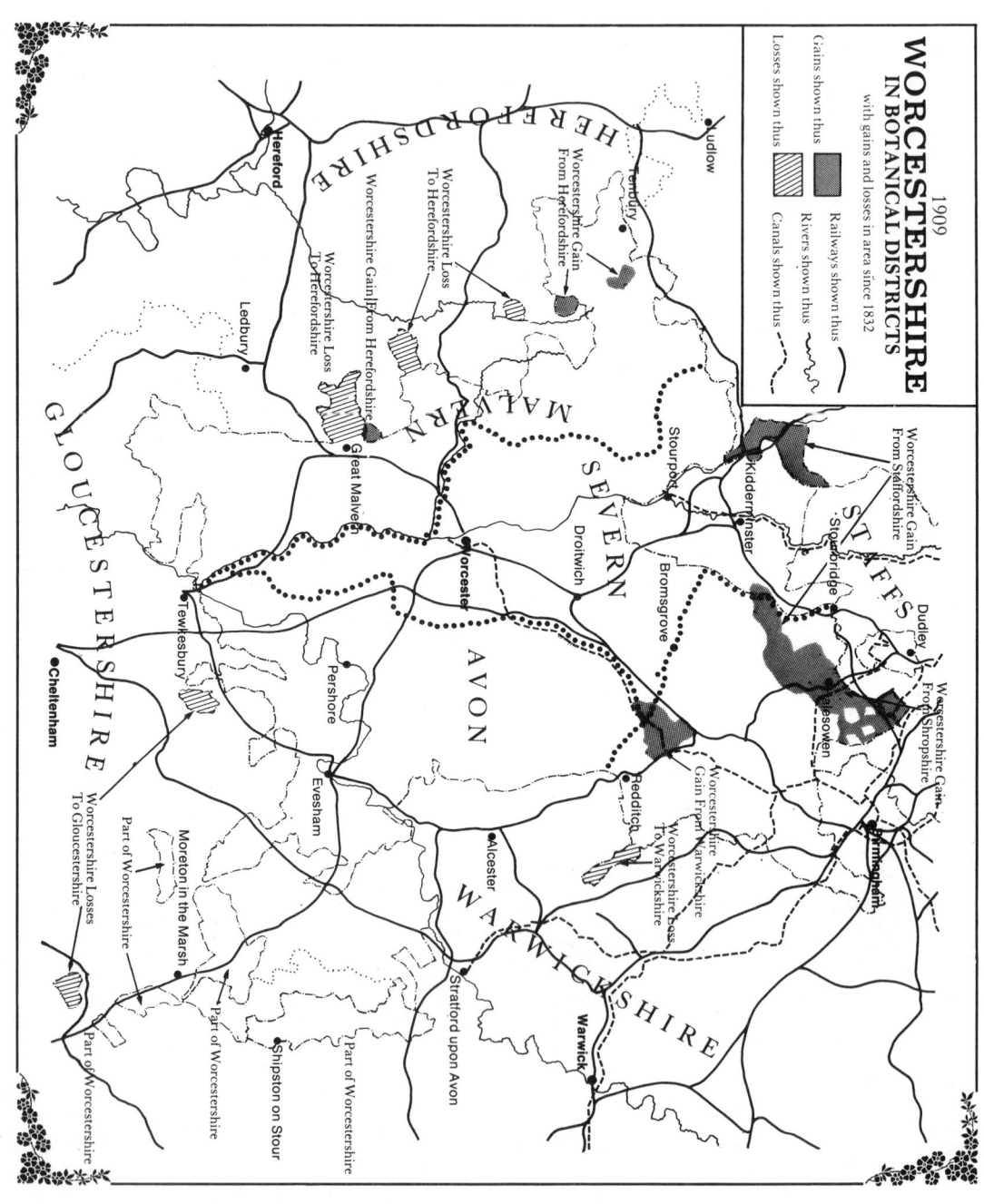

1909
WORCESTERSHIRE
IN BOTANICAL DISTRICTS
with gains and losses in area since 1832

Gains shown thus
Railways shown thus
Losses shown thus
Rivers shown thus
Canals shown thus

HEREFORDSHIRE

GLOUCESTERSHIRE

MALVERN

SEVERN

STAFFS

AVON

WARWICKSHIRE

Ludlow
Tenbury
Hereford
Ledbury
Great Malvern
Worcester
Droitwich
Stourport
Kidderminster
Stourbridge
Dudley
Bromsgrove
Halesowen
Redditch
Birmingham
Tewkesbury
Cheltenham
Pershore
Evesham
Alcester
Moreton in the Marsh
Shipston on Stour
Stratford upon Avon
Warwick

Worcestershire Gain From Herefordshire
Worcestershire Loss To Herefordshire
Worcestershire Gain From Herefordshire
Worcestershire Loss To Herefordshire
Worcestershire Gain From Staffordshire
Worcestershire Gain From Shropshire
Worcestershire Gain From Warwickshire
Worcestershire Loss To Warwickshire
Worcestershire Losses To Gloucestershire
Part of Worcestershire
Part of Worcestershire
Part of Worcestershire
Part of Worcestershire

PREFACE

THE purpose of this book is to provide for the students and field botanists of Worcestershire an account of those Flowering Plants and Cryptogams which they may meet with in the county, both in the relation of such plants to the county itself, and in some of the wider aspects connected with them. In endeavouring to do this, the writers have not thought it necessary to append descriptions to the plants, such as are found in the Botanies; nor have they done more than sketch in a general way the geology, climatology, or petrology of the county. Disquisitions on these subjects would take up space already almost too limited for the real object of the work. They have selected the London Catalogue as their standard of nomenclature, since whatever Botany a student may use, a London Catalogue will be in his hands as well; while by adding the best known synonyms to the names there given, they hope to have made their book useful in any circumstances, and to have banished with regard to the less advanced botanist some of the ever-present difficulties of the modern system of Priority. The writers cannot expect to have eliminated errors in their work. In the first place the Flora was primarily written in accordance with the 9th edition of the London Catalogue, and after it was completed the appearance of the 10th edition of the Catalogue necessitated a change in the nomenclature throughout, the erection of some varieties into species or the converse depression of species into varieties. To do this in the case of critical genera or species was a matter of considerable difficulty. Again, there is a peculiar danger in a work of this character, a work which involves at the same time

compilation and generalization; that there should be an absolute absence of discrepancies is not to be hoped for. The continuous mental application necessary to bring into orderly reference and to correctly appreciate many thousands of facts referring to the various plants is very apt to blunt the sense of relative value or of due proportion; and it is certain that some mistakes will have been fallen into, or some fallacious data allowed to pass unchallenged. For such failures and errors the writers ask for sympathy and deprecate blame. They have added no list of errata; such lists, in their experience, are seldom consulted. Instead, they trust that every reader who should discover an error will mark it in the margin of his book, pass on, and extend his pardon to the writers.

Such a book as this could not be written without the generous help and sympathy of fellow workers. Such the writers have abundantly received, and to one and all, whether help has been great or small, they give their best thanks, and not only their own, but those of all who are interested in Worcestershire Botany.

CONTENTS

INTRODUCTION

Botanical Worcestershire, in the eyes of the writers of this book, consists of all the area which ever has belonged to the county ; including that which is now transferred to others, and that which has been added to it in modern times. To exclude the former, if the botanists of adjoining counties took the same view, would cause many interesting plants to be unnoted in County Botanies ; while to exclude the latter would end in cutting out places like Halesowen, Tardebigge, and Clent, which were brought into Worcestershire so long ago as 1835. There is no logical difference between their case and that of Arley and Dowles and Cowleigh Park, which were made part of the county only a few years ago. So many alterations have been made in the boundary of Worcestershire that they would take long to describe in words, but they are all indicated on the accompanying map.

Even with all these alterations the outline of Worcestershire is extremely irregular, and has been compared to that of a vine leaf, of which fragments have been broken off to form islands in the neighbouring counties. Except in the extreme north-east of the county where the streams run into the Rea and so into the Trent, and the detached portions in the south-east where the streams run into the Evenlode and so find their way into the Thames, the whole of the county is in the basin of the Severn. Geologically it consists of a central undulating plain of Triassic measures overlaid in places with gravelly drift, and intruded upon from the south-east by a wedge-shaped patch of Lower Lias, its point lying northwards. Its western boundary is the range of Malvern, rising near that town to a height of 1,395 feet, and extending as a whole north and south at lesser heights, from nearly the south-west corner of the county for a considerable distance towards its northern boundary, some ten miles out of thirty. The range is composed of Archaean or Pre-Cambrian rocks with Silurian deposits on their western flanks, against which the Devonian of Herefordshire rests ; the north-west area of Worcestershire is the only portion of the county upon the Old Red Sandstone. In the north of the county the coal measures are met with forming two patches, Wyre Forest on the west, and the manufacturing districts of Dudley, Halesowen, and Oldbury on the east, separated by a tract of light sandy country resting on the Bunter pebble beds of the New Red Sandstone. Stretching out towards Birmingham the county again meets the New Red measures over which the eastern boundary runs until it reaches the Lias of the south-east, and in its extremity and

detached portions, the Inferior Oolite. The boundary from the east of the county to the Severn at Tewkesbury is on the Lias, and from the Severn to the south-west boundary across the Marl.

Towards the north-east of the county an interesting range of hills known as Bromsgrove Lickey rises to a height of 956 feet. It consists of the worn-down stumps of a range of mountains showing remnants of deposits of many periods from the Archaean to the coal measures, and against it on both sides the newer measures have been deposited. In the south-east the county includes Broadway Hill, an outlier of the Cotswolds, 1,024 feet high, and the detached hill of Bredon, 961 feet high, both belonging to the Inferior Oolite formation. Many of the heights in the northern part are covered with Permian Breccia, angular or sub-angular rocks cemented together with a kind of calcareous marl, the red remains of the screes on the sides of a vast range of mountains.

The larger portion of the county, however, lies upon the Keuper and Bunter measures of the Trias. The Bunter measures lie at the north of the county and consist of the pebble beds well developed on the south side of Clent Hills, where they are found at an elevation of 800 feet ; and the soft red sandstone, largely developed in the district between Stourport and Stourbridge. The soil of this district is light and comparatively sterile, much of it uncultivated, and is described in the speech of the neighbourhood as ‘ blow-away ’ land. As a band to the south of this formation occurs the Keuper sandstone, affording a much more fertile surface, rich brown arable fields diversified with woodland and fruit-trees, and with the marl, the favourite home of the small-leaved elm. The Keuper marl stretches from the sandstone to the south till it meets the Lias, forming a district of orchards and meadows, pasture and plough land, where the apple and pear are ordinary trees of the hedge-rows and the mistletoe flourishes, seldom seen off the formation. In the south-east of the county the Lias is met with, and upon the Lower Lias Clay rests the vale of Evesham. The surface is scattered with deposits of sand and gravel, which ameliorate the stiffness of the clay, and conduce to the fertility exemplified in the market gardens and fruit orchards in the neighbourhood of Evesham and Pershore. Still further to the south, Bredon and Broadway hills are formed of Inferior Oolite rising out of a bordering of the Upper Lias, in this respect entirely similar to the rest of the Cotswold chain to which they belong.

The county is watered by three main streams, the Severn, the Avon, and the Teme. The Severn flows nearly due south, and cuts the county into two parts, the eastern part considerably the larger. Below Worcester, which is situated about midway on the course of the Severn through the county, the river is joined by the Teme flowing from the north-west ; and at Tewkesbury, where the Severn leaves the county, the Avon flows into it from the north-east. The Avon differs from the other

two rivers in being always bank high, while they, unless in flood, run between high banks cut in the marl. The chief brooks that flow into the Severn on its right bank, or western side, in the northern part of its course are Dowles Brook, draining Wyre Forest, Gladder Brook, rising on the south of the Forest, and Dick Brook, running down from the high land of Rock and Mamble. Just before it reaches the Severn the Teme receives from the north Laughern Brook ; below the mouth of the Teme the proximity of the Malvern Hills running nearly parallel with the river prevents the water courses reaching any size before they run into the main stream. On its east side the Severn receives two considerable streams, the Stour at Stourport coming down from the high land at Halesowen ; and the Salwarpe, rising in Bromsgrove Lickey, passing by Droitwich and falling into the Severn at Hawford. Into the Avon near Defford flows Bow Brook, coming down from the north, in a course nearly parallel to that of the Severn, whose ultimate origin is nearly as far north as Redditch. Piddle Brook drains a similar line of country further to the east and joins the Avon at Pershore. There are no large natural sheets of water in the county. The largest streams are the lake in Westwood Park some sixty acres in extent, and Pirton Pool near Wadborough, about half that size ; and there are ornamental waters of some extent at Croome Park, Hewell Grange, New Pool, Malvern Wells, Glasshampton, Witley, Thorngrove, Stanford Park, and Spetchley Park. Near Northfield is a group of reservoirs connected with the Birmingham Canal ; at Harborne is a large reservoir forming the boundary between Worcestershire and Staffordshire ; at Frankley is one constructed in recent years connected with the water supply of Birmingham, and a similar one to serve the Malverns is situate at the base of the Herefordshire Beacon. A feature of the hilly district to the north of the county is that nearly every stream valley is formed into a chain of pools, while in many little folds of the surface small pools lurk, not visible till closely approached, which form feeders in dry times to the larger ones. Perhaps the greater number of these pools were made to supply power for the manufacture of the iron of the neighbourhood into various articles.

Very little of the county is below 100 feet above sea level, only that part indeed in the immediate neighbourhood of the three chief rivers, and in the very south of the county, where in olden days were large marshes at Longdon and other places, now drained. The main part of the county is between 100 and 200 feet, but on all sides, except where rivers enter or the Severn leaves it, the ground rises considerably. No stream that enters the county except the Severn leaves it, and with the exception of the very small portions drained by the Rea and the Evenlode, all water that falls on its surface flows into the Severn. The whole of the north-east of the county, fronted by the Lickey range, reaches a height of between 400 and 600 feet or more ; the south-east is entangled in the high levels of the Cotswold district; while on the west the high land connected with the Malvern range stretches from its northern extremity, with the exception of a narrow break forming the valley of the Teme, where it enters the county, as far as Dowles Brook, the northern boundary against Salop.

The meteorological features of Worcestershire in the main do not differ from those of the rest of the valley of the Severn. The mean annual rainfall is 23·18 inches, and the wettest months are the three last of the year. Taking Malvern and Birmingham as representative of the western and eastern portions of the county, their rainfall is heavier than that of the central plain, which may be represented by Worcester. The mean annual rainfall of Malvern is 24·88 inches, Worcester 20·47 inches, and of Birmingham 24·21 inches. The temperature varies considerably over the area of the county. No trustworthy figures for the whole of it are available, but comparing Birmingham with Malvern, the former is the warmer of the two localities. Vegetation on the high land of the Lickey district is a fortnight later than in the centre of the county, and in the low lands near the rivers Dahlias and Kidney Beans are blackened by frost as a rule a fortnight earlier than in the higher districts. The prevailing winds are from the south-west, and are sometimes of considerable force, in the north of the county bending over the outlines of the trees, and causing the stems of the taller of them, such as poplars, to take a permanent slant to the northward. The easterly winds sometimes last for a considerable period, but are powerless to counteract the effect of those from the west, which are assisted by the general drift of the atmosphere even when the wind is not high.

For the purpose of studying the botany of the county it has been divided into four districts, two named after rivers, Avon and Severn, and two after ranges of hills, Malvern and Lickey.

The **Avon** district comprises the south-eastern portion of the county. Its northern limit on the county boundary is reached at Headless Cross, near Redditch, and it continues north-westward along the Bromsgrove road to the point where this crosses the Birmingham Canal near Tardebigge, when the boundary follows the canal in a south-westerly direction as far as Oddingley. From Oddingley the line of demarcation between it and the Severn district stretches nearly due south, roughly following the western edge of the Lias formation as far as Tewkesbury; but actually following in its middle portion the bounds of the various parishes encountered, so as to leave them wholly on one side or the other of the line. Boundaries cannot exactly follow formations, are always more or less artificial, and are not evident on the face of the country to the botanist when in the field; but a man can always find out what parish he is in. Though this system makes a boundary somewhat irregular, it is more convenient to divide by parish boundaries for purposes of record

than by road, railway, or canal, or even by water-parting. Rivers of importance usually do divide parishes. In the south part the boundary when passing through Croome Park, and a short distance north and south of it, follows a small brook till the limits of Strensham and Ripple are reached, whence it proceeds along the eastern margin of the Gloucestershire parish of Twyning to the close neighbourhood of Tewkesbury. The southern base of the district is that of the county stretching from Tewkesbury to the east, and several detached parts of the county are included. The district itself falls into three divisions, the valley of the lower course of the Avon, which is upon the Lower Lias formation; the northern part, where the red marl is met with in the neighbourhood of Redditch; and the hills of Bredon and Broadway, capped with Inferior Oolite. At its northern extremity a very small portion of the district runs off the marl on to the Keuper sandstone, and here is a large reservoir at Tardebigge with a somewhat remarkable flora, which though artificially in the Avon district should from its characteristics more properly belong to the Lickey district which it adjoins. In some low places on the Lias formation, saline springs yet occur, the remains possibly of some salt marsh of earlier times.

The **Severn** district is a strip down the centre of the whole county, broad in the north where it extends from the western limit of the Worcestershire portion of Wyre Forest, bounded by Lemp Brook, and on the east reaches the point at which the main road crosses the Stour at Stourbridge; from this point it abuts on the Lickey district as far as Tardebigge, and then turning south it follows the boundary of the Avon district as far as Tewkesbury, opposite which town it ends in a narrow point between the Gloucestershire parish of Twyning and the Severn. On its west side the boundary is irregular until it meets the Teme at Broadwas; after which it follows the course of that river south-eastward to its junction with the Severn, and thence the Severn itself. While by far the greater part of the district is upon the Red Marl, on which formation one of the writers rediscovered on the railway-banks near the tunnel at Worcester *Senecio squalidus*, a plant only previously recorded as occurring on the old wall near the Cathedral Ferry, the soft red Bunter sandstone is largely developed in the north of it, from Hartlebury Common by Wolverley, Kidderminster, and Hagley Brake to Stourbridge. The formation carries a Flora so distinct from that of most of the district, that it might well form one of itself. If so, a convenient boundary for its southern part would be the Severn from its entry into the county to Hampstall Ferry where the southern limit of Hartlebury parish meets the river, which limit it would follow east and north till the boundary of Stone was reached, and keeping by this as far as Muster Green, it would meet the boundary of the Lickey district where it suddenly bends to the north-east to follow the main road to Stourbridge. This line would approximately mark off from the Red Marl the light sandy land which nourishes a typical Flora

of which *Arabis glabra*, *Hypericum humifusum*, *Ornithopus perpusillus*, *Senecio sylvaticus*, which occurs in extraordinary abundance and of great size, and *Verbascum nigrum* are conspicuous members; while *Chrysanthemum segetum*, very rarely met with in other parts of the county, sometimes makes arable fields sheets of gold with its yellow flowers. This proposed boundary is shown on the map accompanying this book as a thin red line across the upper part of the Severn district. The proposed new districts might be called North Severn and South Severn respectively, which would cause no confusion in the Severn district records of the older botanists, but would only entail their being sorted out into two classes. On the whole, however, it has been thought better not to attempt this division in the present book. The line between the Severn and the Lickey districts follows the main road from Lower Hagley to the Staffordshire boundary just to the north of Stourbridge, from which cause it is sometimes difficult to know whether to locate 'near Stourbridge' records in Staffordshire or Worcestershire, or if in the latter county, in the Severn or Lickey districts. There are several interesting botanical localities in the district. The Droitwich Canal nourishes several marine plants which survive probably on account of the salt, derived from the neighbouring brine springs, with which it is impregnated. Hartlebury Common, near Stourport, is a wild space of land, sandy bluffs in the higher parts, a peaty, moor-like surface with cracks filled with stagnant water at a lower level. Fenny Rough is a thick dingle, filling a ravine in the Red Sandstone, which contains some rare plants; and Habberley Valley is an interesting spot. All these would be in the suggested North Severn part of the district. In the suggested South Severn section is contained Wyre Forest, a large tract of native woodland, from the earliest period the haunt of the botanist, the only locality in the county where many plants and mosses are to be found, and where some rare species have not been met with for some years, notably *Thalictrum minus* and *Spiranthes aestivalis*, the latter found only in Hampshire of other English counties.

The **Malvern** district is a band of uneven width comprising the greater part of the western margin of the county, swelling out to the west in the north to contain the valley of the Teme after it enters the county, till that river is met by the Severn division on the left bank at Broadwas. The eastern limit of this district is the western boundary of the Severn district for the whole of its length. It comprises, as well as the Malvern Hills in the south, the high land of Ankerdine and Woodbury, and the range of Abberley, quite subalpine in appearance, but curiously bare from a botanical point of view. To the county and to this district has lately been added a portion of the Herefordshire parish of Cradley on the west side of Malvern, including Cowleigh Park, where several rare brambles have a home; while from it has lately been taken the

parish of Acton Beauchamp and the greater part of Mathon, from which latter parish several plants are recorded in this book, many of which, therefore, have now become Herefordshire records, as well as, in our view, still remaining Worcestershire ones. In the higher part of the valley of the Teme the formation is Old Red Sandstone, and near Eastham and Stanford are deposits of Travertine of which Southstone Rock is a fine example; and owing probably to the calcareous nature of the soil, many orchids are found in the neighbourhood in an abundance unknown elsewhere in the county. The surface of the outcrop of coal measures near Pensax and Mamble is peculiarly unfertile botanical ground. In the south of the district the low land at the base of the Malvern Hills, known as Malvern Chace, was in former years unenclosed, and though now for the most part brought into cultivation, much of it still remains wet and waste. The southern portion of this tract, Longdon Marsh, near Upton-on-Severn, which in rainy seasons used to assume the appearance of a vast lake, was a locality where several good marsh-loving plants were found; but now, as from several similar places in the county, they have vanished.

The **Lickey** district comprises the north-eastern corner of the county, and is bounded, except in the north and east where it abuts upon Staffordshire and Warwickshire, entirely by two main roads. The southern boundary follows the main road from Headless Cross, near Redditch, to Bromsgrove, and through Bromsgrove towards Kidderminster; but some four miles before reaching that town it turns suddenly to the north at Muster Green, along the road from Worcester to Stourbridge, which town it bisects before it reaches the border of Staffordshire, at the bridge over the Stour at the north end of the town. The north-eastern part of this district is markedly bleaker and colder in aspect than other portions of the county. It contains the Lickey Hills, where many good plants now extinct in the locality formerly flourished; and the Clent Hills, the next highest land in the county after Malvern, are comprised in it. This part is peculiarly bare of showy wild flowers after the first burst of spring is over; the hedgebanks and roadside wastes have little floral decoration. The island of the county on which is situated the town of Dudley lies in this district separated from the main body of Worcestershire by a narrow band of Staffordshire. This part of the district is typical Black Country, rugged with pit mounds of all ages, black with the perpetual smoke from the chimneys of countless collieries and iron works. Yet on these pit mounds is found a showy if not extensive flora in the late summer, and a gayer nosegay could be gathered on many of the banks than could be found by the roadside for a dozen miles further south. *Epilobium* flushes the mounds with a rosy hue; Hawkweed clothes hundreds of acres of them; Toadflax, Thistles, and Groundsels add to their gaiety, and in some places the banks are grey with *Artemisia*

Absinthium. On them is found Bracken, and in one place *Calluna vulgaris*, unlikely plants one would think to choose such a habitat; and in an oasis of woodland, on the shaly sides and ballast of a railway which traverses it, has lately been rediscovered for the county *Senecio viscosus* growing in great abundance. In the north part of the island of Dudley a narrow intrusive arm of Staffordshire stretches down for half a mile from the north, robbing this district of *Atropa Belladonna*, which is always to be found in the Castle yard, and causing a similar confusion as to Dudley records to that previously noted in regard to records near Stourbridge. In the vicinity of Birmingham was Moseley Bog, which used to afford many marsh-loving plants, including *Osmunda regalis*; but the site of it is now nearly covered with extending Birmingham, and the plants are gone. In the south part of the Lickey district are the Randans Woods, with their adjuncts of Chaddesley and Pepper Woods, natural woodland, the relics of Feckenham Forest of early times, probably untouched, except to lop and crop and fell, by the hand of man. There are no streams of any size in the district, it being the source whence many streams spring, nor any large sheets of water except reservoirs, which however have been formed for such a length of time that they have taken upon themselves the characteristics of nature, and good plants are found in their vicinity. Deep ravines have been cut in many places in the softer measures by the streams flowing from the hills, for the most part clothed with belts of woodland which shelter shade-loving plants.

There are five counties which actually touch the main body of the county of Worcester; a sixth, Oxfordshire, borders on two sides the little island of Worcestershire lying furthest south-east, Daylesford, and a portion of one side of another detached parish, Evenlode. The entire distance of this community of boundary is so small that as a bordering county Oxfordshire may be disregarded. The other five are Shropshire and Staffordshire on the north, Herefordshire on the west, Warwickshire on the east, and Gloucestershire on the south. Worcestershire and these five counties, with the addition of Monmouthshire, which nowhere approaches Worcestershire, make up the Severn province of Watson's *Topographical Botany*; and Worcestershire, Herefordshire, and Warwickshire form the sub-province of Mid-Severn. Gloucestershire has been divided by Mr. Watson into two vice-counties, East and West Gloucestershire, both of which impinge upon Worcestershire and form its southern boundary, West Gloucestershire however for a comparatively short distance. In Mr. Watson's scheme of numbering, Worcestershire is No. 37, East and West Gloucestershire Nos. 33 and 34, Herefordshire No. 36, Warwickshire No. 38, Staffordshire No. 39, and Shropshire No. 40. The margin of our county touched by West Gloucestershire is so comparatively small that both divisions of that county are grouped

together, in noting the occurrence of plants, and the county as a whole called Gloucestershire.

'The Authors of the best local Floras of modern date have deemed it a useful and honourable task to trace the history of the discovery of the indigenous vegetation of the districts upon which they have written, and to describe the successive contributions made by our predecessors to our knowledge of the present day. This task has not yet been adequately performed for the County of Worcester. I have ventured to attempt it, and hope by so doing, to lay the foundation for a new Flora of the County, whenever, and by whoever, it may be undertaken.'

So William Mathews opens his 'History of the County Botany of Worcester', at page 85 of vol. x of the *Midland Naturalist*, in 1887. And upon his work this book in the main rests. The First Records given in the present book previous to 1893, are taken from the series of papers which followed the above paragraph, and his observations upon them are incorporated throughout. But his method, while admirably adapted to prevent anything of importance escaping the sieve of research, is not one which leads to ease of reference. Even were his work in book form, instead of scattered through a periodical for seven years and deeply buried in its pages, it could not easily be consulted. Its form is far from convenient. But it is a work of extreme thoroughness, and considered as a foundation, which he hoped it might become in the future, it is broad and safe, for which his followers must stand deeply indebted to him.

Mr. Mathews' method was to take the work of every writer from the earliest time in order of date, making lists of the plants he found noticed in the books; and since this led to constant reiteration, he marked with an asterisk those he had met with before. There was no index, and no means, except by looking at every page, of finding the record of any particular plant. For the earlier books he depended upon the MSS. of the Rev. W. W. Newbould, preserved in the Natural History Museum at South Kensington, who made notes of the records referring to each county that occur in them. It has not been thought necessary to examine these books again. The first record of any plant attributed to Worcestershire is in Leland's *Itinerary*, 1549, where, in Hearne's 2nd edition, vol. vi, p. 71, folio 80, the juniper is mentioned as growing at Tetbury Castle, two miles from Tewkesbury. This locality, however, is not in Worcestershire, but in the Gloucestershire parish of Twyning, an area entirely surrounded by Worcestershire, with the exception of a third of a mile on its extreme south where it joins its parent county. Tetbury Castle is now called Towbury Hill and is situated upon the Worcestershire boundary near Ripple; but in spite of its proximity to the county, this notice must be abandoned as a Worcestershire record. The first indubitable county Record occurs in the first British Flora, William How's *Phytologia Britannica*, 1660, and is a notice of *Ranunculus*

Ficaria. In Merrett's *Pinax*, 1666; Ray's *Catalogue*, 1670; that part of Camden's *Britannia* relating to Worcestershire, 1695; and in other books, further records are found. But up to 1732, a date which Mr. Mathews considered to be the end of the pre-Linnaean era, only eleven records are to be found.

The first English Flora on the Linnaean system was William Hudson's *Flora Anglica*, the first edition of which appeared in 1762. It yields only one new record of a Worcestershire plant, that of *Scirpus romanus*, now known as *S. Holoschoenus*, Linn.; a record which has been discredited, possibly without sufficient reason. Worcestershire has been fortunate in being near the homes of several eminent British botanists in an early period. William Withering passed the greater part of his life in Birmingham, and there he died in 1799. The first edition of his well-known *Botanical Arrangement of British Plants*, in two volumes, was published in 1776, and contains no Worcestershire records. It was not till the second edition was published in 1787 that Worcestershire plants are mentioned, and before that was issued another author requires to be dealt with. The second edition was anticipated by the appearance of Dr. Treadway Nash's *History of Worcestershire*, which, published in 1781, contains in an introduction the first list of Worcestershire plants as such, containing forty-three names. It is not known who supplied Dr. Nash with this list, which appears to be the work of a competent botanist, and affords thirty-nine new records. In 1787 the second edition of Withering's *Botanical Arrangement* was published. It was edited by Dr. Jonathan Stokes, who first resided at Kidderminster, and afterwards removed to Chesterfield, where he died in 1831, in his seventy-seventh year. In this edition Dr. Stokes gives many records for Worcestershire, Warwickshire, and Staffordshire, partly from his own observations, and partly from those of his correspondents, among whom was a Mr. Ballard, a surgeon residing at Malvern Wells. At this time the large parish of Halesowen, now in Worcestershire, was in Shropshire. In the book 105 new records are given. A third volume was published in 1792 which adds four new records to the previous list; and in 1796 a third edition of the *Arrangement* was issued containing nineteen new county records. Among the names of correspondents who furnished them are those of the Rev. Mr. Baker of Stout's Hill, Gloucestershire, Miss Read, who probably resided at Bromsgrove, and William Pitt, author of a series of Reports upon the Agriculture of English counties. Three years after the appearance of this work, Dr. Nash published a Supplement to his History with, at p. 96, an additional list of plants, from the observations of Dr. Sheward, a Worcester medical man, which contains twenty-two new county records. This list marks the close of the eighteenth century, and at this time 201 Worcestershire records had been published.

Shortly after the new century began, in 1801, a fourth edition of

Withering's *Botanical Arrangement* was published, two years after the author's death, edited by his son. In this edition there are five new records. The next new records, four in number, are to be found in Turner and Dillwyn's *Botanists' Guide*, published in 1805. In 1810 appeared William Pitt's *General View of the Agriculture of the County of Worcester*, one of a series of such reports which he prepared at the order of the Government. He lived first at Pendeford, near Wolverhampton, and afterwards at Birmingham. In a list of plants growing in Worcestershire, at p. 317, seventy-six species are mentioned for the first time. The next book to be noticed is the *Midland Flora*, by Thomas Purton. He was a surgeon living at Alcester, and of his book the first two volumes were published at Stratford-on-Avon in 1817, and the third in London in 1821. He notices plants in all the neighbouring counties reaching as far as Derbyshire on the one hand and Monmouthshire on the other, and it is often difficult to disentangle the Worcestershire records from those of Warwickshire. Among his correspondents were the Rev. W. S. Rufford of Badsey, near Evesham, and William Scott of Stourbridge, author of a book hereafter to be noticed. In the first two volumes of this book are eighty-four new records, and in the third sixteen that are new. Part 2 of the third volume contained at p. 335 a list of 'Additions and Corrections', which contained seven new records. Between 1824 and 1828, the first edition of Sir J. E. Smith's *English Flora* was published in four octavo volumes, the first two of which contain two Worcestershire records. There is no new record in the two succeeding volumes.

In 1825 a *Descriptive and Historical Account of Dudley Castle*, written by the Rev. Luke Booker, LL.D., vicar there, was published at Dudley. The book contains, at page 107, a list of plants, which contains thirteen new records, growing 'near the Castle'. Here the difficulty of county boundaries is met with. Dudley Castle is actually in Staffordshire, but its surroundings are in Worcestershire, and it must remain a matter of doubt how close to the Castle the phrase 'near the Castle' meant. However, the list may be taken to give the number of new records mentioned above.

We now come to consider the work of Edwin Lees, a name more intimately connected with the botany of Worcestershire than that of any other writer. He was born on May 12, 1800, and died in his eighty-eighth year on October 27, 1887. In 1828 he was a printer and stationer at 87 High Street, Worcester, and in that year he published under the name of Ambrose Florence, a *Strangers' Guide to the City and Cathedral of Worcester*. At page 152 is a 'Catalogue of Plants growing wild in the vicinity of Worcester'. This was the first utterance of Mr. Lees on the botany of a county on which he was to work so much. The list contains the names of 106 plants, of which twenty-seven are new records. In this list no distinction is made between the plants observed by Mr. Lees himself and those recorded on the authority of other observers; nor does it contain a single Composite. In 1828 also, in Jameson's *Edinburgh Philosophical Journal* for that year, there is at page 99 a list of thirty-six Malvern plants contributed by William Ainsworth, M.R.C.S., of which six are new records.

The first volume of the *Midland Medical and Surgical Reporter* was published in Worcester in 1828–9. The first number, for August 1828, contains an article on the 'Medical Topography of Worcestershire', which enumerates some of the plants growing in the neighbourhood of Malvern, and gives a short list, mainly copied from previous writers. The author is unknown, and does not appear to have been an accurate botanist. This was followed in No. 2 for November in the same year by an essay entitled 'Some observations on the Coal District of Worcestershire, and on the Botanical and other Peculiarities of the Malvern Hills', by J. K. Walker, M.D., of Huddersfield, in which he mentions the list of plants in the previous number, and begs to add an additional one, the plants given in which were all collected within four miles of the Holy Well. It contains five new records. The next year, in March 1829, Mr. Lees started a serial publication entitled *The Worcestershire Miscellany*. Five numbers appeared, the last in March 1830, which were afterwards collected and reissued, with a preface and a supplement, in 1831. In the first number is a list of plants from Malvern and from Perry Wood, near Worcester, which contains two new records. In the same year, 1830, Mr. Lees published in Loudon's *Magazine of Natural History*, vol. iii, page 160, a catalogue of the 'Plants of the Malvern Hills'. Mr. Lees' botanical district of Malvern did not regard county boundaries, and his list contains Herefordshire records which it is in some cases difficult to separate from those relating to Worcestershire. The list is dated September 18, 1829, and contains nineteen new records which may be taken to belong to the latter county. The next volume of Loudon's *Magazine*, vol. iv, also contained, at page 450, a list of the rarer plants of Worcestershire, communicated by Mr. W. G. Perry, a bookseller of Warwick, which he stated contained only the names of plants observed and gathered by himself. He mentions eighty plants, of which twenty-four constitute new records.

In 1832 William Scott of that town published a *History of Stourbridge*, and at page 540 of that book is a 'Select descriptive Botanical Catalogue'. The difficulty of locating plants in this list has already been mentioned, and it is added to by the fact that some of the localities he mentions are divided by the line separating Worcestershire from Staffordshire. Added to this are several obvious errors of identification, and some cases in which specific names are omitted. With all its defects, the list is an interesting one; it has preserved for us the Flora of Cradley Park, woodland in Scott's time, but now an integral part of the adjoining Black Country, covered with collieries and brick fields. His list contains

sixty-one records that are new to the county. Two years after the appearance of Scott's book, Sir Charles Hastings, M.D., of Worcester (after whom is named the Hastings Museum, now housed at the Victoria Institute in that city), published his *Illustrations of the Natural History of Worcestershire*. Appendix D, contributed by Mr. Lees, contains notices of the principal authorities for the Botany of the county, and a list of 388 species. Many of these records were those of predecessors, but others were supplied by contemporary botanists, among whom Dr. R. J. N. Streeten, a physician practising in Worcester, was a large contributor. Mr. Lees has included some Herefordshire and Gloucestershire records without any notice that the localities are not in Worcestershire. There are sixty plants mentioned for the first time in this list as occurring in Worcestershire.

The next new records for the county are to be found in Hewett Cotterell Watson's *New Botanists' Guide*, published in 1835. The list of plants for Worcestershire was nearly entirely founded on Mr. Lees' materials, and he also supplied him with a checked catalogue distinguished by Mr. Watson by the words 'Lees' Cat.'. Thirty-nine plants are mentioned in it for the first time. One new record is to be credited to William Addison, F.L.S., surgeon to H.R.H. the Duchess of Kent, who was in practice at Malvern previously to 1852. To the fourth volume of *The Transactions of the Provincial Medical and Surgical Association* he contributed in 1836 a list of rarer plants indigenous to Malvern, chiefly plants already noticed by Mr. Lees. One however was new; *Poa nemoralis*. A serial called the *Naturalist* was started in 1837 by Benjamin Maund, F.L.S., and W. Holl, F.G.S. To Mr. Holl are due, with the Rev. J. H. Thompson, most of the records of Lichens given in this book.

We are now taken to the neighbourhood of Birmingham. Dr. William Ick, Curator of the old Philosophical Institution of Birmingham, contributed to the *Analyst*, vol. vi, for 1837, page 20, a paper entitled 'Remarkable Plants found growing in vicinity of Birmingham in the year 1836'. In the same volume, at page 293, is a list of Moseley plants contributed by Miss M. A. Beilby, afterwards Mrs. Avery of Birmingham. Dr. Ick also contributed a list of plants to the *Midland Counties Herald* for August 5, 1838. In this some of Miss Beilby's records are repeated. Two other papers were contributed to the *Phytologist*, one by Samuel Freeman to the issue for July 1842, first Series, vol. i, p. 261; and the other by Edward Newman, the editor, to the number for March 1843, vol. i, p. 512. This was a list of Worcestershire Ferns. In these papers in the *Analyst* and in the *Midland Counties Herald*, twenty plants are mentioned for the first time.

In the year 1842, Mr. Edwin Lees published the first edition of his *Botanical Looker-out among the Wild Flowers of the Fields, Woods, and Mountains of England and Wales*, of which a second edition appeared in

1851. Five new records, all of trees, are to be found in the first edition. The next year he published the first edition of his *Botany of the Malvern Hills*. Of this book there were three editions, none of which bears a date on the title-page; but the prefaces are dated respectively May 12, 1843, August 1852, and July 31, 1868. Mr. Mathews remarks that this work is of some importance in the history of Worcestershire Botany. It is the first in which all the plants are recorded, and the first also in which any attempt is made to discriminate the Brambles. Mr. Lees' Malvern district did not regard county boundaries, as Herefordshire and Gloucestershire localities are admitted. Few localities are given in the first edition; more in the second and third. In the first edition there are no less than 190 first records of Worcestershire plants. In the second edition of the *Botanical Looker-out*, 1851, *Geum intermedium* is recorded for the first time. In the second edition of the *Botany of the Malvern Hills* there are forty-three new records. In this book many names of gentlemen who contributed to the knowledge of the botany of the county are mentioned. They are the Rev. Dr. Cradock, Principal of Brasenose College, Oxford; the Rev. J. H. Thompson, vicar of Cradley, near Halesowen; Thomas Baxter, second master of the College School at Worcester; Thomas Westcombe; and Thomas Reece, curator of the Hastings Museum at Worcester.

The fourth volume of the *Phytologist*, part 2, 1852, contains several mentions of Worcestershire plants, five of them new. One of these deserves special mention; it occurs at page 1142, and is a record of *Thymus Chamaedrys*, by George Jorden. Mr. Jorden was butler to James Fryer of Bewdley, and possessed an extraordinary knowledge of the plants of Wyre Forest. He was the first botanist to recognize the two forms of British Thymes, which he pointed out about 1844 to his fellow botanists, and which were distinguished by Professor Babington at a meeting of the Botanical Society of Edinburgh on April 14, 1853. In 1853 Stanley's *Worcester and Malvern Guide Book* was published, and it contained a list of the more remarkable Flowers and Ferns indigenous to the neighbourhood of Worcester, contributed by Mr. Thomas Baxter, assisted by Edwin Lees, Thomas Westcombe, and the Rev. J. H. Thompson. The list contains 227 species, but only three are mentioned for the first time. A second series of the *Phytologist* was begun in 1855, edited by Alexander Irvine, and the six volumes that appeared before it came to an end in 1863 contain many notices of Worcestershire plants. In the first volume is an account of a 'Visit to Wyre Forest', by T. W. Gissing. In fact Mr. Gissing made two visits, in July 1854, and June 1855, and both times he was accompanied by George Jorden. It was on the first visit that Mr. Jorden discovered a single plant of *Spiranthes aestivalis* at the great bog on the Worcestershire side of the Forest. The botanical history of the county from this time forward till 1867 is mostly contained

in successive volumes of the *Phytologist*. In vol. ii, page 244, 1857, is a notice of Mr. Irvine's *Illustrated Handbook of the British Plants*, in which some Clent plants are mentioned, and the discovery of the two *Elatines* published; and at page 385 of the same volume is a paper 'On the Botany of the Clent Hills', by Mr. Irvine. In volume iv, under date June 20, 1862, is an account by Mr. Jorden of the destruction by fire of the old Sorb Tree of Wyre Forest, *Pyrus domestica*, Sm. In these volumes there are thirteen new records. But at the same time were being published *Transactions of the Malvern Naturalists' Field Club*, Part I appearing in 1855. Part II contains, at page 19, a list of plants added to the Malvern Flora since the publication of the second edition of Mr. Lees' *Botany of the Malvern Hills*. In the list are two fresh records.

The *Botany of Worcestershire*, by Mr. Edwin Lees, appeared in 1867, and is the first attempt to record all the plants growing in the county. It begins with an introduction, pp. i to xci; there are then 147 pages devoted to the Local Distribution of the Plants of Worcestershire. In this book the first division of the county into Botanical Districts was effected, and these he named: I. The Severn Valley; II. The Malvern Hills and Valley of the Teme; III. The Avon and Lias Country; IV. Bromsgrove Lickey and the intervening district to Birmingham, Halesowen, Stourbridge, and Dudley. He gives a map showing the limits of his districts, but, as in the case of his books on Malvern, he does not confine himself to county boundaries, but colours his districts from point to point, and so as to include the portions of an adjoining county in which islands of Worcestershire might be situated. After this section of the book comes a Table of Plants, separately paged from 1 to 36, page 37 beginning an appendix of twelve pages devoted to Worcestershire Brambles; and then come three unnumbered pages of Additions and Corrections. The Table of Plants consists of a continuous list, against which are four columns, one for each district, in which the occurrence of the plant is noticed with descriptive marks and abbreviations in each case. In the book are 113 additional records for the county, of which forty-three are aliens, and a few are varieties of types previously noticed. Mr. Lees' book lacks method; it has three separate sets of pages, and there is no semblance of an index. The third edition of the *Botany of the Malvern Hills* was published in 1868, and was Mr. Lees' last serious work, though he lived for nearly twenty years afterward. There is a notice of his Life and Work in the *Journal of Botany*, vol. xxv (1887), page 384.

In 1868, William Mathews, of Edgbaston, published a local Flora, consisting of the more remarkable plants and ferns to be found in the county of Worcester within a radius of six and a half miles from the summit of Clent Hill, and in 1881 a new edition was published, in which Mr. Mathews extended his boundary to that of Mr. Lees' Lickey District, and, following him in not regarding the county boundary, he took into his district that part of Staffordshire in the north of the Severn district which lies between the Stour and Worcestershire. Mr. Mathews' lists formed supplements to *Clentine Rambles*, published by Messrs Mark and Moody, Stourbridge.

The two parts in which H. C. Watson's *Topographical Botany* was first issued are dated respectively 1873 and 1874, but as they were printed in a very limited number, and for private circulation only, the second edition dated 1883 is the one usually met with. Although the *Topographical Botany* is a valuable book of reference, it is not entirely trustworthy. Mr. Watson was capricious; he quoted no books; he refused to insert some of Mr. Lees' records, while he took the records of other correspondents on trust, without any real knowledge of their botanical attainments. Mr. Lees was not at all times wise enough to conceal his annoyance with Mr. Watson. The editors of the second edition inserted (E. Lees) wherever they could find that Mr. Lees had recorded the plants enumerated. The value of the work consists in establishing a system of numbers by which to compare the rarity or commonness of plants throughout Britain. Lately, an addendum has been published in the pages of the *Journal of Botany* for 1905, by Mr. Arthur Bennett.

The Botanical Record Club was established in 1873 for the purpose of continuing county records. None were admitted except those personally vouched by authentic specimens submitted to eminent botanists called the Referees. As the *Topographical Botany* was taken as the criterion of previous records, several plants well known in the county for many years were published as new records. The editor was F. Arnold Lees, M.R.C.S., who resided during part of the time at Kidderminster. Reports were issued up to 1886, and in them, besides the repetitions mentioned, there are several new records. The work of the club has since been carried on by other agencies, and new records, though of course in diminishing frequency, in spite of the division of aggregates so much in vogue, are made from day to day.

We are now approaching modern times. In 1887, Mr. Mathews began his series of papers in the *Midland Naturalist*, in which he listed the first record of each plant noticed by Mr. Lees, and upon which the present book is founded. His last paper is dated at foot December 17, 1892. He died in September, 1901, within five days of completing his seventy-second year. Besides being an able and painstaking botanist, whose researches extended over much of Europe and North Africa, he was a man eminent in many ways. He took high mathematical honours at Cambridge, and was one of the original members of the Alpine Club, becoming an early President of that body. The main part of his Herbarium is at Kew, but his Worcestershire plants are in the Hastings Museum at Worcester. In 1901, the first volume of the *Victoria History of Worcestershire* was issued, containing an article on the botany of the county by one of the writers of the present book, which it was hoped

contained a complete list of Worcestershire plants up to that date, localized in their several districts. Worcestershire Botany is by no means neglected at the present day. A Bibliography of works relating to the Botany of Worcestershire, compiled by Mr. John Humphreys, was published by the Worcestershire Historical Society in 1907. The Worcestershire Naturalists' Club is actively at work, and under the care of one of the writers responsible for this book a complete set of its Transactions from the earliest times has been published, and is continued year by year; while several botanists have various parts of the county under continual observation.

The Mosses and Hepatics of Worcestershire until late years have not received much attention. Mr. J. E. Bagnall has kindly written out for this book a list compiled by himself, with Messrs. E. Cleminshaw, M.A., and J. B. Duncan, which appeared in the *Journal of Botany*, vol. xli, Nos. 491 and 492, November and December, 1903. But since the compilation of this list much work has been done, and for subsequent additions and many additional localities we are indebted to Mr. Duncan; while Mr. H. H. Knight, M.A., of Cheltenham, has supplied records of several interesting plants, and localities for the rarer species. With the exception of the *Sphagna*, which follow Warnstorf's system, the Mosses are arranged according to Dixon's *Student's Handbook of British Mosses*, 2nd ed., 1904, and the Hepatics follow the *Census Catalogue of British Hepatics*, by Mr. S. M. Macvicar, 1905. There is still much to be done before a perfect survey of the distribution of these mosses in the county can be obtained, and this may be said in a greater degree with regard to the Hepatics. Many causes throughout the county contribute to form an environment unfavourable to these plants; the comparatively low rainfall and dry atmospheric conditions of the Midlands, the absence of hills sufficiently high and extensive to provide moorland, the improvement of land by agriculture, the drainage of bogs, the cutting of timber, and last, but not least, the spread of the manufacturing areas in the north and east, all combine to impoverish the cryptogamic flora of the county. The older botanists have recorded many plants from the Malvern Hills which are now either very rare or extinct. Possibly some of these records are unreliable, but there can be little doubt that the drainage of these hills for water supply and other purposes has been responsible for the disappearance of several interesting species. Still, as some compensating influence must be reckoned the very varied geological features of the county, each formation of which furnishes its quota of characteristic species. For some reason difficult to ascertain, the south-east of the county is much more favourable to the truly arboreal species than the north or west; but the wooded areas of Wyre Forest and the hilly country lying between the Severn on the east and Herefordshire on the west have formed a retreat for many of the rarer Mosses and Hepatics which are scarcely met with outside

these areas. There is quite a trace of a subalpine character about the Mossflora of this district; and some of the plants survive only in one, or at most two or three spots, the only localities for them in the county. On the whole, however, the Moss-flora of Worcestershire compares favourably with that of neighbouring counties.

The list of Lichens has been compiled from specimens preserved in the Hastings Museum at Worcester, chiefly collected by Dr. Holl and the Rev. J. H. Thompson. Mr. Lees, in his third edition of the *Botany of the Malvern Hills*, at page 139, gives a somewhat full account of the Lichens of that locality, but the method he adopted there does not allow his work easily to be incorporated in a book on the present plan. In some cases he gives mere lists of names; and the localities he mentions are very general. Still, as far as regards First Records many no doubt are to be found there, if it were necessary to search them out. The writers of this book have thought a reference to this work of Mr. Lees' sufficient.

The portion of this book allotted to Fungi is the work of Mr. Carleton Rea, who has given especial attention to this tribe of plants, and it may be taken as an accurate list of all Fungi known at the present time to inhabit the county. Among those recorded are many new to the county and several new to Britain, including *Amanita muscaria* Linn. var. *aureola* Kalchbr., *Lepiota submarasmioides* Speg., *Lepiota leucothites* Vitt., *Tricholoma squarrulosum* Bres., *Tricholoma glaucocanum* Bres., *Inocybe corydalina* Bres., *Agaricus perrarus* Schulz., *Coprinus squamosus*, Morgan., *Coprinus Patouillardii* Quél., *Merulius Guillemoti* Boud., *Lycoperdon hyemale* (Bull.) Vitt.=(*depressum* Bon.), *Lysurus australiensis* Cke. and Mass., *Leptosphaeria vagabunda* Sacc.; while the six new to science are: *Collyba eriocephala* Rea=(*veluticeps* Rea), *Pholiota grandis* Rea, *Inocybe duriuscula* Rea, *Flammula rubicundula* Rea, *Coprinus roseotinctus* Rea, *Clavaria luteo-alba* Rea. Mr. Rea's work on these plants is well known to Fungologists not only in our own country but on the Continent also, and in America.

The Fresh-Water Algae and the Desmids and Diatoms of the county have received but little attention. The list in this book is compiled from the observations of Mr. W. J. Farthing of Worcester, and the records given by Mr. Lees in the book alluded to above. The Algae are arranged according to *Introduction to Fresh-Water Algae*, by M. C. Cooke, 1890.

The writers must express their great indebtedness to their many friends who have rendered them assistance, and especially to Mr. J. E. Bagnall, Mr. John Humphreys, and Mr. R. F. Towndrow, under whose eyes all the manuscript of the Flora has passed. To these gentlemen and to many others in a lesser degree, this book owes very much.

J. A.

CLENT, *January* 1909.

PLAN OF THE FLORA

THE arrangement and nomenclature adopted are those of the *London Catalogue*, tenth edition, varied by additions, both as regards plants which are not mentioned in it, and the distinguishing marks used in this book.

The names of the Natural Orders are printed in Roman Capitals.

Beneath the name of the order comes that of the Genus in heavy Capitals. The usually accepted derivation of the Generic name is given in brackets.

The Specific name of native plants is printed in Clarendon type; of not-native, but well-established plants, aliens happy in their environment, in Clarendon type with the addition of an asterisk. It seems to the writers that the distinction between Colonist and Denizen is one hard to draw, and of no practical use. Colonists and Denizens are regarded by them as 'Established Aliens'. The names of not-native plants, not yet established, or not able to maintain themselves for any length of time, are regarded as 'Casual Aliens', and their names are printed in Small Roman Capitals. Many such not usually recorded are given in this Flora, since at any time one of these may burst out as some of their predecessors have done and overrun the countryside, and such records would then become valuable. The specific name is followed by the English 'Book-name', the name usually given to the plant in the Botanies.

Before the Specific name are placed two numbers, one in Italics, and one in black figures. The number in Italics corresponds to the one attached to the plant in the *London Catalogue*, tenth edition, by which means the Catalogue is made of use as an index to the Flora; and the number in black figures is the order of the plant in the list of Worcestershire plants recorded therein, which are numbered consecutively throughout; Casual Aliens, in Small Roman Capitals, are not always numbered.

After the Specific names, are given in Italics the more common synonyms, by which the plant was known to other Botanists.

In a fresh paragraph comes its character as native, alien, or casual, followed by the habitat of the plant, its rarity or scarceness in the county, and a letter which denotes annual, biennial, or perennial, or a word denoting tree or shrub. These details are followed by the time of flowering, and the index number of its scarcity or rarity throughout Britain, as given in the *London Catalogue*, tenth edition, preceded by the abbreviations *Top. Bot.*

A fresh line is devoted to giving the First Record of the occurrence of the plant in Worcestershire, the authority for it, and the year of the record in brackets.

Underneath the First Record are given localities for the plant in those Botanical districts of the county, Avon, Severn, Malvern, and Lickey, in which it occurs, with the authority for some of the records, and sometimes the date. The writers have omitted the marks of exclamation by which botanical writers usually signalize their own observations ; they desire to take upon themselves the responsibility for all the records in this Flora. They have put none in these pages of the trustworthiness of which they have any doubt without notifying that doubt, advising caution, or pointing out error. Of the Records without authority by far the larger number are made from specimens in the Hastings Museum at Worcester, the Herbarium in which, containing the collections of many early Botanists, has been carefully collated by the writers.

After this follows a paragraph giving any particulars about the plant which the writers think may be useful or interesting, and on its place in rural life, its use in mediaeval medicine, or observations on its characters or distribution. These are concluded by mentioning those of the adjoining counties in which the plant occurs ; or sometimes this fact is put negatively by stating those of the neighbouring counties in which the plant does not occur.

It is hardly possible, in view of the discoveries which are continually being made, that this comparison should be in all cases correct. The comparisons are made for Warwickshire and Staffordshire from a *London Catalogue*, ninth edition, marked by Mr. J. E. Bagnall ; and for Herefordshire from a *London Catalogue*, ninth edition, marked by Mr. Bagnall, and from his MS. notes, with references to the Rev. A. Ley's *Flora of Herefordshire*. Shropshire comparisons are made from a *London Catalogue*, ninth edition, marked by the late Mr. R. de G. Benson of Pulverbatch, Shrewsbury, supplemented by a few records from the *Transactions* of the Caradoc Field Club. For Gloucestershire the comparisons are not so satisfactory ; the *Topographical Botany* has been chiefly relied upon.

The following distinctions have been used :—

Native plants ; Clarendon type.

Established Aliens, Clarendon type with an added asterisk.*

Casual Aliens, Small Roman Capitals.

[ ] Wrong records, whether of locality or of scientific nomenclature, are placed in square brackets.

{ } Braces are used to enclose the articles on plants which are believed to have become extinct.

The Index comprises the Latin names of the Genera, and English names, both book-names sufficient to guide the inquirer, and also the names used by the rural community.

LIST OF ABBREVIATIONS

AND BOOKS TO WHICH REFERENCE HAS BEEN MADE IN THE FLORA

A. Flor. Strang. Guide. The Stranger's Guide to the City and Cathedral of Worcester, by Ambrose Florence, printed by Edwin Lees, 87 High Street (Worcester), 1828.

Analyst. A Magazine containing in vol. vi records by William Ick, Ph.D., and Miss M. A. Beilby, 1837.

Booker, Dud. Cast. A descriptive and historical Account of Dudley Castle and its surrounding scenery, by the Rev. Luke Booker, LL.D., Vicar of Dudley. Dudley, 1825.

Bot. Exch. Club, Rep. Reports of the Botanical Exchange Club of the British Isles.

Bot. Guide. Turner and Dillwyn's Botanists' Guide, two vols., 1805.

Bot. Looker-out. The Botanical Looker-out among the Wild Flowers of the Fields, Woods, and Mountains of England and Wales, by Edwin Lees, 1st ed., 1842, 2nd ed., 1851.

Bot. Malv. Hills. Botany of the Malvern Hills. Three editions. Prefaces only dated May 12, 1843 ; August 1852 ; and July 31, 1868, respectively.

Bot. Rec. Club Rep. Reports of the Botanical Record Club, established, 1873.

Dill. Hort. Eltham. Dillenius's Hortus Elthamensis, 1732.

E. B. English Botany, in thirty-six volumes, by Sir J. E. Smith. First volume published 1790.

Ed. Phil. J. Jameson's Edinburgh Philosophical Journal, containing a paper by W. Ainsworth, M.R.C.S., entitled A Sketch of the Physical Geography of the Malvern Hills, 1828.

Florence, Strang. Guide. See *A. Flor.*, above.

Hast. Mus. Worc. The Hastings Museum, Worcester, housed at the Victoria Institute in that city.

How, Phyt. Phytologia Britannica, 1650. No name of author mentioned, but known to have been written by William How, M.D.

Hudson, Flor. Angl. Flora Anglica, by William Hudson. First edition, 1762 ; second edition, 1778.

Illus. Nat. Hist. Worc. Illustrations of the Natural History of Worcestershire, by Sir Charles Hastings, M.D. Appendix D is a Catalogue of Worcestershire Plants, by Edwin Lees, 1834.

J. Linn. Soc. Journal of the Linnean Society.

J. of B. Journal of Botany.

Lees' Bot. Worc. The Botany of Worcestershire, by Edwin Lees, 1867. The localized *Table of Plants* at the end of this book is referred to sometimes without mentioning the book in which it occurs.

Lees, M. N. H. Lees, in Loudon's Magazine of Natural History, vol. iii, p. 160, 1829.

Lees, Worc. Misc. The Worcestershire Miscellany, edited by Edwin Lees. Published 1829-30. Collected and reissued 1831.

Malv. Advert. The Malvern Advertiser, a local newspaper.

Malv. Field Handb. The Malvern Field Handbook, and Naturalist's Calendar, by the Rev. G. E. Mackie, M.A., Malvern Advertiser Office, 1886.

Mat. Clent and Lickey Hills. A list of plants contributed to Clentine Rambles (Mark and Moody, Stourbridge), by William Mathews. First edition, 1868, second edition, 1881.

Merrett, Pinax. Pinax rerum naturalium Britannicarum, by Christopher Merrett, 1666.

Mid. Count. Her. The Midland Counties Herald, a newspaper published at Birmingham.

Mid. Nat. The County Botany of Worcestershire, by William Mathews, published in the Midland Naturalist (Birmingham, Cornish Brothers), from vol. x, p. 85, to vol. xvi, p. 156, 1887-93.

Mid. Surg. Rep. The Midland Medical and Surgical Reporter, vol. i, 1828-9 ; vol. ii, 1830-1 ; vol. iii, 1831-2. Published at Worcester.

Nash, Worc. Collections for the History of Worcestershire by Dr. Treadway Nash. The first volume contains, Introduction, p. lxxxix, a list of plants, 1781 ; a supplement, with a further list, was published in 1799.

New Bot. Guide. The New Botanists' Guide, by Hewett Cottrell Watson, 1835.

Perry, M. N. H. A list of plants in Loudon's Magazine of Natural History, vol. iv, 1831, p. 450, by W. G. Perry, bookseller, Warwick.

Phyt. N. S. The second series of the Phytologist, edited by Alexander Irvine. Six volumes, bearing date from 1855 to 1863.

Phyt. O. S. The first series of the Phytologist, June 1841 to 1854. The date on the title-page of the last volume is in error 1856.

Pitt, Agric. Worc. General View of the Agriculture of the County of Worcester, by William Pitt, 1810. One of a series of reports published by order of the Board of Agriculture.

Pitt, Phil. Trans. Philosophical Transactions No. 139, 1678, containing an extract from a letter from Edmund Pitt, of Worcester, on the Sorb Tree in Wyre Forest.

Purton, Add. Additions and Corrections, forming Part 2 of the third volume of Purton's Midland Flora, 1821.

Purton, App. Appendix to Purton's Midland Flora, forming vol. iii. 1821.

Purton, Midl. Fl. The Midland Flora, or Botanical Description of British Plants, by Thomas Purton, vols. i and ii published at Stratford-on-Avon, 1817, and the third in London in 1821.

Ray. Cat. Catalogus Plantarum Angliae, by John Ray. First edition, 1670.

Ray, Gibs. Camden. A catalogue of plants contributed to Edward Gibson's translation of Camden's Britannia, 1695.

Scott, Stourb. Stourbridge and its Vicinity, by William Scott, 1832.

Sm. Engl. Fl. English Flora, by Sir J. E. Smith. First edition in four volumes ; vols. i and ii, 1824 ; vol. iii, 1825 ; vol. iv, 1828.

Southall, Descrip. Malv. A Description of Malvern, by Mary Southall. Second edition, 1825.

Stan. Worc. Malv. Guide. Stanley's Worcester and Malvern Guide Book. Worcester, printed by John Stanley, Sidbury. Date, probably, 1853.

Stokes in With. Dr. Stokes's contributions to the third volume of the second edition of Withering's Botanical Arrangement, 1792.

Stokes's With. Bot. Arr. The second edition of Withering's Botanical Arrangement, edited by Dr. Jonathan Stokes, 1787.

Top. Bot. Topographical Botany, by H. C. Watson. In this Flora these abbreviations are placed before the comital number of the plant, taken from the tenth edition of the London Catalogue, which number is primarily founded on the record in Watson's book.

Trans. Malv. Nat. Club. Transactions of the Malvern Naturalists' Field Club Three parts, respectively dated 1855, 1858, 1870.

Trans. P. Med. Surg. Ass. The fourth volume of The Transactions of the Provincial Medical and Surgical Association, London and Worcester, 1836. Containing an Article by William Addison on the Medical Topography of Malvern.

Vict. Hist. Worc. The Victoria History of the County of Worcester, 1901.

W. N. C. Worcestershire Naturalists' Club, the *Transactions* of which, in three volumes, and still in issue, are frequently referred to.

With. Bot. Arr. Botanical Arrangement of British Plants, by William Withering ; first edition in two volumes published 1776 ; the second in three volumes, 1787 ; the third edition in four, 1796. After his death, five further editions were issued.

With. 3rd ed. The third edition of the Botanical Arrangement.

Worc. Misc. The Worcestershire Miscellany, edited by Edwin Lees. See *Lees*, above.

c

DICOTYLEDONES

1. RANUNCULACEAE, Jussieu

CLEMATIS, Linn. 1. (κλῆμα, a vine-shoot.)

1. 1. **C. Vitalba**, Linn. *Traveller's Joy.*

Native. Hedgerows, wood-borders, and thickets. Locally plentiful. Climbing shrub. June to September. *Top. Bot.* 49.

First Record. Hedges near Malvern and north of Evesham, *Pitt, Agric. Worc.*, p. 317 (1810).

Avon. Confined to calcareous soil, *Lees.* Craycombe. Bredon Hill, very abundant at the quarries. Broadway. Trench Woods. Hedge near Eckington Churchyard. Berrow Hill. Feckenham, in profusion.

Severn. Battenhall, *Lees.* Hallow, *Lees.* Wyre Forest, *Jorden.* Not confined to calcareous soil, *Lees.* Dam at Churchill Pool.

Malvern. Near Pendock. Ankerdine Hill. Madresfield. Abberley Hill. Near Tunnel Hill, Upton-on-Severn, *Rev. W. S. Symonds.* Abundant near fifth milestone on the Worcester to Hereford road, near Leigh Sinton.

Lickey. Roadside near Dordale Brook, Belbroughton. Hedge of Gunstool Meadow, Halesowen, *Mathews.*

Usually fairly abundant where the plant occurs at all, it is very nearly confined to calcareous soil, though it sometimes strays on to neighbouring marl. The Belbroughton plants, luxuriant enough, and on the marl, are several miles from even a garden specimen, but the Halesowen plant was afterwards discovered to be a garden escape. The plant is an irritant, but the young shoots are edible; in some places the branches are used for making baskets, and the leaves as fodder for cattle; while boys smoke the dried stems. The plant is locally called Honesty, and Old Man's Beard. It occurs in all the neighbouring counties.

THALICTRUM, Linn. 2. (θάλλω, I flourish, ἴκταρ, close together.)

{4. 2. **T. minus**, Linn. *Lesser Meadow Rue.*
(*T. collinum*, Wallr.)

Native. Stony pastures, thickets, lane-sides. Infrequent generally; one record only for Worcestershire, which requires confirmation. P. June and July. *Top. Bot.* 36.

First Record. As *T. minus.* Bewdley Forest, *Jorden, Phytog.*, N.S., vol. i, p. 281 (1855).

Severn. Bewdley Forest, as above.

B

In the article in the *Phytologist*, from which this record is taken, written by Mr. George Jorden himself, no definite locality for this and many other plants is given. About half the area of the forest, called also Wyre Forest, is in Shropshire, and until it is confirmed this record must remain doubtful. The plant is recorded for Gloucestershire, Hereford, and Salop.}

7. 3. **T. flavum**, Linn. *Common Meadow Rue.*

Native. Banks of rivers, ditches, and moist meadows. Not common except in the Severn district. P. June and July. *Top. Bot.* 71.

First Record. Meadows and banks of rivers; meadows on Severn, *Pitt, Agric. Worc.*, p. 317 (1810).

Avon. Not common, *Lees.* Banks of Avon near Evesham and Eckington Bridge.

Severn. Very fine at Grimley brickfields. Kepax Ferry. Meadow near Bewdley.

Malvern. Longdon Marsh, *Lees.* Near Teme's mouth. Bransford, Powick, Upton-on-Severn, *Towndrow.*

Lickey. Moseley, *Lees,* now gone. Meadows by the Rea, *Miss M. A. Beilby,* possibly in Worcestershire.

A variable plant, not only in the size of its leaflets, but in the laxness of its panicle and the shape of the fruit. It is recorded for all neighbouring counties.

ANEMONE, Linn. 3. (ἄνεμος, the wind; the plant flowers when the spring winds prevail.)

[8. 4. **A. Pulsatilla**, Linn. *Pasque Flower Anemone.*

Native. Grassy calcareous downs. One record only for Worcestershire. P. April and May. *Top. Bot.* 18.

First Record. *Lees, Bot. Worc.*, p. 104 (1867).

Avon. Specimen in Herb. Hast. Mus., Vict. Inst. Worc., 'near Broadway,' *Rev. H. Roberts.*

Although given by Mr. Lees in his table of Worcestershire plants, the locality he quotes is Snowshill, which is a parish adjoining Broadway, but in Gloucestershire, and there is possibly no reason why the plant should not also occur in Worcestershire. Mr. Roberts's specimen is probably from the same locality as that of Mr. Lees. There is no indubitable record for this county. It has been recorded in Gloucestershire, but no other neighbouring county.]

A. FULGENS, G. G.

First Record. At the top of an orchard at the back of Albion house, Ombersley, *W. N. C.*, vol. ii, p. 7.

Severn. As first record.

Some authorities make this only a variety of *A. hortensis.* It is a well-known garden border plant, with brilliant scarlet flowers.

9. 5. **A. nemorosa**, Linn. *Wood Anemone.*

Native. Woods, hedgebanks, and meadows, generally distributed throughout the county, and especially plentiful in the Lickey district. P. March to May. *Top. Bot.* 108.

First Record. Near Dudley Castle, *Booker, Dudley Castle,* p. 107 (1825).

No special localities are necessary in the case of so widely distributed a flower. It is often found more or less deeply flushed with pink. Var. *rubra,* Pritzel, is recorded by Mr. Towndrow from Leigh Sinton in the Malvern district, and it also occurs at Middleyards and Knightwick. The Wood Anemone is found in every neighbouring county.

11. 6. **A. *apennina**, Linn. *Blue Mountain Anemone.*

Alien. Certainly introduced. P. April and May.

First Record. Tunnel Hill, Upton-on-Severn, *Rev. W. S. Symonds,* in *Lees' Bot. Worc.*, p. xxviii (1858).

Malvern, as above. But in *Lees' Table of Plants* this locality is given as in the Severn district. Orchard at Malvern Link, *Towndrow.*

The discovery of this plant was first announced by Mr. Symonds in his second anniversary address to the Malvern Naturalists' Field Club in 1858. In Mr. Towndrow's Herbarium is a specimen labelled in Mr. Lees' handwriting, which he says was from a plant in his garden originally growing in a hilly wood at Shelsley Walsh, and found there by Mr. J. S. Haywood, who said there were several plants in the wood. It has, however, no claim to be a British plant; it has been mentioned as occurring in Salop and Stafford.

ADONIS, Linn. 4*. ("Αδωνις, a youth who was killed by a wild boar and whose blood is fabled to have stained the flowers.)

{12. 7. **A. *annua**, Linn. *Pheasant's Eye.*
(*A. autumnalis*, Linn.)

Colonist. Cornfields. The plant is not now found in Worcestershire. A. May. *Top. Bot.* 8.

First Record. Not localized, *Lees, Bot. Worc.*, p. lviii (1867).

Malvern. One place, now lost, *Lees.*

A plant of doubtful nativity in England, though frequent as a garden flower. Watson's *Top. Bot.* gives no Worcestershire record, but mentions it as occurring in Gloucestershire with other counties; the greater part of these records, he says, 'really illustrating the proneness of botanists to swell out their local lists on the slightest possible pretence'. When seen it is no doubt merely an escape from cultivation.}

MYOSURUS, Linn. 5. (μῦς, mouse, οὐρά tail, from the shape of the fruit.)

13. 8. **M. minimus**, Linn. *Mousetail.*

Native. Very local, but sometimes in plenty where it occurs. A. April to July. *Top. Bot.* 45.

First Record. Malvern Chase, *Ballard,* in *Stokes's With. Bot. Arr.*, 2nd ed., p. 336 (1787).

Severn. Helbury Hill, *Lees,* 1836. Porter's Mill, Claines, *Rev. J. H. Thompson,* 1852.

Malvern. Malvern Chase, as above. Powick Ham, *Westcombe,* 1856. Abundant in a field near the Old Hills; also at Madresfield, *Towndrow.*

Lickey. Harborne Reservoir, *Garner,* 363.

Mr. Lees makes a note, *Bot. Worc.*, p. 39, that this plant was most abundant at Porter's Mill at the date mentioned, but fifteen years afterwards few plants remained ; nor could he afterwards find it at Helbury Hill. It is recorded for all neighbouring counties except Salop.

RANUNCULUS, Linn. 6. (*Rana*, a frog.)

14. 9. R. circinatus, Sibth. *Water Crowfoot.*
(*R. divaricatus*, Schrank.)
Native. Rather uncommon throughout the county, and especially so in the Avon district. P. June to August. *Top. Bot.* 62.
First Record. As *R. aquatilis* β, *pantothrix*. In little pools on Welland Common, *Lees*, *Bot. Malv.*, Hills, 1st ed., p. 23 (1843). First record as *R. circinatus*, Pools, frequent, *Mathews*, *Clent and Lickey Hills* (1881).
Avon. Very uncommon, *Lees*. Dunhampstead. Ferry at Eckington.
Severn. Lady Pool, Blakedown, *W. N. C.* iii. 29. Churchill and Stakenbridge Pools, *Mathews*. Stanklin Pool, *Humphreys*. Pond, Spetchley Park. Near Camp Weir. Ditch, north side of Hartlebury Common.
Malvern. Rather uncommon, *Lees*. Hanley Castle, *Towndrow*.
Lickey. Harborne Reservoir, *Mathews*. Harvington Hall Moat, *Mathews*. Pools at Halesowen, *Mathews*. Birchill Pool, Halesowen. Canal by Bittell Reservoir.
The Water Crowfoot is remarkable for its freedom from the acridity so characteristic of Buttercups in general. In some parts of England it is used as food for horses and cattle, which eat it greedily. In *Top. Bot.* it is recorded for every neighbouring county except Salop, but it is found there ; *R. de G. Benson* (MS.).

15. 10. R. fluitans, Lam. *Water Crowfoot.*
(*R. fluviatilis*, Wigg.)
Native. Occurs in most of the streams throughout the county. P. June to August. *Top. Bot.* 57.
First Record. As *R. aquatilis* δ. In shoals in the Severn, *Stokes*, *With. Bot. Arr.*, 2nd ed., p. 578.
Avon. In the rivers, *Lees*. Bredon, *W. N. C.* ii. 61.
Severn. Abundant in the Severn, *Lees*. Bewdley. Hoo Brook, *Mathews*.
Malvern. In the Teme at Powick, *Lees*.
Lickey. Moat Mill Pool, Bromsgrove, *Humphreys*. Drayton near Belbroughton. Brook near Catshill.
Recorded in all the neighbouring counties.
Var. **Bachii**, Wirtg.
First Record. As here.
Severn. Severn, near Dowles Brook, *Rea*.
This variety is characterized by having shorter, almost sessile, leaves.

16. 11. R. trichophyllus, Chaix. *Water Crowfoot, Water Fennel.*
Native. Very local. P. May to August. *Top. Bot.* 60.
First Record. Pond near Saldons (1856), *Mathews*, *Worc. Nat. Club Trans.* ii. 67 (1899).
Avon. The Wildmoors, Feckenham, *Mathews*. Pond near Saldons, *Mathews*.

Severn. Priest's Pool, Grafton, Bromsgrove, *Humphreys*. Pond in field behind Holt Lock House. Pond near Moseley.
Malvern. 'I have never seen this in the Malvern district,' *Towndrow*, MS.
Lickey. Chadwich Upper Pool, *Humphreys*.
This Crowfoot has been noticed in every bordering county.

17. 12. R. Drouetii, F. Schultz.
Native. Ponds, ditches, streams. Rare. P. May and June. *Top. Bot.* 71.
First Record. Not localized, but given for 'Worcester, Bagnall MS.', *Bagnall*, *Top. Bot.*, 2nd ed., p. 7 (1883).
Avon. Brook at Rush Farm, Feckenham, *Mathews*.
Malvern. Welland, Leigh Sinton, Upton-on-Severn, *Towndrow*. Castle Morton Common.
Lickey. Upper Pool, Chadwich, *Humphreys*. Hopwood Dingle, 1871, *Bagnall*.
This plant was recorded by Mr. Towndrow, *J. of B.*, vol. xxi, in the same year, 1883, as Mr. Bagnall's record. But the preface to the *Top. Bot.* is dated May, and Mr. Towndrow's plant was gathered later in the year, so Mr. Bagnall is entitled to first record. The plant is recorded in *Top. Bot.* for all neighbouring counties except Salop, but it is found there also, *R. de G. Benson* (MS.).

18. 13. R. heterophyllus, Weber. *Water Crowfoot.*
(*R. aquatilis*, Linn., and of Index Kewensis.)
Native. Rivers, ponds, and brooks, local. P. May to July. *Top. Bot.* 45.
First Record. Pools, frequent, *Mathews*, *Clent and Lickey Hills* (1881).
Severn. Ditch side of lane near Coningree Wood. Pond, New Road, Worcester. Pond, Hallow.
Malvern. In the district, *Towndrow* (MS.).
Lickey. As above. Uffmoor Farm and Bittell Reservoir, *Mathews*.
Mr. Mathews adds to the first record above, 'This is the form also known as *R. diversifolius*, var. *radians*,' but this in *Hooker's Stud. Flor.* is given as a form of *R. trichophyllus*. Mr. Watson says in *Top. Bot.*, 'In books this name has been so much used for any segregate of *R. aquatilis* that it has become impossible in most cases to know which of the recent segregates is intended by the name.' There is a good note on Batrachian *Ranunculi* in Arnold Lees' *Flora of West Yorkshire*, p. 115 ; while Lees in his *Bot. Worc.* washes his hands of the whole tribe of segregations, and places them all under *R. aquatilis*. In his *Bot. Malv. Hills*, 3rd ed., 77, 1868, he takes the same course. 'I cannot endorse,' he says, 'the quasi-species into which the Water Crowfoot has been cut up by modern botanists.' See also Druce, *Flora of Berkshire*, p. 16.

19. 14. R. peltatus, Schrank. *Water Crowfoot.*
Native. Ponds, ditches, and streams. Common. P. April to August. *Top. Bot.* 81.
First Record. Pools, frequent, *Mathews*, *Clent and Lickey Hills* (1881).
Severn. Near Dodderhill, *W. N. C.* iii. 26. Pool near Monk Wood.
Malvern. In the district, *Towndrow*, *Malv. Advert.*, Nov. 5, 1892. Pond, Bransford.

Lickey. Pool at Coleford Priory, *Mathews*. Pool near Wychbury Cottage. Bittell Reservoir, *Humphreys*.
Var. d. **floribundus** (Bab.).
First Record. Not localized, *Mathews*, in *Mid. Nat.* xvi. 68 (1893), but referring to the Lickey district.
Severn. The common pond form, *Rea*.
Malvern. The common pond form, *Rea*. Sherridge, *Towndrow*, *Malv. Advert.*, Nov. 17, 1894.
Lickey. Wychbury, Pedmore, *W. N. C.* iii. 259. Harborne Reservoir, *Mathews*. The Nimmings Farm, Clent.
Var. e. **penicillatus** (*Hiern.*).
(*R. pseudofluitans*, Dumort.)
First Record. As *R. penicillatus*, River Cole, Yardley, *Mathews*, *Bot. Rec. Club Rep.*, 1887.
Severn. In the river near Arley, 1900, *Bagnall*.
Lickey. Harborne Reservoir and stream near Harborne Reservoir, Northfield, *Mathews*.
These are the only localities in Worcestershire from which this plant has been recorded. It was first seen in the Cole by Mr. Mathews, June 11, 1884, who says it occurs 'at Hall Green and up to Coleford Priory', *W. N. C.* ii. 101 ; while a specimen of his in Mr. Towndrow's Herbarium of that date is labelled, 'by bridge under Birmingham and Stratford Road, Yardley, Worcestershire.' The plant has been observed in all bordering counties except Herefordshire.
R. peltatus, in one or another of its many forms, is probably the most widely distributed of Water Buttercups. Both H. and J. Groves, in *Bab. Manual*, p. 8, and Syme, record this plant as *R. peltatus*, Fries, not Schrank. This plant is recorded from all bordering counties.

20. 15. R. Baudotii, Godr. *Water Crowfoot.*
Native. Usually in brackish water. P. May to July. *Top. Bot.* 56.
First Record. Madresfield, Malvern, *Towndrow*, *Bot. Exch. Club Rep.* 1887.
Malvern. The Stews and another small pond at Madresfield Court.
Lickey. Bittell Reservoir, *Humphreys*. Pepper Wood, Belbroughton, 1900, *Humphreys*.
Not given for Worcestershire in *Top. Bot.*, where it is recorded only in Gloucestershire of the neighbouring counties.

[**21. 16. R. tripartitus**, DC. *Water Crowfoot. Top. Bot.* 2.
First Record. Near Dunhampstead, *Worc. Nat. Club Trans.* i. 36 (1856).
This has possibly been recorded in error ; nor is it certain if this is the plant meant. At all events recent confirmation is required. Mr. Bagnall says only var. β, *intermedius*, not given in 10th ed. *Lond. Cat.*, has been recorded (MS.).]

23. 17. R. Lenormandi, F. Schultz. *Water Crowfoot.*
Native. Wet places and ditches. P. June to August. *Top. Bot.* 57.
First Record. As *R. coenosus*, *Lees*, in his tabulated list of plants, *Bot. Worc.* 3. As *R. Lenormandi*, Pedmore Common, *Mathews*, *J. of B.*, ix (1871).
Severn. Pond near the Birches at Hagley, *Mathews*. Pedmore, as above.

Lickey. *Lees*, as above, unlocalized. Near Uffmore Wood, *Bagnall*.
This Crowfoot and the following one usually possess no divided submersed leaves ; and the present plant is distinguished from the next by having five instead of three veins in the petals. This plant is recorded for Gloucester, Warwick, Hereford, and Stafford, and has been met with in Salop, *H. de G. Benson* (MS.).

24. 18. R. hederaceus, Linn. *Ivy-leaved Water Crowfoot.*
Native. Shallow ditches, margins of pools, wet places. Not uncommon. P. February to August. *Top. Bot.* 109.
First Record. In a swampy place on Abberley Hill, above the Hundred House, *Perry*, *M. N. H.*, vol. iv. 450 (1831).
Avon. Not uncommon, *Lees*.
Severn. Not uncommon, *Lees*. Hartlebury Common, *Mathews*. Roadside leading to Monk Wood.
Malvern. Abberley. Not uncommon, *Lees*. Side of old Malvern Road near Powick. West side of Worcestershire Beacon.
Lickey. In most of the rills and streams running down from the higher land. Chadwick. Burcott, *Humphreys*.
This plant is recorded from all the neighbouring counties.

25. 19. R. sceleratus, Linn. *Celery-leaved Crowfoot.*
Native. Pools, muddy ditches, damp places. Abundant in the greater part of the county, not uncommon anywhere. *Top. Bot.* 102.
First Record. Not localized, *Lees*, *Bot. Malv. Hills*, 1st ed., p. 28 (1843).
Avon. Norton near Evesham. Fladbury. Shell. Pond near Goosehill Wood. Flyford Flavell. Near Inkberrow.
Severn. Near Droitwich ; Brake Mill Pool, Hagley, *Mathews*. Hartlebury Common. Whittington. Claines. Hallow. Grimley. Holt. Battenhall. Kempsey.
Malvern. As above. Powick. Bransford. Madresfield. Near Upton-on-Severn. Leigh. Knightwick.
Lickey. Broadmarsh, Hagley. Near the gasworks, Cradley. Moat, Harvington Hall. Near Fockbury Mill.
This plant is one of the most acrid of the Buttercups, and in former times was used as a blister. It is found in nearly every part of the world and extends very far towards the north. It occurs in every neighbouring county.

27. 20. R. Flammula, Linn. *Lesser Spearwort.*
Native. Wet places. General throughout the county. P. April to September. *Top. Bot.* 112.
First Record. Not localized, *Lees*, *Bot. Malv. Hills*, 1st ed., p. 28 (1843).
Avon. Dodderhill Common. Trench Woods. The Slads. Churchill Wood.
Severn. Moseley Green, near Worcester. Hartlebury Common. Wyre Forest. Ockeridge Wood. Monk Wood.
Malvern. Malvern Hills. Alfrick. Welland Common. Castle Morton Common. Malvern Link Common.
Lickey. Ell Wood. Randans. Chaddesley Wood. Meadows on Clent Hills.

This is another very acrid Buttercup, its distilled water being a speedy emetic. Blisters, which are readily caused by it, are very difficult to heal. It occurs in all the bordering counties.

Var. b. **radicans**, Nolte.

First Record. As in this book.

Severn. Hartlebury Common.

Malvern. Malvern Link Common.

This is a small form rooting at the joints.

{30. 21. R. Lingua, Linn.   *Great Spearwort.*

Native. River banks, pools, and ditches. Perhaps now extinct. P. July and August. *Top. Bot.* 85.

First Record. Bogs on Malvern Chase, *Ballard*, in *Stokes's With. Bot. Arr.*, 2nd ed., 572 (1787).

Avon. One locality, *Lees*, Table of Plants, *Bot. Worc.*, p. 3.

Severn. Ockeridge Wood, near Holt, *Lees*, in *A. Florence's Strangers' Guide*, 132 ; found by Mr. Stretch. Pool at Spetchley, *Rea.*

Malvern. As above.

This handsome Buttercup has not been seen in the county for many years. It was at Spetchley in 1879, *Towndrow* (MS.), but had gone from Malvern Chase in 1855, *W. N. C.* i. 17. The Avon locality referred to by Mr. Lees is nowhere mentioned in his books. The plant occurs in all the neighbouring counties except Gloucestershire.}

31. 22. R. auricomus, Linn.   *Wood Crowfoot.*

Native. Woods and hedges. Locally common and fairly widely distributed. P. April and May. *Top. Bot.* 89.

First Record. Not localized, *Lees, Bot. Malv. Hills*, 1st ed., 28 (1843).

Avon. Rather uncommon, *Lees*. Fladbury. Badger's Bank. Crowle. Trench Woods. Pinvin. Churchill.

Severn. Bridge's Stone Mill. Perry Wood. Salwarpe. Claines. Hallow. Grimley. Holt. Kempsey. Monk Wood. Nunnery Wood.

Malvern. *Lees*, as above. Bransford. Powick. Leigh. Crews Hill. Alfrick. Dripshill. Copse between Lower Howsell and Braces Leigh.

Lickey. Wychbury, Pedmore. About Halesowen. Abundant in lanes about Belbroughton and the Randans Wood. Frankley. Uffmoor Wood. Lutley near Halesowen. Barnt Green. Northfield.

This is the earliest to flower of all the Buttercups, appearing with the primroses and violets in the first Spring sunshine. The flowers are usually imperfect, one or more of the petals, and sometimes all, being wanting. Nor is it so acrid as the rest of its tribe with yellow flowers. A common country name for it is 'Goldilocks'. It is recorded from all neighbouring counties.

32. 23. R. acris, Linn.   *Upright Meadow Crowfoot.*

Native. Pastures and meadows, abundant and everywhere. P. April to August. *Top. Bot.* 112.

First Record. Not localized, *Lees, Bot. Malv. Hills*, 1st ed., 28 (1843).

This is one of the common Buttercups of the meadows, perhaps more conspicuous than the two following kinds, since it holds its flowers well above the springing grass on its upright stems. As its name imports, it is very acrid, and cattle will not eat it ; but it loses its acridity when made into

hay. It has many old English names, Kingcup being one of them. Possibly it is the 'Tufted Crow-toe' of Milton's *Lycidas*, and certainly it or one of the following species is the 'Cuckoo bud of yellow hue', and the 'Winking Mary-bud' that begins to 'ope its golden eye' at the coming of Phoebus, of Shakespeare. It assumes a variety of aspects according to the situation in which it grows, some of which are made into varieties, none of which, however, have been recorded for Worcestershire. All neighbouring counties possess the plant.

33. 24. R. repens, Linn.   *Creeping Crowfoot.*

Native. Pastures and roadsides, wherever there is grass, and in gardens ; abundant everywhere. P. May to August. *Top. Bot.* 112.

First Record. Not localized, *Pitt, Agric. Worc.*, p. 317 (1810).

A troublesome weed this, in spite of its glossy, rich, golden flower ; for its creeping shoots run everywhere, rooting at every joint, and they everywhere characterize the plant. A double-flowered form has been found at Stanklin, *W. N. C.* iii, p. 233 ; and a very hairy form of this species occurs on the marl at the cutting of the road at Pickersleigh near Malvern, *Towndrow* (MS.). Country people know it as 'Crazy'. In all bordering counties.

34. 25. R. bulbosus, Linn.   *Bulbous Crowfoot.*

Native. Meadows and pastures throughout the county, common everywhere. P. March to July. *Top. Bot.* 106.

First Record. Not localized, *Lees, Bot. Malv. Hills*, 1st ed., p. 28 (1843).

This is the first to flower of the Buttercups of our meadows, and can easily be distinguished from its fellows by the sepals of its flower cup, which turn downwards. In old times it was called 'St. Anthony's Turnip', but it would be an acrid morsel to eat, and disastrous in its effects when eaten—at all events if raw ; but the acridity is entirely dispelled by boiling. The expressed juice of the root of the plant was used by our forefathers to raise blisters in cases of gout. The name Buttercup is not met with before 1777, and is probably a mixture of the earlier popular names—Butterflower and Goldcups or Kingcups. It was a common idea that cows ate these plants, and that this gave to butter its yellow colour. In all bordering counties.

35. 26. R. sardous, Crantz.   *Hairy Buttercup.*

(*R. hirsutus*, Curt.)

Native. Pond sides and watery places. Rare. A. or B. May to August. *Top. Bot.* 75.

First Record. As *R. hirsutus*, not localized, *Lees, Bot. Malv. Hills*, 1st ed., 28 (1843). Localized in the 2nd ed. of the same book, p. 611, 'I have only gathered *R. hirsutus* on a barren pasture bordering on Longdon Marsh,' *Lees*.

Malvern. Longdon Marsh, as above.

Although the number 75 is given for this plant in *Top. Bot.* it is decidedly rare in the Midland counties. It is recorded in that book for Gloucester, Hereford, Stafford, and Salop, of the neighbouring counties, and from Worcestershire as 'Buckman, sp.', which means that a specimen of the plant in Mr. Watson's Herbarium was so labelled from the county. Possibly the older records are in error. Sometimes the upright hairy form of *R. repens* is mistaken for it.

36. 27. R. parviflorus, Linn.   *Small-flowered Buttercup.*

Native. Dry sunny banks. Local and rather rare. A. May and June. *Top. Bot.* 59.

First Records. Malvern Hill, *Ballard* ; Worcester, *Stokes, Withering Bot. Arr.*, 2nd ed., p. 577 (1787).

Avon. One locality, *Lees*, is near Evesham. Pirton. Hanbury.

Severn. Portfields Road and Norton Road, Worcester, *Lees*. Stagbury Hill, *Thompson*. Near Dodderhill. Canal side below Salwarpe Church. Hallow and Cotheridge, *Lees* in *Loudon's Magazine*, and also 'under hedges by the roadside near the Virgin's Tavern, Worcester'. Kidderminster Railway Station. Doddenham Dingle. Pixham. Opposite Teme's Mouth.

Malvern. Alfrick and Malvern, *Lees*, as above. Bridge's Stone Mill. Neighbourhood of Tenbury, *Lees*. Not very uncommon about Barnard's Green and Powick, *Lees*. Newland Common. Alfrick Pound.

Lickey. Lanes near Hagley, *Scott*. Plentiful in fields on the Clent Hills. The Odnalls, Clent. Broom.

Easily recognized from its prostrate habit and little yellow blossoms, which are often imperfect. Recorded from all neighbouring counties.

37. 28. R. *arvensis, Linn.   *Corn Buttercup.*

Colonist. Cornfields. Generally distributed. A. April to August. *Top. Bot.* 69.

First Record. Not localized, *Lees, Bot. Malv. Hills*, 1st ed., p. 28 (1843).

Avon. Bredon Hill. Trench Woods. Norton near Evesham. Sheriff's Lench. Crowle. Hanbury. Libbery. Upton Snodsbury.

Severn. Northwood near Bewdley. Claines. Hallow. Kempsey. Ladywood. Cotheridge. Grimley. Holt. Hartlebury. Ombersley.

Malvern. *Lees*, as above. Powick. Bransford. Madresfield. Malvern Wells. Leigh. Witley. Martley. Clifton-on-Teme.

Lickey. Cornfields near the Upper Stour. Plentiful on the eastern side of the hills.

This plant is one of the most poisonous of its tribe ; it is easily known from all the other species by its large prickly carpels. It occurs in all the neighbouring counties.

39. 29. R. Ficaria, Linn.   *Lesser Celandine ; Pilewort.*

(*Ficaria verna*, Huds.)

Native. Damp places, woods, gardens, fields, under trees. To be found everywhere. P. March to June. *Top. Bot.* 112.

First Record. 'Chelidonium minus flore pleno nondum descriptum. Pilewort with a large flower. In Worcestershire.' [*How's*] *Phyt. Brit.* (1650).

This widely spread plant is one of the harbingers of spring, starring banks and meadows with gold before the winds of March have blown themselves away. 'There 's a flower that shall be mine, 'Tis the little Celandine,' sang Wordsworth. Celandine is the name of two distinct plants bearing yellow flowers, by the old herbalists regarded as species of the same plant. They stated that the flower appeared when the swallows came, which is the case with the Great Celandine, but not with this flower, which forestalls their arrival by a month or more. The name Pilewort has obtained from its

supposed efficacy against a painful ailment, no doubt on account of the shape of its tuberous roots. From the supposed resemblance of these also to a fig, it received the name *Ficaria*, and because of the difference of the shape of its blossoms from other Buttercups, it was placed in a separate genus with that name. Among country people it shares with *R. repens* the name of 'Crazies'. It is very variable both in leaf and flower. It is found in all the bordering counties.

Var. b. **incumbens**, F. Schultz.

First Record. Bransford, *Towndrow, Malv. Advertiser*, 1892.

This variety has since been observed in all the divisions of the county. Plants with axillary bulbils have been noted at Bevere, Crookbarrow, and Martley. In the bordering counties it appears to have been noticed only in Warwickshire.

CALTHA, Linn. 7.   (κάλαθος, a cup.)

40. 30. C. palustris, Linn.   *Marsh Marigold.*

Native. Wet places. Generally distributed. P. March to June. *Top. Bot.* 112.

First Record. Not localized, *Lees, Bot. Malv. Hills*, 1st ed., p. 29 (1843).

This is one of the best known flowers of the countryside, combining as it does the conspicuous nature of the flowers with universal distribution, and flowering at a time when everybody is ready to welcome the floral signs of coming summer. In all lands it has many popular names ; in England, Waterdragon, Horse Blobs, Water Blobs, and in Worcestershire, May Blobs, Butter Boats, King Cups, Water Bubbles, and also sometimes Crazies. Where the custom of carrying garlands about the villages on May-Day still survives, the one flower that is certain to be conspicuous among their constituents is the Marsh Marigold. With the Meadow Buttercups it shares the name of King Cups, and as in their case, the yellowness of butter is attributed to cows feeding upon it. But cattle only eat it when urged to do so by hunger, and then the consequences are usually disastrous. It is found in all the adjoining counties.

Var. b. **Guerangerii** (Boreau).

First Record. As below.

Lickey. 'Used to be abundant on the marshy ground at the foot of Bilberry Hill,' *Bagnall* (MS.).

The flowers of this variety are smaller, the sepals remote when expanded, and the follicles spreading. It has been suggested that it is the origin of the double-flowered form of gardens, which is very different in appearance and habit from the single-flowered type. The variety has been noticed in every bordering county except Salop.

TROLLIUS, Linn. 8.   (*Trol*, a globe, in old German.)

{42. 31. T. europaeus, Linn.   *Mountain Globe-flower.*

Casual. Moist meadows. Once recorded. P. June to August. *Top. Bot.* 65.

First Record. Moist meadows at the foot of Bredon Hill, *Lees, Illus. Nat. Hist. Worc.*, p. 167 (1834).

Avon. As above.

This, the only record for the county, was not acknowledged afterwards by

Mr. Lees, and in *Top. Bot.* the record has a mark of interrogation after it. Yet it seems to have been taken as unquestionable in the account of the distribution of the plant in Hooker's *Students' Flora*. It is recorded for Stafford and Salop, and, as likely to have been introduced, in Herefordshire.}

HELLEBORUS, Linn. 9. (ἑλλέβορος, the Greek name ; ἑλεῖν, to injure, βορά, food ; that is, poison.)

{*43.* 32. H. [viridis], Linn.   *Green Hellebore.*
Var. b. occidentalis (Reuter).
Denizen or possibly native. In woods, local and rare. P. February to April. *Top. Bot.* 32.
First Record. Orchard near Mr. Ballard's, Robinson's End, Malvern Chase, *Mr. Welles, Ballard,* in *Stokes's With. Bot. Arr.,* 2nd ed., p. 581.
Avon. Near Woodgate, Bentley, *Humphreys.*
Severn. One spot, lost, *Lees.* Warndon Wood, *Herb. Hast. Mus.*
Malvern. Two localities, *Lees.* Eardington near Tenbury, *Miss Ada Ballard,* possibly gone.
Lickey. Long Sawcroft Wood, Frankley, *Mathews.*
This plant has not been seen wild in Worcestershire for several years, the last time possibly at the Bentley locality, and requires recent confirmation. It maintains itself freely in shrubberies and gardens. Watson, *Top. Bot.,* does not give it for Worcestershire, but says it has been reported for the county, as it has been reported from all counties bordering upon it.}

44. 33. H. foetidus, Linn.   *Stinking Hellebore.*
Denizen or native. Woods and thickets. Rare. P. February to April. *Top. Bot.* 16.
First Records. Southstone Rock, *Mr. Gardner* ; Hagley, *Mr. Hickman, Purton,* vol. iii, pt. ii, p. 335 (1821).
Severn. Near Cotheridge, *Walcot.* The Old Hyde Farm, Elmley Lovett.
Malvern. Three localities, *Lees.* Lane near Bransford Chapel, *Baxter.* Southstone Rock, still there, 1907, *Rea.* Near Leigh Sinton. Leaves on a hedgebank at Lulsley, 1879, *Towndrow.*
Lickey. Hagley, as above. Lost, *Lees.*
Watson, *Top. Bot.* 17, gives this plant as native from Gloucester and Hereford, and reported from the other neighbouring counties. Popular names for this plant are Bear's-foot, Setterwort, and Oxhead. The well-known Christmas Rose is *H. niger,* the latter name being used to specify it, not from its white flowers, but from its black root.

ERANTHIS, Salisb. 10*. (ἦρ, spring, ἄνθος, flower.)

{*45.* 34. E. *hyemalis, Salisb.   *Winter Aconite.*
Alien. Plantations and parks. Rare. Not a native plant. P. January to March. Not in *Top. Bot.*
First Record. Planted in shrubberies, sometimes increasing, *Lees, Bot. Worc.,* p. xxviii.
Mr. Lees, in his *Table of Plants,* marks this as occasional in all the districts, but gives no definite locality in the body of his book. It has not lately been seen and requires recent confirmation.}

AQUILEGIA, Linn. 11.   (*Aquila,* an eagle, from the form of the petals.)

46. 35. A. vulgaris, Linn.   *Columbine.*
Native. Woods and dingles. Local. P. May to July. *Top. Bot.* 62.
First Record. Souston's (Southstone) Rock near Shelsley, *Ballard* ; *Stokes, With. Bot. Arr.,* 2nd ed., p. 562 (1787).
Severn. Several places, *Lees.* Shrawley Wood. Habberley Valley. From Furnace Coppice to Bewdley, *Jorden.* Road above Ribbesford Wood. On the bank of Dowles Brook in Wyre Forest. Near Lickhill, Bewdley, *Hickman.* Fenny Rough.
Malvern. Silurian eminences, *Lees.* Crews Hill Wood, Knightwick. Malvern Hills, *Lees.* Leigh Sinton.
Lickey. One place, *Lees.*
With white flowers this plant is found in Wyre Forest, Shrawley Wood, and near Habberley Valley, and also in the Malvern district. This plant is better known in the garden border than as a wild flower, but in places it occurs abundantly enough, especially in Wyre Forest, where it is one of the characteristic flowers. It is a well-known flower all over Europe, and the Italians give it the name of *Perfetto Amore,* which may be Englished into ' True Love '. The name Columbine may be found in the earliest literature, and is supposed to have arisen from the resemblance of the inverted flower to five pigeons clustered together ; and the horned nectaries acquired for it in past ages certain ribald allusions. Chapman, in 1605, calls it a ' thankless flower '. The name was applied by Parkinson to some of the Thalictrums, which he called ' Tufted Columbines ' ; and also in much earlier times to Vervain, among which Gerarde says in his *Herbal,* ' pigeons are delighted to be '. The plant occurs in all neighbouring counties.
[A. ALPINA, Linn.
This plant is recorded to have been gathered at the Lickey by Dr. Streeten. It has no claim to be a British plant, and *A. vulgaris,* which it much resembles, though it has larger flowers, no doubt was mistaken for it. The Columbine has not been seen at the Lickey for many years.]

DELPHINIUM, Linn. 12*.   (δελφίν, the dolphin, from the form of the flower.)

47. 36. D. AJACIS, Linn.   *Larkspur.*
D. Consolida, Sibth.
Casual, or at most colonist. Cornfields and waste places. A. July and August. In *Top. Bot.* the plant is not mentioned.
First Record. ' Consolida regalis, Tab. Delphinium, Gesn., Flos Regius, Dod., Larksheele. In a cornfield by Pershore in Worcestershire and thereabout frequently.' [*How's*] *Phyt. Brit.* (1650).
Avon. One place, extinct, *Lees.* East side of Craycombe Hill, sp. *Herb. Hast. Mus. Worc.,* gathered by *Rev. J. H. Thompson,* who added ' apparently wild '. Pershore, as above.
Severn. Holt, *Thompson,* sp. Churchill Viaduct, *Thompson,* sp. East side of Hartlebury Common, *Thompson,* sp. Two places, extinct, *Lees.* Near Grimley, *Edmunds.* Hartlebury Common, August 3, 1906.
Malvern. Two places, extinct, *Lees.* Rare, *Lees, Bot. Malv. Hills,* 1st ed. On the site of a former rubbish heap near Malvern, 1880, *Towndrow.*

Lickey. Chadwick Manor, Bromsgrove, 1805, *Carpenter.* One place, abundant, *Lees.*
The Lickey record occurs in the *Agriculture of Bromsgrove,* by J. Carpenter, who lived at the Manor, published soon after the enclosure of land in that part of the Lickey. He mentions that the newly enclosed fields were infested with the plant to which he gives the name ' Stavesacre ', and recommends its destruction as a most pernicious weed, while in his frontispiece he gives a picture of the plant. Mr. Lees had never heard of the plant as a cornfield weed in Worcestershire. The Lickey record given by him in his *Table of Plants* no doubt refers to the above locality. Although the plant is well established in Cambridgeshire, it is a doubtful native. Gerarde, besides Larkspur, gives it the names of Lark's heel, Lark's toes, Lark's claws, and ' Munkes hoode ' ; and the name Larkspur has been applied to some kind of ' Indian Cress ', probably the Nasturtium of our gardens. The Larkspur has occurred in the counties of Stafford, Salop, and Warwick.

ACONITUM, Linn. 13.   (ἄκων, a dart, from its use to poison such weapons.)

48. 37. A. Napellus, Linn.   *Monkshood, Wolfsbane.*
Alien. Meadows and hedgerows. Certainly introduced. P. July and August. *Top. Bot.* 7.
First Record. Meadow at Northfield, *Ick,* in *Mid. Counties Herald,* August 5, 1838.
Malvern. Spout Farm, Eastham, still flourishing, *Rea.* Near Tenbury, *Lees.* Brookside near Tenbury, *Thompson.* Apparently wild near Shelsley Church. Field on the Tanhouse Farm, Knighton-on-Teme. Teme side opposite Rochford Church. Brookside near Bayton.
With regard to the Eastham locality Mr. Lees says, ' it grows in a marshy spot among alders at the top of Eastham Hill, more than 500 feet in height, where it has as natural an aspect as it could have on the Continent, nor was any house near.' The Rev. Augustine Ley says of the plant, ' It was first detected as a native of Britain in this county at Little Hereford on the Salop border. It is now known to occupy the bushy banks of streams at many stations both in Herefordshire and Monmouthshire and can claim to be a truly native plant,' *Vict. County Hist., Hereford,* Botany, p. 411. But it has been a common garden flower from the earliest times, and botanical opinion is as a whole against its nativity ; there is no record of it as a wild plant until about ninety years ago. It is extremely poisonous, and fatalities have occurred from the root being mistaken for horse radish. It has occurred in all neighbouring counties.

2. BERBERIDACEAE, Vent.

BERBERIS, Linn. 16. (*Berberys* is the Arabic name meaning ' shell ', from the form of the flower.)

51. 38. B. vulgaris, Linn.   *The Barberry.*
Alien or denizen. Hedges. Spiny shrub. Local and rare. May and June. *Top. Bot.* 86.
First Record. In a hedge by the side of Comer Lane, *Lees, Illus. Nat. Hist. Worc.,* p. 160 (1834).

Severn. Very uncommon, *Lees.* Hedgerow near Upton. Titton Lane. Hartlebury Common. Comer Lane. Dynes Green Lane. Hedge about one mile from Bewdley on the Tenbury Road.
Malvern. Roadside four miles towards Leigh Sinton, *Westcombe.* Kent's Green, Powick, *Towndrow.*
The Barberry suffers from the enmity of farmers, who ruthlessly eradicate it from the hedgerows of their fields, for it has long been supposed to exercise a deleterious influence upon wheat. Modern investigation has proved the truth of this supposition and has discovered why it is so. The Barberry is the host in one of its stages of *Puccinia graminis,* Pers., which in another stage infests and destroys members of the grass tribe. The Barberry is found in most European countries, and has established itself on the shores of New England across the Atlantic to such an extent as to cause great annoyance to agriculturalists. Birds refuse to feed upon the sour berries, which, however, are sometimes made into conserves by housewives, and their rich red colour causes them to be frequently used as a garnish for dishes. The Barberry occurs in all neighbouring counties.

EPIMEDIUM, Linn. 17. (Name of doubtful origin, said to be from ἐπί, akin to ; μήδιον, a plant said to grow in Media.)

{*52.* 39. E. ALPINUM, Linn.   *Barrenwort.*
Alien. Woods and thickets. P. May and June.
First Record. In shrubberies near Tewkesbury, *Lees, Bot. Worc.* xxvii (1867).
Severn. One locality, *Lees.*
It is doubtful if this locality is in Worcestershire. Tewkesbury itself is in Gloucestershire, divided from Worcestershire by the Avon ; but on the other side of the river is our county, and in the close locality the southern parts of the three botanical districts of Avon, Severn, and Malvern converge. But ' near Tewkesbury ' is too indefinite a record upon which to lay absolute claim to the plant as a denizen of Worcestershire, and until recent confirmation is obtained the matter must remain doubtful. It certainly has not been seen in modern times. It is not mentioned in *Top. Bot.,* and has not been found in any neighbouring county unless indeed, the above record relates to Gloucestershire.}

3. NYMPHAEACEAE, De Cand.

NYMPHAEA, Linn. 18. (νύμφη, a water-nymph.)

53. 40. N. lutea, Linn.
(*Nuphar lutea,* Smith.)
Native. Rivers, ditches, and ponds. Fairly common, especially in the Avon and Severn districts. P. June to August. *Top. Bot.* 93.
First Record. As *Nuphar lutea,* River Avon at Pershore, *Ballard* ; *Mr. Waldron Hill* ; *Stokes, With. Bot. Arr.,* 2nd ed., p. 554 (1787).
Avon. Between Stoulton and Pershore on the Avon. Near Tewkesbury. Near Trench Woods. Abundant in the tributaries of the Avon, *Lees.* Brook at Huddington. Crowle Brook. Shell Brook.

Severn. Shrawley Wood. Blakedown. Broadwaters, near Kidderminster, *Scott.* Pansington Mill Pool, Hartlebury. Pools at Witley.

Malvern. Not in the streams near the Hills, *Lees.* Stanford Bridge. Hanley Castle, probably planted, *Towndrow.*

Lickey. Pools in Hagley Park; the Leasows, Halesowen; pools in the Cole at Yardley, *Mathews.* Blakedown Pool. Vine Mill Pool, Clent.

This plant Mr. Lees marks as 'not common' in the Malvern and Lickey districts; and as regards the first, the above is the only mention he makes of it, except to say that it occurs in the Leadon at Ledbury, which is in Herefordshire. A common name for the flower is 'Brandy Bottle', arising probably from the fact that its seed vessels are somewhat bottle-shaped, and the flower has a strong scent of brandy. The seeds are eaten in some countries, and taste like those of the poppy. This plant occurs in all neighbouring counties.

CASTALIA, Salisb. 19. (*Castalia*, a fountain of Parnassus, sacred to the Muses.)

55. 41. C. alba, Wood. *White Water Lily.*

(*C. speciosa*, Salisb. *Nymphaea alba*, Linn.)

Alien. Ponds, everywhere introduced. P. May to August. *Top. Bot.* 90.

First Record. As *Nymphaea alba*. In the Avon under Littleton Bank, according to Mr. George Perrott, Lees, *Illus. Nat. Hist. Worc.*, p. 166 (1834).

Avon. Elmley Castle, *Lees.*

Severn. Planted in pools at Glasshampton, *Lees.* Westwood Park. Shrawley Wood.

Malvern. Little Malvern, *Lees.*

Lickey. Two places, introduced, *Lees.*

Though common enough as a wild plant in several parts of England, and especially plentiful so near as Oxfordshire, where it abounds in the still tributaries of the Thames, the White Water Lily is nowhere found truly native in our county, which cannot lay claim to this queen of wild flowers. With our great grandmothers, in the days when modelling flowers in wax was an elegant pastime, this flower was most frequently taken as a model, and, with its green leaves placed on a sheet of looking-glass under a round glass shade, often seen as an ornament in living-rooms. It is a favourite with the poets. Cowper, longing for one of these flowers out of the river Ouse, had his wish gratified by his dog 'Beau', who dashed into the stream and brought one out, his efforts forming the subject of well-known stanzas. It grows only in still waters, and has been recorded as occurring in all neighbouring counties.

4. PAPAVERACEAE, Juss.

PAPAVER, Linn. 20. (Celtic *papa*, pap; because it is administered with pap to induce sleep.)

{**56. 42. P. *somniferum**, Linn. *Opium Poppy.*

Alien. Waste places. Rare. A. June to August. Not mentioned in *Top. Bot.*

First Record. Shores of the Severn near Worcester, Lees, *A. Florence, Strangers' Guide* (1828).

Severn. As above.

This is the only record for this alien in the country. Mr. Lees amplified it in Hastings's *Illustrations Nat. Hist. Worc.* by adding the words, 'below Worcester Bridge.' It has occurred in all the neighbouring counties except Gloucestershire.}

{Var. *glabrum, H. C. Wats.

First Record. As in this book.

Severn. Kepax Ferry, sp. *Herb. Vict. Inst.* Worcester. Several times observed, but regarded as an escape. Not given in 10th ed. *Lond. Cat.*}

57. 43. P. *Rhaeas, Linn. *Common Red Poppy.*

Colonist. Cornfields and waste places. Abundant, but chiefly limited to arable fields. A. May to August. *Top. Bot.* 106.

First Record. Not localized, Pitt, *Agric. Worc.*, p. 317 (1810).

This well-known plant is a glory of the countryside when it makes a whole field a sheet of scarlet, or when among its numerous flowers the golden blossoms of Corn Marigold (*Chrysanthemum segetum*) are intermingled, as they frequently are about Kidderminster and to the north of it. Sometimes country people call the flower Corn-rose, or Red-weed, Headache, or Cheesebowl. This last word is a variant of the old name for the plant, Chesboll, found as early as 1410, which possibly has nothing to do with either cheese or a bowl, but is connected with the obsolete name for an onion, Chiboll, which is given in a fifteenth-century vocabulary as 'poppy', while 'Chesbolle' is given for onion. The poppy is seldom found outside the cultivation of the soil; but it has followed our countrymen to our colonies, where its scarlet flowers are as conspicuous in the cornfields as with us. It is found in all neighbouring counties.

Var. b. *strigosum (Boenn.).

First Record. Hanley Castle and Malvern Wells, Towndrow, *Malvern Advert.* (1892).

Severn. Churchill, Kidderminster.

Malvern. As first record.

This variety differs from the type only in having the hairs on the peduncles all adpressed instead of spreading.

Var. c. *Pryorii, Druce.

First Record. Malvern Wells, Towndrow, *Malvern Advertiser*, 1892.

Malvern. Malvern Wells, Towndrow. Near Malvern, Townsend (MS.).

In this plant the hairs on the peduncle are more or less crimson in colour.

58. 44. P. *dubium, Linn. *Long Smooth-headed Poppy.*

Colonist. Cornfields and waste places, chiefly in arable fields. A. May to August. *Top. Bot.* 105.

First Record. Not localized, Pitt, *Agric. Worc.*, p. 317 (1810).

This plant is as widely distributed as is the foregoing, from which ordinary observers do not discriminate it. It is recorded from all neighbouring counties.

C

Var. *Lamottei, Bor.

First Record. Stakenbridge, Hagley, Mathews, *Trans. Worc. Nat. Club*, ii. 80 (1874).

In this variety the milky sap does not turn yellow on exposure to the air. It is general throughout the Malvern district, Towndrow (MS.), and in other parts of Worcestershire. It is not mentioned in *Lond. Cat.*, 10th ed.

59. 45. P. *Lecoqii, Lamotte.

Colonist. Fields and waste places, but chiefly on cultivated ground. A. May to August. *Top. Bot.* 25.

First Record. Old limestone wall near the Berrow Church, Lees, *Bot. Worc., Add. and Cor.* (1867).

Malvern. As first record. Barnard's Green, Towndrow.

This plant is distinguished by its sap being always yellow. Of bordering counties it has been observed in Gloucestershire and Warwickshire.

60. 46. P. *Argemone, Linn. *Long Rough-headed Poppy.*

Colonist or native. Cornfields and waysides. Not uncommon and widely distributed. A. May to July. *Top. Bot.* 88.

First Record. Not localized, Lees, *Bot. Malv. Hills*, 1st ed., p. 28 (1843).

Avon. Norton, near Evesham. Near Trench Woods.

Severn. Waste ground at Lower Hagley, W. Whitwell. Fields near Hartlebury Common. Warndon. Lower Wick, near Worcester. Stakenbridge. Grimley. Hartlebury Common.

Malvern. Ankerdine Hill. Malvern Wells.

Lickey. Cornfields, Mathews.

This is the only one of the scarlet poppies that can lay any claim to be a native plant, being sometimes, though seldom, found in an uncultivated district. The flowers are neither so large nor so brilliant as those of other field poppies, and there is a black spot at the base of their petals. It occurs in all the neighbouring counties.

61. 47. P. *hybridum, Linn. *Round Rough-headed Poppy.*

Colonist. Cornfields, very local. A. June to September. *Top. Bot.* 41.

First Record. Doubtful, in Pitt, *Agric. Worc.*, p. 317 (1810). First certain record, In a calcareous cornfield at Tibberton, about three miles on the eastern side of Worcester, 'where only I have seen it in the district,' Lees, *Bot. Worc.*, p. 39 (1867).

Severn. Tibberton, as above. Near Warndon, Westcombe.

This is much the rarest of the four red poppies. Possibly the calcareous cornfield spoken of by Mr. Lees as the home of this and several scarce flowers has not disappeared, but certainly modern research has failed to discover any locality corresponding to it, where the plants he mentions are growing in company at the present time. This poppy is more frequent in the south than elsewhere in England; it occurs in Gloucester and Hereford of the neighbouring counties.

GLAUCIUM, Hill 22. (γλαυκός, pale blue; from the hue of the foliage of the plants.)

{**63. 48. G. flavum**, Crantz. *Yellow Horned Poppy.*

Casual. Waste places. Once recorded. A. June and July. *Top. Bot.* 52.

First Record. As *G. luteum*. Once found on Liassic debris at Shipston-on-Stour, Lees, *Bot. Worc.*, p. xxviii.

Avon. As above.

Malvern. Once observed one hundred yards above Saint Anne's Well, Malvern, but regarded as an escape, *Rea.*

This is a plant of the seashore. Mr. Lees, in his *Table of Plants*, marks it as suspicious, and now lost. It is recorded for Gloucestershire only of the neighbouring counties, and that only for its western portion where the salt of the sea comes up the estuary of the Severn.}

G. phoeniceum, Crantz.

Casual. Once found. A. June and July. Not mentioned in *Top. Bot.*

First Record. Railway embankment, Malvern Common, August 3, 1899, Towndrow (MS.).

The petals of this flower are scarlet, not yellow. It has not occurred in any neighbouring county.

CHELIDONIUM, Linn. 24. (χελιδών, the swallow; because it is in bloom whilst the swallow is here.)

65. 49. C. majus, Linn. *Great Celandine.*

Denizen or native. Weedy banks and waste places, nearly always near houses. Generally distributed. P. April to September. *Top. Bot.* 96.

First Record. Hedges in Shrawley, Pitt, *Agric. Worc.*, p. 317 (1810).

Avon. Several places, Lees. Himbleton. Huddington. Crowle. Near Goosehill Wood. Norton, near Evesham.

Severn. Comer Lane, Worcester, Lees. Lincombe. Churchill, near Kidderminster. Kepax. Dines Green. Bromyard Road. Ombersley.

Malvern. Bransford. Witley. Madresfield. Old Hills. Newland.

Lickey. Clent. Danesford. Tardebigge. Harvington.

This is one of the plants which follow man. In damp spots near houses; on the banks against which the household debris is piled; or where some unconsidered drain conducts its contents to the air, there grows the Celandine. Where ruined walls and deserted buildings show where man once has dwelt, there also it is to be found. This is one of the numerous plants called Crazy by country people, and it is also called Lively Sally. The thick yellow juice in the brittle stems is a popular cure for warts, and diluted with milk is supposed to be good for some affections of the eye. Pliny says that the virtues of the plant in this direction were discovered by the swallows, who anointed the eyes of their young ones with its juice; whence is derived the plant's Greek name. It is recorded in all neighbouring counties.

Var. b. *laciniatum, Mill.

First Record. As in this book.

Severn. Lane at Sandford, sp. *Herb. Hast. Mus.*, Worc.

This variety is characterized by the segments of the leaves being deeply pinnatifid and the lobes incise serrate.

5. FUMARIACEAE, De Cand.

NECKERIA, Scop.　(Named in honour of the botanist Necker.)

N. BULBOSA, N. E. Br.

(*Corydalis*, DC. *Capnoides solida*, Moench. *Fumaria solida*, Miller.)

Alien. A garden escape. P. April and May. Not in *Top. Bot.*

First Record. As *Fumaria solida*. Abberley Woods, *Mr. Hickman*, Surgeon. Ludlow, *Purton, App.*, vol. iii, p. 58 (1821).

Malvern. Abberley, as above. Near the Rhydd, *Towndrow* (*MS.*).

This plant has occurred in Staffordshire, Shropshire, and Warwickshire.

CORYDALIS, Vent. 25.　(κορυδός, the crested lark, from the appearance of the spur of the flower.)

66. 50. C. *lutea, DC.　*Yellow Fumitory.*

(*Fumaria lutea*, Linn. *Neckeria lutea*, Scop.)

Alien. On walls and rocks. P. May and June. Not in *Top. Bot.*

First Record. As *Fumaria lutea*. Found by the Rev. W. S. Rufford on Broadway Hills, Gloucestershire (error for Worcestershire), *Purton, Midl. Fl.*, p. 327 (1817).

Avon. Broadway, as above.

Severn. On sandstone, the Worcester side of Holt. Redstone Rock.

Malvern. Limestone wall near the church at Abberley, *Perry, Nat. Hist. Mag.*; persistent there, *Lees.* Malvern Wells.

Lickey. Old wall in Edgbaston lane, Moseley, *Ick.* This may be in Warwickshire, as the lane is in both counties.

Neither of the above plants have any claim to be considered native. The latter is a strong grower, frequently found in gardens, and would have little difficulty in maintaining itself if it escaped. It is a native of rocky hills in Southern Europe; and it is recorded from all adjoining counties.

67. 51. C. claviculata, DC.　*Climbing Fumitory.*

(*Fumaria claviculata*, Linn. *Neckeria claviculata*, N. E. Br.)

Native. Woods and coppices. Local and rare. A. June to August, and sometimes later. *Top. Bot.* 89.

First Record. As *Fumaria claviculata*. In rough stony places by the side of Malvern Hill above Great Malvern Town, *Nash, Hist. of Worc., Int.*, p. lxxxix (1781). 'North Hill, Malvern,' is given as a first record for *C. claviculata*, in *Bot. Rec. Club Rep.* for 1883 (pub. 1884); but it was not new.

Severn. Two localities, *Lees.* Winterdyne, Bewdley, *Jorden.* Habberley Valley. Pettiford Lane, near Park Hall, Kidderminster. Hurcott Wood, *Mathews.*

Malvern. Malvern Hills, as above. British Camp. North Hill, Malvern to the British Camp.

Lickey. Dingle near the Lye, *Scott.* Lower Lickey. Rubery Hill. Wood on Clent Hill, *Amphlett.* Alvechurch, *Mathews.* Rednall Hill. Church Wood, Lickey.

This plant, though rare in England, is quite common in Scotland. It is long and straggling, delicate in appearance, and its flowers are pale yellow. Of the neighbouring counties it has occurred in all.

FUMARIA, Linn. 26.　(*Fumus*, smoke; but the allusion is doubtful. The ancients thought the smoke of this plant dispelled evil spirits.)

68. 52. F. capreolata, Linn.　*Ramping Fumitory.*

(*F. pallidiflora*, Jord.)

Colonist. Fields and waste ground. Rather uncommon. A. July to October. *Top. Bot.* 40.

First Record. Dingle near Lye, *Scott, Stourbridge*, p. 544 (1832).

Avon. Norton, near Evesham.

Severn. Many places, *Lees.* In a hedge at Shrawley and near Abberley, *Lees.* Upper Wick near Worcester. Bewdley. Near Perdiswell Lodge. Fields south of Hartlebury Common.

Malvern. Powick.

Lickey. Sandy lanes near Yardley, *Miss M. A. Beilby.* Possibly in Worcestershire. Hedgebanks, rare, *Mathews.*

The word Fumitory means 'earth smoke', and the plant has an equivalent name in most European countries. Why this name became attached to the plant is doubtful; probably because it comes up out of the earth like smoke, covering it closely as smoke does. No particular superstition is attached to the plant, nor is there any popular or ancient use of the smoke of the plant known. Nor is its odour particularly like smoke. In early days, however, it was supposed to have some medicinal virtue, and was used in the medicine of the village. *F. capreolata* has occurred in all neighbouring counties except Stafford.

F. confusa, Jord.

Colonist. Waste ground. Rare. A. July to September. *Top. Bot.* 32.

First Record. Malvern Link, *Towndrow, J. of B.*, vol. xxii, p. 39.

Malvern. Malvern. Barnard's Green. Malvern Wells, *Towndrow* (*MS.*). Malvern College Grove. Malvern Link.

This sub-species of *F. capreolata* is eliminated from *Lond. Cat.*, 10th ed.

71. 53. F. Boraei, Jord.

First Record. *Towndrow, Sup. to Top. Bot.*, p. 3, from *J. of Bot.*, 1905.

Malvern. Allotments, Pickersleigh, Malvern, October 7, 1907, *Towndrow.* The plant found by Mr. Towndrow was var. d. *serotina*, Clavaud.

72. 54. F. muralis, Sonder.

Colonist. Hedgebanks and arable land. Rare. A. July and August. *Top. Bot.* 5.

First Record. Arable land at Malvern; Dripshill Wood, *Towndrow* and *Rendall, J. of B.*, Dec., 1900.

Malvern. As above. Malvern Link, *Towndrow.*

Lickey. Recorded in error by Mr. Mathews in his *Bot. Clent and Lickey Hills*, in 1881. See *Mid. Nat.* xv. 277, where he says, 'No form of Capreolate Fumitory has been found within this district.'

The plant has occurred in Gloucestershire and Warwickshire.

74. 55. F. officinalis, Linn.　*Common Fumitory.*

Colonist. Cultivated fields, garden ground, anywhere where the soil is turned. Common. A. May to October. *Top. Bot.* 106.

First Record. Not localized, *Lees, Bot. Malv. Hills*, p. 34 (1843).

This is one of the common cornfield plants of Europe, being found from

Lapland in the North to Greece in the South. In England it is a weed of very long standing, as its seed was discovered by Mr. Clement Reid along with those of other weeds of cultivation among Neolithic remains in Scotland. This plant has bitter and tonic qualities, and was used in cases of jaundice and indigestion. It is sometimes called 'Devil's Bit'. It occurs in all neighbouring counties.

76. 56. F. Vaillantii, Loisel.

Colonist. Cultivated fields and waste places. Very rare. August and September. *Top. Bot.* 15.

First Record. As here below.

Malvern. Arable land, Pickersleigh, Malvern, *Towndrow*, July 16, 1896. Recorded only for Staffordshire of the neighbouring counties.

77. 57. F. parviflora, Lam.

Colonist. Cultivated fields. Very rare. A. June to September. *Top. Bot.* 22.

First Record. Ballast hole, Ripple, Worcestershire, collected by Mr. C. C. Waterfall, *Towndrow, J. of B.*, 1905.

Severn. Ripple, as above. A weed in White's Nurseries, July 16, 1896. Of the neighbouring counties this plant has occurred only in Gloucestershire.

6. CRUCIFERAE, Juss.

CHEIRANTHUS, Linn. 28*.　(*Keiri*, Arabic for Wallflower; or from χείρ, the hand, and ἄνθος, flower.)

80. 58. C. *Cheiri, Linn.　*Wallflower.*

Denizen. More or less naturalized on old walls. P. April to June. Not in *Top. Bot.*

First Record. The ruins of the old church and abbey walls of Evesham, *Purton, Midl. Flor.*, p. 311 (1817).

Avon. Walls at Evesham, *Lees.*

Severn. Walls at Worcester, *Lees.* Bewdley.

Malvern. Little Malvern Priory, *Lees.* Abberley.

Lickey. The Manor Abbey, Halesowen, *Mathews.*

The Wallflower is indigenous among rocks in the Mediterranean region. Elsewhere in Europe it is a denizen, and more or less naturalized, the result of its general abundance as a garden flower. With the carnation and the white garden stock it shares the common name of Gillyflower, which possibly in its case alone is shortened to 'Gillies'. But Gillyflower as a name does not rightly belong to it, being properly that of the Clove Pink, Clove alluding to the nail-like shape of the flower, adopted from the French *clou*; while Gilofre, from which Gillyflower is a later corruption, being cognate with the Latin name *Caryophyllum*, means a Clove, the spice, also so called from its resemblance to a nail. Having thus been corrupted, the name has passed on to various highly scented flowers, which have no connexion either with the Spice or the 'Clove Pink'. As in the case of nearly every plant that grows, various virtues were attributed to it by our forefathers; and the flowers were supposed to be good for apoplexy and palsy, and singularly efficacious in cases of gout and pains in the joints and sinews. The plant, more or less naturalized, occurs in all the neighbouring counties.

RADICULA, Hill 29.　(Diminutive of *radix*, a root.)

81. 59. R. Nasturtium-aquaticum, Rendle and Britten.　*Watercress.*

(*Sisymbrium Nasturtium*, Linn. *Nasturtium officinale*, R. Br.)

Native. Marshes, brooks, and slow streams. General. P. April to October. *Top. Bot.* 112.

First Record. As *Sisymbrium Nasturtium*. Watercress, Vale of Severn, *Marshall, Pitt, Agric. Worc.*, p. 317 (1810).

This plant is far better known upon the breakfast table than in its native home, for the flowers are inconspicuous and its habit lowly, and with it usually grows lush vegetation of several kinds. It may be found over a great portion of the Northern hemisphere; it is at home among the hills of India and in the streams of Northern Africa, and it is a denizen of North America, and many of our colonies. It is largely cultivated near towns, its pungent leaves having long been used for salad, and *Poulet au Cresson*, among other things. It is known among country people as Water Cissies. Its leaves contain a notable amount of sulphur, and turn brown in the sun. It occurs commonly in all neighbouring counties.

Var. b. **siifolia**, Rendle and Britten.

First Record. Hanley Castle, *Towndrow, Malvern Advertiser*, 1892.

Malvern. Hanley Castle.

This is a luxuriant state of the type, with a stout stem, and its terminal leaflet is lanceolate, and not conspicuously broader than the lateral ones.

82. 60. R. sylvestris, Druce.　*Creeping Yellow Cress.*

(*Sisymbrium sylvestre*, Linn. *Nasturtium sylvestre*, R. Br.)

Native. Wet meadows, river-banks. Local. A. May to October. *Top. Bot.* 64.

First Record. As *S. sylvestre*. On the banks of the Severn near Worcester, *Stokes, With. Bot. Arr.*, 2nd ed., p. 691 (1787).

Avon. One place, *Lees.* Badsey, *Rufford.*

Severn. Abundant on the banks of the river at Worcester, *Lees.* Severn at Lincombe. Northwick Brick Yard. Eyemore. Corndon Bridge.

Malvern. Many places, *Lees.* Banks of Teme. Near Upton-on-Severn.

This plant varies in the length of the pods and their pedicels, and sometimes the leaves are more deeply cut than in the ordinary type. It occurs in all neighbouring counties.

83. 61. R. palustris, Moench.　*Yellow Cress.*

(*Sisymbrium terrestre*, Sm.)

Native. Rivers, ditches, meadows, and ponds. Local. A. May to October. *Top. Bot.* 84.

First Record. As *Nasturtium terrestre*. On the banks of the Lodge Pools, Kidderminster, *Perry, M. N. H.*, vol. iv, p. 450 (1831).

Avon. Bredon Hill, *Mathews.* One place, *Lees.*

Severn. Meadow at Ribbesford, *Mathews.* Kidderminster, as above. Near Diglis Lock. Field above Pitchcroft. Side of Bromyard Road. West bank of Severn, below Worcester Bridge. Brake Mill Pools, Hagley. Sutton Pool.

Malvern. Many places, *Lees.* Longdon Marsh. New Pool, Malvern Wells.

Lickey. Harborne Reservoir, *Mathews*. Occasional, *Lees*. Lodge Ford. Cradley. Birchell Pool. Halesowen. Bittell Reservoir.

This plant may be known from the last by its smaller flowers and more swollen pod. It occurs in all neighbouring counties.

84. 62. R. amphibia, Druce. *Great Yellow Cress.*

(*Sisymbrium amphibium*, Linn. *Armoracia amphibia*, Koch.)

Native. Banks of rivers, ditches, wet places. Not very common. P. May to September. *Top. Bot.* 48.

First Record. As *Sisymbrium amphibium*. Side of the Severn near Worcester, *Stokes, With. Bot. Arr.*, 2nd ed., p. 692 (1787).

Avon. Several places, *Lees*. Cleeve Hill.

Severn. Ditches by the side of Pitchcroft, Worcester, *Lees*. Lincombe, *Mathews*. Northwick Brickfields. Viaduct Pool, Blakedown. Banks of Canal near Worcester. Banks of Severn above Pitchcroft.

Malvern. Banks of Teme near to its junction with the Severn. Abundant in the Teme at Bransford, *Towndrow*.

Lickey. Bittell Reservoir. Tardebigge, *Humphreys*. Common along the Cole, Yardley, *Mathews*.

This plant varies extremely in the shape and cutting of its leaves. It is remarkable for the long stringy roots which spring from the lower joints of the stem. It is found in all adjoining counties.

BARBAREA, B. Br. 30. (Dedicated to St. Barbara.)

85. 63. B. vulgaris, Ait. *Winter Cress.*

Native. Hedgebanks, watersides, and damp places. General. B. April to July. *Top. Bot.* 102.

First Record. Not localized, *Lees, Bot. Malv. Hills*, 1st ed., p. 33 (1843).

This generally distributed plant is often called Hedge-Mustard, and sometimes Yellow Rocket and Herb St. Barbara. Its leaves are bitter and nauseous, yet were eaten in winter and early spring by our forefathers. Perhaps in those days lettuces and radishes were not easily come by! Cows will eat it, but it is refused by horses. It is recorded from all neighbouring counties.

86. 64. B. arcuata, Reichb.

First Record. In this book.

Avon. Roadside between Pershore and Evesham, *Westcombe*.

Severn. Along the river from Kempsey to Upton, *Towndrow* (*MS.*). Porter's Mill.

Malvern. Upton-on-Severn, *S. H. Bickham*.

This plant is with difficulty distinguished from the above. See Bab., *Manual*, 9th ed., p. 27. It has larger flowers, laxer racemes, more persistent petals, and longer styles, and is yellower green in colour. This plant has been found also in Warwickshire.

87. 65. B. stricta, Andrz.

Native. Damp places. Rare. B. May to July. *Top. Bot.* 12.

First Record. Near the Severn, Worcester, *Towndrow, J. of Bot.*, vol. xviii, p. 374, 1880.

Severn. Not uncommon along the line of the Severn, extending apparently but a short distance west of that river, *Towndrow*. Railway embankment Wyre Forest. Near the Ferry, Arley. Bromwich. Pixham.

Malvern. Abundant at Pool Brook, Upton-on-Severn.

Lickey. The Manor Abbey, Halesowen, *Mathews* (1893).

This plant is, like the above, a segregate of *B. vulgaris*, from which it differs in its much smaller flowers, and in the smaller size of the upper pair of lobes of the lyrate lower leaves, while its flowering raceme is close instead of lax. It has occurred in all neighbouring counties except Staffordshire.

88. 66. B. *intermedia, Boreau.

Colonist or casual. Cultivated fields. B. April to June. *Top. Bot.* 14.

First Record. Newland, *Towndrow, Malvern Advertiser*, 1892.

Severn. Severn bank, Grimley, *Towndrow*.

Malvern. Malvern. Newland. Blackmore Park. Malvern Link, *Towndrow*.

Lickey. Manor Abbey, Halesowen.

This is distinguished by its all pinnatifid upper leaves. It is intermediate between *B. stricta* and *B. verna*, the following plant. It has occurred in the counties of Gloucester, Hereford, and Warwick.

89. 67. B. verna, Aschers. *American Cress.*

(*B. praecox*, R. Br.)

Alien. Waste ground. Rare. B. April to October.

First Record. Roadside by New Pool. Gathered by Mr. T. Westcombe, 1846, *Lees, Bot. Malv. Hills*, 2nd ed., p. 64 (1852).

Severn. Broadheath. Near Worcester, *Lees*. Brake Lane, Hagley, 1904, *W. Whitwell*. Broadheath. Ombersley. Habberley Valley. Sandy field east of Kidderminster Waterworks. A dozen or more plants between the bridge over the canal and the bridge over the Salwarpe at Hawford, 1907.

Malvern. Welland. Malvern. Abundant on the railway at Malvern, Malvern Link, and Malvern Wells. Hedgebank by roadside from Malvern to Welland Common.

This plant is everywhere a garden escape, being an excellent salad, and formerly widely cultivated ; and in spite of its name is as alien in America as it is with us. It is a native of damp grassy places in Southern Europe. It has been observed in all bordering counties.

ARABIS, Linn. 31. (Named from Arabia.)

94. 68. A. hirsuta, Scop. *Hairy Rock-cress.*

Native. Walls and dry banks. Rare. B. May to August. *Top. Bot.* 98.

First Record. Worcester, *Watson's Top. Bot.* 2nd ed., p. 42 (1883).

Malvern. On the railway at Malvern Link, *S. H. Bickham*.

Lickey. In the district, *Bagnall*.

Although this plant has the high comital number of 98, until its discovery at Malvern Link, it had been mentioned as occurring in this county only in *Top. Bot.* In that book Worcester is given without any authority, and therefore its occurrence must have been personally brought to the notice of the compiler of that book. Mr. Lees, in his *Table of Plants*, does not notice it, nor are there any specimens in the Herbaria of the older local botanists. The plant has been recorded from all neighbouring counties.

96. 69. A. glabra, Bernhardi.

(*A. perfoliata*, Lam. *Turritis glabra*, Linn.)

Native. Common in the sandy districts of the north of the county ; elsewhere rare. B. May to July. *Top. Bot.* 38.

First Record. Near Stourbridge, *Rev. W. Wood, Bot. Guide*, 1805, vol. ii, p. 656.

Severn. About Kidderminster ; Habberley Valley ; Lea Castle, *Lees*. Hagley Brake Lanes near Stourbridge, *Purton*. Wollaston, *Scott*. Bewdley. Hartlebury Common. Barnett Hill near Kidderminster. Pedmore. Roadside north of Stourport. Sandstone Quarry, Bishop's Wood, Hartlebury. Park Hall. Danesford.

Malvern. One place, *Lees*.

Lickey. Chaddesley Corbett. Clent, sometimes plentiful.

This unmistakable plant is capricious in its habitats. Somewhere in the North Severn district it is abundant every year, seldom is it abundant in the same place a second year. It will be perhaps represented by one or two spikes, and afterwards disappear, only to spring up again in quantity on some hedgeside-bank half a mile away. In the closing years of last century it studded the western bank of the railway cutting just outside Kidderminster Station towards the north. Not a plant is to be seen there now. The above records, to which numbers could be added, can only be taken to show where it has been seen, not where it is likely to be found at the present time, but in the neighbourhood of which some year or other it will probably recur. The plant is not recorded from Herefordshire, only of bordering counties.

CARDAMINE, Linn. 32. (From κάρδαμον, a kind of cress.)

97. 70. C. amara, Linn. *Large Bitter-cress.*

Native. River-banks. Osier beds. Marshy places by pools. Locally common. P. March to June. *Top. Bot.* 76.

First Record. On the banks of the Avon below Great Comberton, plentifully, *Nash, Worc., Int.*, p. lxxxix (1781).

Avon. Rather uncommon, *Lees*. Norton, near Evesham. Fladbury.

Severn. Wyre Forest, *Jorden*. Lincombe. Mill near Hartlebury Common. Stanklin. Redstone. Severn side, Stourport. Brookside, Cotheridge. Willow bed below Worcester. Fenny Rough. Blakedown. Severn side, Hallow.

Malvern. Rather uncommon, *Lees*.

Lickey. Wychbury Wood. Moseley, Halesowen Abbey, *Lees*. Twyland Wood, Uffmoor Wood, *Mathews*. Belbroughton. Kingsnorton. Northfield. Bittell Reservoir, *Humphreys*. Manor Farm, Halesowen. Alvechurch.

Larger than the next species, with white flowers and purple anthers. It is exceedingly bitter in taste. It is found in every neighbouring county.

98. 71. C. pratensis, Linn. *Meadow Bitter-cress.*

Native. Damp meadows, common everywhere. P. July. *Top. Bot.* 112.

First Record. Var. 2, double-flowered, in a field S.W. of the Taphouse at Hagley, *Withering*, 4th ed., p. 568 (1801).

This common plant, called by country people Lady's Smock, Smock Frock, Butcher's Smock, Lady's Petticoat, or Cuckoo Flower, from its blossoming

when the cuckoo is heard, is very plentiful in moist meadows and damp woods. The flowers are pale lilac, and sometimes double. The double form is still to be found in the locality indicated in the first record. It is widely distributed through the northern regions of Europe, Asia, and America, and a similar plant occurs in Tasmania. It occurs in every neighbouring county.

99. 72. C. hirsuta, Linn. *Hairy Bitter-cress.*

Native. Walls, ditch-banks. Common and widely distributed. A. March to August. *Top. Bot.* 110.

First Record. On the banks of the Avon below Great Comberton, *Nash, Hist. Worc., Int.*, p. lxxxix (1781).

This plant is a common weed, and found everywhere, sometimes small and depauperated, and at other times growing to some size. It is not, however, so common a plant in Ireland. It occurs in every bordering county.

100. 73. C. flexuosa, With.

(*C. sylvatica*, Link.)

Native. Damp shady places and brooksides. A., B. or P. March to August and in the autumn. *Top. Bot.* 104.

First Record. As *C. sylvatica*. Not localized, *Lees, Bot. Malv. Hills*, 1st ed., p. 32 (1843).

This plant in the earlier botanies was placed as synonymous with the preceding one. It possesses six stamens instead of the usual four of *C. hirsuta*, and is altogether larger and laxer. It occurs in all bordering counties.

101. 74. C. impatiens, Linn. *Narrow-leaved Bitter-cress.*

Native. Moist banks, shady places. Very local. A. May to August. *Top. Bot.* 28.

First Record. On loose earth thrown up from a quarry above Lench Ford, nearly opposite Shrawley, and in Cliffy Wood near Hanley, *Stokes, Stokes's With. Bot. Arr.*, 2nd ed., p. 685 (1801).

Avon. Norton, near Evesham. Crowle. Fladbury. Pershore.

Severn. Shrawley Wood. Severn banks, Lincombe. Wyre Forest, *Jorden*. Powick Bridge, *Lees*. Fenny Rough. Near the mouth of the Teme. Willow bed, Diglis. Shrawley Wood. Hallow. Lincombe. Near North Wood.

Malvern. Malvern Hill, *Lees*. Near Ankerdine. Longdon Marsh. Knightsford Bridge. Powick. Bransford.

Lickey. Wychbury Wood. Brinks Pool dam, Clent.

The plant is not at all common, though usually found in quantity at its localities. It is easily recognized by the stipule-like appendages which embrace the stem at the base of the leaf stalk. It appears to have been recorded from all neighbouring counties.

ALYSSUM, Linn. 33*. (ἀ, not, λύσσα, frenzy, because of its curative qualities.)

103. 75. A. incanum, Linn.

Alien. Waste places. Very rare and not permanent. A. or B. July to September.

First Record. In this book.

Severn. Sandy field, Leap Gate, Hartlebury, *Miss Ladbury*, 1896.

Malvern. Cherkenhall, Leigh, *Towndrow, Malvern Advertiser*, January 20, 1894.

A native of Central Europe and Siberia. In this country it shows a tendency to spread along railways and roads. This plant has been met with in Warwickshire.

104. **76. A.** ALYSSOIDES, Linn.

(*A. calycinum*, Linn.)

Colonist or alien. Waste places. Rare. A. May to August.

First Record. Railway embankment at Astwood near Worcester, *Lees, Bot. Worc.*, p. xxix (1867).

Severn. As first record. Near Kidderminster, 1858. Rubbish heap at Hoo Mill, *Thompson*. Leap Gate, Hartlebury, *Miss Ladbury*. Sutton Common, near Kidderminster.

Malvern. Field near Halfway Inn, between Worcester and Malvern. Mill Meadow, Malvern Link, *Towndrow, Malvern Advertiser*, November 14, 1896.

A native of dry stony ground in Central Europe. In this country it is most frequently seen in clover or cornfields, sometimes acquiring some degree of permanence. It appears to have been met with in every neighbouring county except Staffordshire.

105. **77. A.** MARITIMUM, Lam.

(*Koniga maritima*, Br.)

Alien. Waste places, usually near the sea. A. May to August.

First Record. As *Koniga maritima*. Once found growing on the west bank of the Severn below Worcester Bridge by Mr. James Goodman, *Lees, Illus. Nat. Hist. Worc.*, p. 169 (1834).

Severn. As first record. Sutton Common, near Kidderminster, *Rev. J. H. Thompson.*

Malvern. Near Malvern, *Lees.*

This is a favourite in gardens, whence it has sometimes escaped and established itself as a weed of cultivation. Its native home is around the Mediterranean, where it is very abundant. It does not appear to have been seen outside a garden in any neighbouring county.

EROPHILA, DC. 35. (ἔαρ and φιλῶ, from flowering in spring.)

110. **78. E.** VERNA, E. Meyer. *Vernal Whitlow-grass.*

(*E. vulgaris*, DC. *Draba verna*, Linn.)

Native. Walls, dry banks, gravelly fields. Many places throughout the county, abundant in the Avon district. A. February to May. *Top. Bot. 109.*

First Record. As *Draba verna*, not localized, *Lees, Bot. Malv. Hills*, 1st ed., p. 32 (1843).

Avon. Plentiful on walls on Bredon Hill. Blockley.

Severn. In a garden at Lower Hagley, 1903 and 1904, *W. Whitwell.* Fenny Rough. Park Hall. Redstone. Hartlebury Common.

Malvern. Abundant on the hills, *Lees.* Ankerdine Hill. Abberley Hill. Malvern Wells Common.

Lickey. On walls and banks, very common, *Mathews.*

This plant takes many various forms, which may be divided into broad-fruited and long-fruited plants. M. Jordan described fifty-three species, and many of them kept constant under cultivation, even for twenty years. The children in the neighbourhood of Stoulton call it Draba as the popular English name. It is found in all bordering counties.

111. **79. E.** PRAECOX, DC.

Native. Walls and dry places. Very rare. A. February and May. *Top. Bot. 37.*

First Record. In *Worc. Nat. Club Trans.* ii. 7, as being found by Mr. Towndrow on Malvern Common, 1897.

Malvern. As first record.

This plant is recorded only from Warwickshire of our neighbouring counties. It differs from the preceding plant by its more orbicular pod.

COCHLEARIA, Linn. 36. (*Cochlear*, a spoon, from the shape of the leaves.)

[**118.** **80. C.** GROENLANDICA, Linn. *Top. Bot.* 10.

This plant of the seashore is not likely to occur in Worcestershire. Yet it was recorded among Malvern plants by Lees, *Worc. Misc.*, 1829-30. Its locality was said to be a marshy tract, somewhere on the North Hill ; it was never afterwards acknowledged by Mr. Lees.]

120. **81. C.** *ARMORACIA, Linn. *Horse-radish.*

Armoracia rusticana, Gaertn. Mey and Scherb.

Denizen. Riversides, waste places. Several places throughout the county. P. May to September.

First Record. On the banks of the Severn, truly wild, *Lees, Illus. Nat. Hist.*, p. 169 (1834).

Everybody knows the Horse-radish, and in spite of Mr. Lees' assertion above, it is not 'truly wild' anywhere in England, though often seen in apparently wild conditions. Once introduced into a garden the exceeding vitality of its root makes it most difficult to eradicate, and often the site of an old garden long converted to other use is betrayed by the leaves of this plant. It is a native of Eastern Europe, and in this country hardly ever ripens seed. It has been observed in apparently wild conditions in all neighbouring counties.

HESPERIS, Linn. 37*. (ἕσπερος, the evening, when the scent is powerful.)

121. **82. H.** *MATRONALIS, Linn. *Dame's Violet.*

Alien. Riversides and waste ground. Rare. P. May to July.

First Record. Occurring sometimes, but obviously a garden outcast, *Lees' Cat., New Bot. Guide*, 1835.

Severn. Bevere Island, *Rea.* Roadside west of Stourport.

Malvern. Between Bransford and Powick.

This native of Southern Europe and Central Asia has often been considered a British plant. It is such in so far as it is a familiar garden flower, but no further. It has been noticed as escaped in every bordering county.

SISYMBRIUM, Linn. 38. (Etymology doubtful. σισύμβριον, a species of Cress.)

122. **83. S.** THALIANUM, Gay. *Common Thale-cress.*

(*Arabis thaliana*, Linn.)

Native. Sandy fields, heaths, and walls. General. A. April and May. *Top. Bot.* 105.

First Record. As *A. thaliana*. Not localized, *Lees, Bot. Malv. Hills*, 1st ed., p. 33 (1843).

This little plant is so generally distributed that no localization is necessary. It seldom grows more than six inches high. It is found in every bordering county.

123. **84. S.** OFFICINALE, Scop. *Common Hedge-mustard.*

Native. Roadsides, waste places, and arable ground. Common everywhere. P. April to September. *Top. Bot.* 111.

First Record. Not localized, *Lees, Bot. Malv. Hills*, 1st ed., p. 33 (1843).

Everybody knows this plant, which, stiff and angular, may be seen all the summer long, grey with the dust of the road beside which it loves to grow. Its flowers are small and yellow, at the ends of its rigid branches, against which its seed pods lie closely adpressed. It has had much repute as a remedy for hoarseness and weak lungs ; but its flavour is extremely disagreeable, as well as biting, as all mustards are. It is found in all neighbouring counties.

Var. b. leiocarpum, DC.

First Record. As in this book.

Malvern. Malvern Link, North Malvern. Near Upton-on-Severn, 1897, *Towndrow.*

Severn. Near Hoo Mill. Ombersley. Kempsey.

This variety is known by its glabrous pods. It has not been recorded from neighbouring counties.

124. **85. S.** PANNONICUM, Jacq.

Alien, casual. Waste places. Very rare. A. or B. July.

First Record. One plant in a garden, Malvern Link, *Towndrow, Malv. Advert.*, December 17, 1892.

Malvern. As first record.

Severn. Hartlebury Common, *Humphreys.*

The home of this plant is probably Eastern Europe and about the Caspian Sea, where it is very plentiful. It has of late years been seen by Mr. Humphreys on Hartlebury Common, where it appears to be increasing, but as yet it has not been noticed in any of the bordering counties.

125. **86. S.** SOPHIA, Linn. *Flix-weed.*

Native. Waysides, sandy fields. Local and rare. A. June to August. *Top. Bot.* 64.

First Record. Near Stourbridge, *Ick, Mid. Count. Herald*, Aug. 5, 1838.

Severn. As first record. About Kidderminster, 'almost unknown elsewhere in the county,' *Lees.* Habberley Valley. Wyre Forest. By the roadside between Habberley and Kidderminster. Northwick,

Worcester, *Westcombe.* Hartlebury Common, *Humphreys.* Common Hill near Worcester. Banks of Severn.

Malvern. Leigh Mill.

A characteristic of this plant is that the leaves are divided into narrow segments unusual in the tribe to which it belongs. In old days it had some medicinal virtue, at the present day unknown, as it was called 'Sophia Chirurgorum', the Wisdom of Surgeons, from which appellation it derives its specific name. Since it delights in sand the Stourbridge locality may be claimed for the Severn district, which in the neighbourhood of Stourbridge is of that formation. It is found in all bordering counties.

126. **87. S.** *IRIO, Linn. *London Rocket.*

Alien. Waste places. Very rare. A. May and June. *Top. Bot.* 1.

First Record. In *Lees' Catalogue, New Bot. Guide*, 1835. First certain record, Midland Siding, *Towndrow*, 1896.

Malvern. As first certain record.

Mr. Mathews considers Lees' record to be an error, and it is not mentioned in his Botany. Considering, however, that the plant is a weed of the waysides throughout Europe, it is possible that Mr. Lees may have seen a casual plant. It gains its name from its appearance about the streets of London after the great fire of 1666 ; but it had been seen in the suburbs in preceding years, for both Merrett and Ray so mentioned it. Parkinson in 1640 did not know it as an English plant. It has not been observed in bordering counties.

127. **88. S.** ALLIARIA, Scop. *Hedge Garlic, Jack by the Hedge, Sauce Alone, Snakes' Food.*

(*Alliaria officinalis*, Andrz. *Erysimum Alliaria*, Linn.)

Native. Hedgebanks, gardens. Common and widely distributed. B. March to June. *Top. Bot.* 99.

First Record. As *Erysimum Alliaria*. Not localized, *Lees, Bot. Malv. Hills*, 1st ed., p. 33 (1843).

In Mr. Lees' *Table of Plants* this is mentioned as *Alliaria officinalis*. It is well-named after the onion tribe, for its odour when bruised is as powerful as garlic. Cows and sheep will eat it, but horses and goats refuse it, and it imparts its flavour to milk. It had some repute for medicinal virtues in old times, and even yet in villages its leaves are used as an external remedy for sore throats, and also for wounds ; and sometimes it is used as a salad. It is as common in all bordering counties as it is in Worcestershire.

ERYSIMUM, Linn. 39. (ἐρύω, I cure, from its supposed medicinal virtues.)

128. **89. E.** CHEIRANTHOIDES, Linn. *Treacle-mustard.*

Native. Cornfields, garden ground, waste places. Local and rare. A. April to October. *Top. Bot.* 38.

First Record. Caledonia, near Stourbridge, *Ick, Mid. Counties Her.*, August 5, 1838.

Avon. One place, *Lees.* Near Evesham. Near Nafford's Mill.

Severn. Grimley. Fenny Rough. Stakenbridge. Hagley. Hallow. Wilden, near Stourport, *Miss Ladbury.* Blakedown and Lower Hagley,

1902-4, *W. Whitwell.* Severn banks above Kepax. Iverley, near Stourbridge. Severn banks, Grimley.

Malvern. In an uncultivated garden at Malvern, *Towndrow, Malvern Advertiser,* Nov. 5, 1892.

Lickey. Potato field on Walton Hill, *Mathews.*

A weed of cultivated ground throughout Europe. It is recorded from all bordering counties.

129. 90. E. ORIENTALE, Mill.   *Hare's-ear.*

(*E. perfoliatum,* Crantz.)

Alien. Waste places. Very rare. A. Spring and Summer.

First Record. *Towndrow, Malvern Advertiser,* Nov. 14, 1896.

Malvern. Rubbish heap near Malvern Link; Garden at Lower Howsell, *Towndrow,* 1896.

A foreign weed probably introduced with corn or fodder. It has been recorded from Shropshire only of the neighbouring counties.

CAMELINA, Crantz 40*. (χαμαί, on the ground (dwarf), λίνον, flax.)

130. 91. C. SATIVA, Crantz.   *Common Gold-of-Pleasure.*

Alien. Cornfields, waste places. A. June and July.

First Record. Not localized, *Table of Plants, Lees, Bot. of Worc.,* 1867.

Severn. Several places, *Lees.* Churchill, *Mathews.* Stourbridge. Garden at Bewdley. Severn side below Worcester. Broadwas.

Malvern. Leigh Mill.

Lickey. One place, *Lees.*

This plant is a native of the south-east of Europe; in other regions where it occurs it has been introduced with corn. It does not appear to have been met with in Gloucestershire of our neighbouring counties.

Var. FOETIDA (Fr.).

First Record. In this book.

Malvern. Newtown, Malvern, *Towndrow, Malvern Advertiser,* Nov. 5, 1892.

This variety differs in having obovate pods, truncate at the apex.

C. SYLVESTRIS, Wallr.

This plant is not given in the *Lond. Cat.* It was recorded as having occurred at Malvern in *Mid. Nat.* vi. 117 (1883). It is a greater stranger even than the last plant, and has not been seen in Gloucester many times.

BRASSICA, Linn. 42. (From the Celtic *bresic,* cabbage pottage.)

133. 92. B. NAPUS, Linn.   *Rape,* or *Cole-seed.*

(*B. campestris,* Ind. Kew.)

Alien. Sides of fields, waste places. Not uncommon. A. or B. May to September.

First Record. In the common fields around Bredon Hill, *Pitt, Agric. Worc.,* p. 317 (1810).

Avon. About Bredon. Banks of Avon, from seeds taken to the oil mills, *Rev. Winnington-Ingram.* Wyre Piddle. Fladbury.

Severn. Oldington, *Hickman.* Pitchcroft, *Westcombe.* Teme side. Boreley. Kempsey. Diglis. Hagley Brake. South of Kidderminster.

Malvern. Banks of Teme, occasionally, *Lees.* Ankerdine Hill. Bransford. Stockton. Stanford Bridge.

Lickey. Occasional, *Mathews.*

This plant is a cultivated form, and is hardly permanent in its localities. It is recorded from all counties except Herefordshire, where no doubt it also occurs.

135. 93. B. Rapa, Linn.   *Common Turnip.*

(*B. campestris,* Ind. Kew.)

Native or denizen. Occasional throughout the county. B. or P. April to August.

First Record. Not localized, *Lees, Bot. Malv. Hills,* 1st ed., p. 33 (1843).

These cultivated varieties of the old Linnaean *B. campestris* are a little confused. Bentham (*Brit. Flor.,* 3rd ed., p. 37) calls *Napus* the turnip, and *Rapa* the Rape; Babington (*Manual,* 9th ed., p. 32), on the other hand, calls *Rapa* the turnip, and *Napus* the Rape. Hooker (*Stud. Fl.,* 3rd ed., p. 32) follows Babington, while H. C. Watson (*Stud. Fl.,* above) takes the other side. Whatever they may be, they are survivors more or less established of former cultivation. This plant does not appear to be recorded for either Hereford or Salop of the adjoining counties.

Var. b. **sylvestris,** H. C. Watson.

This variety has been gathered by Mr. Towndrow on the banks of the Teme.

Var. c. **Briggsii,** H. C. Watson.

First Record. Malvern Link, *Towndrow, Malv. Advert.,* Nov. 5, 1892.

Malvern. As first record, *Towndrow.*

{**136. 94. B. monensis,** Huds.   *Isle of Man Cabbage.*

Casual. Very rare. B. or P. June to August. *Top. Bot.* 19.

First Record. Discovered by John Fraser, Esq., of Wolverhampton, on Sutton Common, near Kidderminster, *W. Mathews, Bot. Rec. Club Report,* 1873.

Severn. As first record.

This is a rare plant of the seashore, and usually found in the north-west of our island. It has not been seen in any bordering county. Specimens gathered at this locality were first labelled *Sinapis Cheiranthus,* Koch, which plant has a branched leafy stem and is more luxuriant than *B. monensis,* of which the stem is nearly simple, and the leaves chiefly radical. }

138. 95. B. nigra, Koch.   *Black Mustard.*

(*B. sinapioides,* Roth. *Sinapis nigra,* Linn.)

Native. Waysides and fields. General, except in the Lickey district, where it has not been recorded. A. or B. May to September. *Top. Bot.* 63.

First Record. As *S. nigra.* On the banks of the Severn, *Stokes, Stokes's With. Bot. Arr.,* 2nd ed., p. 714 (1787).

Avon. Norton, near Evesham. Fladbury. Pershore. Nafford's Mill. Sheriff's Lench.

Severn. Banks of river at Bewdley. Upper Arley. Hallow. Claines. Diglis. Kempsey. Upton-on-Severn. Ombersley.

Malvern. Madresfield. Teme side. Cotheridge. Bransford. Brockamin. Powick.

D

A rough tall plant, and commonly cultivated for the mustard of commerce in the east of England and for an early salad in gardens. Old writers had the highest opinion of its invigorating powers. The mustard seed of the parable is not the seed of this plant, but is supposed to be that of the tall *Salvadora Persica,* which grows abundantly on the banks of the Jordan and is there used for a condiment, and has very small seeds and develops numerous branches. *Sinapis nigra* is found in all bordering counties.

140. 96. B. arvensis, O. Kuntze.   *Charlock.*

(*B. sinapistrum,* Boiss. *Sinapis arvensis,* Linn.)

Native. Cornfields, very common on arable land. A. May to September. *Top. Bot.* 112.

First Record. As *Sinapis arvensis.* Cornfields and turnip grounds, *Pitt, Agric. Worc.,* p. 317 (1810).

This well-known plant is a pestilent weed in arable land, and in various parts of the country has popular names of somewhat similar sound, Garlock, Chadlock, Cadlock, and in North Worcestershire, Ketlock, being among them, whilst in the south it is known as Wild Mustard and Farmers' Enemy. It is in all our bordering counties.

141. 97. B. *alba, Boiss.   *White Mustard.*

(*Sinapis alba,* Linn.)

Colonist. Cultivated fields. Rare. A. May to September. *Top. Bot.* 83.

First Record. As *Sinapis alba.* By the bank of the Leominster Canal, by the roadside near Tenbury, *Pitt, Agric. Worc.,* p. 317 (1810).

Severn. Blakedown. Stone.

Malvern. Knightwick. Doddenham. The Gritt Farm, Upper Howsell, *Towndrow, Malvern Advertiser,* Nov. 20, 1897.

This plant, which is by no means common in our county, is abundant on the chalk in Berkshire, where it replaces the common Charlock. Its native home is the south of Europe and the north of Africa; in Northern Europe it is only a weed on waste and cultivated ground. It does not appear to have been seen in Hereford or Salop of neighbouring counties.

DIPLOTAXIS, DC. 43. (διπλοῦς, double, τάξις, a row, from its two rows of seeds.)

143. 98. D. tenuifolia, DC.   *Wall Rocket.*

Native. Old walls and dry places. Very rare. P. June to September. *Top. Bot.* 41.

First Record. Localized in the Severn district, *Table of Plants, Vict. Hist. Worc.* 1901.

Severn. Embankment near Hartlebury Station, 1896, *Towndrow.* Hartlebury Common, 1900, *Humphreys,* 'first appearance.'

This plant has been observed in all bordering counties except Herefordshire. 'It is *D. muralis,* not *D. tenuifolia,* that occurs in Malvern, and I suspect in Severn also,' *Towndrow (MS.).*

144. 99. D. *muralis, DC.   *Sand Rocket.*

Denizen. Waste ground, rubbish heaps. Rare. A. and B. May to October. *Top. Bot.* 54.

First Record. Stourbridge Goods Sidings, *Rev. J. H. Thompson, Mathews, Clent and Lickey Hills,* 1881.

Severn. In a garden at Lower Hagley, 1904, not to be accounted for, *W. Whitwell.* Hartlebury Common, May 29, 1906.

Malvern. Malvern Wells, *Towndrow.*

Lickey. As first record.

This plant is a native of the Mediterranean area, but of late years has taken to spread itself along the ballast of railways even so far north as Scotland. It has been observed in every neighbouring county.

CAPSELLA, Medic. 44. (*Capsula,* a little box.)

145. 100. C. Bursa-pastoris, Medic.   *Shepherd's Purse.*

(*Bursa pastoris,* Weber.)

Native. Waysides, cultivated ground, wall-tops. Abundant everywhere. A. March to November. *Top. Bot.* 112.

First Record. Not localized, *Lees, Bot. Malv. Hills,* 1st ed., p. 89 (1843).

This plant is one of the most abundant weeds in all localities frequented by man in all parts of the globe except the Tropics. It varies much in size, on the wall-top or dry bank being only an inch or two high, but among the weeds of the deserted garden growing sometimes nearly to the height of a couple of feet. Its flat heart-shaped pouches well distinguish it. It is called Pickpocket from its bad qualities, and was termed Poor Man's Parmacety from its supposed good ones, though no medicinal virtues are now known to belong to it. It is also called 'King Herod and his family' in the neighbourhood of Stoulton. It occurs in all bordering counties, as well as practically all over the world.

CORONOPUS, Hall. 45. (κορωνίς, crooked, and πούς, a foot. The name is used by Theophrastus.)

146. 101. C. *didymus, Sm.   *Lesser Wart-cress.*

(*Senebiera didyma,* DC.)

Colonist. Waste ground and gardens. Rare. A. or B. July to September. *Top. Bot.* 45.

First Record. Not localized, ' Worcestershire, E. Lees,' *Watson's Top. Bot.,* p. 28.

Severn. Hartlebury Common, 1900, *Humphreys.* Hartlebury Common, in great abundance, 1907. Just above Camp Lock, *Jeffery.*

Malvern. Midland Siding, Malvern Common, *Towndrow.*

Lickey. A casual in Clent cottage garden, 1879, *Amphlett.* Goods Station, Stourbridge.

This plant is supposed to be a native of temperate South America, but American botanists are inclined to assign it to Europe. At all events it is a weed of roadsides and waste ground, and is especially fond of the neighbourhood of the sea. It is plentiful in Kew Gardens, and has been noticed in Gloucestershire, Salop, and Staffordshire. In spite of its appearance in *Top. Bot.* with Mr. Lees' name, that gentleman does not mention it in his botany of the county.

147. **102. C. procumbens**, Gilib.   *Common Wart-cress, Swine's Cress.*
(*C. Ruellii*, All. *Senebiera Coronopus*, Poir. *Cochlearia Coronopus*, Linn.)
Native. Waysides, bare places. Not uncommon except in the Lickey
district. A. or B. May to October. *Top. Bot.* 81.
First Record. As *Cochlearia Coronopus*, Heath Roadside, *Scott, Stourbridge*,
p. 541 (1832).
Avon. Bredon Hill.
Severn. As first record. Top of Pitchcroft. Bromyard Road, near
Worcester. Upper Hop Yard, Lower Wick. Outside the Hawford
Inn, Ombersley Road, *Jeffery*. Whittington. Canal side, Astwood,
Jeffery.
Malvern. Newland, Malvern. Near Tenbury. Road to Powick Mill.
This plant has been recorded from all neighbouring counties.

LEPIDIUM, Linn. 46. (λεπίς, a scale, from the shape of the fruit.)

148. **103. L. latifolium**, Linn.   *Dittander.*
Denizen. Usually in salt marshes round the coast. In one or two places
only. P. July and August. *Top. Bot.* 19.
First Record. Close to the river Salwarpe, near Droitwich, *Rev. J. H.
Thompson, Phyt.* O. S., vol. iv, pt. 2, p. 970 (1853).
Severn. As first record. Droitwich Canal. Bullow Ferry.
This plant, otherwise confined to the neighbourhood of the sea, is still
flourishing in some quantity where Mr. Thompson discovered, it, and doubtless draws its vigour from the brackish nature of the water in its vicinity
caused by the manufacture of salt in the town. It was at one time grown in
gardens, and here is without doubt a relic of cultivation, maintaining itself
by reason of the conditions of its environment. The name Dittander, which
is found as early as 1265, has the same origin as Dittany, the English name
of several plants, chiefly labiates, and especially of *Dictamnus Fraxinella*, the
Burning Bush, a well-known plant of the flower border. The ultimate
origin no doubt is the Greek δίκταμνον, from the mountain Dicte in the island
of Crete, where a herb so called grew. The origin of the final syllable, however, cannot be explained. The plant is not found in any bordering county.

149. **104. L. *ruderale**, Linn.   *Bowyer's Mustard.*
Casual. Waste ground. Rare. A. July to August. *Top. Bot.* 38.
First Record. Rubbish on the side of the Severn above Worcester, *Stokes.
Stokes, With. Bot. Arr.,* 2nd ed., p. 672 (1787).
Avon. Tardebigge Reservoir.
Severn. As first record. Tardebigge, *Humphreys.* Droitwich Canal, rare
except at one place. St. Peter's, Droitwich, *Rev. J. H. Thompson,* 1850.
Near the Sewage Farm, Droitwich. Near Hawford Bridge. Near
Oddingley. Tagwell Green, Droitwich.
Malvern. Midland Siding, Malvern Common, also on the Common,
Towndrow, Malvern Advertiser, November 14, 1896.
This plant is found generally in the north temperate zone, but almost
always in places frequented by man. The word 'Bowyer' in its English
name has nothing to do with archery, but is a corruption of Boor,
a peasant or countryman, and it was given to this plant by Gerarde. By

other herbalists it has been applied to *Thlaspi arvense.* This plant has been
recorded from Gloucester, Stafford, and Warwick.

150. **105. L. sativum**, Linn.   *Garden Cress.*
Casual. Waste places. Not permanent. A. April to August.
First Record. Naturalized in many spots, *Lees, Bot. Malv. Hills,* 2nd ed.,
p. 64 (1852).
This plant is found only near gardens in Western Europe ; in Eastern
Europe and Western Asia it becomes a weed of cultivated fields. Until the
nineteenth century the name of this class of plants was almost always used
in the plural. Bailey, in 1736, gives this definition : *Cresses,* 'an Herb used in
Sallets ; it has no singular number.' This plant does not appear to have
been observed except cultivated in any bordering county besides Staffordshire.

151. **106. L. campestre**, R. Br.   *Mithridate Pepperwort.*
(*Thlaspi campestre,* L.)
Native. Cornfields, waysides, open places in woods. Widely distributed.
B. April to September. *Top. Bot.* 86.
First Record. Not localized, *Lees, Bot. Malv. Hills,* 1st ed., p. 32 (1843).
No special localities need be given for this widely distributed and not
uncommon plant. It is to be found in all bordering counties.

152. **107. L. [heterophyllum**, Benth.]
Var. b. **canescens**, Gren. and Godr.
(*L. Smithii*, Hook. *L. hirtum,* Index Kew.)
Native. Hedgebanks, commons, roadsides. Local and rare. B. or P.
July and August. *Top. Bot.* 89.
First Record. As *L. Smithii.* Near Stourbridge, *Ick, Mid. Count. Her.*,
August 5, 1838.
Avon. Sheriff's Lench. Pershore. Norton, near Evesham. Crowle.
Cleeve.
Severn. Hagley Brake. Abundant by footpath near Pedmore Quarry,
1904, *W. Whitwell.* Helbury Hill, *Lees.* Near Bewdley. Roadside,
Crown East. Ombersley. Kempsey Common.
Malvern. Welland Common, *Lees.* Cowleigh Park. Malvern Wells.
Malvern. Hanley Castle.
Lickey. Lower Clent. Monument Hill, Hagley. Fields on Clent
Hill, *W. Whitwell.* Furnace Forge, Halesowen. Wychbury Wood.
In this plant the anthers are violet, instead of yellow as in the preceding
one, and the style is longer than the notch, which it equals in *L. campestre.*
It occurs only in Western Europe. It has been observed in all bordering
counties.
L. virginicum, Linn.
Casual. Waste places and near Flour Mills. Uncommon. A. May to
July.
First Record. At Leigh Mill, *Jeffery, Transactions Worcestershire Naturalists'
Club,* vol. iii, p. 90.
Malvern. Leigh Mill.
A North American weed increasingly frequent in connexion with town
and mill rubbish.

L. perfoliatum, Linn.
Casual. Waste places and near Flour Mills. Uncommon. A. June to
August.
First Record. At Leigh Mill, *Transactions Worcestershire Naturalists' Club,*
vol. iii, p. 45, *Parkinson,* determined by G. Claridge Druce.
Malvern. Leigh Mill.
A native of Eastern Europe and Western Asia, and a characteristic weed
of the cornfields of that part of the world, and so found in England where
grain siftings have been thrown.

153. **108. L. *Draba**, Linn.   *Whitlow Pepper-wort.*
Denizen. Waste places, embankments. Very local. P. May to July.
First Record. Near the bridge over the Teme at Powick, 1843, *Lees,
Phyt.,* O.S., vol. i, July 12, 1843.
Severn. Rifle Range, Worcester, *Rea.* Hartlebury Common.
Malvern. Near Powick Bridge. Willow-bed near the Rhydd Ferry.
This plant is especially abundant in the deserts in the Caspian region, and
appears to have been introduced into Britain not more than one hundred
years ago. It has now, however, become a weed on waste banks in the neigh-
bourhood of towns. It has been observed in Gloucester and Warwick also.

THLASPI, Linn. 47. (θλάω, I flatten, from its compressed seed vessels.)

154. **109. T. *arvense**, Linn.   *Field Penny-cress, or Mithridate Mustard.*
Colonist. Cultivated and waste ground. On arable fields throughout.
A. April to October. *Top. Bot.* 84.
First Record. Lickhill, *Hickman, Purton,* vol. ii, pt. 2, p. 335 (1821).
Avon. Norton, near Evesham. Bengeworth. Fladbury. Pershore.
Eckington. Bredon.
Severn. In gardens at Kidderminster, *Perry.* Hartlebury, *Perry.* Har-
tlebury Common. Churchill, Kidderminster. Blakedown. Hurcott.
Lincombe. Claines, opposite Camp Ferry. Old Rifle Range, Worcester.
Bilford, Worcester. Stagbury Hill. Rainbow Hill. Warshill.
Malvern. In cultivated fields, *Lees.* Elmbank, Powick. Blackmore
Park. Malvern Wells. Abberley. Knightwick. Bransford.
Lickey. Kendal End, Lickey. West Slope, Walton Hill. Clent Hill,
Mathews. The Lye, *Mathews.* Thicknall.
A common cornfield weed throughout Europe and Western Asia. Its
English name comes from the large flat seed vessels, which are the size of
silver pennies, while its flowers are quite small. When rubbed it possesses
a faint odour of garlic. It is recorded from all neighbouring counties.

{ *155.* **110. T. perfoliatum**, Linn.   *Perfoliate Penny-cress.*
Casual. Grassy banks. Once only recorded. A. March and April.
Top. Bot. 4.
First Record. At Evenlode, an isolated portion of Worcestershire, *Cheshire,
Lees, Bot. Worc.,* p. 105 (1867).
Avon. As first record.
This very local plant is wild only in the counties of Oxford, Wilts, and
Gloucester. In Gloucester only of those bordering Worcestershire has it been
observed.}

IBERIS, Linn. 48. (From Iberia, Spain, the home of several species.)

158. **111. I. amara**, Linn.   *Bitter Candy-tuft.*
Colonist. Cultivated and waste ground. Rare. A. May to September.
Top. Bot. 14.
First Record. About Malvern, here and there, sparingly, *Lees, Bot. Worc.,*
xxix (1867).
Severn. One place, suspicious, *Lees.* Severn side above Kepax.
Malvern. As first record, *Lees.* Among Clover, Malvern Link, *Towndrow.*
Stanbrook.
Mr. Lees, in his *Bot. Worc.,* says the plant was lost in Malvern, and he does
not mention it in his *Bot. Malv. Hills,* 3rd ed., 1868. The English name
of this plant has nothing to do with sugar, but is derived from Candia
or Crete, from which island one of the genus, *Iberis umbellata,* was originally
brought. The plant has been observed in Gloucester, Stafford, and Warwick
of bordering counties.

TEESDALIA, R. Br. 49. (Named in honour of Mr. Robert Teesdale,
a Yorkshire botanist.)

159. **112. T. nudicaulis**, R. Br.   *Naked-stalked Teesdalia.*
(*Iberis nudicaulis,* Linn.)
Native. Bare spots on heaths, pastures. Local and rare. A. May to
August. *Top. Bot.* 73.
First Record. As *Iberis nudicaulis.* In some old stone or gravel-pits by the
side of Pensham field on the footway, *Nash, Worc.,* Int., p. lxxxix (1781).
Avon. One place, *Lees.*
Severn. Falling Sands Common, *Perry.* Hartlebury Common. Near
Broom Pool, 1904, *W. Whitwell.* Habberley. Roadside beyond Shraw-
ley Wood. Near Bewdley. Cookley.
Malvern. Road to the Wyche. Rosebury Rock. Knightsford Bridge, *Lees.*
The genus *Teesdalia* is confined to two species, natives of Europe. This
plant has been recorded from every neighbouring county.

ISATIS, Linn. 51*. (Ισάτις was the name given to this plant
by the Greeks.)

[*161.* **113. I. *tinctoria**, Linn.   *Dyer's Woad.*
Denizen. Marl cliff. One locality only. B. or P. June and July. *Top.
Bot.* 2.
First Record. On a marl cliff close to the Severn, near the Mythe Toot
Hill, Tewkesbury, where the Severn divides Worcestershire and Glou-
cestershire, *Lees' Cat., New Bot. Guide,* 1835.
This locality is actually in Gloucestershire, though geographically in
Worcestershire. The Gloucestershire parish of Twyning projects balloon-
shaped to the north into Worcestershire, its neck taking up not half a mile
of the southern boundary of that county. In this short distance grows
Isatis tinctoria, on a high cliff by Severn side, where the width of the river,
which forms one side of the balloon, separates it from Worcestershire.
Here some botanists consider it to be truly wild ; 'Wild on cliffs by Severn,'
Bab., Manual, 9th ed., p. 40, with Hooker as the authority. ' Where indeed it

appears to be indigenous,' Benth., *Brit. Flor.*, 8th ed., p. 48. ' Wild on cliffs by the Severn, Tewkesbury,' Hooker, *Stud. Flora*, 3rd ed., p. 42. With the dye derived from this plant the ancient Britons are supposed to have stained their bodies. In old days Woad was cultivated, until it was displaced by Indigo ; now Indigo itself is displaced by a product of gas-tar. Woad is at the present day cultivated in some of the eastern counties, but it is not now grown for the beautiful blue colour which is so characteristic of genuine old tapestries, but as the mordant for the best class of clothes. It occurs in no neighbouring county except Gloucestershire.]

RAPHANUS, Linn. 54. (*ράφανος*, a cabbage.)

164. **114. R. *Raphanistrum,** Linn. *Wild Radish.*
Colonist. On arable fields. Common on sandy soil. A. May to September. *Top. Bot.* 111.
First Record. In the common fields around Bredon, *Pitt, Agric. Worc.*, p. 317 (1810).
This plant hardly requires localization ; it is of widespread occurrence. The flowers are pale straw colour veined with purple ; sometimes so pale as to become very nearly white. It is a native of the Mediterranean region, and a common weed elsewhere. Of neighbouring counties, this plant does not appear to have been recorded for Salop, but it must occur there.

7. RESEDACEAE, DC.

RESEDA, Linn. 55. (*Resedo*, I calm, from its supposed sedative quality.)
166. **115. R. ALBA,** Linn. *Shrubby-based Dyer's Rocket.*
(*R. suffruticulosa*, Linn.)
Casual. Cultivated and waste places. Rare. B. or P. July and August.
First Record. As *R. suffruticulosa*. Waste ground, Britannia Square, Worcester, *Lees, Bot. Worc.*, p. xxix.
Severn. Two places, *Lees.* Near Collins's Factory, Northwick Lane, Worcester.
Lickey. One place, *Lees.*
Both suspicious, says Mr. Lees. This plant is a native of Europe, but is only known in Britain as a weed of garden or grain origin. It has been observed in Warwickshire, but in no other neighbouring county.

167. **116. R. lutea,** Linn. *Wild Mignonette.*
Native. Fields and waysides. Local and rare. P. May to October. *Top. Bot.* 53.
First Record. Not localized, *Lees' Cat. New Bot. Guide*, 1835.
Avon. Bredon Hill. Alderminster, *Cheshire.*
Severn. Brickyard, Hartlebury Station, 1895.
Malvern. Leigh Sinton, *Towndrow.* Railway at Malvern Wells, *Towndrow.*
This plant is much like the garden Mignonette, but its odour instead of being pleasing is rather disagreeable. Nor does it possess the reddish-brown stamens of that flower ; they are yellow green. It appears to have been observed in all bordering counties.

168. **117. R. Luteola,** Linn. *Dyer's Rocket, Weld.*
Native. Waysides, quarries, fields, pit-mounds. Common. B. May to September. *Top. Bot.* 95.
First Record. About the ruins of Dudley Castle, *Withering. Stokes, Withering Bot. Arr.*, p. 492 (1787).
Avon. Norton, near Evesham. Evesham. Pershore. Fladbury.
Severn. Near Churchill Station, Grimley. Newtown, Worcester. Diglis. Rainbow Hill. Blakedown. Lincombe. Ombersley. Kempsey. Bewdley.
Malvern. Near Holywell, *Walker.* Leigh Sinton. Ankerdine Hill. Abberley Hill. Alfrick.
Lickey. Hasbury Quarry. Coal-pit Mounds near Dudley and Oldbury.
As its name imports, it was in former times much used in the dyeing trade, and from its juice the colour known as Dutch pink is manufactured. It is one of the first plants to appear on spoil heaps thrown out from coal-pits. It is recorded from every bordering county.

8. CISTACEAE, Juss.

HELIANTHEMUM, Mill. 56. (*ήλιος*, the sun, *άνθεμον*, a flower.)
171. **118. H. Chamaecistus,** Mill. *Dwarf Cistus, Rock Rose.*
(*H. vulgare*, Gaertn. *Cistus Helianthemum*, Linn.)
Native. Hilly pastures, roadsides. Locally common. P. April to September. *Top. Bot.* 92.
First Record. As *C. Helianthemum.* Unlocalized, in a list of Malvern plants, *Ainsworth, Edin. Phil. J.*, p. 99 (1828).
Avon. Bredon Hill, abundant. Trench Woods. Crowle. Cleeve Banks.
Severn. Near Droitwich. Cruckbarrow Hill. Roadside, Crown East. Near Conigree Wood.
Malvern. Plentiful on the Ridgway, *Lees.* Alfrick. Clifton-on-Teme. Ankerdine Hill. Knightwick. Walsgrove. Gadbury Banks. Sarn Hill. Martley. Ombersley.
This plant is recorded from all bordering counties.

{*172.* **119. H. polifolium,** Mill. *White Rock Rose.*
Casual. Stony places. Once recorded, possibly in error. P. May to July. *Top. Bot.* 2.
First Record. Unlocalized, in a list of Malvern plants, *Ainsworth, Edin. Phil. J.*, p. 99 (1828).
Malvern. As first record.
Mr. Mathews says this record must be an error. It has never been seen in the county since. In Watson's *Top. Bot.*, p. 53, the plant is given for Worcestershire in brackets, therefore supposing the record to be an error. In the same way he records it for Gloucestershire, the only adjoining county in which it has been recorded, even doubtfully.}

9. VIOLACEAE, DC.

VIOLA, Linn. 57. (The Latin name.)
173. **120. V. palustris,** Linn. *Marsh Violet.*
Native. Marshes, bogs, swamps, wet commons. Local. P. April to June. *Top. Bot.* 105.
First Record. On Hartlebury Common, *Lees, Illus. Nat. Hist. Worc.*, p. 155 (1834).
Severn. Near the Birches, Hagley, 1904, *W. Whitwell.* Brake Mill Pool, Hagley. Wyre Forest. Stanklin.
Lickey. Bog on Bromsgrove Lickey, 1850, *Mathews.* Alvechurch, *D. Mathews.* Little Farley Wood, 1901, *Humphreys.* Pool opposite Independent College, Moseley, 1858, *Mathews.* Rednal Hill. Coppice near Harborne, *H. S. Thompson.* Rowley Green, Moseley. Cofton Hill. Chaddesley Wood.
Like other violets in form, this flower is paler than most of the species, and the petals are marked with darker veins. It occurs in all bordering counties.

174. **121. V. odorata,** Linn. *Sweet Violet.*
Native. Hedgerows, borders of woods, coppices. Abundant except in Avon, where it is less common. P. March to May and sometimes in the autumn. *Top. Bot.* 80.
First Record. Blue, purple, white, common, *Scott, Stourbridge*, p. 540 (1832).
The violet is widely spread in all the countries of Europe and Asia as far as the Himalayas, and everywhere and all down the ages has been a favourite with their peoples, the subject of poems, partner in their joys of life, and finding for them also medicine in their sickness. The flower is sold in the markets of Persia and Arabia as commonly as it is in the streets of London, or as it was in Athens in classical times ; prized always for its odour and its associations. Pliny said the scent of the violet cured headache ; the old herbalists considered that great healing virtue existed both in leaf and flower, and even in later times a syrup of sweet violets has been considered a suitable medicine for a child. The flowers take many colours ; in hedge-banks and woody places they are perhaps most usually found white. It grows in every neighbouring county.

Var. b. **imberbis.**
First Record. As herein.
Severn. Ombersley Banks.
Malvern. Madresfield, *Towndrow, Malvern Advertiser*, January 26, 1895.

Var. **barbata.**
First Record. As herein.
Malvern. Madresfield, *Towndrow.*
This variety is not given in *Lond. Cat.*, 10th ed.

175. **122. V. hirta,** Linn. *Hairy Violet.*
Native. Hedgebanks, borders of woods, grassy places. Locally plentiful. P. February to May. *Top. Bot.* 73.

First Record. Lane leading from Kempsey to Green Street, *Dr. Streeten, Lees' Illus. Nat. Hist. Worc.*, p. 155 (1834).
Avon. Craycombe. Trench Woods, with white flowers, *Rea.* Fladbury. Crowle. Bredon. Broadway.
Severn. Whittington. Henwick. Witley, with white flowers, *Rea.* Powick. Kempsey Grove. Devil's Catchem Meadow between Ockeridge and Monk Woods. Doddenham. Spetchley. Leopard Hills. Cruckbarrow Hill. Dynes Green. Shrawley. North Wood, Bewdley.
Malvern. Hollybush Hill. Ankerdine Hill. Malvern Link. Gadbury Banks. Sarn Hill. Martley.
This flower much resembles that of the sweet violet, but the bract is lower on the peduncle, and unlike that it is quite odourless, nor does it possess creeping shoots. The rough hairs on the leaf-stalks and leaves usually distinguish it, but Mr. Lees remarks that the plant sometimes becomes smooth on marl or sand. The plant has been recorded from every neighbouring county.

× **odorata.**
(*Viola permixta*, Jord.)
First Record. Not localized, *Mackie, Malv. Field Handb.*, 1886.
Malvern. Madresfield, *Towndrow, Malvern Advertiser*, November 12, 1892. West Malvern.
Mr. Lees, in the 3rd ed. of *Bot. Malv. Hills*, p. 50, recognized this plant, doubtfully assigning it to *V. permixta*, Jord.

177. **123. V. sylvestris,** Kit.
(*V. Reichenbachiana*, Jordan.)
Native. Woods, thickets, shady banks. Local. P. April and May. *Top. Bot.* 55.
First Record. As *V. Reichenbachiana.* Frequent on limestone, *Towndrow, Malv. Advertiser*, October 22, 1892.
Avon. Trench Woods. The Slads. Fladbury. North Piddle. Croome.
Severn. Fenny Rough. Ombersley. Perry Wood. Boreley. Shrawley Wood. Wyre Forest. Spetchley.
Malvern. As first record. Middleyards Coppice. Knightwick. Aileshurst Coppice. Alfrick. Martley. Tiddesley Wood. Dripshill Wood.
Lickey. With white flowers near the Pool House, Belbroughton, 1907, *Amphlett.*
This plant is recorded from every bordering county. Extreme varietal forms occur at Boreley Banks, Ombersley.

178. **124. V. Riviniana,** Reichb. *Dog Violet.*
(*V. canina*, Curt. *V. sylvatica*, Fries.)
Native. Woods, hedges, heathy places. Common and widely distributed P. April to June. *Top. Bot.* 112.
First Record. As *V. canina.* Woods about Cradley, *Scott, Stourbridge*, p. 540 (1832).
This is a form of *V. canina*, Linn., in which the flowering branches are all lateral. It is abundant everywhere, coming into flower as the sweet violet

is fading away, paler in colour, and though scentless a flower of no little charm. In old times it was considered good for cutaneous disorders. The plant is found in all neighbouring counties.

f. villosa, Neum., Wahlst. and Murb.
First Record. In this book.
Malvern. Long Coppice, Leigh, 1898, *Towndrow*. Croft Banks, West Malvern, *Towndrow*.

b. nemorosa, Neum., Wahlst. and Murb.
First Record. In this book.
Malvern. Long Coppice, Leigh, 1898, *Towndrow*.

× sylvestris.
First Record. In this book.
Malvern. Stocks Lane, Newland, *Towndrow*.

180. 125. V. canina, Linn.   *Dog Violet.*
(*V. flavicornis,* Sm.   *V. ericetorum,* Schrad.)
Native. Sandy heathy ground. Local. P. April to June. *Top. Bot.* 88.
First Record. As *V. flavicornis,* not localized, *Lees, Bot. Malv. Hills,* 1st ed., p. 18 (1843).
Avon. Pirton. Cleeve Banks. Rous Lench. Dodderhill Common.
Severn. Hartlebury Common. Stagbury. Holt Bank. Habberley Valley. Wyre Forest. Blackstone Rock. Redstone.
Malvern. Malvern Common, *Westcombe.* Malvern Link. Old Hills. Worcestershire Beacon. Midsummer Hill. Abberley Hill. Berrow Hill. Ankerdine.
Lickey. Walton Hill. Near the Monument, Bromsgrove Lickey.
This plant has been recorded from every neighbouring county except Gloucestershire.

126. V. tricolor, Linn., Sp. Collect.   *Wild Pansy.*
Native. Cultivated ground. Common. A. B. or P. January to November. *Top. Bot.,* Sp. Coll. 112.
First Record. As *V. tricolor, β.* Blossoms blue, blue and yellow, or blue and white, about Stourbridge, *Stokes, Stokes's With. Bot. Arr.,* 2nd ed., p. 957.
Viola tricolor of Linnaeus is now divided into several segregates, of which *V. arvensis,* Murr., the next plant, is the more common form, in which segregate the petals are usually shorter than the sepals; but there is such a range of variability in the species that some botanists think the division unsatisfactory; see Druce, *Flor. Berks.,* p. 79. This aggregate is not numbered in *Lond. Cat.,* 10th ed. The Wild Pansy has several local country names, Kit-run-the-street and Heartsease among them, the latter name being also applied by mediaeval herbalists to the Wallflower. The florist's Pansy is generally allowed to have been derived from this plant, though of late years hybridization has been so largely introduced that it is doubtful if any garden pansies are of pure descent. Lately, too, a race of plants called Violas or Tufted Pansies have been introduced into gardens for bedding purposes, which are hybrids between garden pansies and various alpine species. They are more truly perennial than the Pansy of the florist, which usually dies after flowering. Few flowers have been so transformed out of their normal wild condition

by the management of the gardener. *V. tricolor* occurs in all bordering counties.

183. 127. V. arvensis, Murr.   *Wild Pansy.*
Native. Cultivated ground. General. A. or B. January to November.
First certain Record. Not localized, *Lees, Bot. Malv. Hills,* 1st ed., p. 89 (1843).
This is the commonest of the forms into which the preceding plant has been divided. It does not appear to have been recorded from Herefordshire, but this is possibly because no local botanist has recognized the division, for it must occur in that county as in all others which border Worcestershire.

10. POLYGALACEAE, Juss.

POLYGALA, Linn. 58. (πολύς, much, γάλα, milk; because the plants make cows give plenty of milk.)

187. 128. P. vulgaris, Linn.   *Milkwort.*
Native. Pastures, boggy places, grassy slopes. Abundant in Avon and Lickey, not uncommon elsewhere. P. May to July. *Top. Bot.* 83.
First Record of aggregate. With white flowers at Picket (Pecket) Rock near Kidderminster, *Perry, M. N. H.,* vol. iv, p. 450 (1831). First record as a segregate, Railway bank near Kidderminster, *F. A. Lees, Bot. Rec. Club Rep.,* 1883 (pub. 1884).
Avon. Bredon Hill, plentiful. Sheriff's Lench. Cleeve Bank. Stoulton.
Severn. *P. eu-vulgaris.* Railway cutting in Wyre Forest. Near Kidderminster, *Arnold Lees.* Bewdley. Spetchley Common. Ombersley. Holt.
Malvern. With white flowers, near Bridges Stone, Alfrick. Near Alfrick Church. Storridge Common. Martley. Abberley Hill. Malvern Hills. Ankerdine Hill.
Lickey. Pepper Wood, Belbroughton, *Humphreys.*
The flowers of this plant vary much in colour, and are found purple and lilac, white, reddish purple, and blue. In old days it had the special names of Rogation Flower, or Gang or Procession Flower, intimating its use at the old ceremony of beating the bounds of a parish during Rogation Week, in which connexion it is mentioned by Bishop Kennet, 'of which the maids make garlands and use them in these solemn processions.' Gerarde, in his *Herbal,* says it is so called 'because it doth specially flourish in the Crosse or Gang weeke, or Rogation weeke'. At first the ceremony of beating the bounds was strictly religious, and was an opportunity of public praise and thanksgiving; but, like many ancient customs, it became in course of time perverted into a season of revelry not without buffoonery and the inordinate consumption of beer. Its origin was possibly the old Roman ceremony of *Frimitiae,* or First Fruits, transformed into a religious service by the early Christians. This service is still indicated in many country districts by the word 'Gospel' being prefixed to some tree or other natural object. 'Gospel Oaks' are not uncommon, and were places where the procession stopped and offered prayer; not necessarily on the exact bounds of the parish.

In later days these Gospel, or as they were sometimes called 'Bannering', places were usually a neighbouring farm-house where refreshment for the body, and not for the soul, could be found. It is called Hedge Hyssop in the vicinity of Stoulton. The plant occurs in all neighbouring counties.

188. 129. P. oxyptera, Reichb.
Native. Commons and grassy banks. Rare. P. May to July. *Top. Bo!.* 46.
First Record. Malvern, *Towndrow, Malvern Advertiser,* November 12, 1892.
Malvern. Malvern Link Common, *Towndrow.*
This plant is recorded from Hereford, Salop, and Warwickshire.

189. 130. P. serpyllacea, Weihe.
(*P. depressa,* Wend.)
Native. Dry pastures and commons. Widely distributed. P. April to September. *Top. Bot.* 100.
First Record of segregate. As *P. depressa.* Dry banks and heaths, *Mathews, Clent and Lickey Hills,* 1881.
This is the ordinary plant usually met with. It has been recorded from all bordering counties.

12. CARYOPHYLLACEAE, Juss.

DIANTHUS, Linn. 60. (δῖος, divine, ἄνθος, flower, from the beauty of the flowers.)

194. 131. D. Armeria, Linn.   *Deptford Pink.*
Native. Sandy fields and hedgebanks. Very local. A. June to August. *Top. Bot.* 48.
First Record. Upon banks under hedges on a clayey soil about Pershore, Eckington, Great Comberton, and many other places, *Nash, Hist. Worc., Int.,* p. lxxxix (1781).
Avon. As first record.
Severn. Wyre Forest. Clarkton Leap (Clerkenleap), Worcester, *Stokes.* About Cotheridge; Shrawley, near the church; near Mudwall Mill, *Lees.* Battenhall, *Baxter.* Shrawley Wood. Laughern Brook, a little below Stoulton Bridge.
Malvern. Powick, *Westcombe.* Alfrick, *Lees.* Pastures below the Priory Church, *Lees.* Meadow at Kempsey, *Lees.*
Although this plant is no doubt a true native, it has been so long cultivated in gardens that in many instances it has escaped into conditions more or less resembling those of a wild plant, and as such is usually observed. It has been observed in all bordering counties except Gloucestershire.

195. 132. D. deltoides, Linn.   *Maiden Pink.*
Native. Grassy banks. Very rare. P. July to September. *Top. Bot.* 58.
First Record. Blackstone Rock, near Bewdley, *Sheward, Nash, Hist. Worc. Sup.,* p. 96 (1799).
Severn. Blackstone Rock, *Sheward* and *Purton.* Rediscovered 1851, *Westcombe.* Canal side, Cookley, west of the church, 1852, *Mrs. B. Williams.*

Lickey. Warwick Hall, Bromsgrove, *Humphreys.*
This plant has not been observed at Blackstone Rock, in spite of close search, in recent years, though the environment remains the same as it was in the time of the older botanists. The plant has been recorded from every adjoining county.

200. 133. D. prolifer, Linn.   *Proliferous Pink.*
Casual in the county. Gravelly places and fields. Rare. A. June to October. *Top. Bot.* 8.
First Record. In a marl pit, Landridge Hill, Hanley Castle, *Ballard, Stokes's With. Bot. Arr.,* 2nd ed., p. 441 (1787).
Malvern. As first record. Clover field at Leigh, 1895, *Miss B. Norbury.*
This plant was given from the Hanley Castle locality in Mr. Lees' time. In Watson's *Top. Bot.,* p. 62, the Worcestershire record is in brackets. The plant has not been observed in any bordering county.

SAPONARIA, Linn. (*Saponarius,* soapy, because of the use to which the plants are applied.)

S. vaccaria, Linn.
Casual. Waste places, roadsides, railway ballast. A. July to August.
First Record. As below.
Severn. Roadside near Hartlebury Station, July 13, 1899. *Transactions Worcestershire Naturalists' Club,* vol. iii, p. 2.
This plant, given in the 9th edition of the *London Catalogue,* is omitted from the 10th edition. It is native of oak woods in Asia Minor, and one of the most frequently introduced grain aliens in Britain. On the Continent it is a very common cornfield weed. It has been observed as a casual in Warwickshire.

201. 134. S. *officinalis, Linn.   *Soapwort.*
Denizen. Hedges, waste places, usually near houses. P. July to September.
First Record. Hedges near Hanley, *Ballard, Stokes's With. Bot. Arr.,* 2nd ed., p. 438 (1787).
Severn. Wyre Forest, *Jorden.* Near Dodderhill Church, *Humphreys.* Severn Banks, Bewdley and Lincombe. Also at Hallow. Grimley. Railway banks. Droitwich. Cotheridge, *Lees.* Lincombe. Eyemoor Wood. Ribbesford.
Malvern. Hedges near Hanley, *Ballard.* Alfrick. Severn side below Teme's mouth.
Lickey. Banks of Stour at Lye, *Scott.*
This plant is full of mucilaginous juice, which will lather with hot water, and hence its name. In Spain the plant is used for washing clothes; in Italy for cleaning wool and cloth; and in Switzerland sheep before they are shorn are washed with a decoction of this plant. It is doubtfully a native of Britain. It has long been cultivated in gardens, and the creeping nature of its roots, difficult to eradicate if the plant is once established, tends to make it persistent wherever it has effected a lodgement. In gardens it is frequently seen with double flowers. It is recorded from every bordering county.

SILENE, Linn. (Σειληνός, the oldest and most famous of the satyrs, from the shape of the calyx.)

202. 135. **S. latifolia,** Rendle and Britten.   *Bladder Campion.*
(*S. Cucubalus,* Wibel. *S. inflata,* Sm. *S. angustifolia,* Guss.)
Native. Waysides, cultivated and fallow fields. Not uncommon except in Avon, where it is scarce. P. May to August. *Top. Bot.* 104.
   First Record. As *S. inflata,* not localized, *Walker, Mid. Med. Surg. Rep.,* vol. i, no. 2, November, 1828.
   Avon. Bredon Hill. Stoulton.
   Severn. Rainbow Hill. Near Bewdley. Park Hall. Blackstone Hill. Hartlebury Common.
   Malvern. Malvern Wells.
   Lickey. Rubery, *Humphreys.* Clent Hills, *Amphlett.*
Easily recognized by its thin globular flower-cup, which with the foliage is of a pale sea-green colour. The flavour of the young shoots is reminiscent of green peas, and they have been used as a vegetable, but even when boiled the plant is too bitter to be pleasant to eat. It is called in Worcestershire Catch Fly, Titty Bottle, Brandy Bottle, Cockles, Snake Bottle, and Wild White Buttons. It is found in all neighbouring counties.

Var. **puberula** (Jord.).
   First Record. Malvern Wells, *Towndrow, Malv. Advertiser,* November 12, 1902.
   Avon. Close to Corncockle Quarry, Bredon, 1902, *Bickham.*
   Malvern. As first record.
This variety is downy, not glabrous, as is the typical form. It has been observed in all bordering counties except Staffordshire.

205. 136. **S. conica,** Linn.   *Striated Corn Catch-fly.*
Denizen. Pastures, sandy commons. Local. A. May to July. *Top. Bot.* 12.
   First Record. One field, Iverley, now extinct, *Scott, Stourbridge,* p. 540 (1832). First certain record. Hartlebury Common, *W. N. C. Trans.* iii. 236 (1904).
   Severn. Hartlebury Common, 1900, *Humphreys.*
It is doubtful in which county, Worcestershire or Staffordshire, Scott found this plant, as they both meet in the locality of Iverley, to the south-west of Stourbridge. Unless Scott's record relates to Staffordshire, the plant has not been observed in any bordering county.

206. 137. **S. anglica,** Linn.   *English Catch-fly.*
Colonist. Cornfields, waysides. Local. A. June to August. *Top. Bot.* 57.
   First Record. Areley, near Stourport, *Mrs. Gardner,* late of Stourport; *Purton, App.,* vol. iii, p. 37.
   Severn. Areley, as first record. Churchill, Kidderminster, 1879, *Mathews.*
Lees says, *Bot. Worc.,* p. 12, that Scott mentions this plant in his *History of Stourbridge,* but this is not the case. It has been observed in Herefordshire, Staffordshire, and Warwickshire of bordering counties.

S. DICHOTOMA, Ehrh.
Casual Alien. A.
   First Record. Among wheat, High Mill, Leigh, *Miss B. Norbury,* 1900; and at Woodfield, near Malvern, June 21, 1898, *Towndrow;* *Malv. Advert.,* December 22, 1900.
   Malvern. As first record.
A common weed on cultivated land in Central Europe, in England usually occurring in the neighbourhood of mills.

209. 138. **S. nutans,** Linn.   *Nottingham Catchfly.*
Casual. Dry places and walls. Once recorded. P. June and July. *Top. Bot.* 18.
   First Record. Cornfield at Upton Warren, *Humphreys, W. N. C. Trans.,* vol. iii, p. 45 (1901).
   Severn. As first record.
This is the only record for this plant in the county. In the *Table of Plants, Vict. Hist. Worc.,* vol. i, p. 48, it is marked as occurring in the Lickey district, which is an error. One of the chief homes of this plant is on the cliffs near Dover, where its powerful odour scents the evening air. Its profusion in this locality formerly gave the plant the name of Dover Catchfly. It has received the name of Nottingham Catchfly, because it is common in the neighbourhood of that town. Of our bordering counties the plant has been observed in Salop, Stafford, Hereford, and Warwick.

212. 139. **S. noctiflora,** Linn.   *Night-flowering Catchfly.*
Colonist. Cornfields, especially on sandy soil. Local. A. July to November. *Top. Bot.* 46.
   First Record. In a sandy field behind Birchengrove, Broad Heath, *Lees, Illus. Nat. Hist. Worc.,* p. 163 (1834).
   Avon. Near Corn Cockle Quarry, Bredon Hill.
   Severn. As first record.
   Malvern. A garden weed at Sherridge, *Miss B. Norbury,* 1895; *Towndrow.* In a field of Clover at Madresfield, *Towndrow, Malvern Advertiser,* Jan. 20, 1894.
Although the first record is Mr. Lees', the plant is not mentioned in his *Botany of Worcestershire.* The plant is not unlike that of the much commoner *Lychnis alba,* but it may be readily distinguished from that by its smaller flowers, with more deeply divided petals, which are yellowish on the outside and white on the rose-coloured inside. It has been observed in all bordering counties except Gloucestershire.

LYCHNIS, Linn. 64. (λύχνος, a lamp, from the shape of the capsule.)

214. 140. **L. alba,** Mill.   *White Campion.*
(*L. vespertina,* Sibth.)
Native. Hedges, cultivated fields, commons. Common and generally distributed. B. or P. April to October. *Top. Bot.* 103.
   First Record. As *L. dioica.* Red and White Campion, *Lees, Bot. Malv. Hills,* 1st ed., p. 23 (1843). First record as *L. vespertina,* not localized, *Lees, Bot. Malv. Hills,* 2nd ed., p. 46 (1852).
E

This is a common plant, getting its name (*vespertina*) from the fact that it is not till evening that it gives out its scent. This and *Lychnis dioica* have been considered varieties of the same plant, and certainly they easily hybridize; plants with petals flushed with pink are frequently seen, usually when *L. dioica* is not far away, though possibly change of colour is not always due to hybridization. The two plants, apart from the colour of their flowers, are very similar in appearance, the distinction depending upon the shape of the calyx teeth, the shape of the capsule, and whether the ten teeth of the capsule are straight as in this plant, or recurved, as in the next. It is a widespread plant in Northern Europe and Russian Asia as far east as Lake Baikal. It is recorded from every neighbouring county.

215. 141. **L. dioica,** Linn.   *Red Campion.*
(*L. diurna,* Sibth.)
Native. Woods, hedges, railway-banks, damp, and shady places. General. P. April to September. *Top. Bot.* 111.
   First Record. As *L. dioica.* Red and White Campion, *Lees, Bot. Malv. Hills,* 1st ed., p. 23. First record as *L. diurna, Lees, Bot. Malv. Hills,* 2nd ed., p. 46 (1852).
This is a more elegant plant than the relative it so much resembles, and it is less viscid. Worcestershire children know the plant by the names Billy Buttons and Cockles. It is equally common in and recorded from all bordering counties.

216. 142. **L. Flos-cuculi,** Linn.   *Ragged Robin.*
Native. Wet meadows, woods, osier beds. Common and generally distributed. P. April to October. *Top. Bot.* 112.
   First Record. Not localized, *Walker, Mid. Med. Surg. Reporter,* vol. i, no. 2 (1828).
This well-known plant, with the preceding ones, is too widely distributed to require localization. It is one that comes with the cuckoo in spring, a fact recognized in the Latin of its specific name, sharing its association with that bird with many another country favourite. Its rose-coloured jagged petals grow on a stem the upper part of which is clammy and the lower portion hairy. In some parts of the county it is called Hens' Feathers, Cuckoo's Flower, and Batchelor's Buttons, so we may suppose that buttons were at one time made in the somewhat eccentric shape of the flower. It occurs in every bordering county.

219. 143. **L. *Githago,** Scop.   *Corn Cockle.*
(*Agrostemma Githago,* Linn.)
Colonist. Cornfields. General, but sparingly. A. June to August. *Top. Bot.* 100.
   First Record. Not localized, *Lees, Bot. Malv. Hills,* 1st ed., p. 23 (1843).
This plant is never seen out of the cornfield, where its handsome flower is occasionally to be observed among the stalks of the growing corn. The name Cockle has been applied to this plant from early times, but appears in the first instance to have been that of an equally pestilent cornfield weed of southern and eastern regions, *Nigella arvensis,* Linn. In the north, however, that plant was unknown, and the name became transferred to this one,

of the same habits and equally undesirable. The seeds are black and glossy, and getting among flour, fill it with black specks. In America, a special machine is used to separate these seeds from wheat, and also in France. It is not the 'Cockle' of the Bible, as the plant is not known in Palestine. Nor does the name appear to have any reference to the numerous other things to which the word is applied in the English language. The specific name is formed from the old word 'gith', of unknown origin, which first belonging to the herb *Nigella,* became in time transferred to this plant. 'Cockle,' says Gerarde, 'is called Gith, yet not properly.' The Corn Cockle is recorded from all neighbouring counties.

CERASTIUM, Linn. 66. (κέρας, a horn, from the horn-like capsule.)

221. 144. **C. tetrandrum,** Curtis.   *Mouse-ear Chickweed.*
Casual. Sandy places. Rare. A. April to October. *Top. Bot.* 78.
   First Record. Given for Worcestershire in *Top. Bot.,* 2nd ed., p. 81, with E. Lees in brackets for authority (1883).
   Severn. Hartlebury Common, *W. N. C. Trans.* iii. 236.
This plant is a lover of sea-air. It differs but little from the two following plants, but may be distinguished by its four stamens, which, however, is not quite a constant character. It has been noted in Gloucestershire and Herefordshire.

{*222.* 145. **C. pumilum,** Curtis.   *Mouse-ear Chickweed.*
Native. Dry banks. Very local. A. April and May. *Top. Bot.* 12.
   First Record. Worcester, *E. Lees, Top. Bot.,* 2nd ed., p. 81 (1883).
   Severn. Crookbarrow. Hartlebury Common, Herb. Hastings Museum, *Thompson.*
This plant, like the preceding one, is recorded in *Top. Bot.* with E. Lees in brackets, as the authority. Except as above, no one else has seen it in the county. 'I am afraid it will have to be dropped,' Bagnall (*MS.*).}

223. 146. **C. semidecandrum,** Linn.   *Little Mouse-ear Chickweed.*
Native. Walls, dry fields, heaths. Local. A. March to June. *Top. Bot.* 90.
   First Record. Not localized, *Lees, Bot. Malv. Hills,* 1st ed., p. 23 (1843).
   Avon. South Littleton. Pinvin. Crowle. Pirton.
   Severn. Hartlebury Common. Harborough Hill, Hagley, *Mathews.* Hagley Brake. Bewdley. Stourport. Highwoods Hill. Bromsgrove. Entrance of Fenny Rough, Harvington.
   Malvern. Malvern Hills. Witley. Knightwick. Alfrick. Malvern Link Common.
   Lickey. Winwood Heath, *Mathews.*
This is an early plant to flower. It usually has five stamens, but sometimes becomes 'tetrandum'. It is recorded from all neighbouring counties.

224. 147. **C. viscosum,** Linn.   *Broad-leaved Mouse-ear Chickweed.*
(*Cerastium vulgatum,* Sm. *C. glomeratum,* Thuill.)
Native. Walls, waste places, fields, woods. Widely distributed. A. or B. April to September. *Top. Bot.* 112.
E 2

First Record. As *C. vulgatum*. Broad-leaved Mouse-ear Chickweed, *Lees, Bot. Malv. Hills*, 1st ed., p. 23 (1843).

This common plant occurs everywhere, and of course in all neighbouring counties.

Var. b. apetalum (Dum.).

First Record. In this book.

Severn. On the rock, south side of Hartlebury Common, *Rea*.

This plant, besides being without petals, is usually much more slender than the type.

225. 148. C. vulgatum, Linn. *Narrow-leaved Mouse-ear Chickweed.*
(*C. triviale*, Link. *C. viscosum*, Sm.)

Native. Meadows, walls, banks, and heaths. Common and widely distributed. B. or P. April to September. *Top. Bot.* 112.

First Record. As *C. viscosum*. Not localized, *Lees, Bot. Malv. Hills*, 1st ed., p. 23 (1843).

This plant is distinguished from the former one, with which it is equally general, by its sepals and bracts, which are membranous at their margins, and have a glabrous tip, while in *C. viscosum* they are hairy throughout. It is recorded from all neighbouring counties.

228. 149. C. arvense, Linn. *Field Mouse-ear Chickweed.*

Native. Sandy fields, dry banks. Local. P. April and May. *Top. Bot.* 69.

First Record. On Broadway Hills, *Rufford*; *Purton, Midl. Flora*, p. 220 (1817).

Avon. Broadway. Bredon.

Severn. Barnett Hill, Chaddesley. By roadside at Blackstone. Sutton Common, Kidderminster. Stagbury Hill. Railway-bank, Bewdley.

Malvern. Not met with in this district, *Towndrow*.

Lickey. In the district, *W. N. C. Trans.* iii. 45. Near Gooschill Wood.

This is by far the most conspicuous of the Mouse-ear Chickweeds, having flowers as large as those of the common Stitchwort of the hedgerows and of *C. tomentosum*, which commonly adorns our gardens. It is as widely distributed over the world as any of its congeners, occurring from Arctic Europe to Siberia and from temperate Europe to the Himalayas, being found also in North Africa, North America, and Fuegia and Chili. It is almost bathos after this to say that it is recorded from every neighbouring county !

MOENCHIA, Ehrh. 67. (Name given in compliment to Conrad Moench, Professor of Botany at Hesse-Cassel.)

230. 150. M. erecta, Gaertn. *Upright Moenchia.*
(*Sagina erecta*, Linn. *Cerastium quaternellum*, Fenzl.)

Native. Grassy ground. Local. A. April and May. *Top. Bot.* 53.

First Record. As *S. erecta*. Malvern Hill, *Purton, Midl. Flora*, p. 104 (1817).

Avon. Craycombe. Madams Hill. Bredon. Broadway. Crowle. Sheriff's Lench.

Severn. Near Spring Grove, Bewdley, *Jorden*. Wyre Forest. Near Kidderminster. Stagbury. Helbury Hill. Hartlebury Common.

Malvern. Malvern Hills, from the Worcestershire Beacon to Keysend, *Lees*. Round Abberley Flagstaff. North Hill, Malvern. Old Hills. Castle Morton Common. Abberley Hill. Woodbury. Martley. Malvern Link Common.

Lickey. Clent Hill, Walton Hill, and other eminences of the range. Lower Lickey.

Probably this species occurs more frequently than is known. It is a small insignificant early-flowering annual, imitating with great success the short grass among which it grows, and is very easily overlooked. After flowering the plant dries up, and becomes even less conspicuous. On the hills of the Clent range it occurs on the highest parts of the lower ridges, in compact plentiful patches, which, however, are not commonly met with. It occurs in all neighbouring counties.

STELLARIA, Linn. 68. (*Stella*, a star, in allusion to the shape of the flowers.)

231. 151. S. aquatica, Scop. *Water Chickweed.*
(*Cerastium aquaticum*, Linn. *Malachium aquaticum*, Fries.)

Native. Ditches, river-banks, wet places. Not common. P. May to October. *Top. Bot.* 59.

First Record. As *C. aquaticum*. On rubbish near Moseley Park, *Ick, Analyst*, vol. vi, p. 22 (1837).

Avon. Several places, *Lees*. Norton, near Evesham. Fladbury.

Severn. Stanklin Pool. Wannerton Downs, *Lees*. Hurcott Wood. Near Camp. Holt. Falling Sands. Crown East. Severn side near Worcester. Near Upton Warren.

Malvern. Several places, *Lees*. Dripshill. Cotheridge. Bransford.

Lickey. Moseley. Slideslow Pool, near Bromsgrove.

This plant is recorded from all bordering counties.

232. 152. S. nemorum, Linn. *Wood Stitchwort.*

Native. Bushy places, damp hedgebanks. Rare. May to July. *Top. Bot.* 51.

First Record. Not localized, *Lees, New Bot. Guide Catalogue* (1835).

Severn. One place, *Lees*. Wyre Forest, *Lees*. Holtbank, near Lenchford.

Lickey. One place, gone, *Lees*. Field in Hob Lane, Yardley, *Miss M. A. Beilby*. (Mr. Mathews stigmatizes this record as an error.)

This is a plant whose home is Northern Europe, not reaching to the south-east of that continent. It is somewhat similar to the preceding *S. aquatica*, but easily distinguished therefrom by its lower leaves being stalked and the stigmas three instead of five. Of neighbouring counties it is not recorded for Gloucestershire.

233. 153. S. media, Vill. *Common Chickweed.*
(*Alsine media*, Linn.)

Native. In all kinds of localities. Everywhere. A. Flowers throughout the year. *Top. Bot.* 112.

First Record. Form with ten stamens in hedges near Worcester, *Stokes, With. Bot. Arr.*, 2nd ed., p. 324 (1787).

Everybody knows this common weed. It is a valuable plant for the feathered race, as its name indicates, and our forefathers considered it equally valuable for humans, as an effectual remedy against cramps, convulsions, and palsy. It is a very variable plant, according to the circumstances in which it finds itself, but it may always be recognized by two lines of hairs on the stems which cross over to alternate sides at every node. It is, of course, found in all neighbouring counties.

235. 154. S. neglecta, Weihe.
(*S. media*, var. *major*, Koch.)

Native. Damp places and banks. Not uncommon. A. Flowers throughout the summer.

First Record. In this book.

Avon. Crowle. Sheriff's Lench. Pershore.

Severn. Lane near Worcester. Near Hallow Ford. Claines. Ombersley. Holt. Lincombe. Perdiswell. Witley.

Malvern. Old Hills. Bransford. Madresfield. Leigh.

Lickey. Reservoirs at the base of the Lickey Hills, *Humphreys*.

This plant, formerly considered a variety of *C. media*, is characterized by its leaves with longer stalks, its ten stamens, and its seeds with prominent rounded tubercles.

Var. b. umbrosa (Opiz).

Native. Shady hedgebanks. Common. A. or B. June to September. *Top. Bot.* 22.

First Record. Malvern, *Towndrow, Malv. Advertiser*, 1892.

Avon. Norton, near Evesham. The Slads. Crowle.

Severn. Fenny Rough. Hartlebury Common. Vallambrosa.

Malvern. Madresfield. Knightwick. Bransford. Malvern Link.

This plant, says Mr. Mathews (*Mid. Nat.* xvi. 154), is 'common throughout the county'; but it has received but little notice. Mr. Bagnall (*MS.*), however, doubts the accuracy of this statement. It is frequently regarded as a variety of *S. media*, and is recorded for all neighbouring counties except Gloucestershire.

236. 155. S. Holostea, Linn. *Greater Stitchwort.*

Native. Hedgebanks and wood-borders. Generally distributed. P. April to June. *Top. Bot.* 109.

First Record. Not localized, *Lees, Bot. Malv. Hills*, 1st ed., p. 22 (1843).

A great adornment of hedgebanks in spring. It is a rigid and brittle plant, so much so that it is nearly impossible to pull it up by the root. Our ancestors called it 'All-bones', perhaps because of its brittle and numerous joints; and of this word its specific name is a Greek translation, which was used for the plant in classical times. Other country names for it are Milkmaid and Dicky Snaps. It is found in all neighbouring counties.

237. 156. S. palustris, Retz. *Glaucous Marsh Stitchwort.*
(*S. glauca*, With.)

Native. Bogs, wet ditches, and meadows. Local. P. April to July. *Top. Bot.* 54.

First Record. As *S. glauca*. Lickhill Lane, Worcestershire, *Mr. Hickman, Purton, App.*, vol. iii, p. 36 (1821).

Severn. Lady Pool, Blakedown, *Mathews*.

This plant is included in Mr. Lees' Table of Plants, p. 7, but is not assigned to any of the districts. It is not recorded for Gloucestershire.

238. 157. S. graminea, Linn. *Lesser Stitchwort.*

Native. Heaths and pastures. General. P. April to September. *Top. Bot.* 109.

First Record. Not localized, *Lees, Bot. Malv. Hills*, 1st ed., p. 22 (1843).

This plant is not so frequently met with as the Greater Stitchwort, nor is it so remarkable a flower. Its blossoms are much smaller, and the petals more deeply cleft. It is recorded from all bordering counties.

239. 158. S. uliginosa, Murr. *Bog Stitchwort.*

Native. Wet places. General. P. April to August. *Top. Bot.* 110.

First Record. Rivulets on the side of Malvern Hills, and on the side of the hill at west end of Powick Ham, *Stokes, With. Bot. Arr.*, 2nd ed., p. 457 (1787).

This plant is common enough, but those who would find it must look for it by ditches and in bogs, places not sought after as a rule except by the botanist. Its white flowers are small and inconspicuous among the lush vegetation amidst which it thrives. It is recorded from all adjoining counties.

ARENARIA, Linn. 69. (*Arena*, sand ; many species grow in sandy ground.)

243. 159. A. tenuifolia, Linn. *Fine-leaved Chickweed.*

Native. Dry sunny places on wall-tops and stony ground. Rare. A. May to July. *Top. Bot.* 34.

First Record. Malvern Hills, *Ballard*; *Stokes, With. Bot. Arr.*, 2nd ed., p. 461 (1787).

Avon. Broadway, *Cheshire*.

Severn. Lane near the Toot Hill, behind the Virgin's Tavern, Worcester, *Lees*. (Not acknowledged by him afterwards.)

Malvern. Malvern, *Ballard*. Midland Siding, Malvern Common, *Towndrow, Malv. Advert.*, November 14, 1896.

Lickey. Gravelly fields, Yardley, *Miss M. A. Beilby*.

This plant has occurred in all neighbouring counties.

244. 160. A. trinervia, Linn. *Plantain-leaved Chickweed.*
(*A. trinervis*, Sm. *Moehringia trinervia*, Clairv.)

Native. Woods, shady hedgerows, thickets. Generally distributed. A. April to June. *Top. Bot.* 100.

First Record. As *A. trinervis*. Shady lanes about Worcester. Ankerdine Hill, *Lees, Illus. Nat. Hist. Worc.*, p. 163 (1834).

This common plant might easily be mistaken for Chickweed, and, indeed, usually is so mistaken by the unbotanical observer. Close inspection, however, reveals the fact that its white petals are not cleft like those of the Common Chickweed, while the thin nerves in its leaves are always distinct, and a distinction. It occurs in all adjoining counties.

245. 161. A. serpyllifolia, Linn.　*Least Chickweed.*

Native. Walls, cornfields, sandy heaths. Common. A. March to October. *Top. Bot.* 112.

First Record. Not localized, *Lees, Bot. Malv. Hills,* 1st ed., p. 22 (1843).

Rarely exceeding five or six inches in height, this plant is too inconspicuous to attract attention from any one but a botanist, yet every one's eyes must fall upon it from time to time, it is so very common. It occurs in every neighbouring county.

246. 162. A. leptoclados, Guss.

Native. Woods, shady banks. Uncommon. A. April to June. No number in *Top. Bot.*

First Record. Not localized, *Towndrow, Malv. Advertiser,* November 12, 1892.

Severn. About Churchill and Kidderminster, *Mathews.* Hagley Brake. Park Hall. Bissell. Kendal Walk, Worcester.

Malvern. *Towndrow,* as above. Welland.

Usually considered a variety of *A. serpyllifolia,* this plant differs in its lanceolate sepals, its narrower capsules, and its spreading pedicels. It has been recorded from Salop, Stafford, and Warwick.

SAGINA, Linn. 70.　(*Sagina* signifies 'meat that fattens'; not very applicable to these small plants.)

253. 163. S. apetala, Ard.　*Annual Small-flowered Pearlwort.*

Native. Walls and dry sunny places. Not common. A. May to August. *Top. Bot.* 76.

First Record. On a wall belonging to the Almshouse near St. Oswald's, Worcester, *Stokes, Stokes's With. Bot. Arr.,* 2nd ed., p. 812 (1787).

Avon. Sheriff's Lench. Inkberrow. Harvington.

Severn. Northwood, near Bewdley, *Mathews.* St. Oswald's, Worcester, *Stokes.* Saint George's Square and Britannia Square, Worcester. Kempsey. Ombersley.

Malvern. Malvern Hills. Malvern Link. Powick. Stanford Bridge.

Lickey. Rowney Green Bog, Alvechurch. Sandstone Cutting, Burcot Lane, Bromsgrove.

This insignificant little plant is distinguished from the commoner *S. procumbens* by the fact that the central stem lengthens and flowers, which it never does with that; and its branchlets spread upwards, while in *S. procumbens,* as its name denotes, they are always prostrate; and it is slightly hairy. It is found in all adjoining counties.

Var. b. **prostrata,** Bab.

First Record. On the platform of Malvern Railway Station, *Towndrow, Malvern Advertiser,* 1892.

Malvern. As first record, and near the station, *Towndrow.* Also at Newtown Nursery, Malvern.

In this variety the branches are procumbent, but the central stem, bearing flowers, shows its connexion with *S. apetala.*

× REUTERI.

First Record. As here.

Malvern. Malvern Railway Station, *Bickham* and *Towndrow.*

This hybrid was growing in company with both its parents.

254. 164. S. ciliata, Fr.　*Ciliated Pearlwort.*

Native. Dry sandy fields, roadsides. Rare. A. May to August. *Top. Bot.* 68.

First Record. Blakedown, *Lees, Bot. of Worc.,* p. 12 (1867).

Severn. Hoo, near Kidderminster. Stakenbridge, Hagley. Blakedown, *Mathews.* Wyre Forest. Brake Mill, Hagley.

Malvern. Malvern Link Common. The Rhydd. Malvern Wells. Welland.

This plant was given as a new record for the county by Dr. F. Arnold Lees in *Bot. Rec. Club Rep.* for 1883 (1884); but it was not new. The plant occurs in all neighbouring counties except Salop.

255. 165. S. REUTERI, Boiss.　*Top. Bot.* 4.

Alien. A new introduction. A. May to August.

First Record. Great Malvern Railway Station, *Towndrow, Worc. Nat. Club Trans.,* ii, p. 7, June 17, 1897.

Severn. Foregate Street Railway Station.

Malvern. As above. Cotheridge Court.

Since Mr. Towndrow discovered this plant at Malvern it has been found in other localities in the neighbourhood. Its only other known station was in Central Spain, whence, it would be supposed, it was hardly likely to have been introduced. Mr. Towndrow, *J. of B.,* 1897, p. 409, summarizes all that is known of this species as a British plant. He has found it in Herefordshire also and at Tenby. In *Lond. Cat.,* 10th ed., it bears no marks to indicate that it is an alien, and a comital number is given to it. It has not been recorded from other neighbouring counties.

256. 166. S. procumbens, Linn.　*Procumbent Pearlwort.*

Native. Moist places, fields, heaths. General. P. April to September. *Top. Bot.* 112.

First Record. Not localized, *Lees, Bot. Malv. Hills,* 1st ed., p. 28 (1843).

The little blossoms of this tiny plant are the smallest of our wild flowers. It is a common weed in gravel garden paths where they are a little damp, and its rooting stems make it very difficult to eradicate. It is to be found in every place where a little moisture exists to encourage its growth. It is recorded from all neighbouring counties.

[Var. b. **spinosa,** S. Gibs.

This variety has been recorded by Mr. Lees as growing on gravelly ground at Forthampton, *Bot. Malv. Hills,* 3rd ed., p. 17. Forthampton is in Gloucestershire, and this record, even if correct, cannot be claimed for Worcestershire.]

[260. 167. S. subulata, Presl.　*Awl-shaped Spurrey.*

Native. Ditch banks and commons. Rare. A. or P. June to August. *Top. Bot.* 64.

First Record. Not localized, *Lees, Bot. Malv. Hills,* 1st ed., p. 23 (1843).

Although this plant is given by Mr. Lees in all three editions of *Bot. Malv. Hills* as a Malvern plant, it is not localized in that or any other district of the county in the same writer's *Bot. of Worc.,* and no one else has seen it.]

261. 168. S. nodosa, Fenzl.　*Knotted Spurrey.*

(*Spergula nodosa,* Linn.)

Native. Damp meadows and commons. Not common. P. July and August. *Top. Bot.* 99.

First Record. As *Spergula nodosa.* Hanley, near Malvern, *Rufford; Purton, Midl. Fl.,* p. 223 (1817).

Malvern. Western side of the Worcestershire Beacon; near the Wych; Welland Common, *Lees.* Rill between the Worcestershire Beacon and Sugarloaf Hill. Malvern Common.

As to this plant Mr. Watson remarks, *Top. Bot.,* 2nd ed., p. 73, that while it would appear to rank among 'common' plants by its wide and nearly general distribution, its localities are rather dotted over the island than continuous, like those of the Daisy and other common species. It is recorded from all bordering counties.

SPERGULA, Linn. 71.　(*Spargo,* I scatter, from scattering its seeds.)

262. 169. S. arvensis, Linn.　*Corn Spurrey.*

Native. Cornfields, locally plentiful. A. February to September. *Top. Bot.* 112.

First Record. Not localized, *Walker, Mid. Med. Surg. Rep.,* vol. i, no. 2, November (1828).

This common cornfield weed has several names among country people, one of them, Pick-pocket, denoting their conception of the quality of the plant. Yet although a troublesome weed in the wrong place, in some parts of the continent it is cultivated as food for cattle, which eat it with avidity; and sheep eat it freely also. Usually its stems are tinged with red. Its range on the earth is wide, reaching from Europe to North-west India; and it has been introduced into North America, where in some parts it is very abundant. It is recorded from all counties adjoining.

Var. **vulgaris,** Boenn.

First Record. In this book.

Severn. Fields at Lincombe.

Malvern. Common in cornfields, *Towndrow, MS.* (1897).

In this variety, which is not given in *Lond. Cat.,* 10th ed., the leaves are grass green and the seeds are obscurely margined and covered with club-shaped papillae.

263. 170. S. sativa, Boenn.

First Record. In this book.

Malvern. Mathon, *Towndrow.* Malvern Wells, *Towndrow, MS.* (1897).

This is characterized by its grey green leaves and seeds with a narrow margin covered with minute elevated points. This plant was treated as a variety, *Lond. Cat.,* 9th ed.

SPERGULARIA, Presl 72.　(Named from its resemblance to *Spergula.*)

264. 171. S. rubra, Pers.　*Purple Sandwort.*

(*Buda rubra,* Dumort. *Arenaria rubra,* Linn. *Lepigonum rubrum,* Fr.)

Native. Sandy bare ground. Locally plentiful. A. June to September. *Top. Bot.* 100.

First Record. As *Arenaria rubra.* Not localized, *Walker, Midl. Med. Surg. Rep.,* vol. i, no. 2, p. 100 (1828).

Avon. Norton, near Evesham. Pershore. Defford.

Severn. Between Hagley and Kidderminster, *Mathews.* Frequent in sandy places, north of the district, *Mathews.* Hartlebury Common. Harberrow Common, *Mathews.* Ombersley. Grinley. Glashampton.

Malvern. Malvern Hills, abundant on the rocks; also near the Wych, *Lees.* Old Storridge. Malvern Link, *Towndrow.*

Lickey. Winwood Heath, Frankley. Clent Hills, plentiful; also on the gravel of Clent Grove Drive, a weed, *Amphlett.* Adam's Hill, Clent. Rubery Station. Rednall. Cofton Hill, near the Rose and Crown Inn.

This pretty little flower is frequent enough on the sandy land between Clent and Kidderminster, and on Clent Hills occurs on the beaten earthy tracks made by traffic on them. Mr. Lees says it is 'very uncommon' in the Lickey district; he made a wrong estimate of its rarity. The plant occurs in every adjoining county.

266. 172. S. salina, Presl.　*Seaside Sandwort.*

(*Arenaria rubra,* var. *marina,* Linn. *Lepigonum salinum,* Kindeb. *Buda marina,* Dumort.)

Denizen. Very local. P. May to July. *Top. Bot.* 50.

First Record. As *Arenaria marina.* Defford Common, between Pershore and Upton, *Rufford; Purton, Midl. Fl.,* p. 216 (1817).

Avon. Defford Common.

Severn. Rubbish heap at Hoo Mill, Kidderminster, 1875, fide *C. C. Babington, Mathews.* By the side of the Droitwich Canal.

This seaside plant occurs at that home of such, the Droitwich Canal, but statements regarding it must be received with caution. If this is the plant, how it got there is a matter of speculation; probably the seeds were brought there through the agency of natural dispersion, and finding a congenial locality, owing to the saline conditions of the place, have reproduced their kind. It is hardly likely that they should be survivors of a time so long past as that in which the salt Severn Straits stretched up the valley, although this view has been held. The locality on Defford Common is where a salt spring once existed. Mr. Mathews remarks of this record that it is almost certainly *S. neglecta,* Syme; and, considering the Droitwich Canal plant to be var. *media,* claims the first record for the type at Hoo Mill. It has been found in Gloucester and Stafford.

Var. **media,** Fr.

If the Droitwich Canal plant be *Lepigonum medium,* Fr., its first record would appear to be in a paper read by Professor Buckman at the meeting of the British Association in 1847, referred to by Lees, *Bot. of Worc.,* p. 37. *L. medium* has been recorded in the Avon district from Saldon near Himbleton, Rev. W. Lea (1856), and seems the common form to be found in this Canal at the present time.

Var. **neglecta,** Syme.

The first record of this plant would be that of *S. salina* above, if that has been mistaken in its identity.

13. PORTULACEAE, Juss.

CLAYTONIA, Linn. 74*. (Named after John Clayton, who collected plants mostly in Virginia, and sent them to Gronovius, who published them in his *Flora Virginica*.)

271. 173. C. *perfoliata, Donn.
Alien. A North American plant. A. May to July. Not in *Top. Bot.*
First Record. In this book.
> Severn. Near Worcester, *Rea.* Kidderminster. Lansdown Crescent, Worcester.
> Lickey. Broome House Garden, on the flower beds, abundant, *Amphlett.* Clent Cottage Garden, 600 feet above the sea, 1905, *Amphlett.*

This plant has spread rapidly over England since its first introduction, not much more than a hundred years ago, and is found even on sandy fields and commons. It has been found in Stafford, Salop, and Warwick.

MONTIA, Linn. 75. (Named in honour of J. de Monti, an Italian botanist at Bologna.)

272. 174. M. fontana, Linn.   *Blinks, Water Chickweed.*
Native. Wet places. Locally common. A. or P. April to September. *Top. Bot.* 108.
First Record. Malvern, in a bog on the west side of the hill, *Purton, Midl. Fl.,* p. 91 (1817).
> Avon. Bredon Hill, near Woolas Hall.
> Severn. Pedmore Common. Abberley Hill, *Perry.* Areley Common. Wyre Forest. Titton Brook, near Hartlebury Common.
> Malvern. Common near Malvern, *Lees.* Malvern Link. Malvern Hills. Old Hills. Castlemorton Common. Ankerdine.
> Lickey. Stony Lane, Moseley, *Ick.* Bromsgrove Lickey. Clatterbatch, Clent. Streams at Wildmoor, Catshill.

This succulent little plant scarcely ever expands its flowers, whence its name of 'Blinks', as fearing to meet the sun. It is found in all neighbouring counties.

Var. erecta, Pers.
First Record. In this book.
> Severn. Dowles Brook, *Bagnall (MS.).* Areley Common.

This variety, given in *Lond. Cat.*, 9th ed., is not mentioned in 10th ed., where its place is taken by the segregates *minor* and *major* of Allhusen.

15. ELATINACEAE, Camb.

ELATINE, Linn. 77. (ἐλάτη, a pine, because of the appearance of the shape of the leaves of some of the species.)

{274. 175. E. hexandra, DC.   *Hexandrous Water-wort.*
Native. Borders of pools. One record only. A. July and August. *Top. Bot.* 24.
First Record. In a mill pond near Churchill (Kidderminster), Worcestershire, *A. Irvine, Phyt.,* vol. ii, p. 755 (1857).}
> Severn. As above.

{275   176. E. Hydropiper, Linn.   *Small Octandrous Water-wort.*
Native. Borders of pools. One record only. A. July and August. *Top. Bot.* 4.
First Record. In a mill pond near Churchill Railway Station, with *E. hexandra, A. Irvine, Phyt.,* vol. ii, p. 755 (1857).
> Severn. As above.

Nothing has been seen of these plants since the date of these records, although every pool near Churchill Station, and they are not a few, has been repeatedly searched by competent botanists. Irvine's specimens of *E. Hydropiper,* the rarer of the two plants in Britain, are in the Watsonian Collection in the Herbarium at Kew, but there are none of *E. hexandra.* This latter plant is recorded from Warwick and Salop of the neighbouring counties, and the former in Staffordshire, *teste* J. E. Bagnall.}

16. HYPERICACEAE, Benth. and Hook.

HYPERICUM, Linn. 78. (ὑπέρικον, the old Greek name, used by Dioscorides.)

276. 177. H. Androsaemum, Linn.   *Common Tutsan.*
Native. Woods and hedges. Local and rather rare. P. June to August. *Top. Bot.* 80.
First Record. Lanes at the foot of Malvern Hill, *Stokes ; Stokes's With. Bot. Arr.,* 2nd ed., p. 812 (1787).
> Avon. Bredon Hill. Broadway. Above Elmley Castle.
> Severn. Woods between Wolverley and Cookley. Between Worcester and Tewkesbury, *Withering.* Pecket Rock, Kidderminster, *Perry.* Upper Arley. Ribbesford Woods. Seckley Wood. Pedmore.
> Malvern. At the foot of the Hills, as above. West base of Worcestershire Beacon. Purlieu Lane, *Lees.* Abberley Hill. Witley Court. Old Storridge. Guarlford.
> Lickey. Lutley, near Halesowen, *Lees.* Frankley, 1854, *Mathews.* Randans Wood, *Humphreys.*

This is a handsome shrubby plant, and the flowers are numerous and showy. It once was much esteemed for the healing of wounds, and its common English name is a corruption of the French *Toute-saine,* All Heal. Another local name for it is Park-leaves. It occurs in all neighbouring counties.

279. 178. H. calycinum, Linn.   *Rose of Sharon.*
Alien. Shrubberies and plantations. P. June to October. Not in *Top. Bot.*
First Record. In a copse at Little Malvern, but doubtful if truly wild, *Lees, Illus. Nat. Hist.,* p. 173 (1834).

The handsome blossoms of this plant, three to four inches across, with tufted bunches of innumerable stamens in the centre of the golden cup joined by the petals, are often seen at the shrubbery border of some garden, where its creeping habit leads it to form dense masses of dark shiny leaves on stems a foot or more high. It was introduced into England as a garden plant so long ago as 1676, from its native habitat in the East, by Sir G. Wheeler ; and being hardy, evergreen, and insidiously creeping, it has had

plenty of time to make itself at home in a semi-wild state in very many places in England. The plant is recorded from Gloucester, Salop, and Warwick.

280. 179. H. perforatum, Linn.   *Perforated St. John's Wort.*
Native. Open woods, bushy places, waysides. Common and generally distributed. P. June to September. *Top. Bot.* 101.
First Record. In a record of *H. dubium,* 'often mixed with *H. perforatum,*' *Sheward, Nash, Hist. of Worc.,* Sup., p. 96 (1799).

This showy plant of the late summer is well distinguished from others of its tribe by its two-edged stems, and the clear dots, not forming a network, to be seen in the leaves if we hold them up to the light. A considerable amount of folk-lore has gathered about this plant, As its English name betokens, it was intimately connected with the old customs and practices which took place on St. John's Day, that is, Midsummer Day. And besides this, leaves, flowers, and roots were held to have great medicinal virtue ; and even the seeds also were supposed to have marvellous effect in a varied selection of complaints. Especially was a salve of its flowers considered good for wounds. The plant occurs in all neighbouring counties.

Var. angustifolium, DC.
First Record. Malvern Link, *Towndrow, Malvern Advertiser,* November 12, 1892.
In this variety the leaves are linear-oblong instead of elliptic, and the sepals are finely serrated. It has been found in Staffordshire and Warwickshire.

281. 180. H. maculatum, Crantz.   *Imperforate St. John's Wort.*
(*H. quadrangulum,* Linn. *H. dubium,* Leers.)
Native. Coppices, hedgerows, and moist places. Very common. P. July to September. *Top. Bot.* 76.
First Record. Discovered first as an English plant by Dr. Seward, of Worcester, growing plentifully about Sapey in that county, *With.,* 3rd ed., p. 665 (1796).
> Avon. Tiddesley Wood. Sheriff's Lench. Norton, near Evesham. Nafford's Mill. Bredon. Pershore. Fladbury. Crowle. Trench Woods.
> Severn. About Hagley and Churchill, *Mathews.* Wyre Forest, *Jorden.* Banks of the Severn. Witley. Stagbury Hill. Shrawley Wood. Hurcott Wood. Ombersley. Kempsey. Cotheridge. Nunnery Wood.
> Malvern. Martley. Sapey. Malvern Wells. Middleyards Coppice. Knightwick. Clifton-on-Teme. Abberley. Madresfield. Dripshill Wood. Leigh Sinton.
> Lickey. Halesowen Abbey. Kingsnorton. Near Bromsgrove. Fockbury.

The leaves of this plant show few if any dots, and their place is taken by a pellucid network of veins, while the showy flowers are usually strongly marked with black streaks and dots. It has much the same appearance as the last species, and no distinction would be made between them by the ordinary observer. It occurs in every neighbouring county.

Var. b. Babingtonii, H. and J. Groves.
First Record. In this book.
> Severn. Northwood, Bewdley, *Mathews.* Wyre Forest, *Mathews.*
In this variety the leaves are narrower, and the sepals minutely denticulate. It has been noted in Gloucester, Salop, and Warwickshire.

282. 181. H. quadrangulum, Linn.   *Square-stalked St. John's Wort.*
(*H. acutum,* Moench. *H. tetrapterum,* Fries. *H. quadrangulare,* Stokes. *H. quadratum,* Stokes.)
Native. Wet ditches and damp places. Abundant. P. July to September. *Top. Bot.* 104.
First Record. As *H. quadrangulum.* Lane at Kempsey, and many marshy spots in this county, *Lees, Illus. Nat. Hist. Worc.,* p. 173 (1834).

This species is easily distinguished by the four slightly winged angles to its stems, its sturdier habit, and its somewhat closely compacted cymes of yellow flowers. It is recorded from all neighbouring counties.

284. 182. H. humifusum, Linn.   *Trailing St. John's Wort.*
Native. Heaths, dry pastures, sandy fields. Rather uncommon. P. June and July. *Top. Bot.* 100.
First Record. Ronk's Wood, near Worcester, *Stokes.* Malvern Common, *Ballard, Stokes's With. Bot. Arr.,* 2nd ed., p. 814 (1787).
> Avon. Rather uncommon, *Lees.* Sheriff's Lench. South Littleton. Pershore. Defford. Norton, near Evesham. Trench Woods.
> Severn. Blakedown, *Mathews.* Churchill, near Kidderminster, *Mathews.* Webb's Hole, Hagley. Wyre Forest. Shrawley Wood. Hartlebury Common. Lincombe. Ombersley. Kempsey. Earl's Croome, Ladywood, Salwarpe. Broad Heath.
> Malvern. Malvern Hills, *Lees.* Leigh. Alfrick. Bransford. Martley. Stanford Bridge. Powick. North Hill, Malvern.
> Lickey. Little Farley Wood, *Humphreys.* Pepper Wood. Bromsgrove Lickey, *Mathews.* Campion's Pool, Catshill, *Humphreys.* Clent Hill. Twylands. Frankley. Wildmoor, Catshill. Barnt Green. Blackwell.

This is a pretty little prostrate plant. Its thin stems bordered with little leaves trail among the grass or on the surface of the sandy field, unlike any of those of its kindred except the winter rosette of *H. perforatum,* which it somewhat imitates. It is recorded from every neighbouring county.

286. 183. H. pulchrum, Linn.   *Upright St. John's Wort.*
Native. Heaths, dry places in woods. Not uncommon. P. June to August. *Top. Bot.* 111.
First Record. Not localized, *Walker, Midl. Med. Surg. Rep.,* vol. i, no. 2, p. 100 (1828).
> Avon. Trench Woods. Norton, near Evesham. Crowle. Fladbury. Pershore. Sheriff's Lench. Sharpway Gate. Near Inkberrow. Near Redditch.
> Severn. Crown East. Wyre Forest. Ombersley. Claines. Kempsey. Spetchley. Monk Wood. Shrawley Wood. Nunnery Wood. Arley.
> Malvern. North base of the End Hill, *Lees.* Warren Hill, *Lees.* Ankerdine Hill. Clifton-on-Teme. Abberley. Old Hills. Middleyards. Leigh. Alfrick. Malvern Link Common.

Lickey. Frequent on dry banks, *Mathews.* Meadow near Moseley Park, *Ick.* Frankley.

The flowers of this plant are deeper yellow than those of others of the genus, and with the stem and young buds often more or less tinged with red. The stem is slender, the leaves few, and the flowers abundant. It is found in all neighbouring counties.

287. 184. H. hirsutum, Linn.   *Hairy St. John's Wort.*

Native. Woods, coppices, hedgerows. Abundant. P. June to September. *Top. Bot.* 90.

First Record. Not localized, *Walker, Midl. Med. Surg. Rep.*, vol. i, no. 2, p. 100 (1828).

Here again we find a characteristic, which at once marks off this plant from all others of the commoner St. John's Worts, in the round hairy stems and pubescent leaves. The plant prefers a calcareous soil, and is not so common in the Lickey district as in other parts of the county. It is recorded from all bordering counties.

288. 185. H. montanum, Linn.

Native. Wood and hedgebanks. Local. P. July to September. *Top. Bot.* 46.

First Record. Upon banks under hedges, and by wood-sides about Pershore and on Bredon Hill, *Nash, Hist. of Worc., Int.* p. lxxxix (1781).

Avon. Pershore; Bredon Hill, as above. Tiddesley Wood.

Severn. Wyre Forest, *Jorden.* Blackstone Rock, near Bewdley, *Perry.* Wood west of Abberley Church. Pecket Rock, Kidderminster, *Perry.* Shrawley Wood, *Lees.*

Malvern. Southern part of the range, *Westcombe.* Hollybush Hill, sparingly, *Lees.* Silurian eminences, *Lees.* Keysend. Quarry near the Hundred House. Midsummer Hill.

The calyx and bracts of this species, as the last, are fringed with short-stalked glands, but this plant is smooth while that is hairy. It also is a lover of calcareous soil. It has not been recorded from Warwickshire.

{**289. 186. H. elodes,** Linn.   *Marsh St. John's Wort.*

Native. Wet places and bogs. Very local, and now lost. P. June to September. *Top. Bot.* 62, in which it is not recorded for this county (p. 9a).

First Record. In a drained mill pool on Moseley Common, *Miss M. A. Beilby, Analyst*, vol. vi, p. 294 (1837).

Lickey. Moseley, as above. Bromsgrove Lickey, *Lees.*

Whether Mr. Lees saw this plant growing at the Lickey or no, one cannot say. At p. 135 of his *Bot. of Worc.* his words are: 'The plants in the Bromsgrove Lickey division [which do not occur elsewhere in the county] are entirely made up of species which, like the Cranberry and *Hypericum elodes*,' &c. At all events it is not there now, nor is it at Moseley, which for many years has been a suburb of Birmingham. It may be regarded as extinct. The plant has been recorded for every neighbouring county except Gloucestershire.}

to cure coughs and colds. But this was only one of the cases in which the mallow used to be held efficacious. It was good for agues, cured all affections of the eye, took away the pain of the sting of a wasp, and last, but not least, prevented the hair falling off. The plant is certainly very mucilaginous. It is recorded from all neighbouring counties.

296. 189. M. rotundifolia, Linn.   *Dwarf Mallow.*

Native. Roadsides, dry waste places, under walls in villages, and near farm buildings. Not common. P. May to August. *Top. Bot.* 83.

First Record. Query as *M. parviflora.* Roadsides, and often near buildings, *Pitt, Agric. Worc.*, p. 317 (1810).

Avon. Not common, *Lees.* South Littleton. Churchill, near Worcester. Harvington. Fladbury. Pershore. Crowle.

Severn. Various localities, *Lees.* Hartlebury. Chatley. Barbourne Lane. St. John's, Worcester. Kempsey. Near Droitwich. Salwarpe. Ombersley. Hallow. Grimley. Holt. Near Bewdley.

Malvern. Various localities, *Lees.* Bransford. Leigh. Martley. Clifton-on-Teme. Powick. Guarlford. Malvern Link Common.

Lickey. Not common, *Lees.* Catshill, near Bromsgrove. Clent. Finstall.

This species is not nearly so frequent as the last, and rare in Scotland. It is distinguished from it by its prostrate stems and smaller flowers. Both this and the Common Mallow are used in the East as a vegetable and eaten with meat. Job, in his distress, says 'they cut up the Mallows by the bushes for their meat'. What plant was meant has been a matter of discussion. A well-known garden shrub, *Kerria japonica*, whose abundant yellow blossoms, usually in the double form, ornament many a garden shrubbery in the spring, is called Jews' Mallow, but is not a Mallow at all, but a member of the Rose tribe. The Dwarf Mallow occurs in all neighbouring counties.

18. TILIACEAE, Juss.

TILIA, Linn. 82. (Name of obscure origin, perhaps from the Celtic; in modern Gaelic the Lime is called *Teile.*)

300. 190. T. platyphyllos, Scop.   *Large-leaved Lime.*

(*T. grandifolia*, Ehrh.   *T. europaea*, Linn.)

Alien. A planted tree. June to July. *Top. Bot.* 3.

First Record. As *T. grandifolia.* In a field near the Priory Farm, Little Malvern, *Lees, Bot. Malv. Hills*, 1st ed., p. 28 (1843).

Mr. Watson, *Top. Bot.*, p. 87, allows the true nativity of this tree only in three counties, Hereford, Radnor, and York. It is often confused with *T. vulgaris*, from which, however, it is easily distinguished when in fruit. In *Worc. Nat. Club Trans.* i. 194, are some remarks by Mr. Lees on the Lime as indigenous in Worcestershire. In *Lond. Cat.*, 10th ed., its nativity in Britain is assumed. It is recorded, more or less doubtfully, for Hereford, Salop, and Warwick of neighbouring counties, besides Hereford as above.

17. MALVACEAE, Juss.

MALVA, Linn. 81. (μαλακός, soft; from the emollient properties of the mucilage which it contains.)

294. 187. M. moschata, Linn.   *Musk Mallow.*

Native. Field borders, waysides, dry pastures. Locally common. P. May to October. *Top. Bot.* 91.

First Record. Frequent in moist meadows and amongst bushes in rough grounds, *Nash, Hist. of Worc., Int.*, p. lxxxix (1781).

Avon. With white flowers at the Slads and Sheriff's Lench, *Worc. Nat. Club Trans.* iii. 125. Crowle. Norton, near Evesham. Near Pershore. Fladbury. Tiddesley Wood.

Severn. West of Abberley Church, *Perry.* Bewdley, *Perry.* Henwick; Helbury Hill, *Lees.* Hagley Brake. Shrawley Wood. Hurcott Wood. Harvington. Hartlebury. Ombersley. Kempsey. Cotheridge. Wyre Forest. Monk's Wood.

Malvern. North Hill, *Lees.* Martley. Bransford. Leigh. Ankerdine Hill. Powick. Near Southstone Rock. Alfrick. Madresfield (with white flowers). Malvern Link.

Lickey. Moseley, *Ick.* Bilberry Hill. Clent. Lickey Hill. Tardebigge Reservoir. Blackwell.

This handsome flower, when it occurs, is a pretty feature, its large pink flowers crowded together at the extremities of the stems, of a hue somewhat uncommon among our herbaceous plants, and standing out conspicuously among its neighbours. A white form, which is stouter and stronger and more floriferous than the wild plant, is a common garden flower, its purple stamens forming a fine contrast with its showy petals. The flower is slightly scented with musk, more powerfully in the evenings. The plant occurs in all neighbouring counties.

Var. **heterophylla,** Lej.

First Record. As in this book.

Severn. Near Pope Iron Inn, Worcester, *Herb. Hast. Mus.*

Lickey. Clent, *Herb. Hast. Mus.*

In this variety the lower leaves are roundish entire, and the upper leaves are deeply lobed. It is not given in *Lond. Cat.*, 10th ed.

295. 188. M. sylvestris, Linn.   *Common Mallow.*

Native. Waysides, waste ground. General throughout the county. P. May to October. *Top. Bot.* 96.

First Record. Vars. 2 and 3. Near Worcester, *Stokes*; *Stokes's With. Bot. Arr.*, 2nd ed., p. 739 (1787).

This is a common wayside plant, so fond of houses and the neighbourhood of man that doubts have been cast upon its true nativity in Britain. It is a favourite with village children, who call the flattened round seeds 'Cheeses'. Poodle Pink, Bread and Cheese, Cheese Cakes are also country names by which it is known. The plant, like many others, was supposed to have marvellous virtues in many various diseases; even to-day mallow is used

F

301. 191. T. vulgaris, Hayne.   *Common Lime.*

(*T. europaea*, Linn.   *T. intermedia*, DC.)

Alien. A plentifully planted tree. June. Not in *Top. Bot.*

First Record. As *T. europaea.* Not localized, *Lees, Bot. Looker-out*, p. 94 (1842).

Not until about the middle of the seventeenth century was this tree called the Lime. Previously it was called the Lind, a word now obsolete, but in a softened and poetic form still surviving as Linden. Blithe, in 1649, says 'The Lime Tree is also newly discovered as useful in our English plantations'. What the use was is doubtful, for the timber of the tree is comparatively worthless, and at the present day it is never planted except as a quick-growing tree for ornament, in avenues, parks, and garden walks, where it makes a good feature, when its heads are trimmed into shape; and in this condition it is frequently seen in churchyards. John Evelyn in his *Sylva* recommended its culture for avenues, and it is largely used for this purpose on the Continent. The drawback is that it is the first of our trees to shed its leaves in autumn, and on the outskirts of a town looks sere and yellow quite early in the season. The blossoms are very profuse and fragrant, and the hum of bees among them is frequently audible at a considerable distance away; and the leaves are especially liable to be attacked by aphids, to the exudation from which clings every sort of dust and common particle borne on the summer breezes, making the leaves black and dirty. The bark of the tree is largely exported from Russia in the shape of the well-known garden-matting called Bass Mats; and it has been used as a material on which to write. The flowers and bracts are dried and used for coughs, and the young buds of spring are full of mucilage. The common Lime is to be seen in all bordering counties.

302. 192. T. cordata, Mill.   *Small-leaved Lime.*

(*T. europaea*, Linn.   *T. parvifolia*, Ehrh.)

Native. In several localities. June. *Top. Bot.* 18.

First Record. As *T. parvifolia.* Woods at the north end of the range (Malvern), *Lees, Mag. Nat. Hist.*, vol. iii, p. 160 (1830).

Avon. Trench Woods.

Severn. Eymore Wood, Wolverley, *Mathews.* Shrawley Wood. Wyre Forest, *Lees.* Monk's Wood. Ockeridge Wood. Hawford. Walnut Tree Farm, Ladywood.

Malvern. Old Storridge; Clifton-on-Teme, *Lees.* Rosebury Rock. Woods at Berrow and Castle Morton, plentiful, *Lees.* Mathon. Lane between Acton Beauchamp and Suckley. Half Key. Near Knightsford Bridge.

Lickey. Recorded by *Mathews*, with *T. grandifolia*, for this district; acknowledged by him to be a mistake, *Mid. Nat.* xv, p. 278. Redditch, *D. Mathews.*

The Small-leaved Lime is undoubtedly indigenous, and especially at Shrawley Wood, where the undergrowth is largely made up of it. It is recorded for all neighbouring counties.

19. LINACEAE, DC.

RADIOLA, Hill 83. *(Radius, a ray; from the radiating nature of the branches.)*

303. 193. R. linoides, Roth.   *All Seed, Rupture Wort.*
(*Linum Radiola,* Linn. R. *Millegrana,* Sm.)
Native. Damp heaths, grassy sides, places where herbage is short. Not common. A. June to September. *Top. Bot.* 85.
First Record. As *Linum Radiola.* Astwood Heath, *Purton, Midl. Flora,* p. 165 (1817).
Avon. One place, *Lees.*
Severn. Hartlebury Common, *Lees.* Harborough Pools. Pedmore Common, *Scott, Thompson.*
Lickey. Astwood Heath, as above. Moseley, *Lees.* Moseley Common. Moseley Wake Green, *Miss M. A. Beilby.*
Radiola has been recorded for Warwick, Stafford, and Salop.

LINUM, Linn. 84.   (λίνον, thread; *Lin,* in Celtic and Modern Gaelic.)

304. 194. L. catharticum, Linn.   *Purging Flax.*
Native. Dry pastures and waysides. General. A. May to October. *Top. Bot.* 112.
First Record. Near Dudley Castle, *Booker, Dudley Castle,* p. 107 (1825).
This graceful little plant is universally distributed, but not very conspicuous, as its white flowers are very tiny, and it rarely grows more than six inches high. Country people say it is a cure for rheumatism, and sometimes call it Mill Mountain, a name of obscure origin under which it used to be sold in the seventeenth century. It occurs in all the neighbouring counties.

306. 195. L. angustifolium, Huds.   *Narrow-leaved pale Flax.*
Native. Sandy and calcareous places. Rare. P. May to September. *Top. Bot.* 37.
First Record. As in this book.
Avon. Sheriff's Lench, August 5, 1890, *Towndrow, Rea.* The Slads.
This differs from the next plant in having obovate obscurely five-veined sepals instead of ovate three-veined ones. Of neighbouring counties it has not been found in Salop.

307. 196. L. usitatissimum, Linn.   *Common Flax, Linseed.*
Casual. Waste places, rubbish heaps, near flour mills. Always introduced.
A. June to September. Not in *Top. Bot.*
First Record. Astwood, *Purton, Midl. Flora,* p. 164 (1817).
Avon. Between Evesham and Elmley Castle, *Cheshire.* Norton, near Evesham.
Severn. Fields near Hartlebury, *Perry.* Between Dowles Brook and Bewdley. Roadside leading to Nunnery Wood. Bath Road near Worcester. Merryman's Hill. Wolverley.
Malvern. Near Little Malvern, *Lees.* Cowleigh Park. Alfrick. Malvern Common. Malvern Link.
Lickey. Field Lane, Clent, 1904, *W. Whitwell.* Hunnington, near Halesowen. Blackwell Incline.

The plant is probably a native of North-East Africa; it has not the slightest pretension to be considered a native of Britain. It is probably one of the earliest cultivated plants in the world. From it is obtained the fibre which has been made into linen by people in all ages, and continued to be so until its universal use was superseded by the cotton of America. When it was introduced into England is doubtful; the plant is not mentioned in literature before the year 1000. However that may be, by Elizabeth's time it had been recognized as so useful that laws had been passed enforcing its culture in this country. It was macerated by being steeped in ponds and brooks, a noxious proceeding; and in the Manor Court Rolls of these times, numbers of instances occur of penalties inflicted on villagers for this practice, and of injunctions to preserve this or that stretch of water from such contamination. Cattle were seriously affected if they obtained access to flax-poisoned water. In England, Flax culture is almost obsolete; it still obtains in the North of Ireland and some parts of Scotland. Nor is it the fibre alone which is valuable. From the seeds is obtained the well-known Linseed oil, and the refuse forms the also well-known cattle food Linseed Cake, now like Flax fibre, largely deposed in favour of cotton, from the seeds of which a feeding-stuff is made. Flax has been recorded from Stafford, Warwick, and Salop, and without doubt has escaped from cultivation in the two other bordering counties, but has not been noticed.

20. GERANIACEAE, Juss.

GERANIUM, Linn. 85.   (γέρανος, a crane; the fruit resembling the beak of the bird.)

308. 197. G. sanguineum, Linn.   *Bloody Crane's-bill.*
Native. One locality only. P. July and August. *Top. Bot.* 63.
First Record. Wyre Forest, *Gissing, Phyt., O.S.,* vol. i, p. 151 (1855).
With regard to this plant, Mr. Lees makes two statements in his *Bot. of Worc.* Dowles Brook, running through Wyre Forest, divides Worcestershire from Salop, and in his time the plant was fairly plentiful on the Salop side. On p. 6 he says 'a stray plant has been found occasionally over the border', while on p. 7 his words are: 'Mr. Jorden [observes] that scarcely a stray plant ever gets on the Worcestershire side of the brook. This county has, therefore, hardly a colourable claim to the plant.' Nevertheless a few plants can generally be found on the Worcestershire side of the brook, especially after a recent fall of timber. It is recorded from Gloucestershire, Herefordshire, and of course as above from Salop.

309. 198. G. *versicolor, Linn.
(*G. striatum,* Linn.)
Alien. Always a garden escape, not recognized in most Botanies. P. June and July. Not in *Top. Bot.*
First Record. Severn Esplanade, Worcester, 1852, *Lees, Bot. of Worc.,* p. 40 (1867).
Though this garden plant is naturalized in several localities, especially in the south-west of England, it is always traceable to garden culture. Its garden name in Worcestershire, possibly a local one, is 'Lady Coventry's

Needlework'. It is recorded with the following plant, neither being localized, as 'new' to the county, *Worc. Nat. Club Trans.* iii, p. 45 (1903). It has been found 'wild' enough to be recorded in Herefordshire.

310. 199. G. *nodosum, Linn.
Alien. A garden escape. P. June and July. Not in *Top. Bot.*
First Record. Severn Esplanade, after the new walk was made, *Lees, Bot. of Worc.,* p. xxix (1867).
What is said of the status of the last plant may be said also of this. As Mr. Lees records this from the same locality as that possibly he means the same plant. Mr. Towndrow has a specimen, however, which was sent to him by the late Mr. T. R. A. Briggs, and labelled, 'Abundantly naturalized near Malvern, Worcestershire, July, 1877. Coll. *J. Cosmo Melville.*' He remarks that he never met either with this or *G. striatum.* Nor has it been met with in any neighbouring county.

311. 200. G. *phaeum, Linn.   *Dusky Crane's-bill.*
Colonist. Hedgerows and banks. Rare. P. June and July. Not in *Top. Bot.*
First Record. Near Cradley, Worcestershire, *With.,* 3rd ed., p. 605 (1796).
Severn. Grimley, *Lees.* Redstone Rock, plentiful. Near Hartlebury Church. Near Bromsgrove, *Humphreys.*
Malvern. Abberley, *Rev. Mr. Severn, Rufford.* Well Meadow, below, Abberley Village, *Sheward.*
Lickey. Cradley Park, *Scott.* Near Broom Hill, Belbroughton, *Amphlett.* Wood near Hewell.
This plant has better claim to be a native British plant than either of the foregoing, since it is found in a native state on the Continent as far north as Belgium; but most authorities consider it only naturalized in this country. It is frequently cultivated in gardens, where it is remarkable for its deep claret-coloured, almost black, flowers; and it seeds about freely on the borders. It occurs in all the neighbouring counties.

312. 201. G. sylvaticum, Linn.   *Wood Crane's-bill.*
Native. Moist meadows and coppices. Rare. P. June and July. *Top. Bot.* 56.
First Record. Near Halesowen, *Withering, Stokes's With. Bot. Arr.,* 2nd ed., p. 727 (1787).
Severn. One locality, abundant, *Lees.* Wyre Forest, near to Dowles Brook, plentifully, *Walcot* and *Lees.* Roadside near Croome Park, 1908, *Jeffery.*
Lickey. Furnace Coppice, Halsowen, *Mathews.* Forty yards north of the Stourbridge road at the side of the path Oldnall to Cradley, *Rev. J. H. Thompson.*
The flowers of this plant are somewhat smaller than those of the next, from which it may be distinguished also by its deflexed fruit-stems, and its netted instead of dotted seeds. It is recorded from all neighbouring counties.

313. 202. G. pratense, Linn.   *Meadow Crane's-bill.*
Native. Meadows, roadside wastes, and thickets. General in the south and west of the county. Not common in the Lickey district. P. May to August. *Top. Bot.* 92.

First Record. Frequent in moist meadows, and amongst bushes in rough grounds, *Nash, Hist. of Worc., Int.,* p. lxxxix (1781).
Avon. Not uncommon, *Lees.*
Severn. General, *Lees.* An ornament to the bank of the Severn and of many minor brooks, *Lees.*
Malvern. General, *Lees.* Near Tenbury.
Lickey. Near Offad's Well, Bromsgrove.
This is a very handsome plant, growing to a height of a couple of feet, with large blossoms of a bluish-purple colour. In gardens a white form is sometimes seen, all of which, it is believed, originated from a plant with white flowers found by the *Rev. C. Lunn,* just outside Pershore, July 13, 1883. It occurs in all neighbouring counties.

314. 203. G. pyrenaicum, Burm. fil.   *Mountain Crane's-bill.*
(*G. perenne,* Huds.)
Colonist. Roadsides and railway embankments. Local. P. May to September. *Top. Bot.* 65.
First Record. Not localized, *Lees, Cat., New Bot. Guide* (1835).
Avon. One locality. Near Pershore.
Severn. Abundant in the neighbourhood of Droitwich Railway Station. Catchem's End, Bewdley. With white flowers near Warshill Top Wood, Bewdley, *Rea.* Dodderhill Church. Near Hartlebury, *Miss Ladbury.* Mount Pleasant, Stoke Prior, *Humphreys.* Near Cotheridge, *Lees.* Pathside, Droitwich Road to Claines Church. Boughton. Hedge-bank between the Bromyard and Bransford Roads. In a field between Areley Kings and Abberley Hill.
Malvern. Bransford Lane. Near Crews Hill Wood. Saint Mathias's Churchyard, Malvern Link. Hedgerow, Malvern Wells, *Towndrow, Malv. Advert.,* Nov. 20 (1897). Drakelow.
Lickey. Bromsgrove Station.
This plant was unnoticed until the time of Hudson (1762), and is of doubtful nativity in Britain. However that may be, it is abundant at Droitwich Railway Station on the banks behind the platforms, and on the embankments north of it, conspicuous with its purple flowers. It is found in all neighbouring counties.

315. 204. G. molle, Linn.   *Dove's-foot Crane's-bill.*
Native. Abundant in cornfields, in pastures, on waysides, and in waste places. A. April to December. *Top. Bot.* 112.
First Record. Not localized, *Lees, Bot. Malv. Hills,* 1st ed., p. 33 (1843).
One of the most generally distributed of wayside and waste ground plants, with soft leaves grey-green by nature, and often made greyer by the dust of the highway. A country name for it is Dragon's Blood. The leaves spreading out from the centre, on stems of varying length, form sometimes large circular clumps. It occurs in each bordering county.

316. 205. G. pusillum, Linn.   *Small-flowered Crane's-bill.*
(*G. parviflorum,* Curt.)
Native. Cultivated ground and waysides. Except in the Avon district abundant. A. May to September. *Top. Bot.* 81.
First Record. Not localized, *Walker, Med. Surg. Rep.,* vol. i, no. 2, Nov., 1828.

Mr. Lees records the plant with white flowers from sandy ground at Henwick Hill. It is very much like the Dove's-foot Crane's-bill, and difficult of distinction to the unpractised eye, but it has unwrinkled pubescent capsules, instead of glabrous capsules wrinkled transversely. It is recorded from all neighbouring counties.

317. 206. G. rotundifolium, Linn.   *Round-leaved Crane's-bill.*
Native. Dry hedgebanks and waysides. Local. A. or B. May to October. *Top. Bot.* 21.
First Record. On a wall at Hartlebury, *Purton, Midl. Fl.,* p. 321 (1817).
    Avon. Defford Common. Between Tiddesley Wood and Pershore. Norton, near Evesham. Fladbury.
    Severn. Clerkenleap. Near Bewdley. Ombersley.
    Malvern. 'Uncommon, and I feel somewhat uncertain of its locality,' *Lees.* Martley.
The leaves of this plant are of a paler yellow tint than those of any other British species. This also is much like the Dove's-foot Crane's-bill, but is not nearly so common ; and it may be distinguished by its entire petals, and netted, and not smooth, seeds. It is recorded from Gloucester and Salop of the bordering counties.

318. 207. G. dissectum, Linn.   *Cut-leaved Crane's-bill.*
Native. Cornfields, cultivated ground, hedgebanks. General. B. May to August. *Top. Bot.* 110.
First Record. Near Dudley Castle, *Booker, Dudley Castle,* p. 107 (1825).
Next to *G. molle* this is the commonest species of Crane's-bill met with. The leaves are hairy, not soft and downy, and the flowers are on short stalks. It is recorded from all neighbouring counties.

319. 208. G. columbinum, Linn.   *Long-stalked Crane's-bill.*
Native. Fields and hedgebanks. Rather local. A. May to October. *Top. Bot.* 76.
First Record. Roadside toward Broadwaters, Kidderminster, *Perry, Mag. Nat. Hist.,* vol. iv, p. 450 (1831).
    Avon. Bredon Hill. Broadway, *Lees.* Craycombe Hill. Trench Woods.
    Severn. Shady walk beyond the Old Waterworks, Worcester ; lanes about Worcester and Hallow, *Lees.* Roadsides near Stone. Fenny Rough, *Mathews.* Hagley. Roadside near Henwick. Ombersley Road. Near Spetchley. Habberley Valley. Pedmore Lane, Stourbridge. Wyre Forest.
    Malvern. On Abberley Hill, *Lees.* Roadside near Ankerdine Hill. Malvern Link. Newland. Malvern Wells.
    Lickey. Not common, *Mathews.* Kendal End, Lickey.
This graceful plant is easily recognized. The flowers are placed on long slender stalks, which jut out from the main stem. The flowers also are larger than those of the species last described. The plant is recorded from all neighbouring counties.

320. 209. G. lucidum, Linn.   *Shining Crane's-bill.*
Native. Hedgebanks, old walls. Local, and uncommon in the Lickey district. B. May to August. *Top. Bot.* 93.

First Record. In the lanes about Wolverley and on the rocks at Great Malvern, *Purton, Midl. Fl.,* p. 320 (1817).
    Severn. Between Birchwood and Drakelow. Wyre Forest. Barbourne Lane, Worcester. Lincombe. Near Broadwaters, Kidderminster. Wolverley. Near Virgin's Tavern, Worcester. Blackstone. Lane near Rushock.
    Malvern. Plentiful on the rocks. Back lane, North Malvern. Bransford Railway Station. Little Malvern Churchyard. Malvern Hills. Malvern Link.
    Lickey. Blackwell. Bromsgrove Lickey, *Humphreys.* Clent, extinct. *Amphlett.* Near Monks, Chaddesley Corbett, *Amphlett.* Halesowen Hill, *Scott.* Road above Brewins, Halesowen, *Mathews.* Northfield.
The foliage and stems of this plant are smooth and glossy, whence its specific name, and the lower parts of the plant especially in dry places are often tinged with bright red. In the *Bot. Rec. Club Rep.* for 1880 this plant is given, unlocalized, as a new record for the county by Mr. Bagnall. But it was not so. It is recorded from every neighbouring county.

321. 210. G. Robertianum, Linn. *Stinking Crane's-bill, Herb Robert.*
Native. Hedges, woods, wall-tops. General. B. April to September. *Top. Bot.* 112.
First Record. Not localized, *Walker, Med. Surg. Rep.,* vol. i, no. 2 (1828).
The odour of this plant is strong and disagreeable, and believed to be obnoxious to many and certain kinds of insects, and therefore by cottagers it used to be placed near beds to repel them. The English name has been variously supposed to refer to Robert, Duke of Normandy, St. Robert, and St. Rupert, in the one case because it bloomed about the 29th of April, the day of St. Robert, the founder of the order of Carthusians. The plant is called by country people Smelling Robert and Betty Pink. It contains tannin, and by the old herbalists was supposed to be good for wounds. In North Wales it is said to be very efficacious in cases of gout. It has been found with white flowers near Monk Wood, at Holt Mill, Eyemore Wood, and on the Vine Pool dam, Clent ; and also at Hopton Court, Leigh, and Leigh Sinton. The plant occurs in all neighbouring counties.

ERODIUM, L'Hérit. 86.   (ἐρωδιός, a heron ; the fruit resembling the beak of the bird.)

322. 211. E. cicutarium, L'Hérit.   *Hemlock Stork's-bill.*
Native. Sandy fields and heaths, wall-tops, and waysides. Local. A. or B. April to September. *Top. Bot.* 105.
First Record. By the side of the road near Hallow, abundantly, *Lees, Illus. Nat. Hist. Worc.,* p. 170 (1834).
    Avon. Rather uncommon, *Lees.* Bredon Hill. Broadway.
    Severn. Hallow, as above. Near Stourport. Henwick Hill. Pasture at Kempsey. Hagley Brake. Blakedown. Churchill by Kidderminster. Worcester Court Farm. Near Bewdley. Oldbury Road. Upper Wick. Holt. Devil's Spittleful. Lincombe. Hartlebury Common.

    Malvern. Little Malvern. Martley. Malvern.
    Lickey. Sandstone Rock at Catshill, *Humphreys.* Fields at Thicknall, Clent.
This plant occurs in all neighbouring counties.

323. 212. E. moschatum, L'Hérit.   *Musky Stork's-bill.*
(*Geranium moschatum,* Burm.)
Casual in the county. Waste places. Very rare. A. or B. June and July. *Top. Bot.* 12.
First Record. As *Geranium moschatum.* Near Stourbridge, *Stokes ; Stokes's With. Bot. Arr.,* 2nd ed., p. 725 (1787).
    Avon. One place, suspicious, *Lees.*
    Severn. Stourbridge, as above. *Mr. Lees* evidently considered this locality to be in the Lickey district, in which he records the plant as suspicious. About Bewdley, *Purton.* Sandy fields about Stourbridge, *Scott.* 'A large and robust form near Fenny Rough,' *Worc. Nat. Club Trans.* iii, p. 234.
    Malvern. On Lumbertree Bank, Welland Common, *Lees,* who adds in the 3rd ed. of his *Bot. Malv.,* 'I fear lost by enclosure of the waste.'
    Lickey. Near Dudley Castle, *Booker.*
This plant, like the former one, is fonder of the neighbourhood of the sea than inland localities, but it is a larger and handsomer flower. The foliage, which is somewhat clammy, leaves a faint odour of musk when passed through the hand, whence its name. Watson, in *Top. Bot.,* records it only in Gloucestershire and Worcestershire of the West Midland group of counties, but it has been found in Staffordshire by Mr. Bagnall, and in all other counties on the Worcestershire borders.

324. 213. E. maritimum, L'Hérit.   *Sea Stork's-bill.*
Native. Sandy places. Rare. A. or B. May to September. *Top. Bot.* 33.
First Record. Sandy commons between Enville and Bewdley, *Mr. J. A. Hunter, Stokes's With. Bot. Arr.,* 2nd ed., p. 725 (1787).
    Severn. Habberley Valley, more than once, *Miss Ladbury* (MS.). Roadbank, Hartlebury Common. Trimpley. Pedmore Common, 1835, *Rev. W. L. Baynon.* Roadside between Stourport and Bewdley. Button Oak.
    Malvern. Plentiful at the eastern base of North Hill before the path turns to the Ivyscar Rock, *Lees,* who says later, 'since obliterated at this spot.' It was still there in 1888, *Towndrow.* Near Barnard's Green, *Lees.*
    Lickey. Clent, opposite the church at the foot of the hills, *Irvine.*
This plant, as its name denotes, is still more strongly a lover of the sea than either of the preceding, but yet more often appears inland than the latter one, from which it is distinguished by its simple ovate cordate instead of pinnate leaves, and the fewer flowered peduncles. The flowers are small and inconspicuous and the petals often wanting. Except by Mr. Irvine, the plant has never been seen in the locality at Clent. It is recorded from all neighbouring counties.

OXALIS, Linn. 87.   (ὀξύς, acid.)

325. 214. O. Acetosella, Linn.   *Wood Sorrel.*
Native. Woods, thickets, damp banks. Generally distributed and abundant in the Lickey district. P. April to August. *Top. Bot.* 111.
First Record. Not localized, *Walker, Med. Surg. Rep.,* vol. i, no. 2, November, 1828.
This plant is sensitive to the weather, its flowers closing at the approach of rain, while its leaflets close themselves together under the same circumstances, and at the approach of night. All the family of Wood Sorrels exhibit this tendency in a greater or less degree. The acidity of the plant, formerly called Wood-sour, is due to pure oxalic acid, which was formerly obtained from the leaves, a process which is now superseded by the easier one of treating sugar with nitric acid. Oxalic acid is well known, being the familiar Salts of Lemon by which housewives remove the stains of ink or iron-mould from linen. Another old name for the plant was 'Alleluia', either because, as Gerarde says, it flowers when Alleluia was wont to be sung in the churches, or from its Italian name of *Juliola,* whence at all events came its medicinal name of *Luxula.* Gerarde also says it is called 'Cuckowe's meat, either because the cuckowe feedeth thereon or by reason that it springeth forth and flowereth when the cuckowe singeth most '. The root of the plant is curious, being like a string of beads, and sometimes deeply red. The plant is considered a candidate for the honour of being the Shamrock of Ireland, but is popularly deposed in favour of one of the Trefoils. It is found in all neighbouring counties.

326. 215. O. corniculata, Linn.
Casual. Garden ground and garden refuse heaps. Rare. P. June to September. Not in *Top. Bot.*
First Record. One plant on a rubbish heap at Malvern Link, *W. J. Rendall ; Towndrow, Malvern Advertiser,* Dec. 22, 1900.
    Malvern. As first record.
This plant is probably a native of tropical South America, and in Britain is cultivated in gardens, in which it is so much at home that it maintains itself and becomes a weed. It sometimes escapes, and maintains itself outside them. It has been noticed in this condition in Staffordshire and Salop.

IMPATIENS, Linn. 88.   (The name is from the quality the ripe capsules have of bursting when touched.)

I. Roylei, Walp.   *Snapping Balsam.*
Casual Alien. Waste ground. B. June to September. Not in *Top. Bot.,* nor in the *Lond. Cat.,* 10th ed.
Var. macrochila, Lindl.
First Record. On the bank of a brook adjacent to a coppice near to Madresfield, *Rea* and *Edwards, Worc. Nat. Club Trans.* iii, pp. 6, 46 (1903).
    Malvern, as above.
This plant is a native of the Himalayas, and cultivated as a garden plant. Its history as an English plant is given by Mr. Britten, *J. of B.,* 1900, p. 50.

21. AQUIFOLIACEAE, DC.

ILEX, Linn. 89. (Etymology of the name doubtful. Perhaps the same as *Ulex*, Celtic *uile*, all, *ec* or *ac*, a sharp point. It is the Latin name of the holm, or evergreen, oak.)

332. **216. I. Aquifolium,** Linn.   *The Holly.*
Native. Woods and hedges. Locally plentiful, and generally distributed. Tree. *Top. Bot.* 105.
First Record. Almost covering the hills in the southern part of the chain (Malvern), *Lees, Mag. Nat. Hist.*, vol. iii, p. 160 (1830).
The name Holly is a modification of an even older name for the plant, Hollin or Hollen, a word now, except in Scotland, nearly forgotten. It has no connexion with 'Holy', in spite of the association of the tree with the decking of churches or its use for domestic decoration at Christmas, though from this use it has obtained in many places, especially in North Worcestershire, whence it is taken by cartloads into the neighbouring Black Country, the name of 'Christmas'. The Holly plays many parts besides this best-known one. It bears cutting well, and like the Yew can be trimmed to a number of shapes. Especially does it form fine hedges, and many fine trimmed hedges are to be found in all parts of the country, some of them of enormous size. Its wood is firm and white, and is used for many purposes, one being to make the Tunbridge ware, which delighted our grandparents, for which purpose it was stained many colours; but the curiously stained verdigris Tunbridge ware was originally made from oak wood infected with the mycelium of the fungus *Chlorosplenium aeruginosum*. The tree also is especially grown to form whip handles. From the bark is made bird-lime; from the leaves used to be made a decoction which was of service in intermittent fevers. Many foreign species produce leaves that are used for tea, the Jesuits' Tea, or 'Yerba Maté', of Paraguay, being the leaves of *I. Paraguensis*; and it of it millions of pounds are exported from that country annually. Sheep browse on the leaves of the plant, and rabbits attack them ruthlessly in winter on the slightest provocation. In Worcestershire it is especially abundant on parts of the Malvern Hills, one of the eminences of which has obtained from this fact the name of the Hollybush Hill. In the Lickey district it is also abundant on the higher parts, forming in many places great lengths of roadside hedgerows. In the garden are found many varieties in the form and colour of the leaf, but such variations are very scarce in a wild state; variegated Holly has been reported about Oxford, and a form without spines on its leaves has received the name of *f. inermis*. A form *ferox* with spines all over the surface of the leaf has bordered the Malvern Road and Claines Road for many years. When the tree grows to any size, the leaves of the upper portion are always less spiny than those of a smaller or younger plant. The Holly occurs in all neighbouring counties.

22. CELASTRACEAE, R. Br.

EUONYMUS, Linn. 90. (*εὐώνυμος*, the Greek name of some plant, probably this.)

333. **217. E. europaeus,** *Linn.*   *The Spindle-tree.*
Native. Hedges, woods, and bushy places. General in the Avon and Severn districts; less common in Malvern and Lickey. Shrub. May and June. *Top. Bot.* 74.
First Record. Blackstone Rock, near Bewdley, *Scott, Purton*, vol. iii, pt. ii, p. 335.
Avon. Bredon Hill. Trench Woods. Oddingley. Sheriff's Lench. Near Tiddesley Wood.
Severn. Near Dodderhill. Fenny Rough, *Miss Bickmore*. Hagley, *Scott*. Cotheridge, *Lees*. Near Bewdley, *Gissing*. Hedges near Elmbridge and Droitwich. Hawford. Lane near Redstone Rock. Peachley. Rainbow Hill, Northwick. Old Malvern Road, Worcester. Meadow below Bevere Green. Monk's Wood. Wyre Forest.
Malvern. Near Malvern, *Lees*. Little Shelsley. Rock Wood, Shelsley. Knightsford Bridge. Bransford. Mathon. Ankerdine. Alfrick. Malvern Link. Madresfield.
Lickey. Hedges near the Lye, *Scott*.
The fruit of this plant is much more conspicuous than the insignificant greenish flowers, the capsules being a rich pink, which, when they burst open, disclose the brilliant orange seeds within. To the old English herbalists the plant was known as Prickwood. The use from which it takes its English name has died out in these modern days, but the wood is yet used for making skewers and for the pegs of boots. It is very poisonous, and the shrub is generally eradicated by the farmers. Most animals refuse to eat the berries; and they were formerly used in dyeing, affording a good yellow colour. The plant does not occur in France. The Spindle-tree has been observed in all bordering counties.

23. RHAMNACEAE, Juss.

RHAMNUS, Linn. 91. (*ῥάμνος*, a branch.)

334. **218. R. catharticus,** Linn.   *Common Buckthorn, Dog Wood Blossom.*
Native. Hedgerows and coppices. Not uncommon. Shrub. May and June. *Top. Bot.* 58.
First Record. Side of a brook near Hanley Castle, *Ballard*. Near Worcester, *Stokes, Stokes's With. Bot. Arr.*, and ed., p. 239 (1787).
Avon. Alderminster, *Cheshire*. Broadway, *Lees*. Trench Woods. The Slads. Tiddesley Wood. Woodgate, near Bentley.
Severn. Near Worcester. Wyre Forest. Lincombe, near Stourport. Habberley Valley. Laughern Brook. Perry Wood, *Lees*. Field near Whitford, Bromsgrove. Near Stoke Prior Church. Near Dunhampstead. Battenhall Wood. Sinton, near Hallow. Ockeridge.

Malvern. Madresfield, *Lees*. Middleyards Coppice. Hanley Castle as above. Ankerdine. Rosebury Rock. Powick. Lord's Wood. Leigh. Malvern Link.
Lickey. Illey Mill, Halesowen, *Lees*. Yardley Bridge, *Ick*. Stour, above Halesowen.
This shrub possesses glossy dark green leaves, and the berries are purple; they were formerly much used medicinally. Their juice affords the pigment known as Sap Green, and a good yellow dye was formerly produced from the bark. The aecidial form of the common wheat rust, *Puccinia coronata*, Corda has been many times observed on this plant and its co-species *Rhamnus Frangula*. The plant occurs in all the bordering counties.

335. **219. R. Frangula,** Linn.   *Berry-bearing Alder.*
Native. Bushy moist places. Local. Shrub. May to September. *Top. Bot.* 66.
First Record. In hedges north of Evesham, *Pitt, Agric. Worc.*, p. 317 (1810).
Avon. Near Evesham, as above. Trench Woods. Above Wollashill. North Littleton. Norton, near Evesham. Copse by the river.
Severn. Wyre Forest. Monk Wood. Harvington, *Purton*. Ockeridge Wood. Birchin Grove, *Lees*. Hartlebury Common, one plant, *Rea*.
Malvern. Given for Malvern, *Lees*, 1st ed., but omitted in 2nd and 3rd. Wood near fifth milestone beyond Cotheridge.
Lickey. Moseley, *Lees*. Doctor's Coppice, Frankley, *Mathews*. Uffmore Wood, *Lees* and *Mathews*. Whiteford, Bromsgrove. Near Redditch. Chaddesley Wood.
Like the last, this plant possesses medicinal properties, and the bark affords a yellow dye. The foliage is not very abundant, and the berries are deep purple. It is readily distinguished from the last by its alternate branches; in that the branches are opposite. It occurs in all neighbouring counties.

24. ACERACEAE, Juss.

ACER, Linn. 92. (*Acer*, sharp; on account of the hardness of the wood.)

336. **220. A. *Pseudo-platanus,** Linn.   *The Sycamore.*
Denizen. Woods and hedges. Common. Tree. May and June. Not given in *Top. Bot.*
First Record. Malvern, *Med. Surg. Rep.*, vol. i, no. 1, August, 1828.
This well-known tree is so frequently planted, and seeds itself about so freely, that it is often regarded as a native tree, and in Scotland two enormous ones are now shown as the last survivors of Birnam Forest. On the sandy ground north of Hagley Brake it is extraordinarily abundant. The hedgerows are composed entirely of it, groves of it shroud the farm houses and cottages, and in all stages of growth it forms the thickets along the roadside wastes. The timber is useful, being used for calico printing, and wooden platters were made of it. Needless to say the Sycamore in which Zacchaeus hid himself is not this plant, but a species of fig. Very frequently, in autumn, the leaves are disfigured by the black blotches caused by the fungus *Rhytisma acerinum*, but this can be easily eradicated by collect-

ing fallen leaves and burning them. It is found in all neighbouring counties.

337. **221. A. campestre,** Linn.   *The Maple.*
Native. Hedges. Common. Small tree. May and June. *Top. Bot.* 62.
First Record. Malvern, *Med. Surg. Rep.*, vol. i, no. 1 (1828).
This seldom grows to the size of a tree in this county, though in others it can be observed of fair dimensions as a forest tree. In the autumn none of our trees exhibits a brighter foliage than this in English hedgerows, but it is not common in Scotland. The wood is said to be superior to that of the Sycamore for the purposes for which that is used, and from it were made in mediaeval times the drinking-vessels called Mazer-bowls, so called from the Dutch name of the tree, *maeser*, or British *masarn*. The knobs of old maple trees are curiously knurled and veined, and were much in request in earlier days for carving and works of art. A common name of the tree in the county is the Spigot or Spiggot Oak, because the wood is used to make skewers for butchers. In *Worc. Nat. Club Trans.* ii, *App.*, p. 24, are given some pictures of distorted old Maples in Worcestershire and elsewhere. The tree occurs in all bordering counties.
Var. b. leiocarpon, Wallr.
First Record. Malvern Link, *Towndrow, Malv. Advert.*, January 20, 1894.
Malvern. As above.
In this segregate the ovary of the fruit is glabrous, and not downy.
Var. collinum, Wallr.
This is a segregate of the foregoing, and is the typical form. It is not given in *Lond. Cat.*, 10th ed.
Var. laciniatum.
First Record. In this book.
We have observed a variety which we should call var. *laciniatum* at Alfrick, which is characterized by the deeply cut leaves, and the channelled stems of the branches.

25. LEGUMINOSAE, Juss.

GENISTA, Linn. 94. (Etymology obscure; *gen* is the Celtic for a shrub.)

339. **222. G. anglica,** Linn.   *Needle Furze, Petty Whin.*
Native. Heaths. Very local. Shrub. April to July. *Top. Bot.* 86.
First Record. Broadmore, near Birmingham, *With.*, 3rd ed., p. 625 (1796).
Severn. Broad Heath, *Lees*; *Rea*. Moseley Common. Near Monk's Wood, *Lees*. Stourbridge, *Purton* (*Mr. Mathews* believes this to have been in Staffordshire). Hallow. Birchin Grove, *Lees*. Hartlebury Common, in the Bog.
Malvern. Scattered about the hills, *Ainsworth*. Welland Common, *Lees*. Malvern Link Common. Near British Camp Hotel. Mathon.
Lickey. Broadmore, as in first record.
Of the Stourbridge record Mr. Mathews is doubtful. Scott, in his list of plants in the neighbourhood of Stourbridge, says of this, calling it *G. spinosa*, 'it grows at Whittington Common plentifully, but has been observed nowhere else.' Whittington is in the parish of Kinver, in Staffordshire. The plant has been found in all neighbouring counties.

{*340*. 223. **G. pilosa**, Linn.   *Hairy Green-weed.*

Native. Sandy commons. Very rare. Shrub. May. *Top. Bot.* 6.

First Record. Not localized, *Lees' Cat. in New Bot. Guide* (1835).

Malvern. Between Little Malvern and Malvern Wells, Worcestershire, *Borrer, Hook. and Arn., Brit. Flora*, 6th ed., p. 73 (1850).

Mr. Lees, at p. 73 of his *Bot. of Worc.*, suspects some error of substitution in Mr. Borrer's record ; no one else has seen the plant in the locality. It is not recorded for any bordering county ; and in *Top. Bot.*, p. 105, it is marked 'Worcester. Extinct'.}

341. 224. **G. tinctoria**, Linn.   *Dyer's Green-weed.*

Native. Dry pastures. General. Shrub. May to August. *Top. Bot.* 76.

First Record. Coal-pit banks in various directions (some probably in Worcestershire), *Scott, Stourbridge*, p. 540 (1832).

Avon. Meadows near Himbleton, abundant, *Lees.* Near Goosehill Wood. Near Trench Woods. Near Pershore.

Severn. Pepper Wood, *Humphreys.* Bredicot. Crookbarrow Hill. Monk's Wood. Coningree Wood. Near Crown East. Broad Heath. Ombersley.

Malvern. Plentiful in pastures or waste ground, *Lees.* Ankerdine Hill. Martley. Powick. Bransford. Malvern Link. Malvern Wells.

Lickey. Near Harris's Wood, Hunnington, *Mathews.* Abundant at Frankley and Romsley, *Mathews.* Selly Hall Park, *Ick.* Near Stourbridge, *Scott.* Halesowen. Rowney Green, Alvechurch. Chaddesley Woods. Dodford.

Other English names for this plant are Dyer's Whin and Woad-waxen. The milk of cows feeding on this plant is said to acquire a bitter taste, but cattle will not eat it unless they are pressed by hunger. Both the Latin and English names of the plant refer to its use as a dye-stuff. It yields a yellow colour. A decoction of the seeds was formerly used medicinally. It is found in all neighbouring counties.

ULEX, Linn. 95. (Probably from a similar derivation to that of *Ilex*, which see ; but others derive it from *uligo*, marshy.)

342. 225. **U. europaeus**, Linn.   *Gorse, Furze, Whin.*

Native. Commons, roadside wastes, and pastures. Widely distributed. Shrub. January to April, and in the autumn. *Top. Bot.* 112.

First Record. In a list of Malvern plants, by *Ainsworth*, in *Edin Phil J.*, p. 99 (1828).

This gorse seldom ascends to the highest land in the county. In these localities its place is taken by *U. Gallii* ; but it occurs at an elevation of 800 feet on Romsley Hill. Although its Latin name is *europaeus*, its range is extremely limited, and out of England a glorious gorse-clad common is rarely seen. In Russia it is a greenhouse plant. But in England and most parts of Scotland, the case is different ; 'when the gorse is out of blossom, kissing is out of fashion,' says a homely proverb ; and this persistence of blossom is aided by the fact that *U. Gallii* flowers in the autumn, for the rustic observer would not distinguish the two plants. The gorse is one

---

of the few Leguminous plants known to have double flowers, and the double-flowered form is a cultivated shrub for the sake of its gorgeous masses of golden bloom. In olden time Gorse was considered valuable in cases of jaundice, probably from some kind of homœopathic idea. It occurs in all adjoining counties.

343. 226. **U. Gallii**, Planch.   *Autumn Gorse.*

(*U. nanus*, var. *major*, Bab.)

Native. Commons. Except in the Avon district, general on high ground, and in the Lickey district abundant. Shrub. July to October. *Top. Bot.* 59.

First Record. Under its real name, Great Autumnal Furze, *Lees, Bot. Malv. Hills*, 2nd ed., p. 66 (1852).

Avon. Pirton. Crowle. Rouse Lench. Craycombe. Sheriff's Lench. In the Avon district Mr. Lees marks this in his *Table of Plants* as occurring in two localities. There is no need to localize it in the other districts. It is the common gorse of the Lickey, and no other gorse grows on the Clent Hills. It is abundant on the Malvern Hills, from which locality Mr. Lees at first treated it as *U. minor*, correcting his error afterwards. Small stunted bushes of *U. Gallii* are easily mistaken for that plant. It is recorded from all neighbouring counties.

344. 227. **U. minor**, Roth.   *Dwarf Furze.*

(*U. nanus*, Forster.)

Native. Commons and heaths. Unknown in the Avon district, and uncommon elsewhere. Small shrub. July to October. *Top. Bot.* 27.

First Record. Astley Common, *Purton, App.*, vol. iii, p. 59 (1821). (Probably an error for *U. Gallii*, Mathews.)

Severn. East of Iverley Hill, *Scott.* Ockeridge Wood. Wyre Forest. Holt Bank.

Malvern. On the Hills, *Lees* ; afterwards corrected by him. 'Not a Malvern plant,' *Towndrow* (MS.). Welland Common. Ankerdine.

This plant is very like the last, only dwarfer, and differing in its more slender spines, and in the wings of the petals being shorter than the keel. Owing to its early confusion with *U. Gallii*, the records are unsatisfactory. It has been found only in Salop of the neighbouring counties, but is not recorded in *Top. Bot.* even for that.

CYTISUS, Linn. 96. (κύτισος, a kind of clover. Etymology obscure, or, according to some, from Cythnus, one of the Cyclades, where some of the species were first found.)

345. 228. **C. scoparius**, Link.   *Broom.*

(*Spartium scoparium*, Linn. *Sarothamnus scoparius*, Koch.)

Native. Fields, coppices, commons. Local down the centre of the county, abundant north-east and west. Shrub. May to July. *Top. Bot.* 109.

First Record. As *Spartium scoparium.* Little Malvern, *Ainsworth, Edin. Phil. J.*, p. 99 (1828).

This well-known plant, the 'Planta genista' of olden days, is a feature in many parts of the county. In olden days every part of the plant, seeds and root, flowers and leaves, was held to have peculiar virtues, while the juice

---

expressed from the green stalks was pronounced an infallible cure for toothache ; and in an industrial point of view, its name betokens its use. Cattle reject the branches, but goats browse freely on the young shoots. Its connexion with the Royal House of Plantagenet is well known. It is found in all bordering counties.

ONONIS, Linn. 97. (Probably ὄνος, an ass, by which animal the plant is eaten.)

346. 229. **O. repens**, Linn.   *Rest Harrow, Wild Liquorice.*

(*O. inermis*, Huds. *O. arvensis*, Linn.)

Native. Dry fields, roadsides. Not uncommon. P. May to August. *Top. Bot.* 100.

First Record. In a list of Malvern plants, *Ainsworth, Edin. Phil. J.*, p. 99 (1828).

Avon. Near Evesham, common, *Scott.* Defford Common. Norton, near Evesham. Crowle. Hanbury. Fladbury. Pershore. Littleton Banks. Huckers Hill, Stoke.

Severn. Roadside, Crown East. Spetchley. Cruckbarrow Hill. Rainbow Hill. Gregory's Mill. Ombersley. Kempsey. Pirton.

Malvern. Sides of lanes, *Lees.* Bransford. Ridge Hill, Martley. Alfrick. Near Stanford Bridge.

Lickey. Western side of the Clent Hills, *Mathews.* Tardebigge. Dodford.

This plant has been found with white flowers near Evesham, and on Defford Common. The name Rest Harrow has an equivalent in France, where the plant is called *Arrête Bœuf*, both referring to implements of agriculture. The roots of the plant have the sweet flavour of liquorice, and old herbalists said, that among other virtues, it would cure delirium. The leaves of the plant are somewhat sticky. It is found in all neighbouring counties.

Var. **inermis**, Lange.

This is a segregate of the foregoing, and is the typical form. The plant has been reported as var. *inermis* from several localities. It is not given in *Lond. Cat.*, 10th ed.

347. 230. **O. spinosa**, Linn.   *Thorny Rest Harrow.*

(*O. antiquorum*, Linn. *O. campestris*, Koch.)

Native. Barren pastures, borders of fields, roadsides. General throughout, but scarce. P. June to September. *Top. Bot.* 71.

First Record. Roadsides, heaths, and rough ground, *Pitt, Agric. Worc.*, p. 317 (1810).

Avon. Trench Woods. Cleeve Bank. Broadway. Bredon Hill, Sheriff's Lench. Near Wadborough. Flyford Flavell, Wyre. Churchill. Grafton Flyford. Great Comberton. Littleton Banks. Defford Common.

Severn. Martley Quarries. Kempsey. Ombersley. Near Monk's Wood. Spetchley. Hartlebury. Holt.

Malvern. Longdon, *Lees.* Near Welland.

Lickey. Abundant at Frankley and Romsley, *Mathews.*

This is a more erect plant than the last, and not stoloniferous, as that is. A white-flowered form has been observed at Flyford Flavell, Grafton Flyford, and Great Comberton. It is recorded from all neighbouring counties.

TRIGONELLA, Linn. 98. (τρεῖς, three, γόνυ, a knee, in allusion to the form of the corolla.)

349. 231. **T. ornithopodioides**, DC.   *Bird's-foot, Fenugreek.*

(*Trigonella purpurascens*, Lam. *Trifolium ornithopodioides*, Linn.)

Native. Grassy commons. Rare. A. April and May. *Top. Bot.* 29.

First Record. *Watson, Top. Bot.*, 2nd ed., p. 109, 'Fraser, sp.' (1883).

Malvern. Malvern Common, abundant, May, 1895, *Towndrow* (MS.) ; also *Rea.* Railway embankment at the Link, *Westcombe*, 1867. Hanley Castle. Barnard's Green. Near the Wych. Opposite the British Camp Hotel.

This plant is given in Mr. Lees' *Table of Plants*, without being localized in any of the districts, as *Trifolium ornithopodioides.* This is a very tiny plant. It is recorded only from Gloucestershire of the neighbouring counties.

MEDICAGO, Linn. 99. (μηδικὴ, a kind of clover introduced into Greece from Media.)

350. 232. **M. *sativa**, Linn.   *Lucerne.*

Denizen. Fields, waysides, railway-banks. Occasional. P. May to September. Not in *Top. Bot.*

First Record. Lucerne, Cleve, Worcs., *Purton, Midl. Flora*, p. 347 (1817).

Avon. Cleeve, as above. Sheriff's Lench. Defford. Crowle. Hanbury. Norton, near Evesham. Lenchwick. Near Nafford's Mill.

Severn. Powick, *Lees.* Roadside, Pendock. Wyre Forest. Ombersley. Holt. Kempsey. Severnstoke. Near Ribbesford. Droitwich. Ladywood.

Malvern. Near Powick New Bridge. Bransford Bridge. Leigh. Alfrick. Knightwick. Near Upton-on-Severn. Malvern Link. Newland.

This plant is always an escape from, or remainder after, cultivation. The English name has nothing to do with 'Lovely Lucerne', of modern co-operative travel, but is derived from its name in the South of France, *Lauserda*, of unascertained etymology. Agricultural writers of the seventeenth and eighteenth centuries, before which time the name is not found, invariably placed the French article before the word, and called the plant 'La Lucerne'. In neighbouring counties it does not appear to have been met with except in Herefordshire, where it has been observed on the roadside near Colwall and Ledbury.

353. 233. **M. lupulina**, Linn.   *Black Medick, Hop Medick.*

Native. Fields, waysides, wall-tops, commons. Not uncommon. A. or B. May to August. *Top. Bot.* 105.

First Record. Not localized, *Lees, Bot. Malv. Hills*, 1st ed., p. 36 (1843).

This plant is very much like the Common Yellow Trefoil. It was considered in former times the most useful to the farmer of all the 'artificial grasses', but it is not now cultivated. In some parts of the country it is called Black Nonsuch and Shamrock. It is recorded from all neighbouring counties.

Var. **Willdenowiana**, Koch.
First Record. In this book.
Malvern. Half Key. Upton-on-Severn. Near Malvern, *Towndrow*.
The pods of this variety possess yellowish granular hairs.

354. 234. M. denticulata, Willd.
Casual in the county. Sandy ground. Very rare. A. May to July. *Top. Bot.* 22.
First Record. In a field belonging to Mr. Bickley, Moseley Road, *Mr. J. Morley, Lees, Bot. of Worc., Add. and Corr.* (1867).
Severn. Rubbish heap at Hoo Mill, Kidderminster, *Thompson*, 1876. Upper Hop-yard, Lower Wick. Rubbish heap by Toll-bar, Bath Road, Worcester.
Malvern. Bransford. Madresfield. Knightwick.
Lickey. As first record.
This plant differs from any other wild kind of Medick in its pods, which are deeply netted. It is recorded in *Top. Bot.* doubtfully from Worcestershire and Herefordshire, but it has also been found in Stafford and Warwick.

Var. b. **apiculata**, Willd.
First Record. In this book.
Malvern. Brickyard at Malvern Link, *Towndrow*.
This variety is distinguished by its very short, almost tuberculous, spines.

356. 235. M. arabica, Huds. *Spotted Medick.*
(*Medicago maculata*, Sibth.)
Colonist. Waste places. Very local. A. or B. June to August. *Top. Bot.* 46.
First Record. As *M. maculata*. This has been gathered at Hawford, near Worcester, by my friend *Mr. Abraham Edmunds*, junior, *Lees, Bot. Malv. Hills*, 1st ed., p. 36 (1843).
Avon. Norton, near Evesham. Fladbury. Defford.
Severn. Porter's Mill, Claines, *Lees*. Hop-yard at Lower Wick, *Lees*. Near the Ferry, Old Waterworks, Worcester, *Lees*. Roadside near Hawford Ferry. Northwick. Rubbish heap at Hoo Mill, Kidderminster. Hartlebury Station. Barbourne. Rainbow Hill. Droitwich Canal, near Salwarpe Church.
Malvern. Powick. Knightwick. Bransford. West Malvern.
This plant has been found in Gloucester, Warwick, Hereford, and Salop.

{**357. 236. M. minima**, Desr.
Casual. Sandy ground. Very rare. A. May. *Top. Bot.* 10.
First Record. In this book.
Severn. Cornfield near Ombersley, May 30, 1848, *G. Reece*. These specimens are in the Herbarium at the Hastings Museum, but they are so immature that it is impossible to say whether they are correctly determined.}

MELILOTUS, Hill 100. (*Mel*, honey, and *lotus*, a bean ; from the Greek λωτός.)

358. 237. M. altissima, Thuill. *Melilot.*
(*M. officinalis*, Linn.)
Native. Fields, hedgebanks, waste places. Not common. A. or B. June to October. *Top. Bot.* 73.
First Record. Cleve, Worc., *Purton, Midl. Flora*, p. 346 (1817).
Avon. Craycombe. Trench Woods. Defford Railway Bridge. Bredon Hill. Roadside, Fladbury.
Severn. Railway embankment at Worcester, *Lees*. Near Porter's Mill. Near Stourport. Shrawley. Holt. Kempsey. Hartlebury Common.
Malvern. Waste spots, *Lees*. Middleyards Coppice. Bransford Bridge. Bransford Court. Croft Quarry. Near Longdon Marsh. Witley. Leigh Sinton.
Lickey. Fields near Moseley, *Ick*. Uffmore Wood, *Mathews*. Blackwell Incline.
Although the name of this genus would imply the sweetness of honey, this member of it possesses a strong and disagreeable odour while growing, and it is not till it is drying that its scent becomes sweet, like that of new-mown hay ; and this it retains for a considerable time. It owes this to the same volatile principle that gives the Tonquin Bean its powerful scent. The plant has been used for a variety of maladies. It occurs in all bordering counties.

359. 238. M. *alba, Desr. *White Melilot.*
(*M. leucantha*, Koch. *M. vulgaris*, Willd.)
Casual. Fields. Rare. A. or B. May to August. *Top. Bot.* 40.
First Record. Clover Field, Leigh, *Miss B. Norbury* ; *Towndrow, Malv. Advert.*, November 23, 1895.
Severn. Near Oldington. Pedmore.
Malvern. As first record.
This plant is not native further north than Central France, but shows a tendency to follow man, being especially fond of spreading along railway lines. It is reported from Gloucestershire, Salop, Stafford, and Warwick.

360. 239. M. *officinalis, Lam. *Field Melilot.*
(*M. arvensis*, Wallr.)
Colonist. Rubbish heaps, waste ground. Rare. A. or B. July to September. Not in *Top. Bot.*
First Record. Gathered in an arable field at Wolverley in 1853 by *Mr. W. Mathews, Lees, Bot. of Worc.*, p. 12 (1867).
Avon. Trench Woods. Broadway Hill. Near Tiddesley Wood.
Severn. Wolverley, as above. Mildenham Mill, Claines. Stourport Railway Station. Rubbish heap, Hoo Mill. Above Redstone Rock. Barbourne. Kempsey. Gravel-pit near Fever Hospital, Worcester.
Malvern. Sherrard's Green, *Towndrow, Malv. Advert.*, November 12, 1892.
Lickey. Frankley, *Mathews*.
This plant was recorded as new, which it was not, in *Bot. Rec. Club Rep.*, 1883 (pub. 1884). Mr. Lees, gives for it, in his *Table of Plants*, two localities,

noting that both are lost. This also is an introduced plant, and has been found in all neighbouring counties except Salop.

361. 240. M. INDICA, All.
(*M. parviflora*, Desf.)
Casual. Waste ground. Rare. A. July to August. Not in *Top. Bot.*
First Record. As *M. parviflora*. One plant in an orchard at Malvern Link, *Towndrow, Malv. Advertiser*, 1892.
Severn. Hartlebury Common, *Miss Ladbury*. Arable field at Hartlebury.
Malvern. As above.
This plant is even a worse stranger than the last. It has been observed in Staffordshire.

M. SULCATA, Desv.
Casual. Waste ground. Rare. A.
First Record. Mildenham Mill, *Jeffery, W. N. C. Trans.* iii, p. 185 (1904).
Severn. As above.
This is a weed of cultivation in the Mediterranean region, and its presence in England might probably be traced to the importation of merchandise from that area. It is not given in *Lond. Cat.*, 10th ed.

TRIFOLIUM, Linn. 101. (Name τρίφυλλον, in allusion to its three leaflets.)

363. 241. T. pratense, Linn. *Red Clover.*
Native. Meadows, waysides, banks. Common and generally distributed. B. or P. May to October. *Top. Bot.* 112.
First Record. Not localized, *Lees, Bot. Malv. Hills*, 1st ed., p. 36 (1843).
This plant is too common to require the mention of localities. A white-flowered form has occurred in a meadow near to the mouth of the Teme. The honey the flowers contain is very sweet, and they are the haunts of innumerable bees and butterflies. Shakespeare called them Honeystalks, and children sometimes dub them Honeysuckles. The leaves fold themselves together at night and even in wet weather. In all neighbouring counties Red Clover is found.

Var. SATIVUM, Schreb.
This is the cultivated form of the plant, which maintains itself on the borders of fields after they are devoted to other crops.

Var. c. **parviflorum**, Bab.
First Record. As in this book.
Malvern. Hanley Castle, *Towndrow, Malv. Advertiser*, 1892.
A form in which the teeth of the calyx exceed the corolla, which is smaller than usual.

364. 242. T. medium, Linn. *Zigzag Clover.*
(*T. flexuosum*, Jacq.)
Native. Rough borders of meadows and woods. General. P. June to September. *Top. Bot.* 108.
First Record. As *T. flexuosum*. Worcestershire, *Stokes, With.*, 2nd ed., vol. iii, p. cxxvi (1792).

Avon. Trench Woods. Norton, near Evesham. Pershore. Crowle. Hanbury.
Severn. Northwood, Bewdley. Warshill. Lark Hill, Worcester. Near Astwood. Foot of Helbury Hill. Ombersley. Stourport. Kempsey. Earl's Croome.
Malvern. Dry pastures, *Lees*. Martley. Near Malvern. Middleyards Coppice. Powick. Alfrick. Knightwick. Malvern Link. Madresfield.
Lickey. Harris's Wood, Hunnington. The Whetty, *Humphreys*. Abundant on upland fields, from Clent to the Lickey, *Mathews*. Tardebigge. Barnt Green.
Trifolium medium is recorded from all neighbouring counties. It is much like the preceding, but is well marked by its zigzag stem.

365. 243. T. ochroleucon, Huds. *Sulphur-coloured Trefoil.*
Casual. Dry gravelly soil. Rare. P. June and July. *Top. Bot.* 11.
First Record. On the Link Common (Malvern), *Lees, Mag. Nat. Hist.*, vol. iii, p. 160 (1830).
Severn. Cotheridge, *Baxter, Stanley's Worc. and Malv. Guide*, 1853. Broadway, *Miss Moseley, Lees, Bot. Worc.*, p. 89. Well-established, second pasture field from Worcester, between the old and new Broadwas roads, 1907.
Malvern. As first record.
A plant of the east of England, only casual elsewhere. Mr. Mathews apparently was aware of the Malvern record only, and was inclined to consider it as an error. The heads of flowers, at first hemispherical, become at length oval, when the cream-coloured flowers turn brown. It has been noticed in Gloucestershire and Staffordshire.

367. 244. T. incarnatum, Linn. *Crimson Clover.*
Alien. On the borders of fields. An escape from cultivation. Local. A. April to August. Not in *Top. Bot.*
First Record. In *Lees' Table of Plants, Bot. of Worc.*, p. 9 (1867).
Avon. Norton, near Evesham. Sheriff's Lench. Pershore. Near Trench Woods.
Severn. Near Bewdley. Ombersley. Grimley. Kempsey. Railway-bank at Shrub Hill. Barbourne.
Malvern. Powick. Knightwick. Martley. Malvern Common.
A field of this clover in full flower is a very beautiful sight. The plant is a native of Southern Europe, and has no pretension to nativity in Britain. It has been noticed in all neighbouring counties except Hereford, but possibly has escaped into more or less freedom there also.

370. 245. T. arvense, Linn. *Hare's-foot Trefoil.*
Native. Sandy fields and commons. Local. A. May to September. *Top. Bot.* 94.
First Record. On sand in the neighbourhood of Kidderminster, Mitton, and Stourport, *Pitt, Agric. Worc.* (1810).
Avon. Rather uncommon, *Lees*. Cleeve Bank.

Severn. Hartlebury Common. Above Blakedown, Churchill, and Kidderminster, *Mathews*. Hagley Brake, *W. Whitwell*. Rock Dwellings, Wolverley. Sutton Common. The Devil's Spittleful, Bewdley. Redstone Rock. Bissell. Diglis. Aston Fields, Bromsgrove.

Malvern. Not common, *Lees*. Powick. Knightwick.

Lickey. Hill Top, Bromsgrove. Saltwells Wood near Dudley, *Amphlett*.

This is very distinct from any other British species, its pale pink blossoms just peeping through the soft grey down which surrounds them. It is recorded for all neighbouring counties.

372. 246. T. striatum, Linn.   *Soft Knotted Trefoil*.

Native. Dry pastures and commons. Except in the Malvern district, rather uncommon. A. June to August. *Top. Bot.* 77.

First Record. At Picket Rock, near Kidderminster, and on the side of the Kidderminster Road near Bewdley, *Perry, Mag. Nat. Hist.*, vol. iv, p. 450 (1831).

Severn. Spetchley Common. Hartlebury Common. Near Churchill (Kidderminster). Near Kidderminster, as first record. Roadside near Hawford Bridge. Near Bewdley. Sutton Common. Redstone Rock. Near Kepax. Oldbury Farm.

Malvern. Plentiful at the base of the Hollybush Hill, *Lees*. Ankerdine Hill. Herefordshire Beacon. Malvern Link.

A silky looking little plant, loving the neighbourhood of the sea. It has been recorded from all neighbouring counties.

Var. b. **erectum**, Leight.

First Record. As in this book.

Severn. Sutton Common. Oldbury Farm. Hawford Park.

This variety differs in its erect habit and shortly stalked subconical heads, while the corolla exceeds the calyx in length.

373. 247. T. scabrum, Linn.   *Rough Rigid Trefoil*.

Casual. Dry fields and commons. Rare. A. May to August. *Top. Bot.* 50.

First Record. *Worcestershire Victoria County History*, vol. i, p. 49 (1901).

Severn. Hartlebury Common.

This plant has been noticed in Gloucestershire, Salop, and Warwick.

377. 248. T. hybridum, Linn.   *Alsike Clover*.

Alien. Roadsides, fallow, and cultivated fields. Local. B. June to August. Not in *Top. Bot.*

First Record. Introduced, *Mathews, Clent and Lickey Hills*, 1881.

Avon. South Littleton. Norton, near Evesham. Sheriff's Lench. Pershore. Crowle.

Severn. Near St. Peter's, Droitwich, *Worc. Nat. Club Trans.* iii, p. 179. Claines. Sutton Common. Kempsey. Holt. Hallow. Grimley. Shrawley. Wyre Forest.

Malvern. Common, *Towndrow, Malvern Advertiser*, Nov. 12, 1892.

Lickey. Barnt Green.

This plant was formerly much cultivated, but is now less so. It takes its English name from a town near Upsala, in Sweden, one of the localities

mentioned for it by Linnaeus. It is recorded from all the neighbouring counties.

Var. **elegans** (Savi).

First Record. Railway Cutting at Barnt Green, *Worc. Nat. Club Trans.* ii, p. 83, *Mathews* (1899).

Severn. Merryman's Hill.

Lickey. As above.

This variety differs from the type in possessing a decumbent solid stem, while the heads are smaller. It has been recorded from Warwickshire.

378. 249. T. repens, Linn.   *Dutch Clover, White Clover*.

Native. Meadows, pastures, waysides. General throughout the county. P. May to October. *Top. Bot.* 112.

First Record. *T. repens*, var. *proliferum*, Worcestershire, *Stokes, Stokes's With. Bot. Arr.*, 2nd ed., p. 793 (1787).

This plant is in much repute for pastures, and is everywhere met with; it is as general throughout Europe as it is in our own country. Each flower is on a foot stalk which bends down after flowering, when all the legumes are covered with brown withered flowers. It has been held that this is the true Shamrock of Ireland, though the Wood Sorrel has been asserted to have a claim to that distinction. The Irish names for this clover are words analogous to Shamrock, but usually some of the smaller leaved trefoils are selected by Irishmen to represent the national emblem on St. Patrick's day. It is held lucky to find a leaf with four leaflets; and a plant was supposed to be a protection against witches. For a wonder it does not seem to have obtained any reputation for medicinal virtues, as is the case with nearly every common plant. It is recorded from all neighbouring counties.

379. 250. T. fragiferum, Linn.   *Strawberry-headed Trefoil*.

Native. Roadsides, damp meadows. Not uncommon, except in the Lickey district. P. July to September. *Top. Bot.* 72.

First Record. Cleve, Worc., *Purton, Midl. Flora*, p. 346 (1817).

Avon. Craycombe. Evesham on the Stratford Road, *Lees*. Avon Meadows, Pershore, *Lees*. Huddington. Tardebigge. Common by the side of the Crowle Roads. Fladbury. Abbot's Lench. Eckington. Pershore. Norton, near Evesham.

Severn. Croome Perry Wood. Besford. Tibberton. Warndon. Hagley, *Scott* (not seen at Hagley in recent years, *Mathews*). On the Spetchley Road, *Lees*. Porter's Mill. Broadwas.

Malvern. Side of the road to Welland, *Lees*. Powick. Alfrick. Knightwick. Martley. Stanford Bridge. Malvern Link. Mathon.

This plant is recorded for all neighbouring counties except Staffordshire.

382. 251. T. procumbens, Linn.   *Hop Trefoil*.

Native. Cornfields, dry pastures, railway-banks. General. A. April to August. *Top. Bot.* 105.

First Record. Not localized, *Lees, Bot. Malv. Hills*, 1st ed., p. 36 (1843).

This common plant may be known at a glance by its oval yellow heads. It

could only possibly be mistaken for the Hop Medick (*Medicago lupulina*), and from this it is well distinguished by its straight pod. It is recorded from all neighbouring counties.

383. 252. T. dubium, Sibth.   *Lesser Yellow Trefoil*.

(*T. minus*, Relhan.)

Native. Meadows and pastures, on sandy soil. General. A. April to August. *Top. Bot.* 109.

First Record. Not localized, *Lees, Bot. Malv. Hills*, 1st ed., p. 36 (1843).

This common little trefoil is seen frequently in dry grassy places, its stems prostrate, its yellow flowers small. It is recorded from all bordering counties.

384. 253. T. filiforme, Linn.   *Slender Yellow Trefoil*.

Native. Heathy commons, peaty places. Local. A. June to August. *Top. Bot.* 65.

First Record. Astwood Common, &c., *Purton, Midl. Flora*, p. 345 (1817).

Avon. Norton, near Evesham. Bredon. Pershore. Crowle.

Severn. Spetchley Common. Chaddesley Wood. Henwick. Habberley Valley. Wyre Forest. Hartlebury Common. Pitchcroft. Near the Devil's Spittleful. Bewdley.

Malvern. On the turf of the hills, *Lees*. Malvern Link. Barnard's Green. Ankerdine Hill. Martley. Abberley.

Lickey. Foot of the hills, *Humphreys*. Chaddesley Wood.

This plant has been recorded from all bordering counties.

ANTHYLLIS, Linn. 102.   (ἄνθος, a flower, ἴουλος, a beard, in allusion to the pubescence of the calyx.)

385. 254. A. Vulneraria, Linn.   *Kidney Vetch, Lady's Fingers*.

Native. Roadsides, gravel-pits, field-borders. Nearly confined to calcareous districts. P. May to October. *Top. Bot.* 105.

First Record. Limestone pits, Cradley, near Malvern Hills, Worcestershire, *Ballard, Stokes's With. Bot. Arr.*, 2nd ed., p. 765 (1787).

Avon. Craycombe. Bredon Hill. Broadway. Trench Woods. The Slads. Cleve, *Purton*.

Severn. 'The right bank of the Severn above Kepax Ferry was a golden glory with the flowers of this species in 1907, but they turned out on investigation to have been planted by Mrs. W. Lea to bind together the soil of the recently made banks,' *Rea*. Whittington.

Malvern. Ridge Hill, Martley. Martley Quarries. Abundant on the western flanks of the Malvern range, *Lees*. Castlemorton, on red marl, *Lees*.

Lickey. Banks of Dudley Canal between Haywood and Lappal only, *Scott*.

The distribution of *Anthyllis* is very similar to that of *Clematis*, except that this plant likes sunny situations. It is a characteristic plant of calcareous soils. The plant in old times was used to staunch the bleeding of wounds;

in rural districts it is often called Lamb's Toes. The leaves remain green for some time after the flower is dead. It is recorded for all neighbouring counties.

Var. b. **coccinea**, Linn.

First Record. In this book.

Severn. Whittington, near Worcester, *Towndrow*. By the Railway at Battenhall, Worcester.

The flowers in this variety are more or less tipped with red. At the Battenhall locality they are pure red.

LOTUS, Linn. 103.   (From the λωτός of the Greeks.)

386. 255. L. corniculatus, Linn.   *Bird's-foot Trefoil*.

Native. Pastures, roadsides, commons. Very common. P. May to September. *Top. Bot.* 112.

First Record. Not localized, *Walker, Med. Surg. Rep.*, vol. i, no. 2, p. 100 (1828).

This plant, abundant in our pastures in May and June, has numerous country names, Lady's Slipper, Shoes-and-Stockings, Hen and Chickens, Wild Vetch, Butter-jags, and Cross-toes being among them. The leaves turn blue when drying. No medicinal virtue appears to have been attributed to it. The name Lotus is probably of Egyptian origin, but the etymology of it is unknown. The plant occurs in all neighbouring counties.

387. 256. L. tenuis, Waldst. and Kit.   *Slender Bird's-foot Trefoil*.

Native. Roadsides and grassy places. Local and scarce. P. June to August. *Top. Bot.* 67.

First Record. As *L. corniculatus*, var. *tenuissimus*. In the neighbourhood of Worcester, *Stokes, Stokes's With. Bot. Arr.*, 2nd ed., p. 805 (1787).

Avon. Trench Woods. Craycombe. Particularly remarkable about Crowle and Huddington. Tardebigge.

Severn. Leopard Hill. Cornfields near Elmley Lovett. Monk's Wood. Upton-on-Severn.

Malvern. Longdon Marsh. Brace's Leigh. Stocks Lane, Leigh Sinton. Old Hills. Ankerdine. Martley. Shelsley Beauchamp. Knightwick. Powick. Mathon.

Lickey. Tardebigge Reservoir and Canal, *Humphreys*. Webheath, *D. Mathews*.

L. tenuis occurs in all neighbouring counties.

388. 257. L. uliginosus, Schkuhr.   *Greater Bird's-foot Trefoil*.

(*Lotus major*, Scop.)

Native. Marshes, brooksides, wet meadows. Occasional throughout the county. P. June to September. *Top. Bot.* 100.

First Record. In the Gullet, a woody glen of the Malvern Range, and at the base of Abberley Hill. Tiddesley Wood, Pershore, *Lees, Illus. Nat. Hist. Worc.*, p. 173 (1834).

Avon. Trench Woods. Croome Perry Wood. Tiddesley Wood. Littleton. Pershore. Crowle.

Severn. Pools at Hagley Brake. Below Hallow. Bishop's Wood, Stourport. Churchill Viaduct, near Kidderminster. Monk's Wood. Park Hall. Harvington. Eyemore Wood. Claines. Ombersley. Grimley. Kempsey.

Malvern. The Gullet and Abberley Hill as above. Alfrick. Powick. Martley. Stanford. Leigh. Upton-on-Severn. Pyxham. Malvern Link.

Lickey. Hunnington. Near Halesowen. Lower Lickey.

By many botanists this is considered to be a form of *L. corniculatus*, consequent upon growing in damp spots. But more than this, in the present plant the calyx teeth are divergent, spreading like a star, in the bud, while in the plant referred to they are adpressed. It occurs in all neighbouring counties.

ASTRAGALUS, Linn. 102. (ἀστράγαλος, an ankle bone, from the knotted root.)

392. **258. A. danicus,** Retz. *Purple Milk Vetch.*

(*A. hypoglottis*, Linn. *A. arenaria*, Huds.)

Native. Grassy places. Very local and rare. P. May and June. *Top. Bot.* 43.

First Record. As *A. arenaria*. On the south side of Bredon Hill, below the Camp, *Nash, Hist. of Worc., Int.*, p. lxxxix (1781).

Bredon Hill, in the Avon district, is the only locality in Worcestershire where this plant occurs. It is found in two places, one an abandoned quarry and its vicinity, and the other on the verge of the hill towards Bredon's Norton. In the east and south of England it is more plentiful. The heads of the flowers are large in proportion to the plant, which is seldom more than six inches high. It is recorded from Gloucestershire and Herefordshire.

393. **259. A. glycyphyllos,** Linn. *Sweet Milk Vetch.*

Native. Hedgesides, banks, and thickets. Local, on calcareous soil. P. May to September. *Top. Bot.* 66.

First Record. On a marl bank by Gregory's Mill, *Sheward, Nash, Hist. of Worc., Sup.*, p. 96 (1799).

Avon. Craycombe. Trench Woods. Cleeve Banks. Field south of the Camp, Bredon, *Mathews.* Berrow Hill, Feckenham. Trench Woods, 1908, *Humphreys.*

Severn. Wyre Forest, *Jorden.* Helbury Hill. Thrift Wood, *Lees.* One mile and a half up river from Worcester, *Rea.* Severn Stoke. Monk's Wood. Fenny Rough. Harvington.

Malvern. Silurian eminences, *Lees.* Martley Quarries. Mathon, *Lees.* Cradley. Hanley Castle. King's Hill, Powick. Brockhill. Old Hills. Shelsley.

The leaf of this plant is much larger than that of any of our native vetches. It is called Sweet Milk Vetch from the sweetness of its leaves and roots, which, however, though pleasant at first, leave a bitter flavour on the tongue. The plant is left quite untouched among pasture by cattle. It is sometimes called Wild Liquorice. It is recorded from all neighbouring counties.

CORONILLA, Neck. 106. (κορώνη, anything crooked, like a crow's bill; from the shape of the flower.)

396. **260. C. *varia,** Linn.

Colonist. River-banks. P. Summer. *Top. Bot.* 2.

First Record. Growing on the banks of the Severn in a naturalized state below the mouth of Dowles Brook, *W. N. C. Trans.* i. 8 (1853).

Severn. As above, and still continuing at this station, 1907.

Malvern. One plant on the Railway at Malvern Link, 1893.

A native of woods and limestone hills from Normandy and Belgium to Persia. It is frequent as a waste ground plant over most of its range, and perhaps occurs as native at one station in Kent. It occurs in Warwickshire as a casual.

ORNITHOPUS, Linn. 107. (ὄρνις, a bird, πούς, a foot; from the fruit resembling bird's claws.)

397. **261. O. perpusillus,** Linn. *Common Bird's-foot.*

Native. Heaths, sandy places, roadsides. Locally common. A. May to July. *Top. Bot.* 84.

First Record. Abberley Hill, *Sheward, Nash, Hist. of Worc., Sup.*, p. 96 (1799).

Severn. Hartlebury Common. Horseley Bank, Wolverley. Hagley Brake, abundant. Wannerton Down, *Mathews.* Sutton Common, Kidderminster. Kempsey. Blakedown.

Malvern. On the Hills, plentiful, *Lees.*

Lickey. Blakedown. Winwood Heath, Frankley, *Mathews.* Clent Hills. Moseley Wake Green, *Miss M. A. Beilby.* Rowney Green. Lickey Hills. Beacon Hill.

This pretty little plant frequently occurs very abundantly in some congenial spot, chiefly in light sand. It is usually very small, though sometimes its prostrate stems extend for a distance of eight or ten inches. Its tiny flowers are cream-coloured, veined with crimson, and its pods ripen into curves like the foot of a bird, from which fact it gets its name. The leaves consist of numerous close-set leaflets, in pairs, with one terminal one. The herbalists said the plant had a binding quality, good both for internal and external application in the case of wounds. It occurs in all neighbouring counties.

HIPPOCREPIS, Linn. 108. (ἵππος, a horse, κρηπίς, a shoe; from the form of the joints of the seed-pod.)

399. **262. H. comosa,** Linn. *Horse-shoe Vetch.*

Native. Calcareous soil. Local and rare. P. May to July. *Top. Bot.* 45.

First Record. On the south side of Bredon Hill, below the Camp, *Nash, Hist. of Worc., Int.*, p. lxxxix (1781).

Avon. Broadway, *Lees.* The Slads, *Rea.* Bredon, on the sides of a large quarry, three-quarters of a mile south of the Camp, and the side of an escarpment towards Bredon's Norton, *Mathews.* Bredon Hill above Wollas Hall. On the King and Queen, Bredon, 1907.

This is a common plant on chalky soils in many parts of England, but Bredon Hill and the Slads are the only localities where it now occurs in this

county. The singular pods resemble a tuft of horse-shoes united together by their extremities. The plant is recorded from all neighbouring counties, in Warwickshire as a casual.

ONOBRYCHIS, Linn. 109. (ὄνος, an ass, βρύχω, I bray; because the smell was said to excite braying.)

400. **263. O. viciaefolia,** Scop. *Sainfoin.*

(*O. sativa*, Lam. *Hedysarum Onobrychis*, var. α. Linn.)

Colonist. Pastures and cultivated ground. Several places. P. May to August. *Top. Bot.* 30.

First Record. On the limestone rocks, Malvern, *Lees, Mag. Nat. Hist.*, vol. iii, p. 160 (1830).

Avon. Craycombe. Bredon Hill, south of the Camp, *Mathews.* Norton, near Evesham. Sheriff's Lench. Cleeve Bank. Crowle.

Severn. Tibberton. Ombersley. Grimley.

Malvern. Ridge Hill, Martley. Abberley Hill, *Perry.* Silurian eminences, *Lees.* Gadbury Banks, Eldersfield, *Lees.* Croft Quarry. Coningree Hill. Hanley Castle.

With white flowers this plant occurs at Croft Quarry, Malvern, which is a Worcestershire locality now lost to Herefordshire, and near Martley. Its English name is derived from the French, *sain*, healthy, *foin*, hay, alluding to its properties as a cultivated plant, though sometimes the first syllable is referred to the also French *Sainte*. A popular name is 'Go-to-sleep'. It is especially well-suited for cultivation on poor and calcareous soils, as its long roots penetrate deeply, and find sustenance where the roots of no other plant could reach. Possibly it is a native of Britain; but it has long been cultivated, formerly more than at present, and in many places it is a survival of these conditions. Fuller, in his *Worthies*, says it was 'first fetched out of France about Paris, and since is sown in divers places in England', and that 'it will last seven years by which time the native grasse of England will prevail over the foreigner if it be not sown again'. Mr. Lees, in his *Bot. Worc.*, p. xxxi, note, quotes a letter by Mr. Penyston Hastings, an antiquary of Daylesford, to Dr. Thomas, of Worcester, dated Dec. 11, 1732, in which he states that 'at this place (Daylesford) was first introduced the cultivation of Saintfoin, a French grass brought into England by John Hastings in 1650'. It is recorded from all bordering counties.

VICIA, Linn. 110. (The old Latin name.)

401. **264. V. hirsuta,** Gray. *Common Tare.*

(*Ervum hirsutum*, Linn.)

Native. Hedges, bushy ground, sandy fields. Not uncommon and widely distributed. A. May to August. *Top. Bot.* 109.

First Record. As *Ervum hirsutum*. Not localized, *Lees, Bot. Malv. Hills*, 1st ed., p. 36 (1843).

This weed is common in the fields and hedgerows in England, though less so in Scotland. In wet seasons it will sometimes endanger whole crops by entwining itself among them, and hence it has received the country name of 'Strangle Tare'. It is also called 'Fetches and Blows'. Its little reddish

seeds are dotted with black, and formerly were said to produce debility if ground up among flour. At all events they impart to it a disagreeable flavour. It is recorded from all neighbouring counties.

402. **265. V. tetrasperma,** Moench. *Smooth Tare.*

(*Ervum tetraspermum*, Linn. *V. gemella*, Crantz.)

Native. Hedges, sandy fields, waste places. General. A. May to July. *Top. Bot.* 75.

First Record. As *Ervum tetraspermum*. Not localized, *Lees, Bot. Malv. Hills*, 1st ed., p. 36 (1843).

This plant also is a common weed. It is distinguished from the former plant by its short-stalked smooth seed-pods, while that has sessile hairy ones. It is reported from all bordering counties.

403. **266. V. gracilis,** Loisel.

Colonist. Grassy ground, waysides. Very rare. A. May to July. *Top. Bot.* 25.

First Record. Near Worcester, *Lees, Bot. Worc.*, p. 41 (1867).

Avon. Craycombe Hill. Near Trench Woods, not wild, *Mathews.* Flyford Flavell. Near Gooshill Wood. Hanbury.

Severn. Tibberton, *Lees.* Near Coningree Wood. Warndon. Peachley Road.

Malvern. One place, *Lees.* Powick, *Lees.* Knightwick. Martley. Stockton-on-Teme. Malvern Link. Middleyards Coppice. Siding at Bransford Station.

This plant was reported as a new record, which it was not, from 'gravelly ground near Worcester', by John Fraser, M.D., in *Bot. Rec. Club Rep.*, 1875. It differs from the preceding tare in its flowers, which are twice as large, and their peduncles exceed the leaves and its linear acute leaflets, while its seeds are smaller; but there are very many intermediate forms. It has been observed in Gloucestershire and Warwickshire.

404. **267. V. Cracca,** Linn. *Tufted Vetch.*

Native. Hedges, pastures, banks. General. P. May to September. *Top. Bot.* 112.

First Record. Not localized, *Walker, Med. Surg. Rep.*, vol. i, no. 2, November, 1828.

The blue flowers of this most conspicuous of the vetches are a handsome ornament to the hedges of many a country lane-side during the summer months. In the meadow it is welcome as being a most nutritious fodder plant. It is widely distributed, occurring all over Europe to India, North Africa, and in North America and Greenland. So of course it occurs in each of our neighbouring counties.

406. **268. V. sylvatica,** Linn. *Wood Vetch.*

Native. Woods and thickets. Local and rare except in the Severn district. P. June to August. *Top. Bot.* 80.

First Record. In a thicket on the north side of Breedon Hill, *Nash, Hist. Worc., Int.*, p. lxxxix (1781).

Avon. Bredon Hill. Button's Wood on Bredon Hill.

Severn. Warshill Wood. Habberley Valley. Shrawley Wood. Wyre Forest, *Jorden.* Eyemore. Near Bewdley.

Malvern. Clifton-on-Teme, *Stokes ; Rufford.* Worcestershire Beacon, *Lees.* Near the Spout, Malvern. Lower Sapey in great luxuriance (1834), *Lees.* Sapey Brook. Mathon Coppice.

Lickey. Nuns' Walk, Dodford.

This is our handsomest vetch, but not common enough to become a feature with us among summer flowers. The flowers are numerous, white, and streaked with bluish veins. It is recorded from all neighbouring counties.

407. 269. V. sepium, Linn.   *Bush Vetch.*

Native. Hedges, woods, and thickets. General. P. April to September. *Top. Bot.* 112.

First Record. Not localized, *Walker, Med. Surg. Rep.*, vol. i, no. 2, November, 1828.

This plant has been found with white flowers on Old Storridge, Clap Hill Lane, Bromyard Road, near the bridge over Laughern Brook, and at Cotheridge. It is very common in woods and under hedgerows, but its flowers are of a somewhat dull tint. It is one of the earliest to bloom of our vetches, and remains green far into the winter. It is found in each neighbouring county.

411. 270. V. sativa, Linn.   *Common Vetch.*

Casual. Cornfields, waste places. In several places, always a relic of former cultivation. A. May to August. Not in *Top. Bot.*

First Record. Not localized, *Walker, Med. Surg. Rep.*, vol. i, no. 2, November, 1828.

This plant is a native of the Mediterranean region, widely cultivated in all temperate regions for spring fodder, the only species of the genus except the Bean that is grown to any extent. For the well-known Bean of the fields, which scents the whole neighbourhood of its place of growth, of which the Broad Bean of our gardens is a more highly cultivated form, is a Vetch, *V. faba*, Linn. Pythagoras forbade his disciples to eat beans ; why, has been a matter of speculation. There were many superstitions connected with them, one of which was that they were the retreat of the soul after death. Hippocrates thought that beans weakened the eyesight ; and other evil effects have been attributed to them, for which their black-stained flowers have no doubt been considered a kind of prophetic mourning. *V. sativa* has been recorded, in a doubtful condition, from all our neighbouring counties.

412. 271. V. angustifolia, Linn.   *Wild Vetch.*

(*V. Bobartii*, Forst.)

Native. Dry banks, roadsides, short turf. Widely distributed. A. April to July. *Top. Bot.* 92.

First Record. In a swamp on the north side of Falling Sands Common, near Kidderminster, and in a field between Kidderminster and Picket Rock, *Perry, Mag. Nat. Hist.*, vol. iv, p. 450 (1831).

This plant is not mentioned in Lees, *Bot. Worc.*, under any name. It is a very variable plant, and two extreme forms have been described as species, but the characters by which they have been distinguished pass insensibly

---

into each other. It is a near relative of *V. sativa*, being by some considered a variety of that plant, differing in its leaflets, which are linear lanceolate instead of ovate oblong. White-flowered forms have been observed at Sutton Common near Kidderminster, near Bridges-stone, and near Malvern Wells. The plant is recorded from all neighbouring counties.

Var. **Bobartii**, Koch.

First Record. About the bases of the hills and on the commons, Malvern, *Towndrow, Malv. Advertiser*, Nov. 12, 1892.

Avon. Sheriff's Lench. Crowle. Trench Woods. Fladbury. Goosehill Wood. Flyford Flavell. Bradley.

Severn. Hartlebury Common. Lanes near Harborough, Hagley, running towards *V. Bobartii*, 1904, *W. Whitwell*, but this ' cannot be placed under *V. Bobartii*', *H. and J. Groves*, 1905. Wyre Forest. Blackstone. Ombersley. Kempsey. Severn Stoke.

Malvern. Malvern Link, and as in first record. Knightwick. Storridge. Bridges-stone. Stanford. Martley. Old Hills.

Lickey. Barnt Green. Near Great Farley Wood.

In this form the flowers are solitary, the leaflets of the upper leaf linear, the pods patent, and the stems prostrate. It appears to occur in every neighbouring county except Hereford.

413. 272. V. lathyroides, Linn.   *Spring Vetch.*

Native. Dry sandy places. Local and rare. A. April to June. *Top. Bot.* 54.

First Record. Battenhall Lane, Craycombe Hill, and various other places in the county, *Lees, Illus. Nat. Hist. Worc.*, p. 172 (1834).

Avon. Craycombe Hill.

Severn. Battenhall Lane, Hartlebury Common, *W. N. C.* iii. 236. Sandy fields east of Kidderminster.

Malvern. On the hills small, very rare, *Lees.* ' I believe an error for *V. Bobartii*, which is not mentioned in *Bot. Malv. Hills*,' *Towndrow (MS.)*.

The language of Mr. Lees in the first record hardly accords with the localization of this plant in his *Table of Plants, Bot. Worc.*, where he marks it as occurring in one place only in the two districts of Severn and Malvern, in the latter saying it was lost. This is a very minute decumbent plant, which on Hartlebury Common, where is locally abundant, flowers early, dries up, and soon disappears. Its general aspect renders it quite a distinct species, not easily confused with any other vetch. This, like the last, appears not to be recorded for Herefordshire only of neighbouring counties.

414. 273. V. bithynica, Linn.   *Rough-podded Purple Vetch.*

Native. Bushy places. Local. A. May and June. *Top. Bot.* 19.

First Record. Woods near Clifton-on-Teme, *Stokes, Stokes's With. Bot. Arr.*, 2nd ed., p. 779 (1787).

Avon. Broadway, plentiful.

Severn. Bromyard Road, Worcester.

Malvern. Below the Admiral Benbow, Malvern Wells. Between Alfrick and Hopton Court, *Lees.* Clerkenhill, Leigh, *Dr. Abbot*, in *Eng. Botany.* Between Leigh and Alfrick, *Rea.* Clifton-on-Teme, *Stokes.* Sherridge, Leigh. Powick.

H

This vetch loves the neighbourhood of the sea, but in some few places is found in inland localities. It is recorded only for Gloucestershire, and that only in its western parts, of our neighbouring counties.

LATHYRUS, Linn. 111. (λάθυρος, a plant-name used by Theophrastus.)

415. 274. L. Aphaca, Linn.   *Yellow Vetchling.*

Native. Hedgebanks. Local and rare. A. June and July. *Top. Bot.* 27.

First Record. Cleeve. Littleton, *Purton, Midl. Fl.*, p. 339 (1817).

Avon. Crowle, *Sheppard* (1834). Roadside under Craycombe Hill. Roadside near Evesham, *May*, in his *Hist. Evesham.* Norton, near Evesham. Near Trench Woods. Broadway.

Severn. Norton Field, Worcester, *Lees.* Grove Coppice, Stourport, *Hickman* (1821). Hatfield, near Norton, *L. Sutton* (1845). Roadside near Spetchley Railway Station. Kempsey. Broadwas.

Malvern. Powick, *Baxter* (1847). Quarry at Pendock. Upton, *Julian Kent.* Powick. Near Leigh Mill. Berrow.

This plant is unmistakable in appearance. It has no leaves, properly so-called. They are transformed into tendrils, and replaced by the stipules, which become large, leaf-like appendages. It does not appear to have been recorded from Hereford or Salop of our bordering counties.

416. 275. L. Nissolia, Linn.   *Crimson Vetchling.*

Native. Grassy banks. Several places throughout the county, except in the Lickey district. A. May to July. *Top. Bot.* 41.

First Record. On banks by the sides of woods between Pershore and Eckington, *Nash, Hist. Worc., Int.*, p. lxxxix (1781).

Avon. Littleton Banks. Craycombe. Bredon Hill. Broadway. Between Hadsor and Hanbury. Between Evesham and Church Lench, *Mathews.* Tardebigge. Morton-under-Hill, near Inkberrow.

Severn. Norton Field, Worcester. Temple Laughern. By the roadside near Cotheridge. Broadwas. Henwick. Helbury Hill. Hawford. Oldbury Road. Battenhall Lane. Leopard Farm. Near Spetchley Railway Station. Near Porter's Mill.

Malvern. Middleyards Coppice. Madresfield. Doddenham. Abundant by the roadside between Broadwas and Knightsford Bridge. Longdon. Newland. Madresfield. Welland. Powick.

This grass vetch is hardly to be distinguished from the grass among which it grows except when it is in flower, so grass-like are its phyllodes. Like the last plant, it is not recorded for Hereford or Salop of our bordering counties.

419. 276. L. pratensis, Linn.   *Meadow Vetchling.*

Native. Pastures, meadows, hedges, banks. General. P. June to September. *Top. Bot.* 112.

First Record. Not localized, *Walker, Med. Surg. Rep.*, vol. i, no. 2, November, 1828.

This plant may be seen in most bushy, grassy places in July and August, in greater or less luxuriance, according to the nature of the locality in which it grows. Its stipules are large, in the young state sometimes imitating

---

those of *L. Aphaca.* The stems are angular and climbing, often two or three feet long, and in the flower garden, if it finds an entrance, it is a difficult weed to eradicate, its roots running for a considerable distance underground. It is found in all neighbouring counties.

420. 277. L. latifolius, Linn.   *The Everlasting Pea.*

Alien. Bushy places ; a garden escape. P. July and August. Not in *Top. Bot.*

First Record. Severn Stoke Copse, *Ballard, Stokes's With. Bot. Arr.*, 2nd ed., p. 772 (1787).

Severn. As above. Perry Wood.

Malvern. Guarlford.

This plant was lost in the Severn Stoke locality in Mr. Lees' time. It is a well-known and showy climber, adorning cottage porch or garden alcove with its profusion of foliage and purple and pink flowers. It is a native of the woods of Southern Europe, and seems to have become established in several places ; it often enlivens the banks of some railway where it has escaped from a neighbouring little station garden. The well-known Sweet Pea is *L. odoratus*, Linn., and is a native in the woods of Italy and Sicily ; and another, *L. sativus*, Linn., is largely cultivated in Europe as fodder, and at one time the seeds of the plant were made into bread, but with such deleterious effect that its use was forbidden, as it caused, so it was said, a rigidity of the limbs, which was incurable. The Everlasting Pea, in an escaped condition, is reported as occurring in Salop and Warwick.

422. 278. L. sylvestris, Linn.   *Narrow-leaved Everlasting Pea.*

Native. Woods, thickets, hedges. Local. P. June to September. *Top. Bot.* 62.

First Record. By a woodside going from Pershore to Eckington, *Nash, Hist. Worc., Int.*, p. lxxxix (1781).

Avon. Tiddesley Wood. Defford. Hadsor. Woods near Pershore, *Purton.* Woods about Bredon Hill, *Lees.*

Severn. Helbury Hill ; Perry Woods, *Lees.* Near Lenchford. Hazlewood.

Malvern. Pendock Portway. Gadbury Banks, Eldersfield, *Lees.* Roadside between Broadwas and Doddenham.

This is not a very frequent flower in our woods and thickets. The blossoms are large, of a dull purple colour, tinged with green, and marked with purple veins. It might pass for the Everlasting Pea, but its leaflets are linear lanceolate, instead of, as in that, elliptic and pointed, and the seeds of it are smoother. It is recorded from all bordering counties.

{**423. 279. L. palustris**, Linn.   *Marsh Chickling Vetch.*

Native. Boggy meadows. Very rare. P. June and July. *Top. Bot.* 20.

First Record. Only in a marshy meadow on the western side of Longdon Marsh, *Lees, Bot. Malv. Hills*, 1st ed., p. 35 (1843).

Malvern. Longdon Marsh (1857), *W. N. C.* i. 443. Castlemorton, *Lees.*

Probably the two localities refer to the same spot. Mr. Lees says, 3rd ed., *Bot. Malv. Hills*, p. 87 (1868), that it is constant at the locality. It has now long disappeared. It is not recorded for any neighbouring county.}

H 2

425. 280. L. montanus, Bernh.   *Bitter Vetch.*

(*Orobus tuberosus,* Linn. *L. macrorrhizus,* Wimm.)

Native. Open woods, coppices, bushy places, banks. General. P. April to July. *Top. Bot.* 107.

First Record. As *Orobus tuberosus,* not localized, Lees, *Bot. Malv. Hills,* 1st ed., p. 35 (1843).

This plant, sometimes called Peaseling or Wood Pea, has clusters of pink and purple flowers on long stalks in the axils of the leaves, and possesses no tendrils. The tuberous roots of the plant are used in many ways, from the casual wayside meal of the village child to being roasted like chestnuts, and brought to table in Holland ; thus treated they can scarcely, indeed, be distinguished from chestnuts. They were famed for curing ' hunger's gnawing pangs', under the name of Carmele or Cormeille, a modification of the name in Gaelic. The plant is found in all neighbouring counties.

Var. b. **tenuifolius,** Roth.

First Record. As *O. tenuifolius.* Seats Common, Malvern, Lees, *Bot. Malv. Hills,* 2nd ed., p. 35 (1852).

Severn. Wyre Forest, *Jorden.* Monk's Wood.

Malvern. Little Malvern, *Lees.* Berrow Hill. Ankerdine.

Lickey. Dodford.

This variety differs from the type in possessing linear instead of oblong or lanceolate blunt leaflets. It has been found in Staffordshire, Shropshire, and Warwickshire.

26. ROSACEAE, Benth. and Hook.

PRUNUS, Linn. 112. (The old Latin name.)

427. 281. P. spinosa, Linn.   *Blackthorn, Sloe.*

Native. Woods, coppices, hedges. Abundant in all the districts. Shrub. March to May. *Top. Bot.* 108.

First Record. Not localized, Lees, in his *Bot. Looker-out,* p. 94 (1842).

Everybody knows the Sloe, and welcomes the white early blossoms that crowd its black and thorny leafless branches. Its name has passed into a proverb with regard to the seasons, and the term ' Blackthorn winter ' means a cold March. Its delicate leaves come out after the blossom has disappeared, and at one time they were dried and largely used to adulterate tea. The practice was carried on to such an extent that it received the attention of Parliament, and was suppressed. The dark purple fruit is known to every schoolboy, and one well-known comparison is to say a lady has ' eyes as black as sloes '. The fruit has been largely used to adulterate port wine, and in a more legitimate way ' Sloe Gin ' is manufactured from it, like Cherry Brandy from cherries. The straighter stems, usually thornless at their lower parts, are used for making walking-sticks, for which the words a ' stout blackthorn ' often stand in metaphor. The bark is said to possess some of the qualities of Quinine. The plant is found in all neighbouring counties.

Var. b. **macrocarpa,** Wallr.

(*P. fruticans,* Weihe.)

First Record. Malvern, *Towndrow, Malv. Advertiser,* Nov. 12, 1892.

Malvern. As first record.

In this plant the flowers and leaves are found at the same time, and the leaves are pubescent on the veins beneath.

Var. **coëtanea.**

First Record. Malvern, *Towndrow, Malv. Advert.,* Nov. 12, 1892.

Malvern. As first record.

This variety is not mentioned in *Lond. Cat.,* 10th ed.

428. 282. P. insititia, Huds.   *The Bullace.*

Denizen or native. Hedges. Not common. Shrub. April and May. *Top. Bot.* 67.

First Record. Bullace Plum, Badsey, *Purton, Midl. Fl.,* p. 233 (1817).

Avon. One place, *Lees.* Trench Woods. Cleeve Banks. Sheriff's Lench. Craycombe. Tiddesley Wood.

Severn. Wyre Forest, *Jorden.* Hedge at Battenhall, *Lees.* Whittington. Woods Green. Droitwich Road, five miles from Worcester. Leopard Hills. Kempsey.

Malvern. Barnard's Green ; Welland Common ; Castlemorton, *Lees.* Newland. Near Old Hills. Bransford. Knightwick. Ankerdine. Martley. Malvern Link.

Lickey. Dudley Castle Hill, *Lees.*

This is really a Mediterranean plant, now found in various places in English hedgerows ; at one time it was much planted in this country. The fruit is larger than the sloe, and is sometimes yellow, though generally purple. The yellow-fruited variety has been observed at Malvern Link. It is recorded from all neighbouring counties.

429. 283. P. domestica, Linn.   *The Plum.*

Denizen. Hedges. Occasional. Shrub or small tree. April and May. Not in *Top. Bot.*

First Record. Common Plum, Badsey, *Purton, Midl. Fl.,* p. 234 (1817).

Avon. Occasional, *Lees.* Near Oddingley. Tiddesley Wood.

Severn. Wyre Forest, near the old Sorb tree, Lees, *Bot. Worc.,* p. 5, note. Hedges near Battenhall. Leopard Hills. Droitwich Road, five miles from Worcester. Martley Road. Astley.

Malvern. Occasional, *Lees.*

Lickey. Rather uncommon, *Lees.*

Whatever may be the case with *P. insititia,* this plant, if it be really distinct, has no claim at all to be considered British. In fact, several botanists consider this and the two former plants all one, *P. communis,* Huds., and that it is only by rejecting intermediates that we get three fairly marked species, *spinosa, insititia,* and *domestica.* This plant is the parent of all our orchard plums, and of that useful fruit of Worcestershire cottage gardens, the Damson. The plum blossom of the Vale of Evesham, when fully out, is a sight unequalled in our country, or elsewhere. Plums found wild, seedlings usually from cultivated ones, are far from equalling in quality their parents, and in succeeding generations approach more nearly their ancestral type. The Plum is found undomesticated in all neighbouring counties.

---

430. 284. P. avium, Linn.   *Wild Cherry,* or *Gean.*

Native. Coppices, woods, and hedges. Not common, except in the Severn and Malvern districts. Tree. April and May. *Top. Bot.* 97.

First Record. (According to Mathews) as *P. cerasus,* Wild Cherry tree, near Gregory's Mill, &c., *A. Florence, Stranger's Guide* (1828).

Avon. Tiddesley Wood. Trench Woods.

Severn. Wyre Forest, as *Cerasus avium, Jorden.* Woods opposite Holt Castle, *Lees.* Ombersley. Perry Wood, *Lees.* Blackstone Rock. Stagbury Hill. Dowles Brook. Boreley Coppice. Landmore Coppice, Broadwas.

Malvern. In most of the hilly woods, *Lees.* Ankerdine Hill.

Lickey. Woods at Hagley, Clent, and Frankley, *Mathews.* Wychbury Wood, Pedmore.

From this tree are descended all the tribe of garden and orchard cherries, except the Morello, which is said to be derived from the next plant, the chief difference of which from this one is that it is always dwarfer and never a tree, while its always red fruit does not possess a staining juice. The wild cherry is a great ornament to the woods in which it occurs, especially if they are situated on some sloping hillside ; and not only in spring, when its white flowers are conspicuous, but in autumn also, when its leaves become a rich red colour. From the stems often exudes a sticky gum, said to be exceedingly nutritious, though the bark is astringent. The wood is tough and red in colour, and will take a high polish. Cherries are extensively planted along roadsides in Germany, and from them is made an insidious drink called *Kirschwasser.* Cherry orchards in our county are especially a feature in the neighbourhood of Bewdley and Tenbury, and in the valley of the Teme, and no more beautiful sight can be imagined than a cherry orchard in full bloom. As far as the cultivated cherry is concerned, it is said to have been brought to Europe from the city of Pontus in Asia, now called Kerasoun, from which the Latin name *Cerasus* is derived ; and from *Cerasus,* probably, through many stages, Cherry. That there is no native name in Celtic or Teutonic for the cherry confirms the opinion of botanists that the tree is not indigenous in Western Europe. A mist of folklore has gathered about the cherry, and, however derived, the name has firmly established itself in many associations in our language. The form with double flowers is an ornament to many a garden shrubbery. The wild cherry is recorded for all neighbouring counties.

431. 285. P. Cerasus, Linn.   *Dwarf Cherry.*

Native. Hedges and woods. Rare. Shrub or small tree. April and May. *Top. Bot.* 36.

First Record. Probably as *Cerasus austera.* Hedges, Barnard's Green, rare, Lees, *Bot. Malv. Hills,* 1st ed., p. 23 (1843).

Severn. Hillditch Coppice, Hartlebury, *Miss Ladbury.* Helbury Hill, *Lees.* Near Bewdley. Near Hartlebury Common. Wyre Forest.

Malvern. Barnard's Green, as in first record. Rosebury Rock, Knightwick ; Ankerdine Hill, *Lees.*

Mr. Mathews thinks that all the above records of Mr. Lees refer to *P. avium, Mid. Nat.* xi, p. 159. In fact he seems to doubt Mr. Lees' knowledge of *Prunus* altogether ! Mr. Towndrow observes that he has never seen this plant in

---

the Malvern district, and is inclined also to doubt Mr. Lees' records (MS.). In his *Table of Plants,* Mr. Lees assigns to this plant only two localities, one in each of the districts Severn and Avon. It has been reported from all neighbouring counties.

[*Prunus Padus,* the Bird Cherry, with the number in *Top. Bot.* 70, and occurring in every bordering county, has not been recorded in Worcestershire.]

SPIRAEA, Linn. 113. (Supposed to be the σπειραία of Theophrastus, which is derived from σπείρα, a coil, from the twisted shape of the fruit.)

433. 286. S. salicifolia, Linn.   *Willow-leaved Spiraea.*

Alien. Bushy places. Four localities only, at two extinct. Shrub. July and August. Not in *Top. Bot.*

First Record. Welland Common, perhaps naturalized there, as a garden was not far from the spot, Lees, *Mag. Nat. Hist.,* vol. iii, p. 160 (1830).

Severn. By the side of Dowles Brook, Wyre Forest, *Jorden,* extinct. Snuff Mill Valley, Bewdley, 1900, *Miss Ladbury.*

Malvern. Welland Common, *Lees,* extinct.

Lickey. Red Lane, Redditch.

This plant, commonly seen as an ornamental shrub in gardens, is a native of wet river-banks and bushy places in Southern Europe, and the greater part of the north temperate zone, but further north than Central France it is only known in a naturalized condition. In that condition it has been observed in Stafford and Salop.

434. 287. S. Ulmaria, Linn.   *Meadow Sweet.*

Native. Moist meadows, damp places in woods, sides of ditches. Common. P. May to September. *Top. Bot.* 112.

First Record. Not localized, *Walker, Med. Surg. Rep.,* vol. i, no. 2, November (1828).

This common ornament of wet meadows and stream-sides throughout the county needs no localization ; it is everywhere to be met with where such conditions exist. It was much used for medicinal purposes in mediaeval times, and was supposed to cure nearly every ailment that flesh was heir to. The fragrance of the flower is due to the presence of prussic acid, and though delicious in the open air is not so pleasant in a closed room. The whole plant is bitter and astringent. It is also called Queen of the Meadows, and Meadow Queen, and about Stoulton, ' Coddled Apples.' It is found in all neighbouring counties.

Var. b. **denudata,** Boenn.

First Record. The specimens found this year were clearly within this county, *Towndrow, Malv. Advert.,* Nov. 14, 1896.

Malvern. Near Malvern, as first record. Bransford.

This variety occurs in all the districts, and no doubt in all the neighbouring counties also, though it does not appear to have been observed in any of them. It is distinguished by the stem leaves being glabrous instead of downy beneath.

435. 288. S. Filipendula, Linn.   *Dropwort.*
Native.   Dry pastures.   Abundant in the calcareous districts.   P.   May to August.   *Top. Bot.* 65.
First Record.   On Bredon Hill, above Overbury, plentifully, *Nash, Hist. Worc., Int.,* p. lxxxix (1781).
Avon.   Bredon Hill.   Himbleton.   The Slads.   Defford.   Broadway. Trench Woods.   Alderminster.   Very abundant in a field near Croome Perry Wood.   Tiddesley Wood.
Severn.   Spetchley Common.   Frequent in the neighbourhood of Worcester, *Stokes.*   West side of Perry Wood, *Lees.*   Brookend, near Kempsey, *Lees.*   Near Nunnery Wood.
Malvern.   Old Hills.   Between Madresfield and Powick.
This is a smaller plant than the last, with finely divided dark green leaves. Before the buds expand they are tinged deep rose-colour.   A double form is a handsome and frequent garden flower, and if allowed to establish itself, spreads itself everywhere about the borders.   It is recorded from every bordering county, doubtfully from Salop.

RUBUS, Linn. 114.   (Celtic *reub,* to tear ; but some people derive it from *ruber,* red, from the colour of the fruit.)

This intricate tribe of plants has of late years received much attention at the hands of many botanists, especially in this country from the Rev. W. M. Rogers, whose *Handbook of British Rubi,* 1900, may be considered the latest expression of the knowledge of these plants in Britain.   But nearly every botanist arranges them in a different order, employs a different nomenclature, and raises or depresses plants into species, sub-species, or varieties, as his own observations guide him.   Consequently, it is difficult frequently to recognize the species indicated by a certain name, or to be sure that the plant intended is being dealt with.   It is to be feared that many otherwise enthusiastic botanists are, to use a slang term, 'put off' by this extreme specialization, and so assume the old attitude of Linnaeus, dubbing the bramble *R. fruticosus,* and passing it by.

436. 289. R. idaeus, Linn.   *The Raspberry.*
Native.   Coppices, heaths, roadside wastes.   Abundant in the Severn and Lickey districts, not so common elsewhere.   Shrub.   June and July. *Top. Bot.* 111.
First Record.   In woods abundantly near Kidderminster ; and in hedge-rows, Chaddesley Corbett, &c., *Purton* ; *Purton,* vol. iii, pl. ii, p. 335 (1821).
The fruit of the wild raspberry is not so large as that of the cultivated form, which is much better known than the wild plant ; and the domestic uses its fruit is put to are a matter of common knowledge.   An old name for the plant was Hindberry, a word which is found in use as early as the eighth century, and is still locally current in Cumberland.   Probably it got the name from forming a fruit that hinds could find and eat without question from their masters.   The specific name is given to it from Mount Ida, a mountain range in Mysia in Asia Minor, well known to readers of Tennyson's *Œnone.*   There is another mountain called Ida, in Crete, but it is not

the one whose name is perpetuated in the word *Idaeus.*   The Raspberry is abundant in the North of Europe, especially in Sweden.   In our county it ascends to 900 feet, being found forming thickets on roadside wastes on the Lickey at that height.   It is found in all neighbouring counties.

437. 290. R. fissus, Lindl.
(*R. plicatus,* Leighton.   *R. nessensis,* Hall.)
Native.   Heathy woods and open spaces.   Local.   Bush.   July to September. *Top. Bot.* 57.
First Record.   Mentioned by Lees as a variety of *R. suberectus, Bot. Worc., App.,* p. 47 (1867).
Severn.   Hartlebury Common, *Westcombe.*   Oldington, *Mathews.*
Mr. Lees says this is a more prickly and hairy variety of the following plant, *R. suberectus,* Anders.   Mr. Rogers says it is only quite characteristic in sunny spots, becoming more like that plant in damp shady places.   It is recorded in Salop, Stafford, and Warwick.

438. 291. R. suberectus, Anders.
(*R. nessensis,* W. Hall, which is applicable to this plant as well as the one above.   *R. fastigiatus,* Rub. Germ.)
Native.   Woods and bushy places.   Local.   Shrub.   July and August. *Top. Bot.* 42.
First Record.   Not localized, *Lees, New Bot. Guide* (1835).
Severn.   About Kidderminster.   Hartlebury Common.   Falling Sands Lock, Stourport.   Blakedown Pool and Hurcott Wood, sparingly, *Mathews.*   Wyre Forest.   Birchin Grove.
Malvern.   'In its normal form I have not met with it in the Malvern district,' *Lees.*
Lickey.   Moseley.   Bromsgrove Lickey.   Near Headless Cross, *Lees.* Near Old Rose and Crown Inn, *Rev. J. H. Thompson.*
This is a tall large plant, the former plant being comparatively short and small, and the stem is sharply angled, with the prickles confined to the angles.   It is a well-distinguished species, says Mr. Lees, and its stems never root.   It is found in all neighbouring counties.

441. 292. R. plicatus, Wh. and N.
(*R. corylifolius, affinis,* or *nitidus* of many authors.)
Native.   Commons and heathy woods.   Not common.   Shrub.   May to September.   *Top. Bot.* 69.
First Record.   Birchen Grove, Worcester, *Lees, Bot. Malv. Hills,* 2nd ed., p. 57 (1852).
Severn.   Birchen Grove.   Hodge Hill, Kidderminster, *Mathews.*   Road to Shatterford.   Warshill, Bewdley.   Crown East, *Westcombe.*
Malvern.   In most thickets below Worall's Well, West Malvern.
Lickey.   On the lower Lickey range at about 800 feet, *Lees.*
This plant forms a tall shrub, the stem often remaining suberect the second year, and throwing out flowering shoots from the summit after the manner of the Raspberry ; it very rarely bends over and roots.   It has not been recorded for Gloucestershire except in the western vice-county.

[442. 293. R. nitidus, Wh. and N.
This plant, the *Top. Bot.* no. of which is 24, has not been recorded in this county, but Mr. Mathews says, *Mid. Nat.* xiv, p. 91, that it, and its white-flowered var. *hamulosus,* Lefèv. and Muell., are certain to occur within its boundaries.   Another *nitidus* is mentioned below, and there may be confusion.]

443. 294. R. affinis, Wh. and N.
Native.   Hilly situations on sandy soil.   Local.   Shrub.   July to September.   *Top. Bot.* 20.
First Record.   Not localized, *Lees, Bot. Malv. Hills,* 1st ed., p. 27 (1843).
Severn.   Kidderminster.   Bewdley.   Habberley Valley.
Malvern.   Forming thickets among waste pastures below Malvern Wells, but rare, *Lees.*
Lickey.   Near St. Kenelm's.   Offmoor Wood.
This bramble, Mr. Lees says, he has never seen in hedges.   Easily recognized by its marked long and narrow prickles, and its very gradually narrowed terminal leaflet.   It has not been recorded for Warwickshire or Staffordshire.

450. 295. R. carpinifolius, Wh. and N.
Native.   Heathy places, hedges, and open woods.   Local.   Shrub.   July and August.   *Top. Bot.* 38.
First Record.   Not localized, *Lees, Bot. Malv. Hills,* 1st ed., p. 27 (1843).
Severn.   Wannerton.   Bissell ; Hodge Hill, Broadwaters : ? near Kidderminster.   Near Bewdley.
Malvern.   West Malvern, *Lees.*
Lickey.   Occurs in the Lickey district, *Bagnall.*
Mr. Mathews says, *Mid. Nat.* xiv, p. 91, 'This name has been given to many forms of *Rubus,* and it is not certain that the plant so named by Mr. Lees is the *R. carpinifolius* of Wh. and N.   It is probable that Mr. Lees' plant is *R. Maassii,* Focke.'   Mr. Bagnall also says that Mr. Lees' plant is not correct.   Mr. Rogers says that he has seen no British bramble, labelled *R. Maassii,* Focke, which he would not venture to call *R. pulcherrimus,* hereafter treated of, or *R. Lindebergii,* which does not occur in Worcestershire.   The hornbeam-leaved bramble is usually pale and very prickly, and the typical plant is well-marked.   It occurs in all neighbouring counties.

451. 296. R. incurvatus, Bab.
Native.   Hedges and heathy ground.   Rare.   Shrub.   July and August. *Top. Bot.* 36.
First Record.   Thickets between Cowleigh and Worcester, *Lees, Bot. Malv. Hills,* 2nd ed., p. 55 (1852).
Malvern.   As above.
This bramble has not yet been discovered on the Continent.   It was not accepted for Worcestershire by Dr. A. Lees, but it has been observed in Hereford, Salop, and Stafford.

452. 297. R. Lindleianus, Lees.
(*R. nitidus,* Bell Salt., Leighton, Bloxham.)
Native.   Hedges and thickets, less frequently in woods, generally distributed.   Shrub.   June to August.   *Top. Bot.* 77.

First Record.   As *R. nitidus.*   Not localized, *Lees, Bot. Malv. Hills,* 1st ed., p. 27 (1843).   Established by *Mr. Lees, Phyt.,* O.S., vol. iii, p. 361, November 2, 1848.   In *Bot. Malv. Hills,* 2nd ed., p. 57, and 3rd ed., p. 73, termed *R. Lindleianus.*
Severn.   Common about Kidderminster, *Mathews.*   A form with large cordate leaves, abundant about the Brake Mill, Hagley, *Mathews.* Birchen Grove, *Westcombe.*   Churchill, near Kidderminster.
Malvern.   West Malvern, August 5, 1893, *Towndrow,* fide *Rev. W. M. Rogers.*   Malvern Wells.
Lickey.   Warstone Farm, Frankley, *Mathews.*   General in the district, *Mathews.*
This is one of the few most generally distributed and best known British and Irish brambles, though apparently confined to Germany on the Continent. It is, when typical, easily discriminated by the lustrous stem, very marked leaves, and the cylindrical truncate panicle with remarkably small fruit.   It occurs in all neighbouring counties.

453. 298. R. argenteus, Wh. and N.
(*R. erythrinus,* Genev.)
Native.   Bushy places.   Not common.   Shrub.   July to September.   *Top. Bot.* 31.
First Record.   As *R. fruticosus.*   Not localized, *Lees, Bot. Malv. Hills,* 1st ed., p. 27 (1843), which plant in the 2nd ed., p. 56, and 3rd ed., p. 74, is identified with *R. argenteus.*
Severn.   Cotheridge, *Lees.*   Blackstone.   Bewdley, *Westcombe.*
Malvern.   Southern end of Malvern Hills, *Lees.*   Madresfield, July 18, 1895, *Towndrow,* fide *Rev. W. M. Rogers.*   Abundant in Cowleigh Park, Malvern, *Rogers* and *Ley, Woolhope Club Trans.* 1893-1894, *Additions to the Flora of Herefordshire,* p. 60.
Much like *R. Lindleianus,* but distinguished by the lax, interrupted, round-tipped panicle, and the larger fruit.   The plant does not appear to have been noticed in Gloucestershire.

455. 299. R. rhamnifolius, Wh. and N.
(*R. cordifolius,* Blox.)
Native.   Hedges and wood-borders.   Generally distributed ; very common, *Lees.*   Shrub.   June to September.   *Top. Bot.* 70.
First Record.   Not localized, in *Lees' Cat., New Bot. Guide* (1835).   Recorded as *R. cordifolius, Lees, Bot. Malv. Hills,* 2nd ed., p. 57 (1852).
A common and generally well-marked thicket bramble, well distinguished by its smooth barren skin, and its closely-downy cylindrical panicle.   The terminal leaflet is frequently on a much elongated foot-stalk.   It is found in all neighbouring counties.

456. 300. R. nemoralis, P. J. Muell.
(*R. umbrosus,* of British authors, in part.)
Native.   Hedges and wood-borders.   Rare.   Shrub.   *Top. Bot.* 12.
First Record.   In this book as below.
Severn.   In this district, *Bagnall (MS.).*
Lickey.   In this district, *Bagnall (MS.).*
This bramble has been observed in Hereford and Salop of bordering counties.

459. 301. R. pulcherrimus, Neum.

(*R. polyanthemus*, Lindeb. *R. Neumani*, Focke. *R. umbrosus*, of British authors, in part. *R. Maassii*, Bab., in part. *R. carpinifolius*, Blox., in part.)

Native. Hedges, thickets, woods. Shrub. July to September. *Top. Bot.* 79.

First Record. In this book, as below.

    Severn. In this district, *Bagnall (MS.)*.

    Lickey. In this district, *Bagnall (MS.)*.

This bramble has not been recorded from Gloucester.

463. 302. R. villicaulis, Koehl.

Native. Heaths, open places in woods, wood-borders. Shrub. July to September. *Top. Bot.* 31.

First Record. Not localized, *Lees, Bot. Malv. Hills*, 1st ed., p. 27 (1843).

It is doubtful if Mr. Lees intends this plant in his first record. In the second edition of his book he records a bramble of his own, *R. pampinosus*, Lees, which by Mr. Babington, *Brit. Rubi*, p. 141, is named as *R. villicaulis*, Wh. and N. In the supplement to his *Bot. of Worc.*, p. 44, Mr. Lees localizes his *R. villicaulis*, saying it is rather uncommon, at Rough Wood Dingle, near Cowleigh Park ; woods at Alfrick ; Broadheath, and in Wyre Forest. And his *R. pampinosus*, which he calls 'a very remarkable Worcestershire bramble', he says occurs in dense thickets in Cowleigh Park, and in Wyre Forest.

Var. b. **calvatus**, Blox.

First Record. As *R. calvatus*. Wyre Forest, *Lees, Bot. Worc., App.*, p. 45. *Top. Bot.* 13.

'A large, remarkably savage-looking and strong bramble,' says Mr. Lees, 'whose stem becomes in age quite denuded' of the hairs which give rise to the name of its type, *R. villicaulis* ; while the leaves are very thinly hairy beneath. Mr. Bagnall has found this bramble in the Malvern district, as well as that of Severn, and it has been noted in Stafford, Warwick, and Salop.

464. 303. R. Selmeri, Lindeb. *Top. Bot.* 81.

(*R. affinis*, Blox.)

First Record. In this book.

This plant has been found by Mr. Bagnall in the Severn and Lickey districts. The whole plant is conspicuously more glabrous than *R. villicaulis*, with rounder leaflets conspicuously more shining above. The variety has been noted in all neighbouring counties except Gloucestershire.

470. 304. R. Godroni, Lecoq and Lamotte.

(*R. argentatus*, P. J. Muell.)

Native. Hedges. Very local. Shrub. July to September. *Top. Bot.* 31.

Observed at Malvern Link by Mr. Towndrow. Cowleigh Park, abundantly, Rogers and Ley, *Woolhope Club Trans.* 1893–1894, *Additions to the Flora of Herefordshire*, p. 64. A handsome bramble, coming into flower late. It has been observed in Herefordshire and Warwickshire.

471. 305. R. rusticanus, Merc.   *Blackberry.*

(*R. discolor*, of many authors. *R. ulmifolius*, Schott, in part.)

Native. Hedges and waste spots, and often an epiphyte upon willows. General everywhere. Shrub. June to September. *Top. Bot.* 74.

First Record. As *R. discolor*. Not localized, *Lees, Bot. Malv. Hills*, 1st ed., p. 27.

This plant and *R. corylifolius*, Sm., hereafter to be treated of, make up the rank and file of the great army of brambles met with throughout the country, and commonly known as 'Blackberries'. What exists so 'plentiful as blackberries' ? In the north of the county every lane is haunted during the Blackberry season by people from the neighbouring Black Country, who come miles to gather the fruit to sell to their neighbours at home. Among other virtues attributed to every part of the plant by the herbalists of old, the fruit and flowers were considered efficacious against serpent bite, and it was believed that eating the young shoots as a salad would fasten teeth that were loose ! The long shoots are used to keep down thatch, and in many a country churchyard they used to be seen on the newly made grave, binding down the sods. Numerous superstitions prevail with regard to the fruit. 'Old Harry' jumps into it on the 1st of October, making it unfit to eat. In Brittany, Blackberries ripen and drop off the hedges for want of picking ; the peasants will not touch them, as they believe the crown of thorns was made from bramble branches. The green boughs produce a black dye ; and silkworms thrive on the leaves. It is hardly necessary to state that the Blackberry is found in all the counties that border Worcestershire.

472. 306. R. pubescens, Weihe.

(*R. thyrsoideus*, Wimm.)

Native. Hedges. Rather rare. Shrub. June to September. *Top. Bot.* 15.

First Record. Wyre Forest and Seckley Wood, near Bewdley, *Lees, Bot. Worc., Sup.*, p. 41 (1867).

    Avon. Tardebigge, *Bab., Flora of Herefordshire*, p. 87.

    Severn. As in first record.

    Malvern. In this district, *Bagnall (MS.)*. Mentioned but not localized, *Lees*. Cowleigh Park, 1887, *Ley* (teste Rogers and Focke), *Woolhope Club Trans.* 1893–1894, *Additions to the Flora of Herefordshire*, p. 66.

This bramble has not yet been detected in Scotland, except perhaps in Aberdeenshire, or in Ireland. Under this Mr. Lees gives a variety, *macroacanthus*, which he localizes at Rough Hill Wood, near Cowleigh Park, which locality is in Herefordshire. *R. macroacanthus*, Blox., in part, is synonymized by Mr. Rogers with the variety *robustus*, P. J. Muell., of *R. argentatus*. It is not clear if Mr. Lees intends this plant. *R. pubescens* has not been recorded for Shropshire in the neighbouring counties.

476. 307. R. macrophyllus, Wh. and N.

Native. Woods and hedges. Local. Shrub. June to September. *Top. Bot.* 51.

First Record. Only in the upper part of Cowleigh Park, Malvern, *Lees, Bot. Worc., Sup.*, p. 46 (1867).

    Malvern. Cowleigh Park, as above.

    Lickey. The Randans, 'I think,' *Bagnall (MS.)*.

'A truly forest plant,' says Dr. Focke, 'loving fresh somewhat moist soil and moderate shade.' A much branched bramble with very conspicuous leaves, green on both sides. It is recorded in all neighbouring counties.

Var. b. **Schlechtendalii**, Weihe.

First Record. As of Wh. and N., not Weihe. Cowleigh Park, in a low part, *Lees, Bot. Worc.*, p. 71, note (1867). *Top. Bot.* 50.

    Severn. In this district, *Bagnall (MS.)*. Monk's Wood, *Reece.*

    Malvern. Cowleigh Park, as above, and Ley, *Woolhope Club Trans.* 1893–1894, *Additions to the Flora of Herefordshire*, p. 67.

    Lickey. In this district, *Bagnall (MS.)*.

Mr. Lees says, 'a singular bush of this form has existed in Cowleigh Park, Malvern, to my knowledge above thirty years, having enormously developed panicles and with very paniculate branches. This shrub extends itself proliferously by annual shoots, not rooting, proceeding from the axils of the leaves,' *Bot. Worc., Sup.*, p. 45. His first record, without doubt, refers to the same plant, and here he says it 'has grown for fourteen years to my knowledge, its barren stems excessively thick'. This is usually the form of *R. macrophyllus* most abundantly met with, but does not appear to be such in Worcestershire. It has not been met with in Gloucestershire of the neighbouring counties.

Var. d. **amplificatus**, Lees.

First Record. *Lees, in Steele, Handb. Field Bot.*, p. 58 (1847). *Top. Bot.* 23.

    Severn. In this district, *Bagnall*. In most of the woods about Malvern and Worcester, *Lees*. Wyre Forest.

    Malvern. As above.

    Lickey. Offmoor Wood, *Lees*. Woods, occasional, *Mathews.*

Near to var. *Schlechtendalii*, but differing from it by its long more pyramidal panicle, with a very prickly central stem, while the leaflets have more deeply incised compound teeth. Mr. Lees says it occurs generally throughout England in woods. It has not been observed in Stafford or Salop.

481. 308. R. Sprengelii, Weihe.

Native. Open woods and heathy places. Local. Shrub. July to September. *Top. Bot.* 47.

First Record. Bromsgrove Lickey, *Lees, Phyt.*, O.S., vol. iv, pt. 2, p. 817.

    Lickey. On the Lickey, about 800 feet above sea-level. Moseley. Redditch, *Mr. Lees.*

Mr. Lees includes in this *R. Borreri*, Bell Salt. Mr. Rogers says the *Borreri* is that of Babington, and places Bell Salter's elsewhere, *Handb. Brit. Rubi*, pp. 46, 61. On the Lickey this bramble creeps among the Bilberry bushes, and presents a very characteristic appearance. It rarely becomes stout or very strong, but varies according to habitat ; it can usually be recognized by its loose panicles, the long branches of its flower-stalks, and erect fruit sepals. It is found in all neighbouring counties.

486. 309. R. pyramidalis, Kalt.

(*R. villicaulis*, Blox., and of many British authors.)

Native. Heathy places, open woods. Local. Shrub. July to September. *Top. Bot.* 63.

First Record. Shrawley Wood and Wyre Forest, *Lees, Phyt.*, vol. iv, pt. 2, p. 917 (1852).

    Severn. As in first record.

    Lickey. In this district, *Bagnall.*

It is doubtful if Mr. Lees intends this plant. He refers to Babington, and describes it as a variety of *R. Menkii*, now *foliosus*, of W. and N. Mr. Mathews remarks that it is not so considered by other botanists. Moreover, Mr. Lees does not mention it by this name either in his *Bot. Worc.* or in the last edition of his *Bot. Malv. Hills*. The plant of Kaltenbach seems to have occurred in all bordering counties except Gloucestershire.

Var. **eglandulosa**.

First Record. Abundant in Cowleigh Park, *Rogers and Ley, Woolhope Club Transactions* 1893–1894, *Additions to the Flora of Herefordshire*, p. 70.

    Malvern. As first record.

This variety is not mentioned in *Lond Cat.*, 10th ed.

487. 310. R. leucostachys, Schleich.

(*R. vestitus*, Wh. and N.)

Native. Woods, hedges, and heathy places. General throughout the county. Shrub. July to September. *Top. Bot.* 71.

First Record. In a hedge on the north side of Worcester, *Lees, Mag. Nat. Hist.*, vol. iii, p. 160 (1830). In his *Bot. Malv. Hills*, 2nd ed., p. 55, he calls it *R. vestitus*.

    Severn. As first record. Near Bewdley, *Mathews.* Horseley Bank, Kidderminster, *Lees.* Redstone, *Reece.*

    Malvern. As *R. vestitus.* Woods about Malvern. Abberley Hills, *Lees.*

    Lickey. Offmoor Wood, *Lees.* King's Heath.

R. Grabowskii, Weihe, is mentioned by Mr. Lees as a form of this bramble, but no bramble of this name is recognized by Mr. Rogers. 'It was one of Mr. Bloxam's imaginary species,' says Mr. Bagnall (MS.). Although variable, *R. leucostachys* is seldom difficult to recognize. It has a densely hairy-felted stem and long straight prickles with roundish softly-felted leaflets. The flowers are both red and white, the white usually on limy soil. The plant is recorded in all bordering counties.

494. 311. R. mucronatus, Blox.

(*R. mucronulatus*, Boreau.)

Native. Bushy places, heaths, and hedges. Local, confined to forest ground. Shrub. July to September. *Top. Bot.* 67.

First Record. As *R. mucronulatus*. Not localized, *Lees, Bot. Worc., Sup.*, p. 42 (1867).

This bramble is recorded in all neighbouring counties.

497. 312. R. anglosaxonicus, Gelert.

Native. Hedges and commons. Local. July to September. *Top. Bot.* 23.

First Record. In this book, as under.

    Lickey. In this district, *Bagnall.*

This bramble has been found in all bordering counties except Gloucestershire.

499. 313. R. infestus, Weihe.

Native. Heaths, dry woods, and hedges. Local. Shrub. July to September. *Top. Bot.* 40.

First Record. In this book, as under.

Severn. River-side Bewdley, *Westcombe.*

Lickey. In this district, *Bagnall.* Lower Lickey, *Westcombe.*

Reported for all neighbouring counties.

501. 314. R. Borreri, Bell Salt. *Top. Bot.* 18.

Var. **virgultorum,** A. Ley.

First Record. Hanley Heath in Worcestershire, *Woolhope Club Trans.* 1893–1894, *Additions to the Flora of Herefordshire,* p. 73.

This plant, which Mr. Rogers considers a variety of *R. infestus* above, and gives in his *Handbook,* p. 60, as locally abundant in Worcestershire, does not appear to have been noticed by any Worcestershire botanist. Mr. Rogers says it occurs in Hereford and Salop also. This species is not given in the 10th ed. of the *Lond. Cat.*

503. 315. R. radula, Weihe.

Native. Rough bushy places, hedges. Locally plentiful. Shrub. July to September. *Top. Bot.* 55.

First Record. Not localized, *Lees, Bot. Malv. Hills,* 1st ed., p. 27 (1843).

Severn. Broadwaters, Kidderminster, *Lees.* Near Bewdley.

Lickey. Offmoor Wood, *Lees.* In the district, *Bagnall (MS.).*

In dealing with this plant, which Mr. Lees attributes to Wh. and N., and calls 'a fine straggling thicket bramble', *Bot. Worc., Sup.,* p. 43, he mentions a variety to which in Leighton's *Flora of Shropshire* he gave the name *R. Leightonii,* and which he says is local. Under what name this bramble is now described is not certain. The type has been noticed in all neighbouring counties.

Var. b. **anglicanus,** Rogers. *Top. Bot.* 25.

First Record. As in this book.

Lickey. In the district, *Bagnall (MS.).*

This variety is less stout than the type, and has the prickles not entirely confined to the angles of the stem.

504. 316. R. echinatus, Lindl.

(*R. rudis* of British authors up to 1886.)

Native. Hedges, commons, and open woods. Local. Shrub. July to September. *Top. Bot.* 57.

First Record. As *R. rudis.* Not localized, *Lees, Bot. Malv. Hills,* 1st ed., p. 27 (1843).

Avon. Trench Woods.

Malvern. As first record, *Lees.*

Severn. Helbury Hill, *Westcombe.* Near Bewdley, *Jorden.*

Lickey. In this district, *Bagnall (MS.).* Lickey, *Westcombe.*

Mr. Rogers says this is one of the most frequent of British brambles. The plant has been observed in all neighbouring counties.

512. 317. R. Babingtonii, Bell Salt.

Native. Hedges and heathy places. Shrub. July and August. *Top. Bot.* 28.

First Record. Severn Banks, *Mathews, Worc. Nat. Club Trans.* ii, p. 57 (1899).

Severn. Severn Banks, Bewdley, *Mathews.* Wyre Forest, *Mathews.*

R. Babingtonii has been noticed in Stafford, Warwick, and Herefordshire.

[**513. 318. R. Lejeunei,** Wh. and N. *Top. Bot.* 4.

As, according to Mr. Rogers, this plant occurs only in Breconshire woods, at Llanwrtyd, and near Builth, it cannot be the one intended by Mr. Lees at p. 52 of the 2nd ed. of his *Bot. Malv. Hills.* The plant intended was probably the following one.]

514. 319. R. ericetorum, Lefv. *Top. Bot.* 19.

First Record. As in this book.

Severn. Wyre Forest, *Westcombe.*

This plant, by Mr. Rogers, is considered the prevailing form of *R. Lejeunei* in England. Mr. Lees mentions this bramble at Rough Hill, and includes it in his *Bot. Malv. Hills,* and other books, but the locality is in Herefordshire. It has been noticed in Stafford, Gloucester, Hereford, and Warwick.

[**516. 320. R. mutabilis,** Genev. *Top. Bot.* 6.

Var. b. **nemorosus,** Genev. *Top. Bot.* 1.

This plant, for which, however, the *Top. Bot.* number is only 1, may be the plant intended by Mr. Mathews as gathered by him at Bredon Hill, 1850, *Worc. Nat. Club Trans.* ii, p. 62; but it is hardly likely that this is so. Mr. Lees deals with a plant which he calls *R. nemorosus,* Hayne, and equivalent to *R. dumetorum,* Wh. and N., *Bot. Worc., App.,* p. 39, and probably it is the one intended by Mr. Mathews.]

517. 321. R. Bloxamii, Lees.

Native. Heathy places. Local. Shrub. *Top. Bot.* 22.

First Record. Wyre Forest, *Lees, Bot. Worc., App.,* p. 42 (1867).

Severn. Wyre Forest, as above.

Lickey. In this district, *Bagnall (MS.).*

Of this plant Mr. Lees says (l. c. above), 'I have only observed it on the borders of Wyre Forest, and in Warwickshire, where it was detected by my observant friend, the Rev. Andrew Bloxam'. It has been noticed in all neighbouring counties.

518. 322. R. fuscus, Wh. and N.

Native. Dry woods, borders, and hedges. Local. July and August. *Top. Bot.* 28.

First Record. Not localized, *Lees, Bot. Malv. Hills,* 1st ed., p. 27 (1843).

Severn. In this district, *Bagnall.*

Malvern. As first record, *Lees.* Whippet Brook, North Malvern, *Reece.* Cowleigh Park, *Ley, Woolhope Club Trans.* 1893–1894, *Additions to the Flora of Herefordshire,* p. 79.

Lickey. In this district.

I

Mr. Lees' plant was referred by Professor Babington to *R. glandulosus,* var. *hirtus.* Possibly Mr. Lees did not intend *R. fuscus,* Wh. and N. That plant has been noticed in all neighbouring counties except Gloucestershire.

520. 323. R. scaber, Wh. and N.

Native. Woods and thickets, preferring sandy soil. Rather rare. Shrub. July and August. *Top. Bot.* 25.

First Record. Woods on the Old Storridge, *Lees, Bot. Malv. Hills,* 2nd ed., p. 53 (1852).

Severn. Brake Mill, Hagley, *Mathews.*

Malvern. Between Malvern and Alfrick, *Lees.* Abberley Hill, *Lees.*

An excessively prickly form, and easily known, says Mr. Lees. To this plant he attributes a bramble he records from Bromsgrove Lickey, calling it var. *verrucosus.* This, however, is referred to *R. fusco-ater,* Weihe, by Mr. Rogers, *Handb. Brit. Rubi,* p. 82. Of the neighbouring counties, *R. scaber* has not been observed in Gloucester.

523. 324. R. longithyrsiger, Bab.

Native. Damp hilly shady woods. Rare. July to September. *Top. Bot.* 13.

First Record. As *R. Menkii,* Wh. and N. Not localized, *Lees, Bot. Worc., Sup.,* p. 41 (1867), where he refers to an article in *Phyt.* iv, p. 920 (1853), which may forestall the above.

Severn. Shrawley Wood, *Lees.* Wyre Forest, *Lees.*

Mr. Mathews, and also Mr. Bagnall (MS.) considers that Mr. Lees' plant might be *R. flexuosus,* P. J. Muell., but Mr. Rogers identifies the *R. longithyrsiger* of Mr. Lees' MS., 1849, as the above, *Brit. Rubi,* p. 77. 'A prostrate bramble,' says Mr. Lees, which 'I have noted may be traced to the beautiful variety called *pyramidalis* by Professor Babington'. This variety is referred to *R. longithyrsiger* by Mr. Rogers. *R. flexuosus* he refers to the following plant *R. foliosus,* Wh. and N. It appears that *R. longithyrsiger* has been noticed only in Hereford of the bordering counties.

525. 325. R. foliosus, Wh. and N.

(*R. Guntheri,* Bab. *R. flexuosus,* P. J. Muell.)

Native. Woods on sandy soil. Locally common. Shrub. July and August. *Top. Bot.* 35.

First Record. As *R. Guntheri.* Trench Woods, *Worc. Nat. Club Trans.* i, p. 5 (1853).

Avon. Trench Woods, as first record ; also *Mathews.*

Severn. Wyre Forest, *W. N. C. Trans.* i, p. 8. Abberley, *Lees.* In great abundance, Ribbesford Wood, *Mathews.* Crow's Nest Wood, St. John's, *Lees,* in profusion.

Malvern. Little Storridge Wood, Alfrick, *Lees.* Clifton-on-Teme. Sapey Brook Glen, *Lees.* Cowleigh Park, *Ley, Flora of Herefordshire,* p. 104.

A widely distributed plant, says Mr. Rogers, in south-west and middle England, but unknown in Scotland. It has been recorded from Gloucestershire and Herefordshire.

526. 326. R. rosaceus, Wh. and N.

Native. Hedges, woods, and heathy places. Local. Shrub. July and August. *Top. Bot.* 28.

First Record. Wyre Forest, *Worc. Nat. Club Trans.* i, p. 8 (1853).

Severn. In this district, *Bagnall (MS.).* Wyre Forest, as first record.

The type is only yet known with certainty from Warwickshire and a few south-western counties, says Mr. Rogers, but this plant appears to have been noticed in all neighbouring counties except Gloucestershire. Mr. Lees' plant of this name was referred by Mr. Babington (3rd ed., *Man.*) to *R. pallidus,* Wh. and N. ; *Bot. Worc., Sup.,* p. 42.

Var. b. **hystrix,** Wh. and N. *Top. Bot.* 56.

First Record. As in this book.

Severn. In this district, *Bagnall (MS.).*

To this species Professor Babington referred Mr. Lees' *R. pallidus, Bot. Worc., Sup.,* p. 42. But see *R. dasyphyllus,* below. This variety has not been noticed in Salop.

Var. c. **infecundus,** Rogers. *Top. Bot.* 38.

(*R. thyrsiflorus,* Bab., in part. Var. *hystrix,* Bab., in part. *R. radula,* var. *hystrix,* Bloxam and Coleman.)

First Record. As in this book, below.

Malvern. In the district, *Bagnall (MS.).* North Malvern, hillside above the Hotel, August 4, 1893. Wood at West Malvern, August 27, 1896, *Towndrow,* fide *W. M. Rogers (MS.).*

This variety differs from the type in its smaller prickles and yellowish leaflets, which are very softly hairy beneath, not only on the veins, while the panicle is lax and pyramidal, not broad and diffuse. It has been noted in all bordering counties except Gloucestershire.

Var. e. **adornatus,** P. J. Muell. *Top. Bot.* 21.

Native. Woods and thickets. Locally abundant. July and August. *Top. Bot.* 21.

First Record. As below.

Lickey. In this district, *Bagnall (MS.).*

This bramble in Mr. Rogers's *Handbook,* 1900, is placed as a sub-species of *R. rosaceus,* Wh. and N. It has been noticed in Herefordshire, Staffordshire, and Warwickshire of the neighbouring counties.

529. 327. R. fusco-ater, Weihe.

(*R. verrucosus,* Lees.)

Native. Bushy places. Local. July and August. *Top. Bot.* 11.

First Record. Not localized, *Lees, Bot. Malv. Hills,* 1st ed., p. 27 (1843).

Severn. Abundant at the Brake Mill, Hagley, *Mathews.* Habberley Valley, *Mathews.* Birchin Grove, *Reece.*

Lickey. In the district, *Bagnall (MS.).* Form like this, with white flowers, Bromsgrove Lickey, *Mathews.* Plentifully distributed throughout the hill district from Clent to the Lickey, *Mathews.*

To this was referred a plant Mr. Lees called *R. scaber,* var. *verrucosus,* which he gathered on Bromsgrove Lickey ; perhaps the one that he gathered in company with Dr. Fraser and Mr. Thompson of which he says, *W. N. C. Trans.* i, p. 246, 'they had found the rare *Rubus pygmaeus* on the summit of the Upper Lickey, in full flower in October, and forming low but dense thickets.' *R. fusco-ater* appears not to have been noticed in the counties of Salop and Stafford.

530.   328. R. Koehleri, Wh. and N.
 Native.   Hedges, roadsides, woods.   Common.   Shrub.   July and August.
   Top. Bot. 27.
 First Record.   Not localized, *Lees' Cat., New Bot. Guide* (1835).
   Lickey.   In this district, *Bagnall* (MS.).
 Mr. Lees says that this bramble is widely distributed throughout the county, generally growing solitarily, and need be confounded with no other. On the other hand Dr. Focke has said, *J. of B.,* 1890, p. 134, 'in dry specimens it is difficult to trace the limits between *R. hystrix* and *R. Koehleri.*' It is recorded from all neighbouring counties.

531.   329. R. dasyphyllus, Rogers.   *Top. Bot.* 76.
   (*R. Koehleri,* var. *pallidus,* Bab.)
 First Record (possibly).   Not localized, *Lees, Bot. Malv. Hills,* 1st ed., p. 27 (1843).
   Avon.   Tiddesley Wood, *Lees.*
   Severn.   In this district, *Bagnall* (MS.).   Wyre Forest, *Lees.*   Bewdley, *Westcombe.*   Birchin Grove, *Westcombe.*
   Malvern.   North Malvern, August 4, 1893, *Towndrow,* fide *W. M. Rogers.* Cowleigh Park.   Croft Woods.
   Lickey.   In this district, *Bagnall* (MS.).   As *R. Koehleri.*   Hedges, general, *Mathews.*
 This bramble, no doubt, is the plant to which the records of *R. pallidus* by Mr. Lees and his contemporaries must be referred.   *R. pallidus,* Wh. and N., is very rare.

537.   330. R. Bellardii, Wh. and N.
   (*R. glandulosus,* Bell.)
 Native.   Chiefly in moist woods.   Local.   July and August.   *Top. Bot.* 17.
 First Record.   As *R. glandulosus.*   Bromsgrove Lickey, *Lees, Mag. Nat. Hist.,* vol. iii, p. 160 (1830).   As *R. Bellardi.*   Not localized, *Lees, Bot. Malv. Hills,* 2nd ed., p. 51 (1852).
   Severn.   Wyre Forest, *Lees.*   Seckley Wood, *Babington.*
   Malvern.   Cowleigh Park.   Abberley, *Lees.*
   Lickey.   Near Halesowen, *Lees.*   Bromsgrove Lickey.
 Mr. Rogers, who does not record the plant from Worcestershire, says this is a constant and easily recognized species, from its ternate leaves with sub-equal evenly-toothed leaflets, and its remarkably short few-flowered panicles. A local species fond of hilly woods, forming extensive thickets.   It has not been noticed in Gloucestershire.

539.   331. R. hirtus, Waldst. and Kit.
   (*R. glandulosus,* Bell, var. *hirtus.*)
 Native.   A forest bramble.   Local.   Shrub.   July to September.   *Top. Bot.* 31 (?).
 First Record.   Not localized, *Lees, Bot. Malv. Hills,* 2nd ed., p. 51 (1852).
   Severn.   Wyre Forest, *W. N. C. Trans.* i, p. 8.   In this district, *Bagnall* (MS.).

   Malvern.   Priory Grove, Little Malvern, *Lees.*
 Mr. Rogers cannot define this as a species, but gives Dr. Focke's description. Mr. Lees says little about it, but remarks, 'this is a very hairy form.'   It has not been noticed in Gloucestershire.
 Var. b. **rotundifolius,** Bab.   *Top. Bot.* 13.
 First Record.   In this book.
   Lickey.   In this district, *Bagnall* (MS.).

541.   332. R. saxicolus, P. J. Muell.
 Native.   Woods and bushy places.   Rare.   Shrub.   July and August. *Top. Bot.* 3 (?).
 First Record.   In thick woods, rare, *Lees, Bot. Worc., Sup.,* p. 41 (1867).
 To this bramble Mr. Mathews refers Mr. Lees' *R. humifusus,* Weihe. Mr. Rogers refers *R. humifusus,* Bab., to *R. pallidus,* Wh. and N.   This plant does not appear to have been noticed in any neighbouring county, and the above record is good only if Mr. Mathews's identification is correct. Mr. Rogers observes, *Handb. Brit. Rubi,* 1900, p. 90, 'that he cannot write confidently of this species or say positively whether it is British or not.'

544.   333. R. velatus, Lefv.
 Native.   In woods and thickets.   Rare.   Small shrub.   July.   *Top. Bot.* 8.
 First Record.   Thicket in Cowleigh Park, 1887, 1893, *Rogers* ; *Ley, J. of Bot.,* 1893, p. 7 ; 1895, p. 104.
   Malvern.   As first record.
 'First named as a British plant in 1889 or 1890 by the late Professor Babington from Cowleigh Park specimens,' *Woolhope Club Trans.* 1893-1894, *Additions to the Flora of Herefordshire,* p. 86.

545.   334. R. dumetorum, Wh. and N.
 Native.   Hedges.   Locally abundant.   Shrub.   June to September.   *Top. Bot.* (sp. coll.) 68.
 First Record.   Not localized, *Lees, Bot. Malv. Hills,* 1st ed., p. 27 (1843).
   Severn.   Near Bewdley, *Mathews.*
   Malvern.   As in first record, *Lees.*
 Mr. Lees refers his *R. nemorosus,* Hayne, to this plant in *Bot. Worc., Sup.,* p. 39.   A very variable plant.   It has been noticed in all bordering counties.
 Var. a. **ferox,** Weihe.   *Top. Bot.* 43.
 First Record.   Abundant near Worcester in hedges, *Lees, Bot. Worc., Sup.,* p. 40 (1867).
   Severn.   Near Worcester, as first record.
   Lickey.   Roadsides around Halesowen, *Lees.*
 This 'very rough and glandulose' variety, says Mr. Lees, has been 'elevated into a species' by Professor Babington under the name of *R. tuberculatus* ; Mr. Mathews, however, says it is *R. diversifolius,* Lindl.   Both these forms are now considered varieties of the type.   Worcestershire brambles here become a little mixed !   At all events, the variety has been noticed in all neighbouring counties.

 Var. c. **diversifolius,** Lindl.   *Top. Bot.* 56.
 First Record.   Not localized, *Lees, Bot. Malv. Hills,* 1st ed., p. 27 (1843). Not mentioned in the 2nd or 3rd edition of the book.
   Malvern.   As first record.
   Lickey.   In the district, *Bagnall* (MS.).
 The note above would apply also to this bramble.   It is found in Salop, Stafford, Hereford, and Warwickshire.

 Var. f. **tuberculatus,** Bab.   *Top. Bot.* 37.
 First Record.   As *R. nemorosus,* b. *ferox.*   Abundant near Worcester, in hedges, *Lees, Bot. Worc., Sup.,* p. 40 (1867).
   Severn.   As first record.
   Lickey.   In this district, *Bagnall* (MS.).
 Also in this case there is doubt.   Mr. Mathews, *Mid. Nat.* xiv, p. 92, gives this as an equivalent of *R. scabrosus,* P. J. Muell., and says it is one of the forms of *R. nemorosus,* according to Mr. Lees.   Above it would seem that Mr. Lees himself considered it equivalent to the variety he called var. *ferox.* Mr. Rogers, however, confirms Mr. Mathews's opinion.   This variety has been noticed in Gloucester, Stafford, Hereford, and Warwick.

 Var. g. **triangularis,** Ley.
 First Record.   Very abundant in the valley of the Teme both above and below Stanford Bridge, *Ley, J. of Bot.,* Feb., 1902.
   Malvern.   As first record.
 Near vars. *ferox* and *britannicus,* from the latter of which it differs in the crowded, unequal, very stout, straight thorns, and short-stalked glands of stem and rachis ; in the leaves being nearly always ternate or ternate-lobate, not quinate ; their leaflets shorter, broadly triangular-ovate, acute or shortly acuminate, with shallow crenate-lobate serration, and with their under surface more constantly felted ; in the panicle with long straight divaricate lower branches, often forming a triangular figure.   Sepals broadly triangular, short, at length clasping.   The *triangular* aspect of the very numerous broad-based thorns, of the sepals, of the spaces between the panicle branches, and of the whole panicle ; and to a less degree of the leaves, their leaflets, and the leaf serration, suggests the proposed varietal name as appropriate.

 Var. j. **fasciculatus,** P. J. Muell.   *Top. Bot.* 25.
 First Record.   In this book, as below.
   Severn.   In this district, *Bagnall* (MS.).
   Lickey.   In this district, *Bagnall* (MS.).
 That this form is included in the Worcestershire list is due to the records of Mr. Bagnall above.   No other Worcestershire botanist has noticed it, but it is recorded for Herefordshire.

546.   335. R. corylifolius, Sm.
 Native.   Hedges.   Common and widely distributed.   Shrub.   June to September.   *Top. Bot.* (sp. coll.) 91.
 First Record.   Not localized, *Lees, Bot. Malv. Hills,* 1st ed., p. 27 (1843).
   Avon.   Bredon Hill, *Mathews.*   Trench Woods, *Mathews.*
   Severn.   Rather frequent in hedges west of Hagley village, *Mathews.*   Near Bewdley, *Mathews.*   Abundant about the Brake Mill, Hagley, *Mathews.*
 In Mr. Rogers's *Handb., R. sublustris,* Lees, is constituted a variety of this

species, and Mr. Druce, in his *Fl. of Berks.,* says it is commoner than the type.   One 'remarkable' variety of his bramble Mr. Lees called *R. coenosus,* having its stem and panicle very white with close down, studded with white glands, and the prickles themselves hairy.   This occurs near Worcester.   In his *Bot. Malv. Hills,* 2nd ed., p. 50 (1852), Mr. Lees records a variety which he calls *R. Wahlbergii,* Arrh., which Mr. Mathews equates with *R. corylifolius,* var. *purpureus,* Bab.   But this plant Mr. Rogers in his *Handb.* identifies with the variety *fasciculatus* of the preceding species.   *R. corylifolius,* Sm., including varieties, appears to have been noticed in all surrounding counties.

 Var. a. **sublustris** (Lees).   *Top. Bot.* 67.
 First Record.   Not localized, *Lees, Bot. Malv. Hills,* 2nd ed., p. 51 (1852).
   Malvern.   Malvern Link, *Towndrow* (MS.).
 One would suppose that a botanist knows his own bramble, and Mr. Lees knew what plant he intended ; but if, as he says, his *R. sublustris* is identical with the type, it is difficult to see how it has become a mere variety of it. Mr. Rogers even says it fairly describes *R. corylifolius,* Sm., the type in question.   Mr. Lees says his bramble is very general throughout Worcestershire ; Mr. Rogers does not locate it in the county, but marks it as occurring only in East Gloucester, Hereford, and Warwick.

547.   336. R. Balfourianus, Blox.
 Native.   Hedges.   Local.   Shrub.   July and August.   *Top. Bot.* 44.
 First Record.   As *R. tenui-armatus.*   Not localized, *Lees, Bot. Malv. Hills,* 2nd ed., p. 51 (1852).   As *R. Schleicheri.*   Not localized, *Lees, Bot. Malv. Hills,* 1st ed., p. 27 (1843).
 R. Balfourianus has been noticed in all adjoining counties.

549.   337. R. caesius, Linn.   *The Dewberry.*
 Native.   Hedges, ditches, and stream-sides.   Many places, except in the Lickey district.   Shrub.   May to August.   *Top. Bot.* 78.
 First Record.   Not localized, *Lees, Bot. Malv. Hills,* 1st ed., p. 27 (1843).
   Avon.   Bredon Hill, *Mathews.*   Craycombe, *Lees.*   Sheriff's Lench.   Norton, near Evesham.   Crowle.   Crowle Banks, Pershore.
   Severn.   Banks of the river near Bewdley, *Mathews.*   Railway cutting at Defford, profusely, *Lees.*   Claines.   Ombersley.   Kempsey.   Severn Stoke.   Wyre Forest.   Clarkenleap.   Dynes Green.
   Malvern.   In low, shady places, *Lees.*   Leigh Sinton.   Ankerdine. Knightwick.   Powick.   Martley.   Stanford Bridge.
   Lickey.   Illey Mill, *Mathews.*
 This is a trailing bramble, with round stems that root profusely, but like most of its congeners, it is exceedingly variable, and hybridizes freely with many of them.   The fruits are few and large, more juicy than the Blackberry, and grow either singly or two or three together, not in clusters.   It is recorded from all adjoining counties.   Mr. Lees mentions a variety *pseudo-idaeus,* which has an erect, barren stem, much resembling a Raspberry, and which occurs in a coppice at Rushwick, near Worcester.

 Var. **aquaticus,** Wh. and N.
 First Record.   As below.
   Malvern.   Near Malvern, *Towndrow* (MS.).

A form with a very slender stem and few small prickles. In 9th ed., *Lond. Cat.*, not 10th, and marked *Top. Bot.* 9.

Var. tenuis, Bell Salt.
First Record. As below.
    Malvern. Near Malvern, *Towndrow* (MS.).
A form with a very slender stem, but the small prickles are many and strong. In 9th ed., *Lond. Cat.*, not 10th, and there marked *Top. Bot.* 12.

Var. arvensis, Wallr.
First Record. As below.
    Malvern. Near Malvern, *Towndrow* (MS.).
This also appears in the 9th ed., *Lond. Cat.*, and is not given in the 10th ed. There it is marked *Top. Bot.* 13.

550. 338. R. saxatilis, Linn.
    Native. Moist woods and thickets. Stem annual. Only in one locality. July and August. *Top. Bot.* 70.
    First Record. Bewdley Forest, *Gissing, Phyt.*, N.S., vol. i, p. 151 (1855).
    Severn. Wyre Forest.
This is a very different plant from any of the preceding ; it possesses a herbaceous stem, or very nearly so. Though it was observed in the locality by all the earlier botanists, it only now lingers on the railway-banks at the Bewdley end of Wyre Forest. It has not been recorded from Hereford or Salop.

GEUM, Linn. 116. (γεύω, to season, from the aromatic roots.)

553. 339. G. urbanum, Linn.   *Common Avens, Herb Bennet.*
    Native. Woods, thickets, hedges, banks. Common and widely distributed. P. May to August. *Top. Bot.* 107.
    First Record. Not localized, *Walker, Med. Surg. Rep.*, vol. i, no. 2, November (1828).
The small bright yellow flowers of this plant, on straight stiff stalks, are well known in our hedgerows ; and following them, the round spiny rich brown-coloured ball of fruits, each armed with its little hook. The plant is known throughout the Continent by some name analogous to the second one above. There was a plant in the Middle Ages of such potency that when the root was in the house the devil could do nothing, and fled from it, and so it was considered blessed among herbs ; but what plant this was, or how our Common Avens became identified with it, cannot be determined. At all events the plant was considered a remedy for a considerable number of ailments ; and besides these, the root in spring-time, steeped in wine, gave the liquid ‘ an excellent savour and taste, and being drunk fasting every morning comforteth the heart ’. It is still used in country places to put in homemade wines, and, in the spring, put into ale, it is said to prevent its turning sour. But it was not all good, for mingled with water and given to sick people, it was apt to cause delirium. Herb Bennet occurs in all adjoining counties.

554. 340. G. rivale, Linn.   *Water Avens.*
    Native. Meadows, bushy places, damp hedgerows. Not common. P. May to July. *Top. Bot.* 94.
    First Record. Near a flight of steps leading from the Hope Farm to Sapey Church, *Sheward, Nash, Hist. Worc.*, Sup., p. 96 (1799).
    Avon. One place, *Lees*.
    Severn. Abberley, *Hickman*.
    Malvern. Sapey Brook.
    Lickey. Illey Mill, *Lees*. Bromsgrove Lickey, *Mathews*. Rather common in the woody dingles of the tributaries of the Stour, *Mathews*. Field below St. Kenelm's, *Amphlett*. Frankley. Halesowen. Near Rubery Station.
This is a very different looking plant from the last. Its flowers are purplish red, and nodding, and instead of being a glossy green the plant is more or less hairy and pubescent. It is altogether a shorter, stouter, plant. It is widely distributed on the face of the globe, occurring throughout Europe, North and West Asia, North and South America, and even in Australasia. In North America the plant is much used medicinally. A proliferous example was found at Halesowen. The plant has occurred in all the neighbouring counties.

× urbanum.
    (*G. intermedium*, Ehrh.) *Top. Bot.* 60.
    First Record. As *G. intermedium*, Sapey Brook, *Lees, Bot. Looker-out*, 2nd ed., p. 181 (1851).
    Malvern. Sapey Brook, near the bridge, *Lees*.
    Lickey. Above Illey Mill. Rubery. Woods about Halesowen. Frankley.
This hybrid has been noticed in all bordering counties.

FRAGARIA, Linn. 117.   (*Fragrans*, fragrant, from the perfume of the fruit.)

555. 341. F. vesca, Linn.   *Wood Strawberry.*
    Native. Woods and hedgebanks. Generally distributed. P. April to July. *Top. Bot.* 111.
    First Record. Not localized, *Lees, Bot. Malv. Hills*, 1st ed., p. 27 (1843).
The Wood, or Wild Strawberry, like many other berries, is more frequent in the woods of the North of Europe than it is with us. The name is probably a corruption of Strayberry, and has nothing to do with the straw frequently laid under them in gardens, or the bent of grass on which in olden times country people used to thread them. Most of our present cultivated kinds are derived from American species ; the plant does not appear to have been much grown in gardens before the beginning of the seventeenth century. Mention of the fruit rarely occurs in early household accounts. But wild strawberries were appreciated. Ben Jonson talks of a ‘ pot of strawberries gathered in the wood to mingle with your cream ’. Earlier than this the historic scene at Ely Place, Holborn, dramatized by Shakespeare, may be remembered. There is not much medicinal lore attached to the strawberry. The fruit was perhaps too pleasant to the taste ; with the old herbalists it would seem that the nastier the taste of the remedy the more efficacious it was. The plant occurs in all neighbouring counties.

556. 342. F. *moschata, Duchesne.   *Hautboy Strawberry.*
    (*F. elatior*, Ehrh.)
    Alien. Woods. Rare. P. June to September. Not in *Top. Bot.*
    First Record. Redway on the Ridgeway, almost at the eastern verge of the county, *Cheshire, Lees, Bot. Worc.*, p. 96 (1867).
    Severn. Oldbury Road.
    Malvern. Near the Rhydd, *Towndrow, Malv. Advert.*, Nov. 12, 1892.
The English name is a corruption of Hautbois, from its growing in the high woods of some parts of Central Europe. It has no claim to be indigenous in Britain, but it has been noticed in Herefordshire, Gloucestershire, and Warwickshire, having escaped from some place of cultivation.

POTENTILLA, Linn. 118.   (*Potens*, powerful, from the powerful medicinal effects attributed to some of the species.)

558. 343. P. sterilis, Garcke.   *Strawberry-leaved Cinquefoil.*
    (*P. Fragariastrum*, Ehrh.   *Fragaria sterilis*, Linn.)
    Native. Woods, hedgebanks, commons, wastes. Common and generally distributed. P. January to November. *Top. Bot.* 106.
    First Record. Unlocalized, *Lees, Bot. Malv. Hills*, 1st ed., p. 27 (1843).
This is one of the earliest flowers to peep forth among the herbage of the roadside bank when winter is loosening its hold on the year, well known and welcome everywhere. Country people know it as the Wild Strawberry, from the likeness of its flowers and leaves to that genus ; but the petals are notched instead of rounded at the apex, and it bears no succulent fruit. Linnaeus, while assigning it to brotherhood with the strawberry, emphasized that fact by giving to it the specific name *sterilis*. It is found in all bordering counties.

559. 344. P. verna, Linn.   *Spring Cinquefoil.*
    Native. Hilly, rocky places. Very local. P. April to June. *Top. Bot.* 22.
    First Record. (If the locality was in Worcestershire), Limestone Rock on the western side of the hills, in *Southall's Descrip. Malvern*, p. 215 (1825). First certain record in Worcestershire, Rocks and summit of the Malvern Hills, *Lees' Cat., New Bot. Guide* (1835).
    Malvern. As first records. Wind's Point. Little Malvern, *Towndrow*. Turnpike Road above Little Malvern Church, *Westcombe*.
This plant was given as a new record in *Bot. Rec. Club Rep.*, 1887, by Mr. Towndrow, but it was not so. The writer in the *Midl. Med. Surg. Rep.*, vol. i, no. 1, would assign to this plant a wider range than the above, for he includes the hilly limestone country generally in this locality, in some parts of which the plant may occasionally be met with. In the 3rd ed., *Bot. Malv. Hills*, p. 75, Mr. Lees gives a list of several localities, mostly in Herefordshire, for the plant. The plant has been recorded in Gloucester, Hereford, and Stafford, and has been noticed in Shropshire.

561. 345. P. erecta, Hampe.   *Common Tormentil.*
    (*P. tormentilla*, Neck.   *Tormentilla officinalis*, Linn.   *P. sylvestris*, Neck.)
    Native. Dry pastures, hedgebanks, commons. General. P. April to September. *Top. Bot.* 112.
    First Record. As *Tormentilla officinalis*. Not localized, *Lees, Bot. Malv. Hills*, 1st ed., p. 27 (1843).

The yellow flowers of this little plant are commonly to be seen among short turf during the summer months. The root-stock is large and woody, and its astringency is such that these roots were used for tanning leather. With oak bark, however, they have been superseded in this use by modern chemicals. Sheep are very fond of the plant. It occurs in all neighbouring counties.

562. 346. P. procumbens, Sibth.   *Creeping Tormentil.*
    (*Tormentilla reptans*, Linn.)
    Native. Woods, heaths, hedgebanks. Not common. P. June to September. *Top. Bot.* 83.
    First Record. As *Tormentilla reptans*. Edgebaston Lane, near Avern's Mill (possibly in Worcestershire, *Mathews*), *Ick, Analyst*, vol. vi, p. 22 (1837).
    Avon. Cleeve Banks. Tiddesley Wood.
    Severn. Wyre Forest. Droitwich Canal. Hartlebury Common. Pixham.
    Malvern. Welland Common, *Lees*. Middleyards Coppice. Malvern. Mathon. Ankerdine.
    Lickey. As first record. Bromsgrove Lickey, *Mathews*. Frankley.
This plant, by some botanists considered a variety of the above, differs from it in its stalked leaves and larger, generally solitary flowers. Its creeping stem does not root at the joints as in the following plant. This one has been recorded from all neighbouring counties except Gloucestershire.

563. 347. P. reptans, Linn.   *Common Cinquefoil.*
    Native. Hedgebanks, roadsides, cornfields, pastures. Common. P. May to September. *Top. Bot.* 99.
    First Record. Not localized, *Lees, Bot. Malv. Hills*, 1st ed., p. 27 (1843).
This yellow flower is seen by every wayside, and country people sometimes call it Yellow Strawberry. The plant is astringent and bitter, and has not escaped the attention of the herbalist. Fevers and agues, inflammations, palsy, as well as diseases of the lungs, were held to be amenable to its virtues. Except in the extreme North, it ranges over Europe, North and West Asia, and is found also in the Canaries and the Azores. Throughout Europe, as in our own country, the form of the leaf gives it its name. It occurs in all bordering counties.

564. 348. P. Anserina, Linn.   *Silver-Weed, Goose Grass.*
    Native. Waysides, commons, fields. Common and widely distributed. P. May to September. *Top. Bot.* 112.
    First Record. Not localized, *Lees, Bot. Malv. Hills*, 1st ed., p. 27 (1843).
This is as common as, if not commoner than, the preceding plant, and much more conspicuous by reason of the shining silvery down which clothes the under side of its pinnate leaves, and sometimes encroaches upon the upper side also. The foliage is so much appreciated by geese that through this it has acquired its specific name ; and the leaves are sometimes boiled for a cottage meal. The roots are sweet, and eaten by children. It occurs in all bordering counties.

566. 349. P. argentea, Linn.   *Hoary Cinquefoil.*
    Native. Roadsides and hedgebanks, on sandy soil. Local and rare. P. June and July. *Top. Bot.* 57.

First Record. Side of the turnpike road in the Parish of Holt Castle, *Ballard, Stokes's With. Bot. Arr.*, 2nd ed., p. 532 (1787).

Severn. Frequent about Kidderminster, *Mathews*. Astley. Holt. Between Bromsgrove and Droitwich, *Lees*. Between Worcester and Ombersley, near the second milestone from Worcester. Hartlebury Common, abundant, *Humphreys*. Hagley Brake; Churchill; Blakedown; *Mathews*. Fenny Rough. Oldington. Near Mitre Oak. Near Bewdley. Bissell.

Malvern. Worcestershire Beacon, and the North and Raggedstone Hills.

Lickey. Slideslow, near Bromsgrove.

This plant is nearly confined to the light sandy districts of the north of the county which nourish a distinct flora of their own, among which are *Arabis glabra*, *Spergularia rubra*, *Ornithopus perpusillus*, and *Senecio sylvaticus*. Yet, although many of these plants stray into the bordering Lickey district, this one has only once been recorded in it. Nor is it recorded from Wolverley in the Severn district, though this parish would seem to be a typical locality for its occurrence. It is found in all neighbouring counties except Gloucestershire.

568. 350. P. palustris, Scop. *Purple Marsh Cinquefoil.*
(*Comarum palustre*, Linn. *Potentilla Comarum*, Nestl.)

Native. Boggy places, shallow pools. Local and rare. P. June to August. *Top. Bot.* 100.

First Record. As *Comarum palustre*. In boggy places on the Lickey, near Bromsgrove, *Nash, Hist. Worc., Int.*, p. lxxxix (1781).

Severn. Pedmore Common, *Scott*, gone. Oldfield near Ombersley, *Perry*. Stanklin Pool; Harberrow Pool, *Mathews*. Hartlebury Common, plentiful. Worcester and Birmingham Canal. Oldington.

Lickey. Bromsgrove Lickey, *Nash*, gone. Pool at Moseley, *Mathews*, gone.

This is one of those plants which are steadily disappearing through the drainage of bogs and the advance of modern conditions. The pool at Moseley was opposite the Independent College, and there the plant was gathered by Mr. Mathews in 1858. The locality is now purely urban. Still, however, a small pool on Hartlebury Common is filled with a mass of its leaves, and in a few other places it yet exists. In some countries it is abundant enough to come into the common uses of life. Its roots will dye wool yellow; and in Ireland, it is said, the insides of milking-pails are rubbed with this plant to make the milk appear richer and thicker. Of neighbouring counties, it has not been recorded in Gloucestershire.

ALCHEMILLA, Linn. 119. (The Arabic for *Alchemy*, from the numerous experiments that chemists have tried on this plant.)

570. 351. A. arvensis, Scop. *Field Lady's Mantle, Parsley Piert.*

Native. Dry sandy fields, banks, pastures, wall-tops. Generally distributed. A. May to October. *Top. Bot.* 112.

First Record. Lanes about Henwick and Malvern Hill, *Lees, Illus. Nat. Hist. Worc.*, p. 153 (1834).

This is a common but insignificant weed, the greenish flowers being in little bunches. The name Parsley Piert is derived from the French *Perce-pierre*,

break-stone, a Latin translation of which is the generic name of the Saxifrages. The name possibly came to be applied to all plants popularly supposed to be akin to the Saxifrages; and this was the cause, probably, of the name Parsley Piert being also given, but erroneously, to the Knawell, *Scleranthus annuus*. Parsley Piert occurs in all neighbouring counties.

571. 352. A. vulgaris, Linn. *Common Lady's Mantle.*

Native. Pastures, grassy places in woods. Not uncommon. P. May to August. *Top. Bot.* 107.

First Record. Near Dudley Castle, *Booker, Dud. Cast.*, p. 107 (1810).

Avon. Norton, near Evesham. Pershore. Crowle. Pinvin.

Severn. Grimley. Lane near Henwick Mill. Wyre Forest. Areley Wood. Eymore. Hallow. Ombersley. Kempsey.

Malvern. Alfrick. Malvern Hills. Powick. Southstone's Rock. Near Stanford Bridge. Malvern Link.

Lickey. Beyond Vaughton's Hole, Moseley, *Ick*. Dudley Castle, *Booker*. Clent. Lickey Hill. Upland pastures occasional, *Mathews*. Lutley, near Halesowen. Lower Lickey.

An elegant but not conspicuous plant, with corymbose yellow flowers, and reniform leaves, slightly hairy sometimes beneath. The name, of course, refers to the Virgin Mary; but elsewhere in Europe, except in Sweden, its name appears not to refer to her. It is given to it from the beautiful silkiness of the under side of the leaves of some of the species, especially noticeable in the variety *conjuncta* of *A. alpina*, Linn., a plant that does not occur in our country, a silkiness even more marked than that of the leaves of *Potentilla Anserina*. Formerly the plant was much prized for its medicinal virtues, and formed an ingredient in several of the concoctions of the alchemists. It occurs in all neighbouring counties.

AGRIMONIA, Linn. 120. (Etymology obscure; the word is said to be a transformation of the Greek ἀργεμώνη, a plant which, according to Dioscorides, cured affections of the eye.)

573. 353. A. Eupatoria, Linn. *Agrimony.*

Native. Roadsides, hedgebanks, wood-borders. General; but less so in the Lickey district. P. June to September. *Top. Bot.* 107.

First Record. Roadsides, *Pitt, Agric. Worc.*, p. 317 (1810).

The tall yellow spikes of this flower are a familiar object in the grassy margins of roadsides during summer, and possess a faint odour of lemon if they are bruised. 'Sweethearts' is a country name for it. The Agrimony is still in repute for village doctoring, and used to be considered an especial remedy for snake-bite, which uncommon circumstance in England would appear in olden times to have occurred much more frequently than now. It is an ingredient in herb teas and other concoctions of that class which are made on the countryside; and affections of the liver and throat are supposed to be wonderfully ameliorated by its use. Luckily such a potent weed occurs in all neighbouring counties.

574. 354. A. odorata, Mill. *Sweet Agrimony.*

Native. Woods and grassy places. Local and rare. P. July and August. *Top. Bot.* 46.

First Record. Note by Mr. *Mathews, Phyt.*, 2nd S. i, p. 192 (1855).

Avon. Tiddesley Wood, *Lees*. Near Crowle. Norton, near Evesham. Trench Woods.

Severn. Near Hartlebury, *Miss Ladbury*. Summer Hill, near Hanbury. Wyre Forest. Hagley. Churchill. Stanklin Pool. Shrawley.

Malvern. Powick. Bransford. Knightsford Bridge. Martley. Stanford Bridge.

Lickey. Hunnington, *Mathews*.

More branched and larger in every respect than the previous plant, of which it is sometimes considered a sub-species. It possesses a resinous odour. In spite of the note by himself in 1855, Mr. Mathews, *Mid. Nat.* xiv. 41, gives Mr. Lees as the first recorder at p. 94 of his *Bot. Worc.* The plant was again recorded as new, *Bot. Rec. Club Rep.*, 1883. It has been noticed in Gloucester, Hereford, Warwick, and Stafford.

POTERIUM, Linn. 121. (ποτήριον, a drinking-cup, from its use in preparing 'cool tankard'.)

575. 355. P. Sanguisorba, Linn. *Salad Burnet.*

Native. Dry calcareous pastures, fields. Abundant in the Avon district; less common elsewhere. P. May to August. *Top. Bot.* 74.

First Record. On Bredon Hill; on very barren waste land at Church Lench; on rich red loam near Inkborough (Inkberrow); and in a meadow near Tenbury; not yet common to be found in the county, *Pitt, Agric. Worc.*, p. 317 (1810).

Avon. Craycombe; Ald2erminster; Bredon Hill; Broadway; *Lees*. Norton, near Evesham. Sheriff's Lench. Defford. Crowle.

Severn. Wyre Forest. Near Crookbarrow Hill. Cotheridge. Spetchley. Kempsey.

Malvern. Near Tenbury, *Pitt*. Near Knightsford Bridge. Alfrick. Martley. Many places around Malvern, chiefly upon calcareous soil.

Lickey. Northfield. Tardebigge. Dodford.

This plant owes its English name to the leaves which crowd around its base, which have the flavour of cucumber, and on this account were eaten in the salads of our forefathers. With the Borage the plant was infused in the liquor of the 'cool tankard'; and the herbalists delighted in it. 'A herb the sun challengeth dominion over,' says one; 'and a most precious herb. It is a friend to the heart and liver.' It has sometimes been cultivated as a forage plant, but not very successfully; sheep liked it, but cattle were not fond of it. The plant occurs in all neighbouring counties.

576. 356. P. polygamum, Waldst. and Kit. *Muricated Salad Burnet.*
(*P. muricatum*, Spach.)

Colonist. Calcareous ground and fields. Rare. P. June to August. Not in *Top. Bot.*

First Record. As *Poterium muricatum*. Worcestershire, no locality, *Mr. Thomas Westcombe, Phyt.*, O.S., vol. iv, pt. 2, p. 724 (1852).

Avon. East side of Trench Wood, copious, 1855, *Lees*.

Severn. Wyre Forest.

Malvern. The Wych, 1853, *Thompson*. If, as is possible, Mr. Westcombe's locality is in Hereford (*Bot. Malv. Hills*, 3rd ed., p. 96), the first record

is this locality, *Thompson, Phyt.*, N.S., vol. v, p. 219 (1861). West Malvern; Brace's Leigh; Malvern Link; *Towndrow*.

Lickey. Banks of Railway between the Canal and Furnace Forge, Halesowen, 1860, *Mathews*.

This plant has its home in the dry grassy places in the East, and as a native approaches us no more nearly than Central Europe; but has been sown as a fodder plant in England, and is met with as a survival of such cultivation, and sometimes as a weed in other crops. It has been seen in this condition in Staffordshire and Gloucestershire.

Var. stenolophum, Jord.

First Record. In this book.

Malvern. This variety has occurred at the Malvern Station, but there only. The variety does not appear in *Lond. Cat.*, 10th ed.

577. 357. P. officinale, A. Gray. *Great Burnet.*
(*Sanguisorba officinalis*, Linn.)

Native. Damp meadows. More abundant in the Avon district than elsewhere. P. May and August. *Top. Bot.* 64.

First Record. As *Sanguisorba officinalis*. In meadows, *Pitt, Agric. Worc.*, p. 317 (1810).

Avon. Fladbury.

Severn. Wyre Forest, *Jorden*. Pasture next to Nunnery Wood. Between the second milestone on the Spetchley road and Nunnery Wood.

Malvern. Longdon, *Lees*. Malvern, *Walker*. Madresfield.

Lickey. Beoley, *Lees*. Halesowen; Frankley; Alvechurch; *Mathews*. Northfield Station.

Burnet, the name of this tribe of flowers, is derived from *brun*, brown, being a quasi-diminutive of that word, and therefore the same as 'Brunette'; it is used in the names of several other natural objects. The herbalists used to confound this plant with the Burnet Saxifrage (*Pimpinella Saxifraga*), from the similarity of the leaves, and it had the name of Blood-wort. It also opened 'the stoppings of the liuer' and helped 'the Jaundies'. It was a frequent salad herb, and was used in 'cool tankards'. It is found in all the counties bordering Worcestershire.

ROSA, Linn. 122. (The old Latin name.)

578. 358. R. spinosissima, Linn. *Burnet Rose.*
(*R. pimpinellifolia*, Linn.)

Native. Open sandy places. Local. Small shrub. June and July. *Top. Bot.* 94.

First Record (probably). 'Rosa *pimpinellae fol. fl. rubro*. In some barren fields near Worster, Mr. Brown, and in a barren field at Churchlench four miles beyond Evesham on great plenty,' *Merrett's Pinax* (1666).

Avon. Craycombe. Spetchley Common.

Severn. Wyre Forest, *Jorden*. Hodge Hill, between Kidderminster and Park Hall, *Mathews*. About Kidderminster, *Lees*. About Worcester, *Stokes*. In the sandy country about Bewdley, *Withering*. Blakedown, *Scott*. Newtown, Worcester. Leopard Hill. Cookley, near Kidderminster. Crookbarrow.

Malvern. Leigh Sinton ; Bransford Chapel ; *Lees.*

Lickey. Booley, *Lees.*

This very thorny rose is fond of sand, and it often flourishes on the wind-blown sand-heaps of the seashore. The fruit is black, and, when ripe, very juicy. A number of varieties of this plant, under the name of Scotch Roses, are grown in gardens with flowers of pink, white, and yellow. It is met with in all neighbouring counties.

579. 359. R. involuta, Sm. *Top. Bot.* 60.

The typical form of this rose does not appear to have been recorded in Worcestershire, but the following varieties of it have been observed. The typical plant has been noticed in Stafford, Hereford, and Warwick.

Var. b. **Sabini,** Woods.

First Record. Very rare, *Mathews, Mid. Nat.* xiv, p. 92.

Severn. Wyre Forest, *Jorden* ; very rare, *Lees.* Near Spetchley, three or four miles from Worcester. Crookbarrow.

Lickey. Occurs rarely, *Lees.*

In Warwickshire and Staffordshire.

Var. c. **Doniana,** Woods.

First Record. Craycombe, Crookbarrow Hill, and Battenhall, *Lees, Illus. Nat. Hist. Worc.* (1834).

Avon. Craycombe, *Lees.*

Severn. Crookbarrow Hill ; Battenhall ; *Lees.* Warndon Wood, *Lees.*

Malvern. Side of a wood near Cradley ; near Alfrick ; *Lees.* Cradley Road, seven miles from Worcester, *Westcombe.* Very uncommon, *Lees.* Cowleigh Park.

In Gloucestershire, Herefordshire, and Warwickshire.

Var. d. **gracilis,** Woods.

First Record. Thicket near the rill that runs from Battenhall Farm, *Lees, A. Florence, Stran. Guide* (1828).

Severn. As above.

This rose occurs in no neighbouring county.

582. 360. R. mollis, Sm.

(*R. villosa,* Linn. *R. mollissima,* Willd.)

Native. Hedges and thickets. Not common in the Avon district ; many places elsewhere. Shrub. June and July. *Top. Bot.,* agg., 71.

First Record. Dudley Castle, *Booker, Dud. Cast.,* p. 107 (1825).

Severn. Helbury Hill. Battenhall. Near Kidderminster. St. Peter's. Droitwich. Bromsgrove. Hindlip Wood. Wyre Forest, *W. N. C. Trans.* i, p. 8. Churchill, near Kidderminster.

Malvern. Cowleigh Park, *Ley.* Clifton-on-Teme. Martley.

Lickey. Hedges, general, *Mathews.* Offmoor Wood, *Mathews* and *Bagnall.* Dudley Castle, *Booker.* Between Bromsgrove and Hagley. Shortwood Coppice, near Tardebigge. Halesowen. Frankley. Illey.

Of this rose Mr. Towndrow says (MS.) : 'An error for Malvern, and probably elsewhere in the county, *R. tomentosa,* var. *pseudo-mollis,* being very frequently

mistaken for it. All the specimens I have seen are more correctly referred to *R. tomentosa.*' As *R. mollissima,* it was given from Worcester as a new record, which it was not, by Dr. Arnold Lees in the *Bot. Rec. Club Rep.* for 1874. This rose has occurred in every neighbouring county.

585. 361. R. Andrzeiovii, Steven. *Top. Bot.* 19.

This type does not occur in Worcestershire, but the two following varieties of it have been seen in the county.

Var. b. **pseudo-mollis,** Ley. *Top. Bot.* 21.

First Record. West Malvern, *Towndrow, J. of B.,* vol. xxx, p. 341 (1892).

Mr. Towndrow ascribed his rose to var. *pseudo-mollis,* Baker, according to the description given by Mr. Baker, *J. of B.,* 1892, p. 361, and Mr. Ley, in his paper on the *mollis-tomentosa* group, *J. of B.,* 1907, p. 206, quotes this ascription. Probably in *Lond. Cat.,* 10th ed., the plant is assigned to the wrong authority. This rose has not been noticed in any neighbouring county.

Var. c. **Sherardi** (Davies). *Top. Bot.* 19.

(Var. *subglobosa,* Sm.)

First Record. As *R. Sherardi.* Thicket beyond Battenhall Lane, *Lees, A. Florence, Strang. Guide* (1828).

Severn. Ribbesford Wood, *Mathews.* Lane east of Hartlebury Station. Bewdley.

Malvern. Powick.

This variety has been observed in Salop, Stafford, and Warwickshire.

587. 362. R. scabriuscula, Sm. *Top. Bot.* 20.

First Record. Cowleigh Park and Cradley, *Lees, Bot. Malv. Hills,* 1st ed., p. 24 (1843).

Malvern. Cowleigh. Mill Coppice. West Malvern.

Lickey. Bromsgrove Lickey, fide *J. G. Baker, Bot. Rec. Club Rep.,* 1887.

This rose has been met with in Hereford, Stafford, and Warwickshire.

588. 363. R. tomentosa, Sm., pro parte.

Native. Hedges, thickets, woods. Several places throughout the county. Shrub. June and July. *Top. Bot.,* aggregate, 111.

First Record. Bransford, *Lees, A. Florence, Strang. Guide* (1828).

Avon. Broadway. Trench Woods.

Severn. Between Kidderminster and Blakedown. Helbury Hill. Battenhall. St. Peter's, Droitwich. Bromsgrove. Fernhill Heath. Hartlebury Station. Near Hadsor. Croome Perry Wood. Shrawley. Ribbesford Wood.

Malvern. Below Great Malvern, towards Welland, *Lees.* Between Madresfield and Powick, *Lees.* Bransford. Leigh Sinton. Road below New Pool. Powick. Malvern Link.

Lickey. Hedges, general, *Mathews.* Halesowen Abbey, *Lees.* Frankley.

This plant is recorded from every neighbouring county.

593. 364. R. Eglanteria, Huds.   *The Sweet Briar.*

(*R. rubiginosa,* Linn.)

Native. Hedges and woody places. Not common. Shrub. June and July. *Top. Bot.* 63.

K

First Record. 'Rosa sylv. odore flore duplici, Double Eglantine, G. 1270, invenitur duplici & triplici serie petalorum in sepibus praesertim prope Wigorniam,' *Mr. Brown, Merrett's Pinax* (1666).

Avon. Craycombe. Bredon Hill. The Slads. Tiddesley Wood.

Severn. Gravel-pit near Claines Church (1787), *Stokes.* Foot of Cruckbarrow Hill, *Lees.* Between St. John's and Pitmaston, Worcester. Hedge between seventh and eighth milestone on the Suckley Road from Worcester, just beyond the old toll-gate.

Malvern. Ankerdine, *Lees.* Malvern Hills, *Lees.* Shelsley. Bransford.

Lickey. Between Dudley and Tipton (1787), *Stokes.* Near Bilberry Hill, *Mathews.* Lane in Chaddesley Wood, *Humphreys.*

This is far from being a common plant in this county except in gardens, where it is well known. In early times it was a favourite plant, the Eglantine of the old poets ; and though Milton gave this name to the Honeysuckle, he seems to have been alone in doing so. The sweetness of the leaf sometimes perfumes the air in the neighbourhood of the bush. It is a native of Europe and Asia as far as North-West India. In Tasmania it is largely used as a hedgerow plant, the fences formed of it being ten or twelve feet high and nearly as wide. It is recorded from all neighbouring counties.

A curious variety having long, straight prickles, much like *R. spinosissima* but stronger, occurs at Bransford, *Towndrow, Malvern Advertiser,* Nov. 12, 1892.

× **canina.**

First Record. In this book.

Malvern. Crump End, Malvern, *Baker, Sp. Hast. Mus. Worc.*

This hybrid is not given in 10th ed. *Lond. Cat.* The specimen in the Herbarium of the Hastings Museum is named and signed in Baker's handwriting.

594. 365. R. micrantha, Sm.

Native. Hedges. Bushy places. Not common. Shrub. May to July. *Top. Bot.* 59.

First Record. Bransford, *Lees,* in *A. Florence, Strang. Guide* (1828).

Avon. Norton, near Evesham, in a copse by the river Avon. Sheriff's Lench. Littleton Banks. Tiddesley Wood. Croome Perry Wood.

Severn. Railway-side, Wyre Forest. Bransford. Near Bewdley. Near Pitmaston.

Malvern. Thickets, side of the Warren Hill, *Lees.* Cowleigh Park. Rock Wood, Shelsley. Martley. Beyond Leigh Sinton. Malvern Wells. Woodsfield.

This rose does not appear to have been noticed in Shropshire, but occurs in other bordering counties.

595. 366. R. agrestis, Savi.

[(*R. sepium,* Thuill.)

Native. Hedges. Rare. Shrub. June to August. *Top. Bot.* 17.

First Record. Not localized, *Lees' Cat., New Bot. Guide* (1835).

Malvern. Little Malvern, *Lees.*

Mr. Lees says of this rose, calling it *R. sepium, Bot. Malv. Hills,* 3rd ed., that Mr. Bloxam doubted if it was true *R. sepium,* and that he himself considered

it identical with a rose Mr. Baker named *R. Blondaeana,* which is a form of *R. canina,* and he gives the neighbourhood of Little Malvern as the locality, *Bot. Worc., Add. and Cor.,* 1867. But this cannot be considered a record. *R. sepium* has been noticed only in Warwickshire of neighbouring counties.]

Var. d. **inodora,** Hook. fil.

First Record. In bushy pastures below Malvern Wells, westward, *Lees, Bot. Malv. Hills,* 1st ed., p. 25 (1843).

Avon. Himbleton.

Severn. Perry Wood. Wyre Forest, *Jorden.*

Malvern. Not uncommon, *Lees.* Leigh Sinton. Cradley Wood, near county boundary.

Lickey. A characteristic plant, *Lees.* Offmoor Wood. Halesowen.

Mr. Lees does not appear to have been very sure about this rose. In the 1st ed. he says it is *R. Borreri,* Woods, and in the 3rd, *R. tomentella,* Léman. This rose has been noticed in Warwickshire.

596. 367. R. Borreri, Woods. *Top. Bot.* 23 (?).

First Record. In the woods (Malvern), *Lees, Mag. Nat. Hist.,* vol. iii, p. 160 (1830).

Severn. Perry Wood, *Lees.*

Malvern. As first record. Malvern Wells, *E. F. Linton, Towndrow, Malvern Advertiser,* Nov. 12, 1892.

This rose has been noticed in Staffordshire by Dr. Fraser ; and in Warwickshire.

Var. b. **tomentella** (Léman).

First Record. Specimens of a rose, with hairy and setose petioles, the leaflets hairy and silvery beneath, which I had referred to *R. inodora,* Mr. Bloxam says is truly the above, *Lees, Bot. Worc., Add. and Cor.* (1867).

Severn. Grows in this district, *Lees.*

Malvern. Grows in this district, *Lees.*

This variety has occurred in Hereford, Stafford, and Warwick.

Var. f. **arvatica** (Baker).

First Record. Cowleigh Park, *Flora of Herefordshire,* p. 119.

Malvern. Cowleigh Park, *Ley, Towndrow.*

Lickey. Near Billesley and Hewell Grange, *Bagnall* (MS.).

This variety occurs in Staffordshire and Warwickshire.

597. 368. R. canina, Linn.   *Dog Rose.*

Native. Hedges and thickets. Generally distributed. Shrub. May to July. *Top. Bot.,* aggregate, 112.

First Record. Near Dudley Castle, *Booker, Dud. Cast.,* p. 107 (1825).

The Dog Rose is common nearly everywhere, enlivening roadsides, coppices, and woods in early summer with its long sprays of blush-pink flowers. The name Dog Rose, English though it looks, is not of English origin. It is a translation of the Latin name, which was the one Linnaeus gave to the plant as above, which in its turn was a translation from the Greek κυνόροδον, a name given it because the root was thought to cure the bite of a mad dog. Several varieties of garden roses have originated from this plant, but the immense number of garden roses in cultivation spring from many kinds.

All nations have loved the rose. Wreaths of its flowers to crown triumph; heaps of its petals to adorn the banquet; a plant of it placed on the tomb of a loved one; its uses, symbolism, and folklore among all peoples would fill books upon books. In the East as well as in the West the rose is prized above all flowers, and no flower has entered so intimately into the life and association of people of all nationalities. Of course it has not escaped in our country the attention of the herbalist. Its red fruits, called Hips, were supposed to strengthen the heart, and improve the memory; they were made into 'pleasant meates and banketting dishes'; they form pretty playthings for the children of the countryside. Every one who has noticed this plant has seen here and there green and red mossy tufts on its branches, which country people call Robin's Pincushions, though needless to say the robin has nothing whatever to do with them. Probably their ruddy flush has brought to mind the Robin's breast. These originate from the puncture of an insect *Cynips Rosae*, and were formerly called Bedeguars, a word not yet obsolete. This is an Oriental word made up of the Persian *bād*, wind, and the Arabic *ward*, rose, and at one time was applied to the Milk Thistle, *Silybum Marianum*. The Dog Rose occurs in all neighbouring counties.

R. canina has been broken up into numerous varieties, tending to make the study of the plant as difficult to the ordinary botanist as that of brambles. These varieties depend chiefly upon the nature of the leaves, whether they are entirely smooth on both sides, are more or less hairy on the under surface, or are more or less hairy both above and below. And as in the case of the brambles, it is difficult to be sure what plant is really meant by the names used by various botanists.

Var. a. lutetiana (Léman).

First Record. As *R. glaucophylla*, Winch (?). Between Cowleigh and Cradley, *Lees, Bot. Malv. Hills*, 1st ed., p. 26 (1843).

Avon. South Littleton. Sheriff's Lench. Norton, by Evesham. Trench Woods.

Severn. Kempsey. Claines. Ombersley. Holt. Shrawley. Bewdley. Wyre Forest. Hallow. Clerkenleap. Lincombe. Severn Stoke.

Malvern. Near the Cradley turnpike gate, *Lees*. Powick. Bransford. Alfrick. Knightwick. Martley. Stanford. Knighton-on-Teme.

Mr. Lees does not appear to be certain whether his rose is this variety, or var. l. *andegavensis*, following. In the Supplement to his *Bot. Worc.*, as below under that variety, he states that he had sent some Worcestershire roses to Mr. Bloxam, and in the result he came to the conclusion that *andegavensis* or *verticillacantha* his rose must be referred. *R. lutetiana*, Léman, he says, 'must now designate our old friend the common glabrous-leaved *canina*.' This without doubt is the commonest form throughout Worcestershire, and occurs in all bordering counties.

Var. g. dumalis (Bechst.).

First Record. On and about Rough Hill, near Cowleigh Park and its vicinity, but not very general, *Lees, Bot. Malv. Hills*, 3rd ed., p. 65 (1868).

Severn. Near Norton by Kempsey. Spetchley.

Malvern. Woodsfield, *Towndrow*. Alfrick.

This is one of the roses which were determined for Mr. Lees by the Rev. Andrew Bloxam. Rough Hill Wood is in Herefordshire, near Cowleigh Park, which, now in Worcestershire, is divided from it only by a road, so that half at least of the 'vicinity' is in our county; and this may stand for the first record. It is not an infrequent rose, and is reported from all neighbouring counties.

Var. h. biserrata (Mérat).

First Record. Cowleigh Park, *Ley, Flora of Heref.*, p. 118.

Malvern. As above.

This variety, hardly if at all more than a mere form, has been observed in Warwickshire by Mr. Bagnall.

Var. l. andegavensis (Bast.).

First Record. Not localized, *Lees, Bot. Worc., Sup. at end* (1867).

Avon. Sheriff's Lench. Trench Woods.

Severn. Hallow. Claines. Kempsey. Broad Heath. Birchen Grove.

Malvern. Powick. Bransford. Alfrick. Cowleigh Park, *Ley*.

Mr. Lees says, *Bot. Malv. Hills*, 3rd ed., p. 65, 'I have in my herbarium specimens (in which) the leaflets more resemble those of *R. verticillacantha*, but the peduncles, sepals, and fruit are very setose. It agrees best, therefore, with *R. andegavensis*.' He gives as localities for it between Malvern and Worcester, and the southern side of Birchin Grove. This variety is frequent in Warwickshire, and must occur in other bordering counties, but does not appear to have been recorded from them.

Var. o. surculosa (Woods).

First Record. On the side of a lane leading from Welland Common to Castlemorton and Longdon, *Lees, Bot. Malv. Hills*, 1st ed., p. 26 (1843).

Severn. Near Spetchley, *Herb. Vict. Inst. Worc.*

Malvern. Powick, *Herb. Vict. Inst. Worc.* Malvern Link. Woodsfield, *Towndrow*.

This rose has been recorded from all bordering counties except Salop.

Var. p. verticillacantha (Mérat).

First Record. Barnt Green, *Bot. Rec. Club Rep.* (1887).

Malvern. Malvern, *Ley*. Common, *Towndrow*.

Lickey. As first record.

Mr. Lees had been doubtful of several of his roses, this one among the number, as noted above under var. *lutetiana*. He calls it the 'pale thin-leaved rose'. The sparingly setose peduncles, he says, distinguish it from *R. dumalis*, which it very nearly approaches. This rose has been noticed in all bordering counties except Salop.

Var. r. latebrosa (Déségl.).

First Record. Malvern, *Towndrow, Malv. Advertiser*, 1892.

'I have met with it once, I think, no more,' Towndrow (MS.). This variety has been observed in Warwickshire and Staffordshire. In the 9th ed. *Lond. Cat.*, this rose is given as a form of var. t. *vinacea* of *R. canina*, which variety has not been recorded from Worcestershire.

Var. s. aspernata (Déségl.).

First Record. (Referring to var. *verticillacantha*), 'round Malvern it often puts on the extreme clothing of aciculi by virtue of which it becomes

the *R. aspernata* of Déséglise,' Rev. A. Ley, address by retiring President of the Woolhope Field Club, April 13, 1882.

This does not appear to have been recorded from any neighbouring county, but Herefordshire is very near to 'round Malvern'.

598. 369. R. dumetorum, Thuill. Not numbered in *Top. Bot.*

Native. Hedges and bushy places. Rare. Shrub. May to July. No *Top. Bot.* number, *Lond. Cat.*, 10th ed.

First Record. As *R. dumetorum*. Between Worcester and Malvern, *Lees, Illus. Nat. Hist. Worc.*, p. 166 (1834).

Avon. Norton, near Evesham. Fladbury.

Severn. Chatley. Holt. Kempsey. Cruckbarrow, *Herb. Hastings Museum*.

Malvern. As first record. Cowleigh Park, *Ley*. Bransford.

Lickey. Halesowen Abbey, *Lees*.

This rose is reported from all neighbouring counties except Salop.

Var. b. obtusifolia, Desv.

First Record. Bromsgrove Lickey, *Bot. Rec. Club Rep.* (1887).

Lickey. As first record.

This rose has been observed in Hereford, Stafford, and Warwick.

Var. d. urbica, Léman.

First Record. As *R. Fosteri*. In the woods (Malvern), *Lees, Mag. Nat. Hist.*, vol. iii, p. 160 (1830).

Severn. Bewdley, *Bagnall* (MS.). Monk's Wood. Perry Wood. The Denes, Ombersley.

Malvern. As first record, *Towndrow*. Rough ground between Bromyard and Bransford Road, *Herb. Vict. Inst. Worc.*

Lickey. Plentifully near Northfield, *Bagnall* (MS.). Dayhouse Bank, *Mathews*.

All the bordering counties possess this rose.

Var. e. frondosa (Baker).

First Record. Powick, *E. F. Linton, Malv. Advert.*, 1892.

Malvern. As above.

This variety occurs in Hereford, Stafford, and Warwickshire.

Var. n. caesia, Sm.

First Record. About Leigh Sinton, but rather uncommon, *Lees, Bot. Malv. Hills*, 3rd ed., p. 66 (1868).

Malvern. As record above. Cowleigh Park, *Baker*.

This plant has occurred in Warwickshire and Staffordshire.

599. 370. R. glauca, Vill. *Top. Bot.* 24 ?.

Native. Hedges. Rare. Shrub. May to July. *Top. Bot.* 24.

First Record. South Littleton, *Worc. Nat. Club Trans.* iii, p. 237 (1905).

Avon. As above. The type, *Rea* (MS.).

Rosa glauca has occurred in Herefordshire, Staffordshire, and Warwickshire.

600. 371. R. coriifolia, Fr. No *Top. Bot.* number, *Lond. Cat.*, 10th ed.

First Record. Not localized, *Lees, Bot. Malv. Hills*, 3rd ed., p. 65 (1868).

Malvern. As above.

This has been observed in Hereford, Stafford, and Warwickshire.

601. 372. R. stylosa, Desv.

Native. Hedges and wood-borders. Local. Shrub. May to July. *Top. Bot.* 38.

First Record. As in this book.

Severn. Roadside, Droitwich Road, four and a half miles from Worcester, *Rea*.

Malvern. Cowleigh Park, *Rea*.

This rose is recorded from Gloucestershire and Herefordshire.

Var. b. systyla (Bast.). *Top. Bot.* 38.

First Record. Not localized, *Lees' Cat., New Bot. Guide* (1835).

Avon. Sheriff's Lench. Tiddesley Wood. Defford. Craycombe.

Severn. Broadwas. Ombersley. Sinton Green. Wyre Forest. Kempsey.

Malvern. Malvern Link. Newland. Malvern Wells. Sand-pits, Powick. Half Key. Bransford. Ankerdine Hill. Doddenham. Martley. Knighton-on-Teme. Stanford Bridge. Little Malvern. Welland Common.

Lickey. Butler's Hill, Hewell, *Bagnall*.

R. stylosa seems in the aggregate to be made up of a number of forms, all intermediate between *R. canina* and *R. arvensis*. The variety *systyla* occurs in Warwickshire.

Var. e. pseudo-rusticana, Crépin. *Top. Bot.* 4.

Probably to this variety is to be referred a rose, gathered by Mr. Towndrow at Brace's Leigh, July 27, 1898, though he thinks his plant does not exactly match it. He considered it a hybrid between *R. stylosa*, var. *systyla* and *R. arvensis*.

602. 373. R. arvensis, Huds. *White Dog Rose*.

Native. Hedges, woods, bushy places. Widely distributed. Shrub. May to July. *Top. Bot.* 69.

First Record. Near Dudley Castle, *Booker, Dud. Cast.*, p. 107 (1825).

Avon. Norton, near Evesham. Sheriff's Lench. Fladbury. Pershore. Defford. Crowle. Trench Woods. Craycombe. Flyford Flavell. Tardebigge.

Severn. Bewdley. Droitwich. Crookbarrow. Laughern Brook. Astwood Road. Ombersley. Holt. Grimley. Kempsey. Wyre Forest. Claines.

Malvern. Abberley Hill. Berrow Hill. Martley. Madresfield. Powick. Stanford Bridge. Alfrick. Malvern Link.

Lickey. Clent. Frankley. California, Northfield. Randans Wood.

This rose cannot be mistaken by the most ordinary observer, for its habit or growth, as well as its white flowers, well distinguishes it from those just dealt with. Though it grows freely in the lowlands, in the Lickey district it tends to supplant *R. canina*, and at the higher altitudes occurs much more frequently than that does. It occurs in all the neighbouring counties.

Var. b. gallicoides, Baker.

(*R. arvensis*, var. *setosa*, Bagnall.)

First Record. Near Malvern, *Towndrow, Malv. Advert.*, 1892.

This variety occurs in Warwickshire also.

Var. c. **ovata**, Desvaux.

First Record. As *R. ovata*, Lej. Not localized, *Towndrow, Bot. Exch. Club Rep.* (1887).

  Malvern. Madresfield.

This variety does not appear to have been noticed in any neighbouring county.

Var. **bibracteata**, Bast.

First Record. Malvern, *Towndrow, Malv. Advert.*, 1892.

  Malvern. Malvern. Malvern Link. Pickersleigh. Madresfield.

This rose cannot be discovered in the 10th ed. *Lond. Cat.* In his *Flora Berks.*, p. 206, Mr. Druce varies this spelling, and calls it var. b. *dibracteata*, Bast., which is either an error, or a refinement calculated to add a further puzzle to a puzzling tribe. This variety has been observed both in Herefordshire and Warwickshire.

[**606. 374. R.** cinnamomea, Linn.

First Record. In hedges at Claines, *Lees, Bot. Worc.* xxix.

This rose has no claim to be considered a native of our islands. It belongs to the mountains of Northern Europe and Siberia, and only occurs in England in the neighbourhood of houses. It has been noticed in Gloucestershire.]

[**608. 375. R.** sempervirens, Linn.

First Record. As *R. Melvini*, Towndrow. Madresfield, *Towndrow, Bot. Exch. Club Rep.*, 1885-6.

  Malvern. As above. Leigh Sinton, *Towndrow.* Malvern Wells, 1907.

R. Melvini, as named by Mr. Towndrow, has been referred by Messrs. H. and J. Groves to this plant, *Babington's Manual*, 1904, p. 138. In spite of this, Mr. Towndrow maintains that this rose should undoubtedly come under *R. arvensis*, as M. Déséglise recognized. He believes it to be a sterile cross of varieties of that species, and that var. *bibracteata* is pretty certainly one parent. Mr. Linton suggested to him that it was a hybrid, or more correctly a 'cross' in Darwin's sense (MS.). *R. sempervirens* is not a native of Britain.]

PYRUS, Linn. 123.   (Gaelic *peur*, a pear, or more immediately from πῦρ, flame.)

609. 376. P. torminalis, Ehrh.   *Wild Service Tree, Sorb.*

Native. Woods. Fairly common. Tree. April and May. *Top. Bot.* 50.

First Record. Clerkenleap, &c., *Lees, A. Florence, Strang. Guide* (1828).

  Avon. Sparingly in woods on the Lias, *Lees.* Trench Woods. Tiddesley Wood. Badger's Bank. The Slads.

  Severn. Wyre Forest, *Lees.* Marl cliffs by the Severn, *Lees.* Clerkenleap. Blackstone Rock. Bush Hill, Powick. Perry Wood. Monk's Wood. Ockeridge. Shrawley. Nunnery Wood. Lincombe. Helbury Hill.

  Malvern. Madresfield. Ankerdine Hill. Martley. Crews Hill. Newland.

Lickey. Pepper Wood, Belbroughton, *Humphreys.* Fox Lydiate Wood, Redditch, *Mathews.*

This is a small tree, something like a Hawthorn in appearance, and often bearing abundant fruit of a greenish-brown colour, nearly as large as the hips of the rose, which have a not unpleasant flavour and are sometimes seen in country markets. It is recorded for all neighbouring counties.

611. 377. P. Aria, Ehrh.   *White Beam Tree.*

Native. Woods. Local and rare. Tree. April and May. *Top. Bot.* 50.

First Record. Bewdley Forest, *Jorden, Phyt.*, N.S., vol. i, p. 281 (1855).

  Avon. Bredon Hill above Overbury.

  Severn. Wyre Forest, rare, *Lees.*

  Malvern. Stanford. Hanley Castle, probably planted, *Towndrow.*

This small tree is remarkable for the whiteness of the under surface of the leaves, and the contrast with the rich green upper surface when the wind passes over the foliage is very conspicuous. Mr. Lees records it in his *Table of Plants* as occurring in the Avon district; but it would appear that he was referring to the locality known as Hyate's Pits in the parish of Snowshill, just outside Worcestershire, and in Gloucestershire. It occurs in all neighbouring counties.

{**378. P.** domestica, Ehrh.   *The True Service.*

First Record. In a letter from *Mr. Edmund Pitt*, quoted in *Philosophical Trans.*, no. 189, p. 978, 1678.

This is the first record of the celebrated Sorb Tree of Wyre Forest, possibly the only wild specimen of the plant ever known in Britain. It has attracted the attention of generations of botanists. In the neighbourhood of the Forest it was known as the Witty Pear, and in 1831 it had come to the limit of its age, and was rapidly decaying (*Mag. Nat. Hist.*, vol. iv, p. 450); but it lingered on until 1862, when it was destroyed by fire. To the present generation of Worcestershire botanists the precise locality where it grew even is unknown. Now it is gone from the *Lond. Cat.*, being considered, no doubt, a cultivated variety of the Mountain Ash, planted in the Forest. So our illusions disappear! Two strong descendants of the tree, raised from seed, are flourishing in the grounds of neighbouring Arley Castle. A full history of the tree, mainly from the *Gardeners' Chronicle*, April 13, 1907, is given in *Hortus Arleyensis*, by R. Woodward, Jun. (privately printed, 1907), p. 67. A view of the tree, with figures of the foliage and fruit, are given in Nash's *Hist. Worc.*, vol. i, p. 10 (1781), and a description of it at p. 11. A sketch of the tree, 'taken many years ago,' will be found in Lees' *Bot. Worc.*, p. 4 (1867); and at p. xci of the same book, another sketch of the tree, in the last stage of decay, as it appeared in 1856. 'Its wanton destruction by ruffian hands,' says Mr. Lees, 'was a great source of sorrow to Mr. Jorden, and he gathered up and preserved all the fragments that remained. On October 21, 1862, Mr. Jorden exhibited at a meeting of the *Worc. Nat. Club*, a fragment of the tree; and from this was afterwards manufactured a cup, the silver mounts of which were engraved with representations of parts of the tree; and it was presented to Mr. William Mathews at a meeting of the Club, October 25, 1864. No need to say that no neighbouring county can vie with Worcestershire in ever having possessed a True Service Tree.}

---

617. 379. P. Aucuparia, Ehrh.   *Mountain Ash, Rowan.*

Native. Woods. Local. Tree. May and June. *Top. Bot.* 108.

First Record. Near Bromsgrove Lickey, abundantly, *Purton*, vol. iii, pt. ii, p. 335 (1821).

  Avon. Tiddesley Wood.

  Severn. Wyre Forest, occasional, *Lees.* Shrawley Wood. Hackbury Lane, near Kidderminster. Perry Wood. Ockeridge Wood. Monk's Wood.

  Malvern. Mathon. Cradley. Sparingly, *Lees.*

  Lickey. Lower Lickey, *Lees.* Frankley Wood, *Lees.* Woods at Clent. Bilberry Hill, *Humphreys.* Rare, *Mathews.*

This tree is far more familiar to us from its being planted for ornament than in its wild state. In the autumn its red clusters of fruits are very conspicuous so long as the birds will spare them; and each one of the little berries is like a tiny apple, with core and seeds within. The plant was, and perhaps is, in the north, considered efficacious against witchcraft and the evil eye. Farmers placed branches of it in their cow houses to ward off evil influence, and the milkmaid used to go to her work with a piece in her hand or tied to her pail. The wood is finely grained and hard, and the bark and roots contain a large amount of oil of Almonds. The tree occurs in every neighbouring county.

618. 380. P. communis, Linn.   *Wild Pear.*

Denizen. Hedges. Except in the Avon district. Not uncommon. Tree. April and May. *Top. Bot.* 49.

First Record. Perry Wood, Worcester, *Lees, Worc. Miscell.* (1829-30).

  Severn. Crown East Wood. Perry Wood. Clerkenleap. Atchen Hill Wood. Wyre Forest.

  Malvern. Thorny Pear Coppice, Alfrick.

The Wild Pear is inclined to be a thorny tree, and the fruit of it is hard and harsh. Nevertheless, it is the origin of all the luscious pears that appear upon our table. In a less delicious form its descendants are seen in all Worcestershire hedgerows and orchards, sometimes of very considerable size, and from them is made the perry for which Worcestershire is celebrated, its especial drink, outvying cider as belonging to the county. The most esteemed variety for its manufacture is the Barland Pear. Pears are the charges in the coat-of-arms used for the county, a coat originally belonging to the City of Worcester, and borrowed from it to supply an aching void. Pears have been cultivated from the earliest times; possibly, they were introduced by our Roman conquerors, as very numerous kinds were grown in Italy in classic times, which had names as various as those of to-day. Pears are mentioned in the domestic accounts of the Middle Ages from those of king to those of commoner; and the monks paid especial attention to them. Does not a tradition say that King John was poisoned by a dish of pears prepared therefor by the Monks of Swineshead? And in view of this early introduction the Wild Pear may possibly be the degenerate descendant of the cultivated tree. Whether this be so or not, that a plant may degenerate so much or be brought to such delicious perfection is equally remarkable. It does not appear to have been noticed in Salop.

Var. a. **Pyraster,** Linn.

First Record. Newland, *Towndrow, Malv. Advert.*, December 3, 1898.

  Malvern. As first record.

Mr. Towndrow has observed this variety also near Gritt Farm, Malvern Link. The leaves are downy beneath when young, and the fruit is elongated. It has been noticed in Staffordshire, Herefordshire, and Warwickshire.

Var. b. **Achras,** Gaert.

First Record. Near Malvern, *Towndrow, Malv. Advert.*, December 3, 1898.

  Severn. Clerkenleap. Atchen Hill Wood.

  Malvern. As first record.

In this form the mature leaves are slightly downy, and the fruit is more globose in shape and rounded below. This variety has been observed in Warwickshire.

620. 381. P. Malus, Linn.   *Crab Apple.*

Native. Woods, hedges, thickets. Generally distributed. Small tree. April to June. *Top. Bot.* 89.

First Record. Perry Wood, Worcester, *Lees, Worc. Miscell.*, 1829-30.

As the Wild Pear is the father of the pear of the dessert table, so the well-known Crab is the mother of all apples; and not only mother in the sense of descent, but the support and sustainer from youth to age, being the stock upon which the cultivated kinds are usually grafted. The apple in all its forms is peculiarly a Worcestershire tree, in the south and west of the county decorating every hedgerow; in the north and east, dying away and seldom seen, nor flourishing even to any great extent in the more protected orchard. The northern limit of the hedgerow apple tree in the county is approximately a line drawn from the neighbourhood of Bewdley through Belbroughton, and then turning a little to the south-east towards Redditch. The Crab had its uses; the sour juice was used to curdle milk; vinegar made from the fruit was used to cure sprains, and the juice was thought to have cosmetic qualities. But the uses of the cultivated kinds have been more prominent, and references to them are scattered through literature of all kinds and of all times. Especially in Worcestershire in the apple districts are they used to make cider, the common beverage of that part of the county. Another forgotten use, though still betokened by the name of the modern substitute, pomatum, was to make an ointment for the skin and hair. Apples had their medicinal value in old times, and were especially recommended in diseases of the spleen against melancholy, while an echo of these old ideas still exists in the well-known proverbial saying, 'An apple a day keeps the doctor away.' The Crab Apple occurs in all neighbouring counties.

Var. **sylvestris,** Linn.

(Var. *acerba*, DC.)

First Record. As in this book.

  Avon. Badger's Bank. Tiddesley Wood. Trench Woods.

  Severn. Wyre Forest, sp. *Hast. Mus. Worc.* Gregory's Mill. Monk's Wood. Nunnery Wood. Shrawley Wood.

  Malvern. In the district, *Towndrow* (MS.). Middleyards. Ankerdine Hill.

In this variety the young branches, calyx tube, and underside of the leaves are glabrous. It has been reported from all neighbouring counties except Gloucestershire.

Var. *mitis, Wallr.

First Record. Malvern, *Towndrow, Malv. Advertiser,* November 12, 1892.

Severn. Hedge in lane, Churchill, near Kidderminster. Wyre Forest. Foot of Helbury Hill.

Malvern. As above.

In distinction from the former variety, in this one the parts there glabrous are pubescent or woolly. This variety also does not appear to have been noticed in Gloucestershire.

[*621.* P. *germanica, Hook. fil. *Medlar.*

Mr. Lees goes out of his way in his *Illus. Nat. Hist. Worc.* to record a plant which is neither a denizen of Worcestershire nor quite certainly a native of Britain, which he says is to be found in Deerhurst Lane, opposite the Lower Lode, near Tewkesbury.]

CRATAEGUS, Linn. 124. (κράτος, strength, in allusion to the hardness of the wood.)

622. 382. C. Oxyacantha, Linn.   *Thorn, Hawthorn, White-thorn, May.* (*C. oxyacanthoides,* Thuill.)

Native. Hedges, woods, thickets, commons. Abundant. A small round-headed tree or hedgerow bush. April to June. *Top. Bot.* 30.

First Record. Near Dudley Castle, *Booker, Dud. Cast.,* p. 107 (1825).

The abundance of this plant is greatly added to by its universal use in the formation of Quickset hedges, that is 'living' hedges, for the word has no reference to its rate of growth, which, though rapid when young, is quite slow when it attains any age ; and from this use it is called ' Quick ', a word never used for the plant when it has attained the dignity of a tree or large bush, adorned with flowers. Indeed, its use from the earliest times as a fencing plant gives it one of its names, for Haw is cognate with Hedge, both being derived from the Old English *haga* ; and sometimes Haw by itself has been used to designate the plant. Into the usages and rejoicings of the country-side the plant has entered largely. There are several localities in the county where especially fine thorn-trees are to be found, on the sides of Bredon Hill for instance, in Kyre Park, near Tenbury, The Slads, near Sheriff's Lench, and on some of the declivities of Malvern Hills. Its ragged appearance when old has led to the proverb, ' As bare as a hawthorn.' Some drawings of old thorns in Worcestershire and elsewhere are given in the Supplement to vol. ii of *Worc. Nat. Club Trans.,* p. 20, and the following ones. The reputation of the Glastonbury Thorn, which flowers in winter, is widespread, and the legend which attributes it to the staff of Joseph of Arimathea well known ; but Glastonbury did not possess a monopoly in winter flowering thorns, for in later days several places possessed one. There is one of these trees at the Swan on Newland Common. The deep red berries, bruised and boiled in wine, were held to be a remedy for ' tormenting pains ' ; and their distilled water applied to thorns or splinters in the flesh would ' notably draw them out', showing that homœopathy is no new science, and that a ' hair of the dog that bit you ' is of wider application than

in merely canine matters. The Hawthorn is recorded from all neighbouring counties.

Var. oxyacanthoides, Thuill.

First Record. Lane north of Hartlebury Common, *Lees* and *Thompson, Bot. Rec. Club Rep.,* 1883 (pub. 1884).

Avon. Tardebigge Reservoir. Near Goosehill Wood. Crowle. Fladbury. Norton, near Evesham. Bredon Hill.

Severn. Right bank of Severn, near Diglis Weir. Near Monk's Wood. Kempsey. Wyre Forest. Near Hartlebury Common.

Malvern. Bransford. Newland. Mathon. Martley. Abberley Hill. Powick.

This variety of the aggregate is considered frequently the typical plant, and in several botanies is equated with it, as in the 10th ed. *Lond. Cat.* It differs from the following species, sometimes considered a sub-species or variety, by having usually a glabrous pedicel and calyx, and two or three carpels, while in the following variety the carpel is solitary, and the pedicel and calyx are villose. This variety occurs in Staffordshire and Warwickshire.

623. 383. C. monogyna, Jacq.   *Top. Bot.* 111.

First Record. As in this book.

This plant is abundant in each of the botanical districts of this county, and is really the common 'May', both in it, and all bordering counties.

27. SAXIFRAGACEAE, Juss.

SAXIFRAGA, Linn. 126. (*Saxum,* a stone, *frango,* I break, from this species rooting into rocks and breaking them up.)

630. 384. S. *umbrosa, Linn.   *London Pride.*

Casual. A garden escape. P. June and July. *Top. Bot.,* only native in Ireland.

First Record. In a sloping field a little below Moseley Common, *Mr. W. Evans, With.,* 4th ed., p. 394 (1801).

Malvern. Abberley Hill, garden outcast, *Lees, Bot. Worc.,* p. xxix.

This, the well-known plant of every flower border, has no more claim to a place among the Worcestershire Flora, or the Flora of any other county in Britain, than any casual garden flower, which, thrown on a neighbouring rubbish heap, manages to maintain itself for a longer or shorter period in a more or less flourishing condition. No doubt in every neighbouring county such plants could be and are found, if it were thought necessary to record them.

633. 385. S. tridactylites, Linn.   *Rue-leaved Saxifrage.*

Native. Old-walls, dry ground, roofs of buildings. Common in the Lickey district, less frequent elsewhere. A. March to June. *Top. Bot.* 83.

First Record. Adorning the wall fronting the bank of the Teme at Powick Bridge ; but the river having undermined the wall, it has now fallen down, and the plant has gone with it, *Lees, Illus. Nat. Hist. Worc.,* p. 162 (1834).

Avon. Bredon Hill. Broadway. South Littleton. Walls at Badsey.

Severn. Britannia Square, Worcester. Charlton, near Hartlebury. Near Kidderminster. Hartlebury Common. Upper Areley. Broomhouse, Bromsgrove.

Malvern. Garden wall at Little Malvern. Wall, Powick Bridge, 1907. Roof of Cottage, Old Malvern Road, Powick. Cowleigh Park. Alfrick. Brockamine, Madresfield. Malvern Link.

Lickey. Clent, on every old roof. Halesowen Abbey. Hagley Hall. Harvington Hall. Tutnall.

Of this little plant, as of the above, the true nativity of the plant in England is questioned, but with less chance of truth. It is a tiny plant as a rule, but it grows six or eight inches high in the damp gutters of old tiled roofs. After flowering it turns red, and the whole plant is viscid. It is recorded from all neighbouring counties.

636. 386. S. granulata, Linn.   *White Meadow Saxifrage.*

Native. Hedgebanks, pastures. Locally plentiful. P. May to July. *Top. Bot.* 81.

First Record. Bevere, near Worcester, *Stokes, Stokes's With. Bot. Arr.,* 2nd ed., p. 434 (1787).

Avon. Bredon Hill. Broadway. Near Mill at Churchill, near Worcester.

Severn. Wolverley. Laughern Brook. Canal-side at Astwood. Barbourne. Hartlebury. Harberrow. Hagley, plentiful. Redstone Rock. Blackstone Rock. Holt Mill. Drakelow. Banks of Salwarpe, Stoke Prior. Honeybottom, near Kidderminster. Dunclent. Near the Dog at Dunley.

Malvern. End Hill, *Lees.* Rock. Abberley. Martley. Knightwick.

Lickey. Clent. Belbroughton. Park Gate, near Bromsgrove. Hagley.

This plant in Worcestershire is nearly confined to the new red sandstone in the north of the county, and when that formation comes to the surface it is usually to be found. In some hedgebanks about Hagley it is so plentiful that the banks are as whitened with it in patches as with growths of Greater Stitchwort. The root consists of a number of reddish round little tubers, which on the principle of like cures like, were supposed to be efficacious in gravelly disorders. The plant is recorded from all neighbouring counties.

{*642.* 387. S. hypnoides, Linn.   *Mossy Saxifrage.*

Casual. Probably a garden escape. P. May to July. *Top. Bot.,* no number.

First Record. As in this book.

Avon. Below the Fish, Broadway, *Rev. J. H. Thompson,* sp. *Vict. Inst., Worc.*

This plant is found in rocky mountainous situations in England, Scotland, and Ireland. The Fish Inn stands at the top of Broadway Hill, nearly 1,000 feet above the sea, but even so, its neighbourhood is hardly likely to be a natural habitat for this plant. It has been noticed in Herefordshire, Salop, and Stafford.}

CHRYSOSPLENIUM, Linn. 127. (χρυσός, gold, σπλήν, the spleen, allusion to the colour of the flowers and the shape of the leaves.)

643. 388. C. oppositifolium, Linn.   *Common Golden Saxifrage.*

Native. Damp places in woods, brook-sides, ditches. General. P. February to April. *Top. Bot.* 107.

First Record. Near Sapey Brook, in many places, *Sheward, Nash, Hist. Worc., Sup.,* p. 96 (1799).

This is one of the earliest plants to flower in spring, making masses of bright green leaves, which are flushed over from a little distance with the yellow of its clusters of flowers. It was supposed by the ancients to be a golden remedy for affections of the spleen, which accounts for its generic name ; while its specific name is derived from the fact that its leaves are opposite, and not, as in the following species, alternate. It is recorded from all neighbouring counties.

644. 389. C. alternifolium, Linn.   *Alternate-leaved Golden Saxifrage.*

Native. Damp places in woods, brook-sides, ditches. Not common. P. February to April. *Top. Bot.* 72.

First Record. Purlieu Lane, leading from the Wych to Malvern, *Ballard, Stokes's With. Bot. Arr.,* 2nd ed., p. 404 (1787).

Severn. Holt Mill. Hurcott Wood. Orls Coppice, Rushwick. Rock near Leigh Church. Shrawley Wood. Southstone Rock, *Rea.* Astley Copse. Spout Mill Brook, Hagley.

Malvern. Purlieu Lane. Whippet's Brook, Bridgestone Mill. Sapey Brook. Rosebury Rock Wood. Woodbury Hill.

Lickey. Below Dales Wood. Clent. Frankley. Romsley. Halesowen. Hagley. Alvechurch. Lickey Hill. Dodford. Twilands Wood, Frankley. St. Kenelm's.

So much is this plant like the preceding one, except for the position of the leaves upon the stem, that it is no doubt frequently passed over. In the higher parts of the Clent Hills, this plant is much more frequent than its opposite leaved brother. It is recorded from all bordering counties.

PARNASSIA, Linn. 128. (From Mount Parnassus.)

645. 390. P. palustris, Linn.   *Grass of Parnassus.*

Native. Marshy places. Very rare. P. August to October. *Top. Bot.* 83.

First Record. In some low boggy meadows on the south side of Bredon Hill and eastward of Overbury, *Nash, Hist. Worc., Int.,* p. lxxxix (1781).

Avon. Bredon Hill, gone. Feckenham Moors, *Purton,* gone.

Severn. Stanklin Pool, *Mathews* and *Amphlett.*

Lickey. Moseley, *Miss M. A. Beilby,* gone. Bromsgrove Lickey, *Purton,* gone.

This plant is disappearing from the county. It is still to be found at Stanklin Pool, in one locality on its margin only, in some little plenty. It is a plant of northern climes, frequent in Scotland, where it is seen ornamenting the face of some dripping rock, or decorating with its conspicuous white flowers some piece of boggy ground. It gets its name of Grass, with which tribe of plants of course it is not even distantly related, because Dioscorides called it ἄγρωστις ἐν τῷ Παρνάσσῳ, ἄγρωστις being some kind of grass that mules fed upon. The Grass of Parnassus has been recorded from all neighbouring counties except Herefordshire. From Warwickshire it is disappearing, as in our county.

RIBES, Linn. 129. (An Arabic word, meaning Rhubarb.)

646. 391. R. *Grossularia, Linn.   *The Gooseberry.*

Denizen or native. Hedges, thickets, woods. Several places throughout the county. Shrub. April and May. Not numbered in *Top. Bot.*, 2nd ed., p. 174, where the author says it has been recorded in eighty counties, chiefly, if not solely, the offspring or descendants from seeds conveyed out of gardens.

First Record. A frequent straggler from gardens, and an epiphyte on old willow trees, *Lees, Illus. Nat. Hist. Worc.*, p. 157 (1834).

Although doubts are continually being cast on the nativity in Britain of this well-known plant, it is an undoubted native of most of Europe, reaching the Northern shores of the Continent; and there is nothing in the geographical range of the species to make its nativity improbable. But it is so widely cultivated in England, an inhabitant of every cottage garden, that any locality in a hedgerow is certainly suspicious, however far from any existing habitation. The presumption would be in finding the Gooseberry in such conditions, not that the plant was really wild, but that in former times some human habitation existed in the vicinity which has utterly disappeared. The first syllable of the name undoubtedly means what it says, the Goose. It has been suggested that it is a modification of 'Gorse', nearly as thorny as which shrub the Gooseberry is; or that it is derived from the French *gros*, to which hypothesis the French name *groseille* lends some justification. But there is no trace of the letter 'r' in the name of the plant throughout English literature. The grounds on which plants have received names associating them with animals are so commonly inexplicable, that want of appropriateness in any case is not a sufficient reason for assuming corruption in the etymology of the word. The Gooseberry has been observed more or less wild in appearance in all neighbouring counties.

{**647. 392. R. alpinum**, Linn.   *Alpine Currant.*

Denizen. Woods and hedges. Rare. Shrub. April and May. *Top. Bot.* 35.

First Record. At Northfield, and in hedges at Halesowen, abiding, *Lees, Bot. Worc.*, p. xxix (1867).

  Lickey. As above. Lappal, Halesowen. Near Holt Farm, Northfield. Northfield, *Dr. W. Hinds*.

These are the only records of the occurrence in the county of this plant, whose chief home is among woods in the north of England; and Mr. Lees gave his on the authority of the Rev. J. H. Thompson. The fruit is scarlet, and in racemes like that of the red currant; but the racemes are always upright both in flower and fruit. The plant has been recorded in each neighbouring county except Salop.}

648. 393. R. rubrum, Linn.   *Red Currant.*

Native or denizen. Thickets, woods, borders of streams. Local. Shrub. April and May. Mentioned, but not numbered, in *Top. Bot.*

First Record. In the deep dingle of a wood at Hailstone Hill, near Suckley; also in a ravine at Clifton-on-Teme, between that place and St. Catherine's Well, *Lees, Illus. Nat. Hist. Worc.*, p. 156 (1834).

  Avon. On a willow near Evesham. Near Tiddesley Wood.

  Severn. Laughern Brook. Orl's Coppice, Rushwick; Hurcott Wood, *Lees*. Northwood, Bewdley, *Mathews*. Dunclent. Stourport. Vallambrosa. Boughton. On a pollarded willow near Teme's mouth. Shrawley Wood. Sandford. Broadwas.

  Malvern. Brook-side at Madresfield, *Lees*.

  Lickey. Halesowen Manor, *Mathews*.

Our English name for this well-known garden fruit doubtless originates from the currant of commerce, the dried seedless grape of the Levant. The White Currant is a variety of this plant. Both Black and Red Currants were introduced into English cultivation some time before 1578, when they are called by Lyte the Black and Red 'Beyond the sea Gooseberry'; and one of the names for the currant in France is *Groseille d'outre-mer*. They were at first considered to be, in a degenerate kind of way, the plant that produces the Levantine currant; but both Gerarde and Parkinson protested against calling them 'currants'. It is clear that the cultivation of this fruit did not originate from any British native plant, which is some reason for the pretty general exclusion by botanists of a plant, found growing in England amid natural surroundings independently of cultivation, from the British Flora. The Red Currant is reported from all neighbouring counties.

649. 394. R. nigrum, Linn.   *Black Currant.*

Native or denizen. Hedges, stream-sides, thickets. Local. Shrub. April and May. Mentioned in *Top. Bot.*, but no number given.

First Record. On the banks of the Severn, in several places, *Lees, Illus. Nat. Hist. Worc.*, p. 157 (1834).

  Avon. Norton, near Evesham. Harvington. Defford Brook.

  Severn. Stream by Titton, Stourport; Hartlebury, *Lees*. Blakedown Pools, *Mathews*. Harberrow Pools. Laughern Brook. Severn side, near Camp Weir. Claines. Grimley.

  Malvern. Near the Gullet; banks of the Teme, *Lees*.

  Lickey. Yardley Bridge, *Freeman*. Manor Farm, Halesowen, *Mathews*.

This is a common plant in the woods of Russia and Siberia, where a wine is made of the berries; in England jelly or lozenges made of them are considered valuable for affections of the throat. The leaves have a strong odour, much disliked by many persons, but equally pleasing to others. The plant is reported from all bordering counties.

R. SANGUINEUM, Linn.   *Flowering Currant.*

This well-known occupant of shrubberies and gardens, which has no claim to be included in any British Flora, was reported by Mr. Lees, *Bot. Worc.*, p. xxix, as 'appearing on newly-built thick syenitic walls at Malvern and in hedges near Bromsgrove'.

L

28. CRASSULACEAE, DC.

COTYLEDON, Linn. 131. (κοτύλη, a cup, from the shape of the leaves.)

651. 395. C. Umbilicus-Veneris, Linn.   *Pennywort.*

Denizen or native. Old walls, hedgebanks, and rocks. Local. P. June to August. *Top. Bot.* 54.

First Records. Malvern Hill, *Ballard*. In the clefts of rocks above Great Malvern, *Stokes, Stokes's With. Bot. Arr.*, 2nd ed., p. 464 (1787).

  Severn. Pecket Rock, Habberley Valley, *Lees*. Cliffs near Cookley Wood, *Lees*. Fox-holes, near Kidderminster, *Perry*. Wyre Forest. Wolverley. Hartlebury village.

  Malvern. Malvern Hills, *Lees*. Rosebury Rock, *Lees*. Wall of Bockleton churchyard. Pickersleigh.

  Lickey. Near Clent Church, *Scott*, gone.

This very succulent plant is quite common in the West of England, but there are many districts where it is scarcely ever seen. Doubts even have been cast on its true nativity in Britain. It is recorded from all neighbouring counties.

SEDUM, Linn. 132. (*Sedeo*, I sit, from the squatting habit of the species.)

653. 396. S. Telephium, Linn.   *Live Long, Orpine.*

Native. Dry woods, hedgebanks. Several places but not common. P. June to September. *Top. Bot.* 75.

First Record. Fields about Robinson's End, Malvern Chase, *Ballard, Stokes's With. Bot. Arr.*, 2nd ed., p. 465 (1787).

  Avon. South Littleton, *W. N. C. Trans.* iii. 237.

  Severn. Wyre Forest, *Jorden*. Shrawley Wood. Laughern Brook. Hurcott, Kidderminster, *Lees*. Wet meadow near Bubble Bridge, *Lees*. Blackstone Rock. Redstone Rock.

  Malvern. Rocks on the North Hill, *Lees*. Ivy Scar Rock, Malvern.

  Lickey. Lickey Hills, *Scott*. Tardebigge Reservoir, *Humphreys*.

This plant has a very succulent stem, terminated with clusters of showy red purple flowers. It gets its name of Live Long because it keeps fresh for a long time after it is gathered, and will even continue to grow if it is hung up in a room. Orpine as its name has a longer history. Orpine is another English name for Orpiment or yellow arsenic, used as a pigment under the appellation of King's Yellow. It is conjectured that Orpine, therefore, was first given to some of the yellow stone-crops on account of their colour, and then extended to the race of stone-crops as a whole, afterwards becoming restricted to this plant, in spite of its flowers not being yellow. It was a favourite flower with our predecessors, and is named in all the accounts given of the practices of Midsummer Eve, one of its old names being Midsummer-men. It was considered also a love-charm. Girls used to hang up a piece in their rooms on Midsummer Eve, as the bending of the leaves to right or left was supposed to indicate the constancy or faithlessness of the object of their affections. The plant is recorded from all bordering counties.

In the *Lond. Cat.*, 9th ed., a var. *purpureum*, Linn. (S. *purpurascens*, Koch) was listed; this was recorded by Mr. Lees as growing in a field close to Laughern

Brook, Bubble Bridge, Worcester, *Bot. Malv. Hills*, 2nd ed. (1852). This variety has been noticed also in Salop. Its leaves are larger than those of the type, and often orbicular. It is not given in *Lond. Cat.*, 10th ed.

654. 397. S. purpureum, Tausch.
(*S. Telephium*, var. *purpureum*, Linn. *S. Fabaria*, Koch.)

First Record. In a field close to Laughern Brook, Bubble Bridge, *Lees, Bot. Malv. Hills*, 2nd ed. (1852).

This identification must be received with caution. Neither *S. purpureum*, Tausch. or Linn., is mentioned in *Index Kewensis*, but in *Lond. Cat.*, 9th ed., a var. *purpureum*, Linn., is given. The var. *purpureum*, Linn., has been observed in Salop.

656. 398. S. album, Linn.   *White Stone-crop.*

Alien. Roofs of buildings. Rare. P. June to August. *Top. Bot.* 2.

First Record. On the rocks by the side of Malvern Hill above Great Malvern town, *Nash, Hist. Worc., Int.*, p. lxxix.

  Avon. Evesham, *Lees*. Stone wall, Bredon Hill.

  Severn. Old roof at Churchill (Kidderminster), *Mathews*. Wall at Bewdley, *Mathews*. Habberley.

  Malvern. Roofs at Shelsley. Rocks of the North Hill, *Lees*, seldom flowering. Ivy Scar Rock.

  Lickey. The Leasowes, Halesowen.

Although this is a native plant up to the northern coast of the continent of Europe suspicions are cast at its true nativity in our island. It is not uncommonly seen, however, as an escaped garden plant on a neighbouring wall or roof. It has been noticed in all neighbouring counties.

Var. teretifolium (Haw.).

First Record. A little dwarf Sedum with red, oblong, fleshy leaves on the rocks of the North Hill, Great Malvern, *Lees, Bot. Worc.*, p. 65 (1867).

  Malvern. As in first record.

S. teretifolium, Haworth, is equated by Hooker with *S. album* proper. Mr. Lees submitted his specimens to Mr. Borrer, who stated they were *S. teretifolium*, and Mr. Lees remarks, 'whether that is more than a var. of *S. album* I am unable to say.' Mr. Mathews does not mention this plant. It is omitted as a variety in the *Lond. Cat.*, 10th ed. It has not occurred in any of our bordering counties.

658. 399. S. *dasyphyllum, Linn.   *Round-leaved Stone-crop.*

Denizen. Old walls and about villages. Not common. P. June to August. Not in *Top. Bot.*

First Record. Badsey, near Evesham, *Rufford, Purton, Midl. Fl.*, p. 218 (1817).

  Avon. Badsey. South Littleton.

  Severn. Wyre Forest. Wribbenhall. Blackstone Rock. Kidderminster Road, Bewdley, *Mathews*, perhaps Wribbenhall above. Northwood, near Bewdley, *Mathews*. Habberley Valley. Arley.

  Malvern. Old Storridge, *J. S. Haywood*. Wall near Warner's Farm, Leigh. Millham Farm, Alfrick. Wall near the Chapel, Langley Green. Quarry at Suckley. Leigh Sinton.

This is a small plant, having very fleshy pale green dotted leaves, tinged with red. It is a native of Southern Europe, and into more northern parts it is only introduced. Of adjoining counties it does not appear to have been observed in Staffordshire or Warwickshire.

660. 400. S. acre, Linn. *Wall Pepper, Biting Stone-crop.*
Native. Dry heathy places, wall-tops. Locally common. P. May to July. *Top. Bot.* 108.
First Record. Roofs and walls, *Pitt, Agric. Worc.*, p. 317 (1810).
This plant is very frequently to be seen on the roofs of cottages, pigsties, and farm buildings, less frequently in its native habitats of dry slopes and sandy heaths. Country people call it, besides Wall Pepper, a name it has gained from its acrid juice, Prick Madam and Gold Chain. The herbalists used it, both as an outward application and boiled in beer, for fevers. It stood pre-eminent as a cure for ague. The plant is recorded from all neighbouring counties.

662. 401. S. *reflexum, Linn. *Crooked Yellow Stone-crop.*
Alien. Old walls and dry banks. Several places. P. June to August. Not in *Top. Bot.*
First Record. Malvern Hill, *Ballard, Stokes's With. Bot. Arr.*, 2nd ed., p. 466 (1787).
Severn. Sandstone rocks at Shrawley, *Lees.* Severn Stoke Hill on the Upton side. Bewdley.
Malvern. On Little Malvern Church, *Lees.* On the Abbey Church, Malvern, *Lees.*
This is another garden introduction, which has been recorded from all neighbouring counties.
Var. b. albescens, Haw.
First Record. In this book, as below.
Severn. Bewdley.
This variety is characterized by its lighter yellow flowers, glaucous not reflexed leaves, and its smaller size. The plant does not appear to have been observed in any bordering county except Hereford.

663. 402. S. rupestre, Linn. *Rock Stone-crop.*
Casual. Old walls. Various places. P. June and July. *Top. Bot.* 13.
First Record. Planted on walls in numerous spots, *Lees, Bot. Worc., Int.*, p. xxix (1867).
Mr. Lees gives two localities for this plant, one certainly in Herefordshire, but says that in both cases it was either planted or had escaped from gardens. It has been observed in all adjoining counties.

SEMPERVIVUM, Linn. 133*. (*Semper*, always, τίνο, I live, because the plant is always green despite any drought.)

665. 403. S. tectorum, Linn. *Common House-leek.*
Alien. Old walls, cottage roofs. Scattered throughout the county, but with no claim to be considered a native plant. P. June to August. Not in *Top. Bot.*
First Record. Planted generally about cottages on roofs, *Lees, Bot. Worc., Int.*, p. xxix (1867).

The superstition which caused the House-leek to be extensively planted on buildings has died out in England, and perhaps on the Continent also. For it was deemed a preservative against thunder; in Holland the old name of the flower was the Dutch equivalent of Thunder-bloom. It had its uses in the rural pharmacy. Boiled in milk it quenched thirst in fevers; mixed with honey it allayed inflammation of the throat. The leaves are slightly acid, and will allay the irritation of nettle- or bee-stings; and bound round the forehead 'ease the headache, and distempered heat of the brain in frenzies, or through want of sleep'. It has been observed in all neighbouring counties.

{S. montanum, Linn.
Alien. Old walls. P. June to August.
First Record. As in this book.
Severn. Frog Lane, Worcester, *Lees, Bot. Worc.*, p. xxix (1867).
Mr. Lees says, 'established for half a century or more on an old wall in Frog Lane, but unknown when placed there'.}

29. DROSERACEAE, DC.

DROSERA, Linn. 134. (δρόσος, dew, because the red hairs exude drops of a viscid fluid, especially when the sun is shining.)

666. 404. D. rotundifolia, Linn. *Round-leaved Sundew.*
Native. Bogs and wet heathy ground. Local and rare. P. May to September. *Top. Bot.* 109.
First Record. Malvern Chase, on the side of the rivulet flowing from the Spa, *Ballard, Stokes's With. Bot. Arr.*, 2nd ed., p. 331 (1787).
Severn. Hartlebury Common, plentiful. Falling Sands Common, *Perry.* Devil's Spittleful, near Bewdley, *Perry.* Pedmore Common, *Scott.* Harberrow Pool, *Scott.* Wyre Forest, *Jorden.* Oldington.
Malvern. West side of Malvern Hill, *Purton.*
Lickey. Bromsgrove Lickey, *Purton.* Bogs on Moseley Common, *Miss M. A. Beilby.* Near Alvechurch, *Mr. D. Mathews.* Bog north of Old Rose and Crown, Lower Lickey.
This is one of the plants which are rapidly vanishing from the face of our county. From most of the localities mentioned above it is gone, and at the one locality, Hartlebury Common, where it is still plentiful, in one place it is perilously near—within a few yards of, in fact—mean new red-brick houses; and in another, yet in good quantity, between a road and the hedge of the adjoining cultivated land. It was at Bromsgrove Lickey till after 1870; at Falling Sands, now comprised in Oldington Wood, till 1885; and at the same date it was gathered near the Devil's Spittleful. Moseley is a waste of houses, and nothing has been seen of the plant at Pedmore or Harberrow since Scott's time. Soon, it is to be feared, Worcestershire will know it no more. It is an interesting plant, covered with viscid red-stalked glands, which, when some small insect adheres to them, close in upon the unfortunate victim, while the margins of the leaf curl over it. The plant was unnoticed by the ancients, but in the Latin of the Middle Ages it is called by the translation of the common name, *Ros Solis*; and translations of the same name denote the plant in most European countries, but the

French call it *Herbe aux goutteux*, much more unromantically associating it with the gout, though it is possible to consider this a variation of *herbe de la goutte* and connect the plant with a drop of dew. A celebrated decoction made from it was highly praised by old writers as a remedy for convulsions, and even for the plague. In those places where the plant is plentiful on pastures it is called Red Rot, because it is supposed to deleteriously affect sheep; possibly the damp nature of the soil where alone it could grow in plenty was really responsible. The flowers are not often seen expanded, because as soon as pollination is effected they close up. The flower-stems of the plant rise from three to six inches from the centre of its rosette of battle-dore-shaped leaves, and go black in drying, staining the paper in which they are placed a reddish purple. The plant is recorded from all neighbouring counties.

30. HALORAGACEAE, R. Br.

HIPPURIS, Linn. 135. (ἵππος, a horse, οὐρά, a tail, from the shape of the plant.)

669. 405. H. vulgaris, Linn. *Mare's-tail.*
Native. Slow streams, ponds, ditches. Very local. P. May to July. *Top. Bot.* 90.
First Record. Rare, Clifton-upon-Severn, *Dr. Streeten, Lees' Illus. Nat. Hist. Worc.*, p. 149 (1834).
Severn. Uncless, near Bewdley, *Jorden.* Pirton Pool.
Malvern. Longdon, *Lees.*
This plant could be mistaken for nothing but a Horse-tail, standing up as it does a foot or so above the water. It has flowers of the simplest character, with one stamen, one style, and the fruit with one cell and one seed. It is recorded from all adjoining counties.

MYRIOPHYLLUM, Linn. 136. (μυρίος, a thousand, φύλλον, leaf, from the finely divided leaves.)

670. 406. M. verticillatum, Linn. *Water Milfoil.*
Native. Streams and ponds. Local. P. June to August. *Top. Bot.* 49.
First Record. Not localized, *Lees' Cat., New Bot. Guide* (1835).
Severn. Northwick Pool, Claines, *Lees* (gone; now drained). Northwick Brickyard, Severn side. Below the Copse beyond the Ketch.
Malvern. In the Teme, *Lees.* Teme, near Powick Bridge. Ditch, Upton-on-Severn. Longdon Marsh.
The plant occurs in all the bordering counties.
Var. b. pectinatum, DC.
First Record. As in this book.
Severn. Near Worcester, *Westcombe.* Northwick Brickyard, Severn side.
Although the books say of this that the floral leaves are hardly longer than the flowers, the variety would at once strike the most casual observer.

671. 407. M. spicatum, Linn. *Spiked Water Milfoil.*
Native. Ditches, ponds, streams. Locally common. P. June to September. *Top. Bot.* 80.
First Record. (In the) Rea, near Vaughton's Hole (possibly in Worcestershire), *Miss M. A. Beilby, Analyst*, vol. vi, p. 294 (1837).
Severn. Canal at Worcester, *Lees.* Stanklin Pool, *Humphreys.* Droitwich Canal at Hawford. Northwick Brickyard, Severn side. Spetchley Pool. Pool near Camp Weir. Pool, Ombersley Park.
Malvern. Various pools and marshy spots, *Lees.* Near Clifton-on-Teme. Leigh. Ponds, roadside between Powick and Malvern.
Lickey. Blakedown Pools, *Mathews.* Tardebigge Reservoir, *D. Mathews.* Hewell.
This plant is more frequently met with than the previous one, and forms tangled masses with its slender stems and branches, from which its spikes of tiny flowers rise a few inches above the water. It differs little from the last plant, but its flowers form a leafless spike, and the bracts are smaller than with that. The plant has been observed in all neighbouring counties.

672. 408. M. alterniflorum, DC. *Water Milfoil.*
Native. Ditches and ponds. Rare. P. July. *Top. Bot.* 80.
First Record. In little pools on Welland Common, *Lees, Bot. Malv. Hills*, 2nd ed., p. 73 (1852).
Malvern. Barnard's Green; Welland Common, *Lees.*
On the spike of this plant the sterile flowers are alternate, and the spike nods when in bud, afterwards becoming erect. To the ordinary eye all three Water Milfoils are very much alike. This one has been observed in all neighbouring counties.

CALLITRICHE, Linn. 137. (καλός, beautiful, θρίξ, hair, from the beauty of its capillary ramification.)

673. 409. C. palustris, Linn. *Water Starwort.*
(*C. verna*, Linn. (1753). *C. vernalis*, Koch.)
Native. Streams, ponds. General (*Lees*). A. or P. April to September. *Top. Bot.* 19?.
First Record. Not localized, *Lees, Bot. Malv. Hills*, 1st ed., p. 40 (1843).
Avon. Tiddesley Wood. Bredon Hill. The Slads.
Severn. Ponds and ditches, Ombersley road. Cotheridge. Northwick. Perry Wood. Nunnery Wood. Monk Wood. Ockeridge.
Malvern. Middleyards Coppice. Martley. Abberley.
Lickey. Campion's Pool, Wildmoor, *Humphreys.* Spadesbourne Brook.
The above are aggregate records. The plant is to be found in most of our streams and ditches. Mathews considers Mr. Lees' record to refer to the following plant. Why, he does not say. This plant appears to have been recorded for all bordering counties.

674. 410. C. stagnalis, Scop.
(*C. platycarpa*, Kuetz.)
Native. Ditches, ponds, wet places. General (*Lees*, one locality in the Severn district). P. April to August. *Top. Bot.* 93.

First Record. As *C. platycarpa*, Northwick, *Westcombe*, *Lees*, *Bot. Worc.*, p. 52 (1867).
Avon. Tiddesley Wood. The Slads. Trench Woods. Pirton. Pitcher Oak Wood, Redditch. Crowle. Fladbury.
Severn. Blakedown, Stanklin, Stone and Hurcott Pools, *Mathews.* Churchill, near Kidderminster. Pond, Hawford. Claines. Kempsey. Ombersley. Near Crown East Wood. Near the Ketch. Wyre Forest, Grimley Brickyard.
Malvern. Middleyards. Fries Wood. Martley. Abberley. Powick.
Lickey. Broadmarsh Farm, Hagley. Short Wood, Clent. Bromsgrove Lickey.

The Starwort is well named, for its leaves form a starry rosette just floating on the water, or submersed, or lying on the mud at the margin of a pool; while from the joints of the stem a number of white hair-like roots descend. This plant is recorded from all neighbouring counties.

676. 411. C. intermedia, Hoffm.
(*C. hamulata*, Kuetz. *C. pedunculata*, DC.)
Native. Streams, ponds, and moist places. Not common. A. or P. April to September. *Top. Bot.* 56.
First certain record. As in this book.
Avon. Tiddesley Wood. Trench Wood. Bow Wood.
Severn. Nunnery Wood. Shrawley. Wyre Forest. Ockeridge. Bewdley.
Malvern. Upton-on-Severn. Bransford. Middleyards Coppice. Martley. Abberley. Malvern Link Common; Barnard's Green. Castlemorton Common, *Towndrow.*
Mr. Mathews considers a record by Mr. Lees of *C. autumnalis*, at p. 40 of the 1st ed., *Bot. Malv. Hills*, to be this plant, under the name *C. pedunculata*; because that plant and not *C. autumnalis* is given afterwards in the 2nd (p. 73) and 3rd (p. 95) editions of Mr. Lees' book. Mr. Watson, *Top. Bot.*, p. 169, says under this plant, it 'varies much according to its situations, in deep water, in shallow water, in still or running water, or in places alternately covered with water or left bare. The variations so induced are described as different species; their localities inextricably confused in book records'. Neither this nor the next plant is localized in Mr. Lees' *Table of Plants*. *C. intermedia* has been noticed in Salop, Stafford, and Warwick.

678. 412. C. autumnalis, Linn.
Native. Streams, ponds, marshes. Local. P. May to August. *Top. Bot.* 28.
First Record. In some deep ditches between Barnard's Green and the road to Madresfield, *Lees, Bot. Malv. Hills*, 1st ed., p. 40 (1843).
Mr. Towndrow is convinced that this plant does not occur either in the Malvern district or elsewhere in the county, but that the foregoing plant has been mistaken for it. Further confirmation is required before the plant can be certainly added to the county list. The distinction between the forms of *Callitriche* depend mainly on the character of the fruit, minute differences which may easily be mistaken. Hence the confusion. This plant does not appear to have been noticed in the adjoining counties.

31. LYTHRACEAE, Juss.

PEPLIS, Linn. 138. (πεπλίς, the Greek name for some plant, perhaps from πέπλος, a curtain.)

680. 413. P. Portula, Linn. *Water Purslane.*
Native. Wet places and pond margins. Local. P. July to September. *Top. Bot.* 100.
First Record. At Cookhill, by the side of a pool, *Purton, Midl. Fl.*, p. 181 (1817).
Avon. Cookhill, Inkberrow, *Purton.*
Severn. Hartlebury Common, *Mathews.* Pedmore Common, *Scott.* Broad Heath. Moseley Green, near Hallow.
Malvern. Commons near Malvern, *Lees.* Welland Common. Newland Common. Barnard's Green. Link Common.
Lickey. Bottom of Farley Wood, *Mathews.* Pool, Hagley Hall.
A lowly creeping plant which would be passed by unnoticed by the casual observer. It is recorded from all neighbouring counties.

LYTHRUM, Linn. 139. (λύθρον, blood, probably from the colours of the flowers.)

681. 414. L. Salicaria, Linn. *Spiked Purple Loosestrife.*
Native. Sides of rivers and streams, wet meadows. General, except Lickey district, in which it does not now occur. P. July to October. *Top. Bot.* 92.
First Record. Moist places, *Pitt, Agric. Worc.*, p. 317 (1810).
This is one of the handsomest of our native flowers. The blossoms form tall tapering spikes, sometimes four or five feet high, of a rich purplish red colour, rising up among the vegetation of the borders of rivers and streams. It has had many country names, Long Purples, Purple-strife, and Willow Lythrum among them. In Gerarde's *Herbal* it is called by 'the Dutch name of Grusse Poley, which name we may also very fitly retain in English', he says. The flower is common in the meadows bordering on the Avon. It is gone from the Lickey district, where formerly it occurred at Hay Mill Brook, Yardley, and is so recorded by Miss M. A. Beilby in 1837. It is recorded from all neighbouring counties.

{**682. 415. L. Hyssopifolia, Linn.** *Hyssop-leaved Purple Loosestrife.*
Colonist. Damp places. Very rare. A. June to September. *Top. Bot.* 6.
First Record. Badsey, near Evesham. Stubble fields, Bretforton, Worc., *Purton, Midl. Fl.*, p. 227 (1817).
Avon. Badsey. Bretforton.
The plant has not been seen lately in the county, and probably has gone from it. It is a humble little annual four or six inches high, with small axillary flowers, of a dull purplish lilac colour. No one would suspect its affinity with the last plant, so different is it. Some doubts have been thrown upon its true nativity in England, but it is native in wet places throughout Europe, and so well may be native in our country. It has spread rapidly in waste places far from its native range and has become

locally abundant in Australia and New Zealand. But when it has occurred in England it is usually as a grain-field weed, though near Winchester and the neighbouring chalk downs it flourishes in an apparently wild state. It does not appear to occur in any bordering county except Herefordshire.}

32. ONAGRACEAE, Juss.

EPILOBIUM, Linn. 140. (ἐπί, upon, λοβός, a pod, from the position of the flower upon the long seed-vessel.)

683. 416. E. angustifolium, Linn. *Rose Bay, Willow Herb.*
Native. Woods, bushy places, commons, railway embankments. Locally common. P. June to September. *Top. Bot.* 96.
First Record. Near Bewdley, *Mr. Dyer, Bot. Guide*, vol. ii, p. 656 (1805).
Avon. Craycombe. Tiddesley Wood.
Severn. Pedmore railway cutting, *Mathews.* Wyre Forest, *Jorden.* Croome Perry Wood, *Lees.* Bourne and Grove's wood yard, Worcester. Bridwell, Bewdley. Northwood, Bewdley. Hurcott osier swamp. Hagley. Shatterford.
Malvern. Northern border of Welland Common, *Lees.* An epiphyte near Birtsmorton Rectory, *Lees.* Old Storridge. Mathon. Malvern. Malvern Link. Leigh Sinton.
Lickey. Bromsgrove Lickey, *Mathews.* Alvechurch, *D. Mathews.* Bilberry Hill, *Humphreys.* Barnt Green. Frankley Wood. Near Bittell Reservoir. Saltwells Wood.
This is a handsome plant frequently grown in gardens, but not so frequently to be seen wild. Its seeds are furnished with a little plume, and wind carries it far; moreover the root is especially far-creeping; so that it is well adapted for spreading itself when once established. It is recorded from all neighbouring counties.
Var. brachycarpum, Leighton.
First Record. In this book.
Severn. Wyre Forest. Wribbenhall. Bourne and Grove's wood yard, Worcester.
This variety is characterized by the short capsules and the wide base of the leaves. It is not given in the *Lond. Cat.*, 10th ed.

684. 417. E. hirsutum, Linn. *Great hairy Willow Herb, Apple Codlins.*
Native. By the side of water in most of its conditions. General. P. June to September. *Top. Bot.* 96.
First Record. Not localized, *Lees, Bot. Malv. Hills*, 1st ed., p. 21 (1843).
A common country name for this well-known plant is 'Codlins-and-cream', from the odour of the leaves when bruised, codlins being the term applied in old times to apples which were only fit for cooking, not for eating raw. The plant varies considerably in the hairiness of its leaves and the pubescence of its pods. It has occurred with white flowers at Earl's Croome. It is recorded from all neighbouring counties.
×parviflorum.
First Record. Not localized, *Victoria Hist. Worc.*, i. 61 (1901).
This hybrid has been observed by Mr. Towndrow at Pickersleigh and Woodsfield, Malvern.

685. 418. E. parviflorum, Schreb. *Small flowered Willow Herb.*
Native. Not so dependent on water as the preceding species. Not uncommon. P. June to September. *Top. Bot.* 103.
First Record. Near Kingsnorton, *Ick, Analyst*, vol. vi, p. 22 (1837).
Easily distinguished from the last plant by its unbranched stem and fibrous, not creeping, root. The variety *rivulare*, Wahl., has been recorded by Mr. Towndrow in the *Malv. Advert.*, Nov. 19, 1892. The leaves of the variety are nearly glabrous, instead of hairy as in the type. It occurs in every bordering county.
×roseum.
First Record. Not localized, *Victoria Hist. Worc.*, i. 61 (1901).
This plant has been observed by Mr. Towndrow at Malvern, Pool Brook, Hanley Castle, Madresfield, and Malvern Link.

686. 419. E. montanum, Linn. *Smooth-leaved Willow Herb.*
Native. Woods, banks, ditches, and a garden weed. Everywhere. P. May to September. *Top. Bot.* 112.
First Record. Not localized, *Lees, Bot. Malv. Hills*, 1st ed., p. 21 (1843).
This everywhere meets the eye, tall and luxuriant, perhaps two feet high, among the herbage of the damp shady bank, small and stunted, and bearing perhaps a single flower, in the dry garden border; in gardens it is an annoying weed. A country name for it is Coddled Apples. One wonders why it should have obtained the specific name of *montanum*. There is nothing especially 'mountain' about it; indeed it is a nasty low-land weed. It occurs in every county, neighbouring or otherwise.
×parviflorum.
First Record. Not localized, *Victoria Hist. Worc.*, i. 61 (1901).
Malvern. Purlieu Lane; Malvern Wells; Malvern Link, *Towndrow.*
×roseum.
First Record. Not localized, *Victoria Hist. Worc.*, i. 61 (1901).
Malvern. Malvern Link, *Towndrow.*
Var. verticillatum, Koch.
First Record. In an uncultivated garden at Malvern Link, *Towndrow, Malv. Advert.*, Nov. 19 (1892).
This variety is characterized by having the leaves in threes. It is not in *Lond. Cat.*, 10th ed.

689. 420. E. roseum, Schreb.
Native. Ditches, stream-sides, garden ground. Not uncommon. P. June to September. *Top. Bot.* 46.
First Record. In a garden before a house in Church Street, Kidderminster, 1816, and in Mr. John Lea's drying ground, Mill Street, Kidderminster, 1829, *Perry, Mag. Nat. Hist.*, vol. iv, p. 450 (1831).
Avon. Evesham. Fladbury. Pershore. Defford. North Piddle.
Severn. Wyre Forest, *Gissing.* Kidderminster, as first record. Churchill, near Kidderminster. Bromyard Road near Worcester. A garden weed in Worcester. Near Laughern Brook, Bromyard Road. Claines. Ombersley. Kempsey.
Malvern. Near Barnard's Green, *Bloxam.* Bransford Bridge Mill, *Thompson.* Near the 'Devil's Oak', *Lees.* Little Malvern. Madresfield. Bransford. Malvern. Malvern Link.

By the brook-side this plant will grow more than two feet high ; in garden ground it sometimes hardly reaches six inches, and though essentially a water plant it will bear the smoke of towns well. It is recorded from all neighbouring counties.

690. 421. E. tetragonum, Curt.   *Square stalked Willow Herb.*
(*E. adnatum*, Grisebach.)
Native. Ditches and stream-sides. General, except in the Avon district. P. July to September. *Top. Bot.* 45.
First Record. As *E. tetragonum*, Foot of Malvern Hills, *Sherard, Nash, Hist. Worc., Sup.*, p. 96 (1797).
Avon. Wyre Piddle. Fladbury. Evesham. Pershore. Bredon.
Severn. Witley Court plantations, *Mathews*. Diglis. Wyre Forest. Holt Mill. Shrawley. Glasshampton.
Malvern. Bransford. Newland. Welland. Powick. Mathon, *Town-drow*. Newland. Witley. Malvern Hills in wet places.
Lickey. Occasional, *Mathews*. Clent. Rowney Green. Marsh at Alvechurch. Frankley Beeches.
This plant was given as a new record by Mr. Towndrow, as *E. eu-tetragonum*, in *Bot. Rec. Club Rep.*, pub. 1887 ; but it was not new except as marking the division of the aggregate. The plant is distinguished from the preceding one by the more distinct angles of the stem and its narrower leaves which are without stalks. *E. adnatum*, as a segregate of *E. tetragonum*, does not appear to have been observed in Salop.
× **Lamyi.**
First Record. Bransford, *Malv. Advert.*, Nov. 19, 1892, *Towndrow*.
× **montanum.**
First Record. As above plant.
× **parviflorum.**
First Record. As for the above two plants.

691. 422. E. obscurum, Schreb.
(*E. tetragonum*, Linn. *E. virgatum*, Gren. and Godr.).
Native. Ditches and damp places. Probably common, but not recorded. P. June to September. *Top. Bot.* 100.
First Record. As *E. virgatum*, Mr. *Westcombe, Lees, Bot. Malv. Hills*, 2nd ed., p. 44 (1852).
This is the more common form of the two segregates into which Linnaeus's plant has been divided ; it differs from *E. tetragonum* by producing stolons in summer with pairs of opposite leaves, while that plant produces stolons in autumn with leaves in a ' rosette ' ; the leaves of this, also, are not shining above as those of that are. *E. obscurum* is reported from all neighbouring counties.
× **parviflorum.**
First Record. Not localized, *Victoria Hist. Worc.*, i. 61 (1901).
This hybrid has been observed by Mr. Towndrow in the vicinity of Malvern, at Malvern Link, and Old Storridge.

692. 423. E. Lamyi, F. Schultz.
Native. Woods and roadsides. Rare. P. July and August. *Top. Bot.* 9.
First Record. Field at Malvern, *Towndrow, J. of Bot.*, vol. xxiii, p. 349 (1885).

Malvern. Bransford ; Newland ; Malvern Link.
This plant is nearly allied to the preceding two species, differing chiefly in the glaucous lanceolate acuter, less strongly dentate, leaves, close pubescence, and its larger flowers. This species is entirely ignored in Hooker's *Stud. Fl.*, 3rd ed., 1884, nor does Syme acknowledge it. It appears to have been observed in Herefordshire only of bordering counties.

693. 424. E. palustre, Linn.   *Marsh Willow Herb.*
Native. Marshes, boggy ground, wet places. General. B. June to September. *Top. Bot.* 110.
First Record. Feckenham Bog, *Purton, Midl. Fl.*, p. 190 (1817).
Avon. As first record.
Severn. Stanklin. Churchill, near Kidderminster.
Malvern. Mathon ; Herefordshire Beacon (Worcestershire portion), *Towndrow*.
Lickey. Rowney Green Marsh. Halesowen.
This plant sends out stolons in summer which run underground and end in autumn in scaly buds which become detached. The top of the raceme of flowers is usually nodding. It occurs in all neighbouring counties.
× **parviflorum.**
First Record. As in this book.
Malvern. Mathon, *Towndrow.*

OENOTHERA, Linn. 142*.   (οἶνος, wine, θήρα, a search, it is said because the root searches out the flavour of wine.)

697. 425. Oe. *biennis, Linn.   *Evening Primrose.*
Alien. Railway banks, waste places. Not common. B. June to August. Not in *Top. Bot.*
First Record. In Worcestershire ; *Rev. Mr. Bourne, Withering*, 4th ed., p. 361 (1801).
Avon. Banks of the Arrow (*Purton*).
Severn. Teme side at Powick. Bank of the river at Holt. Droitwich Sewage Farm. Railway Tunnel, Worcester.
This plant is common in North America on river-banks and lake margins, and common in England in gardens. It has long been in the habit of escaping from such, in some counties more frequently and more successfully in the matter of establishing itself than in ours. Evening Primroses were originally given the name *Onagra*, from ὄνος, an ass ; but this word was changed by putting another letter into it and making it οἶνος, and giving it a name equivalent to ' wine trap ', it is said because the roots were formerly eaten as olives are in these days. The tubers of the plant can be used as a substitute for potatoes, if desired. The plant does not appear to have been observed outside the garden fence in Staffordshire or Herefordshire.

CIRCAEA, Linn. 143.   (*Circe*, the enchantress.)

701. 426. C. lutetiana, Linn.   *Enchanter's Nightshade.*
Native. Moist shady woods and banks. General, except in the Avon district. P. June to August. *Top. Bot.* 103.
First Record. South part of Perry Wood, *Lees, A. Florence, Strang. Guide* (1828).

This is a widely distributed plant, but to find it one must seek among the vegetation of the damp wood, or on the lane-side bank where it is deeply shaded by trees. It is frequent in the shrubberies of gardens, where it is a troublesome weed, without anything in its appearance to recommend it. It is supposed that its preference for dark and gloomy spots have been the reason of its name. The German name is the equivalent of Witch-wort ; in the vicinity of Stoulton it is called Dead Man's Grapes. It is recorded from all neighbouring counties.
Var. **intermedia**, *Lond. Cat.*, 9th ed.
First Record. As below.
Severn. Coleridge Wood, Wolverley, *Worc. Nat. Club Trans.* iii. 273.
This is not given by name, *Lond. Cat.*, 10th ed.

33. CUCURBITACEAE, Juss.

BRYONIA, Linn. 144.   (βρύω, I grow quickly.)

703. 427. B. dioica, Jacq.   *Red-berried Bryony.*
Native. Hedges, wood-borders, thickets. Generally distributed, but not so common in Malvern district as elsewhere. P. May to July. *Top. Bot.* 59.
First Record. Hedges on sandy or gravelly soil in the north of the county, *Pitt, Agric. Worc.*, p. 317 (1810).
Although another plant of our hedgerows disputes with this the name Bryony, this is the true and original of the appellation. It is the only native member that Britain can claim of the great tribe of plants which provides cucumbers and melons, and medicines and drugs for suffering humanity. The blossoms are inconspicuous, but the bunches of red berries in the autumn, hung up on its withering stems among the branches of the hedgerow, are often very ornamental ; but they are very poisonous. The root is large, full of nauseous milky juice, and supposed to be a remedy, used externally, for the black eye of the pugnacious. It has been largely used in country medicine, but it is a dangerous plant to meddle with in this way. It is sometimes called Wood-vine, Wild Bryony, Snakes' Food, and Tetterwort ; and in North Worcestershire, Mandrake. It is rare in Scotland. It is recorded from every neighbouring county.

34. UMBELLIFERAE, Juss.

HYDROCOTYLE, Linn. 145.   (ὕδωρ, water, κοτύλη, a cup, from the shape of the peltate leaf.)

704. 428. H. vulgaris, Linn.   *White Rot, Marsh Rot, Pennywort.*
Native. Bogs, marshes. Rather uncommon, except in the Lickey district. P. May to August. *Top. Bot.* 110.
First Record. Feckenham Bog. Astwood Common, *Purton, Midl. Fl.*, p. 153 (1817).
Avon. Pirton Pool. The Slads. Near Feckenham.
Severn. Harberrow Pool, *Mathews*. Stanklin Pool. Pedmore Common, *Scott*. Hartlebury Common. Sutton Common. Wyre Forest. Moseley, near Hallow.

Malvern. Malvern Hills. Castlemorton Common. West side of Worcestershire Beacon. Leigh. Malvern Common.
Lickey. Pool near Independent College at Moseley, 1858, *Mathews*. Bromsgrove Lickey, *Humphreys*. Alvechurch, *D. Mathews*. Moseley Wake Green, *Miss M. A. Beilby*.
This plant is better known by its leaves than its flowers to those whose interest leads them to investigate marshy and wet places ; other people would probably be unaware of its existence. It got its name of White Rot because it was considered prejudicial to sheep, but probably with no more cause than in the case of the Sundew. If sheep found themselves on ground where either of these plants flourish many things would prove more deleterious than they are. But country people are ever ready to find an easy cause for every effect. The plant is in all neighbouring counties.

ERYNGIUM, Linn. 146.   (ἠρύγγιον of Dioscorides.)
[E. campestre, Linn.   *Field Eryngium.*
Worcestershire has no claim to this exceedingly rare plant, whose number in *Top. Bot.* is 7. But it grew tantalizingly near to it, just over the hedge in fact, in a Herefordshire field bordering the county. It was gathered in 1867 by the Rev. Phipps Onslow between 'Tedstone Delamere and Upper Sapey. See *Lees' Bot. Worc., Add. and Cor.* (1867).]

ASTRANTIA, Linn. 147.   (ἀστήρ, a star, from the star-like umbels.)
{A. *MAJOR, Linn.
This plant is mentioned by Mr. Lees in his *Bot. Worc.*, p. xxix, as having been found ' between Whitborne and Malvern ', he says, ' by Mr. Babington from the late Mr. Borrer's information ; but the spot is unknown to any Worcestershire botanist '. It never had any claim to be more than alien ; perhaps if the locality were near to Whitborne it never possessed even this as far as our county is concerned. It is a plant indigenous in Central Europe : in this country it is a garden escape.}

SANICULA, Linn. 148.   (*Sano*, I make whole, from its supposed healing qualities.)
708. 429. S. europaea, Linn.   *Wood Sanicle.*
Native. Woods, thickets, and bushy places. General throughout the county. P. April to August. *Top. Bot.* 109.
First Record. Cradley and other woods, *Scott, Hist. Stour.*, p. 540 (1832).
The flowers of this plant, to be met with in nearly every wood, are not so much an umbel, such as is characteristic of the tribe of plants, as a panicle, being irregularly placed round the top of the stem. It had many healing virtues as its name indicates, which in the various languages is current through Europe. It is found in all neighbouring counties.

CONIUM, Linn. 150.   (κώνειον of Theophrastus, said to be from κῶνος, a top, from the giddiness caused by the juice.)
710. 430. C. maculatum, Linn.   *Common Hemlock.*
Native. Coppices, hedges, river-banks. General, except in the Lickey district where it is uncommon. P. June to August. *Top. Bot.* 104.

First Record. (In) hedges, *Pitt, Agric. Worc.*, p. 317 (1810).

Avon. East side of Trench Woods, 1855. Near Tiddesley Wood. Norton, near Evesham. Fladbury. Pershore.

Severn. Harberrow; Hagley, *Mathews.* Caunsall Bridge, Wolverley. South of Kidderminster. Droitwich. Stourport. Portfields Farm, near Worcester. Hindlip. Battenhall. Ombersley. Claines. Kempsey. Stourport. Bewdley. Severn Stoke. Grafton Manor. Upton Warren.

Malvern. Rampant on hedgebanks, *Lees.* Eight feet high at 'the Bower', Old Storridge, *Lees.* Bransford. Powick. Knightsford Bridge. Stanford. Martley. Madresfield.

Lickey. Lower Clent, *Mathews.* Near Bromsgrove. Gallows Brook, Hagley. I have never seen a plant in the Lickey district, *Humphreys.*

This plant cannot be mistaken. Its spotted stem and its elegant though dull green foliage mark it anywhere. Its dead stalks share with those of many other umbels the name of Kex, really the singular number of Kecksies, and often transferred from the dead stalks of such plants to the living state of most tall growing umbellifers. The deadly nature of this plant is a matter of tradition from the time of Socrates downwards. It has been doubted if the κώνειον of his fatal draught was really made from this plant, as its recorded effects were different; but Hemlock is very plentiful in Greece, while other poisonous plants of the same tribe are much more scarce. In English literature the first mention of its poisonous properties appears in the works of the herbalists of the sixteenth century; it was too deadly for them to play with! It is called also Wild Parsley, and by children Poor Man's Plaything. It occurs in all neighbouring counties.

SMYRNIUM, Linn. 151. (σμύρνα, myrrh.)

{711. 431. S. Olusatrum, Linn.   *Alexanders.*

Denizen. Hedgebanks and thickets. Very local. B. April to May. *Top. Bot.* 63.

First Record. Between Great Comberton and Woollershill, under the hedges of some enclosures near the Avon's side, *Nash, Hist. Worc., Int.*, p. lxxxix (1781).

Avon. As first record. Pirton, plentifully, *Mr. Hollefear, Stokes.* Ditches about Badsey, *Purton.* About Pershore, *Illus. Nat. Hist. Worc.*

Severn. Hill Croome, *Rev. Mr. Welles, Ballard.*

Malvern. Scarce on the red marl, east of the hills, *Lees.*

This plant grows abundantly in the wood by the Severn north of the Mythe Tout, a mile north of Tewkesbury, but in Gloucestershire, though, as Mr. Lees would say, 'locally' in Worcestershire, in company with *Isatis tinctoria.* The English name Alexanders is of very old standing, and has been considered a corruption of *olus atrum*, the black pot-herb, 'as if one would say *olusatres*,' but this derivation is doubtful. In former times the plant was cultivated and eaten like celery, which was not introduced till the seventeenth century, but by which it was quickly displaced. Being a vegetable of fairly pleasant taste, it was not seized upon by the herbalists for medical compounds. It has been observed in Warwickshire, but only as a casual.}

BUPLEURUM, Linn. 152. (βοῦς, an ox, πλευρόν, a rib, from the ribbed leaves of some species.)

712. 432. B. *rotundifolium, Linn.   *Common Hare's Ear.*

Colonist. Cornfields. Local and rare. A. June to August. *Top. Bot.* 39.

First Record. Badsey, Bretforton, *Purton, Midl. Fl.*, p. 148 (1817).

Avon. Craycombe, *Lees.* Shemington Hill, Alderminster, *Cheshire.* Trench Woods. Sheriff's Lench. Norton, near Evesham. Badsey. Near Chadbury Mill.

Severn. Calcareous field, Tibberton, *Lees.* Churchill, Worcester, *Lees.* Bredicot. Barbourne, Worcester.

Malvern. Field at the top of Folly Copse, Alfrick, *Miss Moseley* (1849). Rare in the district, *Lees.*

Lickey. In the district, *Mathews.* Cornfield, Frankley, *Mathews* (1871). Between Oxwood Cottage and Twylands, *Mathews.*

The precise locality of Mr. Lees' 'calcareous field', whence are recorded many good plants, is quite unknown to Worcestershire botanists of to-day. This is a singular looking plant with perfoliate leaves, and its partial involucres are far more conspicuous than the greenish-yellow flowers that nestle among them. An old name for the plant is 'Thorough-wax', said to be given to it because it 'waxes', or grows, 'thorough', or through, the leaves. It is called Hare's-ear, says Gerarde, in his *Herbal*, because of its 'hauing in the middle of the leafe some hollownesse resembling the same'. The plant was formerly much used as a vulnerary. It does not appear to have been observed in Salop of the bordering counties.

{713. 433. B. opacum, Lange.

(*B. aristatum*, auct. angl. *B. odontites*, Sm., not of Linn.)

Colonist. Cornfield weed. Very rare. A. July. *Top. Bot.* 2.

First (and only) Record. Mr. (G. W.) Sandys also stated to me that he had gathered *Bupleurum odontites* at Craycombe, but this was as far back as 1834, and some later confirmation is desirable, *Lees, Bot. Worc.*, p. 94 (1867).

This is a native of the continent of Europe up to the north coast of France, and in the Mediterranean region is a cornfield weed. The only two counties of England for which it is recorded are Devon and Sussex.}

714. 434. B. tenuissimum, Linn.

Colonist. Pastures. Very rare. A. August and September. *Top. Bot.* 24.

First Record. Found by Mr. Addison of Malvern on the common between Malvern and the Rhydd, *Lees, Illus. Nat. Hist. Worc.*, p. 159 (1834).

Avon. Third mile from Evesham on the Worcester Road, *May's Evesham.* Defford Common.

Severn. Nunnery Farm, Worcester, *Lees.*

Malvern. Welland Common, *Lees.* Malvern Link Common. Roadside near the Turnpike beyond Newland Common from Worcester. Old Hills. Roadside near the Rhydd.

Mr. Addison's locality is more closely indicated by Mr. Lees in his *Bot. Malv. Hills*, 1st ed., p. 19 as being the common by the roadside just beyond Garford Court, Barnard's Green. It is really a plant of the salt marshes on the

M

---

south and east coasts of England. It has been recorded in Gloucestershire only of the bordering counties.

B. FRUTICOSUM, Linn.

Alien. Waste places. Very rare. A shrub. P. July.

First Record. As below.

Malvern. Near the Wind's Point, *Rea.*

The occurrence of this species, which is a native of Spain, is due to the fact that it had escaped from the neighbouring garden of Mrs. Otto Goldschmidt, *née* Jenny Lind. It is not mentioned in *Lond. Cat.*, 10th ed.

APIUM, Linn. 154. (*apon*, Celtic for water; *ap, ab*, or *av* meaning water.)

717. 435. A. graveolens, Linn.   *Wild Celery, or Smallage.*

Native. Marshy meadows, water-sides. Local. P. June to September. *Top. Bot.* 58.

First Record. Moors, Sansom Fields, Worcester, *Stokes, Stokes's With. Bot. Arr.*, and ed., p. 326 (1787).

Avon. Common in ditches, *Lees.* Bretforton, *Purton.* Upton Snodsbury, *Purton.* Bredon's Norton. Badsey. Near Wyre Piddle.

Severn. Droitwich Canal. Brookside, Salwarpe, *Sheward in Nash's Sup.* (1799). Ditch, near Cotheridge, *Lees.* Lower Lightwood Lane, Worcester.

Malvern. Longdon Marsh; Pendock Portway, *Lees.*

The strong odour of this plant at once betrays it, alive or dead, green or dry. But however acrid it may be in its native condition, it is pleasant enough when cultivated as its extensive use on our dinner tables testifies. It is very abundant by the side of the Droitwich Canal. It has been observed in all neighbouring counties, more or less truly wild.

718. 436. A. nodiflorum, Reichb. fil.   *Procumbent Marshwort.*

(*Helosciadium nodiflorum*, Koch.)

Native. Ditches, shallow slow streams, wet places. Common and widely distributed. P. June to October. *Top. Bot.* 82.

First Record. As *Helosciadium nodiflorum.* In the *Table of Plants, Lees' Bot. Worc.*, p. 13 (1867).

This is abundant in nearly all streams and rivulets, so much when out of flower imitating another well-known plant to the uneducated eye that it has gained the name of Fool's Water-cress. It has many distinctions from it, however, and in blossom of course could not be taken for it. It is recorded from all neighbouring counties.

Var. e. repens (Koch).

(*Helosciadium repens*, Koch. *Sium repens*, Sm.)

First Record. As *Sium repens.* Bogs on the side of Abberley Hill. Cookhill, near Alcester, *Purton, Midl. Fl., App.*, vol. iii, p. 25 (1821).

Avon. As first record.

Malvern. In marshy ground on the west side of the Worcestershire Beacon, *Lees.* Malvern Hills. Malvern Common.

On 'Malvern Hills among the trickling springs', says Mr. Lees, becoming poetical in the *Illus. Nat. Hist. Worc.* But Mr. Towndrow says, 'True *A. repens*

---

does not occur at Malvern; a small form of *A. nodiflorum* has been passed for it' (MS.). And so in other cases also, no doubt. Yet there are good distinctions. In the type the leaflets are bluntly serrate, and the umbels longer than their peduncles, while in the variety the leaflets are sharply toothed and the peduncles are longer than the umbels. However, *A. repens*, or something mistaken for it, has been noticed in Salop, Stafford, and Warwick.

719. 437. A. inundatum, Reichb. fil.   *Water Honewort.*

(*Helosciadium inundatum*, Koch. *Sison inundatum*, Linn.)

Native. Sides of pools and ditches. Local. P. June to August. *Top. Bot.* 97.

First Record. As *Sison inundatum.* Sides of rivulets on Malvern Chase, *Ballard, Stokes's With. Bot. Arr.*, and ed., p. 275 (1787).

Avon. Pool, Dodderhill Common.

Severn. Hartlebury Common, *Humphreys.*

Malvern. Longdon Marsh. Commons near Malvern, *Lees.* Abberley Hill, above the Hundred House, *Purton.* Newland Common. Barnard's Green. Castlemorton Common. Pond, Castlemorton.

Lickey. Moseley Wake Green, *Miss M. A. Beilby.*

Mr. Lees, in his *Bot. Malv. Hills*, p. 19 (1843), says this plant occurs in the Malvern district, 'only in deep stagnant water holes by the brook on Welland Common'; but at p. 65 of his *Bot. Worc.* he leaves one to infer that the plant had a wider range in the locality. This is a small straggling plant, sometimes floating, but usually submersed, with a few of the upper leaves and the flowers rising above the water. The leaflets of its lower leaves are cut into capillary segments, the upper leaves being pinnate. The umbels are usually of two rays only, with very small white flowers. It does not appear to have been noticed in Gloucestershire of the bordering counties.

CICUTA, Linn. 155. (*Cicuta*, a shepherd's pipe, from the hollow stems.)

[720. 438. C. virosa, Linn. *Top. Bot.* 37.   *Water Hemlock, Cow Bane.*

This plant is recorded, but not localized, in Mr. Lees' catalogue in the *New Bot. Guide*, but was not acknowledged by him in his following publications. 'It is not likely to occur in Worcestershire,' says Mr. Bagnall. It is marked in *Top. Bot.*, and ed., p. 188, for Worcestershire, but without authority. 'From a catalogue sent by Mr. Lees to Mr. Watson,' says Mr. Bagnall (MS.). It is a fine strong plant three to four feet high, with a furrowed hollow stem. In its young state it does not seem to be distinguished by cattle, which eat it among other herbage, with deleterious effects; but when it is well grown they will not touch it. It is marked in *Top. Bot.* for Hereford, Stafford, and Salop among neighbouring counties.]

CARUM, Linn. 156. (Perhaps from the country Caria; or Celtic *carbh*, a ship, from the shape of the carpels.)

722. 439. C. PETROSELINUM, Benth. and Hook. fil.   *Common Parsley.*

(*Petroselinum sativum*, Hoffm.)

Alien. Walls and waste places. Rare. B. or P. July to August. Not in *Top. Bot.*

First Record. On garden walls, *Lees, Bot. Worc.*, p. xxix (1867).

M 2

This alien frequently escapes from gardens, and makes a more or less successful struggle to maintain itself in a state of liberty. The extremely crisped form usually cultivated is var. *crispum*, DC.; but left to itself its leaves quickly come out of curl, and it assumes the form of the type. The plant is usually noticed in some connexion with garden rubbish. It flourishes on the railway cutting near Dodderhill Church, Droitwich, in the Severn district. It is native on the dry hills of Italy and a few neighbouring countries. In many parts of the world it appears as it does in England, in a semi-naturalized manner, being only known as a cultivated plant or an escape. And in this way it has appeared in all neighbouring counties.

723. 440. C. segetum, Benth. and Hook. fil.　　　*Corn Parsley.*
(*Sison segetum*, Linn.　*Petroselinum segetum*, Koch.)
Native. Dry hedgebanks and field-borders. Local and rare. A. or B. July to September. *Top. Bot.* 45.
First Record. As *Sison segetum*. On a marl bank near Gregory's Mill, *Sheward, Nash, Hist. Worc., Sup.*, p. 96 (1799).
Avon. Badsey, *Rev. W. S. Cheshire.* Between Evesham and Elmley Castle, *Lees.*
Severn. Near Dodderhill Church. Between Henwick and Hallow. Spetchley. Broadwas, *Lees.* Rainbow Hill. Near Dodderhill, Droitwich. About 100 yards up a lane terminating at the sixth milestone at Broadwas.
Malvern. Eastward of Castlemorton Church, *Lees.*
A doubtful locality as regards the district is one given by Purton, ' between Hanbury and Droitwich.' The boundary line between the Avon and Severn districts also passes about half-way, along the canal, between Hanbury and Droitwich. It does not appear to have been recorded from Salop.

724. 441. C. Carvi, Linn.　　*Caraway.*
Casual. Waste places, rarely in meadows. Very rare. A. July. *Top. Bot.* 2.
First Record. Meadows near Worcester, *Lees, Illus. Nat. Hist. Worc.*, p. 158 (1834).
Severn. In fields near the Bridge at Powick, *T. Westcombe.* Banks of Kidderminster and Bewdley Railway, *Mathews.* Roadside towards Hallow. Hedge by steep path south-west of Dodderhill Church. Spetchley. Near Broadwas. Rainbow Hill.
Malvern. Malvern Link Common and occasionally elsewhere, but always a casual, *Towndrow.*
This plant is usually considered an alien in Britain, but it is native in the meadows of Central and Northern Europe, reaching Holland and Scandinavia, so there is no reason why it should not be native in Britain also. Moreover, its seeds have been identified by Mr. Reid as occurring in interglacial deposits in our country, which fact would go to prove that it is a very old inhabitant indeed. Perhaps of all the foods, flavours, essences, and medicaments afforded by the umbelliferous tribe of plants to the uses of humanity, the Caraway seed is the best known of all of them. In olden days the root also was in common use, for one old writer said it was

'pleasunt and comfortable, and helpeth digestion'; and another, ' eaten as men eat parsneps, strengthens the stomachs of ancient people exceedingly; and they need not make a whole meal of them neither.' As the last plant, this also does not appear to have been noticed in Salop. It is casual in Warwick and Staffordshire.

SISON, Linn. 157.　(*Sison*, Celtic, a running brook.)
726. 442. S. Amomum, Linn.　　*Hedge Bastard Stone Parsley.*
Native. Hedgebanks. Local. B. July to October. *Top. Bot.* 54.
First Record. Not localized, *Lees' Cat., New Bot. Guide* (1835).
Avon. Cleeve Bank. Craycombe. Alderminster. Bredon Hill; Trench Woods, *Mathews.* Croome. Pirton. Fladbury. Pershore. Stoulton. Norton, near Evesham. Crowle. Bishampton. Throckmorton.
Severn. Crutch Lane, Droitwich. Broadwas. Hartlebury Station. Spetchley. Ombersley. Grimley. Kempsey.
Malvern. Leigh Sinton. Shady lanes, *Lees.* Witley. Bransford. Leigh. Knightwick. Malvern Link.
Lickey. Woodcote Green; Randans, *Mathews.* Tardebigge. Rubery. Park Gate.
The umbels of flowers of this plant are very small, and the whole plant possesses a very offensive odour, especially when bruised. Country names for it are Parsley Breakstone and Honeywort. It is recorded from all neighbouring counties.

SIUM, Linn. 159.　(σεῖω, I brandish; or *siw*, Celtic, water.)
{**728. 443. S. latifolium**, Linn.　　*Broad-leaved Water Parsnep.*
Native. Water-sides. Rare. P. July to September. *Top. Bot.* 42.
First Record. In the Moors near Pitchcroft, Worcester, *Dr. Thompson, Junr., Stokes's With. Bot. Arr.*, 2nd ed. (1787).
Severn. Blakedown Pool, *Scott,* in *Lees' Bot. Worc.*, p. 13. Also *Purton,* with initials *T. P.* Side of Henwick Old Weir Pond, *Lees* (1828).
Malvern. By the weir at Newman's Bridge, near the Devil's Den, Clifton-on-Teme, *Lees.*
This plant has never, since Purton's time, been seen at any of the numerous pools in the vicinity of Blakedown. Mr. Mathews, in *Mid. Nat.*, xi, p. 16, does not mention it as a plant recorded by Scott, in his *Hist. Stour.*; indeed Mr. Scott seems entirely to have passed over umbellifers. In his *Bot. Worc.* (1867), at the reference above, Mr. Lees says the plant ' is now unknown in Worcestershire.' It is a large, strong, conspicuous plant, standing up four or five feet above the water's edge, bearing flat umbels of white flowers, and not likely to be overlooked. Of neighbouring counties it has been recorded in Gloucester erroneously for Stafford, and doubtfully in Salop.}

729. 444. S. erectum, Huds.　　*Narrow-leaved Water Parsnep.*
(*S. angustifolium*, Linn.)
Native. Ditches and streams. Local. P. July and August. *Top. Bot.* 82.
First Record. As *S. angustifolium*. Ditches near Perry Wood, *Sheward, Nash, Hist. Worc., Sup.*, p. 96 (1787).

Avon. One place, abundant, *Lees.* Huddington.
Severn. Wick, *Baxter, Westcombe, Reece.* Opposite ' Camp ', *Gissing.* Stour side at Falling Sands, Kidderminster, *Mathews.* Blakedown Pools. Pool next Wannerton Downs, *Mathews.* Hoo Pool Mill. Fenny Rough. Chawson. Hurcott Bog. Pond near second brickyard from Worcester. Small pond, Hartlebury Common.
Malvern. Powick, *Reece.* Ditch near Teme's mouth, *Mr. T. Baxter.* Longdon. Madresfield; Hanley Castle.
Lickey. Bittell Reservoir, *Humphreys.* Bromsgrove, Charford Mill Pool, and Moat Mill Pool, *Humphreys.* Tardebigge Reservoir. Canal side, Halesowen.
This is a much smaller plant than the last and much more frequently met with. It occurs in all bordering counties.

AEGOPODIUM, Linn. 160.　(αἴξ, a goat, πούς, a foot, from the likeness of the leaf to a goat's foot.)
730. 445. Ae. Podagraria, Linn.　　*Gout Weed, Bishop's Weed, Wild Ash.*
Native or denizen. Hedges, waysides, rubbish heaps, usually near habitations. Generally distributed and common. P. May to August. *Top. Bot.* 100.
First Record. Not localized, *Lees, Bot. Malv. Hills,* 1st ed., p. 19 (1843).
Common as this plant is, the universal inhabitant almost of the untidy bank and the neglected place in the garden, whence it is most difficult of eradication on account of its long running stems, the least bit of which will grow and start a colony of its own, it is not a native of our land. It is at home in woods and meadows in Eastern Europe. It was doubtless introduced as a reputed cure for gout, indicated by its specific name. Such a potent remedy was it considered that one writer says ' the very bearing of this herb about one easeth the pains of the gout '. It has many names besides those given above, such as Herb Gerard, Ash Weed, and Ground Elder, the last very applicable from the similarity of its leaves to those of the Elderberry. It occurs in every neighbouring county.

PIMPINELLA, Linn. 161. (Name altered, Linnaeus tells us, from *bipennula*, in allusion to the twice pinnate leaves.)
731. 446. P. Saxifraga, Linn.　　*Burnet Saxifrage.*
Native. Dry pastures. General; abundant in the Avon district. P. July to September. *Top. Bot.* 102.
First Record. Near Dudley Castle, *Booker, Dud. Cast.* (1825).
This is a common plant on dry pastures. Its root leaves are so like those of the Burnet that its name is quite justifiable, while the leaves of the stem are finely divided. ' Saxifrage' means the ' stone breaker', which, while it may have some meaning in connexion with the regular Saxifrages, many of which delight in rocky habitats, does not appear particularly suited to a plant chiefly found upon turf, and resembling its namesake neither in flower nor appearance. It is supposed to cure the toothache, and a decoction of the plant is used as a cosmetic. All neighbouring counties possess it.

Var. c. **dissecta**, With.
First Record. Malvern, *Towndrow, Malv. Advertiser,* 1892.
Severn. Newtown Road, Worcester, *Rea.*
Malvern. As first record.
In this variety the leaves are more finely cut than in the type. It does not differ from it in any other way, and as the leaflets of the type are very variable, it is possibly a matter of opinion when the type ends and the variety begins. The variety has been recorded in Salop, Stafford, and Warwick.

732. 447. P. major, Huds.　　*Greater Burnet Saxifrage.*
(*P. magna*, Linn.)
Native. Woods, thickets, hedgebanks. Local. P. July and August. *Top. Bot.* 52.
First Record. As *P. magna.* Worcestershire, *Ballard, Stokes;* in Marle, *Stokes; Stokes's With. Bot. Arr.*, 2nd ed., p. 314 (1787).
Avon. Trench Woods.
Severn. Wyre Forest, *Jorden.* Droitwich. Near Dodderhill Church. Hadsor. Barbourne. Helbury Hill. Near Hampton Lovett Church. North of Fernhill Heath Station. Salwarpe. Hadley, near Droitwich.
Malvern. About Cradley, rare, *Lees.* Between Howsell and Brace's Leigh. The Rhydd. Madresfield.
Lickey. Roadside between Longbridge and the Lickey, *Mathews.* Barnt Green.
This is larger in all respects than the last plant and the tuft of root leaves is unmistakable. It grows three or four feet high. The plant is recorded from all neighbouring counties.

Var. b. **dissecta**, N. E. Br.
First Record. As in this book.
Severn. Wyre Forest, *Herb. Hast. Mus. Worc.*

CONOPODIUM, Koch, 162.　(κῶνος, a cone, πούς, a foot, from the conical disk-lobes.)
733. 448. C. majus, Loret.　　*Earth Nut, Pig Nut.*
(*Bunium flexuosum*, With. *B. denudatum*, DC. *Carum flexuosum*, Fries. *Conopodium denudatum*, Koch.)
Native. Woods, thickets, hedges, meadows. General throughout. P. April to July. *Top. Bot.* 109.
First Record. As *Bunium flexuosum.* Pastures, *Lees, Bot. Malv. Hills,* 1st ed., p. 19 (1843).
This is a graceful plant, with a slender stem and a few much-divided leaves. Every schoolboy knows it; the single tuber from which the stem springs is dug up and eaten raw, having a sweetish flavour, which, however, is not very appealing except to early youth. The tubers are eagerly sought for by pigs, whence one of the popular names of the plant. It is found in all neighbouring counties.

MYRRHIS, Linn. 163. (μύρρα, myrrh.)

734. 449. M. Odorata, Scop.   *Sweet Cicely.*

(*Scandix odorata*, Linn.)

Native or denizen. Pastures and hedges, generally near houses. Very local and rare. P. May to July. *Top. Bot.* 65.

First Record. As *Scandix odorata*. In an orchard at the top of Souston's Roche, near Shelsley Walsh, *Ballard, Stokes, Stokes's With. Bot. Arr.*, 2nd ed., p. 303 (1787).

Severn. Side of the Royal Mount, facing the London Road, Worcester, *Lees* (1828). Waste ground between the road and Bewdley Station.

Malvern. Southstone Rock.

Lickey. Near Rubery. By the Sun Inn at Hunnington, *Humphreys* (1907).

This plant, though in Europe it extends as a native to Scandinavia, is not found in natural conditions in Britain, where it has been cultivated from early times. The whole plant is very fragrant, the large seeds especially so. It had its use in salads and its medicinal use as well, being held to be a preservative against the infection of plague. The name Cicely has nothing to do with Cecilia, but is a corruption of Seseli, the generic name of a plant that comes further on in our list. *Myrrhis* is recorded from all neighbouring counties.

CHAEROPHYLLUM, Linn. 164. (χαίρω, I rejoice, φύλλον, a leaf, from the agreeable odour of the leaf.)

735. 450. C. temulum, Linn.   *Rough Chervil.*

(*C. temulentum*, Sm.)

Native. Thickets, hedgebanks. Very common and widely distributed. P. or B. May to July. *Top. Bot.* 99.

First Record. Not localized, *Lees, Bot. Malv. Hills*, 1st ed., p. 20 (1843).

This and *Conium maculatum* are the only two British plants belonging to the umbellifers which have spotted stems, but this plant is hairy while *Conium* is smooth. This plant is conspicuous in hedges and bushy places in June and July, following *Anthricus sylvestris*, an earlier bloomer, in similar situations. As in that plant, the umbels of this are drooping before the flowers expand, but to a more noticeable extent; and this is a softer-looking plant, which does not grow quite so strongly. The stem is hairy throughout, and swells below each joint, while in its predecessor in the hedgebanks the upper part of the stem is smooth. It is recorded from all neighbouring counties.

SCANDIX, Linn. 165. (σκάνδιξ, the Greek name of some plant.)

736. 451. S. Pecten-Veneris, Linn.   *Shepherd's Needle.*

Colonist. Cornfields. General throughout. A. April to October. *Top. Bot.* 93.

First Record. Not localized, *Lees, Bot. Malv. Hills*, 1st ed., p. 20 (1843).

The remarkable thing about this common plant is the extraordinary development of the beaks of the fruit, from which it gets the name given

above and others, such as Venus's Comb and Pucker-needle. The fruits are bright green and sometimes two or three inches long. The plant occurs in all bordering counties.

ANTHRISCUS, Bernh. 166. (Etymology not known; the name is used by Pliny.)

737. 452. A. vulgaris, Bernh.   *Common Chervil.*

(*Cerefolium*, Beck. *Scandix Anthriscus*, Linn.)

Native. Waysides, hedgebanks, waste places. Not common. A. April to June. *Top. Bot.* 80.

First Record. Not localized, *Lees, Bot. Malv. Hills*, 1st ed., p. 20 (1843).

Avon. In this district, *Lees*. Norton, near Evesham. Pershore.

Severn. Brake Mill Farm, Hagley, *W. Whitwell*. In great abundance near Falling Sands Rolling Mill, 1882, *Mathews*. Hartlebury Common. Oldington, Kidderminster, *Mathews*. Stanklin Pool, *Humphreys*. Cookley, 1885, *Mathews*. Hedge in field next to Northwick Hall. Hedge in first pasture by path from Droitwich Road to Claines Church. Northwick. The Moors, Worcester. Near Pope Iron Inn, Worcester.

Malvern. Waste places, *Lees*. Roadside near the Red Lion, Powick. Bransford.

Lickey. Dry banks, rare, *Mathews*.

This plant is much commoner in the hedgerows than the above records would lead one to think, as the very lovely leaves of this plant are soon crowded out and obscured by the ranker vegetation. This plant was published in *Bot. Rec. Club Rep.* for 1873 as a new record, but it was not so. The umbels are stalked and opposite the leaves. The plant does not appear to have been met with in Herefordshire only of neighbouring counties.

738. 453. A. sylvestris, Hoffm.   *Cow Parsley, Hedge Beaked-Parsley.*

(*Chaerophyllum sylvestre*, Linn.)

Native. Hedgebanks, fields, thickets. Abundant. P. March to May. *Top. Bot.* 107.

First Record. As *Chaerophyllum sylvestre*. Pastures, *Pitt, Agric. Worc.*, p. 317 (1810).

This is the first of the umbelliferous plants to expand its flowers in the hedgerow bank or the meadow put up for hay. It is a favourite food for rabbits, and the small boys of urban districts, to whom it is known sometimes as 'Kake', no doubt a form of 'Kex', carry large bundles of it into the towns for their pets. After flowering it goes a dark dull green, and the fruits become nearly as conspicuous along the banks of roadside hedges as its flowers had been. It occurs in all bordering counties.

739. 454. A. Cerefolium, Hoffm.   *Garden Chervil.*

(*Cerefolium sativum*, Lam. *Scandix Cerefolium*, Linn.)

Casual. Waste ground. Rare, not native. A. May. Not in *Top. Bot.*

First Record. As *Scandix Cerefolium*. Found near Worcester in considerable plenty, on the south-east side of the Worcester Road, just beyond the turnpike, May, 1777, and in the hedges in Upper and Lower Old Swinford, though not to be discovered in any of the neighbouring gardens, *Stokes's With. Bot. Arr.*, 2nd ed., p. 304 (1787).

The plant mentioned by Dr. Stokes, just beyond the turnpike gate, remained there in great profusion till 1830. But in that year the road was altered at this point, the bank thrown down, a wall built, and every vestige of the plant destroyed. The English name of the plant is a corruption of *Cerefolium*, itself a corruption of the Greek-derived *Chaerophyllum*. Chervil is a well-known garden pot-herb, and was much esteemed by our ancestors. It was 'verye profytable unto the stomach'; one of the 'necessarie herbes to growe in the garden'; and one, said Evelyn, 'whose tender tops are never to be wanting in our sallets'. And coming down to a modern cookery-book, it is there said 'Chervil is largely used in salads, stuffings, sauces, and omelettes'. Whenever it occurs it is a garden escape of old or new standing. It does not appear to have been noticed in this condition in Gloucester or Stafford.

SESELI, Linn. 167. (σίσελι, a Greek plant-name.)

{**740. 455. S. Libanotis**, Koch.

Casual. A native of chalk hills in Cambridge, Hertford, and Sussex. P. July and August. *Top. Bot.* 3.

First Record. As in this book.

Severn. Severn side, *Reece*.

This plant has not heretofore been recorded for Worcestershire, but in the *Herb. Hast. Mus.*, *Worc.*, is a specimen labelled, 'Severn side, G. Reece'. The name 'Cicely' among English plants is a corruption of the generic name of this plant. It has not been recorded for any neighbouring county.}

FOENICULUM, Hill. 168. (*Foenum*, hay, from its scent or the much-divided leaves.)

741. 456. F. vulgare, Mill.   *Common Fennel.*

(*F. officinale*, All. *Anethum Foeniculum*, Linn.)

Alien. Waste places, railway-banks. Rare. P. June to August. *Top. Bot.* 32.

First Record. As *Anethum Foeniculum*, near Spetchley, *Stokes, Stokes's With. Bot. Arr.*, 2nd ed., p. 311 (1787).

Avon. Under the Keuper Marl Cliff at Crowle, *G. Reece*.

Severn. Spetchley, *Stokes*. Rainbow Hill, Worcester, *Mathews*. Dodderhill Churchyard, 1900, *Humphreys*.

Malvern. At Powick, *G. Reece*. Near Powick Church. Malvern Hill, *Mathews*.

This much-cultivated plant has been reported from various localities all about our island, but if native anywhere with us, it is among the salt-marshes and on the cliffs of the east of England. There is a slight difference between this and the garden plant, *F. dulce*, and it is the latter, probably, which occurs at Dodderhill Churchyard. The root of the Fennel is a common vegetable on the Continent and especially in Italy, but its flavour is not pleasant to the average English taste. The English name for this plant is a corruption of the Latin one—a diminutive of *foenum*, hay. It has been known as a garden herb from the earliest times, and one of its chief uses was as a garnish for fish, a use which survives to the present day in the case

of salmon; and not only as a garnish, but as an ingredient of a fish-sauce also. The odour is not much appreciated at the present day; formerly this was not the case. 'There's fennel for you, and columbines,' said Ophelia; and it was usually among the plants strewn over the pathway of the newly-married. As for its medicinal uses, they were many and various. It relieved those who had eaten poisonous plants; was good for gout and cramp, and yellow jaundice; while it was advised for those who, growing too stout, desired to be lean. The plant does not appear to have been observed out of cultivation in Herefordshire or Staffordshire. In Warwickshire it is a casual.

OENANTHE, Linn. 170. (οἶνος, wine, ἄνθος, flower, from the vinous scent of the blossoms.)

743. 457. Oe. fistulosa, Linn.   *Water Dropwort.*

Native. Wet ditches, marshy fields, pools. Rather local. P. June to September. *Top. Bot.* 68.

First Record. Brookside, Yardley, *Ick, Analyst*, vol. vi, p. 22 (1837).

Avon. Several places, *Lees*. Stream by roadside near Throckmorton. Defford Common. Norton, near Evesham. Fladbury. Dodderhill Common.

Severn. Westwood Park. Pool near Droitwich. Dunhampstead. Bog on right of road near Button Oak, Wyre Forest, *Mathews*. Pool by roadside leading to Monk's Wood. Astwood Road. Bewdley. Kempsey.

Malvern. Several places east of the hills, *Lees*. Malvern Common. New Pool. Pendock. Longdon Marsh. Powick. Knightsford Bridge. Stanford Bishop.

Lickey. One place, *Lees*. Foucher's Pool, Stourbridge.

This is a somewhat remarkable plant, being, so to speak, entirely composed of tubes. The lower leaves are always below the surface of the water, and are flat, but the rest of the plant well merits the name *fistulosa*. The angled corky fruits form dense globular heads, each as large as a marble. The plant is recorded from all neighbouring counties.

744. 458. Oe. pimpinelloides, Linn.   *Knotty-rooted Water Dropwort.*

Native. Meadows and banks. Rare. P. June to August. *Top. Bot.* 17.

First Record. By the side of rills, ascending the north side of Bredon Hill, *Nash, Hist. Worc., Int.*, p. lxxxix (1781).

Avon. Badsey, *Purton*. Bredon Hill, as above. Defford Common.

Severn. Crookbarrow, Worcester, *Lees*. Kempsey.

Malvern. Tewkesbury to Powick, on the west bank of the river, *Lees*. Madresfield. West of Brook House, Powick. East of Powick Church, about a quarter of a mile. Newland Common.

This appears to be observed in no bordering county except Gloucestershire, and there we have seen it at Upton Saint Leonard's.

745. 459. Oe. silaifolia, Bieberstein.   *Sulphur Water Dropwort.*

(*Oe. peucedanifolia*, Pollich.)

Native. Wet, low-lying meadows. Several places in the west of the county. P. June and July. *Top. Bot.* 22.

First Record. As *Oe. peucedanifolia*. On the borders of Longdon Marsh also in a ditch in the lane between Castlemorton and Longdon Marsh, *Lees, Bot. Malv. Hills*, 1st ed., p. 19 (1843).

Avon. The Marsh Common.

Severn. Kempsey Ham. Kempsey Grove. Monk Wood. Meadow west of Kempsey Grove.

Malvern. Longdon, *Lees*. Commons near Malvern, *Lees*. Powick.

This plant is very near *Oe. pimpinelloides*, but is larger and stouter, and the root-fibres are rarely tuberous in the middle. It has been reported from Gloucester and Hereford.

746. 460. Oe. Lachenalii, C. Gmel.   *Parsley Water Dropwort.*

Native. Rough meadows, bogs. Local. P. July to September. *Top. Bot. 72.*

First Record. Welland Marshes, *Lees, Phyt.*, O.S., vol. ii, p. 357 (1845).

Avon. Near the Saline Spring on Defford Common, *Dr. Streeten* (1845). Craycombe, *Lees*. Badsey, *Rev. W. S. Cheshire*. Clay-pits north of Machine Farm, Craycombe. Ditch by roadside, Wyre Piddle. On Defford Common, 1907, *Rea.*

Severn. Ditch by the roadside at Cotheridge.

Malvern. Longdon ; Welland, *Lees*.

This plant is also very similar to *Oe. pimpinelloides*. Dr. Streeten stated that he thought his plant was this when he gathered it. It differs in its root-fibres, which are never tuberous, and the partial umbels are less crowded. It has not been recorded either for Stafford or Salop.

747. 461. Oe. crocata, Linn.   *Hemlock Water Dropwort.*

Native. Sides of ditches, streams, and ponds. Local. P. May to August. *Top. Bot. 92.*

First Record. Frequent in ditches, and by river-sides in many parts of the county, *Nash, Hist. Worc., Int.*, p. lxxxix (1781).

Avon. Norton, near Evesham. Harvington Brook. Fladbury.

Severn. Near Stourport. Lincombe. Severn below Worcester. Ditches near Bewdley, *Mathews*. Grimley Brickfields. Rough, back of White Hall. Hedges near Severn, Ribbesford.

Malvern. Chalybeate Coppice, destroyed 1850, *Lees*. Towards Barnard's Green, *Lees*. Bransford Weir. Knightsford Bridge.

A tall, fine plant, which could hardly escape notice. Its leaves much resemble those of celery, and fatal consequences have frequently resulted from the mistake ; it is poisonous also to cattle. In spite of its virulently poisonous properties the herbalists of old seized upon it, though Gerard uttered warnings against it. An infusion of the leaves has been considered good for leprosy. It occurs in all neighbouring counties.

748. 462. Oe. aquatica, Poir.   *Water Horsebane.*

(*Phellandrium aquaticum*, Linn.   *Oe. Phellandrium*, Lam.)

Native. Pools and ditches of stagnant water. Local. P. June to August. *Top. Bot. 56.*

First Record. As *Phellandrium aquaticum*. Clifton, near Severn Stoke, *Ballard, Stokes* ; *Stokes's With. Bot. Arr.*, 2nd ed. (1787).

Avon. Pershore.

Severn. Near Worcester, *Rufford*. Kempsey Grove, *Lees*. Upton-on-Severn. Pool, New Road, Worcester. Diglis Brick-Pits. Pool near Moseley. Hallow. Grimley Brickfields. Near Northwick Brickyard.

Malvern. Near Southwood, Martley. Longdon Marsh. Stagnant Pools about the Chace, *Lees*. Malvern Link.

Lickey. Meadows near Hagley Hall, *Ballard*.

This plant grows in the water. It is well distinguished from all the preceding Water Dropworts by its roots of slender fibres, never tuberous, and its multifid submersed leaves, while the umbels are lateral, opposed to the leaves. It occurs in all neighbouring counties.

AETHUSA, Linn. 171.   (αἴθω, I kindle, because of its acrid qualities.)

750. 463. Ae. Cynapium, Linn.   *Fool's Parsley.*

Native. Cultivated ground. Universal. A. May to October. *Top. Bot. 96.*

First Record. Not localized, *Lees, Bot. Malv. Hills*, 1st ed., p. 20 (1843).

This weed grows everywhere on cultivated land, being especially fond of garden ground. It is difficult to see how it can be mistaken for parsley, especially the curled cultivated kind, but this has sometimes been the case, with fatal results. Its upright growth, finely-divided leaves, without a curl among them all, and its long, deflexed bracteoles, quite mark it off from its namesake. If it is so mistaken, its English name is quite justified. It occurs in all neighbouring counties.

SILAUS, Bess. 173.   (Etymology unknown ; it was a name given by Pliny to many umbellifers.)

752. 464. S. flavescens, Bernh.   *Pepper Saxifrage.*

(*S. pratensis*, Bess.   *Peucedanum Silaus*, Linn.)

Native. Roadsides, meadows, pastures. General. P. June to October. *Top. Bot. 68.*

First Record. As *Peucedanum Silaus*. A meadow plant, *Pitt, Agric. Worc.*, p. 656 (1810).

Avon. Bredon Hill. Trench Woods. Huddington. Flyford Flavell. Crowle. Tiddesley Wood. Fladbury. Abberton. Norton, near Evesham. The Slads. Hanbury. Near Gooseshill Wood. Inkberrow. Croome Park.

Severn. Monk's Wood. Crutch Lane, Droitwich. Droitwich Canal. Near Nunnery Wood. Kempsey. Ombersley. Holt. Grimley. Spetchley. Cotheridge. Shrawley. Trimpley.

Malvern. In barren fields, *Lees*. Powick. Bransford. Alfrick. Knightwick. Martley. Stanford Bridge. Malvern Link.

Lickey. Romsley and Frankley, *Mathews*. Woodcote Green, Randans, *Mathews*. Field in Hob Lane, Yardley, *Miss M. A. Beilby*. Dodford. Tardebigge Reservoir.

The Pepper Saxifrage occurs in all bordering counties.

ANGELICA, Linn. 177.   (ἄγγελος, a messenger, from its cordial and medicinal properties.)

756. 465. A. sylvestris, Linn.   *Angelica.*

Native. Damp woods and thickets, stream-sides. General. P. July and August. *Top. Bot.* 112.

First Record. Moist hedges, *Pitt, Agric. Worc.*, p. 656 (1810).

This large and handsome plant is to be found in most damp woods, with umbels of pinkish-white flowers. The stem is of a purplish colour, and stout, and downy towards the top, as are the umbels. It was formerly employed as a remedy against the itch. It occurs in all neighbouring counties.

ARCHANGELICA, Hoffm. 178*.   (ἀρχός, chief, ἄγγελος, angel, from its supposed excessive medicinal properties.)

{**757. 466. A. officinalis,** Hoffm.   *Garden Angelica.*

(*Angelica Archangelica*, Linn.)

Alien. Damp places. Twice observed. B. August and September. Not in *Top. Bot.*

First Record. As *Angelica Archangelica*. Recorded as occurring at Broadmoore, about seven miles N.W. of Birmingham, *Withering*, 4th ed., p. 293 (1801).

Severn. Meadow by the Severn, *Herb. Hastings's Museum, Worc.*

Lickey. Probably as first record.

North-west of Birmingham is not in Worcestershire, and, as Mr. Mathews remarks, must be an error for south-west. But *Archangelica* is in no sense a native plant ; its home is in wet woods and on river-banks from Scandinavia to Central Europe. But it was much prized by our ancestors and cultivated to eat as celery, or to be preserved with sugar. Its name is given it because of its angelic, or even archangelic, qualities, which in some countries are thought worthy of an even higher origin of similitude, and the plant is called by a name signifying the Holy Ghost. Archangel, however, is not a name limited in England to this plant. Several species of Dead Nettle, and, as an anti-climax, the Black Stinking Horehound, have been or are called by it. In medicine it was held good for diseases of the chest, would avert hydrophobia, remove the effects of too extensive wine-bibbing, and it possessed various other qualities. It had its superstitious uses also. The roots of the *Archangelica* hung round the neck destroyed the power of witches over the bearer. The plant has been observed in an escaped condition also in Staffordshire.}

PEUCEDANUM, Linn. 179.   (πεύκη, pine-wood, δᾶνος, a gift, because many of the species have a resinous smell.)

761. 467. P. sativum, Benth. and Hook. fil.   *Wild Parsnep.*

(*Pastinaca sativa*, Linn.)

Native. Roadsides, field-borders, hedgebanks. General throughout the county and abundant in the Avon district. B. July and August. *Top. Bot.* 58.

First Record. As *Pastinaca sativa*. Roadside near Stoughton, abundantly. Battenhall, Craycombe, &c., on red marl and lias marl, *Lees, Illus. Nat. Hist. Worc.*, p. 159 (1834).

The name Parsnep, or more rightly Parsnip, is an ultimate corruption of the Latin *Pastinaca*, connected with *pastinare*, to dig with a fork, while the

last syllable, which appears also in turnip, is probably related to the Latin *napus*. Cultivated parsnips are improved descendants of our wild plant, but they are becoming less popular as a vegetable now than in former years, the flavour not being pleasing to modern taste. Parsnip wine was formerly made in village places, like most home-made wines, owing most of its excellences to other ingredients, not to the parsnip. The plant occurs in all neighbouring counties.

HERACLEUM, Linn. 180.   (Ἡρακλῆς, Hercules, who is said to have brought this, or some allied plant, into use.)

762. 468. H. Sphondylium, Linn.   *Cow Parsnep.*

Native. Hedgerows, thickets, pastures. Abundant and generally distributed. P. June to August. *Top. Bot.* 112.

First Record. Not localized, *Lees, Bot. Malv. Hills*, 1st ed., p. 20 (1843).

This, a tall, coarse-looking plant, with thick umbels of white or pinkish-white flowers, and big, rough, hairy leaves, is conspicuous enough in hedge-row and pasture during its period of flowering. It is said to be wholesome and nourishing to cattle ; and it is one of the plants especially sought out by town boys for rabbit-meat. In the North of Europe an intoxicating drink is made from the plant, or one cognate to it. The Cow Parsnep occurs in every bordering county.

Var. b. **angustifolium,** Huds.

First Record. Thorngrove, near Worcester, *R. E. Jeffery, W. N. C. Trans.* iii. 81 (1902).

Severn. Near Bewdley. Meadow by Severn, *Reece*. Shrawley Wood. Claines.

Malvern. Pickersleigh.

Lickey. Round Hill Wood, Hagley.

The leaves of this variety are narrowly pinnatifid. It has been observed in Warwickshire.

CORIANDRUM, Linn. 182*.   (κόρις, a bug, from the smell from the bruised leaves.)

764. 469. C. sativum, Linn.   *Common Coriander.*

Alien. Waste places and arable land. Rare. A. June. Not in *Top. Bot.*

First Record. Not localized, *Lees' Cat., New Bot. Guide* (1835).

Malvern. Arable land at Newland, *Towndrow, Malv. Advert.* Nov. 19, 1892.

This plant, a field weed of Southern Europe and the East, finds its way into England as a grain introduction. It has no claim to be a native of this country. It is most frequently met with in the south and east of England, and is not reported from any neighbouring county.

DAUCUS, Linn. 183.   (δαῦκος of Dioscorides, probably from δαίω, I burn, from the seeds being heating.)

765. 470. D. Carota, Linn.   *Wild Carrot.*

Native. Dry pastures, roadsides, railway embankments. Very common. B. May to August. *Top. Bot.* 109.

First Record. Not localized, *Lees, Bot. Malv. Hill*, 1st ed., p. 20 (1843).

We have now come to that section of the umbelliferous tribe the fruits of which are armed with spines. The fruits of all the members of the tribe consist of two carpels adhering by their faces to a common axis, and each carpel possesses five primary ridges on the outside; while in the substance of the pericarp are usually linear channels containing oil, called vittae. But the forms these universal characters of ridges and vittae take are very varied, and upon them one of the chief distinctions between the genera rest. In Hooker and Arnott's British Flora (6th ed., 1850) is a series of plates giving sections of the pairs of carpels of the several genera, and showing what may almost be called the kaleidoscopic changes of form, character, and position, which these two elements of ridges and vittae assume. In some cases they nearly disappear, at others they are exaggerated in various directions; while the carpels assume all kinds of shape, always retaining some reminiscence of a pentagonal origin. In Daucus and Caucalis, the following genus, the ridges are armed with spines or hooks. The carrot is one of the most frequent of the tribe, characteristic of the red marl of the county, and may always be recognized by the central flower of the umbel, which is red; while the rest of the umbel is usually more or less tinted with pink; and its pinnatifid bracts separate it from all other British Umbellifers. It is a well-known occupant of our gardens, a food for both man and beast. It occurs in all neighbouring counties.

CAUCALIS, Linn. 184. (κεῖμαι, I lie down, καυλός, a stem, from the trailing habit of the plant.)

767. 471. C. latifolia, Linn. *Great Bur Parsley.*
Colonist. Cornfields. Very rare. A. July. *Top. Bot.* 7.
First Record. As in this book.
Severn. Saint John's, Worcester, *Rea.*

768. 472. C. daucoides, Linn. *Small Bur Parsley.*
Colonist. Cornfields. Very rare. A. June and July. *Top. Bot.* 28.
First Record. Calcareous Field at Tibberton, *T. W. Gissing, Lees, Bot. Worc.,* p. 44 (1867).
Avon. One place, lost, *Lees.*
Severn. Tibberton, as above. Near Warndon, *Westcombe.*
This plant is a native of the dry hills of Persia, and has become abundant as a cornfield weed in many parts of Central and Southern Europe. It occurs rarely in England. It has been noticed in Gloucestershire and Warwickshire.

769. 473. C. arvensis, Huds. *Field Hedge Parsley.*
(*Torilis infesta*, Spreng.)
Native. Cornfields. Local. A. June to September. *Top. Bot.* 57.
First Record. As *Torilis infesta.* Not localized, *Lees' Cat., New Bot. Guide* (1835).
Avon. Bredon Hill; Trench Woods, *Mathews.* Aldberminster. The Manor House Farm, Sheriff's Lench. Defford Station.
Severn. Port Fields and other places about Worcester, *Lees.* Roadside between Worcester and Spetchley. Rainbow Hill. Ladywood, Salwarpe. Kempsey.

Malvern. Powick. Leigh. Sherrard's Green.
Lickey. Alvechurch, *D. Mathews.* Near Oldbury, *Ick.*
This plant is recorded from all bordering counties.

770. 474. C. Anthriscus, Huds. *Upright Hedge Parsley.*
(*Torilis Anthriscus*, Gmel.)
Native. Hedgerows, borders of fields, woods, and waste places. Not uncommon throughout the county. A. July to September, *Top. Bot.* 107.
First Record. As *Torilis Anthriscus.* Not localized, *Lees, Bot. Malv. Hills,* 1st ed., p. 20 (1843).
First in the hedgerows to blossom is *Anthriscus sylvestris,* then comes *Chaerophyllum temulum,* and then follows this as the common plant of the waysides in many localities. It is a slender plant, with a stem two or three feet high, solid and rough. The bristles of the fruit are not hooked. It occurs in all neighbouring counties.

771. 475. C. nodosa, Scop. *Knotted Hedge Parsley.*
(*Torilis nodosa*, Gaertn.)
Native. Dry sunny banks, borders of fields. Local. A. May to August. *Top. Bot.* 74.
First Record. As *Torilis nodosa.* Spetchley and Badsey, *Purton, Midl. Fl.,* p. 146 (1817).
Avon. Bredon Hill. Broadway. Badsey.
Severn. Rainbow Hill, Worcester. Henwick. Near the Ketch. On the Spetchley Road. Near Crookbarrow. Near Spetchley House. Droitwich Canal.
Malvern. Not very plentiful, *Lees.* Pixham. Malvern Common.
Lickey. Sidemoor.
This is a prostrate plant, and the umbels are almost globular. The bristles of the fruit are hooked. It is recorded from all neighbouring counties.

35. ARALIACEAE, Juss.

HEDERA, Linn. 185. (Supposed to be from ἕδρα, a seat.)

772. 476. H. Helix, Linn. *The Ivy.*
Native. Woods, hedges, old buildings. Frequent everywhere. Climbing shrub. August to November. *Top. Bot.* 109.
First Record. Covering the Ivyscar Rock (Malvern) most luxuriantly. On Little Malvern Priory, &c., *Lees, Bot. Malv. Hills,* 1st ed., p. 18 (1843).
The Ivy was anciently sacred to Bacchus, whence the placing of a bush of Ivy outside a tavern door to indicate that wine was sold therein. Lyly, in *Euphues,* says, 'Where the wine is neat, ther needeth no Iuie-bush,' a saying that has become proverbial, though the fact that the bush was of Ivy is left out of consideration. Ivy is considered injurious to trees, but this may be disputed. Many timber trees that are covered with it attain a large size, even when the stem of the Ivy is so thick that it must be of equal age with the tree. The natural support of the plant in a condition of nature was no doubt the tree, and it would not be to the advantage of a parasite to kill too early

N

the support on which it depended. The Ivy never twines round a tree like the Honeysuckle, nor does it suck nourishment by means of the root-like fibres by which it climbs; the only harm it can do is by keeping light and free circulation of air from the trunk, or by the weight of the spreading branches of its head. Equally injurious is it often said to be in the case of walls, but possibly with the same lack of reality. It may run its roots into cracks in the masonry, but the masonry must be already perishing if the plant has the opportunity, while the dense covering of leaves is as good nearly as a roof of tiles in preventing moisture from getting beneath them. On the other hand, the Ivy, by the strong framework of its branches, not infrequently supports an ancient edifice and prevents its further decay. Pliny, however, was unsparing in his denunciation of Ivy as a destroyer of plant or building. It was one of the plants the old herbalists considered would prevent the plague, an ailment that loomed large in their eyes, and to be good for the headache; and to drink out of an Ivywood cup would ease those troubled with the spleen. Any one also who had 'a surfeit by drinking wine' would find his speediest cure in drinking some of the same liquor in which Ivy leaves had been steeped. Ivy was largely used in classic times for crowns, garlands, and bedeckings generally. It is not native in Russia, but is often carefully tended in that country for purposes of decoration. It is hardly necessary to say that Ivy grows in all our bordering counties.

36. CORNACEAE, DC.

CORNUS, Linn. 186. (*Cornu*, horn, from the hardness of the wood.)

774. 477. C. sanguinea, Linn. *Dogwood.*
Native. Hedges, thickets, water-sides. General; less common in the Lickey district. Shrub. June and July. *Top. Bot.* 67.
First Record. Hedges near the Lye; Love Lane; Hanbury Hill, *Scott, Hist. Stourbridge,* p. 540 (1832).
The foliage of this plant in autumn becomes more or less tinged with dark purple or red, and when the hedges are leafless the red stems of the shrub are often conspicuous. The wood is hard, and used for making butcher's skewers. The purple-black berries are bitter and astringent; and because they were not fit even for a dog is said to be the origin of the English name. But they were never very likely food for that animal, and there was probably some deeper connexion, as the tree has received several names in which dogs and hounds figure. The plant occurs in all neighbouring counties.

37. CAPRIFOLIACEAE, Juss.

ADOXA, Linn. 187. (ἀ, without, δόξα, glory, in allusion to its insignificance.)

775. 478. A. Moschatellina, Linn. *Moschatel.*
Native. Woods, damp hedgebanks, shady places. Plentiful, less so in Malvern district. P. March to May. *Top. Bot.* 91.
First Record. Purlieu Lane, Mathon, *Ballard.* Between Stone and Mitton, *Stokes, Stokes's With. Bot. Arr.,* 2nd ed., p. 417 (1787).

Avon. Broadway, *Lees.* Tardebigge.
Severn. Laughern Brook, *Lees.* Near Holt Mill. Franche, near Kidderminster. Coney Green. Areley Wood. Stourport. Astley. Near Blackstone Rock. Wood at Rimmels Farm, Crowneast. Bog near Doverdale Church. Above Henwick Mill.
Malvern. Old Storridge; Mathon; by the bank of Teme between Leigh Church and Bransford Bridge, *Lees.* Sapey Brook, *Sheward* in *Nash.* Alfrick.
Lickey. Frankley; Tardebigge; Barnt Green; Alvechurch, *Mathews.* Clent. Offmoor Lane, near Hasbury, *Humphreys.* Bromsgrove Lickey. Between Belbroughton and Drayton; Foxcote; Oldnall, near Stourbridge, *Scott.* Blackwell. Edgbaston Lane, near Moseley Hall, *Ick.* Abundant in the upland country about Halesowen, *Mathews.* Redditch, *D. Mathews.* Cofton Hackett. Halesowen Abbey.
This is a humble little plant, one of the first which blossoms to greet the spring, some six or eight inches high, with pale green leaves and little round heads of five yellow green flowers. It has an odour of musk from which it takes its name. It is recorded from all neighbouring counties.

SAMBUCUS, Linn. 188. (σαμβύκη, a musical instrument, in the making of which the wood was anciently employed.)

776. 479. S. nigra, Linn. *Elderberry.*
Native. Woods and hedges. Common and generally distributed. Small tree. May to July. *Top. Bot.* 109.
First Record. With laciniated leaves, Chester Lane, near Land Oak Turnpike, Kidderminster, *Perry, Mag. Nat. Hist.,* vol. iv, p. 450 (1831).
The Elderberry was a perfect pharmacopoeia to the rustic practitioner of old times; nor was it useful only in medicine. A wine made from its berries is still considered efficacious for sore throats, and its hollow branches had many uses. Traditionally, it is the tree on which Judas hanged himself; and the purplish brown fungus that grows in clusters on its bark, *Hirneola auricula-Judae,* Berk., is to this day called the Jew's Ear, a name if not personal to Judas, at least designating his nation. Cattle are very fond of the leaves. The Elder grows in all bordering counties.
Var. b. **laciniata**, Mill.
First Record. As above.
Severn. As above.
Malvern. Leigh Sinton, *Towndrow.* Stanford Park.

777. 480. S. *Ebulus, Linn. *Dwarf Elder, Danewort.*
Denizen or native. Hedges and banks. Very rare. Shrub. July to September. *Top. Bot.* 77.
First Record. Overend, Cradley, *Scott, Hist. Stourbridge,* p. 540 (1832).
Severn. Ripple, *Ballard* in *Withering.* Lower Wick, *J. S. Haywood.* Between Wallsgrove and Brockhill. Leopard Hills. White's Nursery.
Malvern. Near Doddenham Church. Powick.
Lickey. Cradley, as first record; gone.
The old name for this plant was Wallwort; it is not called Danewort till

the sixteenth century, and the name can hardly have belonged to early tradition. Late tradition has said that it was brought into England by the Danes, or that it sprung up and flourished where slaughter occurred between Danes and Englishmen. Parkinson (1640) has another theory, that it took the name from its strong purging qualities, which were so potent that when a man used it he was said to be 'troubled with the Danes'. The plant is native in bushy places in the Centre and South of Europe, hardly reaching the northern coasts, except as a denizen, so that if the Danes brought it with them they were introducing a stranger to themselves. The plant occurs in all bordering counties.

VIBURNUM, Linn. 189. (Etymology unknown.)

778. 481. V. Opulus, Linn.   *Guelder Rose, Water Elder.*
Native. Moist woods, stream-sides, damp hedges. General. Shrub. May and June. *Top. Bot.* 102.
First Record. Cradley Park, *Scott, Hist. Stourbridge*, p. 540 (1832).
The Guelder Rose is well known in our shrubberies, perhaps better there than it is as a wild plant. In the garden it is conspicuous for its large clusters of white flowers, of which all are sterile, instead of only a few of the outer ones as in the wild kind; and in the hedgerow or thicket for its beautiful branches of ruby red fruit, which adorn the leafless branches in autumn. From the time of its flowering it is called the Whitsun, or Wissanboss. One of our most frequent garden shrubs belongs to this genus, the Laurustinus, which is *Viburnum Tinus*, a native of the South of Europe, which from its evergreen leaves has gained in its popular name the addition of Laurus, with which plant it has nothing to do. Our glossy leaved shrubbery Laurel is a cherry, *Prunus Laurocerasus*. The true *Laurus* is the Bay tree. The Guelder Rose occurs in all neighbouring counties.

779. 482. V. Lantana, Linn.   *Wayfaring Tree.*
Native. Hedges and thickets. Many places in the Avon district, rather uncommon elsewhere. Shrub. May and June. *Top. Bot.* 45.
First Record. Ripple Field, *Ballard, Stokes's With. Bot. Arr.*, 2nd ed., p. 318 (1787).
Avon. Bredon Hill. Tiddesley Wood. Craycombe Hill. Broadway Village. Badger's Bank. Bishampton Banks. The Slads. Dovedale.
Severn. Hedge of a plantation near Berwick's Bridge, *Lees.*
Malvern. Silurian eminences, *Lees.* Near Lord's Wood.
This plant is also called the Mealy Guelder Rose, because the young shoots, petioles, and undersides of the leaves are densely, and the upper sides of the leaves more sparingly, covered with stellate down. It loves the Lias country more than any other part of Worcestershire. It does not appear to have occurred in Stafford or Salop of the neighbouring counties.

LONICERA, Linn. 191. (In honour of Adam Lonicer, a German botanist.)

781. 483. L. *Caprifolium, Linn.   *Pale perfoliate Honeysuckle.*
Alien or denizen. Woods and hedges. Very rare. Shrub. May to July. Not in *Top. Bot.*

First Record. Copse beyond the Ketch, *Lees, A. Florence, Strang. Guide* (1828).
Severn. The Ketch, Worcester, *Lees.* Bridewell, Bewdley, *Jorden; Mathews.* Near Kempsey Grove, *Lees.*
Mr. Lees does not acknowledge his first locality in his after-publications, but gives the Kempsey Grove Station on p. xxix of his *Bot. Worc.* The plant is a native of Southern and Eastern Europe. The upper leaves are connate, from which fact it takes its English name; the rest are distinct. It has been noticed only in Salop of neighbouring counties.

782. 484. L. Periclymenum, Linn.   *Honeysuckle or Woodbine.*
Native. Woods, thickets, hedges. Widely distributed, common everywhere. Shrub. May to September. *Top. Bot.* 112.
First Record. Not localized, *Lees, Bot. Malv. Hills*, 1st ed., p. 18 (1843).
The sweet adornment of our country hedges during the height of summer, dear to all lovers of nature, whose budding leaves at an earlier period are one of the first earnests of coming spring. The twist of the climbing shoot follows the sun, from east to west. The foliage of the Honeysuckle is very agreeable to goats, whence the specific name of the preceding species; and the fact is alluded to in many of the names of the plant in European countries. It occurs in all adjoining shires.

783. 485. L. Xylosteum, Linn.   *Upright Fly Honeysuckle.*
Native or denizen. Hedges and thickets. Very rare. Shrub. July. *Top. Bot.* 1.
First Record. On the eastern side of Longdon Marsh, and at Powick, *Lees, Bot. Malv. Hills*, 1st ed., p. 18 (1843).
Severn. Hilditch Coppice at the end nearest Hartlebury Common, *Miss Ladbury.* Head of Pool near Vallombrosa, Hartlebury Common, *Rea.* Woody border of Lake in Westwood Park, *Lees.* In a coppice close to the South Lodge of Ombersley Park, *Jeffery.*
Malvern. As first record.
Although this plant is excluded by botanists from the list of native plants of Britain, there is a good deal to be said for its claims to be in it. There is nothing against them geographically, for it is native in the woods of Belgium. No doubt in many of its stations it is a bird-sown plant from some neighbouring garden, but in many its appearance is perfectly wild among natural surroundings, and it is not a very frequently cultivated plant, having neither odour nor appearance to recommend it in comparison with others of the genus. The plant has not been observed in Gloucester or Hereford.

SYMPHORICARPUS, Michx. (σύν, together, φέρω, I bear, καρπός, fruit, in allusion to the cluster of berries.)

S. RACEMOSUS, Michx.   *Snowberry.*
This North American shrub is extensively planted in gardens, and well known from its large white berries. It is rapidly advancing in England to become the asterisked established alien instead of the italicized stranger of the books. It is strongly established at Hagley Brake in the Severn district, where its creeping roots in the light soil will probably enable it to maintain itself indefinitely. It is also growing freely on the left bank of the

Severn nearly opposite Gladder Brook. It is not mentioned in *Lond. Cat.*, 10th ed.

38. RUBIACEAE, Juss.

GALIUM, Linn. 193. (γάλα, milk, from some species being used to curdle milk.)

786. 486. G. Cruciata, Scop.   *Crosswort Bedstraw, Mugwort.*
Native. Hedges, woods, thickets. General. P. April to June. *Top. Bot.* 97.
First Record. Plentiful from Newcastle to within a few miles of Worcester, but further south it is scarce, *Mr. Baker, Withering*, 3rd ed., p. 187 (1796).
This dull looking plant, usually covered with roadside dust which clings to its hairy leaves, is a frequent object. The upper blossoms of the clusters are female, the outer ones male. Its leaves, broader than those of its relations, are arranged crosswise in whorls of four. It occurs in all neighbouring counties.

787. 487. G. verum, Linn.   *Yellow Bedstraw.*
Native. Dry pastures and field-borders. Generally distributed. P. May to September. *Top. Bot.* 112.
First Record. In pastures, *Pitt, Agric. Worc.*, p. 317 (1810).
This is conspicuous among our summer flowers. The plant was specially 'Our Lady's Bedstraw' of our ancestors, and from it the name Bedstraw, for no reason in particular, was given generally to other members of the genus by later writers. The plant will curdle milk, and at one time was used in cheesemaking; hence the name *Galium.* The children at Stoulton call all the yellow and white bedstraws Lady's Needlework, Jack-in-the-bed, Wild Needlework, and Lady's Wash, the last from their use in driving away freckles. It is a common European and North Asian plant, extends southward to the Himalayas, and is found also in North Africa. Of course it occurs in all bordering counties.

788. 488. G. erectum, Huds.   *Upright Bedstraw.*
Native or colonist. Meadows and pastures. Very local and rare. P. July to September. *Top. Bot.* 34.
First Record. Near Alfrick, *Lees, Bot. Malv. Hills*, 1st ed., p. 15 (1843).
Avon. Near Alderminster, *Cheshire.* Roadside between Eckington and Evesham.
Severn. Bank of the Bewdley and Kidderminster Railway.
Malvern. Near Alfrick, *Lees.* Malvern Wells, *Towndrow.*
This plant is often considered a sub-species of the following *G. Mollugo*, from which it differs in possessing narrower leaves and cymes with slender ascending branches instead of spreading ones, while the whole plant is suberect, instead of lying on the ground and only rising at the end of the stem. It occurs in all bordering counties.
×verum.
First Record. As here.
Malvern. Malvern Wells, 1907, *Rendall.*

789. 489. G. Mollugo, Linn.   *Great Hedge Bedstraw.*
Native. Hedges and thickets. General. P. May to September. *Top. Bot.* 77.
First Record. Malvern Hills, *Stokes's With. Bot. Arr.*, 2nd ed., p. 155 (1787).
This is a frequent plant in hedgerows and among bushes in most parts of the county, though it is very seldom seen in the Lickey district. It bears its flowers in loose spreading panicles at the end of long slender stems. It occurs in all neighbouring counties.
Var. c. **Bakeri**, Syme.
First Record. On dry banks, Half Key, *Towndrow, Malv. Advert.*, November 19, 1892.
Malvern. As first record.
This variety is characterized by its linear leaves and few-flowered cymes.
Var. **elatum**, Thuill.
(*G. scabrum*, With. Hooker, *Stud. Fl.*, 3rd ed., 194.)
First Record. As *G. scabrum.* In a hedgerow on marly soil on the side of Redhouse Lane, Worcester, *Stokes, Stokes's With. Bot. Arr.*, 2nd ed., p. 155 (1787).
Severn. As above.
Sir Joseph Hooker, at the reference given above, considers this plant to be *G. Mollugo* proper. It is not in *Lond. Cat.*, 10th ed.
×verum.
(*G. ochroleucum*, Syme.)
First Record. Madresfield, *Towndrow, Malv. Advert.*, November 19, 1892.
Malvern. As first record.

790. 490. G. saxatile, Linn.   *Heath Bedstraw.*
Native. Commons and open places. Locally abundant. P. May to August. *Top. Bot.* 111.
First Record. On all the heaths in the vicinity of Kidderminster, *Perry, Mag. Nat. Hist.*, vol. iv, p. 450 (1831).
When the conditions are those congenial to the plant, there it is to be found. Usually a low growing plant, and quite at home in the grassy turf of a sheep-trimmed common, in places where it is protected by some bush or encouraged by some dampness in the soil it will grow a foot or more in height. On the Clent Hills it is very abundant, here and there quite whitening the turf with its flowers. It is found in all neighbouring counties.

791. 491. G. asperum, Schreb.   *Mountain Bedstraw.*
(*G. pusillum*, Sm. *G. sylvestre*, Poll. *G. umbellatum*, Lam.)
Native. Grassy slopes, very rare. A. May to July. *Top. Bot.* 30.
First Record. As *G. pusillum.* East side of Red Hill, *Lees, A. Florence, Strang. Guide* (1828).
Severn. Red Hill, Worcester, as above.
Malvern. Field near the Hornyold Arms Hotel, *Worc. Nat. Club Trans.* i, p. 404.
Mr. Mathews supposes that the first record above refers to the record by Dr. Stokes of *Galium scabrum*, referred to under *G. Mollugo*, var. *elatum* above, 'as growing at this spot,' he says. It must be supposed that he knew the locality, since to the ordinary reader there is nothing common between

the two records except the adjective 'red', and the neighbourhood of Worcester. Mr. Mathews goes on to say that *G. pusillum*, Sm., is a limestone plant of the northern counties not likely to occur in Worcestershire. At all events Mr. Lees records the extermination of the Worcestershire plant at p. 153 of the *Illus. Nat. Hist. Worc.* (1834). The plant has been observed in Staffordshire and Gloucestershire.

792. 492. G. palustre, Linn.   *White Water Bedstraw.*
Native. Meadows, marshes, ditches, pond-sides. Not uncommon. P. June to September. *Top. Bot.* 112.
First Record. *Lees, Bot. Malv. Hills*, 1st ed., p. 15 (1843).
This plant occurs in all neighbouring counties.
Var. c. **Witheringii** (Sm.).
First Record. As *G. Witheringii.* Not localized, *Scott, Hist. Stourbridge*, p. 540 (1832).
Severn. North-east end of Hartlebury Common. Near Shrawley. Hadsor. Shrawley Wood. Poolside by first coppice, Broadheath. Hoo Pool Mill. Near Holt Turnpike.
Malvern. Admiral's Covert, Malvern Wells, *Towndrow, Malv. Advert.*, November 23, 1895.
This is a smaller form of the type, with very rough stems, and ascending panicled branches. The type varies considerably, and doubtless the leaf development varies with the character of the season. The variety has been observed in Salop, Stafford, and Warwick.

793. 493. G. uliginosum, Linn.   *Rough Marsh Bedstraw.*
Native. Wet heaths, marshes, among rushes, borders of ponds. General, except in the Malvern district, where it is rare. P. June to August. *Top. Bot.* 93.
First Record. Feckenham Bog, *Purton, Midl. Fl.*, p. 99 (1817).
This plant and the former one are frequently to be found growing together. Like many of the genus it turns blackish in drying, while *G. palustre* remains green; and it is a rougher plant than that, the leaves being always narrower and more rigid, and mucronate, not blunt. Mr. Towndrow says the plant is rare in the Malvern district. It occurs in all neighbouring counties.

794. 494. G. anglicum, Huds.   *Wall Bedstraw.*
Native. Dry places. Very rare. A. June and July. *Top. Bot.* 10.
First Record. Among clover at Malvern Link, *Towndrow, Malv. Advert.*, November 19, 1892.
Malvern. Arable land, *Towndrow.* Introduced with clover seed.
This plant does not appear to have been observed in any neighbouring county.

795. 495. G. *Vaillantii, DC. *Top. Bot.* 4.
Severn. Stanklin, *Herbarium, Hast. Mus., Worc.*
This plant is given by Hooker, *Stud. Flor.*, 3rd ed., 1884, p. 194, as a sub-species of the foregoing; while Babington, *Manual Brit. Bot.*, 9th ed., 1904, p. 187, gives it as variety β. of *G. spurium*. *G. spurium* was recorded by Pitt, *Agric. Worc.*, p. 317 (1810), calling it Corn Galium or Hairough. Mr. Mathews,

Mid. Nat. x, p. 202, thinks this was possibly an error for *G. tricorne*, and considers it the first record of that plant. The number of *G. Vaillantii* in *Top. Bot.* is 4, and it is not recorded for any neighbouring county.

796. 496. G. Aparine, Linn.   *Goose Grass, Cleavers.*
Native. Hedgerows, waste places, cultivated ground. Very common. A. May to August. *Top. Bot.* 112.
First Record. Marshes, *Scott, Hist. Stourbridge*, p. 540 (1832).
Everybody knows this plant, festooning the hedgerows with its long lank stems, sometimes forming great matted masses; and often intruding itself among the crops of the garden. It is very rough, and clings to everything, while its round seeds, covered with short hooked bristles, cling to the clothing of the passer by most tenaciously, well justifying its country name of Cleavers. Its other common name is derived from the supposed fondness of geese for its herbage. Another wide-spread name for it is Hairiff, an appellation it possessed in Saxon times, frequently in Worcestershire with the 'h' omitted and the word 'grass' added to it. Its juice has been praised as a purifier of the blood, and it was one of the numerous remedies for the bite of 'serpents', which our ancestors seem to have thought lurked in every bush. Also a broth made from it was taken to prevent the 'lank and lean' from growing fat. The plant is found in all neighbouring counties.

797. 497. G. tricorne, Stokes.   *Rough-fruited Corn Bedstraw.*
Colonist. Cornfields. Local. A. June to August. *Top. Bot.* 43.
First Record. In cornfields at the Croft, Mathon, *Lees, Bot. Malv. Hills*, 1st ed., p. 15 (1843).
Avon. Trench Woods. Craycombe Hill. Near Crowle.
Severn. In the neighbourhood of Bewdley, *Jorden.* Tibberton. Field near Spetchley Common.
Malvern. Silurian eminences, *Lees.* Cornfield on Old Storridge Hill. Croft Farm, Mathon; Hanley Castle, *Lees.* Near New Pool. Ridge Hill, Martley.
A wide-spread weed of cultivated ground in Europe possibly native in the East and Palestine. It is not native of our island, being only seen as a cornfield plant. It has been noticed in Gloucester and Warwick only of our neighbouring counties.

ASPERULA, Linn. 194.   (*Asper*, rough, from the rough hairs of many of the species.)

798. 498. A. odorata, Linn.   *Sweet Woodruff.*
Native. Woods, thickets, hedgebanks. General in the Avon and Lickey districts, many localities elsewhere. P. May and June. *Top. Bot.* 106.
First Record. At the Leasows, near Halesowen, *Withering, Stokes's With. Bot. Arr.*, 2nd ed., p. 158 (1787).
Avon. General, *Lees.* Trench Woods. Norton, near Evesham. Tiddesley Wood.

Severn. Wychbury Wood, Pedmore. Wassell Hill, Bewdley. Wyre Forest. Wood at Hawford. Monk Wood. Ockeridge. Shrawley Wood.
Malvern. Leigh Sinton. Old Storridge. Southstone Rock. Stanford. Ankerdine Hill. Clifton-on-Teme. Sapey Brook.
Lickey. Offmoor Wood. Clent Hills. Beoley, *Mathews.* Romsley. Frankley. Tardebigge; Alvechurch, *D. Mathews.* Hagley Wood. Chaddesley Woods. Randams.
The plant when dried is fragrant with the odour of new mown hay, but when growing the scent is scarcely perceptible. It was put in wine 'to make a man merrie', and was thought to be 'good for the heart and liver'. It is eaten by cattle and horses. The Woodruff occurs in all neighbouring counties.

800. 499. A. cynanchica, Linn.   *Small Woodruff, Quinancy-wort.*
Native. Limestone pastures and open places. Very local. P. May to September. *Top. Bot.* 40.
First Record. Broadway Hill, *Sheward, Nash, Hist. Worc., Sup.*, p. 96 (1799).
Avon. Broadway. Bredon Hill.
The one place in Worcestershire where this plant is always to be seen is Bredon Hill. It is very different in appearance from the Sweet Woodruff, and in odour also, for it smells most disagreeably. Its name of Quinancy or Quinsey, wort, refers to its uses in disorders of the throat. It is found also in Gloucestershire and Warwickshire of the bordering counties.

801. 500. A. arvensis, Linn.
Casual alien. Waste ground, shrubberies. Very rare. A. June. Not in *Top. Bot.*
First Record. Near Hartlebury Church, *Worc. Nat. Club Trans.* iii, p. 2 (July 13, 1899).
Severn. As first record.
This plant is a native of rough stony ground at high altitudes in Syria, Persia, and Afghanistan. In Mid and Southern Europe it is a cornfield weed, and has been known in England for many years. The flowers are bright blue. It has not been met with in any bordering county.

SHERARDIA, Linn. 195.   (In honour of James Sherard, an English botanist.)
802. 501. S. arvensis, Linn.   *Field Madder.*
Native. Cultivated fields, roadsides, waste ground. Common in suitable situations. B. February to October. *Top. Bot.* 109.
First Record. Corn and grass fields, *Scott, Hist. Stourbridge*, p. 540 (1832).
This is a small plant, usually decumbent, but the umbel of lilac flowers is a pretty little object, while the structure of the calyx and fruit is very curious. It occurs in all the neighbouring counties.

39. VALERIANACEAE, Juss.

VALERIANA, Linn. 196.   (*Valeo*, I am powerful, from its powerful medicinal properties.)

803. 502. V. dioica, Linn.   *Small Marsh Valerian.*
Native. Marshes, wet places in meadows. General down the centre of the county, not so common elsewhere. P. April to June. *Top. Bot.* 73.
First Record. In moist meadows, hedgesides, &c., *Pitt, Agric. Worc.*, p. 317 (1810).
Avon. Tiddesley Wood. Fladbury. Pershore. Defford Brook.
Severn. Wild in all the bogs in Wyre Forest, and in every osier plantation round Worcester, *Rea.* Blakedown Pool; Stanklin Pool; Fenny Rough; Pepper Wood, Belbroughton, *Humphreys.* Bubble Bridge, *Lees.* Doverdale Bog. Dunclent. Shrawley Wood. Ockeridge Wood. Monk Wood.
Malvern. Boggy places, local, *Lees.* Woodside beyond Old Storridge. Walms Well. New Pool.
Lickey. Bilberry Hill Reservoir; Bittell Reservoir; Rubery, *Humphreys.* Dales Wood, Romsley. Woods in Clent. Illey. Halesowen. Hunnington. Lickey Woods.
A pretty little flower of damp meadows in May, with a corymb of pink flowers of two sizes, the males larger than the females. The cotyledonary leaves resemble those of *Parnassia palustris*. The plant occurs in all bordering counties.

804. 503. V. officinalis, Linn.   *Wild Valerian.*
(*V. Mikanii*, Syme.)
Native. Dry woods. Once recorded, as below. P. May to August. *Top. Bot.* 21.
First Record. Malvern, *Towndrow, Malv. Advert.*, November 19, 1892.
Malvern. As first record.
This plant is the typical form of *V. officinalis*, Linn. It has eight to ten pairs of leaflets, while the following form, *V. sambucifolia*, has four to six. Though only once seen in our county, it occurs in all that border us.

805. 504. V. sambucifolia, Willd.   *Wild Valerian.*
(*V. officinalis*, Linn.)
Native. Watersides, damp hedges, wet woods. General. P. June to August. *Top. Bot.* 111.
First Record. As *V. officinalis.* In moist meadows and hedgesides, &c., *Pitt, Agric. Worc.*, p. 317 (1810). First record of the segregate, Malvern, *Towndrow, Malv. Advert.* 1892.
This is a plant so frequently met with as to require no list of localities. The powerful scent of the Valerian is displeasing to many people; on the other hand it is peculiarly agreeable to cats; they seem nearly to go mad over it! A plant of the genus is supposed to have afforded the Spikenard of the East. The leaves are used by country people as an application to fresh wounds, whence one of its common names, All-heal. Gerarde calls the plant 'Setewall'. Wild Valerian occurs in all bordering counties.

KENTRANTHUS, Neck. 197.* (κέντρον, a thorn, ἄνθος, a flower, from the spurred corolla.)

807. **505. K. *ruber**, DC.   *Red Spur Valerian.*
(*Valeriana rubra*, Linn.)
Alien. Railway-banks, walls. Rare. P. June and July. Not in *Top. Bot.*
First Record. On the old wall of the western entrance to the Cloisters, *A. Florence, Strang. Guide* (1828).
Severn. As above.
Malvern. Rock, Wych Road, *Towndrow, Malv. Advert.*, November 14, 1896.
This showy garden flower is a native of the rocks of Southern Europe. It is thoroughly naturalized in the southern parts of England. It is a sturdy plant, seeding itself freely when in a state of cultivation, in which condition its blossoms are several shades of red, pink, and white. Railway embankments seem a favourite habitat, and sometimes glow crimson with its flowers from a considerable distance away. Lady's Needlework is a country name for it. The plant has been noticed in Gloucester, Salop, and Stafford.

VALERIANELLA, Hill 198. (Diminutive of *Valeriana*.)

809. **506. V. olitoria**, Pollich.   *Common Corn Salad, Lamb's Lettuce.*
(*Fedia olitoria*, Vahl.)
Native. Dry hedgebanks, walls, cornfields. General. Abundant in the Lickey district. A. April to June. *Top. Bot.* 99.
First Record. Not localized, *Lees, Bot. Malv. Hills*, 1st ed., p. 14 (1843).
This is not a very attractive plant; its flowers are very small, usually a dull pale blue, and the leaves pale green. The stem is repeatedly two-forked. As its specific name would betoken, it has its uses in the kitchen, being cultivated for salad, especially on the Continent. Doubts have been thrown upon its true nativity in England, as at one time it was considered to be truly indigenous only in Sardinia and Corsica. It has been noticed in every bordering county.

{*810.* **507. V. *eriocarpa**, Desv.
Colonist. Fields and waste ground. Very rare. A. May to September. *Top. Bot.* 5.
First Record. By the side of the road between New Pool and the Hanley Turnpike Gate below Malvern Wells, *Lees, Bot. Malv. Hills*, 1st ed., p. 14 (1843). 'I have not since found it,' *Lees*, 3rd ed., p. 44 (1868).
Malvern. As first record.
Mr. Lees equates his plant with *Fedia mixta*, Vahl, under which name he first described it, according to a statement in *Bot. Worc.*, p. xlviii. The above first record must be received with caution; more probably it is in its right position under *V. dentata*, var. b. *mixta*, where in *Bot. Worc.*, it is also placed. *Valerianella mixta*, Dufr., is a variety of *V. dentata*. But the distinctions between all these corn salads are rather fine, chiefly depending upon their fruit. This plant is sometimes cultivated in gardens under the name of Italian Corn Salad. It is a native of the South and West of Europe. It has been observed in Salop and Stafford.}

811. **508. V. carinata**, Loisel.   *Keeled Corn Salad.*
Colonist. Dry banks. Very rare. A. May to June. *Top. Bot.* 16.
First Record. Two very local plants Mr. Cheshire sent me from near Alderminster—*Galium erectum* and *Valerianella carinata* (*Fedia carinata*, Stev.), and on examination I find no reason to disprove the designations given, *Lees, Bot. Worc.*, p. 98 (1867).
Avon. Near Alderminster, *Cheshire*.
Severn. North Wood, near Bewdley, *Mathews*, 1850.
Malvern. Near Malvern, 1897, *Towndrow*. Newtown Nursery, Malvern.
This is another doubtful native, though as it is indigenous throughout Europe as far as the north coast of Guernsey, there would seem to be no sufficient reason for excluding it from the list of native plants in our island also. It has been noticed only in Salop of our neighbouring counties.

{*812.* **509. V. rimosa**, Bast.   *Sharp-fruited Corn Salad.*
(*V. Auricula*, DC.)
Colonist. Cultivated fields, very rare. A. June and July. *Top. Bot.* 37.
First Record. Worcester(shire), with (E. Lees) after the record, *Watson, Top. Bot.*, 2nd ed., p. 217 (1883).
This plant is not localized in any district in Mr. Lees' *Table of Plants*, nor mentioned in the body of his *Bot. Worc.* It is another native of the Mediterranean region, which spreads northward through Europe as a weed of cultivated ground. It is recorded also for Gloucester and Hereford of neighbouring counties.}

813. **510. V. dentata**, Poll.   *Narrow-fruited Corn Salad.*
(*Fedia dentata*, Vahl.)
Colonist. Cultivated fields. Local. A. May to September. *Top. Bot.* 8a.
First Record. As *Fedia dentata*. Not localized, *Lees' Cat., New Bot. Guide* (1835).
Avon. Bredon Hill. Broadway. Field near Trench Woods.
Severn. Field near Wyre Forest, Bewdley. Near Coningree Wood. Railway Bank, near Brake Mill, Hagley. Bewdley. Near Woods Green. Warndon.
Malvern. Cornfields and roadside banks, *Lees*. Near New Pool. Malvern Wells. Half Key.
Lickey. Field between Hopwood Wharf and Rowney Green, Alvechurch. Farley Farm, Romsley.
This is a common weed of cultivated ground throughout Europe, spreading from its home, Dalmatia, where it is undoubtedly native. It is recorded from all neighbouring counties.

Var. b. *mixta*, Dufr.
First Record. By the side of the road between New Pool and Hanley Turnpike Gate below the Wells, *Lees, Bot. Malv. Hills.*, 1st ed., p. 14 (1843).
Severn. Warndon, 1850, *Thompson, Towndrow*.
Malvern. As first record.
In this variety the fruit is hispid.

40. DIPSACEAE, Juss.

DIPSACUS, Linn. 199. (δυψάω, I thirst, from the upper connate leaves usually containing water in their hollows.)

814. **511. D. sylvestris**, Huds.   *Teasel.*
(*Dipsacus fullonum*, var. a, Linn.)
Native. Damp hedges and roadsides, wet woods. General, rather uncommon in the Lickey district. B. July to September. *Top. Bot.* 74.
First Record. Moist hedges, *Pitt, Agric. Worc.*, p. 317 (1810).
Avon. Bredon Hill. Fladbury. Trench Woods. Littleton Banks. Norton, near Evesham. Pershore. Tiddesley Wood. Defford Common. The Slads. Craycombe. Besford.
Severn. Hagley Brake (one plant), 1874, *Mathews* and *Thompson*. Shrawley Wood. Wyre Forest. Ribbesford Woods. Monk Wood. Ockeridge Wood. Nunnery Wood. Kempsey. Ombersley. Hadley. Banks of Laughern Brook. Cotheridge. Claines.
Malvern. Abundantly dispersed, *Lees*. Bransford. Powick. Madresfield. Leigh. Knightsford Bridge. Martley. Stanford Bridge. Dripshill Wood. Malvern Link.
Lickey. Occasional on a clay soil, *Mathews*. Feckenham, *Mathews*. Dodford. Randans Woods. Chaddesley Woods. Tardebigge.
The chief difference between this plant and the Fuller's Teasel consists in the spines in the latter being hooked; but as under the influence of poorer soils the hooks tend to disappear there is reason to believe that the plants really are the same. The use of the Teasel in the cloth trade is well known, and one that has existed from classical times. During the winter the chaffy bristly heads stand up conspicuous, and give rise to the country names of Wood Broom and Hair Brushes. The water collected in the leafy cups was considered a cure for inflamed eyes. The larva of a small insect (either *Penthina gentiana* or *Eupoecilia roseana*) that infests Teasel heads was considered to charm away agues, but Gerarde, who had personally tried this and other ' physick charmes', stigmatizes them all as ' foolish toies that I was constrained to take by fantasticke people's procurement', and says 'they did me no good at all'. Old writers also recommended the plant for hygrometrical purposes. The head was said to grow smoother at the coming of wet and windy weather, and ' against rain will close up all his prickles'. The Teasel occurs in all bordering counties.

815. **512. D. pilosus**, Linn.   *Small Teasel, Shepherd's Rod.*
Native. Damp banks and sides of streams. Not common, less rare in the Severn and Malvern districts. P. July to September. *Top. Bot.* 52.
First Record. Below the Abbey, Great Malvern, and most abundant in a lane between the Church and the Priory Farm at Little Malvern, *Lees, Bot. Malv. Hills*, 1st ed., p. 15 (1843).
Avon. The Slads. Near Evesham. Hanbury. Defford Brook. Crowle Brook.
Severn. Bubble Bridge. Orls Coppice, Rushwick. Blackstone Rock. Shrawley Wood. Dick Brook. Wyre Forest. Elmbridge Mill. Astley.

Upper Wick. Kenswick Brook. By the Salwarpe, near Hawford. By the Salwarpe, near Impney.
Malvern. Silurian eminences, *Lees*. Between Powick and the Old Hills; Old Storridge; Mathon Park Copse; Between the Church and Priory Farm, Little Malvern; Below the Abbey, Great Malvern; *Lees*. Newland. Madresfield. Clifton-on-Teme. Witley Court. Stanford Bridge. Martley.
Lickey. Moseley, *Ick*. Halesowen Manor; Illey Mill, *Mathews*.
The small Teasel grows in all neighbouring counties.

SCABIOSA, Linn. 200. (*Scabies*, a scab, alluding to its use in curing cutaneous diseases.)

816. **513. S. Succisa**, Linn.   *Devil's Bit Scabious.*
Native. Pastures, moist woods, commons. Generally distributed. P. June to August. *Top. Bot.* 54.
First Record. Rough pastures, *Pitt, Agric. Worc.*, p. 317 (1810).
The purplish blue flowers of this scabious are commonly to be seen during late summer on pastures and hillsides standing on a stiff stem a foot or so in height. The root, after the first year, is as it were cut or bitten off abruptly. It is said the Devil did it, whence its English name. He did it out of envy, or otherwise 'it would be good for many uses', says an old writer. Another one attributes the Devil's interference to the fact that he 'was not troubled with any disease for which it was proper'. However, what the Devil left of the root was 'very powerful against the plague and all pestilential diseases', and, of course, 'the bitings of all venomous beastes.' Apparently old herbalists lay no stress on its efficacy against the complaints indicated by its generic name, though its common names on the Continent all have reference to this fact. The plant is recorded from all neighbouring counties. It occurs with white flowers at the Ockeridge end of Monk Wood, and, indeed, generally throughout the county.

817. **514. S. Columbaria**, Linn.   *Small Field Scabious.*
Native. Dry pastures. Very rare. P. June to September. *Top. Bot.* 72.
First Record. On Breedon Hill, *Nash, Hist. Worc., Int.*, p. lxxxix (1781).
Avon. Bredon Hill. Shemington Hill, Alderminster, *Cheshire*. Roadside near Craycombe Hill. Below the Fish, Broadway.
Malvern. Near Knightwick, *Rea*.
The purplish flowers of this plant have a more fully expanded appearance than those of the last kind. Its leaves are lighter in colour, the cauline ones finely divided, and the radical on long stalks. It is a plant of calcareous soil, and occurs in all the bordering counties.

818. **515. S. arvensis**, Linn.   *Field Scabious.*
(*Knautia arvensis*, Coult.)
Native. Cultivated fields, hedgebanks. General. P. June to August. *Top. Bot.* 99.
First Record. Not localized, *Lees, Bot. Malv. Hills*, 1st ed., p. 15 (1843).
This is a handsome flower, varying much in colour but generally of a light bluish lilac, with large outer florets, and frequently to be seen among the

ripening corn and on roadsides. Let the smoker puff a little tobacco smoke over the blossoms ; they will assume a rich green colour, and for a time, seem otherwise uninjured. The plant occurs in all neighbouring counties.

41. COMPOSITAE, Juss.

EUPATORIUM, Linn. 201. (*Eupator*, a name of Mithridates the Great, who is said to have brought this plant into use.)

820. 516. E. cannabinum, Linn. *Hemp Agrimony.*
Native. Marshes, wet woods, stream-sides. General. P. July to September. *Top. Bot.* 98.
First Record. Yardley, *Miss M. A. Beilby, Analyst,* vol. vi, p. 294 (1837).
Detailed localities are unnecessary in the case of this widely distributed plant. It is a tall plant, with dense clusters of small, rather dingy looking, flesh-coloured blossoms. Yet when well grown it makes a handsome addition to the autumn garden border, always exciting remark and astonishment among non-botanists that it is a common British wild flower. Ignorant of the etymology of the name, an old herbalist asserts that it is called *Eupatorium*, as if that word was *Hepatorium*, because it strengthened the liver. It is said that if laid near loaves it will prevent them going mouldy. It is recorded from all neighbouring counties.

SOLIDAGO, Linn. 202. (*Solidare*, to unite, from its supposed vulnerary properties.)

821. 517. S. Virgaurea, Linn. *Golden Rod.*
Native. Woods, thickets, heaths. Not common. P. July to September. *Top. Bot.* 110.
First Record. Not localized, *Walker, Med. Surg. Rep.,* vol. i, no. 2 (November, 1828).
Avon. Norton, near Evesham, on the Railway-banks.
Severn. Burnt Wood near Bewdley, *Perry.* Wyre Forest. Coleridge Wood, Wolverley. Shrawley Wood. Lincombe.
Malvern. Malvern Hills, *J. K. Walker.* Near Dripshill Wood. North Hill. Martley. Knightwick. Old Storridge.
Lickey. Bromsgrove Lickey, *Humphreys.* Coppices at Cradley and Stamber Mill, *Scott.* Halesowen Road, *Ick.* On Railway embankment at Barnt Green, and at Blackwell.
Gerarde says this plant was ' extolled above all the herbes for the stopping of blood ', and says at one time it was sold in London for half a crown an ounce, but afterwards when it was found in Hampstead wood, no one would give half a crown for a hundredweight of it. Whereupon he moralizes on the instability of human affairs ! The plant occurs in all neighbouring counties.

Var. b. **cambrica**, Huds.
First Record. As below.
Malvern. Malvern Hills, *Rea.*
The variety is squat in appearance, possesses broader ciliate leaves, and

simple large-flowered cymes. Neither variety has been recorded for any bordering county.

Var. **angustifolia**, Gaud.
First Record. As below.
Severn. Wyre Forest, *Rea.*
This variety is characterized by its narrower leaves, the upper ones often serrate, and its fewer flowers. This variety is not given in *Lond. Cat.*, 10th ed.

BELLIS, Linn. 203. (*Bellus*, pretty.)

822. 518. B. perennis, Linn. *Common Daisy.*
Native. Fields, meadows, everywhere. Abundant and widely distributed. P. January to December. *Top. Bot.* 112.
First Record. ' Bellis flore herbaceo globoso. In Mr. Selden's Cops neer his house in Worcestershire,' *Mr. Morgan, Merrett's Pinax* (1666).
This familiar and favourite flower, not only of Britain but of Europe generally, gets its English name probably not so much from the appearance of the flower, but because in the evening the ray flowerets close over the yellow disk, to open again when morning comes. In sunny places the rays are often tipped with crimson, while in the shade they usually remain white ; if the rays are entirely red, the plant is advanced to the dignity of a variety, var. *coloratus*, Peterm. In France daisies are called Marguerites, a name in England mainly reserved for larger flowers of similar appearance. Several eminent women of the name of Margaret have been connected with the flower, notably Margaret of Anjou, who adopted the daisy as her device ; but the name is taken doubtless from St. Margaret, who was the type of female innocence and meekness. The range of the Daisy is across Central Europe from East to West, and into Asia Minor. It is not known in North Russia nor in Greece. It occurs in all neighbouring counties.

Var. **non-radiata**, Lees.
First Record. Not localized, *Lees, Bot. Malv. Hills,* 1st ed., p. 39 (1843).
Severn. Garden, 34 Foregate Street, Worcester. Hartlebury Common. Wyre Forest.
Malvern. Malvern Wells, *Towndrow.*
This variety is perhaps a passing state, more or less permanent. It is not mentioned in the botanies or catalogues.

ERIGERON, Linn. 205. (ἦρι, early, or perhaps ἔριον, hair, γέρων, an old man, from the early ripening of the grey seed down.)

826. 519. E. canadense, Linn. *Canadian Fleabane.*
Alien. Introduced from North America, occurring on waste ground. A. July to September. Not in *Top. Bot.*
First Record. One plant with other casuals in a hop-yard at Leigh Sinton, *Towndrow, Malv. Advert.,* Dec. 22, 1900.
Avon. Near Pershore Mill.
Severn. Porter's Mill. Bewdley.
Malvern. As above. Leigh Mill.

o

This North American plant has spread over all the temperate regions of the world. It was noticed in France in the seventeenth century, and a few years later in England also. It is common near London, and in the southern and eastern counties. In the midlands, previous to the above occurrence in Worcester, it had been observed in Staffordshire only.

827. 520. E. acre, Linn. *Blue Fleabane.*
Native. Dry pastures, railway-banks, wall-tops. Locally abundant, but absent from considerable parts of the county. A. or B. May to October. *Top. Bot.* 65.
First Record. Tops of walls about the Cathedral, *Sheward, Nash, Hist. Worc., Sup.,* p. 96 (1799).
Avon. Craycombe. Shemington Hill, Alderminster, *Cheshire.* Bredon Hill. Broadway. Sheriff's Lench, *Rufford.*
Severn. About Worcester Cathedral. Hartlebury Common, first appearance, 1900, *Humphreys.* Sutton Common. Dunghill at Hoo Pool. Bewdley. Foley Park. Outside Oldington Wood.
Malvern. Leigh, north of the churchyard, *Lees.* Barnard's Green. Brace's Leigh.
Lickey. Lime Rocks, Dudley, *Withering.*
The whole plant is rough to the touch. It occurs in all neighbouring counties.

FILAGO, Linn. 206. (*Filum*, a thread, the whole plant being covered with cottony hairs.)

829. 521. F. germanica, Linn. *Common Cudweed.*
(*Gnaphalium germanicum*, Linn., Sp. Pl.)
Native. Cultivated fields, heaths, railway-banks. Plentiful, scarcest in the Lickey district. A. June to September. *Top. Bot.* 96.
First Record. As *Gnaphalium germanicum*. Not localized, *Lees, Bot. Malv. Hills,* 1st ed., p. 38 (1834).
A frequent and singular looking plant, having at the top of its cottony stem a globular assemblage of heads, from the base of which rise two or more stalks, bearing the same kind of head. The herbalists called it *Herba impia*, as if the smaller stalks were undutiful in raising themselves above the parent head. In the ' New Locality List ', *Bot. Rec. Club Rep.* for 1883, Dr. F. A. Lees notices under this head a *Filago apiculata*, a remarkable plant growing in sandy fields near Kidderminster (Dr. Fraser) with crimson-tipped involucral bracts. Professor Babington referred it to this plant. The English name of the plant was given it because it was administered to cattle that had lost their cud. The plant occurs in all neighbouring counties.

{**830. 522. F. apiculata**, G. E. Sm.
Native. Dry sandy soil. Very rare. A. July to September. *Top. Bot.* 19.
First Record. Not localized, Worcestershire, *Mr. T. Westcombe, Phyt.,* O.S., vol. iv, pt. 2, p. 715 (1858).
Severn. Near Hartlebury Common, *Westcombe.*
Mr. Lees, *Bot. Worc.,* p. 46, gives the locality as near Hartlebury Common. By some botanists the plant is still ranked as a variety of the foregoing,

from which it differs in being taller, with broader leaves, heads acutely five-angled, involucral bracts purplish and boat-shaped, the tips reddish. It is found chiefly in the eastern counties, and has not been noticed in any of those bordering Worcestershire.}

832. 523. F. minima, Fr. *Least Cudweed.*
(*F. montana*, Sibth. *Gnaphalium montanum*, Huds.)
Native. Dry sandy fields, heaths, and commons. Many places in the Severn and Malvern districts. A. June to September. *Top. Bot.* 91.
First Record. As *Gnaphalium montanum*. On a common between Ombersley and Hartlebury, near the Mitre Oak. Malvern Hill, *Purton, Midl. Fl.,* p. 390 (1817).
Severn. Hartlebury Common. Wannerton Downs, Churchill. Habberley Valley. Sandy fields east of Kidderminster. Wyre Forest. Hagley.
Malvern. Malvern Hills, *Lees.* North Hill. Malvern Common. Malvern Link Common.
Lickey. Dry places, common, *Mathews.*
This little plant is seldom more than six inches high, and the branches are forked ; it is usually well smothered in cottony down. It is recorded from all neighbouring counties.

ANAPHALIS, DC. 208*. (Said by De Candolle to be an ancient Greek name of some Gnaphalioid plant, and that it may be taken as an anagram of the very similar genus Gnaphalus.)

835. 524. A. *margaritacea*, Benth. and Hook. fil. *Pearl Cudweed.*
(*Antennaria margaritacea*, R.Br.)
Alien. Moist meadows. Rare. August. Not in *Top. Bot.*
First Record. By a rivulet in the heart of Wire Forest, *Rev. Mr. Butt, Bot. Guide,* vol. ii, p. 656 (1805).
Severn. By the side of a small stream in a wood about two miles west of Bewdley, *Westcombe.* Wood near Bewdley and Wyre Forest, *Herbarium, Vict. Inst. Worc.*
This plant is a native of dry woods in North America, and is an old favourite in cottage gardens, whence it has sometimes escaped. All the specimens in the herbarium seem to have been gathered at the locality mentioned by Mr. Butt. It has been noticed in Herefordshire, Stafford, and Salop.

GNAPHALIUM, Linn. 209. (γναφάλιον, down, with which the leaves are covered.)

836. 525. G. uliginosum, Linn. *Marsh Cudweed.*
Native. Damp places, roadsides where water has stood, pond-sides, cultivated ground. Common and generally distributed. A. July to October. *Top. Bot.* 112.
First Record. Not localized, *Lees, Bot. Malv. Hills,* 1st ed., p. 38 (1843).
A common little plant, seldom more than three or four inches high, white, like all its relatives, with cottony down. It occurs in all neighbouring counties.

o 2

838. **526. G. sylvaticum,** Linn. *Highland Cudweed.*
Native. Dry pastures and places. Local. P. July to October. *Top. Bot.* 103.
First Record. Plentifully in rough pastures near Fladbury, *Nash, Hist. Worc., Int.,* p. lxxxix (1781).
  Avon. The Ridgway near Cookhill, *Purton.* Charlton.
  Severn. Wyre Forest, *Jorden.* Shrawley. About Kidderminster. Bissell. Near Churchill. Shrawley Wood. Charlton, near Hartlebury. Hagley Brake. Hurcott Wood. Near Holt.
  Malvern. End Hill, *Lees.* North Cottage, Malvern Wells. Base of the North Hill. Near the Quarry, North Hill. Old Storridge.
  Lickey. Bromsgrove Lickey, *Humphreys.* Rowney Green, near Alvechurch.
This is a very common plant in Scotland, whence its English name, but it is not by any means confined to that country. Many of the Everlastings of our gardens are Cudweeds. The Immortelle of France, so largely used for funeral wreaths, is *Gnaphalium orientale.* Another Cudweed is the well-known Edelweiss of the Alps, *Gnaphalium Leontopodium, Scop.,* which grows as well in our English gardens from seed as it does in its native home, though perhaps with some loss of whiteness and woolliness; and is largely grown from seed in Switzerland to sell, with sham alpenstocks, to the ordinary traveller. Everlastings take dye well, but like other dyed flowers, look bizarre and incongruous when so treated. *G. sylvaticum* grows in all neighbouring counties.

INULA, Linn. 210. (Name said to be a corruption of *Helenium,* this plant being supposed to have sprung from the tears of Helen of Troy.)

841. **527. I. *Helenium,** Linn. *Elecampane.*
Denizen or native. Pastures and hedgesides. Rare. P. July and August. In *Top. Bot.* no numbers are given, but a note says that the plant is recorded in more than sixty counties, chiefly in a doubtfully native condition.
First Record. In great abundance on the side of Breedon [sic] Hill, in the ascent from Great Comberton, *Nash, Hist. Worc., Int.,* p. lxxxix (1781).
  Avon. Bredon Hill. At the edge of a pool in the Red Deer Park near Defford, 1857, *Rev. F. K. Clarke; Towndrow, Malv. Advert.,* Nov. 20, 1897.
  Malvern. Western side of Woodbury Hill. Wallsgrove Hill. Berrington, near Tenbury. Stansford, *T. B. Stretch.* Knightsford Bridge, *Newman.* Roadside near a cottage garden, Half Key, *Towndrow, Malv. Advert.,* Nov. 20, 1897. Bransford.
  Lickey. Near Shortwood Coppice. Clent.
Doubts have been cast on the true nativity of this plant in England; it is a handsome flower, frequent in the flower border, and its credit must suffer thereby. For it is undoubtedly native from Spain to the Altai Mountains in Russia, and there seems no reason why it should not be native in Britain also. The English name has a curious history; it is a corruption of *Enula*

campana, in which the first word stands for Inula, the second, while it may mean 'Campanian', most probably is an adjective meaning 'of the fields', as a fourteenth-century writer distinguishes two species of the plant, *campana* and *hortulana.* Elecampane lozenges were made a century ago, often in the form and resemblance of notable persons. The leaves are bitter and aromatic, and the root also, which when it is dry has something of the scent of orris-root, itself the root of an Iris, whence its name. This Inula has been observed in all neighbouring counties.

842. **528. I. squarrosa,** Bernh. *Ploughman's Spikenard.*
(*Conyza squarrosa,* Linn. *I. Conyza,* DC.)
Native. Roadsides and hedgebanks, heathy places. Local, rare in the Lickey district. P. July to October. *Top. Bot.* 58.
First Record. As *Conyza squarrosa.* In a lane near Hartlebury and about Chaddesley Corbett, *Perry, Mag. Nat. Hist.,* vol. iv, p. 450 (1831).
  Avon. Bredon Hill. Broadway. Inner fosse of Elmley Castle. Near Defford Common. Rous Lench. Dunstall Common.
  Severn. Hartlebury, *Perry.* Near Stanklin Pool. Hartlebury Common. Near Droitwich, *Humphreys.* Wolverley. Pedmore. Lincombe. Between Broadheath and Cotheridge. Habberley Valley. Roadside at Holt. Hartlebury Railway Station. Cornsall. Cookley. Spetchley. Coleridge Wood. Wyre Forest. Shrawley.
  Malvern. Base of Raggedstone Hill; The Croft, Cowleigh, *Lees.* Woodbury. Near Ankerdine Hill. Crews Hill Wood. Old Hills. Alfrick. Martley. Great Malvern. Mathon.
  Lickey. Corngreaves, Halesowen; Chaddesley Corbett, *Perry.* Between Bromsgrove Railway Station and Stoke Prior.
The plant has a slightly aromatic odour, hardly perceptible till it is bruised. It is not particularly handsome though of some size—two to three feet high. In its young state, before flowering, it is not unlike the young state of the Foxglove. It is recorded from all neighbouring counties.

PULICARIA, Gaertn. 211. (*Pulex,* a flea, which is supposed to be driven away by its powerful smell.)

846. **529. P. dysenterica,** Gray. *Common Fleabane.*
(*Inula dysenterica,* Linn.)
Native. Marshes, wet ditches, sides of roads. General throughout the county. P. July to September. *Top. Bot.* 79.
First Record. Roadside, on moist ground, common, *Pitt, Agric. Worc.,* p. 317 (1810).
This plant is conspicuous among the green grasses of the damp ditch-side, in its living state the haunt of bees and butterflies, but when dried supposed to have the qualities indicated by both its generic and English name. It is not the only flower to bear that title, *Erigeron, Inula,* and *Plantago Psyllium,* Linn., also possessing it. Sprinkled on beds, besides the dominant insect, a decoction of it was supposed to keep away gnats. Possibly its specific name is given it from its having proved a valuable medicine in the Russian army, a fact which Linnaeus speaks of. It grows in all neighbouring counties.

847. **530. P. vulgaris,** Gaertn. *Small Fleabane.*
(*Inula Pulicaria,* Linn.)
Native. Watery spots, margins of water. Very local. A. July to August. *Top. Bot.* 25.
First Record. In watery spots on Barnard's Green, &c., *Lees, Bot. Malv. Hills,* 1st ed., p. 39 (1843).
  Avon. Trench Woods, *Mathews.*
  Malvern. Commons near Malvern, *Lees.* Newland Common, near the Swan. Near Powick. Longdon Heath. Barnard's Green. Near Malvern Chase.
  Lickey. Tardebigge. Reservoir; Bittell Reservoir; near Chaddesley Woods, *Humphreys.*
This plant is about a foot high, with heads of yellow flowers devoid of the attractive ray-florets of *P. dysenterica.* The stem is much branched and hairy. It is recorded only from Hereford and Warwick of bordering counties.

XANTHIUM, Linn. (*ξανθός,* yellow, from its effects as a hair-dye.)

531. X. spinosum, Linn. *Spring Bur-weed.*
Alien.
First Record. On a rubbish-heap at Hoo Mill, Kidderminster, *Thompson, Worc. Nat. Club Trans.* i. 227 (1876).
  Severn. As above.
  Malvern. Hop-yard at Leigh Sinton, *J. H. White,* 1900.
A species now widely spread along many of the trade routes of the world. Substantial claims have been advanced for countries so widely separated as the Steppes of Russia and South America as its true home. At the present day it seems to occur in England where shoddy manure is employed. It has not been seen in any neighbouring county. The plant is not given in *Lond. Cat.,* 10th ed.

BIDENS, Linn. 212. (*Bis,* twice, *dens,* a tooth, from the two awns which crown the fruit in some species.)

848. **532. B. cernua,** Linn. *Nodding Bur-marigold.*
Native. Pond-sides and wet places. Not common. A. or B. June to September. *Top. Bot.* 82.
First Record. Banks of streams, *Scott, Hist. Stourb.,* p. 540 (1832).
  Avon. Norton, near Evesham.
  Severn. Brake Mill Pool. Boggy Common at Moseley, near Hallow. Hartlebury Park. Wannerton Pool. Blakedown Pool. Ditch near Pitmaston, *Lees.*
  Malvern. Upton-on-Severn. On the commons, *Lees.* Martley. Welland Common.
  Lickey. Tardebigge, *D. Mathews.* The Valley, Bromsgrove, *Mathews.* Canal bank near Bittell Reservoir.
In this species the fruit usually possesses three or four bristles, each thickly set with prickles. The plant has opposite branches, but the leaves are not stalked. It occurs in all neighbouring counties.

849. **533. B. tripartita,** Linn. *Trifid Bur-marigold.*
Native. Pond-sides and waste places. Not uncommon throughout. A. or B. July to September. *Top. Bot.* 84.

First Record. Not localized, *Lees, Bot. Malv. Hills,* 1st ed., p. 38 (1843).
  Avon. Roadside near Throckmorton. Craycombe. Stoulton.
  Severn. Churchill and Blakedown Pools, *Mathews.* Wannerton Swamp. Ladywood, Salwarpe. Near the Bridge over the Severn at Worcester. Hallow. Ombersley.
  Malvern. Wet spots and pools on the commons, *Lees.* Near Powick. Martley. Barnard's Green. Malvern Link.
  Lickey. Harborne Reservoir, *Mathews.* Tardebigge, *Humphreys.* Bittell Reservoir.
This plant is much like the former one, but the leaves are narrowed into winged foot-stalks, and are often tripartite or pinnate, while the fruit-bristles are usually two. The flower-heads also hold themselves up. It is recorded from all neighbouring counties.

GALINSOGA, Ruiz and Pav. 213*.

850. **534. G. parviflora,** Cav.
Alien. Waste places.
First Record. Hewell, *Pettigrew, Worc. Nat. Club Trans.* iii. 61 (1901).
This plant is a native of Central and South America, and was first noticed in Europe about one hundred years ago, but it is now locally abundant in different parts of the Continent and of this country. The introduction of American wheat is doubtless the cause of its appearance in Europe. We have no records of it for neighbouring counties.

ACHILLEA, Linn. 214. (Named after Achilles, who is said to have discovered its healing virtues.)

851. **535. A. Millefolium,** Linn. *Yarrow* or *Milfoil.*
Native. Pastures, meadows, roadsides. Very common and generally distributed. P. May to September. *Top. Bot.* 112.
First Record. Not localized, *Walker, Med. Surg. Report,* vol. i, no. 2, p. 100 (November, 1828).
This is one of the commonest flowers of the countryside. It has had many names: Soldier's Wound-wort, Nose-bleed, and Old Man's Pepper indicate former uses of the plant. It was, and perhaps is, taken in large quantities as a remedy for consumption, and a salve is made of it. The flowers, usually white, are sometimes deeply flushed with purplish pink, and this form in the immediate neighbourhood of Worcester is almost replacing the type. Yarrow grows in all neighbouring counties.

852. **536. A. Ptarmica,** Linn. *Sneezewort.*
Native. River-sides, damp meadows, thickets. Not uncommon. P. June to August. *Top. Bot.* 112.
First Record. Cornfield at the Rhydd, *Dr. Streeten;* also at Battenhall and Little Malvern, *Lees, Illus. Nat. Hist. Worc.,* p. 175 (1834).
This plant is by no means uncommon, though not to be seen, like Yarrow, on every patch of grass. All parts of it are pungent, and it would appear that it obtains its popular name, not like many plants, because it is a remedy for some ailment, but from its being a cause of the condition indicated by its name. The dried leaves are used as a substitute for snuff. It occurs in all neighbouring counties.

ANTHEMIS, Mich. 216. (ἀνθεμίς, a flower, from profusion of blossom and value of the blossoms as a medicine.)

***855*. 537. A. Cotula**, Linn. *Stinking May-weed. Top. Bot.* 76.

Native. Cultivated fields and waste places. Generally distributed. A. May to October.

First Record. Not localized, *Lees, Bot. Malv. Hills*, 1st ed., p. 39 (1843).

This is a common weed. Stubble fields are frequently full of it. It has an offensive odour, and as several of these white-rayed, yellow-disked composites are much alike to the ordinary observer, it is frequently mistaken by him for the sweet-scented chamomile, to his great disappointment. The juice is acrid, and in times when corn was reaped used to blister the hands of reapers. The plant occurs in all bordering counties.

***856*. 538. A. arvensis**, Linn. *Corn Chamomile.*

Native. Cornfields. Local. A. May to September. *Top. Bot.* 73.

First Record. On the Ridgeway, upon the new-made turf-mounds, in great plenty, *Purton, Midl. Fl.*, p. 398 (1817).

Avon. Norton, near Evesham. Sheriff's Lench. Pershore. Earl's Croome. Crowle. Flyford Flavell.

Severn. Blakedown, *Mathews.* Near Camp House. Hartlebury. Near Bewdley. Dunclent. Between Lower Wick and Boughton. Kempsey. Ladywood, Salwarpe.

Malvern. Cowleigh Park; Berrow, *Lees.* Near Old Hills. Knightwick. Martley. Powick. Malvern Wells.

Lickey. Rare, *Mathews.*

The reclining habit of this plant and its more pubescent foliage distinguish this from the preceding. It is recorded from all neighbouring counties.

***857*. 539. A. nobilis**, Linn. *Chamomile.*

Native. On commons. Local and rare. P. June to September. *Top. Bot.* 49.

First Record. Hanley Common and Malvern, *Purton, Midl. Fl.*, p. 398 (1817).

Avon. Defford Common. Dodderhill Common. Feckenham.

Severn. Bliss Gate, Bewdley, *Jorden.* Trimpley. Kempsey Common. Holt. Camp. Stagbury Hill. Hartlebury Common.

Malvern. Barnard's Green and Welland Common, abundant, *Lees.* Seats Common (now enclosed), *Lees.* Newland Common. Malvern Wells. Malvern Link Common. Barnard's Green.

This plant is seen more often as a cultivated herb than as a wild plant, for perhaps no plant has a greater reputation, which survives, in part, to the present day, than the Chamomile, though in the garden the yellow disk is often supplanted by white rays after the manner of a double daisy. The flowers are strongly fragrant and bitter; Chamomile tea is a rustic remedy, well known, and its use penetrates, or did do so before these days of advertisement puff so many medicaments into notice, into all ranks of society. It is cultivated both in England and on the Continent for medicinal purposes, the English growths being considered of better quality than the foreign ones in the Pharmacopœia. It is considered to be improved in quality by being trodden upon, a fact spoken of by Shakespeare. It is recorded from all neighbouring counties.

CHRYSANTHEMUM, Linn. 217. (χρυσός, gold, ἄνθεμον, a flower, many species having yellow flowers.)

***858*. 540. C. *segetum**, Linn. *Corn Marigold.*

Native. Cultivated fields on sandy soil. Locally abundant. June to August. A. *Top. Bot.* 110.

First Record. Cultivated ground, *Pitt, Agric. Worc.*, p. 317 (1810).

This plant is remarkable for its smooth and glaucous foliage, and the flower, all golden yellow, stands on a branched stem with amplexicaul leaves. Usually when it occurs it clothes the fields with gold, as the poppy robes them in scarlet; and frequently both plants grow together, forming a beautiful sight. From the railway between Kidderminster and Hagley fields so decked may often be seen, especially about Churchill and Hagley stations. But being an annual, it does not thus display itself every year. Yet for all its beauty it is a bad weed. As a truly native plant it is quite rare, being only known in pastures in a few localities in the Mediterranean region; but as a weed of cultivation it is common throughout Europe. It is recorded from all adjoining counties.

***859*. 541. C. Leucanthemum**, Linn. *Ox-eye Daisy.*

Native. Grass fields, railway-banks, grassy wastes. Abundant everywhere. P. April to July. *Top. Bot.* 112.

First Record. Not localized, *Lees, Bot. Malv. Hills*, 1st ed., p. 39 (1843).

This flower is one of the most prominent features of our grass fields. It is sometimes called the Moon Daisy, Dog Daisy, Cow Daisy, Queen of the Daisies, or Maudlin Daisy. The juice is acrid, and it has the reputation, though not the name, of being a certain remedy against those annoying insects after which *Pulicaria* is designated; it absolutely destroys them! It occurs in all neighbouring counties.

***860*. 542. C. *Parthenium**, Bernh. *Common Feverfew.*

(*Matricaria Parthenium*, Linn. *Pyrethrum Parthenium*, Sm.)

Denizen. Hedgebanks, walls, waste places. Several places throughout the county. P. May to August. Not localized in *Top. Bot.*, except to say that it is recorded, with a few exceptions, in all counties as a more or less distrusted native.

First Record. As *Matricaria Parthenium.* Hedgesides, Shrawley, *Pitt, Agric. Worc.*, p. 317 (1810).

This is a plant commonly seen in gardens, and nearly as frequently on the neighbouring garden rubbish heap, from which it escapes into more or less wild-looking conditions. Its native home is in the woods of South and South-East Europe, but it is a common weed of cultivated land throughout Europe. The odour of the plant resembles that of chamomile, but it is not so strong. Feverfew is a corruption of the word 'Febrifuge', and it is often still further corrupted into 'Featherfew', to which name its pinnatifid leaves doubtless lead. It is called Curly Jack in some parts of Worcestershire. It was in old times regarded as a specific in cases of fever and ague. It has been noticed in every neighbouring county.

MATRICARIA, Linn. 218. (*Mater*, a mother, from its former medicinal uses.)

***861*. 543. M. inodora**, Linn. *Scentless May-weed.*

(*Chrysanthemum inodorum*, Linn. *Pyrethrum inodorum*, Sm.)

Native. Fields, waysides, waste places. Generally distributed. A. or B. May to November. *Top. Bot.* 111.

First Record. Not localized, *Lees, Bot. Malv. Hills*, 1st ed., p. 39 (1843).

In spite of its Latin name the plant is not by any means odourless, but the scent is very different from that of *Anthemis Cotula* or *Matricaria Chamomilla*; probably the name was given to it as not having the disagreeable stench of the Stinking May-weed, *A. Cotula.* It occurs in every bordering county.

***863*. 544. M. Chamomilla**, Linn. *Wild Chamomile.*

Native. Cultivated fields and waste places. Generally distributed, but less common in the Lickey district. A. May to August. *Top. Bot.* 65.

First Record. On Welland Common and other waste spots about the eastern base of Malvern Hill, *Lees, Illus. Nat. Hist. Worc.*, p. 175 (1834).

The odour of this plant is very like that of the true Chamomile, but fainter. The receptacle is hollow, a mark of distinction from other similar-looking flowers, and the leaves are not channelled on the under side as in *M. inodora.* It occurs in all neighbouring counties.

***864*. 545. M.** SUAVEOLENS, Buchenau.

(*M. discoidea*, DC.)

First Record. As below.

This alien is a native of open ground in Oregon and other parts of the United States, and has become naturalized in several parts of Europe, the turf of Kew Green, near London, for one of them. It was discovered at Bransford, July 28, 1902, by Mr. H. S. Bickham, and is recorded in *Worc. Nat. Club Trans.* iii. 127. It is a stiffer and more bushy plant than *M. inodora*, and has the disk florets four-toothed and not five-toothed as in the latter.

TANACETUM, Linn. 220. (Name altered from Athanasia, ἀ, not, θάνατος, death, that which does not quickly fade.)

***866*. 546. T. vulgare**, Linn. *Common Tansy.*

Native. Roadsides, hedges, river-banks. Rather uncommon in the Avon and Lickey districts. P. June to September. *Top. Bot.* 105.

First Record. Abundant near the Stour and other rivers, *Pitt, Agric. Worc.*, p. 317 (1810).

Avon. Cookhill Farm, Inkberrow, *Mathews.* Norton, near Evesham. Fladbury. Pershore. Bredon. Shell Brook.

Severn. Pettiford Lane, Kidderminster, *Mathews.* Dodderhill Common. Bewdley. Abundant on banks of Severn, *Lees.* Blakedown; Churchill (Kidderminster), *Mathews.* Hagley. Ombersley. Holt. Stourport. Kempsey.

Malvern. Cowleigh, *Lees.* On banks of Teme, *Lees.* Powick. The Rhydd. Upton-on-Severn. Martley. Knightsford Bridge. Stanford Bridge.

Lickey. Lower Clent. Broome. On Clent Hills. Pitcher Oak. Salter's Lane, Redditch. Tardebigge.

The yellow, button-like, flowers of this plant stand out among masses of

dark green feathery foliage. It was a favourite pot-herb with our ancestors, though its aromatic scent is nowadays not pleasing to many persons. Tansy Pudding, probably a kind of savoury omelet, was a favourite dish; cakes were made of it called 'Tansies'; and it was used as a representative of the bitter herbs to be taken with the Paschal Lamb. Gerarde says the 'Tansies' were pleasant in taste and good for the stomach, and that the root preserved with honey or sugar was an 'especial thing against the gout, if every day for a certaine space a reasonable quantitie thereof be eaten fasting'. Tansy wine used to be a favourite village medicine. Horses and cattle dislike it, and leave it in dark green solid tufts about any field it happens to grow in. The Tansy grows in all neighbouring counties.

T. BALSAMITA, Linn. *Costmary* or *Alecost.*

Alien. Waste places and roadsides. Garden escape. P. September and October.

First Record. As below.

Severn. Battenhall Road, Worcester, July 3, 1900, *Rea.*

The first record of this garden escape, much cultivated formerly to give a flavour to ale, whence the second English name. The first name was given it because of its wide association in the Middle Ages with St. Mary, in France, Germany, and elsewhere. As for 'cost', that has an Oriental source, and being the name for the thick aromatic root of *Aplotaxis Lappa*, a native of Cashmere, got itself transferred to other odoriferous plants. The word occurs in an early eleventh-century Saxon manuscript.

ARTEMISIA, Linn. 221. (From Ἄρτεμις, Artemis, the Diana of the Latins.)

***867*. 547. A. Absinthium**, Linn. *Common Wormwood.*

Denizen. Waste places and roadsides. Not common. P. July to September. *Top. Bot.* 74.

First Record. About Malvern and Alfrick, near farm houses, *Lees, Illus. Nat. Hist. Worc.*, p. 174 (1834).

Avon. Trench Woods. Near Huddington, *Mathews.*

Severn. Wyre Forest. Near Stakenbridge.

Malvern. As first record. Near Eldersfield, *Lees.* Near Wych, Malvern Hills.

Lickey. Hunnington, near Halesowen, *Lees.* Abundant on pit mounds near Dudley.

'As bitter as wormwood' is a popular simile. It was used medicinally by country people; but most of these village remedies are now superseded by the widely advertised pill or patent medicine. The cultivated plant is still in the Pharmacopœia, the unexpanded flowers being the useful part of the herb. A liqueur prepared from it is much drunk in France under the name of Absinthe. The common strongly-scented plant of our garden called 'Old Man' is *Artemisia Abrotanum* of the South of Europe; it seldom flowers with us. Tarragon, used to flavour vinegar, is *A. Dracunculus*, our name coming through the French *Estragon*, which leads up to its specific name. The range of our plant on the surface of the earth extends nearly all round the North temperate zone. The leaves are silky on both surfaces. The plant has been observed in all neighbouring counties.

868. **548. A. vulgaris,** Linn.　*Common Mugwort.*

Native. Hedges, field-borders, bushy places. Common and generally distributed. P.　June to September.　*Top. Bot.* 110.

First Record. Not localized, *Lees, Bot. Malv. Hills,* 1st ed., p. 38 (1843).

This is a common but dull plant, with dark green leaves coated with cottony down underneath, the spikes standing up some three or four feet high with few hardly noticeable flowers. Pliny said that if the traveller had some of the herb tied about him, he felt no weariness at all ; and pilgrims used to put it in their shoes. A good deal of superstition clung to the plant. On Midsummer Eve, a 'coal' could be found under its roots which kept people safe from a number of things, lightning and the quartan ague among them. Some ancients, however, threw doubt upon the 'coals', and said they were nothing but the 'old acid rootes' of the plant. Mugwort occurs in all bordering counties.

Var. b. coarctata, Forselles.

First Record.　In this book.

This variety has been observed in the Severn district by Mr. Towndrow. It has much narrower leaves than the type, and the racemes are condensed. But many intermediate forms are to be observed. It has been seen in Stafford and Warwick.

TUSSILAGO, Linn. 222.　(From *tussis,* a cough, for which the plant was held remedial.)

872. **549. T. Farfara,** Linn.　*Colt's-foot.*

Native. On clay soils chiefly. Abundant. P.　March and April. *Top. Bot.* 112.

There is no need to say much of this common plant, which is one of the earliest to flower in spring, before its leaves appear above the ground. It is a horrible weed in many places, nearly impossible to eradicate, for every little bit of root will grow. Colt's-foot juice is made in many districts ; and the plant was supposed to be extremely efficacious in cases of cold and cough. The down of the plant mixed with saltpetre, before matches came into use, was used to make tinder. It occurs in all bordering counties.

PETASITES, Linn. 223.　(πέτασος, a head-covering, from the size of the leaves.)

873. **550. P. *fragrans,** Presl.　*Winter Heliotrope.*

Alien. Garden escape. P.　January and February. Not in *Top. Bot.*

First Record. Established in two or three places at Malvern, *Towndrow, Malv. Advert.* (1892).

Severn. Arley Castle, *Herb. Vict. Inst. Worc.*

A native of the Mediterranean region, naturalized in most of the countries of Europe. It does not appear to have escaped from cultivation in Staffordshire, as it has in other bordering counties.

874. **551. P. ovatus,** Hill.　*Common Butter Bur.*

(*P. officinalis,* Moench.　*Tussilago Petasites,* Linn.　*P. vulgaris,* Desf.)

Native. Sides of rivers, brooks, canals, and in wet places. Locally abundant. P.　March to May. *Top. Bot.* 109.

First Record. As *Tussilago Petasites.* Rather local upon the banks of the Teme ; by the side of the brook at Alfrick ; banks of the Severn, opposite Cleavelode ; skirting the entire course of the brook at Sapey, *Lees, Illus. Nat. Hist. Worc.,* p. 175 (1834).

Avon. Norton, near Evesham. Fladbury.

Severn. Caunsall Bridge, Wolverley. Hartlebury Common. Lincombe. Grimley. Hallow. Near Droitwich. Upton Warren.

Malvern. All the first records. Bridges Stone Mill, Alfrick. Sapey Brook. Banks of Teme. Knightsford Bridge. Powick. Mathon.

Lickey. Meadow near Offmoor Lane. Near St. Kenelm's. River Stour, Cradley. Charford.

The leaves of this plant are so large that it is called in many places Monk's Rhubarb. The stamens and pistils of the plant usually occur in different flowers, the female panicle being long and lax, the male ovoid and dense. The flowers appear before the leaves, usually from ground made bare of vegetation by the dense shade of the foliage for the rest of the year. Its long roots penetrate the soil in every direction, and it is a most troublesome weed. In former times it had great repute against fevers, and was ' a great strengthener of the heart and cheerer of the vital spirits '. It occurs in all neighbouring counties.

DORONICUM, Linn. 224*.　(δῶρον, a gift, νίκη, victory, because it gave the victory over wild beasts.)

876. **552. D. *Pardalianches,** Linn.　*Great Leopard's Bane.*

Alien or denizen. Plantations and roadsides, pool dams. Very local. P.　July and August. Not in *Top. Bot.*

First Record. By a ditch-side a little below Bewdley, *Gissing, Phyt.,* N.S., vol. i, p. 151 (1855).

Severn. Old bridge at Powick. Snuff Mill Valley, Bewdley. Arley Castle. Bewdley. Near Blackstone Rock. Wyre Forest. Roadside between Ribbesford and High Oak. Near Winterdyne.

Lickey. Hagley Park hedge, above Hagley. Vine Pool Dam, and Clent Grove Pool Dam, Clent.

The nativity of this plant in Britain has perhaps unduly been called in question. While it may be a naturalized alien in the South, originating from garden cultivation, it is found in more natural conditions in the North and in Scotland. It is a native in Normandy, Belgium, and Holland, and whatever its rank among our Flora, it is certainly a plant of old standing. And it has always been considered wonderfully potent against wild beasts. Its specific name is derived from πάρδος or πάρδαλις, a panther, and ἄγχω, I strangle ; and as for scorpions and such like, Leopard's Bane 'layd to a scorpione makethe hyr vtterly amased and num'. Conrad Gesner, who made many experiments on his own person with regard to the properties of plants, died after eating some of this plant in 1565 ; but in later times the poisonous nature of the root came to be doubted. Gerarde says a man 'ate very manie of the rootes at sundrie times and found them very pleasant and comfortable. And thus I leave all controversies ', he adds. He did not try it himself ! It does not appear to have been noticed in Warwickshire.

SENECIO, Linn. 225.　(*Senex,* an old man, from its grey seed down.)

878. **553. S. vulgaris,** Linn.　*Common Groundsel.*

Native. A universal weed of cultivated ground and waste places. A. In flower all the year. *Top. Bot.* 112.

First Record. Not localized, *Lees, Bot. Malv. Hills,* 1st ed., p. 39 (1843).

This ubiquitous plant has been used for poultices ; but it does not seem to have entered very much into the medicine or superstitions of our forefathers. If it has any use at all, it is as green food for caged birds. It is as abundant in all neighbouring counties as it is in our own.

879. **554. S. sylvaticus,** Linn.　*Mountain Groundsel.*

Native. Heaths, hedgebanks, and wastes on light sandy soil. Locally common and generally distributed. April to September. *Top. Bot.* 107.

First Record. Astwood Common, *Purton, Midl. Flora,* p. 405 (1817).

This plant is particularly abundant on the sandstone measures of the north of Worcestershire, especially about Kidderminster, Wolverley, and Churchill, and is common on the higher parts of the parish of Clent. It is much larger and paler than the last plant, sometimes growing three feet or more in height ; the leaves are often hoary. It occurs in all neighbouring counties.

Var. b. auriculatus, Meyer.

(*S. lividus,* Sm.)

First Record. As *S. lividus.* Not localized, but near Malvern, *Trans. Prov. Med. and Surg. Assoc.,* vol. iv, p. 141 (1836), *Dr. William Addison of Malvern.*

Severn. Hartlebury Common, *Herb. Hast. Mus., Worc.,* sp.

Malvern. As first record.

In this variety the upper leaves are more distinctly auricled and clasping.

× viscosus.

First record. As below.

Lickey. Saltwells Wood, Dudley, September 13, 1907, *Amphlett.*

The specimen is in the herbarium of Mr. R. F. Towndrow.

880. **555. S. viscosus,** Linn.　*Stinking Groundsel.*

Native. Waste ground. Very rare. A. July to September. *Top. Bot.* 34.

First Record. Badsey, *Rufford, Purton, Midl. Flora,* p. 405 (1817).

Avon. As in first record.

Lickey. Saltwells Wood, near Dudley, abundant, September 13, 1907, *Amphlett.*

Since Purton's time nothing had been seen of this plant, either at Badsey or elsewhere in Worcestershire, till September, 1907, when it was discovered in great abundance on the ballast and embankments of Lord Dudley's private railway, which traverses Saltwells Wood ; and with it grew *S. sylvaticus,* and the hybrid between the two. Saltwells Wood is in the northern island of Worcestershire, which contains the town of Dudley and is mostly purely Black Country ; and in the midst of it the wood is a green oasis. It had been noticed in Gloucester, Hereford, Salop, and Staffordshire.

881. **556. S. *squalidus,** Linn.　*Inelegant Groundsel.*

Denizen. Railway-banks and waste ground. Increasing rapidly along the permanent way of the Great Western Railway, spreading from Oxford, where it was cultivated in Bobart's time. A. B. or P. April to October. Not in *Top. Bot.*

First Record. Walls near the cathedral and adjoining the river at Worcester. Inserted in the list on the authority of the Rev. A. Bloxam, *Lees' Cat. New Bot. Guide* (1835).

Upon this record hung for many years Worcestershire's claim to this plant. The locality was near the old Water-gate, where it still existed in 1853, and where, from notes Mr. Lees had seen, it must have existed at that time for more than half a century. He gives its history, *Bot. Malv.,* 3rd ed., p. 93. But at last it disappeared. In 1905 the plant reappeared in Worcestershire in great abundance on the high marl banks between Shrub Hill Station at Worcester and the tunnel, where the yellow sheets of flowers are very conspicuous ; and it has now been seen also near Helbury Hill, at Badsey Station, and Pershore Station to the south, and under Dodderhill, near Droitwich Station to the north. The plant is probably a native of Sicily, where it grows profusely on the volcanic sands. The plant has yet been noticed only in Warwickshire of adjoining counties.

× vulgaris.

(*S. vernalis,* Boswell).

First Record. Grows among the parent plants at the locality at Worcester Station, *Rea, Worc. Nat. Club Trans.* iii. p. 235 (1905).

Severn. As first record.

882. **557. S. erucifolius,** Linn.　*Hoary Ragwort.*

(*S. tenuifolius,* Jacq.)

Native. Waysides, field-borders, hedges. Not uncommon. P.　July to September. *Top. Bot.* 67.

First Record. Not localized, *Lees, Bot. Malv. Hills,* 1st ed., p. 39 (1843).

Avon. Trench Woods. Craycombe Hill. Dodderhill Common.

Severn. Roadside beyond Red Hill. Gregory's Mill. Pedmore. Severn side, Bewdley. Near Ockeridge Wood. Hawford.

Malvern. New Pool, Malvern Chace, but sparsely, *Lees.* Abberley House.

Lickey. Turnpike roadside at Hunnington, *Mathews.* Illey Mill. Clent Grove. Clatterbach, Clent. Woodcote Green, Randans. Hayley Green. Halesowen. Chaddesley Woods.

The plant is recorded from all neighbouring counties.

883. **558. S. Jacobaea,** Linn.　*Common Ragwort.*

(*Jacobaea vulgaris,* J. Bauhin.)

Native. Pastures, roadsides, heathy places. General throughout but rare near Malvern. P.　June to September. *Top. Bot.* 112.

First Record. Not localized, *Lees, Bot. Malv. Hills,* 1st ed., p. 39 (1843).

This is a common autumn plant, conspicuous with its deep yellow flowers and tall stem, sometimes three or four feet high. The plant, if bruised, has an unpleasant scent, and from this circumstance in Scotland has gained the name of 'Stinking Willie'.

[S. ERRATICUS, Bert.

(*S. barbareaefolius*, Krock.)

This plant is recorded for the New Pool, Malvern Chase, by Mr. Lees in *Bot. Malv. Hills*, 1st ed., p. 39 (1843) and continued in the 2nd ed.; but in the 3rd it is discarded. It is given in Hooker's *Stud. Man.*, 3rd ed., p. 219, as a form of *S. aquaticus* with pinnatifid leaves, but it is not a British plant, nor indeed anywhere much more than a mere state.]

884. 559. S. aquaticus, Huds. *Water Ragwort.*

Native. Marshes, meadows, damp roadsides. Not uncommon, plentiful in the Malvern district. P. July to September. *Top. Bot.* 111.

First Record. Not localized, Lees, *Bot. Malv. Hills*, 1st ed., p. 39 (1843).

This species is very much like the last, differing chiefly in the form of its lower leaves; and in its fruit being all glabrous, instead of only those of the ray; and it is usually a taller plant. It is recorded from all neighbouring counties.

[886. S. paludosus, Linn. *Top. Bot.* 3. *Great Fen Ragwort.*

This plant is recorded in *Pitt, Agric. Worc.*, p. 317 (1810) as 'Bird's Tongue Groundsel', near Malvern Wells by the roadside, and on the road thence to Upton. Mr. Mathews says it is an incredible record; and as the plant is very rare in its home in the Fens of the east of England, with the number 3 in *Top. Bot.*, there is no doubt that the record is an error.]

CARLINA, Linn. 226. (Named after Charlemagne, who used it medicinally.)

891. 560. C. vulgaris, Linn. *Carline Thistle.*

Native. Dry pastures, open heaths, and downs. Abundant in the Malvern district, not uncommon elsewhere, except in the Lickey district. B. June to September. *Top. Bot.* 83.

First Record. Astwood Bank, *Purton, Midl. Fl.*, p. 386 (1817).

Avon. Bredon Hill. Broadway Hill. Craycombe Hill. Dodderhill Common. Defford Common.

Severn. Wyre Forest. Stagbury Hill. Kempsey Common. Hartlebury Common. Ombersley Bank.

Malvern. Abberley Hill. Malvern Hills, to their summits, *Lees*. Malvern Common. Ankerdine Hill. North Hill. Malvern Wells.

Lickey. Roadside above St. Kenelm's Church, on Clent Hill (1908).

The Carline Thistle prefers a calcareous soil, and in its locality at Clent, where it has been for very many years, a vein of Magnesian conglomerate comes to the surface through Permian sandstone. Once recognized, it is an unmistakable plant; and when gathered it will retain much of its beauty for a considerable time, as it is very rigid and spiny. Charlemagne is said to have been conducted by an angel to the Carline Thistle, and cured his army of the plague by its use. The plant is found in all neighbouring counties.

ARCTIUM, Linn. 227. (ἄρκτος, a bear, from its coarse appearance.)

892. 561. A. majus, Bernh. *Great Burdock.*

(*A. Lappa*, Linn.)

Native. Copses, hedges, waste ground, rubbish heaps. General, except in Malvern district, where it is not uncommon. B. June to September. *Top. Bot.* 44.

First Record. As *Arctium Lappa*. Not localized, Lees, *Bot. Malv. Hills*, 1st ed., p. 38 (1843).

Every one knows this large rough looking plant, growing three or four feet high, with leaves larger than any other native to Britain except the Butter Bur, and whose fruit, by means of its hooked bracts, clings to every living thing it touches, man or beast. It has a reputation for curing rheumatism by the application of its leaves like plasters, and the old herbalists attributed to it a number of other virtues. It can be used as a vegetable in the manner of asparagus, the stems being stripped of their rind just before flowering and boiled. It occurs in all neighbouring counties.

893. 562. A. nemorosum, Lej.

Native. Roadsides, hedges, thickets. Local. B. July and August. *Top. Bot.* 40 ?.

First Record. *Towndrow, Malv. Advertiser*, December 3, 1898.

Malvern. Madresfield; Lower Howsell; Leigh; all, *Towndrow*.

Linnaeus knew only one Burdock, which he called *Arctium Lappa*. This aggregate has now been divided into four segregates, the foregoing plant, the present one, and the two that follow, chiefly differing in the characters of their heads, that is, the collection of flowers surrounded by an involucre, which is the mode of inflorescence of Composites. Some botanists, Hooker, for instance, *Stud. Fl.*, 3rd ed., p. 220, consider these segregates as forms presenting no constant characters. This plant has the heads agglomerated and almost sessile at the apex of the principal stems, and is identical, Mr. W. H. Beeby says, *J. of Bot.*, 1908, pp. 380–382, with *A. Newbouldii*, Ar. Benn. This present form has been noticed in all bordering counties except Gloucestershire.

894. 563. A. minus, Bernh.

(*Arctium Bardana*, Willd.)

Native. Roadsides, hedges, thickets. General. B. July to September. *Top. Bot.* 92.

First Record. As *Arctium Bardana*. Not localized, Lees, *Bot. Malv. Hills*, 1st ed., p. 38 (1843).

Mr. Mathews identifies Mr. Lees' *A. Bardana* with this plant. Mr. G. C. Druce, *Fl. Berks.*, p. 297, appears to identify it both with this plant and with another, *A. tomentosum*, Miller. *A. minus* has been observed in all neighbouring counties, and is the commonest roadside form of this genus in Worcestershire.

564. A. intermedium, Lange.

Native. Roadsides, hedges, thickets. Rare. B. July and August.

First Record. On the bank of the Severn near the Ketch, between Worcester and Kempsey, August 7, 1892, *Towndrow, J. of Bot.*, vol. xxxi, p. 56 (1892).

Avon. Tiddesley Wood. Trench Woods. Dodderhill Common.

Severn. As above. Ombersley. Holt Bank. Whitehall. Shrawley Wood. Wyre Forest.

Malvern. Chapel Hill Coppice, Clevelode. Coneygree Coppice, Leigh, *Towndrow, Malv. Advert.*, Dec. 3, 1898.

Lickey. Uffmoor Wood.

P

This plant possesses woollier heads, twice as large as those of *A. minus*, and the stalks of the heads are longer, those of the lower ones longest. This plant, given in the 9th ed., *Lond. Cat.*, is replaced in ed. 10 by a plant called *A. Newbouldii*, Ar. Benn., referred to above. No such name is to be found in the last editions of either of the three leading English botanies, those of Hooker, Babington, or Bentham. Mr. Beeby, in the article above referred to, says this plant 'represents both *A. majus*, Bernh. x *minus*, and *A. minus*, Bernh., var. *purpurascens*, Blytt.' The plant does not appear to have been observed in Gloucestershire.

CARDUUS, Linn. 228. (Etymology doubtful; perhaps from *card*, Celtic and Gaelic, a card for combing wool.)

897. 565. C. nutans, Linn. *Musk Thistle.*

Native. Waysides, pastures, downs. General, except in the Lickey district where it is less common. B. May to September. *Top. Bot.* 75.

First Record. Not localized, Lees, *Mag. Nat. Hist.*, vol. iii, p. 160 (1830).

Avon. Bredon Hill. Littleton Banks. Sheriff's Lench.

Severn. Castle Hill, Wolverley. Hagley Brake. Churchill (Kidderminster). Stagbury Hill. Near Worcester. Upper Wick. Park Hall. Wannerton Downs. Field by Bewdley Railway Station.

Malvern. Abundant on the Hills, *Lees*. North Hill.

Lickey. Clent, frequent. Broom.

This is one of the most spiny, but at the same time one of the most handsome, of the thistles, with a large purplish-red head, on a sparsely branched cottony stem, gracefully bending over to one side. It grows from two to three feet high, and in the evening exhales a slight musky odour. A whiteflowered form has been gathered at Stagbury Hill. It is recorded from all neighbouring counties.

898. 566. C. crispus, Linn. *Welted Thistle.*

Native. Hedges, open woods, waysides, waste places. General. B. May to October. *Top. Bot.* 88.

First Record. As *C. acanthoides* (see below). In fields between Kidderminster and Picket Rock, *Perry, Mag. Nat. Hist.*, vol. iv, p. 450 (1831).

In this plant the wings of the stem are continuous, and the leaves are usually downy beneath. The flowers are often to be seen white, as in Wyre Forest. It is recorded from all neighbouring counties.

Var. c. **acanthoides**, Linn.

First Record. Not localized, *List of Plants, Victoria Hist. Worc.*, vol. i, p. 52.

In this variety the leaves are less downy beneath and are larger than in the type, and the bracts, instead of being slender, have a short spine. This is a less common plant than the type, and does not appear to have been observed in Staffordshire.

Avon. Fladbury. Defford Common. Dodderhill Common.

Severn. Holt, *Towndrow*. Wyre Forest. Bog near Doverdale Church. Ombersley. Kempsey Grove. Besford.

Malvern. Powick. Shelsley Walsh.

Lickey. Salter's Lane, near Redditch.

CNICUS, Linn. 229. (κνίζω, I prick.)

899. 567. C. lanceolatus, Willd. *Plume Thistle.*

(*Carduus lanceolatus*, Linn.)

Native. Hedges, pastures, roadsides, waste places. Very common and generally distributed. B. May to November. *Top. Bot.* 112.

First Record. Not localized, Lees, *Bot. Malv. Hills*, 1st ed., p. 38 (1843).

This is one of our commonest thistles. The thistles seem to have been quite neglected by the mediaeval herbalists. They could not even, on the ground of 'like cures like', find a remedy for the pricks of the teeth of 'serpents' among them! This plant occurs in all neighbouring counties.

900. 568. C. eriophorus, Roth. *Woolly-headed Plume Thistle.*

(*Carduus eriophorus*, Linn.)

Native. Roadsides, waste places, dry pastures. Locally abundant. B. July to September. *Top. Bot.* 48.

First Record. As *Carduus eriophorus*. On Breedon Hill, *Nash, Hist. Worc., Int.*, p. lxxxix (1781).

Avon. Bredon Hill. Near Defford Common. The Slads. Broadway. Cleeve Bank. The Marsh Common. Dunstall Common. Littleton Banks. Craycombe Hill.

Severn. Kempsey; Cotheridge, *Lees*. Road from Worcester to Norton, *Lees*. Broadwas.

Malvern. Powick. Abberley Hill. On the Hills, Malvern.

This plant rarely grows more than two feet high, and is remarkable for the very thick down that is interwoven among the bracts of the involucre, and the leaves are very thorny above and cottony beneath. It is recorded from all neighbouring counties.

901. 569. C. palustris, Willd. *Marsh Thistle.*

(*Carduus palustris*, Linn.)

Native. Meadows, marshes, damp woods, and hedgebanks. Common and generally distributed. B. May to September. *Top. Bot.* 112.

First Record. With white flowers at Picket Rock, near Kidderminster, *Perry, Mag. Nat. Hist.*, vol. iv, p. 450 (1831).

This is a very common thistle and the tallest of all, of which it is hard to say whether white flowers or purple are the typical ones. It occurs in all neighbouring counties.

x **pratensis**. *C. Fosteri*, Sm.

First Record. As *C. Fosteri*. Low flat meadows between Etherwood (probably Netherwood) and Tibberton, Lees, *Bot. Worc.*, p. 47 (1867).

Severn. As above. Half a mile westward of Crowle, *Worc. Nat. Club Trans.* i, p. 37.

903. 570. C. pratensis, Willd. *Meadow Plume Thistle.*

(*Carduus pratensis*, Linn.)

Native. Marshes, bogs, moist meadows. Local. A. B. or P. May to September. *Top. Bot.* 49.

First Record. Swampy meadows near Robinson's End, Malvern Chace, *Ballard, Stokes, Stokes's With. Bot. Arr.*, 2nd ed., p. 877 (1787).

Avon. Alderminster. Feckenham Bog, *Purton*. Upton Snodsbury, *Lees*.

Severn. Between Tibberton and Crowle, *Lees.*

Malvern. Castlemorton, *Lees.* Longdon Marsh.

Lickey. Moseley, *Miss M. A. Beilby.* Hob Lane, Yardley, *Miss M. A. Beilby,* Alvechurch, *D. Mathews.*

This is a smaller plant, with a cottony stem, a foot or more high, and a single flower head. It occurs in all bordering counties.

905. 571. C. acaulis, Willd.   *Dwarf Plume Thistle.*

Native. Dry commons, hilly pastures, roadsides. Locally abundant in calcareous soil, scarcer elsewhere. P. June to October. *Top. Bot.* 44.

First Record. Sapey, *Sheward, Nash, Hist. Worc., Sup.,* p. 96 (1799).

Avon. Bredon Hill, plentiful ; also with white flowers. Broadway. The Slads. Tardebigge Reservoir. Defford Common.

Severn. Near Castle Hill, Wolverley. Spetchley Common.

Malvern. Silurian eminences, *Lees.*

The plant is recorded from all neighbouring counties.

× **arvense.**

(*C. Clarkei,* H. C. Watson.)

First Record. Tardebigge Reservoir, as below, *Trans. Worc. Nat. Club,* vol. iii, p. 60 (1901).

Avon. Almost replaced the type in a field near Tardebigge Reservoir, July 18, 1901. Bredon Hill. The Slads. Broadway.

Malvern. Old Storridge Common.

906. 572. C. arvensis, Hoffm.   *Creeping Plume Thistle.*

(*Carduus arvensis,* Linn.)

Native. Cornfields, roadsides, waste places, rubbish heaps. Abundant and ubiquitous. P. June to September. *Top. Bot.* 112.

First Record. Not localized, *Lees, Bot. Malv. Hills,* 1st ed., p. 38 (1843).

With *Cnicus lanceolatus* this plant disputes the honour of being the commonest of our thistles, and no doubt it is the worse weed, as its long creeping roots make it extremely difficult to eradicate, and the smallest fragment left in the ground is full of vitality, and quite capable of founding a new and extensive colony. Male and female plants often occur in separate large patches. It occurs in all bordering counties.

Var. d. **setosus,** Bess.

First Record. Croft Farm, West Malvern, *Towndrow, Malv. Advert.,* Nov. 19, 1892.

This variety has a less branched stem, and setose margins to the leaves. It has been noticed in Warwickshire, and, as Croft Farm is now in Herefordshire, must be credited to that county also.

ONOPORDON, Linn. 230. (ὄνος, the ass, πέρδομαι, I break wind, from its supposed effect when eaten by the animal.)

907. 573. O. Acanthium, Linn.   *Cotton Thistle.*

Native or denizen. Roadsides, hedgebanks, waste places. Rare. B. July to September. *Top. Bot.* 60.

First Record. Road from Worcester to Droitwich, near Henlip, *Baker, With.,* 3rd ed., p. 704 (1796).

Avon. Craycombe Hill. Dodderhill Common, *Miss Grafton.* Near Lenchwick.

Severn. As first record. Near Wolverley, *Jorden.* Henwick. Claines.

Norton. Hartlebury Common, *Miss Ladbury* and *Edwards.* Near Wilden Pool, Stourport, *Miss Ladbury.* Evesham Road, Worcester, *Purton.* Summerhill, Kidderminster, *Perry.* Near Churchill Farm (Kidderminster), *Mathews.* Boughton. Helbury Hill.

Malvern. Near Welland Church, *Lees.* The Old Hills, Powick, *Lees.*

This handsome thistle is the one claimed by the Scotsman as his badge, and is often cultivated under the name of the Scotch Thistle. Yet it has only disputed claims to being a native of Britain ; it is a weed of waste ground, and especially of roadsides, throughout the whole of Europe and Western Asia ; and if indigenous anywhere it is on the dry sandy hills of the south of France. It was formerly cultivated for its fleshy receptacle, but has been superseded by the Artichoke. It lasts for only a few years as a rule in localities where it may be discovered. It has been observed in all neighbouring counties.

SILYBUM, Vaill. 231*. (σίλυβον, a Greek name applied by Dioscorides to some thistle-like plants.)

908. 574. S. *Marianum, Gaertn.   *Milk Thistle.*

(*Mariana lactea,* Hill. *Carduus Marianus,* Linn.)

Casual. Roadsides, waste ground. Rare. B. May to September. Not in *Top. Bot.*

First Record. As *Carduus Marianus.* Near Worcester, *Purton, Midl. Fl.,* p. 381 (1817).

This plant is a native of the Mediterranean region and the East. In other parts of Europe and in England it is only known as a weed of waste ground, and as such has been observed in several places in this county, but soon disappearing. It is a stately plant, and at once distinguished from all other thistles by its white-veined leaves. The flower is large, of a rich purple colour. Formerly the whole plant had a culinary value. The heads were cooked like artichokes, the young leaves eaten for salad, and the roots were boiled. It appears to have been noticed in all neighbouring counties except Gloucestershire.

SERRATULA, Linn. 233. (*Serrula,* a hand-saw, in allusion to the serrated foliage.)

910. 575. S. tinctoria, Linn.   *Saw-wort.*

Native. Woods, thickets, hedgerows. General, less common in the Lickey district. P. June to September. *Top. Bot.* 64.

First Record. Perry Wood, Worcester, *Lees, Worc. Misc.* (1829-30).

Avon. Between Trench Woods and Himbleton, *Mathews.* Tiddesley Wood, occasionally with white flowers. The same at Abbots Wood Lane.

Severn. Perry Wood, occasionally with white flowers. Rainbow Hill. Nunnery Wood. Wyre Forest.

Malvern. Dripshill. Longdon Marshes. Madresfield.

Lickey. Frankley and Romsley Woods ; Besley ; Ham Dingle, Pedmore, *Mathews.* Alvechurch and Web Heath, *D. Mathews.* Stourbridge neighbourhood, *Scott.* Near the Rea, Balsall Heath, *Miss M. A. Beilby.* Randans Woods.

A stiff slender plant with small terminal heads in clusters. A good yellow dye can be obtained from the plant. It occurs in all bordering counties.

CENTAUREA, Linn. 234. (So named because the Centaur Chiron healed himself with a plant of this genus when wounded by Hercules.)

[**911. 576. C. *Jacea,** Linn.   *Brown-rayed Knapweed.*

First Record. Cradley Park, *Scott, Hist. Stourb.,* p. 540 (1832).

This record is no doubt an error, and was so regarded by Mr. Mathews, *Mid. Nat.,* xi, p. 18, who suggested that the plant was probably *C. nigra,* with radiant heads. It is probably indigenous in Sussex, but allowing this, it is placed by Sir Joseph Hooker in his Appendix of Excluded Plants, *Stud. Fl.,* 3rd ed., p. 532. It is certainly indigenous in Normandy and Belgium. Its specific distinction consists in the appendages of the bracts of the involucre being erect and rounded instead of patent and pectinated, as they are in the next species, *C. nigra.* The plant has been noticed in Staffordshire.]

912. 577. C. nigra, Linn.   *Black Knapweed.*

Native. Meadows, heaths, pastures, railway - banks. Common and generally distributed. P. May to October. *Top. Bot.* 112.

First Record. Cornfields, *Scott, Hist. Stourb.,* p. 540 (1832).

This plant is to be seen almost everywhere. A common country name for the flowers is 'Hard heads'. They are sometimes found rayed, and the plant then becomes the variety *radiata* or *nigrescens,* Willd., which Mr. Lees records from Warndon, *Bot. Worc.,* p. 46 (1867), and Mr. Mathews from the Trench Woods, *Worc. Nat. Club Trans.* ii. 65. 'Knap' in the name is cognate with 'Knob', which is evidently quite descriptive. Mr. G. C. Druce in his *Fl. Berks.,* p. 305, has some excellent observations on this plant and its forms. The plant occurs in all bordering counties.

913. 578. C. Scabiosa, Linn.   *Greater Knapweed.*

Native. Cornfields, hedgesides, field-borders. Many places. P. May to October. *Top. Bot.* 82.

First Record. Cornfields, *Scott, Hist. Stourb.,* p. 540 (1832).

Avon. Trench Woods. Bredon Hill ; also there with white flowers, and at Conderton Hill, July 23, 1903. Rous Lench. Throckmorton.

Severn. Calcareous field at Tibberton, *Lees.* Railway Bank, Hagley Station, *Mathews.* Crutch Lane, Droitwich. Hagley Brake. Blakedown. Hartlebury Station. Ombersley. Kempsey. Rainbow Hill. Canal at Bilford Lane. Claines. Wyre Forest.

Malvern. Longdon, *Lees.* Bransford. Bridges-stone. Martley.

Lickey. Hasbury Quarry, Halesowen, *Mathews.* Rowney Green. Alvechurch.

This is a handsome flower, the flower heads have very wide spreading rays, and the stem is two or three feet high, much branched. An old name for it, or one of the tribe, was Matfellon, the syllable 'mat' representing the 'mate' in 'checkmate', while 'fellon', an obsolete form of 'felon', is used in a meaning also obsolete, brave or sturdy. The herbalists thought highly of it. It occurs in all neighbouring counties.

---

914. 579. C. *Cyanus, Linn.   *Corn Blue-bottle.*

Colonist. Cornfields, chiefly on light soils. Local. A. or B. June to September. *Top. Bot.* 95.

First Record. Not localized, *Lees, Bot. Malv. Hills,* 1st ed., p. 39 (1843).

Avon. Norton, near Evesham. Sheriff's Lench. Near Comberton, *Westcombe.*

Severn. Kempsey. Railway, Worcester. Claines. Ombersley. Grimley.

Malvern. Powick. Welland. Guarlford.

This plant, though from its universality of occurrence in Britain unquestionably admitted into all our Floras, can hardly be considered more than an alien, which is so satisfied with the surroundings in which it finds itself that it lives and thrives. It is a native of Southern Russia and Asia Minor, and in the south of England is plentiful among cereal crops, getting scarcer as the north is approached. It is perhaps better known in the flower border than in the field, for it and cultivated forms of it, of varying tint, have long been known in the garden, where their long roots and spreading tendencies make them sometimes more or less of a weed. It is called Hurtsickle and Knapweed in many parts of this county. It has been recorded from all neighbouring counties.

915. 580. C. *paniculata, Linn.

This plant in *Top. Bot.* is said only to occur in the Channel Islands ; and is a native of Southern Europe. It was observed as a casual in a clover field at Leigh, escaped no doubt from some garden, September, 1895, by Miss B. Norbury. It has not been observed in any of the bordering counties.

917. 581. C. Calcitrapa, Linn.   *Common Star Thistle.*

Colonist. Gravelly and sandy places. Very rare. A. July and August. *Top. Bot.* 17.

First Record. Broadheath, as below.

Severn. Broadheath, September 12, 1893. *Rea.* Wichenford, *Worc. Nat. Club Trans.* iii. 46.

This plant also is a native of the Mediterranean region, but naturalized in many places in the north temperate zone. The long sharp thorns of its flower cup are very conspicuous, and from them it gets its specific name, as they bring to mind the Caltrop, an instrument of warfare armed with spikes so arranged that, however it fell, one of them was standing upright, with a view of annoying an enemy's cavalry. It has not been observed in any neighbouring county.

918. 582. C. solstitialis, Linn.   *Yellow Star Thistle. St. Barnaby's Thistle.*

Casual. Cornfields. Local. A. June to August. Not in *Top. Bot.*

First Record. Fallow field, near Great Malvern, *Miss Dyson, Trans. Malv. Nat. Club,* pt. 2, p. 19 (1858).

Severn. Orchard, a quarter of a mile west of Bishop's Palace Hartlebury. Charlton, Hartlebury. Walnut Tree Farm, Ladywood. Salwarpe.

Malvern. As first record. Powick. Leigh.

Yet another native of the Mediterranean region that has spread itself throughout Southern Europe as a weed of cultivated ground. It has frequently been observed in England among corn crops raised from foreign seed, but has not been received so kindly as its neighbours, and is allowed no place among British plants, though given in the Botanies. It has not been observed in neighbouring counties.

CICHORIUM, Linn. 235.　(A modification of the Arabic name, in Greek κίχορα or κιχόρεια.)

919.　583. C. Intybus, Linn.　　　*Chicory or Succory.*
Native. Waysides. Cultivated fields, field-borders. Not uncommon in the Avon district, less common elsewhere. P. May to October. *Top. Bot.* 65.
　First Record. At Pinvin, north of Pershore, introduced into cultivation by *Mr. Arthur Young, Pitt, Agric. Worc.,* p. 317 (1810).
　　Avon. Trench Meadow. Near Wadborough. The Slads. Lane between Crowle and Upton Snodsbury. Huddington. Grafton Fields.
　　Severn. Near Stourport, *Lees.* Abbots Wood. Huddington. Bredicot. Rainbow Hill, Worcester. Mildenham Mill. Cock's Tannery, Worcester. Park Hall. Bewdley Railway south of Kidderminster.
　　Malvern. Longdon. Castlemorton, *Lees.* Hill End, Longdon. Malvern Link.
　　Lickey. Clent Hill, 1881, *Amphlett.* Casual in fields at Thicknall, Clent, 1904, *W. Whitwell.*
This plant possesses a beautiful flower, but its beauty is somewhat marred by the habit of the plant, which is rough and angular, and the blossoms, which grow close to the stem, come out one by one, one withering as its successor above it expands. Its use to adulterate coffee is well known, and it is a staple article of food in Egypt. The garden Endive is a nearly allied plant, *Cichorum Endivia*, which is alleged to have been imported into Europe from China in the sixteenth century. Both the specific name of the Chicory, and the English name for the garden kind, are derived from the late Latin *intybea*, a feminine adjectival form. This plant is largely cultivated on the Continent for salad, as well as for the sake of the root. It is recorded from all bordering counties.

ARNOSERIS, Gaertn. 236.　(ἀρνός, a lamb, σέρις, succory.)

920.　584. A. minima, Schweigg and Koerte.　　*Dwarf Nipplewort.*
(*A. pusilla*, Gaertn. *Hyoseris minima*, Linn. *Lapsana pusilla*, Willd.)
Colonist. Cornfields. Very rare. A. July and August. *Top. Bot.* 23.
　First Record. As *Hyoseris minima.* In Pensham Field near Pershore in the most barren and gravelly places, *Nash, Hist. Worc., Int.,* p. lxxxix (1781).
　　Avon. As first record.
　　Severn. Railway-bank east of Stour Viaduct, Kidderminster. Bewdley Railway, *Herbarium Hastings Museum.* Catchem's End near Bewdley, *ex relatione Cleminshaw.*

This plant has seldom been seen in Worcestershire in modern times, though the original locality has been several times searched for it, notably on August 29, 1855, when no trace could be found of it. The leaves are all radical, and the stems swell and become hollow upwards. It is a native of Central Europe, and even as a weed of sandy cultivated fields becomes rare to the south and north. It has not been observed in any neighbouring county.

LAPSANA, Linn. 237.　(λαπάζω, I purge, from its qualities.)

921.　585. L. communis, Linn.　　*Common Nipple-wort.*
Native. Hedges, waysides, cultivated ground, open woods. Very common, and generally distributed. A. May to December. *Top. Bot.* 112.
　First Record. Not localized, *Lees, Bot. Malv. Hills,* 1st ed., p. 38 (1843).
This very common plant is everywhere to be seen in the summer months, tall and branching, with stalked and toothed leaves, heart-shaped at the base, small yellow flowers, and fruit without any pappus. It is found in all neighbouring counties.

PICRIS, Linn. 238.　(πικρός, bitter.)

922.　586. P. hieracioides, Linn.　　*Hawkweed Ox-tongue.*
Native. Hedgebanks, field-borders, among clover. Local. B. or P. June to October. *Top. Bot.* 60.
　First Record. Not localized, *Lees, Bot. Malv. Hills,* 1st ed., p. 37 (1843).
　　Avon. Trench Woods. Craycombe. Fladbury. The Slads. Sheriff's Lench. Near Defford Common. Tiddesley Wood. Norton, near Evesham.
　　Severn. Tibberton. Field near Saldon, Droitwich. Wyre Forest. Hawford. Ombersley. Kempsey. Hadley. Severn Stoke. Shrawley. Blackstone Hill. Severn Meadows above Pitchcroft.
　　Malvern. Leigh Sinton. Silurian eminences, *Lees.* Ankerdine Hill. Ravenshill Banks. Knightwick. Martley. Powick. Bransford. Upton-on-Severn. West Malvern.
　　Lickey. Randans Woods, *Mathews.*
This plant is a lover of calcareous soil. It is found in all bordering counties.

923.　587. P. echioides, Linn.　　*Bristly Ox-tongue.*
(*Helminthia echioides*, Gaertn.)
Native. Waysides, hedges, woods. Local. A. or B. July to October. *Top. Bot.* 65.
　First Record. As *Helminthia echioides.* Not localized, *Lees, Bot. Malv. Hills,* 1st ed., p. 37 (1843).
　　Avon. Near Wadborough. Craycombe. Bredon Hill. Near Trench Woods. Sheriff's Lench. Peopleton. Norton, near Evesham. Lenchwick. Grafton Flyford.
　　Severn. Helbury Hill. Bredicot. Roadside near Cotheridge.
　　Malvern. Silurian eminences, *Lees.* Powick. Newland.
While the former plant is found on calcareous soil this one prefers stiff clayey localities; and one would not suppose the 'Silurian eminences',

given for it in the Malvern district by Mr. Lees, particularly suitable for it. The leaves are covered with strong prickles springing from white tubercles. It is found in all the bordering counties.

CREPIS, Linn. 239.　(κρηπίς, a sandal, from the form of the fruit.)

924.　588. C. foetida, Linn.　　*Stinking Hawkweed.*
(*Barkhausia foetida*, F. W. Schmidt.)
Colonist. Woods, rough ground. Rare. A. or B. July and August. *Top. Bot.* 15.
　First Record. As *Barkhausia foetida.* Not localized, *Mackie, Malv. Field Handbook* (1886).
　　Malvern. Cornfield at Malvern Link, *Towndrow, Malv. Advert.,* Nov. 19, 1892.
A foul-smelling plant, which has not been observed in any neighbouring county.

925.　589. C. taraxacifolia, Thuill.　　*Rough Barkhausia.*
(*Barkhausia taraxacifolia*, DC.)
Colonist. Cornfields. Rare, more frequently met with in recent years, and in parts of the county becoming quite common. A. or B. May to August. *Top. Bot.* 31.
　First Record. As *Barkhausia taraxacifolia.* Rail sidings at Wribbenhall, *F. A. Lees* and *Thompson, Bot. Rec. Club Rep.,* 1883 (published 1884).
　　Avon. Near Evesham. Fladbury. Pershore. Defford.
　　Severn. As first record. Railway-bank, Kidderminster. Ombersley. Grimley. Holt. Kempsey. Ribbesford.
　　Malvern. Malvern Link. Newland. Malvern Wells. Powick. Bransford. Martley. Leigh.
　　Lickey. Field near Thicknall, Clent, 1905, *W. Whitwell.*
This plant would seem to be increasing in the country. It is a native in South-east England in meadows and bushy places, especially in the chalk. In other parts it is a colonizing weed. It does not appear to have yet been observed in Gloucestershire or Staffordshire.

× **Taraxacum officinale.**
　First Record. As here.
　　Malvern. Malvern Wells, 1908, *Towndrow.*
This bi-generic hybrid possesses the leaves of *Crepis* and the scape of *Taraxacum.* It was allowed to grow in hope of multiplication.

926.　590. C. setosa, Hall. fil.
Casual. Cornfields. Very rare. A. July and August. Not in *Top. Bot.*
　First Record. Among clover at Malvern Link. *Towndrow, Malv. Advert.,* Nov. 19, 1892.
This plant is a native of meadows in Central and Southern Europe. It has been observed in Herefordshire and Warwickshire.

927.　591. C. capillaris, Wallr.　　*Smooth Hawk's-beard.*
(*C. tectorum*, Huds. *C. virens*, Linn.)
Native. Cultivated fields, meadows, dry banks, sides of roads. Very common and widely distributed. B. May to September. *Top. Bot.* 102.

　First Record. As *Crepis tectorum.* Not localized, *Lees, Bot. Malv. Hills,* 1st ed., p. 37 (1843).
This is one of the commonest plants of the country side, varying much in height and character according to the situation and character of the place of growth. It bears a corymb of little yellow flowers, on a stem sometimes reaching the height of three feet, and sometimes hardly reaching one foot, and is always easily known by its upper linear sagittate leaves. It occurs abundantly in all neighbouring counties.

928.　592. C. nicaeensis, Balb.
Alien. Fields, introduced with seed. Very rare. B. June and July. Not in *Top. Bot.*
　First Record. Malvern Link, *Towndrow, Bot. Ex. Club Rep.,* 1886.
　　Severn. Railway-bank by Stour Viaduct, Kidderminster to Bewdley.
　　Malvern. Between Lower Howsell and Stocks Lane.
This alien from Central and South-eastern Europe has been observed in Hereford and Salop of our bordering counties. It is intermediate between *C. capillaris* and *C. biennis*, and is chiefly characterized by its scabrous fruit.

929.　593. C. biennis, Linn.　　*Rough Hawk's-beard.*
Colonist. Cultivated fields. Rare. B. May to August. *Top. Bot.* 27.
　First Record. Malvern, *Towndrow, Mid. Nat.,* vi, p. 117 (1883), and *Malv. Advert.,* Nov. 19, 1892.
　　Avon. Hill leading up to Dodderhill Common.
　　Severn. West side of Hanbury Railway Station.
　　Malvern. As above. Callow End, Powick. Mathon.
This plant has been observed in all neighbouring counties except Staffordshire.

931.　594. C. paludosa, Moench.　　*Marsh Hawk's-beard.*
Native. Damp woods and shady places. Rare. P. July to September. *Top. Bot.* 62.
　First Record. Not localized, *Victoria Hist. Worc., Table of Plants,* p. 52 (1901).
　　Severn. In the district, *Bagnall.* By the Severn at Bewdley.
This plant differs from all the foregoing Hawk's-beards in having a cylindrical, unbeaked fruit, and a stiff brittle pappus. It has been noticed in Staffordshire and Warwickshire.

HIERACIUM, Linn. 240.　(ἱέραξ, a hawk, as the plant was supposed to strengthen the vision of birds of prey.)

Of this intricate and puzzling genus of plants, which has exercised the patience and tried the skill of many botanists both in this country and abroad, the latest expression of knowledge as regards British species is to be found in Babington's *Manual of Botany,* 9th ed., 1904, p. 232, and Mr. Linton's *The British Hieracia* of 1905. To these accounts, drawn up under the direction of Messrs. F. J. Hanbury and W. Linton, the student desirous of making an intimate acquaintance with the genus may be referred. It contains, according to the former work, ninety-seven species, with numerous varieties in the case of most of them. A Worcestershire botanist is likely to meet with but few of them ; still, if he is anxious to do so, he may amuse himself by bringing home a few sheaves of the flowers of the kind he meets with, and on going through them carefully it is possible that he may discover a plant with the particular bristle that constitutes a variety. To most people life is not long enough for such a diversion. Different authors differ widely as to the value of the specific characters, the extreme forms pass into each other by insensible gradations, and hybridization appears to be common.

932. 595. H. Pilosella, Linn. *Mouse-ear Hawkweed.*
Native. Commons, dry banks, pastures. General throughout. P. May to August. *Top. Bot.* 110.
First Record. Not localized, Lees, *Bot. Malv. Hills,* 1st ed., p. 37 (1843).
This is a well-known plant of the short grass or the dry common, with pale yellow flowers on single stems, and leaves studded on the top with sparse long hairs, and closely felted with short hairs beneath. Its scions creep out from the parent plant on every side, and lying close to the surface of the soil form thick patches of leaves. It is one of the composites, a much neglected race as a whole by the village practitioner of ancient times, which had reported medicinal virtues, its juice taken in wine being considered good for jaundice. It occurs in all neighbouring counties.

935. 596. H. *aurantiacum, Linn. *Orange Hawkweed.*
Alien. Woods, pastures, and waste places, a garden escape. P. June to August. Not in *Top. Bot.*
First Record. An escape. Malvern, *Towndrow, Malv. Advert.,* Nov. 19, 1892.
Malvern. As above. Garden wall, Tenbury. Roadside near Suckley, *Jeffery.*
This plant, a native of the Alps of Southern Europe, has long been a garden favourite, and establishes itself in rockeries and in borders, growing strongly and spreading widely. Its flowers are a deep orange of a tone rare among border flowers, and are carried in bunches at the top of stems black with sticky hairs, which increase in the neighbourhood of the heads. It has been observed out of confinement in Herefordshire and Salop.

597. H. murorum, Linn. *Wall Hawkweed.*
Native. Walls and dry places. Several places. P. May to July. *Top. Bot.* (aggregate) 68.
First Record. In a wood by Picket Rock, near Kidderminster, and in the Rocky Wood, Fenny Rough, near Stone, *Perry, Mag. Nat. Hist.,* vol. iv, p. 450 (1831).
Avon. Norton, near Evesham. Fladbury. Littleton. Pershore.
Severn. Near Bewdley ; Northwood, Bewdley, *Mathews.* Wyre Forest. Crowneast Wood. Fenny Rough.
Malvern. On the rocks, Malvern, *Lees* ; possibly an error for *H. vulgatum,* Fr. Old Storridge Common. Mathon.
Lickey. Offmoor Lane. Dales Wood, *Mathews.*
Of the aggregate, to which these records refer, twenty-five varieties are added in the *London Catalogue,* 9th ed. ; in the *Lond. Cat.,* 10th ed., the name

with all its varieties is gone. The space is taken up by quite a new lot of names, among which, here and there, glimmer like stars a few of the former ones. For the purposes of this book it will be sufficient to retain the name from the 9th ed. Mr. Mathews throws doubts on Mr. Lees' Malvern records. *H. murorum* is not now to be found there, while *H. vulgatum* is common there. *H. murorum* may be distinguished from its numerous relatives by its leaves persisting at the base while the plant flowers, its stem usually with a single leaf, its green thin leaves, and the usually few hairs on its involucre. The plant is found in all neighbouring counties.

1022. 598. H. vulgatum, Fr. *Common Hawkweed.*
(*H. silvaticum,* Lam.)
Native. Woods and hedgebanks on sandy soil. Local. P. May to September. *Top. Bot.* (aggregate) 90.
First Record. As *H. silvaticum.* Dudley Castle Hill, *Withering,* 4th ed., p. 671.
Avon. Bredon Hill. Broadway. Abbots Wood.
Severn. Crowneast Wood, *Lees.* Northwood, near Bewdley, *Mathews.* Summer Hill and Wassell Wood. Near Kidderminster ; Rock Wood and Burnt Wood, near Bewdley, *Perry.* Shrawley Wood. Nunnery Wood. Bewdley Station. Spring Grove. Cathedral walls. Wyre Forest.
Malvern. Malvern Hills. Alfrick. The Wych. Southstone Rock. Martley Hill.
Lickey. The Valley, Bromsgrove, *Mathews.* Dudley Castle Hill, *Withering.* The Manor Abbey, Halesowen. Between Suchern and Weatheroak.
Both *H. murorum* and *H. vulgatum* were given as new records in *Bot. Rec. Club Rep.* for 1883, which they were not. This plant has a leafy stem, its root-leaves usually taper into a foot-stalk, its flower-stalks are downy, and the involucre is covered with down mixed sometimes with black hairs or setae, while the radical leaves persist to the time of flowering. The plant is recorded from all neighbouring counties.

1033. 599. H. sciaphilum, Uechtritz.
Native. Dry heathy places. As in records below. P. June to August. *Top. Bot.* 22.
First Record. Railway cutting near Upton-on-Severn, July 1, 1897 ; Long Coppice, Leigh, August 1, 1895, *Towndrow* ; named by F. J. Hanbury, *Malvern Advertiser,* November 11, 1899.
Malvern. As in first record. Malvern Link.
This plant has been found in Shropshire and Herefordshire.

1044. 600. H. tridentatum, Fries. *Top. Bot.* 6.
(*H. rigidum,* Hartm. Var. *tridentatum* (Fr.).)
First Record. Powick, 1888, *Towndrow, J. of B.,* vol. xxvi, p. 312.
Severn. As first record. Whitehouse Coppice, Powick.
Mr. Towndrow reported in *J. of B.,* December, 1900, a form of *H. rigidum,* determined by Mr. F. J. Hanbury, which he gathered at Leigh Sinton, near Malvern, July 21, 1898, and again in 1900. It is raised to a species in *Lond. Cat.,* 10th ed. The aggregate has been found in Herefordshire.

1062. 601. H. sabaudum, Linn. *Shrubby Broad-leaved Hawkweed.*
Var. b. **boreale** (Fr.).
(*H. sabaudum,* Huds.)
Native. Heaths, dry woods, hedgebanks. Local. P. July to September. *Top. Bot.* (aggregate) 96.
First Record. As *H. sabaudum,* var. 4. Perry Wood, Worcester, *Stokes, Stokes's With. Bot. Arr.,* 2nd ed., p. 850 (1787).
Avon. Rather uncommon, *Lees.* Tiddesley Wood. Near Defford Common. Near Evesham.
Severn. Near Bewdley ; the Valley, near Bromsgrove, *Mathews.* Wyre Forest, *Jorden.* Crowneast Wood, *Lees.* Perry Wood. Churchill, near Kidderminster. Shrawley Wood. Severn, near Bewdley. Chatley.
Malvern. Hanley ; Crump End, *Lees.* Sherrard's Green. Old Storridge.
Lickey. General, *Mathews.*
This plant has no radical leaves at the time of flowering ; the stem is often reddish, very hairy below, stellately downy above, and leafy throughout ; the teeth of the leaves pointing forward, the lowest leaf with villous petiole, peduncles floccose, involucre nearly glabrous, and the ligules glabrous. This plant occurs in all neighbouring counties.

1063. 602. H. umbellatum, Linn. *Umbellate Hawkweed.*
Native. Hedgebanks, heathy places. Local. P. July to September. *Top. Bot.* (aggregate) 88.
First Record. In a wood by Crow's nest (Crown east) Wood, St. John's (Worcester), *Lees, Bot. Malv. Hills,* 2nd ed., p. 70 (1852).
Severn. Wyre Forest, *Jorden.* Hodge Hill, Kidderminster, *Lees.* Habberley. Crowneast Wood. Between Churchill (Kidderminster) Station and the Church. Hagley Brake, *Mathews.* Hodge Hill, near Kidderminster. Near Bewdley. Blakedown. Lincombe.
Malvern. Crowneast Wood is the one locality referred to by Lees in his *Table of Plants* ; 'I have not seen it elsewhere,' he says. This locality is not in the Malvern district. Coneygree Wood. Alfrick. Malvern Wells. Powick.
Lickey. Rare, *Mathews.* Saltwells Wood.
This plant has no radical leaves at the time of flowering, its stem is short and wiry and has many narrow leaves, and its heads are large sub-umbellate glabrous, the ligules glabrous, and its bracts are recurved. The Hawkweeds, besides their virtues relating to the eyes of hawks and for combatting jaundice, shared with numerous other plants in times gone by their efficacy against the bites of the universal 'serpent'. *H. umbellatum* occurs in all neighbouring counties.

HYPOCHOERIS, Linn. 241. (ὑπό, for, χοῖρος, a hog, the roots being eaten by that animal.)

1065. 603. H. glabra, Linn. *Smooth Cat's-ear.*
Native. Open, sunny, gravelly spots. Rare. A. June to August. *Top. Bot.* 48.
First Record. In Pensham Field, near Pershore, in the most barren, gravelly places, *Nash, Hist. Worc., Int.,* p. lxxxix (1781).
Avon. Pensham Field, *Nash,* extinct.

Severn. Hartlebury Common. Fields between Bewdley and Kidderminster. Sutton Common.
Malvern. North Hill, *Towndrow.* Malvern Common. On the Hills, Malvern Wells.
This rare plant has small yellow flowers. It has not been recorded either for Gloucestershire or Herefordshire.

Var. b. **Balbisii,** Lois.
First Record. As under.
Severn. Sutton Common, *Rea.*
In this variety all the fruits are beaked, whereas in the type the marginal fruits are not beaked.

1066. 604. H. radicata, Linn. *Long-rooted Cat's-ear.*
Native. Pastures, meadows, heaths, lawns, waste places. Very common and generally distributed. P. May to September. *Top. Bot.* 111.
First Record. Not localized, Lees, *Bot. Malv. Hills,* 1st ed., p. 38 (1843).
This is a troublesome weed, on account of its long roots, while the rosettes of leaves, which are all radical, press closely to the earth. It occurs abundantly in all neighbouring counties.

LEONTODON, Linn. 242. (λέων, a lion, ὀδούς, a tooth, from the tooth-like margins of the leaves.)

1068. 605. L. nudicaule, Linn. *Hairy Hawkbit.*
(*L. hirtum,* Linn. *Thrincia hirta,* Roth. *Apargia hirta,* Scop.)
Native. Sandy places, heaths, roadsides, pastures. General ; less common in the Lickey district. A. B. or P. May to September. *Top. Bot.* 74.
First Record. As *Thrincia hirta.* Not localized, Lees, *Bot. Malv. Hills,* 1st ed., p. 37 (1843).
This is a frequent plant on commons and heathy places, bearing a yellow flower on each of its purplish hairy stalks. It occurs in all neighbouring counties.

1069. 606. L. hispidum, Linn. *Rough Hawkbit.*
(*Apargia hispida,* Hoffm.)
Native. Meadows, pastures, grassy downs, waysides. Generally distributed and common. P. May to September. *Top. Bot.* 95.
First Record. As *Apargia hispida.* Not localized, Lees, *Bot. Malv. Hills,* 1st ed., p. 37 (1843).
This plant is very common on pastures and grassy places. It bears its flowers on single leafless stalks, which are slender and swollen at the top. It is found in all bordering counties.

1070. 607. L. autumnale, Linn. *Autumnal Hawkbit.*
(*Apargia autumnalis,* Hoffm.)
Native. Meadows, waysides, and waste places. General. P. May to September. *Top. Bot.* 112.
First Record. As *Apargia autumnale.* Not localized, Lees, *Bot. Malv. Hills,* 1st ed., p. 37 (1843).
An abundant plant all over Britain, and ranging throughout Europe and Russian Asia from the Arctic regions to the Mediterranean. It occurs, of course, in all our bordering counties.

TARAXACUM, Hall. 243. (ταράσσω, I stir up, from its medicinal effects.)

1071. **608. T. officinale**, Weber. *The Dandelion.*
Taraxacum Taraxacum, Karsten. *Leontodon Taraxacum*, Linn. *Taraxacum Dens-leonis*, Desf.
Native. Everywhere abundant. P. January to October. *Top. Bot.* 112.
First Record. As *Leontodon Taraxacum*. Not localized, *Lees, Bot. Malv. Hills*, 1st ed., p. 37 (1843).
This yellow flower, whose commonness leads one to despise and overlook its beauty and charm, shares with the daisy universal popularity. It is the plaything of village children, who manufacture chains of it, and imaginatively tell the time by its graceful heads of plumy seeds ; the friend of those who love the countryside ; and useful in modern medicine, besides affording a salad for the table and a substitute for coffee in its baked roots. What need to say more ? The English name is a corruption of the French *Dent de Lion*, of which the generic name given it by Linnaeus is a translation. The plant is abundant all through its range—Europe, Central and Northern Asia, and North America. Bordering counties share with Worcestershire in this abundance. An interesting bi-generic hybrid with *Crepis taraxacifolia* is recorded under that plant.

Var. **Dens-leonis**, Desf.
First Record of segregate, in this book.
 Avon. Norton, near Evesham. Fladbury. Sheriff's Lench. Pershore. Crowle. Defford.
 Severn. Hartlebury. Pitchcroft, Worcester. Ombersley. Kempsey. Crown East. Diglis.
 Malvern. General, *Towndrow* (MS.). Powick Ham. Martley. Bransford. Knightwick. Stanford Bridge. Broadwas.
 Lickey. Halesowen.
The varieties of this plant depend chiefly upon the characters of the bracts of the involucre. This variety is the common form of the plant in cultivated ground, having the outer bracts deflexed or curled downwards, and the inner one simple at the top. This segregate is not given in *Lond. Cat.*, 10th ed. It has been observed in all neighbouring counties.

1072. **609. T. erythrospermum**, Andrz. *Top. Bot.* 17.
First Record. Not localized, *Victoria Hist. Worc.*, p. 53 (1901).
 Avon. Sheriff's Lench. Littleton Banks. Pershore. Crowle.
 Severn. Claines. Hallow. Grimley. Stagbury Hill. Wyre Forest. Kempsey. Hartlebury Common.
 Malvern. Common, *Towndrow, Malvern Advertiser*, Jan. 29, 1894. Malvern Hills. Powick. Bransford.
 Lickey. Barnt Green.
In this plant the outer bracts are patent or adpressed and the inner ones gibbous or with an appendage at the tip, while the seed, instead of being yellow, is bright red or reddish-brown. It has been observed in every neighbouring county except Gloucestershire.

1073. **610. T. palustre**, DC. *Top. Bot.* 80.
First Record. Not localized, *Victoria Hist. Worc.*, p. 53 (1901).
 Avon. The Marsh Common. Defford Common.
 Severn. In the district, *Bagnall* (MS.). Hartlebury Common. Wyre Forest. Kempsey Common.
 Malvern. Hedgebanks, *Towndrow, Malv. Advert.*, Jan. 20, 1894. Midsummer Hill.
In this plant the outer scales are adpressed and the inner ones simple at the top, and the fruit is pale and spinulose upwards. It has been observed in all neighbouring counties except Gloucestershire.

Var. b. **udum**, Jord.
(Var. *laevigatum*, DC.)
First Record. Malvern, *Towndrow, Malv. Advert.*, Nov. 19, 1892.
 Avon. Crowle. Craycombe Hill. Cleeve Banks. Clarke's Hill, Evesham.
 Severn. Crookbarrow Hill. Helbury Hill. Kempsey Common. Holt Bank.
 Malvern. Croft Farm. Wynds-point. Pickersleigh. Ankerdine. Berrow Hill, Martley.
 Lickey. Barnt Green. Northfield.
In this variety the outer scales are erect and patent, and the inner ones gibbous or appendaged at the top, and the fruit is similar to that of *T. erythrospermum* but lighter in colour. It has been noticed in Hereford, Salop, Warwick, and Stafford.

LACTUCA, Linn. 244. (From *lac*, milk, which flows from the stem when broken.)

{*1075.* **611. L. virosa**, Linn. *Acrid Lettuce.*
Native. Hedges, waysides, and bushy places. One record only. A. or B. July to September. *Top. Bot.* 53.
First Record. Not localized, *Lees' Cat., New Bot. Guide* (1835).
 Severn. One locality, *Lees, Table of Plants*, 17.
This plant is in appearance much like the more common *Lactuca muralis*, the distinction depending on the number of florets in the head, the black fruit, and the horizontal position of the upper leaves of this plant. In many cases Mr. Lees was mistaken, and perhaps was mistaken in this. No other botanist has observed it in Worcestershire, but from all bordering counties it is recorded, Mr. Watson, *Top. Bot.*, p. 226, stigmatizing Salop's record as ' bad authority '.}

{*1076.* **612. L. Serriola**, Linn. *Prickly Lettuce.*
(*L. Scariola*, Linn.)
Native. Waste places. One record only. B. July and August. *Top. Bot.* 6.
First Record. At Longdon Hill End, south of the path to Welland, *Lees, Bot. Malv. Hills*, 2nd ed., p. 69 (1852).
 Malvern. As first record.
In *Top. Bot.* the Worcestershire record is fortified by the addition of ' Westcombe, sp.', showing that Mr. Watson himself had seen a specimen of the plant from Mr. Westcombe's herbarium. Although this plant is very

rare in Britain, it is plentiful in many parts of Europe, and it has been thought that the cultivated lettuce of our gardens, *Lactuca sativa*, is a modified condition of this plant. Pliny, however, says that there was no other lettuce known to his ancestors except a plant he calls ' black lettuce ', which might be *L. virosa*, from its black seeds, especially as he dilates upon its narcotic milky juice, a quality more marked in that plant than this. Both plants are closely allied, but this one is prickly only towards the base. But from whatever plant derived, the lettuce has been so long in cultivation that its origin is lost. Many of its European names are derived from its milky qualities. *L. Serriola* does not occur in any neighbouring county.}

{*1077.* **613. L. saligna**, Linn. *Least Lettuce.*
Native. Chalky places. One record only. B. July and August. *Top. Bot.* 9.
First Record. ' Lactuca sylv. laciniata minima. N. D. near Church Lench in Worcestershire in great plenty,' *Merrett's Pinax*, 1666.
 Avon. Church Lench, as above.
Mr. Mathews considers this record to refer to *L. saligna*. Mr. Watson, *Top. Bot.*, p. 226, says : ' not verified of late years ', and so it remains. It is a rare plant of the east and south-east of England, especially near the sea. It is not recorded from any neighbouring county.}

1078. **614. L. muralis**, Gaertn. *Ivy-leaved Lettuce, Wall Lettuce.*
(*Prenanthes muralis*, Linn.)
Native. Woods, walls, and banks. Locally common. A. or B. June to October. *Top. Bot.* 72.
First Record. As *Prenanthes muralis*. Blackstone Rock and Rock Wood, near Bewdley ; in a wood by Picket Rock and Summerhill, Kidderminster ; and Rock Hill, a mile and a half from Bromsgrove on the road to Alcester, *Perry, Mag. Nat. Hist.*, vol. iv, p. 450 (1831).
 Avon. Craycombe. Broadway. Hanbury. Throckmorton. Stoulton.
 Severn. Birchen Grove. Kyre. Bockleton. Shrawley Wood. Hagley. Hadley. Salwarpe. Wyre Forest. Near Hartlebury Common. Woolverley.
 Malvern. Little Malvern ; Cowleigh, *Lees*. Southstone Rock. Alfrick. Half Key. Abberley. Clifton-on-Teme. Dripshill. Malvern. Malvern Link.
 Lickey. Halesowen Abbey. Romsley. Clent. Rubery ; Catshill, *Humphreys*. Stony Lane, Moseley, *Miss M. A. Beilby*. Fairfield. Burcot. Finstall.
This is the commonest of our wild lettuces, preferring the rocks of the red sandstone measure, on which, in shady places, it sometimes occurs plentifully. It is a slender branching plant, with a smooth round hollow stem, sometimes two or three feet high. Each head has five florets, giving it the appearance of a simple flower of five petals. It is recorded from every neighbouring county.

SONCHUS, Linn. 245. (Named σόγχος in Greek, said to be a modification of σομφός, hollow, from the nature of its stem.)

1080. **615. S. oleraceus**, Linn. *Common Sow Thistle.*
Native. Cultivated ground, waste places. Abundant throughout the county. A. April to September. *Top. Bot.* 111.
First Record. Not localized, *Lees, Bot. Malv. Hills*, 1st ed., p. 37 (1843).
This is a well-known weed of cultivated ground, not only in Britain, but of very wide distribution on the face of the earth, occurring not only in Europe but in North and West Asia, India, North Africa, South Australia, and New Zealand. It is a favourite food with rabbits, being one of the plants called rabbit's meat and sought after for his pets by the schoolboy ; and it is so favoured by swine that it has gained a name in reference to these animals not only in England but all over the continent also. The plant is full of milky juice. It occurs abundantly in all neighbouring counties.

1081. **616. S. asper**, Hoffm. *Sharp-fringed annual Sow Thistle.*
Native. Cultivated ground and waste places. General throughout the county. A. May to September. *Top. Bot.* 106.
First Record. Not localized, *Lees, Bot. Malv. Hills*, 2nd ed., p. 69 (1852).
This plant much resembles the preceding one, differing in having rounded appendages at the base of the leaves instead of arrow-shaped ones, while the fruit is not transversely rugose. The leaves, too, are crisped and not so flat. It occurs in all neighbouring counties.

1082. **617. S. arvensis**, Linn. *Corn Sow Thistle.*
Native. Cultivated fields. Generally distributed. B. May to July. *Top. Bot.* 111.
First Record. Not localized, *Lees, Bot. Malv. Hills*, 1st ed., p. 37 (1843).
This is one of our handsomest plants, with elegant leaves and showy flowers, each as large as a half-crown piece, growing three or four feet high. It has not so wide a range as its relative, *S. oleraceus*, but still is found over most of the temperate regions of the old world and India. It is recorded from all neighbouring counties.

Var. b. **glabrescens**, Hall.
First Record. Madresfield, *Towndrow, Malv. Advert.*, Nov. 19, 1892.
This form differs from the type in the absence of the glandular pubescence of the flower-heads.

TRAGOPOGON, Linn. 246. (τράγος, a goat, πώγων, a beard, from the character of the bearded fruit.)

1084. **618. T. pratensis**, Linn. *Yellow Goat's-beard.*
(*Tragopogon minor*, Fr.)
Native. Meadows, waysides, railway-banks. General. B. May to July. *Top. Bot.* 86.
First Record. Vale of Severn, *Pitt, Agric. Worc.*, p. 317 (1810).
 Avon. Craycombe. Norton, near Evesham.
 Severn. Railway-banks, near Harberrow, Hagley, *Mathews*. Spetchley. Diglis. Kempsey. Wyre Forest. Rainbow Hill, Worcester. Grimley Brickyard. Dodderhill, Droitwich.

Malvern. Powick. Bransford. Knightsford Bridge. Martley. Lickey. New road to Kingsnorton, three miles from Birmingham, *Ick*, 1837. Clent. Tardebigge.

In the *Lond. Cat.* the specific name is *pratense*. But the botanies give *pratensis*, which is warranted by the gender of πράτων. The rule is to correct bad spelling in compounds in accordance with good Greek and Latin.

T. minor, Fries, differs from *T. pratensis*, in having the flowers half as long as the scales of the involucre, instead of quite as long, and is the common form. Mr. Lees, *Bot. Malv.*, 3rd ed., p. 90, says that both forms may be seen on the same specimen according to the age of the flower. The long slender leaves of this plant and Salsify distinguish them from all other British composites. The flowers open at sunrise, and close at noon, whence its country name of 'Johnny go to bed at noon'. The head of seeds is very conspicuous after flowering, elevated on a long stalk and very feathery. Its roots are a good vegetable, and possess in addition the qualities of 'warming the stomache', prevailing 'in consumptions', and strengthening 'those that have beene sicke of a long lingering disease'. It is recorded from all neighbouring counties.

1085. 619. T. minor, Mill.
First Record. Bredon Hill, *Mathews, Worc. Nat. Club Trans.* ii, p. 63 (1899).
Avon. Bredon Hill. Trench Woods.
Severn. In the district, *Bagnall* (MS.).

The above appears to be the first definite record of this plant, which in Babington's *Manual*, 9th ed., p. 226, is considered a separate species. It is reported from the Severn district by Mr. Bagnall (MS.), and by Mr. Mathews from the Trench Woods also, and indeed is the common Worcestershire form. It has been noticed in all bordering counties except Herefordshire.

1086. 620. T. *porrifolius, Linn.   *Salsify.*
Alien. An escape from cultivation. B. May and June. Not in *Top. Bot.*
First Record. Not localized, but placed in Malvern district in *Lees, Table of Plants*, p. 17. Mr. Mathews, *Mid. Nat.* xiv. 186, and also *Mid. Nat.* xv. 141, says in error 'Lickey'.
Severn. Mudwall Mill, Worcester. Railway-bank between the Engine House and Tunnel, Worcester.
Malvern. Brickyard, Malvern Link, *Towndrow, Malv. Advert.*, November 19, 1892.

Linnaeus wrote *porrifolium*, but the remarks made in the case of a former plant apply here also. This plant is a native of Southern Europe and has long been cultivated as a garden vegetable. It is very similar in appearance to the common Goat's-beard, but the flowers are dull purple. It has been noticed in Gloucestershire and Staffordshire.

42. CAMPANULACEAE, Juss.

JASIONE, Linn. 248. (ἰασιώνη, a healing plant mentioned by Theophrastus, from ἰάομαι, I heal.)

1089. 621. J. montana, Linn.   *Sheep's Bit, Sheep's Scabious.*
Native. Dry sandy fields, hedgebanks, heaths. Rather uncommon. A. or B. June to October. *Top. Bot.* 80.

First Record. Burcot Lane, near Bromsgrove, and in other places about that neighbourhood, *Purton, Midl. Fl.*, p. 418 (1817).
Severn. Harberrow. Blakedown. Wannerton Downs, *Mathews.* Shrawley Wood. Near Bewdley, *Westcombe.* Hartlebury Common. Areley.
Malvern. Rosebury Rock, *Lees.* Wood adjoining Woodbury Rock. Old Storridge.
Lickey. As first record. The Birches, Hagley. Churchill. Lickey Hill. Dodford.

This plant is very like *Scabiosa Succisa*. It has a disagreeable odour when bruised. The flowers are bright blue in dense hemispherical heads, within a many-leaved involucre. It is recorded from all neighbouring counties.

WAHLENBERGIA, Schrad. 249. (Named after G. Wahlenberg, Professor of Botany at Upsala.)

{**1090. 622. W. hederacea**, Reichb.   *Ivy-leaved Bell-flower.*
(*Campanula hederacea*, Linn.)
Native. Shady boggy places. One locality only, now lost. P. June to August. *Top. Bot.* 46.
First Record. As *Campanula hederacea*, on Hartlebury Common, *Rev. T. Butt, Sm. Eng. Fl.*, vols. i and ii (1824).
Severn. As first record.

This plant has not been seen in that locality by any other botanist. It is a likely enough place for its occurrence; but Mr. Mathews doubts the correctness of the record, which is by the same observer who reported *Anaphalis* (*Gnaphalium*) *margaritacea* from Wyre Forest. The correctness, however, of this determination was verified by Westcombe and from specimens in the Herb. Hast. Mus. Worc. The plant, however, is unmistakable, and it is difficult to imagine error short of fraud. It has been recorded from all neighbouring counties.}

CAMPANULA, Linn. 251. (The Latin for a little bell, from the shape of the corolla.)

1093. 623. C. glomerata, Linn.   *Clustered Bell-flower.*
Native. Calcareous pastures, dry banks. Rare. P. May to October. *Top. Bot.* 51.
First Record. Knightsford Bridge, *Lees*; *A. Florence, Strang. Guide* (1828). 'Must be an error,' *Mathews*; if so, the first true record is Dudley Castle Hill, *Scott, Stourb.*, p. 540 (1832).
Avon. Broadway Hill, purple and white. Bredon Hill. Near Bredon Hill between Elmley and Evesham, *Westcombe.*
Malvern. West Malvern, *Rogers*; *Towndrow, Malv. Advert.*, January 20, 1894.
Lickey. Dudley, *Scott.*

Mr. Lees, in his *Table of Plants*, p. 18, does not localize this Bell-flower in the Malvern division thus himself repudiating the Knightsford record. This handsome flower with its dark purple-blue bells clustered at the top of its angular stems is a frequent inhabitant of the garden border, where it

grows over a foot in height. It has not been recorded for Staffordshire or Herefordshire in our bordering counties; but is frequent in South Warwickshire.

1094. 624. C. Trachelium, Linn.   *Nettle-leaved Bell-flower.*
Native. Woods, thickets, hedges. Local. P. July and August. *Top. Bot.* 59.
First Record. Blackstone Rock, near Bewdley; banks of the Severn near Stourport; and on a steep bank about four miles on the Hereford road from Stourport (near Witley), *Perry, Mag. Nat. Hist.*, vol. iv, p. 450 (1831).
Avon. Broadway. Bredon. Norton, near Evesham. The Slads, Trench Woods. Fladbury. Tiddesley Wood.
Severn. Hadley Road, Droitwich, *Humphreys.* As in first record, *Perry.* Shrawley Wood. Shrawley. North Wood, near Bewdley. Wyre Forest. Fenny Rough. Near Claines Church. Ombersley. Kempsey. Holt. Hadley.
Malvern. Ankerdine. With white flowers at Knightwick, and lane back of Woodbury Hill. Malvern Wells. Clifton-on-Teme. Powick. Abberley Hill. Cowleigh. Witley. Martley. Fetterlocks.
Lickey. A weed in Clent Cottage Garden, Clent, not known to have been introduced, with white and blue flowers, *Amphlett.* Harvington.

This is a rough-looking plant with a stem some feet high. It requires some little imagination to see the likeness of its leaves to a nettle. The top flower of the spike comes out first. The plant is recorded from all neighbouring counties.

1095. 625. C. latifolia, Linn.   *Giant Bell-flower.*
Native. Woods and thickets. Not common except in the Lickey district. P. July and August. *Top. Bot.* 61.
First Record. On the road from Halesowen to Birmingham, on a shivery sand-bank, *Withering*, 3rd ed., p. 243 (1796).
Avon. Near Evesham, *Westcombe.* Tardebigge Reservoir.
Severn. Laughern Brook, near Boughton, Worcester, *Lees.* North Wood, near Bewdley. Cookley. Shrawley Wood. Dick Brook footbridge, Stourport, *Mr. Gardner.* Lincombe Wood, *Hickman.* Arley. Westwood Park. Plantation near Ardwick's Cell. Ardwick's Cell, abundant.
Malvern. Hedges below Malvern Hills, *Nash.* Wood near Cowleigh Park. Cowleigh Park.
Lickey. Illey Mill. Manor Abbey, Halesowen. As in first record. Shutt Mill, Romsley. Alvechurch, *D. Mathews.* Clent. Belbroughton. Yardley Bridge, *Freeman.* Lutley Holloway, *Scott.* Blackwell.

This handsome plant is a well-known occupant of the garden border and shrubbery walk, and shares with *Campanula media*, Linn., a native of Northern Italy and a commonly cultivated biennial, the name of Canterbury Bell, as it is now called; but the name was always in the plural previously to the nineteenth century. Originally this was the name of *C. Trachelium*, being afterwards transferred to *C. media*, which was called also in the sixteenth century, Coventry Bells and Marian's Violet. Then another leap

brought the name to the present plant. Country people sometimes call it Throat-wort. The name Bell-flower is from a fanciful association of the flowers of the tribe with the small bells worn on their horses by pilgrims in pre-Reformation times. The plant occurs in all neighbouring counties.

1096. 626. C. *rapunculoides, Linn.   *Creeping Bell-flower.*
Denizen. Woods, banks, hedges. Rare. P. July and August. *Top. Bot.* 24.
First Record. Discovered by the Rev. G. H. Piercey, of Chaddesley, near Kidderminster, in a lane on a dry bank near to Shrawley Wood, September 30, 1820, *Purton, App.*, vol. iii, p. 18 (1821).
Severn. Between Churchill Station and Clent, *Irvine.* Shrawley Wood, first record. Between Droitwich and Hadsor.
Malvern. Malvern Railway Station, *Towndrow, Malv. Advert.*, November 14, 1896.

This plant is not a native of our island, being found there, as well as in North-western Europe, only in the neighbourhood of a house. Mr. A. Irvine, in his record, carefully notes that he found the plant only in the neighbourhood of houses; and though the Hadsor record states 'in a copse', it may be suspected that some kind of habitation was not far away. It has been noticed only in Gloucestershire and Warwickshire, and in Staffordshire near Tamworth of neighbouring counties.

1097. 627. C. rotundifolia, Linn.   *Harebell* or *Hairbell, Round-leaved Bell-flower.*
Native. Heaths, commons, sandy banks, fields. Common and widely distributed. P. June to September. *Top. Bot.* 111.
First Record. Not localized in a list of Malvern plants, *Ainsworth, Edin. Phil. J.*, p. 9 (1828).

This graceful little flower is an universal favourite, and has received many familiar names. No one seeing its slender lanceolate and linear stem-leaves would consider its specific name apposite. It is only its winter radical leaves that are at all round, and even so they are rather heart-shaped or kidney-shaped than round, and soon vanish in the growth of the plant. It has been a matter of controversy whether this, or the wild hyacinth of the woods, is the Blue-bell of Scotland, but modern opinion seems to claim this little Bell-flower as the plant. In some parts of England it is called Witch's Thimbles, and Harvest-bells. It occurs in all neighbouring counties.

1099. 628. C. *Rapunculus, Linn.   *Rampions.*
Colonist. Grassy fields. Rare. B. June and July. *Top. Bot.* 31.
First Record. Hindlip, Worcestershire, *Withering*, 3rd ed., p. 242 (1796).
Severn. Hindlip, as first record. Bank of road near Hartlebury Railway Station, July 30, 1893.

In his *Table of Plants*, p. 18, Mr. Lees places this plant at a locality in the Malvern district, referring to himself repeating that he found the plant of Bromsberrow; but Bromsberrow is in Herefordshire. This is an alien plant, whose home ranges from Belgium and Northern France to North Africa, Syria, and Western Siberia; it is not found in Greece, and in England it owes its presence to the fact that it was prized by our ancestors as a vegetable, the roots being boiled, or prepared as a salad with vinegar and pepper. It has not been reported for Salop.

1100. 629. C. patula, Linn.   *Spreading Bell-flower.*
Native. Hedge-banks, thickets. Well distributed throughout the county. A. or B. July. *Top. Bot.* 30.
First Record. 'Sponte nascentem reperi inter dumeta collis cuiusdam sylvosi prope Worcestriam' (in a wood called Elbury Hill, about a mile from Worcester), *Dillenius, Hort. Elth.* fo. 69, with a plate, Tab. lviii (1732).
Avon. Bentley Thrift. Hanbury. Bradley Green. Feckenham. Upton Snodsbury. Near Croome.
Severn. Wyre Forest, *Jorden.* Hurcott Wood. Blakedown. Churchill. Hagley. Shrawley Wood. Stanklin. Harvington. Railway Cutting, Pedmore. Crowneast. Holt Mill. Glasshampton. Hartlebury. Near Hadsor. Oldbury Lane. Salwarpe. Helbury Hill. Near Monk Wood. Leopard Hills. Hindlip. Hartlebury Common. Coningree Wood.
Malvern. Near the Rhydd. Dripshill. Hanley. Between Barnard's Green and Madresfield. Abberley Hill. Near Witley Court. Old Storridge. Leigh.
Lickey. Belbroughton. Brook-side near Feckenham, *Mathews.* Bromsgrove ; Woodcote, *Humphreys.* Chaddesley Corbett, *Perry* ; Lutley Holloway, *Scott.* Clent.
In John Nichols's *History of Leicester*, 1795, at p. clxxix, referring to the first record above, Richard Pulteney, who had previously mentioned it in *Phil. Trans.*, vol. xlix, pt. ii, p. 815 (1756-7) says, 'First discovered in England by Mr. Brewer, in 1726, near Worcester, as recorded by Dr. Dillenius.' This plant is not unlike the common Hairbell, but the flowers are deeper bluish purple, and wider open, and the calyx is toothed at the base. It loves the rocks of the new red sandstone, fringing the railway cutting between Hagley and Stourbridge. It has been recorded from all neighbouring counties.

C. Speculum, Linn.   *Venus's Looking Glass.*
Alien. A cornfield casual. A. July to September.
First Record. Field on Bredon Hill, *Miss Woodward, Lees, Bot. Worc.*, p. 107 (1867).
This is the common garden annual, a cornfield weed throughout Europe except in the Mediterranean region, where it is probably native.

LEGOUSIA, Durande. 252.

1101. 630. L. hybrida, Delarbre.   *Venus's Looking Glass.*
(*Campanula hybrida*, Linn.)
Native. Cornfields, chiefly on calcareous soil. Occasional. A. June to September. *Top. Bot.* 48.
First Record. In the calcareous field at Tibberton, *Lees, Bot. Worc.*, p. 48 (1867).
Avon. Broadway Hill. Bredon Hill, *Mathews.* Flowers deep to light purple, even white, at Broadway, 1856, *Lees.*
Severn. Calcareous field at Tibberton, *Lees.* Near Warndon, *Westcombe.* Between Droitwich and Copcut Elm on the roadside. Coningree Wood.

Malvern. Malvern Wells. Malvern Link. Field near New Pool, Malvern Wells.
This plant is recorded from all neighbouring counties.

43. VACCINIACEAE, DC.

VACCINIUM, Linn. 253. (Etymology doubtful ; perhaps altered from *baccinia*, denoting a plant with numerous *baccae* or berries.)

{1102. 631. V. Vitis-Idaea, Linn.   *Red Whortleberry, Cowberry.*
Native. Mountainous heaths. One place. Shrub. June and July. *Top. Bot.* 67.
First Record. Lower Bromsgrove Lickey, *Lees' Cat., New Bot. Guide* (1835). Lickey. As above.
Mr. Mathews thinks this record must be an error ; the plant, at all events, is not given in *Lees, Table of Plants*, p. 19, as occurring in the Lickey or any district. In *Top. Bot.* it is marked for Worcestershire without any authority, which means that Mr. Watson himself thought the 'authority sufficiently reliable, without being too exacting in the test', *Top. Bot., Int.* xxxiii. There is no special reason why the plant should not have grown at the place indicated at the period named and have disappeared shortly afterwards, one of a number of rare plants that have disappeared there and elsewhere in Worcestershire during the last century. The doubt is occasioned mainly by Mr. Lees not localizing it in his *Table of Plants.* It occurs in all bordering counties except Gloucestershire.}

1104. 632. V. Myrtillus, Linn.   *Bilberry.*
Native. Woods and heaths. Locally common. Small shrub. April and May. *Top. Bot.* 101.
First Record. Rocks above Great Malvern, *Ballard, Stokes's With. Bot. Arr.*, 2nd ed., p. 394 (1787).
Severn. Wyre Forest. Stagbury Hill. Fenny Rough. North Wood, near Bewdley. Eymore Wood. Hartlebury Common, *Westcombe.*
Malvern. Worcestershire Beacon, *Rea.*
Lickey. Plentifully on the hills named after the plant on the Lickey. Light Woods ; Warley, sparingly, *Mathews.* Cradley Park, *Scott.* Farley Woods, Romsley. Pepper Wood. Headless Cross, Redditch, *D. Mathews.* The Randans.
Besides Bilberry this plant has several English names, Whortleberry, Whinberry, Blaeberry, and Whorts among them, though the general one in the Midlands is the first, which is apparently of Norse origin ; Blaeberry is the common one in the north and Scotland. The dark, bloom-covered fruits are much sought after in the neighbourhood of Birmingham and in other neighbourhoods where they are plentiful, to be made into puddings and tarts, in which they are usually mixed with black currants, or sometimes with blackberries. The flowers are wax-like and drooping, nearly round balls, greenish-white suffused with red ; and the leaves when young are rosy, passing through a delicate green to a deep rich colour in autumn. The plant occurs in all neighbouring counties.

OXYCOCCUS, Hill. 254. (ὀξύς, sharp, κόκκος, a berry, from the flavour of the fruit.)

{1105. 633. O. quadripetala, Gilib.   *Cranberry.*
(*O. palustris*, Pers.   *Vaccinium Oxycoccus*, Linn.   *Scholera Oxycoccos*, Roth.)
Native, but not now in the county. Peaty bogs. Small shrub. June to August. *Top. Bot.* 68.
First Record. In the boggy parts of the Lickey, near Bromsgrove, *Nash, Hist. Worc., Int.*, p. lxxxix (1781).
Severn. Pedmore Common, *Scott.*
Lickey. Bromsgrove Lickey. Moseley Common, *Miss M. A. Beilby.*
The Cranberry is a well-known fruit, and large quantities are imported into England from America and various parts of the continent. The plant is a lowly, creeping, spreading shrub growing on heathy bogs difficult to walk upon. The flowers are bright rose colour, at the end of long simple stalks, and the deeply divided segments of the corolla are turned backwards. The leaves are very small. The plant has not been recorded from Gloucestershire.}

44. ERICACEAE, Juss.

ANDROMEDA, Linn. 257. (Named after *Andromeda*, who was chained to a rock and exposed to the attack of a sea monster.)

{1109. 634. A. Polifolia, Linn.   *Marsh Andromeda.*
Native. Peat bogs. Once recorded, possibly in error. Shrub. May to September. *Top. Bot.* 29.
First Record. On the Lickey Hill, near Bromsgrove. It is to be feared now extirpated, the road having since been altered, *Lees, Illus. Nat. Hist. Worc.*, p. 161 (1834).
Mr. Mathews says this record must be an error. Mr. Lees did not acknowledge it in his *Bot. Worc., Table of Plants*, p. 19. Yet the same remarks may be made about it as those in the case of *Vaccinium Vitis-Idaea.* Parts of the Lickey, at the beginning of last century, would appear to have been typical localities for plants delighting in moisture and boggy places. They have all gone now, but this is no reason for saying they never existed. This plant is not marked for our county in *Top. Bot.*, p. 269, but is given as occurring in Stafford and Salop.}

CALLUNA, Salisb. 258. (καλλύνω, I adorn or beautify, whether from the beauty of its flowers or its qualities as a broom is uncertain.)

1110. 635. C. vulgaris, Hull.   *Ling, Heather, Heath.*
(*Erica vulgaris*, Linn.   *C. erica*, DC.)
Native. Heaths, commons, dry sandy places, woods. Abundant in the Lickey district, less common elsewhere. Shrub. May to September. *Top. Bot.* 112.
First Record. No other kind of heath grows throughout the Malvern range, *Lees, Mag. Nat. Hist.*, vol. iii, p. 160 (1830).
This is a well-known plant, and in the Lickey district grows not only on commons and roadside wastes or sandy land, but is found, sometimes very attenuated, in nearly every bit of natural woodland also. It forms a thick scrub in the higher part of Hartlebury Common, and covers the lower

portion in a less luxuriant form. It is not common in the Malvern district, and Mr. Lees remarks that here it grows very small ; but it grows profusely on Old Storridge Common. At Saltwells Wood, near Dudley, it grows on an old pit-mound ; and at Warley Park, near Birmingham, it comes up among the turf of the trimmed garden lawn. But though Clent Hills rise in the centre of a district where it is widely distributed, only one struggling plant is known upon them. Nor is it by any means common in the Avon district. White-flowered specimens were observed in Coleridge Wood. It is recorded from all neighbouring counties.

Var. b. **glabrescens**, Koch.
First Record. In this book.
Severn. Hartlebury Common.
In this variety the leaves are glabrous, and it is the common form occurring throughout the county.

Var. **incana**, *Lond. Cat.*, 9th ed.
First Record. In this book.
Severn. Wyre Forest. Near the Devil's Spittleful, Habberley Valley. Blakeshall Common.
Lickey. Lower Lickey.

ERICA, Linn. 259. (ἐρείκη, heath.)

1112. 636. E. Tetralix, Linn.   *Cross-leaved Heath.*
Native. Damp places or heaths. Not uncommon in the Lickey district, rare elsewhere. Shrub. May to September. *Top. Bot.* 110.
First Record. Hartlebury Common, Worcestershire, *Ballard, Stokes's With. Bot. Arr.*, 2nd ed., p. 398.
Avon. One place, *Lees.*
Severn. Hartlebury Common. Broadheath. Wyre Forest, *Jorden.* Oldington Wood. Devil's Spittleful.
Lickey. Bromsgrove Lickey ; extinct, *Humphreys.* Moseley Common, *Miss M. A. Beilby.* Astwood, *Purton.* Stagbury Hill. Rednall Hill.
Mr. Lees notes that no species of *Erica* occurs throughout the Malvern chain of hills. This is a delicate and drooping plant, with its pale pink, waxen-looking flowers in clusters at the ends of its stems. Scott records it with white flowers, as very rare, somewhere near Stourbridge, probably Pedmore Common ; it is found also with white flowers on Hartlebury Common, and with pink and white flowers at Hungery Hill, Wyre Forest. It is recorded from all neighbouring counties, but does not occur in East Gloucestershire.

1114. 637. E. cinerea, Linn.   *Fine-leaved Heath.*
Native. Heathy places, sandy roadsides. Absent from the Avon and Malvern districts ; not common in the Severn district, but abundant in the Lickey district, except on Clent Hills, where it does not occur. Shrub. May to September. *Top. Bot.* 108.
First Record. In the N. of Worcestershire, *Stokes, Stokes's With. Bot. Arr.*, 2nd ed., p. 399 (1787).
This plant and the Ling make up what is usually termed Heather. The people who dwell about the stretches of moorland where this heath is the

prevailing plant use it for numerous purposes : for thatch, for a rough couch, compacted with earth for walls, and it can be twisted into a kind of rope. Its young twigs and flowers afford a yellow dye, and it can be used for tanning leather; while some kind of beer has been made of or flavoured with it. Lastly, it is the haunt of innumerable bees, though heather honey, which is usually dark-coloured, does not find favour with every one. It is said to be narcotic. *Erica cinerea* is recorded from every neighbouring county.

PYROLA, Linn. 264. (*Pyrus*, a pear, from the shape of the leaves.)

1121. **638. P. rotundifolia,** Linn. *Round-leaved Winter-green.*
Native. Moist woods, bushy places. Two records, possibly referring to the same locality. P. July to September. *Top. Bot.* 29.
First Record. In a wood on the Witley side of Abberley Hill, *Nash, Hist. Worc.*, Sup., p. 96 (1799).
Severn. Wyre Forest.
Malvern. As first record. Abberley Hill, *Lees, A. Florence, Strang. Guide.*
This is the rarest of the Winter-greens recorded for this county. Most of these plants are much alike and sometimes mistaken one for another, their distinction chiefly depending on the relative length and character of the style and stamens. In this plant the style is longer than the stamens, bent down and curved upward at the end, and the stigma is annular with five blunt erect points. Of neighbouring counties the plant is recorded only from Salop and Stafford.

1122. **639. P. media,** Swartz. *Intermediate Winter-green.*
Native. Woods. Very rare. P. July and August. *Top. Bot.* 43.
First Record. Cradley Park, near Stourbridge, *Scott, Purton,* vol. iii, pt. ii, p. 335.
Severn. Wyre Forest, *Jorden.* Habberley Valley. Park End Coppice, Bewdley.
Lickey. Cradley Park, extinct.
Scott, in his *Hist. Stourb.,* says there was only one plant at Cradley Park, which in 1817 was transplanted, cultivated, and soon lost—a fate which befalls many rarities! In this plant the style is nearly always straight, longer than the stamens, and the stigma annular with five erect points. The plant has been recorded only from Staffordshire and Warwickshire among neighbouring counties. It was recorded in *Top. Bot.* for Worcestershire on the strength of a specimen from Mr. Prentice.

1123. **640. P. minor,** Linn. *Lesser Winter-green.*
Native. Woods and bushy places. Very rare. P. July and August. *Top. Bot.* 68.
First Record. Abberley, *Rev. Mr. Severn-Rufford, Purton, Midl. Fl.,* p. 732 (1817).
Severn. Abberley, as first record. Wyre Forest, *Jorden.* Shrawley. Wood at Habberley Valley, *Mathews.* Near Bewdley, *Mrs. Carleton Rea.*
In this plant the style is equal in length to the stamens, and the stigma is without a ring, five-lobed, but pointless. This plant also, like the last, was recorded for Worcestershire in *Top. Bot.* on the strength of a specimen of Mr. Prentice's. It is a common plant of the Highlands of Scotland. It occurs in all bordering counties.

45. MONOTROPACEAE, Nutt.

MONOTROPA, Linn. 266. (μόνος, one, τρόπος, a turn, from the curved raceme.)

1126. **641. M. Hypopitys,** Linn. *Bird's-nest.*
(*Hypopitys Monotropa,* Crantz.)
Native. Woods near roots of fir and beech. Rare. P. June to September. *Top. Bot.* 47.
First Record. Shrawley Wood, *Mrs. Gardner, Purton, App.,* vol. iii, p. 36 (1821).
Avon. The Slads. Sheriff's Lench. Middle Hill, Broadway, *Sir T. Phillips, Cheshire.* Dovedale, Blockley, *Cheshire.* Trench Woods, 1908, *Humphreys.*
Severn. Warshill (Wassall), Bewdley, *Jorden.* Hadley Firs, Ombersley.
Mr. Babington, *Manual,* 9th ed., p. 282, says this plant is 'not parasitical'. It bears every characteristic of being a parasite, and, as it no doubt derives its sustenance from the roots of the trees near which it is always found, it is a distinction without much difference to say it is not parasitical. It has no leaves or branches nor any spot of green anywhere about it, and turns black when dry. Mr. Lees, in his *Table of Plants,* p. 19, marked it as extinct in Severn, but it has come to life again at Hadley Firs. The roots of the Pine and Beech on which this parasite or symbiote grows are covered with dense masses of growth termed 'mycorhiza' of a symbiotic fungus, and these function as root-hairs to the tree, in return for which they receive inorganic material worked up by these plants from the soil ; they are probably the mycelium of an *Elaphomyces* commonly known as False Truffles. The plant is recorded from all neighbouring counties.

47. PRIMULACEAE, Vent.

HOTTONIA, Linn. 269. (Named after Pierre Hotton, a Professor at Leyden during the latter half of the seventeenth century.)

1134. **642. H. palustris,** Linn. *Water-violet, Feather-foil.*
Native. Ditches, slow streams. Very rare. P. April to August. *Top. Bot.* 48.
First Record. In a ditch near Crowle, *Lees, A. Florence, Strang. Guide* (1828).
Avon. Crowle, as first record.
Severn. Near Severnstoke. Crowle, as first record, extinct.
Malvern. Near Chaceley, *Lees.* Upton-on-Severn, near the turnpike gate on the Malvern road, *Kent.* Ditches between Upton Bridge and Southend Farm, Beach. Longdon Marsh.
This is a handsome plant, its pink flowers rising in whorls above the water, while its finely divided leaves are always submerged. The plant is recorded from all neighbouring counties except Herefordshire. It may be doubted if Mr. Lees' Chacely locality is in Worcestershire or Gloucestershire : he says, *Bot. Worc.,* p. 35, 'in a marshy spot opposite Forthampton Court ',

and if this be so it is clearly in Gloucestershire, Forthampton being in the latter county ; in *Top. Bot.,* p. 336, the Gloucestershire record is founded on a specimen of Mr. Lees'.

PRIMULA, Linn. (*Primus,* first, because the flowers appear early.)

1135. **643. P. vulgaris,** Huds. *The Primrose.*
(*Primula acaulis,* Linn.)
Native. Woods, thickets, hedgebanks. Common except in the Lickey district, where the plant is scarce. P. January to May, and in the autumn. *Top. Bot.* 111.
First Record. Ditch banks, *Pitt, Agric. Worc.,* p. 317 (1810).
The primrose is a spring flower of the whole of Europe except the north-east portion, but does not extend into Asia, and it is found also in North Africa. Its Linnaean specific name, stalkless, would at first appear a misnomer, for the flowers have stalks, and sometimes fairly long ones, but these are the peduncles, not the true stem, which appears in the cowslip and the oxlip, and is wanting in the typical primrose. It is the umbel that has no stalk, not the flower. The primrose has entered into English poetry in a well-known way, and it is the plaything of country children, who make of them primrose balls. The plant is remarkably scarce in the Lickey district, caused, no doubt, in some respects by the proximity of Birmingham and the Black Country, where the roots not only find a ready sale, but whose inhabitants, strolling out into the surrounding country, cannot let a primrose-root alone when they see one. The plant is recorded from all neighbouring counties.
Var. b. **caulescens,** Koch.
First Record. Not localized, *Lees, Bot. Malv. Hills,* 2nd ed., p. 38 (1852).
Avon. Tiddesley Wood. Trench Woods. Norton, near Evesham. Upton Snodsbury.
Severn. Elmley Lovett Forest, *Miss Ladbury.* Bog near Doverdale Church. Monk Wood. Spetchley. Hadley. Boreley. Kempsey Grove. Shrawley Wood. Wyre Forest. Elmbridge.
Malvern. Coppice near Bransford Chapel, *Lees.* Orchard at Malvern Link Vicarage, *Towndrow.* Very general, *Lees.* Not so ; the variety is very rare in the Malvern district, *Towndrow, Malv. Adv.,* Dec. 22, 1900. Witley. Southstone Rock. Ankerdine Hill. Old Storridge Common. Lord's Wood.
Lickey. In the upland woods and dingles, *Mathews.* Frankley. Tarde-bigge. Dodford.
There is considerable confusion in the records for this plant and hybrids between *P. veris* and *P. acaulis,* and *P. elatior.* True *P. elatior,* Jacquin's Oxlip, is a plant only found in the eastern counties. Mr. Lees' record of the plant at Bransford and Miss Ladbury's at Elmley Lovett are above attributed to this variety. The only character that can be depended upon to distinguish *P. elatior* from the common oxlip of the fields, allowed to be a hybrid production, is the open mouth of the tube of the corolla. All Worcestershire oxlips may confidently be put down as this variety of *P. acaulis,* or more likely as hybrids between the Primrose and the Cowslip. This variety has been noticed in Gloucester, Stafford, and Warwick.

1136. **644. P. veris,** Linn. *The Cowslip.*
Native. Pastures, hedgebanks, borders of woods, railway-banks. General, except in Lickey, where it is not so frequently met with. P. March to May. *Top. Bot.* 90.
First Record. Meadows, *Pitt, Agric. Worc.,* p. 317 (1810).
Perhaps this flower is even more the plaything of village children than the primrose, and by their elders also it was used to make a homely wine, which usually owed its vinous qualities to more potent ingredients ; but home-made wines have gone out of fashion in these days ! It has received many popular names in different parts of the country—Paigle, Petty Mullen, Palsy-wort among them. Curiously enough, the second syllable of the general name does not appear to have anything to do with the 'lip' of a cow. The two syllables rightly divide ' cow-slip ', the second one being the old English word ' slyppe ', a slimy substance, cognate with 'slop'—not at all a poetical association for the name ! By very early writers the flower was usually termed Cowsloppe. Medicinally it was thought efficacious in cases of palsy, for strengthening the brain and nerves, as an application to wounds, and for inducing sleep. The juice of the flowers was used as a cosmetic to cleanse spots or marks on the face, 'whereof some gentlewomen have found good experience ', says Parkinson. The cowslip occurs in all neighbouring counties.

× **vulgaris.**
First Record. Unlocalized, *Victoria Hist. Worc.,* p. 53 (1901).
The remarks under *P. vulgaris* apply to this also, which is common about Malvern, *Towndrow,* and Mr. Rea has noticed it at Ockeridge Wood ; while in the Herbarium at the *Hast. Mus. Worc.* there are specimens from Narrow Lane Wood, Pirton ; Brookside north of Monk Wood ; Porter's Mill ; and Madresfield.

[*1137.* **645. P. elatior,** Jacq. *Jacquin's Oxlip.* *Top. Bot.* 7.
Mr. Lees gives as a record for this East Anglian plant, as Oxlip Primrose, Coppice near Bransford Chapel, where he says there are several curious varieties of Primrose, *A. Florence, Strang. Guide* (1828). It is not likely that it was the true plant which he observed. See the remarks under *P. vulgaris, var. caulescens,* above. Neither in this county, nor in any adjoining one does *P. elatior* occur, though in Gloucester a hybrid of some sort has been mistaken for it.]

LYSIMACHIA, Linn. 272. (λύσις, a dissolving, μάχη, a battle. The English name has a similar meaning.)

1142. **646. L. vulgaris,** Linn. *Yellow Loosestrife.*
Native. River banks, marshy ditches. Plentiful in the Avon district, uncommon elsewhere. P. June to September. *Top. Bot.* 79.
First Record. By the side of the Avon at Pershore, *Ballard, Stokes's With. Bot. Arr.,* 2nd ed., p. 208 (1787).
Avon. Avon banks near Evesham. Nafford Mill. Eckington Bridge.
Severn. Below Holt Lock. Bridewell, Bewdley, *Mathews.* Hampstall, *Mrs. Gardner.* Powick Weir. Below Pixham Ferry. Bewdley. Grimley Brickyard. River-bank near the Cathedral.

Malvern. Longdon. Severn-side near Kempsey. Banks of Teme above Bransford. Near Old Bridge, Powick. Longdon Marsh. Pool Brook, Upton-on-Severn.

Lickey. Stour-side near Birchill, Halesowen, probably a garden escape, *Mathews*. Tardebigge.

The Greek word λυσιμάχιον is found only in Pliny's Latin translation of it. He says that the plant was discovered by Lysimachus, King of Thrace, a more probable origin for the name; and he continues that oxen are made to eat it to be rendered more willing to draw together, so the mistranslation of what was most probably a personal attribution, λυσιμάχιον, is at least ancient. This is a handsome flower growing two or three feet high. In some places it is called Yellow Willow-herb. In former days it was considered to possess medicinal virtues. It occurs in all neighbouring counties.

1145. 647. L. Nummularia, Linn.   *Creeping Loosestrife, Money-wort.*
Native. Wet meadows, stream-sides, damp woods, and ditches. Plentiful in low situations; not so common on the higher land of the Lickey district. P. June to August. *Top. Bot.* 71.
First Record. Cradley Park. Hodge Hill. Wychbury Hill, *Scott, Hist. Stourb.*, p. 540 (1832).

This plant is a frequent inhabitant of the garden rockery, and often decks the margin of the garden fountain. In damp places it increases rapidly by means of its creeping stems. It is called also Creeping Jenny, and in North Worcestershire, Nimble Sailor, though the origin of the latter name would appear to be not a little obscure. It occurs in all neighbouring counties.

1146. 648. L. nemorum, Linn.   *Wood Loosestrife, Yellow Pimpernel.*
Native. Woods and thickets. General throughout. P. April to October. *Top. Bot.* 109.
First Record. In a wood by Picket Rock, Kidderminster, *Perry, Mag. Nat. Hist.*, vol. iv, p. 450 (1831).

The bright yellow blossoms with narrow linear sepals of this straggling little plant, which both in form and flowers brings to mind the Scarlet Pimpernel of our fields, are to be seen in most of the woodland rides throughout the county, and it occurs in all neighbouring counties also.

GLAUX, Linn. 274. (γλαυκός, sea-green, from the colour of the plant.)

1148. 649. G. maritima, Linn.   *Sea Milkwort, Black Saltwort.*
Denizen. Usually found on sea-shores and salt marshes. One place only. P. June to August. *Top. Bot.* 71.
First Record. In the greatest profusion on the side of the Droitwich Canal between Bevereye and Salwarpe, above Worcester, *Buckman, Phyt., O.S.*, vol. iii, pt. ii, p. 540 (1849).
Severn. As first record.

This owes its location in our county to the salt of Droitwich, as was mentioned in the case of *Lepidium latifolium*, which plant see. It was given as a new record, *Bot. Rec. Club Rep.*, 1875, by Dr. F. A. Lees, as occurring by one of the brine pits at Droitwich, he being doubtless then unaware of its abundance in the neighbourhood. It is a succulent little plant, curious

from the fact that it possesses no petals, the flower consisting of the coloured calyx, as is the case with the Marsh Marigold. It is not recorded for Worcestershire in *Top. Bot.*, nor for any neighbouring county, but *Hooker, Stud. Fl.*, 3rd ed., p. 264, says it occurs in the salt districts of Staffordshire as well as of Worcestershire.

ANAGALLIS, Linn. 275.   (ἀνά, again, ἀγάλλω, I adorn, from the plants readorning, every summer, the places where they occurred.)

1149. 650. A. arvensis, Linn.   *Scarlet Pimpernel.*
Native. Cultivated fields. Very common in suitable situations. A. May to November. *Top. Bot.* 99.
First Record. Not localized, *Walker, Med. Surg. Rep.*, vol. i, no. 2, p. 100 (1828).

The country names of Rain and Sunshine, Poor Man's Weather Glass, and Shepherd's Barometer allude to the fact that the flower never opens on a rainy day, and even closes on the approach of a storm, though the two latter must of course be posterior to the invention of the instrument they allude to. The derivation given above for the generic name does not hold the field without dispute. Dioscorides says it is derived from ἀναγελάω, to laugh aloud, from the plant's medicinal virtues with regard to the liver. It also had wonderful power in drawing out thorns, or even arrows, from the flesh, and therefore some have maintained that its name is derived from that quality, ἀνάγω being the Greek for 'I lead up'. And as its scientific name is thus doubtful, so its English name of Pimpernel has its history. Originally in mediaeval Latin *pipinella*, this was considered a corruption of *bipinella*, a diminutive of *bipennis*, two-winged. In the English name from the earliest times the second syllable was *per*, not the *pin* of the Latin. Whatever was the derivation, the name originally belonged to the Burnets, Great and Salad, and later got itself transferred, how or why, there is no explanation, to the Burnet Saxifrage; but it is now become obsolete in all these instances, and belongs exclusively to the little Scarlet Pimpernel. It is known in some parts of Worcestershire as Pheasant's Eye. It was considered a plant of potent medicinal power, peculiarly efficacious in the bites of mad dogs, good for complaints of the eyes, diseases of the brain, and epilepsy and dropsy. The flowers are also found flesh-colour and white. Mr. Lees observed it with white flowers from Astley, near Stourport; the blue form is considered a distinct species, *A. femina*. In old times this was called the Female Pimpernel, in distinction to the scarlet kind, which was called the Male Pimpernel. The plant is recorded from all bordering counties.

Var. **phoenicea,** Lamk.
First Record. In this book.
Malvern. An uncultivated field at Malvern Link, *Towndrow*.
This variety is characterized by having a purple eye to the scarlet, pink, or white corollas. Not mentioned in *Lond. Cat.*, 10th ed.

1150. 651. A. femina, Miller.
(*A. caerulea*, Schreb.)
Colonist. Cultivated fields and waste places. Local. A. July to September. *Top. Bot.* 48.

R

First Record. As *A. arvensis* var. *flore caerulea*. Upon Breedon Hill, in a cornfield at the top of Overbury Wood, *Nash, Hist. Worc. Int.*, p. lxxxix (1781).
Avon. Plentiful between Netherton and Haselor, *Cheshire*. Bredon. Broadway. Trench Woods. Craycombe Hill. Fladbury. The Manor House Farm, Sheriff's Lench. Near Wadborough Railway Station.
Severn. Calcareous field at Tibberton, *Lees*. Astley, with the white and red forms, *Lees*. Coningree Wood. Claines. Ladywood, Salwarpe. Kempsey.
Malvern. Ham Hill. Newland. Malvern.

Besides in colour, this plant differs but little from the last; it is more erect, and the petals are without the minute glandular hairs that fringe those of the last. The plant has a wide distribution, being with the previous one found in Europe, North and West Asia, India, and North Africa. It occurs in all bordering counties.

1151. 652. A. tenella, Murr.   *Bog Pimpernel.*
Native. Peaty Bogs. Local. P. June to August. *Top. Bot.* 97.
First Record. Feckenham Bog, *Purton, Midl. Fl.*, p. 115 (1817).
Avon. Feckenham Bog, *Purton*, extinct.
Severn. Wyre Forest. Stanklin Pool, *Humphreys*. Holy Well, Wyre Forest, *Mathews*. Great Bog, Wyre Forest. Bevere Green and Hartlebury Common, *Westcombe*. Boggy field, near Fenny Rough, *Perry*, extinct.
Malvern. Malvern Hills, *Lees*. Western base of Worcestershire Beacon, *Lees*.
Lickey. Near Alvechurch, *Mr. D. Mathews*. Bromsgrove Lickey up to 1899, *Worc. Nat. Club. Trans.* ii, p. 52. Bilberry Hill, Lickey.

This is a charming little plant, with rose-coloured flowers, large in proportion to the size of the plant, on rather long root-stalks, which creep upon the mossy surface of the bog. It is a vanishing plant in Worcestershire. It is recorded from all neighbouring counties.

A. indica, Sweet.
(*A. latifolia*, Linn.)
Garden escape.
First Record. As below.
Severn. In a wet meadow near the Stour at Wilden; near Stourport, *Miss Ladbury*.

A garden annual, which does not yet appear to have been noticed in an escaped condition elsewhere in England. Its flowers are small and deep blue.

CENTUNCULUS, Linn. 276. (From Latin *cento*, a patchwork or coverlet, from the way it covers the ground.)

{**1152. 653. C. minimus,** Linn.   *Small Chaffweed, Bastard Pimpernel.*
Native. Damp heathy ground, bare places on sandy commons. Very rare. A. June to September. *Top. Bot.* 64.
First Record. Not localized, *Lees' Cat., New Bot. Guide* (1835).
Malvern. Brand Lodge, Colwall, *Lees* (but this is in Herefordshire).

Lickey. Moseley Common, *Lees*. Moseley Wake Green, *Ick*.

Mr. Lees' Colwall locality is in Herefordshire, and as he says it has 'hitherto occurred nowhere else in the district', Worcestershire can lay no claim to the plant from this record. The two Moseley records refer to the same locality, and this district is now urban. It is to be feared this tiny plant is now lost to the county. In *Top. Bot.* it is recorded for Warwick and Salop besides Worcester, for which latter record no authority is given. It has been seen in Staffordshire, and it is recorded for Herefordshire by Rev. A. Ley in the *Victoria History* of that county.}

SAMOLUS, Linn. 277. (According to Pliny, the name of a marsh plant possessing wonderful sanatory properties, perhaps the same as *slan-lus*, in Celtic, 'the healing herb'.)

1153. 654. S. Valerandi, Linn.   *Brookweed* or *Water Pimpernel.*
Native. Ditches and marshy situations. Very local. P. June to August. *Top. Bot.* 82.
First Record. Side of the brook running from the brine pit on Defford Common, *Messrs. Ballard* and *Hollefear, Stokes's With. Bot. Arr.*, 2nd ed., p. 221 (1787).
Avon. Defford Common. Between Tunnel and Craycombe Hill, *Mathews*. Plentiful by the roadside between Craycombe and Evesham, *Gissing*. Avon near Eckington Bridge. Feckenham Moors, *Westcombe*.
Severn. Near Briar Mill, Droitwich. Battenhall, *Lees*. Droitwich Canal.
Malvern. Rare. East of the hills, *Lees*.

Although Mr. Lees mentions the plant as above in his *Bot. Malv. Hills*, 3rd ed., p. 49, in his *Table of Plants* he does not localize it in the Malvern district. The rounded stem grows some twelve inches high, and the foliage is thick and succulent. It was in former ages said to be a complete cure for any malady affecting the pig. The plant is one of the most widely distributed, occurring throughout the northern temperate hemisphere and in the Himalayas. Of the neighbouring counties it does not appear to have been recorded from Herefordshire.

48. OLEACEAE, R. Br.

FRAXINUS, Linn. 278. (φράξις, a separation, because the wood splits easily.)

1154. 655. F. excelsior, Linn.   *The Common Ash.*
Native. Woods, thickets, hedges. General. Tree. April and May. *Top. Bot.* 109.
First Record. Near Dudley Castle, *Booker, Dud. Cast.*, p. 107 (1825).

This is one of the noblest of our trees, its timber useful in all stages of its life, the young copse wood for making hurdles, when it is older for hop-poles, and when adult for implements of husbandry, carts, and carriages. It bursts into leaf late in the spring, usually after the oak, though popular folklore considers them as competitors in a race, and as the oak or the ash comes first into leaf so will the summer be dry or wet. And it drops its leaves all at once at the first spell of really low temperature in the autumn. The flowers are very simple, possessing neither calyx nor corolla, and the winged fruit is well known under its country name of Keys. The roots are spreading and extend a long distance under the surface of the ground. It was largely used in medicine. It was potent against the usual venomous beast, and water distilled from the tender tops was thought to abate corpulence. A curious superstition regarding it was connected with the little shrew-mouse. When the animal crawled over any beast, it afflicted it with cruel anguish, only to be abated by the application of the twigs of a shrew-ash, usually some old pollarded tree of the kind. A North Worcestershire superstition was that if those unblessed with offspring passed between the riven parts of the trunk of an aged ash, the defect would in due course be remedied. In *Worc. Nat. Club Trans.*, vol. ii, p. 9, is the picture of a riven old ash at Birts Morton, where a lady is depicted, no doubt as acting upon this superstition. The ash entered largely into the mythology of our ancestors. Yggdrasil, in Scandinavian mythology, was the ash-tree of Odin, binding together heaven, earth, and hell, though some say this idea was not primitive, but post-Christian. The tree seeds itself freely, and is found in all neighbouring counties.

Var. b. heterophylla, Vahl.
(Var. *diversifolia*, Ait.)
First Record. Sandbourne, near Bewdley, *Jorden, Lees' Bot. Worc.*, p. 6 (1867).
This variety, which possesses simple, or simple and pinnate, leaves, has been observed near Bewdley by Mr. Towndrow also.

LIGUSTRUM, Linn. 279. (From *ligare*, to bind, from the use of its twigs.)

1155. 656. L. vulgare, Linn.   *Privet.*
Native. Thickets, woods, hedges. Not general, except in Severn and Malvern. Shrub. May to August. *Top. Bot.* 83.
First Record. Abundant in the hedges, *Lees, Illus. Nat. Hist. Worc.*, p. 149 (1834).
Avon. Sheriff's Lench. Fladbury. Crowle. Wood Norton. Rous Lench. Tiddesley Wood. Trench Woods.
Severn. Wyre Forest, near the old Sorb tree, *Lees*. Hagley. Dynes Green. Rainbow Hill, Worcester. Roadside, Rushwick. Near Crowneast Wood. Spetchley. Battenhall Lane. Shrawley Wood. Holt. Ombersley. Kempsey. Hazlewood. Ockeridge Wood.
Malvern. Madresfield. Dripshill Wood. Martley. Witley.
Lickey. Randans Wood.
This plant is far better known as the constituent of the trimmed garden hedge than in its wild condition, at all events in our part of England, and it is often planted in shrubberies as it bears the drip of trees well, and in woods as cover for the pheasants. It is recorded from all neighbouring counties.

49. APOCYNACEAE, Juss.

VINCA, Linn. 280. (*Vincere*, to bind, from the use of its stems.)

1156. 657. V. *major, Linn.   *Greater Periwinkle.*
Alien. Hedgerows and banks. Several places throughout the county. Under-shrub. P. June to August. Not in *Top. Bot.*
First Record. Roadside, Little Witley and Clifton Hill, *Sheward, Nash, Hist. Worc., Sup.*, p. 96 (1799).
Avon. Hampton near Evesham. Near Turnpike, Green Hill, Evesham. Crowle Hill. Inkberrow.
Severn. Opposite Rainbow Hill, *Lees*. Roadside between Cotheridge and Broadwas. Near Shrawley. Wyre Forest. Between Oldbury and Broadheath. Whittington. Bank of Severn opposite Grimley.
Malvern. Near Little Malvern Church. Lane near Welland Common Powick, *Lees*. Martley. Broadwas. Woodsfield.
Lickey. Broughton (Belbroughton?), near the bridge, *Scott*.
This is no true native of England, being at home in the woods of South-east Europe. Wherever it occurs it is an escape from a present or past garden. It has been observed at large in all neighbouring counties.

1157. 658. V. minor, Linn.   *Lesser Periwinkle.*
Denizen. Thickets and banks. Several places in Malvern and Avon districts. Under-shrub. January to October. *Top. Bot.* 73.
First Record. Side of the road to Martley, opposite Kempsey (sic) Mill, *Sheward, Nash, Hist. Worc., Sup.*, p. 96 (1799).
Avon. Near Tiddesley Wood. Norton, near Evesham. Near Trench Woods. Huddington.
Severn. Northwood, near Bewdley, *Mathews*. With white flowers at Whittington and Camp. Crookbarrow Hill. Clerkenleap. Laneside east of Porter's Mill. Norton juxta Kempsey. Lincombe. Lane near Hartlebury Common. Woodcote.
Malvern. Silurian eminences, *Lees*. West end of Keysend Hill. Above the lime-kilns at Leigh Sinton. Alfrick. Castlemorton near the Church. Lane between Powick and Bransford. Ankerdine Hill. Clifton-on-Teme. Madresfield.
Lickey. The Valley, Bournheath.
This, though a native, is also much more common as a garden plant than in an indigenous state, and in the majority of its localities is a garden escape. The Order to which the Periwinkles belong is a most poisonous one. The Tanghin poison of Madagascar is derived from a plant belonging to it. Strychnine, and the Woorali poison of South America, are produced from plants belonging to the allied Order Loganiaceae distinguished only by a difference in the shape of the stigma. The flower of this plant is usually blue, but varies into white and purple. The plant has been observed in all neighbouring counties.

50. GENTIANACEAE, Juss.

BLACKSTONIA, Huds. 282. (After a London surgeon named Blackstone.)

1159. 659. B. perfoliata, Huds.   *Yellow Centaury, Perfoliate Yellow-wort.*
(*Chlora perfoliata*, Linn.)
Native. Dry banks and pastures. Several places. A. June to September. *Top. Bot.* 60.
First Record. As *Chlora perfoliata*, in rough pastures of a stiff clayey soil, about Great Comberton and elsewhere, *Nash, Hist. Worc. Int.*, p. lxxxix (1781).
Avon. Tardebigge Reservoir and Canal, *Humphreys*. Craycombe. Bredon Hill. Broadway. Trench Woods. The Slads. Littleton Banks. Wadborough.
Severn. Northwood, near Bewdley, *Mathews*. Wyre Forest. Kempsey. Old Rifle Range, Worcester.
Malvern. Mathon; Leigh Sinton; Cowleigh Farm; The Rhydd; Bush Hill, Powick, *Lees*. Abberley. Clifton-on-Teme. Near Ankerdine Hill. Bransford. Abberley Hill.
Lickey. Romsley; Frankley; Clent Hill, *Mathews*. Wychbury Hill, Pedmore, *Scott*. Cradley Park, *Scott*. Dudley Castle, *Ick*. Field at Ell Wood. Chaddesley Woods.
This is a glaucous plant twelve inches or more high, the stem-leaves connected by their whole breadth in distant pairs, and very bitter. It is recorded from all neighbouring counties.

CENTAURIUM, Hill. 283. (κενταύριον, a plant name used by Theophrastus.)

1160. 660. C. umbellatum, Gilib.   *Common Centaury.*
(*Chironia Centaurium*, Schmidt. *Erythraea Centaurium*, Pers.)
Native. Dry fields, heathy places, open places in woods, railway-banks. Locally common. A. or B. June to September. *Top. Bot.* 102.
First Record. As *Chironia Centaurium*, var. 3, blossoms white, Upper Battenhall, near Worcester, *Stokes, Stokes's With. Bot. Arr.*, 2nd ed., p. 237 (1787).
Avon. Croome Perry Wood. Craycombe. Tardebigge Reservoir, *Humphreys*. The Slads, also with white flowers. Tiddesley Wood. Churchill, near Worcester.
Severn. Battenhall, as first record. Wyre Forest, also with white-flowered forms. Habberley Valley. Monk Wood.
Malvern. Old Hills, abundant with white flowers. Witley. Bransford. Powick. Alfrick. Martley. Ankerdine. Malvern Link. Madresfield. With white flowers at Mathon.
Lickey. About St. Kenelm's. Randans, *Mathews*. Frankley. Chaddesley Woods.
A pretty pink-flowered plant closing its petals in the afternoons or on rainy days. This plant, like most of the Gentian tribe, is very bitter. It had

much medicinal repute, being considered good for agues and jaundice, and useful for removing freckles. It has its name because its virtues were said to have been discovered by Chiron the Centaur, the wisest of all that race, who lives to-day among the stars in the constellation Sagittarius. The plant occurs in all neighbouring counties.

{1163. 661. C. pulchellum, Druce.   *Dwarf-branched Centaury.*
(*Erythraea pulchella*, Fries.)
Native. Dry heaths. Very rare. A. July to September. *Top. Bot.* 44.
First Record. Gathered by Mr. G. Reece, by the side of a lane between Alfrick Chapel and Grimsend House; also near Malvern, Mr. T. Westcombe, *Lees, Bot. Malv. Hills*, 2nd ed., p. 38 (1852).
Malvern. As records above. Malvern Link, *Westcombe*.
This differs from the former plant in having its flowers all stalked instead of nearly sessile, and the calyx only a little shorter than the tube of the opening corolla instead of not half as long. It has been observed only in Herefordshire and Warwickshire of bordering counties.}

GENTIANA, Linn. 285. (From Gentius, King of Illyria, who brought into use the species so much valued in medicine, *G. lutea*.)

1169. 662. G. Amarella, Linn.   *Small-flowered Gentian.*
Native. Pastures, grassy downs. Very uncommon. A. July to October. *Top. Bot.* 81.
First Record. 'Gentiana autumnalis flore albo foliis longis angustis. In old Pastures on the north-west of Church Lench, Wostershire, plentifully,' *Merrett's Pinax* (1666).
Avon. As first record. Bredon Hill. Craycombe Hill. Shemington Hill, Alderminster, *Cheshire*. Broadway. Elmley. Tardebigge Reservoir. Wyre Forest. Badger's Bank.
Severn. Trimpley, *Jorden*.
Malvern. Silurian eminences, *Lees*. Purlieu Lane; Sarn Hill; Mill Copse, Cowleigh, *Lees*.
A formal-looking plant, remarkably erect, varying in size up to twelve inches. It has been recorded from all neighbouring counties.

1172. 663. G. campestris, Linn.   *Field Gentian.*
Native. Dry, open pastures, heathy commons. Very rare, only recorded for Wyre Forest. A. or B. August to October. *Top. Bot.* 88.
First Record. Bewdley Forest, between Furnace Mill and the Eagle's Nest Inn, *Sheward, Nash, Hist. Worc., Sup.*, p. 96 (1799).
Severn. Wyre Forest, as first record. Also *Jorden* and *Lees*. The New Parks and Hitterell Coppice, and other portions of the forest well within the county boundary. In Wyre Forest.
Although the first record is doubtfully in the county, Furnace Mill being in Salop, and Mr. Jorden does not localize the plant, Mr. Lees definitely says, *Bot. Worc.*, p. 5, that on a visit made by the Worc. Nat. Club to the historic Sorb tree, July, 1853, ' *Gentiana campestris* was added to the flora of Worcestershire'; and though he was somewhat loose as to the outer boundaries of the county botanical districts, he here uses the county name and not that of one of the districts; and he gives it as occurring in the Severn district in his

Table of Plants. It may with safety' be considered a Worcestershire plant. The medicinal qualities of these plants are well known. They are febrifuges and vermifuges, and their tonic qualities rank as high as quinine. It is recorded in *Top. Bot.* from Stafford and Salop besides Worcestershire, in the latter case with the authority (E. Lees). The essential distinctions between these two plants are very marked. In *G. campestris*, the lobes of the calyx are very unequal and 4-partite, whereas in *G. Amarella* the lobes of the calyx are equal and 5-partite. The plant appears to have been observed in Gloucester and Hereford as well as the counties mentioned above.

MENYANTHES, Linn. 286. (μήν, a month, ἄνθος, a flower; some say from the duration of the blossom.)

1174. **664. M. trifoliata,** Linn. *Marsh Buckbean, Marsh Trefoil, Bogbean.*
    Native. Marshes, bogs, borders of slow streams and ponds. Locally abundant. P. May to October. *Top. Bot.* 110.
    First Record. Bell's Mill, near Stourbridge, *Scott, Purton, Midl. Fl.,* p. 122 (1817).
        Severn. Hartlebury Common. By the canal, Wolverley side of Kidderminster, *W. W. How.* Stanklin Pool. Fenny Rough, *Perry.* Wilden, Stourport, *Hickman.* Near Button Oak, Wyre Forest, *Mathews.* Lower Harberrow Pool and Brake Mill Pool, Hagley, *Mathews.*
        Malvern. One place, lost.
        Lickey. Moseley Common, *Miss M. A. Beilby.* Campion's Pool, *Humphreys.*
    This is a very pretty flower, the buds before expanding being rosy pink, and the expanded petals being white and beautifully fringed with white filaments on their inner surface. It is a plant of wide range, extending into Arctic Europe. In Iceland it grows plentifully; and so matted are its roots that travellers know that when they see the plant on the morass, then they may safely tread. The bitter root is said to be good for rheumatism and ague, and in Sweden is used as a substitute for the hop. The original name was 'Buckesbeanes', apparently a translation by Lyte, *Dodoens,* IV. lxxviii. 542, of the Flemish *bocks boonen,* 'goat's beans.' Bog-bean appeared later, and is probably a rationalizing alternative of *buck-bean.* It appears to be a diminishing plant in this county. It is recorded from all neighbouring ones.

NYMPHOIDES, Hill. 287.

1175. **665. N. peltatum,** Rendle and Britten. *Nymphaea-like Villarsia.*
    (*Limnanthemum peltatum,* S. G. Gmel. *Villarsia Nymphaeoides,* Vent.)
    Native. In still pools of the Avon. Very local. P. July and August. *Top. Bot.* 10.
    First Record. As *Villarsia Nymphaeoides,* in the River Avon between Pershore and Eckington, June 1850, *Mr. George Reece,* Curator of the Worcester Museum, *Phyt.* O. S., vol. iv, p. 5 (1850).
        Avon. As first record. Near Tewkesbury. The Swan's Neck above Eckington Bridge.
    This is an elegant water-plant, with large yellow flowers curiously plaited, and leaves like those of the water-lily, only smaller. The canals in Holland are in some parts covered with this plant. It is distributed over a great

part of the world, both in temperate and tropical regions. It has been noticed in Gloucester, Salop, and Warwick, but is not recorded in *Top. Bot.,* p. 281, with certainty for any of them, nor indeed for Worcestershire.

51. POLEMONIACEAE, Juss.

POLEMONIUM, Linn. 288. (πόλεμος, war, the plant having, according to Pliny, caused a war between two kings, rival ' first recorders '.)

1176. **666. P. caeruleum.** *Blue Jacob's Ladder.*
    Alien. Hedgesides and waste places. Once or twice observed. P. June to August. *Top. Bot.* 5.
    First Record. Near Bromsgrove, *Lees, Lees' Bot. Worc.,* p. xxx (1867).
        Malvern. Callow End, *Towndrow, Malv. Advert.,* Nov. 19, 1892.
        Lickey. As first record. Meadow near the Leasowes, Halesowen, an escape, *Mathews.*
    Though this plant is truly native of the limestone tracts of North Britain, elsewhere it is certainly an alien. It is a favourite garden flower. It seeds freely and maintains itself easily. In France the plant is called *La Valériane Grecque.* It has escaped from gardens in all neighbouring counties.

52. BORAGINACEAE, DC.

CYNOGLOSSUM, Linn. 289. (κύων, a dog, γλῶσσα, a tongue. From the shape and texture of the leaf.)

1177. **667. C. officinale,** Linn. *Common Hound's-tongue.*
    Native. Waysides, woods, heaths, commons. Not uncommon. B. or P. June to August. *Top. Bot.* 76.
    First Record. Hedgesides, Shrawley and other places; in the town of Evesham on rubbish, *Pitt, Agric. Worc.,* p. 317 (1810).
        Avon. Honeybourne, *Mathews.* Norton, near Evesham. Church Honeybourne. Roadside, Craycombe Hill.
        Severn. Near Hagley, *Scott.* Near Crowneast. Roadside, Astwood. Shrawley. Honey Bottom, north of Kidderminster. Wyre Forest.
        Malvern. Malvern Hills. Near the landslip, Knightsford Bridge. Ridge Hill, Martley. Berrow Hill, Martley. Abberley Hill. Ankerdine.
        Lickey. Clent Hill, *Mathews.* Brettell Lane ; Canal Banks, *Scott.*
    This is a downy plant, soft to the touch, and it possesses an odour which some describe as that of mice, while to others it is frankly fetid. The flowers are purplish red. It is said to cure the bite of a mad dog ; and that if the plant were laid under the foot, no dog would bark at the person so fortified. The plant occurs in all bordering counties.

1178. **668. C. montanum,** Linn. *Green-leaved Hound's-tongue.*
    (*C. germanicum,* Jacq. *C. sylvaticum,* Haenke.)
    Native. Coppices, waste places. Rare. B. June and July. *Top. Bot.* 17.
    First Record. 'Cynoglossa folio virenti, J. B. Cynoglossum minus folio virente, Ger., Cynoglossum sempervirens, G. B. Park, the lesser Hound's-tongue—also in some shady lanes near Worcester,' *Ray's Cat.* p. 90 (1670). In *Ray's Syn.,* p. 75 (1690) is added to the above, 'by Mr. Pitt,' meaning, no doubt, the recorder of the Wild Sorb-tree in Wyre Forest.

        Avon. Fladbury Road, Evesham, *Edward Rudge, F.L.S.,* and *Rev. W. S. Rufford* in *Purton*; no longer met with, *Lees.*
        Severn. As first record. Third milestone from Worcester to Pershore, *Nash.* Elmley Lovett.
        Malvern. Near Longdon Church, *Lees.* Gullet Glen, *Lees.* Near Madresfield, *Towndrow.*
        Lickey. Brettell Lane canal banks, *Scott*; 'Must, I think, be an error,' *Mathews.*
    This plant is readily to be distinguished from the last by its more or less shining and brighter coloured leaves free from pubescence, and their different figure. The plant appears not to have been met with in Hereford or Stafford.

ASPERUGO, Linn. 290*. (*Asper,* rough, from the rough leaves.)

1179. **669. A.** PROCUMBENS, Linn. *German Madwort.*
    Alien. Waste-ground. Rare. A. June to October. No number given for *Top. Bot.* in *Lond. Cat.,* but recorded therein, p. 328, for fourteen provinces and given doubtfully for nine others, Worcestershire among them.
    First Record. Not localized, *Lees' Cat., New Bot. Guide* (1835). ' Must be an error,' *Mathews.*
        Malvern. Weed in the garden at Sherridge, *Miss B. Norbury.* Malvern Link.
    In his *Table of Plants* Mr. Lees does not localize this plant, nor mention it in his *Bot. Worc.* The only part of the world from which it has been recorded in any but artificial habitats is the region between South-east Europe and Afghanistan, but it extends as a weed of waste and cultivated ground over Europe, Western Asia, and North Africa. The original Madwort of the herbalists was an *Alyssum,* but this plant is recognized and quaintly described at length by Gerarde, *Herbal,* II. cxviii. 379. This plant is not recorded for Warwick or Stafford.

SYMPHYTUM, Linn. 291. (συμφύω, I unite, from its supposed vulnerary qualities.)

1180. **670. S. officinale,** Linn. *Common Comfrey.*
    Native. Sides of rivers and brooks, marshes, ditches. Common except in the Lickey district. P. May to August. *Top. Bot.* 86.
    First Record. In great plenty by the Stour, near Kidderminster, *Pitt, Agric. Worc.,* p. 317.
        Avon. Norton, near Evesham. Fladbury. Pershore. Chadbury. Bredon. Harvington. Eckington.
        Severn. Banks of Severn, purple flowers chiefly, *Lees.* Grimley brick-fields. Banks of River Salwarpe, Droitwich. Kempsey. Ombersley. Stourport. Lincombe. Bewdley. Severn Stoke. Hadley. Grimley. Hallow. Shrawley. Boreley. Ribbesford. Cotheridge.
        Malvern. Bransford. Powick. Upton-on-Severn. The Rhydd. Knightsford Bridge. Stanford Bridge. Clifton-on-Teme.
        Lickey. Brook near Oldmill, Clent.

    The flowers of this plant vary a good deal in colour from purple through pink to yellowish-white, the latter form being considered the normal one. The purple form, which usually has rougher leaves, is var. *patens,* Sibth. It is applied to reduce bruises and sprains. The plant is recorded from every neighbouring county.

1181. **671. S. tuberosum,** Linn. *Tuberous Comfrey.*
    Alien. Damp hedgesides, coppices. Twice met with. P. June and July. *Top. Bot.* 30.
    First Record. As below.
        Severn. Lane above Hagley Station, 1904, *W. Whitwell.* Wyre Forest, *J. B. Duncan.*
    This is a common plant in Scotland, but rare in England. The whole plant is smaller and slenderer than the preceding. It has been noticed in Staffordshire.

S. ASPERRIMUM, Donn. *Prickly Comfrey.*
    Alien. Waste places, a relic of cultivation. P. June to August. Not in *Top. Bot.*
    First Record. An escape, Churchill, *Mathews, Worc. Nat. Club Trans.,* ii. 97.
        Severn. Churchill (Kidderminster); Willow holts by the canal at Wolverley, *Mathews.* Rainbow Hill, Worcester. Field near Shrawley Wood. Harberrow.
        Malvern. Madresfield.
    This plant is a native of the Orient, but has been grown to some extent in England for fodder. It roots deeply and is difficult to eradicate when once planted. It is not given in *Lond. Cat.,* 10th ed.

S. ORIENTALE, Linn.
    (*Borago orientale,* Linn.)
    Mr. Mathews notes of this alien, *Clent and Lickey Fl.,* p. 34 (1881), that it ' is established in the hedge at Hagley Park, by the Birmingham road, below the water-trough, where it has been growing for many years in company with *Doronicum Pardalianches*'. The plant is a native of South-east Europe, and has not a shadow of a claim to be native in our island, where it has been used as a fodder plant.

BORAGO, Linn. 292.* (Said to be from *cor,* the heart, and *ago,* I bring, because it was supposed to bring courage ; but this is not good etymology. Probably from some late Latin word analogous to the word *burra,* a shaggy garment, referring to the roughness of the foliage.)

1182. **672. B. officinalis,** Linn. *Common Borage.*
    Alien. Waste places, usually near villages. Occasional. A. or B. May to September. Not in *Top. Bot.*
    First Record. Lane at Bromwich Farm, *Lees, A. Florence, Strang. Guide* (1828).
        Avon. Upton Snodsbury. Evesham. Huddington.
        Severn. Battenhall. Lower Wick. Near Bewdley. Kidderminster. Churchill, near Kidderminster. Ombersley.
        Malvern. 'An agrarian wanderer,' *Lees.*
    This plant, a native of rich damp ground in the Mediterranean region, has been an inhabitant of gardens from a very early period, both on account of

its flowers and its virtues. Its chief use was as an ingredient in the 'cool tankard' of our forefathers, but besides this they made soup or salad of the leaves, preserve of the blue petals, and pickle of the young shoots. 'I Borage bring courage,' said the old adage, whence possibly the idea that the name was a Latin word for courage with the initial changed. Escaped Borage has been noticed in all neighbouring counties.

ANCHUSA, Linn. 293.* (ἄγχουσα in Greek, from ἄγχω, I throttle, because it caused inflammation in the throat if chewed.)

1184. 673. A. *sempervirens, Linn.   *Evergreen Alkanet.*
Casual. Waysides, waste places, shrubberies. Rare. P. June to August. No numbers given in *Top. Bot.*, records usually reporting it as an uncertain native.
First Record. Near Birmingham, on the Alcester road, *Withering*. Near the Blanketts, Worcester, *Ballard*; *Stokes, Stokes's With. Bot. Arr.*, 2nd ed., p. 192 (1787).
Avon. Westmancote.
Severn. Near the Blanquettes, as first record. Osier-bed near Hartlebury Castle, *Miss Ladbury*. In a field by the old waterworks, Worcester.
Malvern. Near the Lodge, Mathon, *Lees*.
Lickey. Near Birmingham, as first record. Near Moseley Hall, *Miss M. A. Beilby.*
This is a plant of western range, and is quite a possible native of England, though usually when found it is a garden escape. This plant, with others of the genus, contains a red colouring matter in its roots, and the English name is ultimately derived from the Arabic *al henna*, through the Spanish *Alcanna*, Henna being a well-known dye. Dioscorides, accounting for the Greek original of the Latin name, says it was given to it because if any one chewed the leaves and spat into the mouth of a viper he would kill it, because of the power of the plant in causing irritation and inflammation in the throat. It is to be supposed that human beings were not considered so susceptible as the snake tribe to this quality. The plant has been observed in all neighbouring counties.

LYCOPSIS, Linn. 294. (λύκος, a wolf, ὄψις, a face, from the fancied resemblance of the flower to a wolf's head.)

1185. 674. L. arvensis, Linn.   *Small Bugloss.*
Native or colonist. Cultivated fields, waste places, chiefly on sandy soil. Not uncommon, except in the Avon district. A. or B. May to October. *Top. Bot.* 105.
First Record. Not localized, *Lees, Bot. Malv. Hills*, 1st ed., p. 16 (1843).
Avon. One place, *Lees*. Stoulton. Churchill, near Worcester. Sheriff's Lench. Near Trench Woods.
Severn. Devil's Spittleful, Bewdley. Near Wannerton, Churchill (Kidderminster). Lincombe. Hartlebury. Hartlebury Common. Northwick. Near Henwick Mill. Wyre Forest. Glashampton. Claines. Kempsey. Stanklin, *Humphreys*.
Malvern. Roadsides, *Lees*. Martley. Abberley. Mathon.
Lickey. Native, locally abundant, *Mathews.*

A very rough and bristly plant, the hairs or bristles being seated on white callous tubercles. In parts of Worcestershire it is called Butcher's Blue. The flowers are small for the size of the foliage. It occurs in all neighbouring counties.

PULMONARIA, Linn. 295. (From *pulmo*, the lungs, from its former use in pulmonary affections.)

1187. 675. P. officinalis, Linn.   *Common Lungwort.*
Denizen. Woods and thickets. Very rare. P. April and May. *Top. Bot.* 2; with the note in 1st ed., 'not indigenous, but reported in about three dozen counties.'
First Record. By the side of a wood at Lower Sapey, apparently wild, in 1834, *Lees' Cat., New Bot. Guide* (1835).
Severn. Whittington.
Malvern. Lower Sapey, *Lees*. Rosebury Rock, *Lees*.
Lickey. Chaddesley Wood, *Humphreys.*
Although the nativity of this plant in England is disputed, it is plentiful in Belgian woods, and is native across Europe to Southern Russia. In Chaddesley Wood it bears all the marks of a native plant, occurring among the undergrowth, and flourishing and spreading. Chaddesley Wood is part of the ancient forest of Feckenham which has never felt cultivation, nor the hand of man more than to cut timber and brushwood. The resemblance of the spotted leaves to lungs when diseased doubtless procured for the plant its familiar name of Lungwort, and because of this likeness it was supposed to cure the disease it imitated. In some places it is called 'Spotted Mary', but rather in its status as a garden plant than a wild one, for it is too rare in that condition to come under the observation of country people. It has been noticed in all neighbouring counties except Herefordshire.

MYOSOTIS, Linn. 297. (μῦς, a mouse, and οὖς, an ear, from the shape of the leaves.)

1189. 676. M. cespitosa, F. Schultz.   *Tufted Water Scorpion-grass.*
Native. Pond-sides, wet places. Not uncommon. B. or P. May to September. *Top. Bot.* 111.
First Record. Not localized, but at Malvern, *Lees, Mag. Nat. Hist.*, vol. iii (1830).
Avon. Norton, near Evesham. Fladbury. Upton Snodsbury. Besford.
Severn. Burnt Wood, Bewdley, with white flowers, *Perry*. Near Droitwich, *Mathews*. Near Dunhampstead. Crowneast. Cockshut Hill, Droitwich. Brickyard, Claines. Churchill, Kidderminster. Hoo Mill Pool. Sutton Pool. Dynes Green. Stakenbridge. Worcester.
Malvern. Fine and tall at Powick, *Lees*. Newland Common. Malvern Hills. Alfrick. Longdon Marsh. Knightsford Bridge. Malvern Link Common.
Lickey. Offmoor Lane. Hagley Park, *Mathews*. Salter's Lane, Redditch, *Mathews*. Chadwick Pools, *Humphreys*. Moseley, *Freeman*. Hewell.

Much like the following plant, the true 'Forget-me-not', but the flowers are smaller, with the lobes of the calyx as long as the tube, and the pubescence of the stem is adpressed, not usually spreading. The plant is recorded from all neighbouring counties.

1190. 677. M. scorpioides, Linn.   *Creeping Water Scorpion-grass. Forget-me-not.*
(*M. palustris*, Hill.)
Native. Sides of rivers, brooks, and ponds. Wet places. Common. P. April to August. *Top. Bot.* 104.
First Record. As *M. palustris*, *Lees, Bot. Malv. Hills*, 1st ed. p. 16 (1843).
Who does not know the Forget-me-not, not only in England, but all over the Continent as well? Or the legend accounting for its name? How the knight and the maiden, wandering by the stream, saw the bright blue flowers, and in endeavouring to pluck them the knight was carried away, tossing the flowers on shore in his last effort, with the words 'Forget-me-not'? The plant increases by underground stolons, and the stem is angular with spreading pubescence. The plant occurs in all bordering counties.

Var. b. **strigulosa**, Reichb.
First Record. Near Droitwich, *Mathews, Worc. Nat. Club Trans.*, ii. 168.
Severn. Oldbury Lane, *Herb. Vict. Inst., Worcester.*
This variety, which has a more slender and not angular stem and the pubescence adpressed, does not appear to have been noticed in Herefordshire among our bordering counties.

1191. 678. M. repens, G. and D. Don.   *Creeping Water Scorpion-grass.*
Native. Wet heathy places, margins of ponds. Rare. P. July to September. *Top. Bot.* 92.
First Record. *Lees*, in his *Table of Plants, Bot. Worc.* p. 20 (1867).
Severn. Wyre Forest, Worcester side, *Mathews*, 1858. Near Bewdley Road, Wyre Forest.
Lickey. One locality only, *Humphreys*. Rowney Green Bog, Alvechurch.
This plant was recorded, not localized, for Worcestershire by Dr. F. A. Lees in *Bot. Rec. Club Rep.* for 1875, as a first record for the county, but it was not so. The plant is much like the true Forget-me-not, but is distinguished by the narrow lanceolate teeth of the fruit-calyx and its shorter style. It has been recorded from all neighbouring counties.

1193. 679. M. sylvatica, Hoffm.   *Upright Wood Scorpion-grass.*
Native. Woods, shady places. General, except in the Lickey district, where it occurs occasionally. P. June and July. *Top. Bot.* 47.
First Record. Not localized, *Lees' Cat., New Bot. Guide* (1837).
Avon. Bredon Hill. Battons Wood, Bredon Hill. Goosehill Wood.
Severn. Opposite mouth of Worcester and Birmingham Canal. Bewdley. Wolverley. Shrawley Wood. Ockridge Wood. Wyre Forest.
Malvern. Ankerdine Hill. Southstone Rock. Brace's Leigh. Mathon.
Lickey. Stour-side at Halesowen Manor and Illey; Twyland Wood; Alvechurch, *Mathews.*
This plant also was recorded as new when it was not so in *Bot. Rec. Club Rep.* for 1883 (pub. 1884). The flowers are large and handsome, and it

belongs to the section of the genus with some, if not all, of the hairs on the calyx hooked. The plant occurs more frequently in the north of England and the Lowlands of Scotland than elsewhere. It appears not to have been noticed in Herefordshire.

1194. 680. M. arvensis, Hill.   *Field Scorpion-grass.*
Native. Woods, cultivated fields, hedge banks, waste ground. Generally distributed. A. April to July. *Top. Bot.* 112.
First Record. Sandy bank between Hagley and Stourport, *Lees, Illus. Nat. Hist. Worc.*, p. 153 (1834).
For a fine spacious localization for a plant the first record would be hard to beat! The two places are ten miles apart, and the whole town of Kidderminster is between them. Still, this does not cause any difficulty in the case of such a frequent plant as this one. It is the most common species, the whole plant is rough with spreading bristles, and it may be found on every sandy bank. It is recorded from every neighbouring county.

Var. **umbrosa**, Bab.
First Record. Not localized, but in error marked in the Severn district instead of Malvern, *Vict. Hist. Worc.*, I., p. 54.
Malvern. In the district, *Towndrow.*
This variety, with a flatter corolla-limb, and a biennial plant, is often mistaken for *M. sylvatica*. The variety has not been noticed in Gloucestershire. It is not given in *Lond. Cat.*, 10th ed.

1195. 681. M. collina, Hoffm.   *Early Field Scorpion-grass.*
Native. Heaths, sandy fields and banks, walls. Rather uncommon. A. or B. April to June. *Top. Bot.* 92.
First Record. Not localized, *Lees' Cat., New Bot. Guide* (1834).
Avon. Bredon Hill. Roadside near Crowle Church. Road to Spetchley.
Severn. Habberley Valley, *Lees*. Barnett Hill, Chaddesley, *Amphlett*. Astwood Road. Sutton Common. Areley Kings Lane, Stourport. Hartlebury Common. Astley. Blackstone. Banks of fence, Perdiswell Park.
Malvern. On some of the highest rocks of the hills, *Lees*. Powick. Ankerdine Hill. Berrow Hill, Martley. Little Malvern.
Lickey. Bilberry Hill, *Humphreys*. Barnett Hill, Chaddesley, *Amphlett*.
Though localized by Mr. Lees in his *Table of Plants* in the Lickey district, Mr. Mathews never saw it there, but makes a note in *Fl. Clent and Lickey Hills*, p. 34, that it 'is likely to occur'. It has since been seen in two localities. Barnett Hill is a cutting in the sandstone rock through which the main road passes, and this main road is the division between the Severn and Lickey districts. It occurs on both sides of the road. The plant is much like the following species, but the flowers are always blue, and the fruit-calyx is open and not closed. It occurs in all neighbouring counties.

1196. 682. M. versicolor, Sm.   *Yellow and Blue Scorpion-grass.*
Native. Heaths, sandy pastures, dry banks, walls. Locally common. A. or B. April to July. *Top. Bot.* 109.
First Record. On the top of Malvern Hill, nearly opposite the village, *Rev. W. S. Rufford, Purton, App.*, vol. iii, p. 16.

Avon. Bredon Hill. Craycombe Hill. Cleeve Banks. Inkberrow.

Severn. Harberrow Hill. Bishop's Wood. Hartlebury Common. Ribbesford. Belle Isle Rock, near Stourport. Spout Mill Swamp, Hagley. Near Bewdley. Stone Quarry, Holt. Shrawley Wood. Blakedown. Stagbury.

Malvern. Malvern Hills. Woodbury Hill. Midsummer Hill. Abberley Hill. Wall by Mrs. Berrington's house, Little Malvern. Malvern Link. New Pool. Malvern Wells.

Lickey. Moseley, *Mathews*. Kendal End. Pepper Wood, *Humphreys*. Clent. Romsley Hill.

This little plant, which varies in height according to the locality in which it finds itself, at the most scarcely exceeding six or eight inches, is easily recognizable from its flowers, which when they first open are pale yellow, and afterwards blue. This plant and all the genus to which it belongs owe their name of Scorpion-grass to the manner in which the spikes of flowers, before unfolding, are borne in curving racemes, like the supposed curve of a scorpion's tail over its back. It is recorded from all neighbouring counties.

LITHOSPERMUM, Linn. 298. (λίθος, a stone, σπέρμα, a seed, from the hard fruit.)

1198. **683. L. officinale**, Linn. *Common Gromwell.*

Native. Woods, hedges, bushy places. Not common. P. May to August. *Top. Bot.* 78.

First Record. Near Battenhall, *Lees, A. Florence, Strang. Guide* (1828).

Avon. Near Fladbury Station. Berrow Hill, Feckenham, *Humphreys*. Craycombe Hill. Mollylocks Hill. Defford Common.

Severn. Wyre Forest. Dowles Brook. Kempsey Grove. Great Bog, Wyre Forest.

Malvern. Black House Hill Wood. Quarry near Fetterlocks, Woodbury. Between Ankerdine and Berrow Hill. Crews Hill. Alfrick. Ridge Hill, Martley.

This is a dingy-looking plant, remarkable for nothing but the hardness of its seeds, which are grey, as bright and glossy as porcelain, and so hard that it is difficult to break them. The English name was originally Grommel, or some combination of letters having a similar sound, and the form Gromwell is comparatively late, the 'w' being due apparently to analogy with Speedwell. The original name is of doubtful origin. The plant is recorded from all bordering counties.

1199. **684. L. arvense**, Linn. *Corn Gromwell. Bastard Alkanet.*

Native or colonist. Cornfields, but not in the Lickey district. A. March to July. *Top. Bot.* 86.

First Record. Not localized, *Lees, Bot. Malv. Hills*, 1st ed., p. 16 (1843).

Avon. Bredon Hill. Feckenham.

Severn. Rainbow Hill, Worcester. Wyre Forest. Claines. Kempsey. Ombersley. Bewdley. Rushwick.

Malvern. Near Malvern Link. Southwood, Martley. Near Alfrick. Powick. Madresfield.

In this plant the flowers are white and the nutlets tubercled, otherwise it is much like the former plant. The roots are red and provide red dye, hence its name of Bastard Alkanet. It occurs in all neighbouring counties.

ECHIUM, Linn. 299. (ἔχις, a viper, from its supposed medical virtue.)

1200. **685. E. vulgare**, Linn. *Common Viper's Bugloss.*

Native. Heaths, dry sandy fields. Not common, except in the sandstone district. A. or B. April to September. *Top. Bot.* 92.

First Record. Roadside in the north of the county, *Pitt, Agric. Worc.*, p. 317 (1810).

Avon. Bredon Hill, above Woolas Hall.

Severn. Abundant about Kidderminster, *Lees*. Wolverley. Wyre Forest, *Rea*. Churchill (Kidderminster); Blakedown; Bissell, *Mathews*. Blakeshall Common. Devil's Spittleful, Bewdley. Areley Kings. Skey's Wood.

Malvern. Scarcely known, *Lees*. Side of railway, Malvern Wells. Ankerdine Hill. Raggedstone Hill. Malvern. Malvern Link.

This is one of the most brilliant of our wild flowers, bearing its blossoms in lateral curving cymes and forming a pyramid of bloom. The plant was formerly considered to be a cure for the bite of the viper, speckled as is its stem, and in most European languages its name has reference to this belief. Bugloss is from the Greek words βοῦς and γλῶσσα, 'ox' and 'tongue'. The Viper's Bugloss occurs in all neighbouring counties.

53. CONVOLVULACEAE, Juss.

CALYSTEGIA, Br. 300. (καλός, beautiful, στέγη, a covering, from the large bracteas.)

1202. **686. C. sepium**, Br. *Great Hooded Bindweed.*

(*Convolvulus sepium*, Linn. *Calystegia sepium*, R. Br.)

Native. Hedgerows and thickets. General, but not so frequent in the Lickey district. P. June to October. *Top. Bot.* 95.

First Record. As *Convolvulus sepium*. In hedges, *Pitt, Agric. Worc.*, p. 317 (1810).

The form with rose-coloured flowers has been recorded from near Pershore. The plant occurs in all neighbouring counties.

{*1203.* **687. C. Soldanella**, Br. *Sea-side Bindweed.*

Denizen. Once seen; usually found on sandy sea-shores. P. June to August. *Top. Bot.* 46.

First Record. This, as a boy, I found on Hartlebury Common, but in recent years I have not observed it, *Rea, Worc. Nat. Club Trans.* iii. 47 (1901).

Severn. Hartlebury Common, June, 1877, *Rea*.

This sea-side plant was seen on a common which, if now many miles from the sea, has in many places all the characteristics of the wind-blown sanddunes of the shore; but its survival is as curious and interesting as those of *Lepidium latifolium* at Droitwich, and *Glaux maritima* by the Droitwich Canal. It is not recorded from any inland county except our own.}

s

CONVOLVULUS, Linn. 301. (*Convolvo*, I entwine.)

1204. **688. C. arvensis**, Linn. *Small Bindweed.*

Native. Cornfields, cultivated ground, waste places, railway-banks. Abundant everywhere. P. June to October. *Top. Bot.* 96.

First Record. Cornfields, *Pitt, Agric. Worc.*, p. 317 (1810).

This is a most troublesome and, though it bears a very pretty flower, usually pink, but sometimes white, in which form it is especially common at Holt and other places on the banks of Severn. Country people know it by several names, Bearbind, Hedgebell, Withy-wind, Wave-wind, Umbrella, Gardener's Enemy, Strawberry Ruiner, and sometimes Bind alone. It has long white roots, most tenacious of life, and in some soils disputes with *Equisetum arvense* the honour of being the most ineradicable weed of the arable field. The flowers, when naturally pink, are frequently bleached nearly white by the sun, and shut up at night or on the approach of rain. The roots of the plant afford a medicinal substance which, if not so powerful as Scammony or Jalap, both products of plants of this genus, has a similar effect. It is abundant in all neighbouring counties.

CUSCUTA, Linn. 302. (Derived probably from the Arabic name, *Kechout*.)

689. C. Epilinum, Weihe. *Flax Dodder.*

Casual. Parasitical on Flax. Rare. A. August. Not in *Top. Bot.*

First Record. I have seen it in a field at Tibberton, *Lees, Bot. Worc.*, p. 24 (1867).

Severn. Tibberton, as first record.

Malvern. One place, suspicious, *Lees*. Near Doddenham Church.

Only where Flax is cultivated or survives from cultivation is this plant likely to occur. It is very destructive to the Flax crop on the continent, and doubtless was imported into England when the Flax industry was introduced. It has not been seen in Stafford or Gloucester. It is not in 10th ed. *Lond. Cat.*

1205. **690. C. europaea**, Linn. *Greater Dodder.*

Native. Hedges and bushes, damp herbage. Rare. A. July to September. *Top. Bot.* 31.

First Record. Badsey; South Littleton, *Purton, Midl. Fl.*, p. 139 (1817).

Avon. Badsey, *Rev. W. S. Rufford*. Shipston-on-Stour, *Rev. Dr. Jones*.

Severn. Cotheridge, *Lees*.

Malvern. Berrow, *Lees*. Malvern Link on Vetches.

Lickey. Pen Orchard Farmhouse, Clent, 1877, *Amphlett*.

Dr. Jones's plant constitutes the record in Smith's *English Flora*, 1824. The seeds of the Dodder germinate in the soil, but if they do not succeed in finding a suitable host they soon perish. If they succeed, they coil round it from west to east against the apparent course of the sun, the root soon dies, and thenceforth the parasite feeds upon the juice of the adopted plant by means of its haustoria or sap-sucking cells. The Dodders have no leaves, consisting entirely of stems and flowers, and they have had several country names, Hell-weed, Tetter, and Strangle-weed among them. This one grows on many kinds of plants, vetches being its commonest host. It has been noticed in Gloucester, Warwick, and Stafford of neighbouring counties.

1206. **691. C. Epithymum**, Murr. *Lesser Dodder.*

Native. Parasitical on Gorse, Thyme, and the Heath tribe. Rare. A. June to October. *Top. Bot.* 47.

First Record. On the North Hill (Malvern), *Lees, Mag. Nat. Hist.*, vol. iii, p. 160 (1830).

Avon. Craycombe, *G. W. Sandys*. 'Nowhere else met with in Worcestershire,' *Lees, Bot. Worc.*, p. 94. Bredon.

Malvern. Powick. Ankerdine Hill.

Mr. Lees apparently forgets his own first record, though possibly he may afterwards have found that he was mistaken in it. The chief victim of this plant is the Gorse, about which it often twines so thickly that hardly anything is to be seen but a confusion of thin red threads. The plant was held to have considerable efficacy against 'fainting and swooning' and 'trembling of the heart'. The plant is reported from Staffordshire.

1207. **692. C. Trifolii**, Bab. *Clover Dodder.*

Denizen. Parasitical on clover chiefly. Rare. A. July to September. *Top. Bot.* 2?

First Record. Near Alderminster, *Mr. W. Cheshire, jun., Lees, Bot. Worc.*, p. 24.

Avon. Between Badsey and Littleton, 1855. Churchill. Alderminster, 1855. Norton near Evesham. Lenchwick. Sheriff's Lench. Near Trench Woods.

Severn. Near Holt, 1870. Waresley. Kempsey. Ombersley. Wyre Forest. Upton Warren.

Malvern. Malvern Link. Little Witley. Ham Hill. Powick. Bransford. Leigh Sinton. Old Storridge.

Lickey. Warstone Farm, Frankley, 1867; The Doweries, Hunnington, 1869; Goodrest Farm, Hunnington, 1869, *Mathews*. Tutnall, *Humphreys*.

This plant, erected into a distinct species by Babington, with a stem clasping like a ring instead of twining irregularly, with shorter distant scales on the corolla tube, is regarded in Hooker's *Stud. Fl.*, 3rd ed., p. 283, as a variety of the preceding plant, and also by Bentham, 8th ed., p. 307. It is supposed to have been introduced from the continent among clover seed, but it has decreased greatly in recent years as seedsmen now have to supply cleaner seeds. It has occurred in every neighbouring county except Salop.

54. SOLANACEAE, Juss.

SOLANUM, Linn. 303. (Supposed to be a perversion of *solamen*, from the *comfort* or solace derived from medicinal use.)

1208. **693. S. Dulcamara**, Linn. *Woody Nightshade, Bittersweet.*

Native. Woods and hedges, damp situations. Abundant throughout the county. Climbing Shrub. May to August. *Top. Bot.* 100.

First Record. Hedges, *Pitt, Agric. Worc.* (1810).

The purple flowers of this plant, with the yellow anthers united into a cone, are frequently to be seen in the shrubby hedgerow, and the scarlet berries that succeed them are conspicuous in the autumn. The stems straggle away from the root sometimes to a distance of eight or ten feet. The root, when chewed, is at first bitter, but afterwards leaves a sweet taste in the mouth. Hence probably the English name. It is also known locally as Touch-me-not and Naughty Man's Plaything, a name similar to one given to Hemlock. The potato is *Solanum tuberosum*, but though grown by the herbalist Gerarde and figured in his *Herbal* in 1597, nearly 200 years passed before its value as a food product was recognized. The also well-known Egg Plant is *Solanum Melongena*; and the Tomato, *Lycopersicum esculentum*, belongs to an allied genus. The Bittersweet occurs in all neighbouring counties.

1209. 694. S. nigrum, Linn. *Common Nightshade.*
   Colonist. Cultivated ground. Local. A. June to October. *Top. Bot.* 64.
   First Record. Roadsides and on dunghills in most parts of the county, *Pitt, Agric. Worc.*, p. 317 (1810).
      Avon. Pensham. Norton, near Evesham. Sheriff's Lench. Badsey. Fladbury. Pershore. Bredon. Crowle. Inkberrow.
      Severn. Wolverley, *Scott.* Hartlebury Common, abundant, 1900, *Humphreys.* Very large near Stourport, 1860. Bewdley. Sutton Common. Barbourne, near Worcester. Kempsey. Claines. Ombersley. Holt. Hallow.
      Malvern. Hanley (Castle), *Lees.* Powick. Bransford. Leigh. Alfrick. Martley. Abberley. Stockton. Barnard's Green.
      Lickey. Several places, *Lees.* Hagley Lanes, *Scott.* A frequent weed, Clent Cottage Garden.
   The specific name of this plant is taken from the fruit, which is black, and not the flowers, which are white. The plant is a weed of such world-wide distribution, occurring in all temperate and tropical regions, that it is difficult to know where its native land is. At all events it is hardly in England. It seldom grows more than a foot or so high. It has been noticed in all neighbouring counties.

LYCIUM, Linn. 304*. (Λύκος, a wolf, from the sharpness of the thorns of some species; or from λύκιον, a name given by Dioscorides to *Rhamnus* as coming from Lycia, in Asia Minor.)

1210. 695. L. *chinense, Mill. *Box Thorn, Tea Tree.*
   (*L. barbarum,* Linn.)
   Alien. Near houses. Only found as an escape. Climbing Shrub. May to August. Not in *Top. Bot.*
   First Record. Malvern, an escape, *Towndrow, Malv. Advertiser*, November 19, 1892.
      Severn. Near Kidderminster Station. Near Dunhampstead.
      Malvern. The Wyche, *Towndrow.*
   A native of the Mediterranean region, usually when recorded an escape from cottage gardens where this plant is sometimes seen on the house-front or forming an arch by the garden gate, and where it makes dense thickets of growth with long straggling shoots. The flowers are small, and inconspicuous among the foliage. There is some amount of it by the railway-side in the neighbourhood of Kidderminster Station, where it has escaped from the gardens of the houses at the top of a bank by a bridge, and grows among

the masonry of the bridge itself nearly to the ground, like a gigantic wall plant. A local name for it as garden inhabitant is Tea Plant. It has been noticed in Salop and Warwickshire.

ATROPA, Linn. 305. (Named after *Atropos*, one of the Three Fates, from its deadly qualities.)

1211. 696. A. Belladonna, Linn. *Deadly Nightshade, Dwale.*
   Denizen. A native of woods on chalk or limestone. Very local. P. June to August. *Top. Bot.* 54.
   First Record. Dudley Castle, *Withering, Stokes, Stokes's With. Bot. Arr.*, 2nd ed., p. 233 (1787).
      Severn. Lincombe, Hartlebury, *Dr. James Nash.* Near Bewdley, *Jorden.*
      Lickey. Dudley Castle yard, as first record. Bell's Mill, Stourbridge, *Scott.*
   Dudley Castle yard is 'locally' only in Worcestershire, the castle being situated on a short narrow spur of Staffordshire which juts out into Worcestershire; but the record all through the books has been treated as a Worcestershire one. There the plant still flourishes. Luckily the county does not depend upon this record only for the possession of the plant. There are two other records of its occurrence, though the plant is not now known at either of them. Dr. James Nash's record was communicated to *Med. Surg. Rep.*, vol. i, no. 2, p. 105 (1828), wherein he says it has 'grown and flourished upon an old wall, formed of broken pieces of red sandstone and iron dross . . . ever since the recollection of the principal inhabitants'. In Mr. Lees' *Table of Plants* he appears to disregard the Dudley and Stourbridge localities, and notes it only at one place, in the Severn district, 'suspicious'; no doubt his Lincombe habitat. The Belladonna is an interesting plant, a deadly poison, and from its quality of inducing sleep and madness comes the name Dwale, meaning dead sleep or torpor, most probably of Scandinavian origin. The black shining fruits are like cherries, but possess a calyx. A tale tells of the poisoning of a Danish army with the plant by Macbeth; and it is supposed to be the 'insane root', which 'takes this reason prisoner', alluded to by Banquo. Its juice used to be mixed with corn to take birds with, for after eating it 'they shall slepe that ye may take them with your handys'. It is used at the present day to cause dilatation of the pupil of the eye in surgical operations upon that organ. The plant has been noticed in all bordering counties.

DATURA, Linn. 306*. (From its Hindi name *dhatura*.)

1212. 697. D. STRAMONIUM, Linn. *Common Thorn-apple.*
   Colonist. Waste ground and gardens. Local. A. June to October. Not in *Top. Bot.*
   First Record. Near Beverye, *Lees, A. Florence, Strang. Guide* (1828).
      Severn. Bevere, as first record. Hartlebury Common, 1900, *Humphreys.* Near Hartlebury, *Miss Ladbury.* Stoke Prior. Stonehall Common. Diglis. Near Bewdley. White's Nursery, Lower Wick. Gardens, Britannia Square, Worcester, *Westcombe.*
      Malvern. Near the Church, *Lees.* Workhouse Garden, Upton-on-Severn, abundant. Lower Howsell. Sherrard's Green.

This plant is probably a native of the Caspian region which was introduced into Europe in very early times, and took so kindly to the west that it has become very abundant in many parts of that continent. Various species of this genus have been used either to produce a kind of intoxication or as a poison in many countries. The plant has been noticed in every bordering county except Gloucestershire.

HYOSCYAMUS, Linn. 307. (ὗς, a hog, κύαμος, a bean. Hogs are said to eat the fruit.)

1213. 698. H. niger, Linn. *Common Henbane.*
   Native. Roadsides, waste places. Rare. B. May to August. *Top. Bot.* 79.
   First Record. Roadsides and amongst rubbish, *Pitt, Agric. Worc.*, p. 317 (1810).
      Avon. Alderminster, *Cheshire.* Bredon Hill. Broadway. On the King and Queen at Bredon. Elmley Castle.
      Severn. Side of the road beyond Spetchley, *Lees.* Droitwich. Stoke. Abundant on Hartlebury Common, 1900, *Humphreys.* Bath Road, Worcester. Hampton Lovett Church. Roadside, Whittington.
      Malvern. Base of the Hills near the Wells, *Lees.* North Hill, *Lees.* Longdon Marsh. Newland.
      Lickey. Cradley Churchyard, *Thompson.* Hagley and Broom Lanes, become very scarce, *Scott.*
   The plant is covered with long clammy hairs, and the flowers are dingy yellow, usually veined with purple. It is very poisonous to man and most animals, though both goats and sheep will eat it, and pigs eat it with impunity. It has been noticed in all neighbouring counties.

NICOTIANA, Linn. (Named after Jacques Nicot, French Ambassador at Lisbon, by whom tobacco was introduced into France in 1560.)

{**N. RUSTICA,** Linn.
   Alien. A. June to August.
   First Record. Naturalized about Bewdley (and growing) in considerable profusion, *Jorden, Lees, Bot. Worc.*, p. 7 (1867).
   This plant is cultivated in Central Asia and in tropical Africa, and has become naturalized in several places in Europe. The tobacco we all know so well is the product of *Nicotiana Tabacum.* This has no claim to be anything more than a casual weed. It is not given in *Lond. Cat.*, 10th ed.}

55. SCROPHULARIACEAE, Juss.

VERBASCUM, Linn. 308. (Name altered from *Barbascum*, from *barba*, a beard, in allusion to the shaggy nature of its foliage and hairy stamens.)

1214. 699. V. Thapsus, Linn. *Great Mullein.*
   Native. Roadsides, banks, hilly slopes, open woods. Local, but widely scattered. B. June to September. *Top. Bot.* 91.
   First Record. Roadside on sandy ground north of Kidderminster, *Pitt, Agric. Worc.*, p. 317 (1810).

      Avon. Tiddesley Wood. The Slads. Crowle.
      Severn. Crowneast. Fenny Rough. About Churchill (Kidderminster). Hagley Brake. Timberhanger. Broom. Blakedown. Malvern Road. Severnside above Holt. Henwick. Near Monk Wood. Wyre Forest, especially on the charcoal heaps. Eymore Wood. Shrawley Wood.
      Malvern. Malvern Hills. Middleyards Coppice. Little Malvern. Near the Rhydd. Dripshill. Ankerdine Hill. Martley. Sherrard's Green.
      Lickey. Clent Hills. Tutnall.
   This is a fine plant locally common on the sandy land north of Kidderminster and especially abundant in the neighbourhood of Hagley Brake. The leaves are woolly on both sides and continued down the woolly stem, and it grows sometimes four or five feet high. The woolly covering of the leaves was used for tinder in the days of flint and steel, and for lampwicks. It is also locally known as Poor Man's Flannel. The plant occurs in all neighbouring counties.

1216. 700. V. Lychnitis, Linn. *White Mullein.*
   Denizen or casual. Waysides, walls. Very rare. B. July and August. *Top. Bot.* 12.
   First Record. Hagley, *Scott, Hist. Stour.*, p. 540 (1832).
      Severn. Hagley, *Scott*, now gone. Near Cookley, locally abundant, *Mathews.* Railway embankment east of Dowles Church. Caunsall, north of Kidderminster. Near Wolverley.
      Malvern. Malvern, *Lees, Illus. Nat. Hist.*, 1834.
   Mr. Lees, in his *Table of Plants*, does not recognize his record for the Malvern district; and Mr. Mathews doubts if the Cookley plants are in the county. But Cookley is in the parish of Wolverley and Staffordshire does not come within a mile of it. The leaves are nearly smooth above, and woolly beneath, and the down of this plant as well as the last has been used for lampwicks, and hence the specific name, from λύχνος, the Greek for lamp. The flowers are cream-coloured or yellow, and are on short stalks. The plant has been observed in Gloucester, Salop, and Stafford.

1217. 701. V. nigrum, Linn. *Dark Mullein.*
   Native. Waysides, banks, borders of fields. Not common. B. or P. June to August. *Top. Bot.* 43.
   First Record. About Stourbridge on the side of the Bromsgrove road, *Purton, Midl. Fl.*, p. 125 (1817).
      Severn. About Kidderminster. Between Kidderminster and Park Hall. Near Bevere. Droitwich Canal at Salwarpe. Harborough Common. Hagley Brake. Churchill, Kidderminster. Roadside near Dowles Church. Stanklin.
      Malvern. Midland Siding, Malvern Common, *Towndrow, Malv. Advert.*, July 18, 1898.
      Lickey. Locally plentiful at Clent.
   The leaves of this Mullein are deep green, not woolly, and the spike of flowers is much like that of *V. Thapsus*, but the hairs of the filaments are purple and it is not so stout a plant. It loves the sandy land of the northern part of the Severn district; elsewhere it is not often met with. It is recorded from all neighbouring counties.

1218. 702. V. virgatum, Stokes. *Large-flowered primrose-leaved Mullein.*
Casual. Waste places and fields. Rare. B. August. Not mentioned in *Top. Bot.*

First Record. First shown me by my late worthy friend Mr. Waldron Hill of Worcester, in a field on the south side of a lane leading from Gregory's Mill to the turnpike road near that town. The side of the Turnpike road from Worcester to Ombersley opposite the road leading to Beverley (Bevere), *Stokes, Stokes's With. Bot. Arr.,* 2nd ed., p. 228 (1787).

Severn. Habberley Valley, *Lees.* Brake Mill Plantation, Hagley, *Mathews.* Roadside near Bevere, as first record. Sandy Fields at High Habberley. Roadside, Perdiswell, Droitwich Road. Roadside and in potato field near ninth milestone towards Bewdley from Worcester, *Westcombe.*

Malvern. One place, gone, *Lees.* Middleyards, Bransford.

This plant differs from the following one in having pedicels shorter than the bracts, one to five together, instead of pedicels nearly twice as long as the bracts, and solitary; in Hooker, *Stud. Fl.,* 3rd ed., p. 293, it is considered a sub-species of it. The two localities mentioned in the first record are not far apart, and the spot is mentioned all through the books as one where the plant is to be found. Mr. Lees, *Bot. Worc.,* p. 9, says that Habberley Valley is 'the only locality in Worcestershire where it can be considered indigenous'. With regard to the Bevere locality Sir James Smith, *E. B.,* vol. viii, opposite to a plate of this plant, no. 550, says the Rev. Mr. Baker took the plant from which the plate was prepared from one of the spots near Worcester mentioned by Withering, and said that 'this Mullein was first observed growing plentifully in a field near Wrexham, by Mr. Nash, who planted it in his garden at Bevere, from whence probably its seeds got to the neighbouring Turnpike road to Ombersley, and from thence to the lane leading from Gregory's Mill'. On this Mr. Mathews remarks, *Mid. Nat.* x. 174, that the plant now grows in several spots in the north of the county, where it is difficult to imagine that it can have spread from Bevere. Scott, in his *Hist. Stourb.,* records the plant from Iverley just outside the county in Staffordshire. It has been observed in Herefordshire, Salop, and Warwickshire.

1219. 703. V. *Blattaria, Linn. *Moth Mullein.*
Denizen or casual. Waysides, waste places. Rare. B. May to September. Not mentioned in *Top. Bot.*

First Record. Roadside, Holt, *Sheward, Nash, Hist. Worc., Sup.,* p. 96 (1799).

Avon. Badsey, *Rufford* in *Purton.*

Severn. Holt, as first record. Sandy ground north of Kidderminster, *Pitt.* Fox holes, Kidderminster, *Perry.* Between Bewdley and Stourport, *Westcombe.* West end of the south side of the railway embankment on the west of the road to Bewdley.

Malvern. Malvern Link, *Towndrow.* Side of road to Worcester, *Lees.* Between Powick and Newland. Near Malvern Chace, *Westcombe.*

Lickey. One place, *Lees.*

Dr. Stokes, *With. Bot. Arr.,* 2nd ed., p. 229, says he had never seen *V. Blattaria* growing wild, and that his specimens were from gardens, so that

he is led to suspect that his *V. virgatum* was often taken for true *V. Blattaria.* Scott, in his *Hist. Stourb.,* records the plant as common about Dunsley and Kinver, whereas Mr. Mathews remarks that it is *V. virgatum* that grows about these places, *Mid. Nat.* x. 223. This plant gets its specific name because it is said to be extremely obnoxious to cockroaches, *Blatta orientalis,* while on the other hand it obtains its English name because moths and butterflies are supposed to be especially fond of it. The plant is widely dispersed both in the old and the new world. It is a native of dry hillsides in Central Europe, and has been recorded from all neighbouring counties.

LINARIA, Hill. 309. *(Linum,* flax, which plant the leaves of some species resemble.)

1220. 704. L. *Cymbalaria, Mill. *Ivy-leaved Toad-flax.*
(Antirrhinum Cymbalaria, Linn.)
Denizen. Old walls, old brickwork. In many places. P. April to November. Not in *Top. Bot.*

First Record. As *Antirrhinum Cymbalaria.* Abbey walls, Great Malvern, *Purton, Midl. Fl.,* p. 288 (1817).

This pretty flower is often to be seen springing from the mortar of old walls, apparently as wild as any flower could be, but a moment's thought would recognize the fact that without the handiwork of man it would not be there! It is found in similar situations in nearly every part of Europe, and in England is supposed to have originated from the Chelsea Botanic Garden, whence Dillenius mentioned its escape. It formerly flourished in great abundance on the wall of old Clent Grove garden next to the road at Clent, but a few years ago the wall was rebuilt and the plant destroyed. It is seen in every neighbouring county.

1221. 705. L. Elatine, Mill. *Sharp-pointed Toad-flax, Fluellin.*
(Antirrhinum Elatine, Linn.)
Native. Arable fields chiefly on light soil. Occasional. A. June to December. *Top. Bot.* 55.

First Record. As *Antirrhinum Elatine.* Cleve, *Purton, Midl. Fl.,* p. 285 (1817).

Avon. As first record, *Purton.* Trench Woods, *Mathews.* Craycombe Hill. Bredon Hill. Near Wadborough station. Badsey.

Severn. Calcareous field at Tibberton. Near Monk Wood. Bewdley.

Malvern. Near Black House Hill Wood, Suckley, 1905. The Croft Farm, Malvern. Brace's Leigh. Near Dripshill. Between hop-yard and lane to Lord's Wood. Near Malvern Wells.

Lickey. Field near Bogs Wood, Halesowen, 1872, abundantly, *Mathews.* A record in *Worc. Nat. Club Trans.* ii, p. 79, Hayley Green, Lutley, 1872, refers to the same locality. Chaddesley Woods.

This plant is not unlike the next, but differs in the shape of its leaves, whence its book name, and in the spur of the corolla being straight and not curved upwards. The plant was formerly used as a remedy in cutaneous disorders. It does not appear to have been noticed in Staffordshire.

1222. 706. L. spuria, Mill. *Round-leaved Toad-flax, Fluellin.*
(Antirrhinum spuria, Linn.)
Native. Arable fields, chiefly on light soil. Occasional. A. June to December. *Top. Bot.* 43.

First Record. As *Antirrhinum spuria.* Cleve and Littleton, *Purton, Midl. Fl.,* p. 287 (1817).

Avon. As first record. Trench Woods. Craycombe Hill. Stoulton. Near Wadborough station.

Severn. Calcareous field at Tibberton, *Lees.* Near Pirton, with one peloria flower, *Rea.*

Malvern. Bransford. Powick.

The English name Fluellin or Fluellen, borne by this and the preceding plant, is derived from the Welsh for 'the herbs of Llewelyn'. Shakespeare uses the name Fluellen as the equivalent of Llewelyn. The name was first used in connexion with *Veronica officinalis,* and afterwards got misapplied to this and the preceding plant. Mr. Druce mentions, *Fl. Berks.,* p. 366, that while he has found *L. spuria* without *L. Elatine* growing with it, he has never found *L. Elatine* without *L. spuria.* The plant does not appear to have been noticed in Salop or Stafford.

1225. 707. L. PURPUREA, Mill. *Purple Toad-flax.*
Alien. Old walls. A garden escape. P. July and August. Not in *Top. Bot.*
First Record. As below.
Malvern. Garden wall at Powick, *Towndrow.*

This plant is a native of the mountains of Southern Europe, and has long been an inhabitant of our gardens. It has been noticed as an escape in Warwickshire.

1226. 708. L. repens, Mill. *Creeping Toad-flax.*
Native. Waysides, cultivated fields, walls, dry ground. Rare. P. June to October. *Top. Bot.* 22.

First Record. As *L. repens* or *Italica.* Roadside near Mitre Oak, *T. Westcombe, Lees, Bot. Worc.,* p. xxx (1867). First certain record. Clent Hill, *Mathews* and *Oliver, Bot. Rec. Club Rep.,* 1883 (pub. 1884). Discovered in this locality by Mr. Amphlett, of Clent, *Mathews, Mid. Nat.* xv, p. 259.

Severn. As first record. Roadside, Common Hill, near Worcester. Stourport.

Lickey. As above.

The Clent Hill station for this plant is the adjoining piece of ground not far from the Four Stones, at an elevation of some 900 feet above the sea. After the plant was first discovered the field was planted with larch; but in 1887 the whole plantation was burnt to the ground. Yet the *Linaria* survived, and was to be seen there during recent years. There are some interesting notes about this plant and its hybrids with *L. vulgaris* in Druce's *Fl. Berks.,* p. 366, one of which hybrids was called *L. Italica,* as in the first record. But there are numerous intermediate forms. The plant has been seen only in Stafford and Warwick of neighbouring counties.

1227. 709. L. vulgaris, Mill. *Yellow Toad-flax.*
(Antirrhinum Linaria, Linn.)
Native. Hedges, waysides, cultivated ground. General. P. June to October. *Top. Bot.* 99.

First Record. As *Antirrhinum Linaria.* In hedges, *Pitt, Agric. Worc.,* p. 317 (1810).

This is one of the most showy flowers of late summer, with a spike of pale yellow flowers, and leaves of pale sea-green hue, standing up in the herbage of the hedge-side, and sometimes among the corn. Country people call the flower Butter-and-Eggs, Pattens and Clogs, Snap Dragons, and Flax-weed. It is recorded from all neighbouring counties.

f. Peloria.
First Record. Hanley Castle; Malvern Link, *Towndrow, Malv. Advert.,* November 26, 1892.

This form, which has an equal and regular corolla and five spurs, is not very frequently met with. The name is derived from the Greek πέλωρ, a monster.

1228. 710. L. minor, Desf. *Small Toad-flax.*
(L. viscida, Moench. *Antirrhinum minus,* Linn.)
Native or colonist. Cultivated ground, waste places. Very rare. A. May to September. *Top. Bot.* 63.

First Record. Broadway Hills, *Rufford, Purton,* vol. iii, pt. ii, p. 335 (1821).

Avon. Near Badsey, *Westcombe.* Bredon Hill. Sheriff's Lench.

Severn. Along the railway in Wyre Forest, and always abundant. Near Monk Wood.

Malvern. Near Knightsford Bridge. Near Witchery Hole. The Croft Farm. Malvern Link.

Mr. Druce remarks, *Fl. Berks.,* p. 370, that this plant is frequent along the permanent way of the Great Western Railway. It is a native of Southern Europe, and a cornfield weed over an extended area. The plant was published as new, which it was not, from the Wyre Forest Station by Mr. F. A. Lees, *Bot. Rec. Club Rep.,* 1883 (pub. 1884). It is recorded in all neighbouring counties.

ANTIRRHINUM, Linn. 310. (ἀντί, in comparison with, ῥίν, a nose, from the mask-like appearance of the flowers.)

1229. 711. A. *majus, Linn. *Snapdragon.*
Alien. On old walls. Occasional, everywhere introduced. Shrubby perennial. June to August. Not in *Top. Bot.*

First Record. Littleton, *Purton, Midl. Fl.,* p. 288 (1817).

Avon. Walls at Evesham, *Lees.*

Severn. Walls at Worcester, *Lees.* With cream-coloured flowers on a wall near the Commandery, Worcester, *Lees.* Bewdley.

Malvern. On old walls, *Lees.* Brockamin. Little Malvern. Powick.

This is well known as a garden flower, having been brought to considerable perfection of form and colour by nurserymen. It is a native of dry pastures in the Mediterranean region. The origin of the English name can be seen by pressing the sides of the corolla when its resemblance to the head of the Dragon of mythology is at once evident, and it can be made to snap its teeth many times. The seeds produce a useful oil. It has been observed in its escaped form in all neighbouring counties.

1230. 712. A. Orontium, Linn.   *Lesser Snapdragon.*

Colonist. Cornfields, arable and garden ground. Rare. A. June to October. *Top. Bot.* 47.

First Record. Worcestershire, *Ballard, Stokes, Stokes's With. Bot. Arr.,* 2nd ed., p. 288 (1817).

Avon. Sheriff's Lench.

Severn. Between Wribbenhall and Spring Grove, *Jorden.* Barbourne, Cotheridge. Near Stourport. Northwick. Near Perdiswell Lodge. Bewdley. Brickyard, Claines. Field between Northwick and Common Hill.

Malvern. Waste Nursery ground, Malvern. Near the Abbey Gateway. Hanley Castle.

This is a smaller plant than the preceding, with dull purple flowers. It is easily known by its long leafy sepals, and the stem is about a foot high. It is recorded from all neighbouring counties.

SCROPHULARIA, Linn. 311.   (From *scrofula,* which disease the roots were supposed to cure.)

1231. 713. S. aquatica, Linn.   *Water Fig-wort.*

Native. Sides of water, wet places, damp woods. General and common. P. June to September. *Top. Bot.* 72.

First Record. Watery places, *Pitt, Agric. Worc.,* p. 317 (1810).

This plant is also called Water Betony, and a country name for it is Fiddle Dock. It is very common by the water-sides, and when in full flower is visited by quantities of wasps, which insects perform the act of fertilization. The name Fig-wort is derived from the supposed efficacy of the plant in curing that painful ailment for which *Ranunculus Ficaria* was recommended by our ancestors. Cream-coloured flowers have been noticed at the North Hill, Malvern, and Alfrick Chapel by the late George Reece, and by Mr. Towndrow at Welland. The plant occurs in all neighbouring counties.

Var. b. **cinerea,** Dum.

First Record. Malvern, *Towndrow, Malv. Advert.,* 1892.

This variety differs from the type in possessing an entire, instead of a reniform scale on the upper lip, and is the common form in this county. It has been observed in Warwickshire.

1232. 714. S. alata, Gilib.   *Ehrhart's Fig-wort.*

(*S. umbrosa,* Dum.   *S. Ehrharti,* C. A. Stev.)

Native. Damp woods, thickets, and hedges. Rare. P. July and August. *Top. Bot.* 21.

First Record. As *S. Ehrharti.* Plentiful in the valley between the Clent Hills (Clatterbach), and is not associated with either of the species to which it is intermediate, *Irvine, Phyt., N.S.,* vol. ii. p. 446 (1857).

Severn. Pixham, *Towndrow.* Severn-side at Shrawley, *Towndrow.*

Malvern. Banks of Teme, *Towndrow.* Near Stanford Bridge, *Ley.* Bransford. Ankerdine. Spout Farm, Eastham.

Lickey. Clent, as first record.

The records for the Severn district are probably the first two certain records for the plant in this county. The former is in *Bot. Rec. Club Rep.,*

1883 (pub. 1884), and the latter in *J. of B.* xxii, p. 38. The plant has been seen at Clent by no other botanist, and Mr. Irvine himself seems to have dropped his record afterwards, as it is not mentioned in another article on Clent Botany in *Phyt.,* April, 1858. It will be remembered that the records of the two *Elatines* depend upon the same gentleman's observations. Whatever may be Mr. Irvine's qualifications as a botanist, no one reading the articles on Clent in the *Phytologist* can come to any other conclusion but that his literary style is exceedingly loose, and shows anything but a trained mind. The distinction of this plant from its brethren depends chiefly on the scale on the upper lip of the flower, which, if we desire to be very scientific, we can call the 'staminode', and which has two diverging lobes ; and the leaves are more toothed. In *Hooker's Stud. Fl.,* 3rd ed., p. 296, the plant is considered a sub-species of *Scrophularia aquatica.* It appears not to have been observed in Gloucester or Stafford of neighbouring counties.

1233. 715. S. nodosa, Linn.   *Knotted Fig-wort.*

Native. Damp woods, thickets, hedges. General and frequent. P. June to August. *Top. Bot.* 109.

First Record. Not localized, *Walker, Med. Surg. Rep.,* vol. i, pt. ii, p. 100 (1828).

The roots of this plant give to it its name, consisting as they do of a number of small white tubers strung together by fibres ; and as they (on the principle of like cures like) were supposed to cure the swellings caused by disease, hence the English name of the plant, and its Latin specific name. Indeed it was a plant of wider repute than this, for Gerarde says that many 'rashly' said that if it was carried about by a man it kept him in health. It is a tall slender plant some three or four feet high ; the flowers are dull purple and green, but sometimes white. It is found with white flowers at Croome Perry Wood. It occurs in all neighbouring counties.

1235. 716. S. *vernalis, Linn.

Alien. Waste places. Rare. P. April to May.

First Record. Waste heaps, Hewell Grange, *Humphreys, Trans. Worc. Nat. Club,* vol. iii, p. 47.

Lickey. As first record.

This plant is a native of Europe as far north as Belgium, but in England, as here, it is more or less associated with garden culture.

MIMULUS, Linn. 312*.   (μμώ, an ape, from the form of the corolla.)

1236. 717. M. *Langsdorffii, Donn.   *Yellow Monkey-flower.*

(*M. luteus,* Linn.   *M. guttatus,* DC.   Mr. Druce says '*M. luteus* of British authors, not of Linn.')

Alien. Brook-sides and marshy places. Thoroughly established in several places. P. June to October. Not in *Top. Bot.*

First Record. Once gathered, but memorandum of spot mislaid, *Lees, Bot. Worc.,* p. xxx (1867).

Severn. Hurcott Wood, Kidderminster. Blakedown Brook. Wannerton Pools. Naturalized near the Birches, Hagley, *W. Whitwell.*

Malvern. Southstone Rock, 1900, *Rea.*

Lickey. Campion's Pool, Wildmoor, *Humphreys.* Bell End, Belbroughton, *Amphlett.*

This plant is a native of stream-sides in Western North America, and was brought into Europe by Langsdorf from Unalaska, one of the Fox Islands, a group of the Aleutian Islands, off the coast of Alaska. It has taken most kindly to its adopted country, and is spreading rapidly in suitable situations. The true *M. luteus,* Linn., is a South American plant. It does not appear to have been yet noticed in Gloucester, Stafford, or Warwick ; but this is only a question of time.

M. moschatus, Douglas.   *Musk.*

First Record. Hurcott Wood, near the brook-side, *Worc. Nat. Club Trans.* iii. 242 (1905).

Severn. As first record.

This, the common Musk of the garden, is also a North American alien. It was discovered as above by Mr. Carleton Rea.

LIMOSELLA, Linn. 313.   (From *limus,* mud, its usual habitat.)

1237. 718. L. aquatica, Gilib.   *Common Mud-wort.*

Native. Muddy ditches. Rare and local. A. August. *Top. Bot.* 45.

First Record. In ditches and roads about Badsey, *Rufford, Purton, Midl. Fl.,* p. 295 (1817).

Avon. As first record.

Severn. Westwood Park. Pools near the Heath, Stourbridge, *Scott.*

Malvern. Welland Common. Newland Common, *Lees.* By the margin of New Pool, *Lees.* Malvern Common. Garrett Pool, *Westcombe.* Newland Common. Barnard's Green.

Lickey. Cofton Reservoir, *Mathews.* Bittell Reservoir, Barnt Green, *Mathews.*

This little plant is one not that would attract the notice of any one but the botanist. The root is creeping and throws up clusters of leaves one or two inches long, and the flowers are minute, and rise from the base of the leaf-stalks. It has been recorded from all neighbouring counties.

DIGITALIS, Linn. 315.   (*Digitale,* the finger of a glove, which its flowers resemble.)

1239. 719. D. purpurea, Linn.   *The Foxglove.*

Native. Heaths, dry open woods, commons, bushy places. Common. B. or P. May to September. *Top. Bot.* 107.

First Record. Blossoms white, Shenstone Lane, near Hartlebury, *Stokes, Stokes's With. Bot. Arr.,* 2nd ed., p. 655 (1787).

This well-known flower is one of the handsomest of our native plants, and its medicinal properties are as well known as its blossoms. None of the old herbalists seem to have known its especial action upon the heart, though they recommended it for 'falling sicknesse' and other ailments. Dr. Withering first made known its modern value. On the Clent Hills the plant is very plentiful, and when the gorse on the hills is burnt, as it often is by naughty boys, the second year after the fire the burnt patch, however large it may be,

is thickly covered with Foxgloves, making a beautiful show. It is frequently plentiful on newly stirred soil, and on Malvern Hills is sometimes in such cases very abundant. It occurs in all bordering counties.

VERONICA, Linn. 316.   (Name certainly from ἱερά εἰκών, 'the sacred picture', in allusion to St. Veronica's handkerchief ; but by some it has been supposed to be a corruption of *Betonica.*)

1240. 720. V. hederaefolia, Linn.   *Ivy-leaved Speedwell.*

Native. Waste and cultivated ground. A plentiful weed. A. February to July. *Top. Bot.* 101.

First Record. Amongst wheat early in the spring, *Pitt, Agric. Worc.,* p. 317 (1810).

This common weed occurs in all neighbouring counties.

1241. 721. V. didyma, Ten.   *Gray Field Speedwell.*

(*V. polita,* Fr.)

Native. Cultivated ground. Uncommon throughout. A. January to December. *Top. Bot.* 89.

First Record. Not localized, *Lees, Bot. Malv. Hills,* 1st ed., p. 14 (1843).

By some botanists this is considered a sub-species of the following plant, from which it differs chiefly in the serratures of the leaves, the shape of the acute sepals, and the number of seeds in each cell, which are eight to ten instead of four to six. Mr. Lees marks it as uncommon in all the districts. Patient search among the weeds of the field would no doubt prove it to be quite as common as the following species. It is recorded from all neighbouring counties.

1242. 722. V. agrestis, Linn.   *Green Speedwell.*

Native. Cornfields, garden ground, waste places, wall-tops. Common everywhere. A. January to December. *Top. Bot.* 112.

First Record. Not localized, *Lees, Bot. Malv. Hills,* 1st ed., p. 14 (1843).

An abundant weed. Recorded from all neighbouring counties.

1243. 723. V. *Tournefortii, C. Gmel.   *Buxbaum's Speedwell.*

(*V. Buxbaumii,* Ten.)

Colonist. Cultivated fields and garden ground. Abundant. A. January to December. *Top. Bot.* 90.

First Record. As *V. Buxbaumii.* On Malvern Link, and near Stanbrook, *Lees, Bot. Malv. Hills,* 2nd ed., p. 34 (1852).

This now universal weed is a native of dry pastures in South-east Europe, and in recent times it has spread all over the continent and all these islands as a weed of cultivated ground. It was first recorded in Britain in 1829, and it is a matter for curious speculation why it had not started on its travels earlier, and why it started upon them when it did ; but the era of railway making seems to agree with its dominant occurrence in these islands. It occurs in all neighbouring counties.

[**1244. V. triphyllos,** Linn.   *Blunt-fingered Speedwell.*

Casual, if the records are correct. Sandy fields in the East of England and Yorkshire. Twice recorded. A. April to June. *Top. Bot.* 7.

First Record. Sandy fields, but rare, *Purton, Midl. Fl.*, p. 53 (1817) ; *Mid. Nat.* x. 224. Also recorded by *Lees*, at the northern extremity of the Link Common (Malvern), *Mag. Nat. Hist.*, vol. iii, p. 160 (1830).

Mr. Mathews to each of these records makes a note that they must be erroneous ; and in the case of the latter one refers to the *New Bot. Guide*, p. 622. The plant occurs in no bordering county. It is not mentioned in *Lees, Bot. Worc.*, so that gentleman had come to the conclusion by 1867 that both Purton and himself were in error. No local specimens are preserved at the Herbarium of the Hastings Museum, Worcester.]

1248. 724. V. arvensis, Linn.	*Wall Speedwell.*
Native. Heaths, walls, cultivated ground, dry banks, turf. Common and generally distributed. A. February to July. *Top. Bot.* 111.
First Record. Not localized, *Lees, Bot. Malv. Hills*, 1st ed., p. 14 (1843).
This is a common plant of fields and old walls, with inconspicuous bright blue flowers almost hidden by the upper leaves. It is downy, and gathers dust freely. It is recorded from all bordering counties.

1249. 725. V. serpyllifolia, Linn.	*Thyme-leaved Speedwell.*
Native. Cultivated ground, roadsides, pastures. Common in every variety of situation. P. March to October. *Top. Bot.* 112.
First Record. Hedges in summer, *Pitt, Agric. Worc.*, p. 317 (1810).
This plant varies much according to its situation. It is a frequent weed in gardens, forming, if allowed, spreading plants rooted at every node, from which the spikes of light blue blossoms ascend. It is a plant of wide distribution, spreading into North Africa and North America. It occurs in all bordering counties.

Var. b. **tenella,** All.
(*V. humifusa* Dickson.)
First Record. As below.
Severn. Hartlebury Common, *Herb. Vict. Inst. Worc.*
This differs from the type in the glandular pubescent capsule, its quite prostrate stems and shorter racemes, rooting at the nodes, and entire leaves. The variety is recorded from Herefordshire.

{**1252. 726. V. spicata,** Linn.	*Spiked Speedwell.*
Casual. Usually on chalky pastures or limestone rocks. Once recorded. P. July and August. *Top. Bot.* 3.
First Record. Near the Severn above Worcester, *T. W. Gissing*, 1852, *Lees, Bot. Worc.*, p. xxx (1867).
Mr. Mathews, *Mid. Nat.* xiv. 189, erroneously says that Mr. Lees did not insert the flower in his *Table of Plants*, but it is there on p. 21, marked as if naturalized, though suspicious. It is a frequent garden plant, where it grows to a much finer state than in wild conditions, breaking into colours, blue, white, and pink. It does not occur out of gardens in any neighbouring county.}

1254. 727. V. officinalis, Linn.	*Common Speedwell.*
Native. Heaths, dry woods, pasture, hedgebanks. General throughout. P. May to August. *Top. Bot.* 112.
First Record. Astwood Bank, *Purton, Midl. Fl.*, p. 51 (1817).

This is a very variable plant, especially in size. The leaves are bitter and astringent, and were formerly extensively used to be made into a kind of tea ; it was supposed also to strengthen the body. The plant occurs in all neighbouring counties.

1255. 728. V. Chamaedrys, Linn.	*Germander Speedwell.*
Native. Hedgebanks, borders of woods. Common and generally distributed. P. March to July. *Top. Bot.* 111.
First Record. Hedges in summer, *Pitt, Agric. Worc.*, p. 317 (1810).
This is the beautiful little blue flower that decks every sunny hedgebank in early summer. Yet the blossoms hold but lightly to the stems, opening in the morning, dropping off at the end of the day. The name Germander is applied to several other plants—those of the genus *Teucrium*, for instance—and is a corruption of the word used as the specific name of this, which itself is derived from the Greek words χαμαί, on the ground, and δρῦς, oak. The plant is frequently the victim of a gall-fly, *Cecidomyia Veronicae*, which causes little round hairy knobs on the leaves and upper part of the stem at the end of summer. Many popular names have been given to the plant— Eye-bright, Bird's-eye, Speedwell, Poor Cuckoo's-eye, Forget-me-not, and in great distinction to Bird's-eye, Cat's-eye also. It is found in every neighbouring county.

1256. 729. V. montana, Linn.	*Mountain Speedwell.*
Native. Shady woods. Local and not frequent. P. April to July. *Top. Bot.* 90.
First Record. Wood at the west end of Powick Ham, near Worcester, *Stokes, Stokes's With. Bot. Arr.*, 2nd ed., p. 16 (1796).
Avon. Bredon Hill. Broadway. Batton's Wood, Bredon Hill.
Severn. Hallow ; Orls Coppice, *Lees*. Harberrow Hill. Shrawley Wood. Near Dyne's Green.
Malvern. Alfrick ; Powick ; Rushwick ; The Gullet, Malvern, *Lees*. Rosebury Rock Wood. Near Ankerdine Hill. Wood, east end of Abberley Hill. Lord's Wood. West of North Hill.
Lickey. Wychbury Wood ; Frankley ; Romsley ; Woods at Hagley Hill, *Mathews*. Dale's Wood, Romsley.
This plant has a weak trailing stem, with a few pale blue flowers, and the leaves are large and all stalked. The seed-vessels are very large and quite flat. It occurs in all neighbouring counties.

1257. 730. V. scutellata, Linn.	*Marsh Speedwell.*
Native. Marshy and peaty places. Several places. P. June to August. *Top. Bot.* 107.
First Record. Broadmoor, near Halesowen, *Withering*, 3rd ed., p. 16 (1796).
Avon. Throckmorton. Pershore. Fladbury. Dodderhill Common. Defford Common. Churchill.
Severn. Ditch on the Crowle Road, *Lees*. Hartlebury Common, *Lees*. Moseley Green, Hallow. Wyre Forest. Hill Ditch, Hartlebury. Brickyard, Claines. Grimley Brick-pits.
Malvern. Western base of the Worcestershire Beacon, *Lees*. Newland

T

Common. Malvern Hills, *Westcombe*. Tinkers Cross. Powick Ham. Southwood. Kyre. Malvern Link Common.
Lickey. Moseley Wake Green, *Miss M. A. Beilby*. Near Halesowen, *Withering*. Alvechurch. Rowney Green, *D. Mathews*. Bittell Reservoir, *Humphreys*.
This plant, like the last, has a long weak stem, and the blossoms are pale flesh-coloured. It is recorded from all neighbouring counties.

Var. b. **hirsuta,** Weber.
First Record. As var. β. *parmularia*. By the little pool on Snead's Green, *Lees, Bot. Malv. Hills*, 3rd ed., p. 43 (1868).
Severn. Hallow ; Northwick, *Lees*. Wyre Forest. Hartlebury Common, 1894, *Rea*.
Malvern. As first record.
This variety has been described by several names besides the name in the *Lond. Cat.* It is *villosa*, Schum., *pubescens*, Hook., *Stud. Fl.*, p. 302, and the name Mr. Lees gives it in the first record. It is the hairy form of a glabrous plant, and both the type and variety grow in the Menyanthes Pool on Hartlebury Common side by side. It does not appear to have been observed in any neighbouring county.

1258. 731. V. Anagallis-aquatica, Linn.	*Water Speedwell.*
(*Anagallis aquatica*, Gerard.)
Native. Ditches, brooks, ponds. Local. Scarce in the Lickey district. P. June to August. *Top. Bot.* 100.
First Record. As *V. anagallis*. Brook above Gregory's Mill, and with pink blossoms, side of Berwick's Pool, *Lees, A. Florence, Strang. Guide* (1828).
Avon. Near Pershore. Near Eckington Bridge. Cleeve Mill. Throckmorton. Fladbury.
Severn. With white blossoms, Hartlebury Common. Duck Brook, near Worcester. Ashmore Common. Laughern Brook. Cotheridge. Henwick Mill. Hadley. Grimley. Doverdale.
Malvern. Near Powick, on the Malvern Road. Southwood. Clifton-on-Teme. Sherrard's Green.
Lickey. Near the bridge at Yardley, *Ick*. Alvechurch, *D. Mathews*.
This plant to the casual observer is not unlike the next and more common one, but may be distinguished from it by its sessile instead of stalked leaves, its angular stem, and its flowers usually of a paler blue. It is recorded from all neighbouring counties.

1259. 732. V. Beccabunga, Linn.	*Brook-lime.*
Native. Ditches, ponds, shallow streams, wet places. Frequent and generally distributed. P. May to August. *Top. Bot.* 112.
First Record. In shallow streams, *Pitt, Agric. Worc.*, p. 317 (1810).
This is a common plant in every watery place, the leaves of it sometimes by the uninitiated being supposed to be those of watercress, and sometimes sold with watercress to be eaten as it is. But they are too pungent to be agreeable. Its stems are prostrate at the base, and root in the mud. It has been seen with white flowers in Crowle Brook, Defford Brook, and near Power's Mill at Ombersley. The plant is as common in all neighbouring counties as it is in Worcestershire.

EUPHRASIA, Linn. 317. (εὐφρασία, joy, from εὖ, well, and φρήν, the mind, from the valuable qualities attributed to it.)

1260. 733. E. officinalis, Linn.	*Common Eye-bright.*
Native. Heaths, pastures, roadsides. Locally abundant. A. May to September. *Top. Bot.* 112.
First Record. Cornfields and pastures, *Pitt, Agric. Worc.*, p. 317 (1810).
Avon. Near Wadborough. Near Croome. Craycombe. Defford. Sheriff's Lench. Pershore. Evesham. Crowle.
Severn. With white flowers near Witley. Hartlebury Common. Wyre Forest. Bewdley. Stourport. Kempsey. Ombersley Hadley. Stanklin.
Malvern. Malvern. Powick. Lord's Wood. Ankerdine Hill. Old Hills. Alfrick. Martley. Stanford Bridge. Knightwick. Bransford.
Lickey. Offmoor Wood. Near Pen Orchard, Clent. Tardebigge. Bell End.
This is a low-growing plant, with white flowers marked with lilac and dashed with yellow, peering out of the short grass of the roadside or the meadow, though sometimes, in favourable situations, it will grow eight or nine inches high. The plant is bitter and astringent, and, as its name denotes, formerly much prized as a remedy in affections of the eyes. Used in any way it was supposed to be efficacious in this respect, and one writer says that if the herb were as much used as it was neglected it would half spoil the spectacle-makers' trade ! Milton represents our first father as purging with it 'his visual orbs, for he had much to see '. And so on down the poets till quite a late time, both actually and metaphorically. Lately the plant has fallen into the hands of the specialists, and Linnaeus's aggregate has been treated like straw in a chaff machine, and chopped into a lot of bits. Out of it have been made at least fifteen segregates, named in the *Lond. Cat.*, 10th ed., with a form, a variety, and numerous hybrids. Such diversions are not for us ! The critical descriptions of these segregates sometimes take up ten lines of small print. The plant is recorded from all neighbouring counties.

BARTSIA, Linn. 318. (Named after *John Bartsch*, a Prussian botanist and a friend of Linnaeus.)

1275. 734. B. Odontites, Huds.	*Red Bartsia.*
(*Euphrasia Odontites*, Linn.)
Native. Cultivated fields, pastures, waysides. Common, and in places abundant. A. June to September. *Top. Bot.* 111.
First Record. As *Euphrasia Odontites*. Var. 2. Flowers white. Sent to me by Mr. Bourne, who gathered it on Northington Farm, Grimley, near Worcester, *Withering*, 3rd ed., p. 543 (1796).
This is a very common plant, with much-branched stem and numerous spikes of dull pink flowers, and there is frequently a pinkish flush over leaves and stem as well. Cattle will not eat it, and are said even to avoid its neighbourhood. The plant has been noticed with white flowers between Crowle and Upton Snodsbury, and in a coppice between Alfrick and Knightwick. It occurs in all bordering counties.

T 2

Var. b. **serotina**, Dum.

First Record. Localized in the Malvern district, *Victoria Hist. Worc.*, vol. i, p. 54 (1901).

Severn. Near Worcester. Monk Wood.

Malvern. Malvern Common, *Towndrow.* Malvern Wells. Suckley Hill.

Lickey. Railway Cutting at Barnt Green, *Mathews.*

Recorded from Staffordshire and Warwickshire.

Var. **verna**, Reichb.

First Record. Localized in the Malvern district, *Victoria Hist. Worc.*, vol. i, p. 54 (1901).

Avon. Broadway.

Severn. Leopard Farm. Oldbury Road, Worcester.

Malvern. Malvern, *Towndrow.*

Lickey. Hunnington, near Halesowen.

This is a variety with leaves rounded below, bracts longer than the flowers, and rather straight ascending branches. This variety is not given in *Lond. Cat.*, 10th ed. Like the former variety it is recorded from Stafford and Warwickshire.

PEDICULARIS, Linn. 319. (*Pediculus*, a louse, from the supposed property of the plant in producing an undesirable condition in sheep.)

1278. **735. P. palustris**, Linn. *Marsh Lousewort.*

Native. Bogs, marshes, wet meadows. Not common. B. or P. May to September. *Top. Bot.* 112.

First Record. Oldfield, near Ombersley ; in Wire Forest ; and in Burnt Wood, near Bewdley, *Perry, Mag. Nat. Hist.*, vol. iv, p. 450 (1831).

Avon. Wyre Piddle, close to the river, *Lees.*

Severn. As first record. Wyre Forest, *Mathews.* Hartlebury ; Stanklin Pool, *Humphreys.*

Lickey. Moseley, *Ick, Miss M. A. Beilby.*

Both this and the following plant have a bad reputation as causing the state referred to in their generic and English names in sheep, but both the plants and the condition are effects of the same cause, the marshy nature of the ground. There are well-marked specific differences between these two plants ; this is larger and stouter than the next, the flowers are not so bright in colour, and the ribs of the calyx are hairy and the lobes of the calyx are not foliaceous. This plant occurs in all neighbouring counties.

1279. **736. P. sylvatica**, Linn. *Pasture Lousewort.*

Native. Wet, heathy, or hilly pasture. General. B. or P. April to July. *Top. Bot.* 112.

First Record. In a swamp on the north side of Falling Sands Common, near Kidderminster ; Oldfield, near Ombersley ; and in high pastures at Trimpley Green, near Kidderminster, *Perry, Mag. Nat. Hist.*, vol. iv, p. 450.

It is hardly necessary to give localities in the case of so common a plant of the damp meadow, which sometimes occurs in considerable quantity. It has been seen with white flowers near Bridges-stone, Alfrick, and near Leigh. It occurs in all bordering counties.

RHINANTHUS, Linn. 320. (ῥίν, a nose, ἄνθος, a flower, in allusion to the beaked upper lip of the corolla.)

1280. **737. R. major**, Ehrh. *Hairy Yellow-rattle.*

Native. Cultivated land. Uncommon. A. July and August. *Top. Bot.* 25.

First Record. As below.

Avon. Feckenham.

Lickey. Halesowen, *Thompson, Herb. Hastings Museum*, 1856.

The difference between this plant and the following one consists chiefly in the lobes of the upper corolla-lip, which in this are short and roundish and in that oblong, while the bracts are green throughout and not yellowish, with green points. It has been observed in Gloucester and Stafford.

1281. **738. R. Crista-galli**, Linn. *Common Yellow-rattle.*

(*R. minor*, Ehrh.)

Native. Meadows, pastures, heaths. Abundant. A. May to July. *Top. Bot.* 112.

First Record. Not localized, *Walker, Med. Surg. Rep.*, vol. i, no. 2, p. 100 (1828).

This is an abundant weed in poor pastures, and the pastures are poor because of its presence. For it is semi-parasitic in its nature, attached by suckers to the roots of grasses. The crested bracts gained for it the name of Cock's-comb, and it bears an equivalent name in many European countries. After the flowers fall off the calyxes become brown and dry, and the loose seeds rattle in them as the wind blows over them. Hence its familiar name. It is also locally known as Rattle Box and Wild Musk. Cattle are not fond of it when growing, but in many cases the hay of the meadows where it grows must largely consist of it. It has a wide distribution north of the tropics extending into North America. It occurs in all bordering counties.

MELAMPYRUM, Linn. 321. (μέλας, black, πυρός, wheat, from the resemblance in shape of its seeds to corn.)

[*1288.* **739. M. cristatum**, Linn. *Crested Cow-wheat. Top. Bot.* 11.

This plant has been recorded thus : Near Ombersley, Worcestershire, *Mrs. Gardner, Purton, App.*, vol. iii, p. 54 (1821). Mr. Mathews says, *Mid. Nat.* x. 256, the record must be an error ; certainly no one has seen it since. Besides, it is found only in eleven counties of England, all of them in the east and south-east. It may be dismissed from the flora of Worcestershire.]

1290. **740. M. pratense**, Linn. *Common Yellow Cow-wheat.*

Native. Woods, bushy places, sometimes in open ground. General except in the Avon district. A. May to August. *Top. Bot.* 108.

First Record. In the woods near the road from Birmingham to Halesowen, *Withering*, 3rd ed., p. 545 (1796).

No plant is named so directly on the principle of *lucus a non lucendo* as this. It is not a plant of pastures, but nearly exclusively found in woods and thickets. It is much relished by cattle, and from this it gains its English name. The plant, with many others of this tribe, goes black in drying,

becoming perfectly inky in the herbarium. It occurs in all neighbouring counties.

Var. d. **montanum**, Johnst.

First Record. Rednal Hill, Bromsgrove Lickey, *Rev. J. H. Thompson, Worc. Nat. Club Trans.* i. 323 (1885).

This plant is smaller in all its parts, with the bracts quite entire and the leaves linear lanceolate. It is recorded from Salop and Stafford.

56. OROBANCHACEAE, Vent.

OROBANCHE, Linn. 322. (ὄροβος, a vetch, ἄγχω, I strangle, from the parasitical habit of the plants.)

1294. **741. O. major**, Linn. *Greater Broom-rape.*

(*Orobanche Rapum*, now *Rapum-genistae*, Thuill.)

Native. Parasitical on the roots of Broom and Gorse. Rare. A. June to August. *Top. Bot.* 63.

First Record. As *O. Rapum*, Shrawley Wood, *Ballard.* On a dry bank near Clifton-on-Teme, *Stokes, Stokes's With. Bot. Arr.*, 2nd ed., p. 658 (1787).

Severn. Warshill, Bewdley, *Jorden.* Shrawley Wood, *Lees.* Stagbury Hill.

Malvern. Abberley, above the Hundred House. Between Suckley and Bear's Wood. Eastern base of Herefordshire Beacon. Hollybush Hill. Near Malvern Wells. Ankerdine Hill. Malvern Hills.

Lickey. Clent Hill, *Mathews.*

The Broom-rapes are all frankly parasitical, and proclaim it in their appearance ; they have retained no green leaf as a shred of respectability, as the more timid sap-suckers *Pedicularis* and *Rhinanthus* have so far succeeded in doing. Each kind seems to have a preference for the juice of one plant or another ; but very possibly as its host varies so may a plant vary, and come to be regarded as several species instead of one. These plants have been used in country medicine to cure ague and toothache, and to remove freckles, which clearly, from their frequent mention in old times in connexion with their supposed remedies, were considered unbecoming by the ladies. This Broom-rape has been recorded from all neighbouring counties.

[*1297.* **742. O. elatior**, Sutton. *Tall Broom-rape. Top. Bot.* 29.

This Broom-rape has been recorded twice in the county, but in each case from a clover-field, which at once throws doubt upon the subject, as this plant is parasitical on *Centaurea Scabiosa*, while the more common *O. minor* is the Broom-rape that preys upon clover. The records are : the first, From a clover-field below the Abbey Church, Great Malvern, *Miss Moseley's Herbarium* ; *Lees, Bot. Malv. Hills*, 1st ed., p. 31 (1843) ; and the other, Tall Broom-rape. Fields of Clover, about the village of Moor and Wyre, *May, Hist. Evesham*, p. 419 (1845). In Watson's *Top. Bot.*, p. 299, from which the number attached to the plant is 28, it is not credited to the account of Worcestershire. Mr. Lees, however, maintained the Malvern record in his three editions of *Bot. Malv. Hills*, justifying it by a description of the plant. It may safely be taken off the list of Worcestershire plants.]

1299. **743. O. Picridis**, F. Schultz. *Top. Bot.* 6.

First Record. As below.

Severn. Hartlebury Common, among *Picris*, near the western boundary, *Humphreys* and *Amphlett* (June 22, 1901).

This plant is not allowed specific rank by all botanists, being considered at most a sub-species of *O. minor*, varied by reason of its irregular host. It has not been observed in any bordering county.

1301. **744. O. minor**, Sm. *Lesser Broom-rape.*

Native. Parasitical, nearly invariably on clover. Locally abundant. A. June to October. *Top. Bot.* 33.

First Record. In a field at Lower Wick, 1847, *Lees, Bot. Malv. Hills*, 2nd ed., p. 63 (1852).

Avon. Two places, abundant, *Lees.* Bredon Hill. Norton, near Evesham. Westmancote.

Severn. Field near Claines Church, *Rea.* Lower Wick, *Lees.*

Malvern. Near Black House Hill Wood, Suckley (1905), *W. H. Edwards.* Birts Morton. Malvern. Malvern Link. Malvern Wells.

Lickey. Thicknall, Clent, 1902-3-4, in a different field each year, following the clover, *W. Whitwell.*

This is the Broom-rape most commonly met with, though it is not by any means a common plant. Although clover-sap is its usual food, it is sometimes found on other plants. In *Bot. Rec. Club Rep.*, 1883 (published 1884), *O. eu-minor* is given as a first record, which it was not, *Mid. Nat.* xv. 259. It appears to have occurred in all neighbouring counties except Staffordshire.

LATHRAEA, Linn. 323. (λαθραῖος, hidden, from the manner of its growth.)

1303. **745. L. Squamaria**, Linn. *Greater Toothwort.*

Native. In damp shady woods, usually among hazel. Local and not common. A. March to May. *Top. Bot.* 63.

First Record. Abberley, *Mrs. Gardner, Purton*, vol. iii, pt. ii, p. 335.

Severn. Wyre Forest, *Jorden.* Between Birchwood and Drakelow. Wolverley. Ribbesford Wood Astley Wood.

Malvern. Rosebury Rock, *Lees.* Little Malvern. Bridges-stone Mill, Leigh. Abberley. Suckley. Malvern. West Malvern.

Lickey. Wychbury Wood, Pedmore. Lutley, near Halesowen. Deep Wood, Clent, *Rev. J. H. Thompson.* Hagley Park. Dales Wood, Romsley, *Mathews.* Beacon Hill, Lickey. Chaddesley Woods. Dordale Brook, Belbroughton, *Amphlett.*

This plant is sometimes fairly abundant in its habitats. It was so at the Dordale Brook locality, but the Hazel Holt in which it grew has been raided by cattle and bitten off nearly to the ground so that stumps only remain, and the plant has not been seen there of late years. It grows freely in Hagley Hall garden. The subterranean stem has upon it a number of scales so like a human tooth that they have given to it both its English and its specific name ; possibly they are transformed leaves. The plant sometimes grows on the elm, and has been recorded from all neighbouring counties.

57. LENTIBULARIACEAE, Rich.

UTRICULARIA, Linn. 324. (*Utriculus*, a little bladder, from the vesicles on the leaves.)

1304. **746. U. vulgaris,** Linn. *Greater Bladderwort.*
Native. Slow streams, ditches, and ponds, preferring stagnant water. Very rare. P. July and August. *Top. Bot.* 88.
First Record. Pools east of the Hills, rare, *Lees, Bot. Malv. Hills,* 1st ed., p. 14 (1843).
    Malvern. As above. Near Chacely, *Lees.*
In his *Table of Plants* Mr. Lees marks the plant as occurring at one locality only, and in the 3rd ed. of *Bot. Malv. Hills,* p. 44, he locates the plant as occurring ' Near Chaceley '; which record, *Bot. Worc.,* p. 35, he further amplifies into ' Between Chaceley and Tirley '. Tirley is in Gloucestershire, a couple of miles from Chaceley, of which distance only about half a mile is in Worcestershire, so that it is open to conjecture whether the station belongs to our county. It has not been seen in any other part of Worcestershire, if this is a county record, but has been noticed in every neighbouring county except Herefordshire.

1306. **747. U. minor,** Linn. *Lesser Bladderwort.*
Native. Pools in heathy situations. Very rare. P. July. *Top. Bot.* 72.
First Record. In the bog on Hartlebury Common, June 14, 1889, *Rea, Worc. Nat. Club Trans.* ii. 8 (1897).
    Severn. As first record.
This is the only Worcestershire record for this plant, and Mr. Rea notes that the bog has largely dried up, and he has not since seen the plant. It differs from the former plant in possessing a nearly flat spreading margin to the lower lip instead of one deflexed at right angles all round. The Utricularias are interesting plants, the little bladders upon the leaves being first filled with water, then with air, when the plant rises to the surface, and when it has flowered they again fill with water; and the plant sinks to the bottom to ripen its seeds. The bladders possess an elastic valve at the orifice opening inwards, through which aquatic insects enter, and by which when inside they are imprisoned. The animals, mostly small crustaceans, die and the products of their decomposition are sucked in by special absorption cells developed within the bladder. The plant has not been noticed in Gloucestershire.

PINGUICULA, Linn. 325. (From *pinguis,* fat, the leaves being greasy in texture.)

1309. **748. P. vulgaris,** Linn. *Common Butterwort.*
Native. Peaty bogs and ditches. Very local. P. May to July. *Top. Bot.* 93.
First Record. Broadmoor, about three miles SW. of Birmingham, *Mr. Brunton.* On the NW. side of the Malvern Hills, but not on S. or SE. sides, *Ballard, Stokes's With. Bot. Arr.,* 2nd ed., p. 16 (1787).
    Avon. Feckenham Bog, *Purton,* lost.

    Severn. Wyre Forest.
    Malvern. Malvern Hills, *Lees.* West side of Worcestershire Beacon, but no other part of the range, *Lees.*
    Lickey. One place, lost, *Lees.*
This is a frequent plant in Scotland and Arctic Europe, but is much rarer in the south. It is used in Lapland to curdle milk, and from this its name of Butterwort is obtained; and the crushed leaves are used as a remedy for bruises. It is recorded from all neighbouring counties.

58. VERBENACEAE, Juss.

VERBENA, Linn. 326. (From the Celtic name *ferfaen,* meaning to drive away stone.)

1313. **749. V. officinalis,** Linn. *Common Vervain.*
Native. Dry waysides and pastures. Local. P. July to October. *Top. Bot.* 67.
First Record. Roadside, Powick. Amongst rubbish in the town of Evesham in great profusion, *Pitt, Agric. Worc.,* p. 317 (1810).
    Avon. As first record. Tredington, *Bagnall.* Crowle. Near Goosehill Wood. Fladbury. Bentley.
    Severn. Kempsey, in quantity. Near Hawford. Elmbridge Mill. Holt. Powers Mill. Top of Newtown Hill, Worcester. Woodcote.
    Malvern. Powick. Near Farm Houses on Barnard's Green. About Hanley, *Lees.* Newland Common. Ankerdine. Malvern Link.
    Lickey. Drayton, Chaddesley, *Scott.*
The Vervain is seldom found very far away from roads and habitations either in England or in Northern Europe, being native probably in pastures in the Mediterranean region. The flowers are very small for the size of the plant. It was a plant of peculiar virtues with our forefathers, not only of medicine, but of magic. In classic times it was well known; Pliny talks of it, Horace mentions it. By the herbalists it was prescribed for something like thirty different maladies; and a superstitious use was against witchcraft, ' much avayling ' in such a case, says Drayton; and yet we moderns can find nothing in it except a slight astringency! It occurs in all neighbouring counties.

59. LABIATAE, Juss.

MENTHA, Linn. 324. (Probably from the Greek μίνθος, which can be left untranslated.)

1314. **750. M. rotundifolia,** Huds. *Round-leaved Mint.*
Native. Damp places. Rare. P. August and September. *Top. Bot.* 54.
First Record. Several places near Sapey Brook, *Sheward, Nash, Hist. Worc., Sup.,* p. 96 (1799).
    Severn. Wyre Forest, *Jorden.* Near Bewdley, *Jorden.* Road near Stone, *Lees.* Pansington, Hartlebury. Hartlebury Common.
    Malvern. Sapey Brook, as first record. Mathon.
The mints are a difficult tribe of plants; the species are often variable,

hybridizing, and hard to discriminate. This plant is covered with soft hairs, and the leaves are shaggy with white down beneath, and it has a strong disagreeable odour. It does not appear to have been noticed in Staffordshire.

1315. **751. M. alopecuroides,** Hull.
Alien. Damp places. Very rare. P. August and September. *Top. Bot.* 15.
First Record. Barnard's Green Common, *Towndrow, Worc. Nat. Club Trans.,* iii. 127 (1903).
    Malvern. As first record.
This is a rare plant in the British Isles, not without suspicion that it is an escape from gardens. Its leaves beneath are hairy, not shaggy. It has been observed in Warwickshire only of neighbouring counties.

1316. **752. M. longifolia,** Huds. *Horse Mint.*
(*M. sylvestris,* Linn.)
Native. Hedges in marshy situations. Rare. P. August and September. *Top. Bot.* 64.
First Record. As *M. sylvestris.* By the side of a rill beyond the yew tree on the Ombersley Road, Worcester, *Lees, Nat. Hist. Worc.,* p. 168 (1834).
    Severn. As first record. Wyre Forest, *Jorden.* Hurcott Brook, near Park Hall, *Lees.* North Wood, Bewdley; Blakedown Pool, *Mathews.* Porter's Mill. Churchill, Kidderminster. Near Kidderminster.
    Malvern. One locality, *Lees.*
The leaves of this plant are silky beneath, and it possesses a pleasant odour. It is recorded from all neighbouring counties.

1317. **753. M. *spicata,** Linn. *Spear Mint.*
(*M. viridis,* Linn.)
Alien or casual. Waste places. Rare. P. August and September. Mentioned in *Top. Bot.,* but not numbered.
First Record. By the side of a stream near Newland, *Lees, Mag. Nat. Hist. Worc.,* p. 168 (1830).
    Avon. Harvington. Pershore. Fladbury.
    Severn. Near the railway viaduct, Blakedown, 1869 and 1872, *Rev. J. H. Thompson.* Near Bewdley. Near Stourport. Ombersley. Kempsey.
    Malvern. Newland, as first record. On the side of a deep ditch by the Link, some years ago (1843), *Lees.* Powick. Knightsford Bridge. Martley.
    Lickey. Harvington Hall Moat, *Mathews.*
This is no native of our country. It is always an escape from the garden, where it is largely cultivated chiefly for use with lamb, in which case it is said to represent the ' bitter herbs of the Passover '; or to be boiled with green peas or new potatoes. And besides this culinary use, it has had its medicinal virtues as well, being one of those useful plants which were ' good for the biting of a mad dog ', besides the lesser evils of wasp and bee stings. This mint does not appear to have been observed outside the limits of the garden in Gloucestershire.

1318. **754. M. piperita,** Linn. *Peppermint.*
Native or denizen. Ponds and wet places. Not common. P. August to October. *Top. Bot.* 77.
First Record. Plentiful by the rills on the Chace, *Lees, Mag. Nat. Hist.,* vol. iii, p. 166.
    Avon. Bredon Hill; Broadway, *Lees.*
    Severn. Wyre Forest, *Jorden.* Dane's Green, Henwick, *Lees.* Spout Mill, Lower Hagley, *Thompson,* 1880. Redstone Rock. Near Bewdley. Field next to Fenny Rough.
    Malvern. Wet places, abundant, *Lees.* Welland Common, *Lees.* Martley. Roadside near Stanbrook. Barnard's Green. Malvern Wells.
    Lickey. Hagley Green, Halesowen, *Mathews,* 1846. Harvington Hall Moat, *Mathews.* Bromsgrove Road. Redditch, *Mathews.*
The essential oil afforded by this species, which exists in glands on the calyx and leaves, is used both in confectionery and for medicinal purposes. It is probably an escape from cultivation in most of the cases where it is found, though in some counties it is asserted that it is native. It is recorded from all bordering counties.

Var. a. **officinalis,** Hull.
First Record. Localized in Malvern district. *Victoria Hist. Worc.,* i, p. 55 (1901).
    Malvern. Hanley Castle; Barnard's Green, *Towndrow.*
This variety has leaves acute or rounded at the base, and elongate spikes. Recorded in Gloucester and Warwick.

Var. b. **vulgaris,** Sole.
First Record. Localized in Malvern District, *Vict. Hist. Worc.,* i, p. 55 (1901).
    Malvern. In the district, *Towndrow.*
This plant has leaves rounded or sub-cordate at the base, and shorter spikes. This variety has been observed in Warwickshire only.

1319. **755. M. aquatica,** Linn. *Water Capitate Mint.*
Native. Sides of water in all states, and in wet places generally. Very common. P. August to October. *Top. Bot.* 112.
First Record. Not localized, as *M. hirsuta, Lees, Bot. Malv. Hills,* 1st ed., p. 29.
The aggregate *Mentha aquatica* is now divided into varieties, hybrids, and varieties of hybrids.

Var. a. **hirsuta,** Huds.
First Record. As segregate, *Mid. Nat.,* xiv. 235, where the name is given as an alternative to the *M. aquatica* of Mr. Lees' *Table of Plants.*
This is the commonest of all the mints, and widely distributed. The flower whorls are in a few axillary and terminal, somewhat globose clusters, and the whole plant usually tomentose. It is recorded from all bordering counties.

Var. b. **subglabra,** Baker.
First Record. Newland, *Towndrow, Malvern Advertiser,* 1892.
In this variety the leaves are narrower, and glabrous, except on the nerves beneath, and the calyx pedicels and corolla hairy. It has been observed in Warwickshire.

Var. c. **citrata**, (Ehrh.). *Bergamot.*
First Record. As below.
Severn. Bewdley.
Lickey. Bell Hall, Belbroughton.
This variety is entirely glabrous. It has been observed in Staffordshire.

× **arvensis.** *Marsh Whorled Mint.*
(*M. sativa*, Linn. *M. verticillata*, Huds.)
Native. Margins of water and wet places. Local. P. July to October. *Top. Bot.* 90.
First Record. Near Worcester, *Herb. Nat. Hist. Soc., Lees, Bot. Malv. Hills,* 2nd ed., p. 60 (1852).
Avon. Tardebigge Reservoir and Canal, *Humphreys.*
Severn. Kempsey. Northwick. Grimley. Doverdale.
Malvern. Leigh.
Nearly all the species of Mints are hairy, with serrated leaves, and they vary chiefly in the direction of becoming more or less smooth, and in the form of their leaves, which are sometimes cut or crisped. When smooth the odour of the species becomes mild and pleasant. This plant varies considerably. It occurs in all neighbouring counties.

Var. b. **paludosa**, Sole.
First Record. Malvern, *Towndrow, Malv. Advert.*, 1892.
This plant is hairy and the upper whorls collected into a spike with smaller bracts. It does not appear to have been observed in Salop or Stafford.

Var. c. **subglabra**, Baker.
First Record. Leigh, *Towndrow, Malv. Advert.*, Dec. 3, 1898.
In this variety the plant is almost glabrous, the whorls all separate, and the bracts all foliaceous. It has been observed in Herefordshire and Warwick.

× **longifolia**, Huds. *Top. Bot.* 9, in *Lond. Cat.*, 10th ed.
(*M. pubescens*, Willd.)
First Record. Banks of Leigh Brook, Alfrick, *Towndrow, Bot. Rec. Club Rep.* (1887). Brockamin.
This is a form with flowers in thick dense spikes with leaves hairy above and woolly beneath. It appears to occur only in Gloucestershire of the neighbouring counties.

Var. a. **palustris** (Sole).
First Record. Leigh Brook, Alfrick, 1884, *Towndrow, J. of B.*, xxii, p. 301.
Of this variety the leaves are tomentose above and woolly beneath. It has not been observed in any neighbouring county.

1320. 756. M. rubra, Sm. *Marsh Whorled Mint.*
Native. Wet places. Very rare. P. August. *Top. Bot.* 14 ?.
First Record. By a pool near the firs at Clifton-on-Teme, *Lees, Illus. Nat. Hist. Worc.*, p. 168 (1834).
Avon. Between Peopleton and Stonebow Bridge, 1846, *Westcombe.*
Severn. Wyre Forest.

Malvern. Clifton-on-Teme.
This is a segregate of *M. sativa*, nearly glabrous, except the teeth of the calyx, and the leaves veined with purple. It has been observed in every neighbouring county.

1322. 757. M. gentilis, Linn. *Whorled Mint.*
Native. Wet places. Very rare. P. August. *Top. Bot.* 10 ?
First Record. Not localized, *Lees' Cat., New Bot. Guide* (1835).
Severn. Blackstone Rock, 1846, *Westcombe.* Powick Weir, *Lees.* Wyre Forest, *Jorden.* Near Bewdley. Northwick Brickyard.
Malvern. Teme-side. Powick Weir, *Towndrow.* Pickersleigh, Malvern.
Lickey. Yardley, *Miss M. A. Beilby.*
This Linnaean species is now sometimes considered a variety of *M. sativa*, or perhaps a sub-species of that plant. It is not mentioned in Lees' *Table of Plants*, p. 22, being no doubt included under *M. sativa.* Its leaves are slightly hairy above, and on the few nerves beneath, and the pedicels and corolla are glabrous. It has been observed in Gloucester, Salop, and Warwickshire. An unnamed variety of the plant has been reported at Pickersleigh, by Mr. Towndrow.

× **arvensis.**
First Record. As below.
Malvern. Growing with *M. gentilis*, and near to *M. arvensis*, at Powick Weir, Sept. 10, 1900, *Towndrow, Malv. Advert.*, Dec. 22, 1900.
Not mentioned in *Lond. Cat.*, 10th ed.

1323. 758. M. arvensis, Linn. *Corn Mint.*
Native. Cornfields, river-sides, damp places. General. P. July to October. *Top. Bot.* 111.
First Record. Cornfields and moist ground, *Pitt, Agric. Worc.*, p. 317 (1810).
This is very commonly to be seen as a weed of cultivated land. It differs very little from *M. sativa*, chiefly in the form of the calyx-teeth; and it varies a good deal in itself. It has rather an aggressive odour, which has been compared to that of decayed cheese. It is recorded from all neighbouring counties.

Var. b. **agrestis** (Sole).
First Record. As *Mentha agrestis.* Near Bewdley, *Jorden, Lees, Bot. Worc.*, p. 6 (1867).
Severn. As first record. Kempsey. Holt.
Malvern. Powick Weir, *Towndrow*, 1900. Alfrick.
This variety is not mentioned in Lees' *Table of Plants.* It has been observed in Warwickshire.

1324. 759. M. Pulegium, Linn. *Pennyroyal.*
Native. Pond-sides, wet heaths. Not common. P. July to October. *Top. Bot.* 54.
First Record. Side of a pool near Robert's End near Hanley Castle, *Ballard, Stokes's With. Bot. Arr.*, 2nd ed., p. 602 (1787).

Severn. Wyre Forest, *Jorden.*
Malvern. As first record. Barnard's Green, *W. Addison.* Newland Common, *Westcombe.*
Lickey. Great Reservoir at the Lickey, *Lees.*
This species is very different in habit from the other species of the genus, and is readily known by its prostrate stems, its frequently recurved leaves, and especially by the hairy throat of the calyx. Its odour when bruised is very powerful. The herbalists all said it was 'good and wholesome for the lungs', and Gerarde tells us that a garland of the plant hung about the head would cure giddiness. The name Pennyroyal seems to be an alteration or corruption of the earlier *pulyole ryale*, the first portion of which is represented by the specific name of the plant, and the second of course is the same as 'royal'; but intermediate stages leading to the present 'penny' have not been found. The French name for the plant is *pouliot.* Pennyroyal occurs in all bordering counties.

LYCOPUS, Linn. 328. (λύκος, a wolf, and πούς, a foot, from a fancied resemblance of the leaf to a wolf's foot.)

1325. 760. L. europaeus, Linn. *Common Gipsy-wort.*
Native. Sides of water, boggy places in woods. General throughout the county. P. June to September. *Top. Bot.* 95.
First Record. Not localized, at Malvern, *Lees, Mag. Nat. Hist.*, vol. iii, p. 160.
Avon. Defford Common. Pirton Pool. Churchill, Worcester. Tardebigge Reservoir.
Severn. Henwick. Hallow. Northwick. Stanklin Pool. Blakedown Pool. Earl's Croome. Diglis Brickyard. Banks of Worcester and Birmingham Canal. Near Crowneast.
Malvern. Brace's Leigh. Roadside, Malvern Road, Leigh. Martley. Sherrard's Green.
Lickey. Moat Mill Pool, Bromsgrove. Pool near Independent College, Moseley, 1858, *Mathews.*
A troublesome weed this, when it becomes established in moist meadows, its creeping roots being full of vitality and difficult to remove. The small flowers are crowded together among the upper leaves and are pale rose in colour. The roots impart a black stain, which was said to be used to darken the complexion of any one who wished to pass for a gipsy, and hence its English name. It is recorded from all neighbouring counties.

ORIGANUM, Linn. 329. (ὄρος, a hill, γάνος, joy, from its ornamenting dry hilly places.)

1326. 761. O. vulgare, Linn. *Common Marjoram.*
Native. Dry sunny banks and pastures. P. July to September. *Top. Bot.* 90.
First Record. Side of the road at the foot of Craycombe Hill, Evesham. About the top of Southstone's Rock and at Lower Sapey, *Lees, Illus. Nat. Hist. Worc.*, p. 168 (1834).
Avon. Craycombe. Bredon. Broadway.

Severn. Wyre Forest, *Jorden.*
Malvern. Southstone Rock. Lower Sapey. Clifton-on-Teme. Witley. Alfrick. Crews Hill. Shelsley.
In chalky districts this is sometimes an abundant flower. It is a frequent ingredient, where it occurs, in the villager's herb-tea, and is hung up to dry for winter use. Members of the genus are inhabitants of the herb garden. *O. Majorana* is Sweet Marjoram, *O. Onites* Pot Marjoram, and *O. heracleoticum* Winter Sweet Marjoram. The etymology of the English name is wrapt in mystery. It has from the earliest times been represented by letters conveying a similar sound. The Latin name was *amaracus*, or said to be so, but this does not take us much further. The plant is recorded from all bordering counties.

THYMUS, Linn. 330. (θυμός, strength, from its odour being supposed to strengthen the spirits.)

1327. 762. T. Serpyllum, Fr. *Wild Thyme.*
Native. Pastures, commons, banks, roadsides. Locally abundant. P. June to September. *Top. Bot.* 112.
First Record. Not localized, *Lees, Bot. Malv. Hills*, 1st ed., p. 29 (1843).
The Thyme is well known. It has been credited with numerous healing virtues, and was as notable a 'strengthener of the lungs' as any that grew. Toothache, too, fled before it, and to the present day the chemist sells Oil of Thyme, which, however, is made from Marjoram, to allay its pains. Sheep have been universally supposed to be benefited by feeding on Thyme, but they do not touch it unless driven by hunger. As the plants that grow in marshy meadows have been said to injure sheep, so the Thyme, which grows in the hilly places suitable for them, has been said to improve them. Thyme is found in all neighbouring counties.

1328. 763. T. Chamaedrys, Fr.
Native. In the same conditions as the above plant, but not so frequent. P. June to September. *Top. Bot.* 40.
First Record. At Bewdley, *Mr. G. Jorden, Phyt. O. S.*, vol. iv, Pl. ii, p. 1142 (1852).
Avon. Craycombe. Sheriff's Lench. Bredon.
Severn. Bewdley, *Jorden.* Wyre Forest. Severn Stoke. Hartlebury Common. Hillside above Ribbesford Church. Near Spetchley. Fenny Rough. Harvington. Hagley Brake.
Malvern. Old Hills. Berrow Hill, Martley. Woodbury Hill. Abberley Hill.
Lickey. Walton Hill, Clent, *Mathews.* Hagley. Romsley Hill.
This is a segregate of the old Linnaean species, the distinction between this and the type being that while that forms a cushion with a fringe of prostrate barren shoots which in the next year produce erect flower-heads and are prolonged at the end, in this the stem of the preceding year ends in a flower shoot. And it is very distinct in appearance, its erect flower-shoots usually flushed in their upper parts as deeply purple as the flowers. The plant does not appear to have been noticed in Salop or Staffordshire.

CLINOPODIUM, Linn. 331. (κλίνη, a bed, πούς, a foot.)

1329. **764. C. vulgare**, Linn. *Wild Basil.*
(*Calamintha Clinopodium*, Spenn.)
Native. Hedges, bushy places, wood borders. Rather local. P. July to
September. *Top. Bot.* 90.
First Record. Not localized, *Lees, Bot. Malv. Hills*, 1st ed., p. 30 (1843).
Avon. Trench Woods, *Mathews.* Fladbury. Pershore. Rous Lench.
Defford. Bredon. Crowle. Hanbury. Craycombe.
Severn. Kempsey. Claines. Ombersley. Stourport. Hartlebury.
Hallow. Grimley. Broadheath. Spetchley. Cookley. Upton
Warren.
Malvern. Powick. Bransford. Alfrick. Knightwick. Martley.
Clifton-on-Teme. Brace's Leigh. Croft Farm.
Lickey. Near Bromsgrove Railway Station.
The name Basil is ultimately derived from the Greek word βασιλικόν,
royal, probably because some 'royal' unguent was prepared for the plant.
In the Latin this word got confused with *basiliscus*, and it came to be supposed
that it was an antidote to the venom of the basilisk. In modern French
both the plant and the serpent are called *basilic*. The plant is recorded from
all neighbouring counties.

CALAMINTHA, Lam. 332. (καλός, good, μίνθος, the word from which
'mint' is derived.)

1330. **765. C. Acinos**, Clairv. *Common Basil Thyme.*
(*Thymus Acinos*, Linn. *Calamintha arvensis*, Lam. *Clinopodium Acinos*, Kuntze.
In several records it is termed *Acinos vulgaris*.)
Native. Dry sandy fields, dry banks, roadsides. Local. A. or B. June
to October. *Top. Bot.* 75.
First Record. As *Thymus Acinos*. Churchill Field Corner, *Scott, Hist.
Stourb.*, p. 540 (1832.)
Avon. Bredon Hill, both purple and white flowers. Broadway. Be-
tween Evesham and Elmley Castle.
Severn. Wyre Forest. Habberley Valley. Blakedown. Near Holt
Mill. Hagley Brake. Churchill, Kidderminster. Near Bewdley.
Hartlebury. Near Stourport. Chaddesley.
Malvern. Abberley Hill. Leigh. Bransford.
This is locally known as Balm in some parts of Worcestershire. The plant
is recorded from all neighbouring counties.

{1331. **766. C. Nepeta**, Savi. *Lesser Calamint.*
(*Melissa Nepeta*, Linn. *Calamintha Nepeta*, Savi. *Thymus Nepeta*, Sm.
Clinopodium Nepeta, Kuntze. *Calamintha parviflora*, Lam.)
Denizen. Dry banks on calcareous soil. Very rare. P. August to
October. *Top. Bot.* 9.
First Record. As *C. Nepeta*. Not localized, *Lees' Cat., New Bot. Guide* (1835).
Mr. Mathews suspected this record to be an error. The plant is included
in a catalogue, attributed to Mr. Herbert New, which is given in *May's
History of Evesham*, 1845, in which it is said to occur in 'The Shrubbery,

Fladbury Rectory'. This record also meets with Mr. Mathew's suspicions.
A form of *C. montana* is very frequently mistaken for this plant, which itself
is by some botanists considered a sub-species. The plant is recorded from
Gloucestershire only of our bordering counties.}

1332. **767. C. montana**, Lam. *Common Calamint.*
(*C. officinalis*, Moench. *Melissa Calamintha*, Linn. *Thymus Calamintha*, Sm.
Clinopodium Calamintha, Kuntze.)
Native. Dry roadsides and hedgebanks. Not very common. P. July to
October. *Top. Bot.* 62.
First Record. As *Melissa Calamintha*. In woods and thickets near Malvern
and elsewhere, not unfrequent, *Nash, Hist. Worc., Int.*, p. lxxxix (1781).
Avon. Hampton Magna, Evesham, *Perry.* Near Croome. Stoulton.
Severn. Cowmore Lane, St. John's, *Lees.* Grimley. West of Hagley
Village, *Mathews.* Near Droitwich. Near Worcester. Hallow. Om-
bersley. Trimpley. Holt. Hill Hampton. Shrawley. Hartlebury.
Malvern. About Malvern, *Lees.* Witley. Abberley. Leigh.
Lickey. Catshill, near Bromsgrove, *Mathews.* Dudley Castle, *Withering.*
Chaddesley Corbett.
This plant is larger altogether than the preceding one, erect and bushy, its
stems pale greyish green, and downy. One of its virtues is to restore tainted
meat to soundness. It is recorded from all neighbouring counties.

{1333. **768. C. grandiflora**, Moench. *Wood Calamint.*
(*C. sylvatica*, Bromf.)
Native. Thickets and woods. Once only recorded. P. August to Sep-
tember. *Top. Bot.* 3.
First Record. Abberley Hill, August 16, 1844, *Dr. J. M. Streeten, Lees, Bot.
Worc.*, p. 50 (1867). Sp., *Herb. Hast. Mus. Worc.*
The principal distinction between this and the last species lies in the
upper lip of the calyx and corolla, and in dried specimens it is almost
impossible to detect these characters. The true plant in *Top. Bot.*, p. 310, is
recorded only from Devon, the Isle of Wight, and Hampshire.}

MELISSA, Linn. 333*. (μέλισσα, a bee, from bees affecting the plant.)

1334. **769. M. *officinalis**, Linn. *Balm.*
Alien. Hedge-sides. A relic of cultivation. P. August. Not in
Top. Bot.
First Record. Hanley, near Farm Houses, naturalized, *Lees, Bot. Malv. Hills*,
1st ed., p. 30 (1843).
Malvern. Hanley, as above. Roadside, Alfrick. Bransford. Abberley
Hill. Guarlford.
This plant is a native of woods in the Mediterranean region. It has been
long somewhat extensively cultivated in gardens, and wherever it occurs in
this country is an escape.

SALVIA, Linn. 334. (*Salvo*, I save or heal, in allusion to its qualities.)

1335. **770. S. Verbenaca**, Linn. *Wild English Clary.*
Native. Dry banks, sides of roads. Not common. P. May to October.
Top. Bot. 64.

U

First Record. By the side of the road at Harvington, leading to the mill,
Purton, Midl. Fl., p. 57.
Avon. Alderminster, *Cheshire.* Harvington, as above.
Severn. London Road and Crowle Road, Worcester, *Lees.* Between
Birchwood and Drakelow. Roadside near Blakedown Viaduct. Near
Spetchley. Red Hill, Worcester, in profusion.
Malvern. Marl banks, *Lees.* Old road, Broadwas, *Jeffery.*
The English name for this plant is not derived, as the herbalists of the sixteenth century supposed, from the use of the seeds in getting rid of substances
that had intruded themselves under the lid. They said Clary meant Clear
eye ; but it means nothing of the sort. It represents the mediaeval Latin
name of the plant, *Sclarea*, though the loss of the initial letter cannot
be explained, nor can the origin of the word itself. Under the
influence of moisture, a mucilage which, says Gerarde, 'if put whole
into the eies it cleanseth and purgeth them exceedingly from redness,
inflammation, and divers other maladies.' Besides its use for the eyes the
wild Clary was recommended for a number of ailments, and its potency
is summed up in the old proverb, 'He that eats sage in May shall live
for aye.' Nothing more need be said ! It is erroneously called Wood-sage in
some parts of Worcestershire. Luckily for their inhabitants, the plant
is recorded from all neighbouring counties.

1337. **771. S. pratensis**, Linn. *Meadow Clary.*
Alien. Grassy fields. Very rare. P. July to September. *Top. Bot.* 3.
First Record. Not localized, *Lees' Cat., New Bot. Guide* (1835).
Severn. Hartlebury. Field beyond Stourport Railway Station.
Malvern. Near the College, Malvern, *Towndrow.*
Lickey. Pasture field near Rubery ; leaves of a sage there, 1908, *Amphlett.*
On the continent this is an abundant plant, especially in the south, and
it is undoubtedly native as far north as Normandy and Belgium ; but it
becomes very rare in those countries. There would therefore seem to be no
special reason why it should not also be rarely native in our island. In *Top. Bot.*,
p. 305, it is only recorded certainly from Kent and Oxfordshire, but it has
been observed in Gloucestershire, Salop, and Warwickshire.

S. verticillata, Linn.
First Record. Well established, side of the railway, Malvern Common,
Towndrow, Malv. Advert., December 3, 1898.
Severn. Hartlebury, 1908.
Malvern. As first record. In a wood at Suckley, 1907.
A native of mountain pastures in the Mediterranean region.

NEPETA, Linn. 335. (Either from Nepi, a town in Italy, or *nepa*,
a scorpion, for the bite of which the plant was considered a cure.)

1338. **772. N. Cataria**, Linn. *Cat-mint.*
Native. Hedges, waysides, borders of fields. Rather uncommon. P.
July to September. *Top. Bot.* 59.
First Record. Dudley Castle, *Withering*, 3rd ed., p. 519 (1796).
Avon. Croome.
Severn. About Kidderminster. Stourport. Witley. Devil's Spittleful,
Bewdley. Near Holt. Near Hallow. Trimpley. Severn Stoke.

Malvern. Great Witley.
Lickey. Thicknall Farm, 1904, *W. Whitwell.* Lane south-east of Hagley
Railway Station.
The stem of this plant sometimes by the friendly help of the hedge in which
it grows reaches a height of three or four feet, and is much branched, and
all white with down. The plant possesses a powerful odour, which is
extremely attractive to cats ; they do not seem to notice the plant when
growing, but appear to be quite intoxicated with it when it is purposely
brought near to them. Dogs, however, show no partiality for it. It was
sometimes used medicinally, and it was said that the roots when chewed
would make the most gentle persons fierce and wrathful. The plant is
recorded from all bordering counties.

1339. **773. N. hederacea**, Trev. *Ground Ivy.*
(*Glechoma hederacea*, Linn. *N. Glechoma*, Benth.)
Native. Woods, thickets, hedges, fields. Abundant and generally distributed. P. March to June. *Top. Bot.* 103.
First Record. As *Glechoma hederacea.* Hedgebanks, *Pitt, Agric. Worc.*,
p. 317 (1810).
'Ground Ivy tea' was in former days a well-known spring medicine ; and
by our forefathers the plant was supposed to possess many medicinal virtues,
and had many popular names ; Gill by the ground, Gill or Jack by the
hedge, Robin run in the hedge, Hay-maid, Ale-hoof, and Tun-hoof among
them, 'hoof' being a variant of Hove, in the earliest times the name of some
Plant, identified with the Ground Ivy. The plant is recorded from all
neighbouring counties.

Var. **parviflora**, Benth.
First Record. In this book.
Severn. Sutton Common.
In this variety the sub-glabrous corolla-tube equals the calyx, whereas in
the type it is twice as long. The variety is not mentioned in *Lond. Cat.*,
10th ed., though given in the 9th ed. It appears to have been observed in
Warwickshire only of neighbouring counties.

SCUTELLARIA, Linn. 336. (*Scutella*, a little dish, which the calyx
with its ear somewhat resembles.)

1340. **774. S. galericulata**, Linn. *Common Skull-cap.*
Native. Sides of water. Not uncommon. P. June to September. *Top.
Bot.* 103.
First Record. Side of the canal between Wolverley and Stourport in
many places, *Pitt, Agric. Worc.*, p. 317 (1810).
Avon. Defford Brook. Bentley Thrift. Badsey. Eckington Bridge.
Bow Brook. Side of brook near Feckenham.
Severn. As first record. Laughern Brook. Kempsey Grove. West-
wood Park. Oldington Wood, Kidderminster. Oldbury Lane, near
Dynes Green. Hartlebury Common. Captain's Pool. Henwick
Mill. Grimley. Near Monk Wood.
Malvern. Danemoor Pool, Welland Common, *Lees.* Ponds on Bar-
nard's Green, *Lees.* Malvern. Leigh. Martley.

Lickey. Tardebigge, *Humphreys*. Bittell Reservoir, *Humphreys*. Pond on Moseley Common, *Miss M. A. Beilby*. Canal-side at California, Northfield.

The English name of this handsome plant is derived from the resemblance of the calyx, when inverted, to a helmet with the visor raised, from which also comes the specific Latin name, derived from *galericulum*, meaning a cap or bonnet; while in its ordinary position the calyx not inaptly gives rise to the generic name. The plant is recorded from all neighbouring counties.

1341. 775. S. minor, Huds. *Lesser Skull-cap.*

Native. Damp heathy places, margins of ponds. Local. P. June to September. *Top. Bot.* 72.

First Record. Hanley Common, *Hickman*, *Purton*, vol. iii, pt. ii, p. 335 (1821).

Avon. Trench Woods. The Marsh Common. The Slads. Thrift Wood. Bentley.

Severn. Wyre Forest, *Mathews*. Hartlebury Common. Rock Wood, near Bewdley, *Perry*. Hurcott Wood, *Mathews*. Great Bog, Wyre Forest.

Malvern. As first record. Middleyards Coppice. Leigh. Martley. Shelsley Beauchamp.

Lickey. Bog on Moseley Common, *Miss M. A. Beilby*. Bittell Reservoir; Halesowen Canal, *Mathews*. A weed in Clent Cottage Garden, Clent. Alvechurch and Redditch, *D. Mathews*.

This plant is not common except in the West of England, though it is recorded from all neighbouring counties.

PRUNELLA, Linn. 337. (Named from the German *bräune*, the quinsy.)

1342. 776. P. vulgaris, Linn. *Common Self-heal.*

(*Brunella vulgaris*, Moench.)

Native. Pastures, heathy woods, in turf. Abundant and widely distributed. P. June to August. *Top. Bot.* 112.

First Record. White flowers, west side of Worcester, *Lees, Mag. Nat. Hist.*, vol. iii, p. 160 (1830).

The English name comes from its supposed efficacy in healing wounds caused by scythes, sickles, and the like; and as its generic name, softened down from the original *Brunella*, would indicate, it was held to be a remedy for the quinsy. It is an abundantly common plant, whose usually dark purple flowers vary sometimes into lilac and white. Besides the locality for white-flowered plants mentioned in the first record, another is near Holt Bank. It is recorded from all neighbouring counties.

MELITTIS, Linn. 338. (μέλισσα, or μέλιττα, a bee, because it yields honey to bees.)

{**1344. 777. M. Melissophyllum,** Linn. *Bastard Balm.*

(*M. grandiflora*, Sm.)

Native. Woods and coppices. Once recorded, now lost. P. May and June. *Top. Bot.* 9.

First and only Record. As *M. grandiflora*. Woods and fields near Halesowen, *Scott, Hist. Stourb.*, p. 540 (1832).

This is a very handsome plant, with purple and white flowers, marked in various ways. When dried, it smells like hay. It is chiefly found in the South of England, and of the neighbouring counties it is recorded only, and that doubtfully, in Gloucestershire.}

MARRUBIUM, Linn. 339. (The origin of the name is not known; some say it comes from the Hebrew *mar*, juice, *ro*, bitter, because of its bitter taste.)

1345. 778. M. vulgare, Linn. *White Horehound.*

Denizen or casual. Roadsides and waste places. Rare. P. May to September. *Top. Bot.* 66.

First Record. Roadsides on sandy and gravelly soils at Shrawley, *Pitt, Agric. Worc.*, p. 317 (1811).

Avon. Craycombe, *Lees*. Nafford Mill. Dodderhill Common.

Severn. Hartlebury Common, *Humphreys*. Mitton, near Stourport. Grimley, *Gissing*. Habberley, Kidderminster, *Perry*. Astley. Pullens Farm, Hartlebury.

Malvern. Clifton-on-Teme. Welland and Castlemorton Commons. Rocks of Malvern above Malvern Chace, *Westcombe*. The Wells Common. Herefordshire Beacon. Near Keys End.

Lickey. Baldwin's Green, Lye Waste, *Scott*.

Any Avon locality is omitted from Mr. Lees' *Table of Plants*, p. 22. This is a bushy plant coated with white down. It is thought to be particularly efficacious in affections of the lungs, and candied Horehound and Balsam of Horehound are still sold by chemists. The plant is recorded from all neighbouring counties.

STACHYS, Linn. 340. (στάχυς, a spike, from the form of the inflorescence.)

1346. 779. S. officinalis, Trev. *Wood Betony.*

(*Betonica officinalis*, Linn. *S. Betonica*, Benth.)

Native. Woods, roadsides, hedgebanks, coppices. Frequent in suitable situations. P. June to November. *Top. Bot.* 82.

First Record. As *Betonica officinalis*. Not localized, *Walker, Med. Surg. Rep.*, vol. i, no. 2, p. 100 (1828).

Pliny gives the specific name as *vettonica*, and says it is the Gaulish name of a plant discovered by a Spanish tribe called the Vettones. It was considered a plant of not only medical but also magical virtue, perhaps more esteemed by our ancestors than any except Vervain. Even wild beasts were said to know of its powers, and healed themselves by it. 'Sell your coat and buy Betony,' was an old proverb; not bad advice, if it really would, as was alleged, cure forty-seven different disorders, and keep off the mischief of witches as well. The powdered leaves have been used in the manner of snuff. Specimens with white flowers have been observed in Wyre Forest, Shrawley and Monk Woods. The plant occurs in all neighbouring counties.

1348. 780. S. palustris, Linn. *Marsh Woundwort.*

Native. Ditches, stream-sides, damp thickets. Frequent. P. June to September. *Top. Bot.* 112.

First Record. Not localized, *Lees, Bot. Malv. Hills*, 1st ed., p. 29 (1843).

This plant is so frequent in suitable situations that it is unnecessary to give localities. Its roots are extensively creeping, and in moist soil are a great nuisance to the agriculturist. It is recorded from all bordering counties.

× **sylvatica.**

(*S. ambigua*, Sm.)

First Record. Malvern Link, *Towndrow, Malv. Advert.*, November 26, 1892.

Avon. Harvington. Wyre Piddle. Defford Brook.

Severn. Roadside between Mopsons Cross and Bewdley, *Westcombe*. Norton juxta Kempsey. Hawford. Wyre Forest. Lincombe. Holt. Dick Brook, Astley. Gladder Brook. Hindlip. Laughern Brook, Hallow.

Malvern. North of Malvern Church. Malvern Link. Leigh. Near Upton-on-Severn. Eastham.

This hybrid or variety has shortly stalked leaves, and the fruit never matures. There are several other hybrids known, some more nearly approaching *S. palustris* than this, which is found in all neighbouring counties.

1349. 781. S. sylvatica, Linn. *Hedge Woundwort.*

Native. Woods, ditches, hedges, shady places. General. P. May to July. *Top. Bot.* 112.

First Record. Not localized, *Lees, Bot. Malv. Hills*, 1st ed., p. 29 (1843).

This is a common plant in woods and hedges. The flowers are reddish purple, the lower lip variegated with white. It is sometimes found with white flowers, and this has been the case at Shrawley Wood and Monk Wood. It is recorded from all bordering counties.

1351. 782. S. arvensis, Linn. *Corn Woundwort.*

Native or colonist. Cultivated fields. Rather frequent. A. March to November. *Top. Bot.* 99.

First Record. Yardley Field, *Miss M. A. Beilby, Analyst*, vol. vi, p. 294 (1837).

This plant is not so frequently met with in the Lickey district as in other parts of the county. It is easily distinguished by its small size and weak stems and its small pale purple flowers four or six in a whorl. White-flowered forms have been observed in the neighbourhood of Worcester. It is recorded from all neighbouring counties.

GALEOPSIS, Linn. 341. (γαλέη, a weasel, ὄψις, aspect, from the appearance of the flowers.)

1353. 783. G. Ladanum, Linn. *Red Hemp-nettle.*

(*G. intermedia*, Vill.)

Colonist. Cultivated ground. General. A. June to August. *Top. Bot.* 3.

First Record. Not localized, *Walker, Med. Surg. Rep.*, vol. i, no. 2, p. 100 (1828).

Avon. Craycombe. Bredon Hill. Broadway. Railway-side near Oddingley. The Slads. Sheriff's Lench. Between Spetchley and Wyre Piddle. White Lady's Aston. Norton, near Evesham. Cornfields about Evesham.

Severn. Bredicot. Ladywood. Salwarpe. Kempsey. Ombersley. Stourport. High Oak. Grimley. Cotheridge. Hallow.

Malvern. Martley Lime Quarry. Between Bransford Bridge and

Leigh Church. Knightwick. Leigh. Malvern Link. Powick. Half Key. Brace's Leigh.

Lickey. General, *Mathews*.

No doubt all these records refer to the aggregate, the old *G. Ladanum*. Babington, *Manual*, 9th ed., p. 333, equates this plant with the next. Bentham, *Brit. Flor.*, 4th ed., p. 356, gives the next as a narrow-leaved form with a longer corolla-tube; while Hooker, *Stud. Flor.*, p. 327, makes it a sub-species. At all events the broad-leaved form of the aggregate, now with the name *G. Ladanum* reserved for it, has only very doubtfully been seen in Worcestershire, and the above records should be placed to the credit of the following plant. The same may no doubt be said of records from neighbouring counties, in all of which *G. Ladanum* has been recorded, while *G. angustifolia* appears to be recorded only from Salop.

1354. 784. G. angustifolia, Ehrh. *Red Hemp-nettle.*

Colonist. Cultivated ground. Little observed. June to October. *Top. Bot.* 71?

First Record. As below.

Malvern. Brace's Leigh. West Malvern; Buckman's near Malvern Link; Half Key, *Towndrow*.

The distinctions between this plant and the preceding one are somewhat fine. It differs chiefly in having narrower leaves connate at the base, approximate upper whorls, and the tube of the corolla much longer than the calyx. Of neighbouring counties it appears to have been observed only in Salop.

1356. 785. G. speciosa, Mill. *Large-flowered Hemp-nettle.*

(*G. versicolor*, Curt.)

Colonist. Cultivated fields and gardens. Rare. A. July and August. *Top. Bot.* 80.

First Record. Hurcott near Kidderminster, 1847, *Mr. G. Reece, in Herb. Nat. Hist. Soc., Lees, Bot. Worc.*, p. 50 (1867).

Avon. Broadway, *Lees*.

Severn. Hurcott, as first record. Barbourne.

Malvern. One plant in arable field, Barnard's Green, *Towndrow, Malv. Advert.*, November 26, 1898.

Lickey. Top of Romsley Hill, *Lees*. By the back Lodge of Clent House, Clent, 1875 and 1880, *Amphlett*. Dayhouse Farm, Romsley, 1858, *Mathews*. Barnt Green Railway Station, 1874, *Mathews*. Two Gates, Cradley, *Rev. J. H. Thompson*.

This showy plant cannot be mistaken or overlooked with its large yellow flowers each with a broad purple spot on the lower lip. In the herbarium, however, it becomes difficult to distinguish from the next species, which in everything but the corolla it much resembles, and with which by some botanists it has been united. It is much commoner in Scotland than in England. It appears to have been recorded from all neighbouring counties except Gloucestershire.

1357. 786. G. Tetrahit, Linn. *Common Hemp-nettle.*

Native. Cultivated and waste ground, hedges, wood-borders. Common and widely distributed. A. June to October. *Top. Bot.* 112.

First Record. On the Halesowen Road, *Ick, Analyst*, vol. vi, p. 32 (1837).

This is a common plant in the cornfields, with yellow, or yellow and purple, or sometimes white flowers. The Hemp-nettles are a tribe of plants in which even the mediaeval herbalists could discover no medicinal virtues. The plant is recorded from all bordering counties.

LEONURUS, Linn. 342*. (λέων, a lion, and οὐρά, a tail, from some fancied resemblance.)

{1358. 787. L. *Cardiaca, Linn. *Motherwort.*
Alien. Waste places and roadsides. Rare. P. August and September. Not in *Top. Bot.*
First Record. Near Malvern, *Purton, Midl. Fl.,* p. 284 (1817).
Severn. Near Bewdley, *Jorden.* Iverley Hills and adjoining fields, *Scott.* Worcestershire adjoins Iverley, which is in Staffordshire, and possibly this is a Worcestershire record. Wilden, near Stourport, *Miss Ladbury.* Old Brake Garden, Hagley, *Mathews.*
Malvern. As first record.
The true native home of this plant is not known; probably it is an Asiatic species. Everywhere it seems to depend more or less upon the work of man. The plant differs from any other of our labiates in the possession of palmately-lobed lower leaves, and it grows three or four feet high. Its specific name was given it because it was supposed to cure heart-burn, and also to give a good heart in another sense. It does not appear to have been noticed in Gloucestershire.}

LAMIUM, Linn. 343. (λαιμός, the throat, from the shape of the flower.)

1359. 788. L. amplexicaule, Linn. *Henbit Dead-nettle.*
Native. Cultivated ground, chiefly on sandy soil. Locally common. A. February to August. *Top. Bot.* 96.
First Record. Not localized, *Lees, Bot. Malv. Hills,* 1st ed., p. 29 (1843).
Avon. Craycombe; Bredon Hill; Broadway, *Lees.* Sheriff's Lench. Crowle.
Severn. Common about Kidderminster. Near Worcester, *Lees.* Hartlebury. Stourport. Lower Wick. Near Bevere. Claines. Bewdley. Kempsey.
Malvern. Near New Pool, Malvern Wells.
Lickey. General, *Mathews.*
This is a brighter looking plant than most of the Dead-nettles. The plant is some six inches high, and as the stem lengthens the floral leaves become more distant. It is recorded from all neighbouring counties.

1361. 789. L. hybridum, Vill. *Cut-leaved Dead-nettle.*
(*L. incisum,* Willd. *L. dissectum,* With.)
Colonist or denizen. Cultivated ground. Not common. A. April to August. *Top. Bot.* 76.
First Record. As *L. incisum.* Not localized, *Lees, Bot. Malv. Hills,* 1st ed., p. 29 (1843).
Avon. Pershore. Badsey.
Severn. Near Earl's Court, St. John's, rare, *Lees.* Lower Wick, Worcester. Crowneast. Ladywood. Kempsey. Between Hartlebury and Pansington, with white flowers.

Malvern. Powick. Bransford. Martley.
No district is assigned to this plant in Mr. Lees' *Table of Plants,* p. 22. Except for the more deeply crenate leaves, and the corolla-tube shorter than the calyx, naked within instead of with a faintly hairy ring, as in *L. molucellifolium,* Fries, a plant which has not been observed in Worcestershire, or a more pronounced ring, as in the next species, the plant is very difficult to distinguish from either of them. It has been recorded in Hereford, Warwick, and Stafford of the bordering counties.

1362. 790. L. purpureum, Linn. *Red Dead-nettle.*
Native. Cultivated ground, hedgesides, and a weed in gardens. Common everywhere. A. January to December. *Top. Bot.* 112.
First Record. Not localized, *Lees, Bot. Malv. Hills,* 1st ed., p. 29 (1843).
This truly common plant, dull individually, forms sometimes when it grows abundantly fine masses of colour, the floral leaves being nearly as red as its flowers. White-flowered specimens have been observed at Lower Wick, Worcester, near Claines Church, and at Hagley. It occurs in all neighbouring counties as abundantly as in our own.
Var. b. decipiens, Sonder.
First Record. As here.
Severn. Hartlebury.
In this variety the bracts and leaves are deeply crenate.

1363. 791. L. maculatum, Linn. *Spotted Dead-nettle.*
Alien. Hedgebanks. An escape from gardens. P. June to August. Not in *Top. Bot.*
First Record. Found near Defford Common by Dr. Streeten, *Lees, Illus. Nat. Hist. Worc.,* p. 168 (1834).
This has no claim to be native. Where it is native is in woods from Southern Belgium to Persia. It is never found in woods in our country; but on the other hand is very often to be seen in the garden border, from whence it makes its escape; and this it appears to have done in every neighbouring county except Gloucestershire. In some parts of Worcestershire it is known as All Heal.

1364. 792. L. *album, Linn. *White Dead-nettle.*
Native. Waysides, waste places, borders of fields, hedgebanks. Plentiful. *Top. Bot.* 102.
First Record. Not localized, *Lees, Bot. Malv. Hills,* 1st ed., p. 29 (1843).
This plant everybody knows. It has come under suspicion as not being truly native, a matter which is interestingly discussed in *J. of B.,* 1902, p. 360. It is recorded from all neighbouring counties.

1365. 793. L. Galeobdolon, Crantz. *Yellow Archangel, Yellow Weasel-snout.*
(*Galeopsis Galeobdolon,* Linn. *Galeobdolon luteum,* Huds.)
Native. Woods, hedges, and the borders of woods. Locally common. P. April to June. *Top. Bot.* 67.
First Record. As *Galeobdolon luteum.* Hedges near Malvern Chace, *Ballard;* Woods near Worcester, *Stokes; Stokes's With. Bot. Arr.,* 2nd ed., p. 611 (1787).
This plant is common enough in suitable situations, but it loves the woodland and is never seen far away from shade. The specific name is from

γαλέη, a weasel, and βδόλος, a bad smell. It is a European plant, not being found outside the limits of the Continent, except in Western Siberia. The tribe of Dead-nettles has been comparatively free from the attacks of the herbalists. The plant occurs in all bordering counties.

BALLOTA, Linn. 344. (βάλλω, I cast away, from its bad smell.)

1366. 794. B. nigra, Linn. *Black Horehound.*
Native. Hedges, waste places, by footpaths and roadsides. P. June to September. *Top. Bot.* 77.
First Record. Hedges in the Vale of Evesham, *Marshall, Pitt, Agric. Worc.,* p. 317 (1810).
A dull-looking plant this, the wrinkled foliage naturally grey-green, and made greyer when growing by the roadside, as it commonly does, by the dust arising from it. The whole plant possesses a disagreeable odour. It is never seen far from human habitation, and where humans go to dwell, if not under a tropic sun, the plant follows them. Specimens with white flowers have been collected at Great Witley and Comer Lane, Worcester, and at Pixham. It is recorded from every neighbouring county.

1367. 795. B. ruderalis, Sw. *Black Horehound.*
First Record. As below.
Severn. Lower Wick, 1888, *Towndrow.*
The calyx teeth of this plant, sometimes considered only a variety of the above, are ovate, gradually acuminate, and erect. Those of the type are broadly ovate, short, and patent or reflexed. It is a very hairy and soft plant, and is a common form in Worcestershire. It has been observed in Herefordshire and Salop.

TEUCRIUM, Linn. 345. (From Teucer, King of Troy, the first to use the plant medicinally.)

1371. 796. T. Scorodonia, Linn. *Wood Germander.*
Native. Dry open woods, commons, banks. General. P. May to September. *Top. Bot.* 110.
First Record. On hedgebanks, *Pitt, Agric. Worc.,* p. 317 (1810).
The Wood Germander, or Wood Sage, is as frequently seen out of the wood as in it. The wrinkled leaves are not unlike those of the garden sage, but a brighter green; the flowers are yellowish-green; and the whole plant smells strongly of hops. Our forefathers called the plant Ambrose, not from the saint of that name, but because they thought it as valuable as the fabled food of the gods, and used it in several ways for medicinal purposes. It is recorded from all neighbouring counties.

AJUGA, Linn. 346. (ἄζυξ, unyoked, in allusion to its supposed properties.)

1372. 797. A. reptans, Linn. *Common Bugle.*
Native. Woods, thickets, wet pastures, damp places. Widely distributed. P. May to July. *Top. Bot.* 109.
First Record. Wollaston Rocks (Stourbridge), *Scott, Hist. Stourb.,* p. 540 (1832).

This plant is very common in damp meadows and woodland rides, varying a good deal in colour, from the normal deep blue to pale lilac or white. With white flowers it has been seen at Ockeridge Wood; Himbleton; Round Hill Wood, Hagley; and several other places. Our ancestors thought highly of it as a remedy for pulmonary affections and as an application for wounds. The derivation of the English name is not known, but it has no connexion with any other word that is spelled in the same way. The plant is recorded from all bordering counties.

60. PLANTAGINACEAE, Juss.

PLANTAGO, Linn. 347. (From *planta,* a plant; but why applied to this genus especially is not known.)

1375. 798. P. major, Linn. *Greater Plantain.*
Native. Pastures, waysides, waste places. Common and generally distributed. P. May to September. *Top. Bot.* 112.
First Record. Roadsides, *Pitt, Agric. Worc.,* p. 317 (1810).
This well-known plant was highly esteemed by our ancestors for its many virtues. Its old name, Way-bread, had no relation to the staff of life, but is a corruption of the Anglo-Saxon Wabron or Wabret. It is also called May Bread, Hard Heads, and Soldiers in some parts of Worcestershire. The juice of the plant was considered a remedy for various ills; the seeds are used as food for caged birds, the long spikes being twined in and out the wires of their cages. The leaves were universally used for application to wounds. It is recorded from all neighbouring counties.
Var. b. intermedia, Gilib.
First Record. As below.
Malvern. In the district, *Towndrow, Malv. Advert.,* Jan. 20, 1894.
In this variety the leaves are downy and coarsely dentate, and the scapes are also downy.

1376. 799. P. media, Linn. *Hoary Plantain.*
Native. Dry meadows, pastures, waysides. General. P. May to October. *Top. Bot.* 82.
First Record. Var. 2. Leaves with straw-coloured stripes. Hawford Bridge in Worcestershire, *Stokes, Stokes's With. Bot. Arr.,* 2nd ed., p. 143 (1787).
This is a common plant on lawns, where the sessile leaves, lying close to the earth, form a close cluster round the root, so that when removed they leave a bare spot. It is said not to be indigenous in Scotland. It is recorded from every neighbouring county.

1377. 800. P. lanceolata, Linn. *Ribwort Plantain.*
Native. Roadsides, pastures, soil cultivated and waste. Ubiquitous. P. April to July. *Top. Bot.* 112.
First Record. In pastures, *Pitt, Agric. Worc.,* p. 317 (1810).
The dark-brown flower-heads of this plant have received several country names, Cocks-and-Hens and Fighters among them, the last one originating from a children's game, where one flower-head is held out to be beheaded

by another. It was also called Hound's Tongue, Fire-grass, and Rib-grass, and it is not unwelcome among the grass of the permanent pasture, if not too abundant. Extreme fasciated forms of this plant were observed by Mr. D. Willis at Hallow, 1908, as many as thirty or forty heads being collected together without much enlargement of the stalk. It is recorded from every neighbouring county.

Var. b. **Timbali**, Reichb. fil.
First Record. In clover fields. An alien, *Towndrow, Malv. Advertiser*, Nov. 26, 1892.
Avon. South Littleton.
Malvern. Malvern Link, *Towndrow*.
The bracts and sepals of this variety have broad silvery margins. It appears to have been observed only in Warwickshire of our bordering counties.

{*1378*. 801. P. maritima, Linn.   *Seaside Plantain.*
Denizen, if not an error. Seashores and mountains. P. June to September. *Top. Bot.* 78.
First Record. Flourishes at Stourbridge, *Scott, Hist. Stourb.*, p. 540 (1832).
Severn. As first record. Hartlebury Common, *Reece*, in the *Phytologist* as below.
The first is certainly an unlikely record ; Mr. Scott has a footnote about it, *Hist. Stourb.*, p. 553 ; the second record occurs in *Phyt.*, vol. iii, pt. ii (1849), p. 13, in a list by Professor Buckman of the plants now growing in the valley of the Severn, where the Professor says the plant is ' recorded in the *Phytologist* as having been found on Hartlebury Common by Mr. Reece of the Worcester Museum '. Mr. Mathews, *Mid. Nat.* xiii. 164, remarks, ' No reference (is given) to the volume and page of *Phytologist* where the record occurs. I have failed to find the record and suspect an error.' No reference to the plant is given in the Bot. Section, *Bibliog. Worc., Worc. Hist. Soc. Pubs.*, 1907. No specimen now has this name in the local collections. Possibly it has been corrected to *P. Coronopus*. Mr. Lees, in his *Table of Plants*, marks it for one place, lost, in the Severn district. At all events the plant is now extinct. So many remarkable plants have been seen on Hartlebury Common that the record is not at first sight impossible. In *Top. Bot.*, 2nd ed., p. 342, the plant is dealt with curtly, and the remark is made that it occurs in some few of the exclusively inland counties without mentioning which they were. It is said to have been observed in Gloucestershire.}

1379. 802. P. Coronopus, Linn.   *Buck's-horn Plantain.*
Native. Dry sandy pastures, commons, roadsides. Locally common. A. May to September. *Top. Bot.* 96.
First Record. On the side of the Bromsgrove Road between Crab's Cross and Headley's Cross, *Purton, Midl. Fl.*, p. 92 (1817).
Avon. As first record.
Severn. Hartlebury Common, *Rea.* Giant's Grave, Habberley, *Lees.* Between Blakedown and Kidderminster, *Mathews.* Lane above Hagley Station, *W. Whitwell.* Kidderminster Common. Kempsey Common.
Malvern. Barnard's Green ; Walks at the Wells ; Common east of the Firs. Near Powick, on the Rhydd road, *Westcombe.* Malvern Common. Malvern Link Common.

Lickey. Moseley Wake Green, *Miss A. Beilby.*
While in everything else in appearance like the other plantains, this one is easily distinguished from them by its cut leaves. Mr. Whitwell submitted his plant, which grows very abundantly in the locality, to Mr. E. G. Baker, who identified it as *f. tenuifolius*, Wirtgen. Buck's-horn Plantain is recorded from all neighbouring counties.

P. arenaria, Waldst. and Kit.
First Record.
Malvern. Clover-field at Leigh, 1895, *Miss B. Norbury.*
This alien plant is a native of sandy pastures in the Mediterranean region, and occurs as a weed, usually of waste sandy ground, throughout the rest of Europe. It is not in the *Lond. Cat.*, 10th ed.

LITTORELLA, Berg. 348.   (*Littus*, the shore, where it grows.)
1380. 803. L. uniflora, Aschers.   *Plantain Shore-weed.*
(*L. lacustris*, Linn.   *L. juncea*, Berg.)
Native. Margins of pools. Very local. P. June to August. *Top. Bot.* 94.
First Record. Marked as ' lost ' in the Lickey district, *Lees, Table of Plants Bot. Worc.*, p. 23 (1867).
Malvern. Newland Green.
Lickey. In some abundance by the side of a pool at Moseley, *W. Southall, Worc. Nat. Club Trans.* i. 217 (1874). Bittell Reservoir, *H. S. Thompson*, 1893.
This plant is given by Scott in his *Hist. Stourb.*, p. 60, from a neighbouring locality in Staffordshire. The plant does not appear to have been recorded from Gloucestershire.

61. ILLECEBRACEAE, Benth. and Hook.

SCLERANTHUS, Linn. 352.   (σκληρός, hard, ἄνθος, a flower, from the indurated perianth.)
1386. 804. S. annuus, Linn.   *Annual Knawel.*
Native. Sandy cornfields and heathy ground. Locally abundant. A. or B. March to September. *Top. Bot.* 100.
First Record. On the rocks of Malvern Hills, *Lees, Illus. Nat. Hist. Worc.*, p. 163 (1834).
Avon. Occasional, *Lees.*
Severn. Hartlebury Common. Near Bewdley. Stone Quarry, Holt. Park Hall. Harberrow. Holt Bank.
Malvern. Between the North Hill and Worcestershire Beacon. Arable land, Malvern Link.
Lickey. Clent. Barnt Green. Blackwell.
This is a prostrate little plant, very inconspicuous, with small green flowers, easily passed over by any one except a botanist. It possibly never possessed any qualities which would lead to its having a popular name, and ' knawel ' is not much more than a book-name, from the German *knawel*, a ball of yarn. ' The base Almaignes cal it knawel, that is to say, knotweed,' says Lyte, in 1578. The plant is recorded from all neighbouring counties.

62. AMARANTHACEAE, Juss.

AMARANTHUS, Linn. 353.*   (ἀ, not, μαραίνω, I fade ; flowers which do not fade. The ' h ' has got in probably from some confusion about ἄνθος, a flower.)
{*1389*. 805. A. Blitum, Linn.   *Wild Amaranth.*
Alien. Waste ground. Once recorded. A. August. Not in *Top. Bot.*
First and only Record. Near Bewdley, *Jorden, Lees, Bot. Worc.*, p. 6 (1867).
This plant is a native of the Orient, a rare casual in England, commoner on the Continent. To this tribe belong the Cock's Combs, Globe Amaranths, and Prince's Feathers of our greenhouses, of which gay company Blite, a name which it used to share with several plants, including garden spinach, is a very humble relative. Nor has it any qualities which make up for its plain aspect. ' Bleets seeme to be dull, vnsauorie and foolish Woorts, hauing no tast nor quickenesse at all,' says Holland, 1601. The plant has been noticed in Salop and Warwickshire.}

A. hybridus, Linn.
First Record. As below.
Malvern. Hopyard at Leigh Sinton, 1900, *Towndrow.*
This plant is not in *Lond. Cat.*, 10th ed.

63. CHENOPODIACEAE, Ventn.

CHENOPODIUM, Linn. 354.   (χήν, a goose, πούς, a foot, from the shape of the leaves of some species.)
1390. 806. C. polyspermum, Linn.   *Many-seeded Goosefoot.*
Native. Cultivated fields, waste ground, hedges, and borders of fields. Local, but in places occurring plentifully. A. June to September. *Top. Bot.* 49.
First Record. Not localized, *Lees' Cat., New Bot. Guide* (1835).
Avon. Norton, near Evesham. Sheriff's Lench. Crowle.
Severn. Claines. Henwick Mill. Bewdley. Dunghill, Droitwich.
Malvern. Uncommon, *Lees.*
Lickey. One place, *Lees.* Lickey Hill.
This plant is remarkable for its seeds, which are very numerous, dark-brown, and shining, in part only enveloped by the perianth. The leaves are sometimes tinged with red. It is not recorded for Staffordshire of our neighbouring counties, but has been observed therein.

Var. b. cymosum, Moq.
First Record. Malvern, *Towndrow, Malv. Advert.*, Nov. 26, 1892.
Severn. Near Pershore.
Malvern. As first record.

Var. spicatum, Moq.
First Record. Malvern, *Towndrow, Malv. Advert.*, Nov. 26, 1892.
Malvern. As first record.
This variety is not given in *Lond. Cat.*, 10th ed.

1391. 807. C. Vulvaria, Linn.   *Stinking Goosefoot.*
(*C. olidum*, Curtis.)
Native. Rubbish heaps, garden ground. Very rare. A. July to September. *Top. Bot.* 37.
First Record. Overend, Cradley, *Rev. J. H. Thompson, Mathews, Clent and Lickey Hills*, p. 39 (1881).
Severn. Bewdley.
Malvern. Rubbish heap, Madresfield, *Towndrow.*
Lickey. As above. Lyde Green. Cradley.
Mr. Thompson's plant was gathered in 1870, and was a small specimen. The foliage of this plant feels greasy to the touch, and is covered with a powdery substance, which smells strongly of bad fish. It is not recorded from any of the bordering counties, but has been observed in Herefordshire.

1392. 808. C. album, Linn.   *White Goosefoot.*
Native. Waysides, cornfields, garden ground, manure heaps. Abundant and generally distributed. A. June to October. *Top. Bot.* 112.
First Record. Not localized, *Lees, Bot. Malv. Hills*, 1st ed., p. 20 (1843).
This plant is the commonest of all the Goosefoots, and the stem is sometimes three feet high, with clustering spikes of greenish flowers. It was formerly used as a vegetable, and a common name for it is Fat Hen. It occurs in all bordering counties.

Var. b. viride, Syme.
First Record. Not localized, *Lees, Bot. Malv. Hills*, 2nd ed., p. 42 (1852).
Malvern. Madresfield, *Towndrow.*
This variety has nearly entire leaves, and the calyx is almost glabrous. It is not mentioned in Mr. Lees' *Bot. Worc.* It has been observed in Herefordshire, Salop, and Warwickshire.

Var. incanum, Moq.
(*C. candicans*, Lamk.)
First Record. Malvern, *Towndrow, Malv. Advert.*, 1892.
Severn. Kempsey.
Malvern. Common, *Towndrow.*
The leaves of this variety are much toothed and the calyx is very mealy. It does not appear to have been noticed in any bordering county. The variety is not given in *Lond. Cat.*, 10th ed.

Var. viridescens, St. Am.
(*C. paganum*, Reichb.)
First Record. Malvern, *Towndrow, Malv. Advert.*, Nov. 26, 1892.
Malvern. Common, *Towndrow.*
In this plant the leaves are sinuate-dentate, and the calyx is only sparingly mealy. It has not been observed in any neighbouring county. It is not given in *Lond. Cat.*, 10th ed.

1394. 809. C. serotinum, Linn.   *Fig-leaved Goosefoot.*
(*C. ficifolium*, Sm.)
Native. Waste and cultivated ground. Once recorded. A. July to September. *Top. Bot.* 18.
First Record. Not localized, *Lees, Bot. Worc.*, p. xlviii (1867).

The distinctions of the Goosefoots depend largely on the characters of the seeds, coupled with the shape of the leaves. In this plant the leaves are three-lobed and shaped somewhat as one of the specific names and the book-name indicate. It has been noticed in Salop and Warwick.

1395. 810. C. murale, Linn.　*Nettle-leaved Goosefoot.*
Native. Waste and cultivated ground, rubbish heaps. One record only.
A. July to September. *Top. Bot.* 42.
First Record. This grows about Worcester, *Lees, Bot. Worc.,* p. 51 (1867).
　Severn. Worcester, *Lees.* The Moors, Worcester. Weed in garden, Saint George's Square, Worcester, *Westcombe.*
　Malvern. Interfields, *Towndrow, Malv. Advert.,* August 26, 1892.
The leaves of this plant are unequally and sharply dentate, and the seeds are minutely granular, and keeled at the edge. It has been observed in Hereford, Salop, and Warwick.

1396. 811. C. hybridum, Linn.　*Maple-leaved Goosefoot.*
Native. Waste and cultivated ground. Very rare. A. August to October. *Top. Bot.* 25.
First Record. Not localized, *Lees' Cat., New Bot. Guide,* 1835.
　Avon. Near Evesham, 1850, *Westcombe.*
This plant has large seeds, minutely pitted, but not keeled at the edge. It has been observed in Gloucestershire and Warwickshire.

1397. 812. C. urbicum, Linn.　*Upright Goosefoot.*
Native. Rubbish heaps. Very rare. A. August to September. *Top. Bot.* 40.
First Record. Not localized, *Lees' Cat., New Bot. Guide* (1835).
　Severn. Bury Hill, Droitwich, 1865, *Lees.* Ladywood. Helbury Hill.
　Malvern. Rubbishy places, not abiding, *Lees.*
The seeds of this plant are nearly as large as Rape-seed. They are very minutely rough, and blunt at the edge. It is recorded from every neighbouring county.

Var. b. intermedium, Moq.
First Record. As below.
　Severn. Gregory's Mill, sp., *Herb. Hast. Mus. Worc.*
In this variety the leaves are deeply sinuately toothed, the spikes are shorter than the leaves, and the panicle is leafy almost to the top.

1398. 813. C. rubrum, Linn.　*Red Goosefoot.*
Native. Cultivated ground. Rather frequent. A. July to September. *Top. Bot.* 66.
First Record. Roadsides, *Pitt, Agric. Worc.,* p. 317 (1810).
　Avon. Pirton Pool. Near Craycombe. Pershore. Pinvin. Stoulton.
　Severn. About Droitwich. Bewdley. Near Gregory's Mill. Dung-hill south of Worcester. Upper Hop-yard, Lower Wick, Worcester. Ombersley. Berwick's Brake. Hartlebury. Lincombe.
　Malvern. Barnard's Green ; Snead's Green ; Mathon, *Lees.* Madresfield. Malvern. Powick. Kyre. Bransford. Leigh.
　Lickey. Halesowen, *Bagnall.*

The flowers of this plant are in dense leafy spikes, and the seeds are very minute and shining, blunt or slightly keeled at the edge. The next plant, which grows as a weed in Hagley Hall gardens, has frequently been mistaken for this. *C. rubrum* is recorded from Gloucester, Hereford, and Warwick, and has been observed in Staffordshire.

1399. 814. C. botryodes, Sm.
(*C. botrys,* Linn.)
Colonist. In Hagley Hall gardens only, till 1907. A. August and September. *Top. Bot.* 8.
First Record. Not localized, *Lees, Bot. Malv. Hills,* 1st ed., p. 20 (1843).
　Severn. Hagley Hall gardens, a weed. Amongst potatoes, Ombersley, 1907.
In the first place some botanists consider this plant a variety of the former one ; in the second place it is not a native of England, being at home on sea-shores and river-banks in Central and Southern Europe; and thirdly the Worcestershire specimens have by several persons been referred to and recorded as *C. rubrum,* because, no doubt, the flowers are red. Irvine, however, in his *Handbook on British Plants,* 1857, rightly diagnosed it, and at p. xxx of *Bot. Worc.* Mr. Lees quotes from him as follows, under *C. Botrys*: ' Comes up every year spontaneously (as I was told, when the plant was shown me) in the gardens of Lord Lyttleton, Hagley.' And it is still as abundant as ever in the same locality. Mr. Lees does not localize the plant in his *Table of Plants.* It differs from *C. rubrum* in its minute shining seeds being keeled at the top, and its spikes being leafless at the top ; but chiefly on account of the extreme succulence of the flower and subsequently of the floral receptacle for the seed, strongly suggestive of the common Strawberry and as difficult as that is to press for a herbarium example of the fruit. It is certainly a most distinct species and no variety of any British Chenopodium. It is not recorded for any neighbouring county, and as regards several of the records which are given in *Top. Bot.,* Mr. Watson says that it is probable that var. *pseudo-botryoides* of *C. rubrum* has been mistaken for it. The plant is not starred as an alien in *Lond. Cat.,* 10th ed. It is not recorded from any neighbouring county.

1401. 815. C. *Bonus-Henricus, Linn.　*Mercury Goosefoot or Good King Henry.*
Native. Waysides near villages. Not common. P. May to September. *Top. Bot.* 100.
First Record. Banks on roadside near Bromsgrove, *Pitt, Agric. Worc.,* p. 317 (1810).
　Avon. Himbleton, *Mathews.* Near Fladbury Church. Roadside near Mere Hall.
　Severn. Tibberton Churchyard, *Lees.* Turquay, Worcester, *Lees.* Near Gregory's Mill, Worcester. Railway, Worcester. Camp Lock, *Jeffery.* Upton Warren.
　Malvern. Church Street, Malvern. Near Little Malvern. Malvern Common.
　Lickey. On Clent Hill in two places, far apart, one over 1,000 feet above the sea, *Amphlett.* Cradley Forge, *Mathews.* Halesowen.

X

This is a dull-looking, dark-green plant, and in old times it was cultivated in gardens and boiled like spinach. Usually it is found near the habitations of man, past or present. On Clent Hill it grows on field borders, and is always cropped close by sheep when they are turned into the fields. It has several popular names, All Good, one of them. It is recorded from every neighbouring county.

BETA, Linn. 355.　(Said to be derived from either the Celtic *bwyd,* food or nourishment, or the Celtic *bett,* red.)

{*1402.* 816. B. maritima, Linn.　*Common Beet.*
Alien. Waste places. On the marl cutting at Shrub Hill Station, Worcester. P. August and September. *Top. Bot.* 8.
First Record. On the railway embankment at Shrub Hill, Worcester. Plentiful for two years, *Lees, Bot. Worc.,* p. xxx (1867).
　Severn. As first record.
This plant is the uncultivated form of *B. vulgaris,* Linn., the garden beet ; they are the same plant. In *Bot. Worc.,* p. 21, Mr. Lees gives an account of the circumstances under which the plant was found. With it was some amount of *Atriplex hortensis,* and on its appearance Mr. Lees wrote a paper in the *Phytologist,* vol. iii, p. 1050. The plant is a native of muddy seashores both of Europe and North America, and besides being the parent stock of garden beet, is probably that of mangold wurzel also. It has not been recorded from any bordering county.}

ATRIPLEX, Linn. 356.　(ά, not, τρέφω, I nourish.)

1404. 817. A. patula, Linn.　*Spreading halberd-leaved Orache.*
(*A. angustifolia,* Sm., and *A. erecta,* Huds., are now usually included under *A. patula,* Linn., of which they are made to form varieties.)
Native. Cultivated ground and waste places. Common and widely distributed. A. July to October. *Top. Bot.* 93.
First Record. Roadsides and upon rubbish, *Pitt, Agric. Worc.,* p. 317 (1810).
There is some misunderstanding in the books as to the Oraches, and it is not easy to unravel the records. For the purposes of the present book we have decided to follow the definition of the species as set out in Hooker's *Students' Flora.* We thus regard the species *A. patula* as those plants having leaves dilated at the base but with decurrent petioles, the lower leaves oblong-lanceolate, the upper linear-pointed, and the whole plant almost glabrous ; whereas in *A. hastata* all the leaves are petioled, the lower and middle leaves hastate and truncate at the base, the upper ones being lanceolate and entire, and the plant mealy. The present plant is an inhabitant of nearly every stubble field in one variety or another, and in one variety or another occurs in all bordering counties.

Var. b. erecta, Huds.
First Record. Not localized, *Lees, Bot. Malv. Hills,* 2nd ed., p. 75 (1852); *Mid. Nat.* xiii. 186.
　Severn. Bromwich Lane, Worcester, *Westcombe.* Helbury Hill. Near Gregory's Mill. Near Stakenbridge.
　Malvern. Arable land, common, *Towndrow* (MS.).
In this variety the stem is erect or ascending, the lower leaves deltoid

serrate, the panicle lax, the spikes short dense, sepals denticulate, usually muricate on the back.

Var. c. angustifolia, Sm.
First Record. Not localized, *Lees, Bot. Malv. Hills,* 1st ed., p. 42 (1843).
　Avon. Norton near Evesham. Sheriff's Lench. Crowle. Pinvin. Bishampton.
　Severn. Kempsey. Helbury Hill. Near Diglis Lock. Merrymans Hill, Kempsey. Claines. Ombersley.
　Malvern. Ladywood, 1888, *Towndrow* (MS.).
In this variety, which is one of the commonest forms in the neighbourhood of Worcester, the stem is weak, procumbent, or ascending, the branches divaricate, leaves sub-entire, the spikes long and lax-panicled, and the sepals usually smooth.

1405. 818. A. hastata, Linn.　*Hastate-leaved Orache.*
Native. Waste ground, rubbish heaps, garden ground. Not uncommon. A. July to September. *Top. Bot.* 60.
First Record. As *A. eu-hastata.* Malvern Link, *Towndrow, Malv. Advert.,* November 26, 1892.
This plant in Hooker's *Stud. Fl.,* 3rd ed., p. 340, is made into a sub-species of *A. patula,* and for the leading distinctions between this species and *A. patula* see our remarks under that species. *A. eu-hastata,* Huds., has the upper leaves not hastate, the spikes lax, terminal of the panicle long, sepals rhombic, much longer than the utricle, and most of the seeds large. It is recorded from all bordering counties.

1406. 819. A. deltoidea, Bab.　*Triangular-leaved Orache.*
Native. Sides of water, damp waste ground. Probably not uncommon. A. August to October. *Top. Bot.* 54.
First Record. Not localized, *Lees, Bot. Malv. Hills,* 2nd ed., p. 75 (1852). *Mid. Nat.* xiii. 186.
　Severn. Brickfields Farm. Side of road to Malvern. Near Britannia Square, Worcester. Chatley. Hadley. Ladywood. Ombersley. Kempsey.
　Malvern. Near Powick. Upton-on-Severn. Bransford.
This plant is the typical *A. hastata* species in this county, and differs from the last in the following features. The upper leaves are usually hastate, the spikes dense, terminal of the panicle short, the sepals truncate but little longer than the utricle, seeds mostly small. It appears to have been observed in Hereford, Salop, and Warwick.

Var. microsperma, W. and K.
First Record. Not localized, *Lees, Bot. Malv. Hills,* 2nd ed., p. 75 (1852).
　Severn. Merrymans Hill, Worcester. Diglis. Near Droitwich.
In this variety the leaves are all opposite, and the valves of the fruit are oval instead of triangular, as in the type *A. deltoidea.*

1407. 820. A. Babingtonii, Woods.　*Spreading-fruited Orache.*
(*A. rosea,* Bab.)
Native. A plant of the seashore. Very rare. A. July to September. *Top. Bot.* 70.

First Record, doubtfully. In a grassy meadow, in a circumscribed bit of boggy ground, the oozy birth-place of a sluggish salt spring, at Saldon, *Lees, Bot. Worc.*, p. 33 (1867).

Avon. Saldon, as above.

Severn. Droitwich Canal, 1904, *Worc. Nat. Club Trans.* iii. 183, and throughout its length.

This plant, in Hooker's *Stud. Fl.*, p. 340, made a sub-species of *A. patula*, is characterized by being usually pale and somewhat mealy, prostrate, branches spreading ascending, leaves mostly opposite, deltoid or rhombic-ovate, entire or sinuate-toothed, upper usually similar, cluster of flowers remote, spikes simple, lax, leafy, sepals connate at the often hardened base or united nearly to the middle, seeds all vertical, large, pale, and rather rough. In *Top. Bot.*, 2nd ed., p. 348, it is recorded only from Gloucester-shire and bordering counties, Worcestershire not being mentioned.

{**A.** HORTENSIS, Linn. *Garden Orache.*

Alien. Gardens and railway-banks. Once recorded before 1906. A. June to August.

First Record. On the railway embankment at Shrub Hill Station, Worcester, with *Beta maritima, Lees, Bot. Worc.*, p. xxx (1867).

Severn. As first record. Ombersley, 1906.

This plant is probably a native of Turkestan, much cultivated in gardens throughout Europe. Mr. Lees does not mention it in his *Table of Plants*, nor of course is it given in *Top. Bot.*}

64. POLYGONACEAE, Juss.

POLYGONUM, Linn. 360. (πολύς, many, γόνυ, a knee, from the many-jointed stems.)

1423. **821. P. Convolvulus,** Linn. *Climbing Bindweed. Black Bindweed.*
Native. Cornfields, garden ground, hedgebanks. Generally distributed. A. June to October. *Top. Bot.* 111.

First Record. Not localized, *Lees, Bot. Malv. Hills*, 1st ed., p. 21.

This is a common plant, especially in cornfields, where it twines round the stems of the wheat, binding them together in masses, sometimes bearing them down with its weight. It is recorded from all neighbouring counties.

Var. b. **subalatum,** V. Hall.

(Var. *pseudo-dumetorum*, Wall.)

First Record. Churchill (Kidderminster), with wires six feet long, *Mathews, Clent and Lickey Hills*, p. 41.

Severn. As first record. Potato ground near Stourbridge Station.

Malvern. Arable land, common, *Towndrow, Malv. Advert.*, December 3, 1892.

This plant is often mistaken for the next species, as no doubt is the case in a record of Mr. Lees' for the following plant, *Bot. Malv. Hills*, 1st ed., p. 21 (1843). To this record Mr. Mathews appends the note, *Mid. Nat.* xii. 185, 'In the 3rd edition this name is queried. Mr. Lees adds "I found either this or a very tall variety of the preceding in a hedge bounding Sarn Hill Wood, Bushley, some years ago, when residing at Forthampton." The true *P. dumetorum* is a plant of the south of England.' Of this variety the seg-

---

ments of the perianth are winged, as in *P. dumetorum*, the nut is opaque, striated with minute points; in that plant the nuts are very smooth and shining. This variety has been observed in Salop.

1424. **822. P. dumetorum,** Linn. *Copse Buckwheat.*
Native. Hedges and borders of woods. Very local. A. July to September. *Top. Bot.* 14.

First certain Record. Hedge in lane between Wannerton and Churchill (Kidderminster), *Mathews, Worc. Nat. Club Trans.* ii. 96 (1899).

Severn. As first record.

Mr. Lees' earlier record of this is mentioned under the last plant, with Mr. Mathews's comments upon it. It is not recorded from any neighbouring county.

1425. **823. P. aviculare,** Linn. *Common Knot-grass.*
Native. Cultivated and waste ground, roadsides, field-paths, gardens. Abundant and universally distributed. A. or B. June to October. *Top. Bot.* 111.

First Record. Not localized, *Lees, Bot. Malv. Hills*, 1st ed., p. 21 (1843).

This is one of the most frequent of our wild flowers. The stems are dark green, furnished with chaffy stipules which, if we wish to be scientific, we can call *ocreae*, bearing all through the summer numerous small pinky-white flowers. A decoction of it was supposed to prevent growth, both in man and animals. It is a favourite with sheep, nevertheless, and it does not affect them at all events in this way. Its specific name refers to the fact that birds feed upon the abundant seeds. It is known by the names of Arsesmart and Pig Grass in many parts of Worcestershire. It is recorded from all neighbouring counties.

Var. a. **agrestinum,** Jord.

First Record. Malvern, *Towndrow, Malv. Advert.*, December 3, 1892.

Severn. Arable fields near Worcester, *Rea.* Severn-side near Worcester. Worcester and Birmingham Canal.

Malvern. As first record.

The common upright robust field form. It has apparently been observed only in Warwickshire.

Var. c. **arenastrum,** Bor.

First Record, doubtful. A variety closely approximating to *P. maritimum* was brought to me by Mr. T. Lewis, a member of the Club, from the soil thrown up at the Diglis cutting, *Lees, Bot. Worc.*, p. 52 (1867). First certain Record. Malvern, *Towndrow, Malv. Advert.*, December 3, 1892.

Severn. Cornfield, Shrawley.

Malvern. *Towndrow*, as record above.

Lickey. Base of the Bilberry Hills, *Mathews.*

This variety is a sand-loving, prostrate form. It has been seen in Stafford-shire and Warwickshire.

Var. e. **rurivagum,** Jord.

First Record, doubtful. Malvern, *Towndrow, Malv. Advert.*, December 3, 1892.

Severn. Brickyard, Claines. Cornfield, Shrawley. Tagwell Green, near Droitwich.

Malvern. As first record.

This is a wayside form, upright, with narrow, very acute leaves. It has been observed in Staffordshire and Warwickshire.

1426. **824. P. Raii,** Bab. *Robert's Knot-grass.*
(*P. Roberti,* Loisel.)
Native. Usually on sandy seashores. Once recorded. A. or P. August and September. *Top. Bot.* 39.

First and only Record. As *P. Roberti.* On a little green or waste spot in a lane near St. Peter's, Droitwich, *Lees, Bot. Worc.*, p. 37.

Severn. As first record.

A plant, says Babington, whose plant it is, resembling *P. aviculare* in habit, but *P. maritimum* in fruit. That is, the fruit is smooth and shining instead of being minutely striate. It has not been observed in any neighbouring county.

1428. **825. P. Hydropiper,** Linn. *Biting Persicaria.*
Native. Wet places, ditches, stream-sides, damp places in woods. Common throughout the county. A. June to October. *Top. Bot.* 105.

First Record. Malvern Chace, *Stokes's With. Bot. Arr.*, 2nd ed., p. 408 (1787).

This plant was formerly held in great repute for its medicinal virtues; its juice is hot and acrid, but not unpleasant to the taste. It was said to be particularly obnoxious to insects, and while it would protect clothing from moths, it would also put to flight the domestic flea. It is recorded from all bordering counties.

1429. **826. P. minus,** Huds. *Small creeping Persicaria.*
Native. Wet meadows, marshy ground. Rare. A. July to October. *Top. Bot.* 52.

First Record. Gravel pit on Malvern Chace, *Stokes's With. Bot. Arr.*, 2nd ed., p. 410 (1787).

Avon. The Marsh Common.

Severn. Ladywood, Salwarpe. Wyre Forest.

Malvern. Barnard's Green; Welland and other Commons, *Lees.* Malvern Common, *Westcombe.*

A small and rather elegant creeping plant, nearly allied to the last species. The fruit is pitchy black and shining, whereas in the former it is punctate. It is recorded only from Staffordshire by old authorities, for Mr. Bagnall has never seen it himself (*MS.*), and Herefordshire of bordering counties, but it has been observed also in Warwickshire.

1430. **827. P. mite,** Schrank. *Lax-flowered Persicaria.*
Native. Sides of water in low-lying situations. Very rare. A. June to September, *Top. Bot.* 23.

First Record. Has been gathered at Boughton, near Powick, *Lees, Bot. Malv. Hills*, 2nd ed., p. 44 (1852).

Severn. As first record. Barbourne Brook. Stanklin.

The stems of this plant are slender and branched, erect, but decumbent

---

and rooting at the base. The chaffy stipules are all fringed, and the sepals are eglandular. It is a species very near to *P. Persicaria*, though suggestive of *P. minus* in appearance. It is not recorded from any neighbouring county with the exception of Herefordshire, and in *Top. Bot.*, p. 355, is said to be insufficiently vouched for Worcestershire.

1431. **828. P. Persicaria,** Linn. *Spotted Persicaria.*
Native. Wet places, cultivated ground, waste places. Common. A. June to October. *Top. Bot.* 112.

First Record. Not localized, *Scott, Hist. Stourb.*, p. 540 (1832).

The leaves of this common plant are rather large and deep green, and usually have a blackish or purplish spot about the middle. It was formerly called Peach-wort, a translation of the specific name, which was given to it, says Gerarde, because of 'the likenesse that the leaues haue with those of the Peach tree'. It was also called by our ruder forefathers Dead Arsesmart in distinction from Arsesmart proper, which was *P. Hydropiper*, given to that plant, says an ancient writer, because if it touches the 'bare skinne, it maketh it smart, as often it doth, being laid into the bed greene to kill fleas'. It occurs in all neighbouring counties.

Var. b. **elatum,** Gren. and Godr.

First Record. Malvern, *Towndrow, Malv. Advert.*, December 3, 1892.

Malvern. As first record.

This is a luxuriant shade form with brighter green, narrower leaves and slender racemes much resembling *P. maculatum.* It has been observed in Warwickshire.

1432. **829. P. lapathifolium,** Linn. *Pale-flowered Persicaria.*
(*P. pensylvanicum,* Huds.)
Native. A weed in fields and on dunghills. General. A. June to October. *Top. Bot.* 103.

First Record. As *P. pensylvanicum*, var. *petechiale.* In a ditch on Stourbridge Common, *Stokes's With. Bot. Arr.*, 2nd ed., p. 412 (1787).

Mr. Lees does not seem to regard this plant as being so common as it really is. He assigns to it only one locality, *Table of Plants*, p. 24, in the Lickey district, against which may be put Mr. Mathews's 'general', in his *Flora* of that district, p. 40. The leaves vary considerably, from glabrous to pubescent or cottony above and beneath, and sometimes they have a dark blotch on the upper surface and the stem is spotted. The plant is recorded from all neighbouring counties.

1433. **830. P. maculatum,** Trim. and Dyer.
(*P. pensylvanicum,* var. *caule maculato,* Curtis. *P. nodosum,* Bab.)
Native. River-banks, newly turned up mud by water, waste ground. Probably not so rare as the records indicate. A. July to September. *Top. Bot.* 34.

First Record. As *P. lapathifolium*, var. *nodosum.* Not localized, *Lees, Bot. Malv. Hills*, 3rd ed., p. 57 (1868).

Severn. Near Bubble Bridge, *Reece*, July 26, 1850.

Malvern. As first record. Pixham; Malvern Wells, *Towndrow* (*MS.*). New Pool, *Westcombe*, 1848, as a var. of *P. lapathifolium.*

This plant is considered a sub-species of *P. lapathifolium* by Hooker, *Stud. Fl.*, 3rd ed., p. 345. It is a smaller plant than the preceding, and the leaves are white and woolly beneath. It has been observed in Warwickshire.

1434. 831. P. amphibium, Linn.

Native. Ponds, rivers, ditches, damp fields. General throughout. P. July to September. *Top. Bot.* 108.

First Record. Not localized, *Lees, Bot. Malv. Hills*, 1st ed., p. 21 (1843).

The pretty pink flowers of the aquatic form of this plant are to be seen in most of the larger streams and the ponds throughout the county, and its leaves float on the surface of the water; but they are very variable, according to its habitation. A terrestrial form is made into the variety following. It is the only perennial species of the *Persicaria* group of Polygonums. The type occurs in all bordering counties.

Var. **terrestre**, Leers.

First Record. Malvern, *Towndrow, Malv. Advert.*, December 3, 1892.

> Avon. Norton, near Evesham. Fladbury. Pershore. Bredon. Nafford Mill. Eckington.
> Severn. Kempsey. Barbourne, Worcester. Ombersley. Mouth of Salwarpe. Stourport. Bewdley. Holt.
> Malvern. Malvern Link; Newland, *Towndrow*.

Of this variety the leaves are narrow-lanceolate, rough with short rigid hairs on both sides, and it is often in a flowerless condition. Neither Babington, in his *Manual*, 9th ed., nor Hooker, in his *Stud. Fl.*, 3rd ed., nor Bentham, *Brit. Flor.*, 8th ed., recognize the variety, which does not appear to have been observed in Herefordshire of our neighbouring counties. The variety is not given in *Lond. Cat.*, 10th ed.

1435. 832. P. Bistorta. *Common Bistort, or Snakeweed.*

Native. Damp meadows and pastures. Not uncommon. Absent from the Lias formation. P. May to August. *Top. Bot.* 74.

First Record. Ham Green, near Mathon, and Martley, *Ballard, Stokes's With. Bot. Arr.*, 2nd ed., p. 406.

> Severn. Severn Banks, Bewdley. Wyre Forest. Near Hagley Station. Near Churchill Station. Trimpley Green. Wichenford. Kempsey. Dunclent. Astley.
> Malvern. Near Rosebery Rock, Knightwick. Meadow near the Spa, 1834, *Lees*. Southstone Rock. Mill Copse, Cowleigh. Side of path by chalybeate Spa, *Westcombe*. Mathon.
> Lickey. Moseley, *Mathews*. Romsley. Frankley. Clent. Near the Cock at Rubery, *Mathews*. Whitehall, Halesowen. Near St. Kenelms. Near Asylum, Rubery.

This plant bears a handsome spike of pink flowers sometimes two feet high or more. The specific name and its equivalent in English come through the French from the Latin *bis*, twice, and *tortus*, turned, in allusion to the twisted form of the large root; and from the contorted form of the root also comes the other English name. The root is black outside and red within, and is a powerful astringent. The old herbalists believed the plant to have a powerful faculty to resist all poison; and it had among them the names of Serpentary, Dragon Wort, Osterisks, and Passions, the last

one now modified into a common name for the plant, Patience Dock. The leaves and young shoots may be used as a vegetable. It is recorded from all neighbouring counties.

[1436. 833. P. viviparum, Linn. *Viviparous Alpine Bistort. Top. Bot.* 30.]

This plant was recorded by Mr. Lees thus: 'On Rosebury conglomerate rock by the Teme, below Knightsford Bridge,' *Illus. Nat. Hist. Worc.*, p. 162 (1834). It was afterwards acknowledged by him to be an error, and it is not localized in his *Table of Plants*, p. 24. Mr. Mathews, *Mid. Nat.* xi. 206, in mentioning the record, adds the note, 'See *New Bot. Guide*, p. 203,' where it is said, 'probably not found on Malvern, as reported in E. Lees' MSS.' It is a plant of the north, being found no further south than Yorkshire and in Carnarvonshire, while in the Highlands it is particularly abundant, but seldom ripens seeds. It is, of course, not found in any neighbouring county.]

FAGOPYRUM, Hall. 361*.   (*Fagus*, Latin, a beech-tree, πυρός, Greek, corn, from the shape of the seeds.)

1437. 834. F. sagittatum, Gilib.   *Buckwheat.*

(*F. esculentum*, Moench.   *Polygonum Fagopyrum*, Linn.)

Casual. Waste places, openings in woods. Occasional. A. July to October. Not in *Top. Bot.*

First Record. As *Polygonum Fagopyrum*. Not localized, *Scott, Hist. Stourb.*, p. 540 (1832).

> Avon. Occasional, *Lees*. Wood Norton.
> Severn. Stanklin Pool, *Humphreys*. Lower Hagley, *W. Whitwell*. Near Barker's Brick Works on Birmingham Canal, Worcester. Habberley Valley. Wyre Forest.
> Malvern. On cultivated land, occasionally.
> Lickey. Cradley, *Rev. J. H. Thompson*. Hunnington Station, near Halesowen.

Apparently a native of Manchuria, Buckwheat was introduced into Europe by the Turks about the thirteenth century. The common name may be a corruption of Beech-wheat, an equivalent for which is its name in Germany and Holland; but it was a familiar name with Turner thirty years before Lyte professed to take it from the Dutch, and it may be English and equal Buck-mast, that is, the seed of the Beech, which the seeds of this plant resemble in shape. However this may be, the plant is in the future likely to be met with much more frequently than in the past, as it is a favourite food in the pheasant-rearing field. It has been met with in every neighbouring county except, apparently, Gloucestershire.

RUMEX, Linn. 363.   (Possibly from *rumex*, a pike, from the shape of the leaves of some of the species.)

1439. 835. R. conglomeratus, Murr.   *Sharp Dock.*

(*R. acutus*, Sm. and of Linn. Herb.)

Native. Ditches, sides of streams, waste ground, roadsides. Common and generally distributed. P. June to August. *Top. Bot.* 97.

First Record. As *R. acutus*. Not localized, *Lees, Bot. Malv. Hills*, 1st ed., p. 21 (1843).

This plant is much like the following, from which it differs in the shape

of the leaves, in possessing a panicle leafy almost to the top, and by having a tubercle on each of the enlarged sepals instead of on only one. It is recorded from all bordering counties.

× crispus.

First Record. Pixham, *Towndrow, Malv. Advert.*, Dec. 3, 1892.

> Malvern. As first record. Malvern Link Common, 1907, *Towndrow*.

The last recorded plant was growing near both parent plants, and showed the characters of each.

1441. 836. R. sanguineus, Linn.   *Bloody-veined Dock.*

Native. Woods, thickets, waysides. Very uncommon. P. June to September. *Top. Bot.*, no number, *Lond. Cat.*, 10th ed.

First Record of the aggregate. Not localized, *Lees, Bot. Malv. Hills*, 1st ed., p. 21 (1843). This was the variety *viridis*; see same book, 2nd ed., p. 43. First Record of type. Under a hedge at Malvern Wells, 1867, *Lees, Bot. Malv. Hills*, 3rd ed., p. 56 (1868).

> Severn. Near Worcester; Bewdley; *Claines, Herb. Hast. Mus. Worc.*
> Malvern. As first record. In this district, *Towndrow*. Roadside, Malvern Wells, 1908.

In spite of Mr. Lees' after-disclaimer of his first record, in his *Table of Plants* he localized the type as 'very uncommon' in the Severn and Malvern districts, the latter on the strength of the Malvern Wells locality, no doubt. He does not give any locality in the Severn district. The variety with green veins is far more frequent. The plant is recorded, probably generally in the aggregate, from all bordering counties.

1441. 837. R. viridis, Sibth.   *Green-veined Dock.*

(*R. nemorosus*, Schrad.)

Woods, thickets, waysides. Common. P. June to September. *Top. Bot.* 90.

First Record. As above.

> Avon. Tiddesley Wood. Deerfold Wood. The Slads. Trench Woods. Bow Wood. Grafton Wood. Croome Perry Wood. Defford Wood.
> Severn. Shrawley Wood. Wyre Forest. Nunnery Wood. Monk Wood. Perry Wood. Ockeridge Wood. Ribbesford Woods. Titton Banks. Hill Ditch, Hartlebury.
> Malvern. Very common on the hills, *Towndrow*. Middleyards Coppice. Crews Hill Wood. Ankerdine Hill. Martley. Wood near Southstone Rock. Kyre Park.
> Lickey. Great Farley Wood. The Randans. Chaddesley Wood.

This plant, a variety in *Lond. Cat.*, 10th ed., is kept a distinct species in *Index Kewensis*, and we, too, think it is distinct, hence we raise it to specific rank. Syme thinks the difference is too slight to constitute a species, but Mr. Druce, *Flor. Berks.*, p. 431, says the red-veined Dock comes true from seed, and it seems contradictory that a type should be so rare but a variety common, as it is in Worcestershire and everywhere. This is an inhabitant of all our woods and shady places, while the other species, *R. sanguineus*, is almost unknown throughout the county. The distinction seems to consist entirely in the absence of red colouring matter from the veins of the leaves, which are generally small, and there is a pale tubercle on the upper fruiting sepal. The plant has been observed in all bordering counties.

1442. 838. R. maritimus, Linn.   *Golden Dock.*

Native. Marshy places, margins of rivers. Uncommon. B. July to September. *Top. Bot.* 39.

First Record. Var. *aureus*, Severn Stoke, *Ballard, Stokes's With. Bot. Arr.*, 2nd ed., p. 371 (1787).

> Avon. South end of Pirton Pool.
> Severn. Below Worcester. Spetchley Park, 1847, *Reece*. Sutton Pool, Kidderminster. Lodge Pools, Kidderminster, 1816, *Perry*. Pool, Grafton Manor.
> Malvern. Longdon. Chalybeate Pool, Malvern, *Lees*. Pond near Ham Castle. Near Clifton-on-Teme.
> Lickey. Harborne Reservoir, 1851, *Mathews*.

This dock can always be distinguished by its narrow leaves and the very long teeth of the inner fruiting sepals, which exceed them in length. It is in its glory in salt-marshes. Its flowers are bright orange-coloured, and the whole plant is yellowish-green. It is recorded from all neighbouring counties except Gloucestershire.

[1443. 839. R. limosus, Thuill.   *Marsh Yellow Dock.*]

(*R. palustris*, Sm.)

Native. Marshy places. Rare. P. July to September. *Top. Bot.* 27.

First Record. As *R. palustris*. Marsh by the Chalybeate Spring (Malvern), *Lees, Mag. Nat. Hist.*, vol. iii, p. 160 (1830).

> Severn. Severnstoke, *Lees*.
> Malvern. As first record.

This plant is very similar to the last, differing in the form and number of its enlarged sepals, and the teeth which border them do not exceed the length of the enlarged sepals. Its nuts also are larger. It is recorded doubtfully from Warwick and Stafford.

1444. 840. R. pulcher, Linn.   *Fiddle Dock.*

Native. Roadsides, churchyards, old pastures. Very local. P. June to October. *Top. Bot.* 43.

First Record. Not localized, *Lees' Cat., New Bot. Guide* (1835).

> Avon. Holy Cross, Pershore.
> Severn. Dodderhill Churchyard.
> Malvern. Longdon, *Lees*. Little Malvern, *Mathews*. Castlemorton Churchyard, and one mile east of Castlemorton Church, *Westcombe*. On the hill north of Longdon Marsh. Malvern Wells.

This plant is not recorded for Hereford or Salop, but has been observed in the former county.

1445. 841. R. obtusifolius, Linn.   *Broad-leaved Dock.*

Native. Pastures, waysides, waste places, orchards. Very common and widely distributed. P. May to September. *Top. Bot.* 112.

First Record. Not localized, *Lees, Bot. Malv. Hills*, 1st ed., p. 21 (1843).

This is a common dock of the wayside and the neglected soil, and may be known by its broad, blunt root-leaves, generally curled at the margin, and the fruiting sepals which are strongly toothed at the base. It is refused by cattle but eaten with avidity by deer. Its most popular surviving use is to allay the smart caused by the sting of the nettle, for which it is an antidote usually ready at hand, for where the nettle grows there grows the Dock. 'Out nettle, in Dock; nettle, nettle stung me,' is the charm the country children use on applying the remedy. It is a difficult weed to eradicate; the roots run deep and any small fragment soon produces a plant. The roots are very astringent. This Dock prefers good soil, and it is said where the Dock will not grow no wheat will thrive. The plant is recorded from all bordering counties.

× **viridis.**
First Record. As below.
Malvern. Madresfield, 1887, *Towndrow, Malv. Advert.*, Dec. 3, 1892.

1446. 842. R. crispus, Linn.   *Common Dock.*
Native. Cultivated ground, pastures, waste ground. Abundant and widely distributed. P. May to October. *Top. Bot.* 112.
First Record. Not localized, *Lees, Bot. Malv. Hills*, 1st ed., p. 21 (1843).
This, like the last, is a common dock; the leaves are lanceolate, wavy, and crisped, and the sub-entire fruiting sepals are broadly ovate or cordate. It is especially common near dwellings; the tubercle is orange-coloured. It is recorded from all neighbouring counties.

× **obtusifo'ius.**
(*R. acutus*, Linn. Sp. Pl. *R. pratensis*, Mert. and Koch.)
First Record. Not localized, *Lees, Bot. Malv. Hills*, 1st ed., p. 21 (1843). In the 2nd ed. the locality is given as Longdon. At the above reference Mr. Lees also records *R. acutus*, intending probably *R. conglomeratus* above.
Severn. As *R. pratensis*. Wannerton Downs, *Scott, Mathews.* Brickyard, Northwick, Worcester. Near Worcester. Roadside, Whittington. Malvern Road. Pixham, *Towndrow, Malv. Advert.*, Dec. 3, 1892.
Malvern. Near Malvern, *Towndrow.* Longdon, *Lees.* Ham Bridge.
This plant is recorded from all bordering counties except Staffordshire, but not from Worcestershire itself. *Top. Bot.* 358.

× **viridis.**
First Record. As below.
Malvern. Madresfield, 1887 and 1889, *Towndrow, Malv. Advert.*, Dec. 3, 1892.

1448. 843. R. Hydrolapathum, Huds.   *Great Water Dock.*
Native. Sides of water. Several places; rare in the Lickey district. P. May to September. *Top. Bot.* 71.
First Record. About Clifton, *Ballard, Stokes's With. Bot. Arr.*, 2nd ed., p. 374 (1787).
Avon. Near Tewkesbury. Near Eckington Bridge. Norton, near Evesham. By the Avon near Evesham. Fladbury. Pershore.
Severn. Crowneast. By Stour, one mile south of Cookley Forges. Henwick. Northwick Brickyard. Grimley Brickyard. Boreley. Near Bewdley. Kempsey.
Malvern. Longdon Marsh. By little pools near the Severn, *Lees.* Bransford. Brockamin. Knightsford Bridge. Southwood. The Rhydd.

Lickey. Bittell Reservoir; Falling Sands, Kidderminster, *Mathews.* As *R. aquaticus*, probably this plant, banks of Stour at Stourbridge, *Scott.*
This is the largest of our docks, sometimes reaching a height of four or five feet. The enlarged sepals are all tubercled, and the panicle is very large. It is a characteristic feature of river-side vegetation in many parts, and is recorded from all neighbouring counties.

1451. 844. R. Acetosa, Linn.   *Common Sorrel.*
Native. Pastures, meadows, woods. Very common and widely distributed. P. April to July. *Top. Bot.* 112.
First Record. Not localized, *Lees, Bot. Malv. Hills*, 1st ed., p. 21 (1843).
This is a common plant in the meadows, their surfaces frequently blushing red with its blossoms. The leaves at the flowering season have a pleasant acid flavour, and in olden times were used with other plants in making verjuice. The juice was also used medicinally and deemed efficacious against agues, jaundice, and other maladies. It is called Sorrow and Cuckoo's Sorrow in some parts of Worcestershire. It is recorded from all neighbouring counties.

1453. 845. R. Acetosella, Linn.   *Sheep's Sorrel.*
Native. Dry fields, waysides, heaths. Locally abundant, preferring sandy soil. P. March to August. *Top. Bot.* 112.
First Record. Not localized, *Lees, Bot. Malv. Hills*, 1st ed., p. 21 (1843).
This is a smaller and more graceful plant than the preceding. The flower-stems rise from a little rosette of halberd-shaped leaves, each rosette connected with a neighbouring one by a thin thread-like root. At the end of summer the foliage becomes very red. It is recorded from all bordering counties.

65. ARISTOLOCHIACEAE, Juss.

ASARUM, Linn. 364. (á, not, σειρά, a band, because it was not bound up in the garlands of flowers by the ancients.)
{**1454. 846. A. europaeum,** Linn.   *Asarabacca.*
Casual. Woods. Very rare. P. May. *Top. Bot.* 6.
First and only Record. Very rare (no locality given), *Purton, Midl. Fl.*, p. 225 (1817).
No other botanist has observed this plant in the county. It is a curious-looking plant throwing up annually a short stem with two evergreen leaves, between which is a drooping dull-green flower. It has been used to remedy the effects of excessive drinking, its aromatic roots were used medicinally, and the leaves, when powdered, to produce the effects of snuff. It is not recorded in *Top. Bot.*, p. 363, except doubtfully from Herefordshire, from any neighbouring county.}

ARISTOLOCHIA, Linn. 365*. (The name originates from its supposed medicinal virtues; ἄριστος, best, λόχιος, belonging to childbirth.)
{**1455. 847. A. *Clematitis,** Linn.   *Birthwort.*
Alien. Waste places, hedges. Very rare. Shrub. June to September. Not in *Top. Bot.*

First Record. This very rare and singular plant was discovered by Miss Mary Anne Rawlins, of Pophills, growing at Chaddesley, near Kidderminster, *Purton, Midl. Fl.*, vol. ii, p. 430 (1817).
Severn. On the site of old gardens at Worcester, *Lees.* Site of the Shire Hall, Worcester, March 9, 1834, *A. D. Gordon.*
Malvern. A weed in a market garden at Upton-on-Severn, *G. A. Clarke.*
Lickey. Chaddesley Corbett, as first record.
This plant is a native of Western Asia, cultivated from ancient times in Europe, being now a common weed in some of the warmer parts of the Continent. This plant, among other virtues, was one of those peculiarly offensive to 'serpents', so much so that they flee from the person who carries a piece in his hand. It was believed that snake charmers availed themselves of this plant to stupefy the animals before they charmed them. The plant has been observed in Herefordshire and Gloucestershire.}

66. THYMELEACEAE, Reichb.

DAPHNE, Linn. 366. (From the nymph Daphne, who was changed into a Laurel.)
{**1456. 848. D. Mezereum,** Linn.   *Common Mezereon.*
Native. Dense woods and thickets. Very rare. Shrub. March and April. *Top. Bot.* 9.
First Record. Eastham and Stanford, Worcestershire, Rev. Edward Whitehead, of Corpus College, Oxon., *Purton, App.*, vol. iii, p. 33.
Avon. Hedge near Wadborough Station. Ditch-side, Pirton end of Croome Perry Wood.
Malvern. Stanford Park; Eastham, as first record. Wood behind Little Shelsley Church, 1854, *Mrs. Smith*, of Shelsley Court, *Lees.*
It is to be feared this plant has been lost to the county, for no trace of it now is to be found in the above localities. The Eastham locality is mentioned in Smith's *English Flora*, 1824. It has been noticed in all bordering counties, 'probably an alien,' *Top. Bot.* 2nd ed., 362. This shrub is a well-known occupant of gardens, the purplish or white flowers expanding early in spring before the leaves appear. The berries, which are showy in autumn, are highly poisonous, but the Robin feeds eagerly upon them. The berries, and the bark as well, are used medicinally in some countries.}

1457. 849. D. Laureola, Linn.   *Common Spurge Laurel.*
Native. Woods, thickets, hedgerows, chiefly on calcareous soil. Local. Shrub. February to April. *Top. Bot.* 52.
First Record. In woods and hedges near Pershore, frequent, *Nash, Hist. Worc., Int.*, p. lxxxix (1781).
Avon. Rather uncommon, *Lees.* Bredon Hill. Trench Woods. Tiddesley Wood. Defford. Sheriff's Lench. Crowle. Huddington. Bentley.
Severn. Witley, *Scott.* Perry Wood, Worcester, *Lees.* Border of lake, Westwood Park. Hadsor Lane, near Droitwich. Shrawley Wood. Ribbesford Coppice. Chatley. Ombersley. Kempsey. Earl's Croome. Ladywood, Salwarpe. Claines. Elmley Lovett. Hallow. Grimley. Lincombe. Hartlebury. Hindlip. Spetchley.

Malvern. Not uncommon, *Lees.* Leigh Sinton. Alfrick. Knightsford Bridge. Abberley Hill. Martley. Bransford. Powick. Hanbury. Malvern Link.
Lickey. One place, *Lees.* Chaddesley Woods.
This is a well-known garden shrub, its sweet bunches of yellow-green blossoms perfuming the air in its neighbourhood. It is a common hedgerow plant for miles round Worcester. The leaves, not unlike small laurel leaves, are chiefly borne at the ends of the branches, somewhat resembling in this respect the Wood Spurge. Therefore the two components of its English name are justified. It is recorded from all neighbouring counties.

68. LORANTHACEAE; Juss.

VISCUM, Linn. 368. (ἰξός, the Greek name for the plant.)
1459. 850. V. album, Linn.   *Mistletoe.*
Native. A parasitic shrub on trees. Abundant in Severn and Malvern districts, less common in Avon, and nearly absent from the Lickey. *Top. Bot.* 40.
First Record. Worcestershire on Apple-trees, sometimes on Limes, and in one instance on a Plane-tree near Lord Coventry's Menagerie, Croome, *Stokes's With. Bot. Arr.*, p. 1112 (1787).
No need to say much about the Mistletoe, except in its botanical aspect; it enters too much into the life of the people at Christmas time, and its ancient and modern history and uses are too well known, to need comment. In *Worc. Nat. Club Trans.*, vol. iii, p. 121, is a paper on the plant, and a list of its Worcestershire hosts is given on p. 122. To those there mentioned may be added a standard rose-tree in Clent Cottage garden at Clent, on which it flourished for many years, becoming a big bush and ultimately killing its host. In Worcestershire its northern limit is reached pretty much where apple and pear trees cease to be seen in the hedgerows. These cease somewhere about a line drawn nearly due east from Kidderminster till the Stourbridge and Bromsgrove road is reached, and then continued to Bromsgrove, and along the boundary of the Lickey district to Redditch. Mistletoe does not extend quite so far as this line; the most northerly plants known are on the poplar at Dordale Brook in Belbroughton, and on apples at Tanwood Farm in Chaddesley Corbett; and it occurs also on the apple at Woodcote Green, all these localities being in the Lickey district, in which alone would there be any need to localize the plant. There are a considerable number of plants in the garden at Ferndale, Pedmore, much further north, where it has been induced to grow on many of the fruit trees; and it grows finely, always affording a bough at Christmas. The history of the English name is curious. In early times there were two plants known as Mistle or Missel: one Basil, called Earthmistel, the other this plant, called Oakmistel. Then there came to be added on to this word the Saxon *tan*, a twig, in two forms, the strong masculine *tan*, and the weak West Saxon feminine *ta*; and from these the two names of the plant, Mistletoe and Misseldine, were derived, the latter finally giving way to the former so lately as the seventeenth century, until at the present day it is obsolete. The Mistletoe is recorded from all neighbouring counties.

70. EUPHORBIACEAE, Juss.

EUPHORBIA, Linn. 370. (Named from *Euphorbus*, physician to Juba, king of Numidia, because he is said first to have brought the plant into use.)

1462. **851. E. helioscopia,** Linn.   *Sun Spurge.*

Native. Waste and cultivated ground. Frequent among crops of turnips and mangolds. Generally distributed. A. February to September. *Top. Bot.* 112.

First Record. Not localized, *Lees, Bot. Malv. Hills,* 1st ed., p. 40 (1843).

The tribe of Spurges is an interesting order, comprising not only herbs, or in the case of the Box, a shrub, as with us, but also large trees. Numerous substances which enter into our common life are derived from members of the tribe. Castor oil is extracted from *Ricinus communis,* Cascarilla is the bark of *Croton Eluteria, Croton Tiglium* affords Croton oil, *Siphonia elastica,* India-rubber, *Hevea braziliensis,* Para-rubber, and *Janipha Manihot,* a most poisonous plant, Tapioca and Cassareep, the latter forming the basis of the West Indian pepper-pot, besides being a main ingredient in Worcestershire Sauce. The Sun Spurge is quite a common weed. The juice is milky and is an old remedy for the cure of warts. It is recorded from all neighbouring counties.

{*1463.* **852. E. platyphyllos,** Linn.   *Broad-leaved warted Spurge.*

Native. Fields and waste places. Once recorded. Annual. June to August. *Top. Bot.* 29.

First and only Record. As *E. platyphylla.* South Littleton, *Purton, App.,* vol. iii, p. 38.

This plant is no doubt extinct in the county, no one has seen it since Purton's time. It is recorded only for Gloucestershire of our bordering counties.}

1469. **853. E. amygdaloides,** Linn.   *Wood Spurge.*

Native. Woods, hedgerows, thickets. General except in the Lickey district, where it is rather uncommon. P. April to July. *Top. Bot.* 52.

First Record. Hedges near Worcester on the Bewdley Road, *Pitt, Agric. Worc.,* p. 317 (1810).

     Avon. Trench Woods. Tiddesley Wood. The Slads. Badgers Bank. Bow Wood. Crowle. Goosehill Wood. Deerfold Wood. Grafton Wood. Churchill Wood. Norton, near Evesham.

     Severn. Wyre Forest. Helbury Hill. Kempsey Grove. Shrawley Wood. Ockeridge Wood. Monk Wood. Astley Wood. Lincombe. Perry Wood. Deadmans Copse. Hadley. Nunnery Wood.

     Malvern. Woody places, abundant, *Lees.* Middleyards Coppice. Aislehurst Coppice. Crews Hill Wood. Ankerdine Wood. Fries Wood. Martley. Stanford Park. Malvern Link.

     Lickey. Alvechurch. Beoley, *Mathews.* Randans Wood. Chaddesley Woods. Lickey Woods.

This is a pretty plant, with greenish-yellow blossoms in spring; and both

in spring and autumn its stem and leaves take on a purplish red tinge. It is seldom seen in the Lickey district. It is recorded from all bordering counties.

1470. **854. E. *Esula,** Linn.   *Leafy-branched Spurge.*

Alien. Waste places. Once met with. P. Not in *Top. Bot.*

First Record. One plant among corn at Bransford, 1892, *Towndrow, Malvern Advertiser,* Dec. 3, 1892.

This plant is a native of woods and meadows in Central and South-east Europe, and is extending along railways in other parts of the Continent. It reappeared at Bransford in 1902. It has not been observed in any neighbouring county.

Var. b. ***Pseudo-Cyparissias,** Jord.

First Record. As below.

     Severn. Ripple, *Towndrow.*

This variety is not mentioned even in the last edition of Hooker, Bentham, or Babington. It is characterized by its strap-shaped leaves, while those of the barren branches are linear.

1474. **855. E. Peplus,** Linn.   *Petty Spurge.*

Native. Cultivated ground. Common. A. May to November. *Top. Bot.* 109.

First Record. Not localized, *Walker, Med. Surg. Rep.,* vol. i, no. 2, p. 100 (1828).

This plant is a common garden weed. It is recorded from all neighbouring counties.

1475. **856. E. exigua,** Linn.   *Dwarf Spurge.*

Native. Cultivated ground, waysides. Common. A. June to November. *Top. Bot.* 83.

First Record. Not localized, *Lees, Bot. Malv. Hills,* 1st ed., p. 40 (1843). This plant is as common in cornfields as the previous one is in gardens. It is found in all bordering counties.

1476. **857. E. Lathyris,** Linn.   *Caper Spurge.*

Denizen or native. Woods. Very rare. P. June to August. *Top. Bot.* 4 ?

First Record. Crow's Nest Woods, *Dr. Streeten, Lees, Illus. Nat. Hist. Worc.,* p. 177 (1834).

     Avon. The Slads. Blockley.

     Severn. Crowneast as above. Railway-bank near Workhouse, Worcester. Bewdley.

     Malvern. Sherrard's Green, *Towndrow, Malv. Advert.,* Dec. 3, 1892.

These are the only records of the occurrence of this plant in the county, but Mr. Lees, *Bot. Worc.,* p. xxx, adds the words, ' and other places where gardens have been.' Under no other circumstances is it likely to appear. It was formerly much cultivated in gardens. Babington and the *Lond. Cat.* spell the specific name *Lathyrus,* but *Lathyris* is the proper spelling. It has been observed in Gloucester, Hereford, and Salop.

{E. CHARACIAS, Linn.   *Red Shrubby Spurge.*

Alien. P. April.

First Record. On Malvern Hill, between the Inn and the Wells, *Withering,* 4th ed. (1801).

               Y

This plant is a native of woody districts in the Mediterranean region. If the record is correct, it was a decided interloper, and has long since taken its departure.}

BUXUS, Linn. 371. (Name altered from πύξος, the Greek name for the plant.)

1477. **858. B. sempervirens,** Linn.   *Common Box.*

Alien. Introduced shrub. April and May. In *Top. Bot.* not mentioned ; *Lond. Cat.,* 10th ed., 3.

First Record. Planted in woods on Bredon Hill, *Lees, Bot. Worc.,* p. xxx (1867).

The Box is a native of Europe as far north as the woods of Belgium and Holland, and since in a few localities in England it grows plentifully and naturally, it is not impossible that it may be a native of England also. But certainly that is not the case in Worcestershire. Wherever it is seen it is a relic of old garden cultivation. Besides being used in a dwarf state in our gardens, the timber is extensively used in the arts and manufactures, and the leaves of the plant have been used for medicinal purposes. The plant occurs in all bordering counties, but in none of them is more than introduced.

MERCURIALIS, Linn. 372. (So called because *Mercury* discovered the virtues of the plant.)

1478. **859. M. perennis,** Linn.   *Dog's Mercury.*

Native. Woods, thickets, hedgerows. General. P. March to May. *Top. Bot.* 107.

First Record. In hedges between Bromsgrove and Feckbury, *Pitt, Agric. Worc.,* p. 317 (1810).

The young stems of this plant are among the first growth to be observed on the hedgerow banks at the approach of Spring. It is very poisonous, and in drying turns blue black. It is recorded from all neighbouring counties.

1479. **860. M. annua,** Linn.   *Annual Mercury.*

Colonist. Fields and gardens. *Top. Bot.* 42.

First Record. Malvern Link, 1905, *Towndrow* (MS.).

     Malvern. As first record.

This plant has been observed in Gloucester, Hereford, and Salop.

71. URTICACEAE, Juss.

ULMUS, Linn. 373. (The Latin name.)

1480. **861. U. glabra,** Huds.   *Wych Elm.*

(*U. campestris,* Linn.   *U. montana,* Stokes.)

Native or Denizen. Woods and hedges. Tree. March and April. *Top. Bot.* 99.

First Record. Near Dudley Castle, *Booker, Dud. Cast.,* p. 107 (1825).

A considerable number of forms of the Elm have been described, but at present there is very considerable divergence of opinion concerning their description and their synonymy. In appearance this tree is lighter

and more branching than the small-leaved kind, but there is infinite diversity of habit in all the tribe. As a hedgerow plant this elm sends out most luxuriant branches during the summer, with large broad leaves. Some sketches of old and distorted Elm trees which have grown in Worcestershire are given in *Worc. Nat. Club Trans.* ii, *Sup.,* p. 11 onwards. This Elm grows in all bordering counties.

1481. **862. U. campestris,** Linn.   *Small-leaved Elm.*

(*U. suberosa,* Ehrh.   *U. sativa,* Miller.   *U. campestris,* Huds., not of Linn. *U. surculosa,* Stokes.)

Denizen. Tree. In every hedge on the sandstone and Marl, but not so common in the Lickey district. *Top. Bot.* 60.

First Record (possibly this plant is intended). Plentiful in Worcestershire, *Withering,* 3rd ed., p. 278 (1796).

This is the Elm so abundant in the county, the 'Worcestershire Weed'; a tree of the open country, seldom to be seen among the trees of the wood except in open hedges. It has been said from early times to be no native of England; it rarely ripens seeds, and seedlings are seldom if ever seen. Yet we have in the county three parishes which take their name from the Elm— Elmley Lovett, Elmley Castle, and Elmbridge, which must have been conspicuous for the number of their Elms when names were being made, while now, at all events, this tree is far more plentiful in these localities than the broad-leaved Elm. The roots of this tree, where once it has grown, throw up quantities of saplings, some of which survive, and in their turn become hedgerow trees. The timber of the tree is one of the most useful we have, and is universally employed to encase poor humanity on its last journey to its final home. The blossoms are very numerous, and in early spring clothe every twig with russet brown. Our forefathers used the leaves and roots for several disorders, and among other purposes, to ' cleanse the skin and make it fair', and also to prevent the hair falling off. *Ulmus campestris* grows in every neighbouring county.

Var. b. **suberosa,** Moench.

First Record. As *U. suberosa,* possibly this variety. Near Dudley Castle, *Booker, Dud. Cast.,* p. 207 (1825).

This variety is considered synonymous with the type; but at all events some young Elms are conspicuous enough from the ridged and corky bark of their branches. This characteristic is not so noticeable when the tree is old, but plants which show it seldom grow up into tall pyramidal Elms. They usually make stumpy bushy-headed trees. This variety does not seem to have attracted notice in any bordering county except Salop.

Var. c. **glabra,** Mill.

First Record. As *U. glabra.* In Crow's Nest (Crowneast) Wood, *Mr. T. Westcombe, Lees, Bot. Worc.,* p. xxx (1867).

This also is said to be the same as the type. Mr. Mathews considers, *Mid. Nat.* xv. 44, Mr. Lees' *U. carpinifolia* to be synonymous with this, and, as the name occurs one line earlier in his *Bot. Worc.* at the above reference, attributes to him the first record. In the former variety the leaves are scabrid above and pubescent beneath, and in this one, nearly glabrous;

               Y 2

many varieties are described, differing in habit and foliage, but their characters do not remain constant. This glabrous variety has been noticed in Salop and Warwickshire.

HUMULUS, Linn. 374.   (*Humus*, rich soil, in which the plant flourishes.)

1482. **863. H. Lupulus**, Linn.   *Common Hop.*

Native. Hedges and thickets. Not uncommon. Climbing perennial herb. May to August. *Top. Bot.* 88.

First Record. Not localized, *Walker, Med. Surg. Rep.*, vol. i, no. 2, p. 100 (1828).

Although the Hop is a native of England it was not until the sixteenth century that its cultivation began to spread. Worcestershire is one of the few counties in which it is largely grown at the present time. No doubt in years gone by, to judge from field names in various parishes, Hopyards extended much farther to the north of the county than they do now, and to long past cultivation many of the plants now to be seen twining in hedgerows doubtless owe their origin. Hops did not escape the notice of the herbalists, who considered them a great purifier of the blood, while in modern days a pillow of Hops is used to induce sleep. Finally, it is said that the young tops of the shoots tied in bundles and boiled, form a pleasant vegetable, and that the bleached shoots served like Asparagus are delicious. The Hop is recorded from all neighbouring counties.

URTICA, Linn. 375.   (From *uro*, I burn, in allusion to its stinging properties.)

1483. **864. U. dioica**, Linn.   *Stinging Nettle.*

Native. Waste places, hedges, woods, meadows. Common everywhere. P. July and August. *Top. Bot.* 112.

First Record. Not localized, *Lees, Bot. Malv. Hills*, 1st ed., p. 40 (1843).

The English name of this plant would obviously appear to be related to the word *needle*; but the relationship is obscure. The plant was much used medicinally by our ancestors, and one of the most curious uses was to stop bleeding of the nose by pressing a leaf against the roof of the mouth. Venomous as the plant is, it can easily be avoided. There is a vast amount of wisdom summed up in the proverb, 'It is better to be stung by a nettle than pricked by a rose.' The young shoots, in spring, when boiled, make an excellent substitute for spinach. The stinging nettle occurs, of course, in all our bordering counties.

1485. **865. U. urens**, Linn.   *Small Nettle.*

Native. Waste ground, manure heaps, gardens. General, but not nearly so frequent as the preceding plant. A. June to September. *Top. Bot.* 108.

First Record. Not localized, *Lees, Bot. Malv. Hills*, 1st ed., p. 40 (1843).

This is a smaller, sturdier plant than the last, never growing so laxly, and not so fond of shade; and, being an annual, is not nearly so noxious a weed, for it does not possess any underground creeping stems, which in the case of its relative spread it far and wide. It is recorded from all bordering counties.

PARIETARIA, Linn. 376.   (*Paries*, a wall, on which it is frequently found.)

1486. **866. P. ramiflora**, Moench.   *Common Pellitory of the Wall.*

(*P. diffusa*, Koch.   *P. officinalis*, Linn., pro parte.)

Native. Old walls, hedgebanks. Not uncommon, except in the Avon district. P. June to September. *Top. Bot.* 94.

First Record. As *P. officinalis*. City walls, &c., *Lees, A. Florence, Strang. Guide* (1828).

Avon. Evesham. Wyre Piddle. Fladbury. Pershore. Defford.

Severn. City Walls, *Lees*. Roadside, Battenhall. Near the Cathedral. Bewdley. Stourport. Hawford. Little Hadley Mill. Salwarpe Mill. Droitwich. Claines. Hallow. Power's Rough. Dodderhill. Grafton.

Malvern. Little Malvern, *Lees*. Powick Church. Martley. Abberley. Upton-on-Severn. Leigh. Malvern Link.

Lickey. Halesowen Abbey, *Mathews*. Chaddesley [Corbett], *Scott*.

The name Pellitory is not found before the sixteenth century; previously the name of the plant was Parietary, of which the present name is a modification. The plant has no beauty to recommend it, the flowers being small and reddish, in the axils of the leaves. It is recorded from all bordering counties.

Var. b. **erecta**, M. and K.

First Record. As here.

Severn. Cathedral Ferry, Worcester. Near Worcester Bridge. Bewdley. Ombersley.

In this variety the stem is erect, and seven or more flowered, whereas the type is prostrate or ascending, and three-flowered. Not given in *Lond. Cat.*, 10th ed., nor in the three leading botanies.

CANNABIS, Linn.   (κάνναβις, the Greek name of the plant used by Dioscorides, which is derived from the Sanskrit *canam*.)

C. sativa, Linn.   *Hemp.*

Alien. Waste ground. Rare. A. August. Not in *Top. Bot.*

First Record. An outcast, *Towndrow, Malvern Advertiser*, 1892.

Severn. Worcester. Kidderminster. Wilden.

Malvern. As above.

Lickey. Hunnington, near Halesowen, *Herb. Hast. Mus. Worc.*

This plant has been of very wide cultivation from prehistoric times, and formerly was a common crop in England. If it had found its environment suitable, it would by this time have been well established in many places. As it is, its sporadic appearance is probably due to the fact that the seed is a favourite food of the feathered race when in captivity, and especially of the Parrot. Its occurrence in the Severn Division is due to its use for this purpose, and the same is probably true of the other divisions.

73. CUPULIFERAE, Rich.

BETULA, Linn. 378.   (Some derive it from *betu*, the Celtic name for the plant, whilst others say it comes from *vetus*, old, because the plant exists for a long time.)

1488. **867. B. alba**, Linn.   *The Birch.*

(*B. verrucosa*, Ehrh.)

Native. Woods and thickets, especially on dry and sandy soil. General. Tree. April and May. *Top. Bot.* aggregate, 112.

First Record. Perry Wood, Worcester, *Lees, Worc. Misc.*, 1829–30.

The Birch is fond of light soil, and in many parts of the sandy district around Kidderminster is the tree that comes up on roadside wastes. The tree is a useful one. The wood is tough and white, and largely used in making brushes. The twigs are made into brooms, and have also acquired a painful significance with schoolboys. From the bark many things are formed, and the oil obtained from the 'white rind' is employed in tanning Russia leather. It is a country remedy for several complaints, and in Scotland a kind of wine is made of the sap. The tree is recorded from all neighbouring counties.

1489. **868. B. tomentosa**, Reith. and Abel.   *The Birch.*

(*B. pubescens*, Ehrh.   *B. glutinosa*, Wallr.)

Native. Heathy woods on clay and heavy soils. Not uncommon. Tree. April to May. *Top. Bot.* 72.

First Record. United with *B. alba*, and localized in all the districts, *Lees, Table of Plants*, p. 26 (1867).

The Birches, like the Elms, are a little bit 'mixed'. The problem may be stated mathematically. *B. alba*, Linn. = *B. glutinosa*, Fr. = *B. pubescens*, Koch., and var. b. of this is *B. pubescens*, Ehrh. So Bab., *Manual*, 9th ed., p. 388. But *B. alba*, Linn. = *B. verrucosa*, Erhr., and a sub-species is *B. glutinosa*, Fries, and of the sub-species *pubescens*, no sponsor, is a variety. Thus Hooker, *Stud. Fl.*, 3rd ed., p. 366. Then take Druce, *Fl. Berks.*, p. 446. *B. pubescens*, Ehrh. = *B. glutinosa*, Wallr. After this, three letters only need be written, Q. E. D. But the leading distinctions between these two species are the following: *B. alba* has the branches drooping, leaves doubly serrate with raised veins above, and the lateral lobes of the trilobed fruit falcate reflexed, whereas *B. tomentosa* has the young branches erect and velvety, leaves simply serrate, long and acuminate with raised veins beneath, and the lateral lobes of the fruit ascending. *B. tomentosa* is met with on wetter, colder land than *B. alba*, which flourishes on dry and sandy ground. This species has been noticed in all bordering counties except Gloucestershire, but in *Top. Bot.* p. 372 Worcestershire is not credited with it.

ALNUS, Adans. 379.   (Possibly *al*, near, *lan*, the bank of a river; the general habitat of the genus.)

1491. **869. A. rotundifolia**, Mill.   *Common Alder.*

(*A. glutinosa*, Gaertn.)

Native. River and brooksides, damp woods. Generally distributed. Tree. February to April. *Top. Bot.* 110.

First Record. Not localized, *Lees, Bot. Looker-out*, p. 94 (1842).

The Alder is a common tree by watersides throughout the county. The foliage is glutinous, a quality our forefathers took advantage of, in the manner of the modern fly-paper, 'to rid the chamber of those troublesome bedfellows,' fleas. The bark was used medicinally; the timber is used for making clogs, and is very durable under water. Alder trees dug out of peat bogs furnish a bog-wood as black as Ebony. The piles on which the lake village of Glastonbury were founded were made of this wood and are in a perfect condition when drawn from beneath the hearthstones at the present day, but of course at once crumble on exposure to the air. It occurs in every neighbouring county.

CARPINUS, Linn. 380.   (Name from *car*, wood, *pin*, head, in Celtic, being used to make yokes for oxen.)

1492. **870. C. Betulus**, Linn.   *Common Hornbeam.*

Native. Hedges, coppices. Not common. Tree. April and May. *Top. Bot.* 37.

First Record. Not common, *Lees' Cat., New Bot. Guide* (1835).

Avon. Wood Norton. Comberton.

Severn. Hadsor. Hartlebury. Vallombrosa. Grimley. Worcester. Bewdley.

Malvern. Madresfield. Witley.

Lickey. Near Redditch, *D. Mathews*. Bell End, Belbroughton, *Humphreys*. Round Bromsgrove.

Probably, if not surely, all these records relate to planted trees; in shrubberies it is much more often seen than in a state of nature, except, for instance, at Epping Forest, which is a forest of Hornbeam. It is doubtless often mistaken for an Elm, but in youth its leaves are very distinct, being prettily folded into plaits. Its catkins too are very different from any others met with in England. Gerarde, not acknowledging the derivation of the English name suggested by the derivation of the Latin one, says the former is derived from the hardness of the wood, which might be compared to horn, and therefore the name, Hardbeam, or Hornbeam. As oxen were common beasts of burden in his day, possibly he would have mentioned them as connected with the name if such had been the case. This tree is well adapted for hedges, and bears trimming well. The tree occurs in every bordering county.

CORYLUS, Linn. 381.   (κόρυς, a casque or cap, from the form of the involucre.)

1493. **871. C. Avellana**, Linn.   *The Hazel-nut.*

Native. Woods, coppices, hedges. General and common. Shrub or tree. January to April. *Top. Bot.* 111.

First Record. Not localized, *Lees, Bot. Looker-out*, p. 94 (1842).

The yellow catkins of this plant are called by country children 'Lambs' tails', and, though so plentiful a plant, the mediaeval herbalist does not seem to have been able to find any virtue in it. It is a plant of wide distribution, occurring all over Europe, Western Asia, and North Africa, and also in all bordering counties.

QUERCUS, Linn. 382. (The Latin name, which some derive from the Celtic *kaër ques*, fine tree, and others from the Greek τραχύς, rough, from the roughness of its bark.)

1494. 872. **Q. Robur,** Linn.   *The Oak.*

Native. Woods, coppices, hedges. Generally distributed. Shrub or tree. January to April. *Top. Bot.* 105.

Var. a. **pedunculata,** Ehrh.

First Record. As *Quercus Robur, β.* Acorns on long fruit stalks. Little Shelsley, *Mr. Hollefear, Stokes's With. Bot. Arr.,* 2nd ed., p. 1084 (1787).

The aggregate, *Q. Robur,* has been divided into three segregates, sub-species *pedunculata,* as above, and the two following sub-species. It is a well-known tree in Worcestershire, usually preferring the red marl and stiff land, while the Elm prefers the red sandstone and Keuper. The finest collection of oaks in the county, without doubt, is the 'Wood-patch' at Kyre Park, some 180 or 200 trees, springing straight from the turf with enormous trunks and fine heads at the top of the fall stems, which here attain a height varying from 100 to 120 feet; but the majority of the Oaks at Kyre Park belong to the *sessiliflora* sub-species. *Q. pedunculata* is the oak generally to be met with. On the sandy land north of Kidderminster and throughout Worcestershire it is frequently abundant in all stages of growth up to a small tree on the wide roadside wastes, in many places forming dense thickets and showing that the countryside, if left to itself, would soon be transformed into an oak forest. The timber of Wyre Forest consists mainly of oaks, but the trees are seldom allowed to grow into any size, being shorn down as soon as they become marketable; but if allowed to grow up they would not attain any remarkable size, seeing that their roots soon strike the carboniferous limestone. The sub-species *pedunculata* is easily distinguished by its sessile leaves and acorns with long peduncles, whereas in *sessiliflora* the leaves are petioled and the peduncles of the acorns very short. The magnificent circle of ten oaks in Kyre Park is one of the finest examples of this tree in the county, and, like the Wood-patch there, belongs to the sub-species *sessiliflora.* The oak is found in all neighbouring counties.

Var. b. **intermedia,** D. Don.

First Record. *Quercus intermedia* seems to predominate [in Wyre Forest], *Lees, Bot. Worc.,* p. 4, note (1867).

Avon. Tiddesley Wood. Bow Wood.

Severn. Wyre Forest. Woods near Bewdley, *Westcombe.* Ribbesford Wood.

Malvern. In the district, *Towndrow.* Martley. Ankerdine. Abberley

This sub-species seems rather to be a variety of *sessiliflora,* from which it is distinguished by the young branches being glabrous instead of downy, as in that species, short petioles to the leaves, and shortly stalked acorns.

Var. c. **sessiliflora,** Salisb.

First Record. Not localized, *Lees' Cat., New Bot. Guide* (1835).

Avon. Stoulton. Tiddesley Wood. Crowle.

Severn. Hallow, *Towndrow.* Kyre Park. Wood between Cotheridge and Broadwas. Ribbesford Wood. Eymore Wood, Wyre Forest. Ombersley.

Malvern. Woods near Ankerdine.

CASTANEA, Hill. 383*. (Named from *Castanea* in Thessaly, where the tree grew finely.)

1495. 873. **C. *sativa,* Mill.**   *Sweet Chestnut.*

(*C. vulgaris,* Lam.   *Fagus Castanea,* Linn.)

Denizen. Plantations, woods, parks. Where introduced. Tree. May and June. Not in *Top. Bot.*

First Record. As *Fagus Castanea.* In Shrawley Wood, apparently wild, *Lees, Illus. Nat. Hist. Worc.,* p. 178 (1834).

Although this tree was a native of England in prehistoric times (*J. of B.,* 1885, p. 253), it is extinct as such now. It very seldom ripens its seed, and is nowhere established. Besides being planted as an ornamental tree, it is largely used to form undergrowth in woods, where it throws out numerous useful saplings from its stools. In the South of Europe, and in Italy especially, the chestnut forms a common food of the peasantry. Chestnuts, roasted on pierced trays over charcoal, are sold in London streets after nightfall, and are used also in several culinary ways, but are said to be difficult of digestion. Mr. Lees, while marking it as general in all other districts in our county in his *Table of Plants,* p. 26, does not give it for the Avon district, where it is now equally common as an introduced tree. Its wood is a very valuable timber, and many of the oak-timbered roofs of our old churches are alleged to be the product of this Chestnut. The tree grows in all neighbouring counties.

FAGUS, Linn. 384. (φηγός, in Greek, from φάγειν, to eat, from the nutritive fruit.)

1496. 874. **F. sylvatica,** Linn.   *The Beech.*

Denizen. Woods, parks, planted in hedgerows. Plentiful throughout the county. Tree. March and April. *Top. Bot.* 87.

First Record. Not localized, *Lees, Bot. Looker-out,* p. 94 (1842).

Some fine Beech-trees may be seen in Kyre Park, Hagley Park, and at Frankley Beeches, which is a well-known landmark in the neighbourhood of Birmingham. Julius Caesar asserted that the Beech did not occur in Britain, but his statement 'requires confirmation'. It is more frequently seen in the south of the county than the north, preferring calcareous measures, but grows into a fine tree everywhere. The wood is used for many purposes. It has been said that our word 'book' is derived from the same source as the name of the tree, since in early times books were bound in beech boards, or inscriptions were cut in the bark of the tree; but there are etymological difficulties in the way. The Beech trunk is still a favourite surface for the holiday-maker to cut his name upon; in Hagley Park is a fine tree on which the words 'Peaceful Silence', misspelt, however, were cut in large letters, with the date, in 1849, and the tree would hardly seem to have increased its girth since. The Beech is to be seen in all the bordering counties.

74. SALICACEAE, Rich.

SALIX, Linn. 385. (The Latin name, from the Celtic *sal lis,* near the water, from the locality of the plants.)

The Willows are a difficult genus. An eminent botanist has said, 'they are now a practically impossible study unless you have a large garden and can grow them for study in all their phases.' There is not, perhaps, in the whole vegetable kingdom a genus more liable to variation at different periods of growth and under different circumstances, especially in the leaves of the fertile plant. In fact, what is the sterile and what the fertile state of the same species can only be determined by growing them from seed. Coupled with this is the common tendency on the part of most of the species to hybridize with others. The genus has baffled the researches of the ablest botanists. A botanical parody runs as follows :

Ranunculi, they puzzle me,
The Brambles make me sad ;
The Hawkweeds rack my very brain,
But the Willows !—drive me mad !

1497. 875. **S. pentandra,** Linn.   *Sweet Bay-leaved Willow.*

Denizen. Watersides. Scarce. Shrub or tree. May and June. *Top. Bot.* 59.

First Record. Teme-side, Powick, *Lees, Bot. Malv. Hills,* 1st ed., p. 41 (1843).

Avon. Alderminster, *Cheshire.*

Malvern. Teme-side, *Lees.*

Lickey. The Leasowes, Halesowen; Westminster and Frog Mill Farms, Frankley; Harborne Reservoir, *Mathews.*

This is perhaps the most ornamental of our willows, and flowers latest of them all ; but it is not a native of the county. The large green leaves are nearly as fragrant, when bruised, as those of the Bay-tree. It has been observed in Salop, Stafford, and Warwick of our bordering counties.

1498. 876. **S. triandra,** Linn.   *Blunt-stipuled triandrous Willow.*

Native. Banks of streams. Rather rare, except in the Severn district. Small tree or shrub. April and May. *Top. Bot.* 66.

First Record. Not localized, *Lees, Bot. Male. Hills,* 1st ed., p. 41 (1843).

Avon. Fladbury. Eckington. Harvington. Near Evesham.

Severn. By the side of Severn, *Lees.* Clerkenleap. Holt. Ombersley. Near Broadwas. Hurcott Pool.

Malvern. Clifton-on-Teme. Stanford Bridge. Bransford.

Lickey. Harborne Reservoir ; Upper Bittell, *Mathews.*

Mr. Lees groups with this in his *Table of Plants, S. amygdalina,* Linn., equivalent with *S. triandra* proper according to *Hooker's Stud. Fl.,* 3rd ed., p. 370 ; variety γ. of it according to *Babington's Manual,* 9th ed., p. 378. Whichever it is, the First Record for *S. amygdalina* is in *Lees' Cat., New Bot.*

Guide (1835). *S. triandra* is recorded for all neighbouring counties as an aggregate.

Var. b. **Hoffmaniana,** Sm.

First Record. Localized in the Severn district, *Lees, Bot. Worc., Table of Plants,* p. 26.

This plant is not mentioned in the text of Lees' *Bot. Worc.* It is recorded from Herefordshire.

1499. 877. **S. decipiens,** Hoffm.   *Top. Bot.* 28.

First Record. Not localized, *Lees, Bot. Malv. Hills,* 1st ed., p. 41 (1843).

'Seems only a slight variety of *S. fragilis,*' Babington. 'A variety of that plant,' Hooker. The plant has been observed in every neighbouring county except Gloucestershire.

1500. 878. **S. fragilis,** Linn.   *Crack-willow.*

Native. Watersides. General. Tree. April and May. *Top. Bot.* 90.

First Record. Not localized, *Lees, Bot. Malv. Hills,* 1st ed., p. 41 (1843).

Avon. Norton, near Evesham. Evesham. Harvington. Fladbury. Pershore. Crowle. Bredon. Eckington. Flyford Flavell. Inkberrow.

Severn. By the side of Severn, *Lees.* Claines. Ombersley. Holt. Grimley. Kempsey. Stourport. Bewdley. Droitwich. Ladywood, Salwarpe. Crowneast. Boreley.

Malvern. General by ponds and brooks, *Lees.* Powick. Bransford. Leigh. Knightsford Bridge. Horsham. Clifton-on-Teme. Stanford Bridge. Shelsby Beauchamp.

Lickey. Damp meadows, general, *Mathews.*

This Willow grows to be a large tree with a bushy head, and the brittleness of the branches originated the English name. It is recorded from all neighbouring counties. A form of this willow, differing merely in the character of its branches and its smooth, glossy leaves, is known as *S. Russelliana,* Sm., the Bedford Willow, and was recorded but not localized by Mr. Lees in his *Catalogue* in the *New Bot. Guide,* 1835. The wood of this variety is the most valuable of any afforded by the Willow tribe, and the only willow from which the best cricket bats are manufactured. *S. fragilis* is recorded from all bordering counties.

Var. b. **britannica,** F. B. White.   *Top. Bot.* 31.

First Record. Malvern district, *Towndrow, J. of Linn. Soc.,* Nov. 13, 1890.

Malvern. As first record.

This variety has been observed in Staffordshire.

1501. 879. **S. alba,** Linn.   *Common White Willow.*

Native. Sides of rivers and pools, damp hedges. Common. Tree. April and May. *Top. Bot.* 92.

First Record. Not localized, *Lees, Bot. Malv. Hills,* 1st ed., p. 41 (1843).

Individuals of this plant form the rank and file of the Willows met with everywhere on the country side, its leaves turning up when the wind blows and showing their silvery undersides. Pollarded, the tree borders many a pond or winding brook, and it is largely planted in such places to furnish wood for poles, fences, crates, and such like. 'Oyle' of willow was a favourite remedy with our forefathers for many ailments, especially cramp, if it came 'of a hot cause'. It is met with in all the adjoining counties.

Var. b. **coerulea**, Sm.   *Huntingdon Willow.*

First Record. In all the districts, *Lees, Bot. Worc., Table of Plants*, p. 25.

This variety is a form with more glabrous leaves, and many of the willows seen about the county belong to it. The plant has been noticed near Malvern by Mr. Towndrow.

Var. c. **vitellina**, Linn.   *Yellow Willow, Golden Osier. Top. Bot.* 14.

First Record. As *S. vitellina.* Not localized, *Lees' Cat., New Bot. Guide* (1835).

Mr. Lees mentions that this tree is frequently seen in or close to cottage gardens. It is easily recognized by the yellow colour of its branches, sometimes deepening nearly into orange. We have no record of this plant from Herefordshire.

× **fragilis.** *Top. Bot.* 21.

(*S. viridis*, Fr.)

First Record. As *S. viridis*, Fr. Malvern district, *Towndrow, J. of Bot.*, vol. xxvi, p. 312 (1888).

Malvern. As first record.

Mr. Towndrow's plant was vouched for by Mr. F. B. White. This willow has not been observed in any bordering county.

× **triandra.**

(*S. undulata*, Ehrh.)

First Record. As *S. undulata.* Not localized, *Lees, Bot. Malv. Hills*, 1st ed., p. 41 (1843).

Avon. Near Evesham.

Severn. Sutton Pool, near Kidderminster, *Mathews.*

S. undulata is not mentioned in the text of Lees' *Bot. Worc.*; but in his *Table of Plants*, p. 20, it is assigned to one locality in the Avon district, and in the Herbarium of the *Vict. Inst. Worc.* there are specimens from near Evesham. Babington makes it a species, equating it with *S. lanceolata*, Sm. It has been observed in Salop and Staffordshire.

1502. 880. S. purpurea, Linn.   *Purple Willow.*

Native. Stream sides. Many places. Small tree. March and April. *Top. Bot.* 76.

First Record. Not localized. *Lees, Bot. Malv. Hills*, 1st ed., p. 41 (1843).

The first record possibly takes no account of segregates, but there is a form called *S. Lambertiana*, Sm., and this Mr. Lees seems to deem equivalent with the type. *S. Lambertiana* is talked of in the text of his *Bot. Worc.*, and is not localized in his *Table of Plants*; while exactly the reverse occurs with regard to *S. purpurea*. Bad as all willows are, this plant seems to be the wickedest of all; its disguises are numerous, and one never knows when one has got it! It occurs in all bordering counties.

× **viminalis.** *Top. Bot.* 36.

(*S. rubra*, Huds.)

(Possibly *S. Helix*, Linn.)

First Record. As *S. Helix.* Astwood, *Purton, Midl. Fl.*, p. 471.

This hybrid has a variety, *S. purpureoides*, Gr. and Godr. This plant was reported by Mr. Towndrow, *Malv. Advert.*, Dec. 3, 1892. Also there is another variety, *S. Forbiana*, Sm. This is included with *S. rubra* by Mr. Lees in his *Table of Plants*, p. 26, as occurring at one place in the Malvern district. When hybrids continue the sins of their parents and take to varying, things have come to a pretty pass! Neither the hybrid in its type nor any of its varieties seem to have been noticed in bordering counties.

1503. 881. S. viminalis, Linn.   *Common Osier.*

Native. Pond-sides, marshes, osier-holts. Common and generally distributed. Shrub or small tree. April and May. *Top. Bot.* 88.

First Record. Not localized, *Lees, Bot. Malv. Hills*, 1st ed. (1843).

The Osier is largely cultivated for making basket-work, and there are extensive holts down the course of the Stour from Kinver to Kidderminster, as well as in other parts of Worcestershire. It is recorded from all neighbouring counties.

× **stipularis**, Sm.

First Record. Not localized, *Lees, Bot. Malv. Hills*, 2nd ed., p. 74 (1852).

This plant has been observed in Salop and Warwick.

× **acuminata**, Sm.

First Record. Not localized, *Lees, Bot. Malv. Hills*, 2nd ed., p. 74 (1852).

This plant has been observed in every bordering county except Herefordshire.

1506. 882. S. caprea, Linn.   *Great round-leaved Sallow.*

Native. Woods, hedges, watersides. Generally distributed. Shrub or small tree. February to April. *Top. Bot.* 106.

First Record. Not localized, *Lees, Bot. Malv. Hills*, 1st ed., p. 41 (1843).

This is the earliest species of Willow to flower at the coming of Spring, the male plants being loaded with bright yellow blossoms before any of its leaves appear; and the twigs with the catkins, gathered at Easter, are called Palm-branches. Later in the season the satin-like shining catkins of the female plant appear. Another name for the plant is Goat Willow, because goats are said to be fond of its catkins. The wood is tough and employed for making handles for agricultural implements, hurdles, and for other rustic purposes. The bark used to be used for tanning leather, and is said to have the medicinal qualities in a less degree of Peruvian bark. It is recorded from all bordering counties.

× **cinerea.** *Top. Bot.* 5.

(*S. Reichardti*, A. Kern.)

First Record. Malvern, *Towndrow, J. of Linn. Soc.*, Nov. 13, 1890.

Malvern. Malvern Link, *Towndrow.*

This hybrid does not appear to have been noticed in any bordering county.

1507. 883. S. aurita, Linn.   *Round-eared Willow.*

Native. Hedges, thickets, damp woods. Locally plentiful. Shrub or small tree. April and May. *Top. Bot.* 106.

First Record. On the Ridgeway Common, *Purton, App.*, vol. iii, p. 76.

Avon. First record as above.

Malvern. At the base of Keysend and the Ragged-stone, *Lees.*

Lickey. Woods and roadsides about Northfield; Frankley; Romsley, *Mathews.* Randans and Chaddesley Woods.

This willow sends out straggling branches, and is well marked for a willow by the deeply sunken veins on the hairy leaves and its large stalked stipules. It is recorded from every neighbouring county.

× **caprea.** *Top. Bot.* 10.

(*S. Capreola*, J. Kern.)

First Record. Malvern, *Towndrow, J. of Linn. Soc.*, Nov. 13, 1890.

Malvern. Malvern Link, *Towndrow.*

× **cinerea.** *Top. Bot.* 30.

(*S. lutescens*, A. Kern.)

First Record. Malvern, *Towndrow, J. of Linn. Soc.*, Nov. 13, 1890.

Malvern. Near Malvern, *Towndrow.*

1508. 884. S. cinerea, Linn.   *Grey Sallow.*

Native. Woods, hedges, stream-sides, damp places. Common. Shrub or small tree. April. *Top. Bot.* 106.

First Record. Not localized, *Lees' Cat., New Bot. Guide* (1835).

This is a common willow, sometimes bordering a stream with its bushy growths, seldom rising into a tree, with rather rusty glittering foliage, and the leaves are somewhat leathery. It cannot be said to possess any beauty, and it is of no use. The Gray Sallow occurs in all neighbouring counties.

f. **aquatica**, Sm.

First Record. Not localized, *Lees, Bot. Malv. Hills*, 1st ed., p. 41 (1843).

This is a plant with broader, larger, and more pliant leaves than the type.

× **viminalis.** *Top. Bot.* 16.

(*S. Smithiana*, Willd.)

First Record. As *S. Smithiana.* Not localized, *Lees, Bot. Malv. Hills*, 1st ed., p. 41 (1843).

Avon. Not common, *Lees.*

Severn. Rather uncommon, *Lees.* Laughern Brook, *Lees.*

Malvern. Banks of Teme, *Lees.*

Lickey. Upper Bittell, planted, *Mathews.*

This plant is recorded from all neighbouring counties.

POPULUS, Linn. 386. (The Latin name, signifying that it was the tree of the people, which it was considered to be at Rome and in France during the revolutions.)

1517. 885. P. *alba, Linn.   *Great White Poplar or Abele.*

Denizen. Woods, hedges, fields. Not very common. Tree. March to April. *Top. Bot.* 70.

First Record. Banks of Stour, *Scott, Hist. Stourb.*, p. 540 (1831).

This is a remarkable tree, with its leaves cottony and snowy white beneath. It is doubtful if it is really an indigenous tree, having been planted in most of the localities in which it is seen. This tree is easily

distinguished from the cross with *P. tremula* when in flower, since the two stigmas in *P. alba* are bifid and placed crosswise, whereas the two stigmas in the hybrid are deeply divided into four linear lobes. *P. alba* is recorded from all neighbouring counties.

× **tremula.** *Grey Poplar.*

(*P. canescens*, Sm.)

Denizen. Wet woods and hedges. Several places. Tree. March and April. *Top. Bot.* 49.

First Record. Not localized, *Lees' Cat., New Bot. Guide* (1835).

This tree is frequently confounded with the preceding one, but may easily be distinguished when in flower as explained above, while some consider it much nearer the next. It is recorded from all bordering counties.

1518. 886. P. tremula, Linn.   *Trembling Poplar or Aspen.*

Native. Woods, hedges. Frequently seen, usually as a planted tree. Tree. March and April. *Top. Bot.* 105.

First Record. Perry Wood, Worcester, *Lees, Worc. Miscell.* (1829-30).

This Poplar is seen about the country occasionally bordering streams, or forming spinneys and plantations. It is a quick-growing tree, and the quiver of its leaves in the slightest breeze is caused by the vertical flattening of the leaf stalk, which counteracts the waving motion induced by the wind, a spring, so to speak, always endeavouring to bring the blade of the leaf back to a position of rest. The wood is not of good quality for ordinary purposes, but it had great reputation, in the days when such things were used, for making pattens. It burns with difficulty, and stands under water well. The tree has entered into poetry from early times chiefly as a simile for tremulousness; and people were so ungallant to liken women's tongues to the leaves, because said Gerarde, 'as the poets and some others' report, these seldom cease wagging. The earliest form of the name was Asp; the second syllable is a later addition. Scott, in his *Hist. Stourb.*, records *Populus communis* as growing on the 'Banks of Stour'. It is impossible to tell what plant he intended by this wide spacious name, but probably this tree, as he had already recorded the White Poplar. The Aspen occurs in all bordering counties.

1519. 887. P. nigra, Linn.   *Black Poplar.*

Denizen. Hedges, plantations. Not uncommon as a planted tree. March. Mentioned in *Top. Bot.*, but not numbered.

First Record. Near the Hayes on the banks of Stour, *Scott, Hist. Stourb.*, p. 540.

The Black Poplar is easily distinguished by its long pointed leaves, which in length exceed their width, and the glabrous scales of the catkins, but in appearance when in full leaf is not very different from the Aspen. It is later, however, in coming into leaf, and the young foliage is orange-yellow instead of yellow-green, and the catkins are of a deep rich red colour. It is a native of Central and Southern Europe, but is extensively planted elsewhere, especially in Normandy and Belgium. The well-known Lombardy Poplar is a fastigiate form of the tree. Apparently it has not been noticed in Gloucestershire, but this can only be from lack of eyes.

76. CERATOPHYLLACEAE, Gray.

CERATOPHYLLUM, Linn. 388.　(κέρας, a horn, φύλλον, a leaf, from the forked leaves.)

1521. **888. C. demersum,** Linn.　*Horned Pond-weed.*

Native. Rivers, ditches, ponds. Not common. P. August and September. Not distinguished in *Top. Bot.*, p. 170, both species being grouped under *C. aquaticum,* aggregate 53.

First Record. In fishponds at W. Rawlins's, Esq., Brockencote, Worcestershire, filling nearly the whole of one pool, *Purton, App.,* vol. iii, p. 70.

Avon. One place, *Lees.*

Severn. As first record. Spetchley, *Lees.* Near Severn Stoke. Harvington Moor. Grafton Manor. Pond on the Spout Farm, Lower Hagley.

Malvern. Longdon Marsh, *Lees.*

This plant grows entirely under water, and the green flowers occur in whorls in the axils of the leaves. It has been noticed in all bordering counties.

1522. **889. C. submersum,** Linn.　*Unarmed Hornwort.*

Native. Ponds. Not uncommon. September. *Top. Bot.,* see preceding species.

First Record. Pools on Welland Common, *Lees, Bot. Malv. Hills,* 1st ed., p. 40.

Avon. Not uncommon, *Lees.* Northwick Park.

Severn. Plentiful, *Lees.* Northwick Pool. Pool at Hallow, by roadside to Shoulton.

Malvern. As first record. Near Malvern Link.

The only distinction between this plant and the last consists in the very short persistent style and the absence of the two spines at the base of the fruit. Of bordering counties the plant appears to have been noticed only in Warwickshire.

77. CONIFERAE, Juss.

JUNIPERUS, Linn. 389. (The classical name, which is derived from the Celtic *jeneptus,* rough, from the taste of the fruit.)

1523. **890. J. communis,** Linn.　*Juniper.*

Native. Downs and woods. Rare. Shrub. May. *Top. Bot.* 78.

First Record. On barren waste land between Evesham and Church Lench, *Pitt, Agric. Worc.,* p. 317 (1810).

Avon. Craycombe Hill. Cleeve Hill, *Purton.* Coldknap Hill, *Lees.* The Slads. Lenchwick.

Severn. Wyre Forest, *Perry.* The New Parks, Wyre Forest.

Malvern. Bush Hill, Powick, *Lees.*

The Juniper appears to be slowly vanishing from Worcestershire, and has quite disappeared from most of the localities spoken of by the older botanists. Some plants are still to be found at Craycombe and the Slads, but in this locality many of them are more or less moribund owing to the attacks of the fungus *Gymnosporangium clavariaeforme.* The plant is useful for its berries, which are about as large as currants, appearing one year and remaining green until the next, when they become dark blue, covered with a whitish powder or bloom. Formerly they were used to flavour gin, but in this useful office they have been superseded by Turpentine. Old writers deemed the berries very remedial. They were 'hot in the third degree' and 'dry in the first'; and of course they were 'good against the bitings of venomous beasts'. Besides use in medicine, superstitious beliefs clung to the tree, and it was burned to expel evil spirits from the dwelling. It occurs in all bordering counties.

TAXUS, Linn. 390. (τόξον, a bow, from the use of the wood, or from τάξις, a row, because the leaves are arranged in two rows.)

1525. **891. T. baccata,** Linn.　*Common Yew.*

Native. Woods, probably planted in other situations. In various places. Tree. February to April. *Top. Bot.,* 2nd ed., p. 380, 69 ; *Lond. Cat.,* 10th ed., 17.

First Record. Numbers scattered over the country between Stourport and Abberley. Clearly an indigenous tree, *Stokes's With. Bot. Arr.,* 2nd ed., p. 1130 (1787).

The wood of the Yew was of considerable importance in England in olden times.

England were but a fling
But for the bow and the grey goose wing,

says an old proverb ; a circumstance put into stirring verse by Sir Conan Doyle in *The White Company.* As the games of one age are often the shadows of the stern business of the preceding ones, the bow still survives as a plaything with modern Archery Societies. 'The Baleful Yew,' says Virgil ; and its noxious character was believed in down the ages. At the present time a controversy rages as to its poisonous effect upon animals. No doubt animals have died after eating it, but whether their death was due to its poisonous nature, or to the mechanical irritation caused by the character of the food, or even caused by it at all apart from local circumstances, is often questioned. Certainly old Yews stand unguarded in many a meadow and pasture where animals have fed from time immemorial without any injurious results. Undoubtedly under some conditions it is poisonous, and not only to animals, but to pheasants ; many experiments, however, have been made, animals have been deliberately fed with yew, but the results have been indeterminate, and the question is still unsolved. The plant has been used medicinally, and one old writer said that wine kept in yew wood was a capital thing to administer to the guest whose removal was desired ! The association of the Yew with the churchyard is well known, but purists keep Yew sprays out of the church at the time of Christmas decoration. The tree bears clipping well. In Worcestershire the 'Twelve Apostles' at Cleve Prior have

Z

several times found a subject for the painter's brush, and in many parts of the county are more or less successful examples of the topiarian art. The Irish Yew is a fastigiate form of the tree. The tree occurs in all bordering counties.

PINUS, Linn. 391. (The classical name.)

1526. **892. P. sylvestris,** Linn.　*Scotch Fir.*

Now denizen, formerly native. Dry heathy woods. Met with throughout the county. Tree. May. *Top. Bot.* 17.

First Record. Planted, *Lees, Bot. Worc., Table of Plants,* p. 26.

The Scotch Fir formerly grew throughout Britain, and its remains are found plentifully in peat mosses and elsewhere. In some places seedling firs spring up in abundance. But whether our present firs are any of them derived from any original wild stock must remain doubtful, as the tree has been and is abundantly planted everywhere. It is a useful tree. The timber when the tree is grown under suitable conditions is good and durable, though as ordinarily grown in England it loses many of its good qualities. The resinous sap finds many uses, the bark is used for lining and covering huts, and the inner bark ground to powder is mingled with flour to make coarse black bread. The tree, of course, is to be seen in all the counties which border Worcestershire.

[Mr. Lees records in his *Table of Plants,* p. 26, *Abies excelsa* and *Larix Europaea* as 'planted'.]

MONOCOTYLEDONES, Juss.

78. HYDROCHARIDEAE, Benth. and Hook.

ELODEA, Michx. 392*. (ἑλώδης, growing in watery places.)

1528. **893. E. *canadensis,** Michx.　*Water Thyme.*

(*Udora canadensis,* Nuttall. *Anacharis Alsinastrum,* Bab.)

Colonist. Streams, ponds, ditches. Widely distributed. P. May to September. Not numbered in *Top. Bot.*

First Record. In the Avon at Evesham, *Mr. Cheshire,* June 1852. *Worcestershire Chronicle,* August 31, 1853.

The history of this plant, which has been seen, one year or another, in nearly every piece of water in the county, is remarkable. Mr. Lees, *Bot. Worc.,* p. 25, gives a summary of the history of this plant, which was first observed in Britain about 1841, having come into England from County Down in Ireland, where it was introduced from America about 1836. At the present time it is showing signs of decrease, perhaps in consequence of its long dependence upon vegetative reproduction in the absence of the male plant, which does not occur in this country. Ducks are very fond of it, and will nearly clear it out of small pieces of water, and Swans have been introduced on many waterways to eradicate this pest. It has been observed in all neighbouring counties.

HYDROCHARIS, Linn. 393. (ὕδωρ, water, χάρις, elegance, from being a showy aquatic plant.)

1529. **894. H. Morsus-ranae,** Linn.　*Common Frog-bit.*

Native. Ditches, ponds, slow streams. Uncommon. P. July and August. *Top. Bot.* 48.

First Record. In a pool by the side of the New Road, Worcester. Also in several other ponds near Powick and Kempsey, *Lees, Illus. Nat. Hist. Worc.,* p. 178 (1834).

Severn. Grimley Brickfield. Kempsey Grove. Brick Pits, Diglis. Pond between Cathedral Ferry and Bromwich. Near Severn Stoke. Pond near the Ketch.

Malvern. Upton-on-Severn, *Lees.* The Rhydd.

The plant is recorded from Gloucester, Stafford, and Salop.

VALLISNERIA, Michx. (Named after A. Vallisneri, an Italian botanist.)

V. SPIRALIS, Linn.

This alien plant, a native of the South of Europe and extending up to the Midi in France, was discovered in the stagnant water of Knapp's Brickyard at Northwick by Mr. George Reece on July 31, 1868, *Worc. Nat. Club Trans.* i. 245. No explanation of the occurrence there was ever forthcoming, as no brooks or drains emptied into the water ; and all that remains of the plant are Mr. Reece's specimens in the Herbarium of the Hastings Museum at Worcester. It is a plant frequently grown in aquariums and bowls by the botanically inquisitive, to see the curious manner in which the female blossoms are fertilized on the surface of the water by detached floating male flowers.

79. ORCHIDACEAE, Lindley.

NEOTTIA, Adans. 398. (νεοττιά, a bird's nest, from the fibrous root.)

1534. **895. N. Nidus-avis,** Rich.　*Bird's-nest Orchid.*

(*Ophrys Nidus-avis,* Linn. *Listera Nidus-avis,* Hook.)

Native. Shady woods, chiefly of beech or hazel. Rare. P. May to July. *Top. Bot.* 87.

First Record. As *Listera Nidus-avis.* In a coppice at Kempsey, *Dr. Streeten, Lees' Illus. Nat. Hist. Worc.,* p. 176 (1834).

Avon. Badger's Bank. Tiddesley Wood.

Severn. Wyre Forest. Habberley Valley. Shrawley Wood. Vallombrosa. Astley Wood. Warshill Wood.

Malvern. Silurian eminences, *Lees.* Sarn Hill. Alfrick. Ankerdine Hill. Malvern Wells. Powick.

Lickey. Ham Coppice, Pedmore ; Dales Wood, Romsley, *W. Mathews.*

The whole plant is dingy brown and withered-looking, much like some of the Broomrapes. It is recorded from all neighbouring counties.

LISTERA, Br. 399. (Named in honour of Dr. Martin Lister, an eminent British naturalist.)

1536. **896. L. ovata,** Br.　*Common Twayblade.*

(*Ophrys ovata,* Linn. *O. Bifolia,* Gerard.)

Native. Woods, pastures, marshes, bushy places. Common, and generally distributed. P. May to August. *Top. Bot.* 109.

Z 2

First Record. As *Ophrys bifolia*, Hurcott Wood, *Stokes, Withering*, ed. 2, vol. iii, p. cxxvii (1792).

This is quite a common plant readily distinguished by the two glossy green, strongly nerved leaves, placed some distance up the stem. The flowers are yellowish in colour and form a long loose spike. It is reported from all neighbouring counties.

SPIRANTHES, Rich. 400.   (σπεῖρα, a coil, ἄνθος, a flower, from the twisted inflorescence.)

1537.   897. S. spiralis, Koch.   *Fragrant Lady's Tresses.*
(*Ophrys spiralis*, Linn.   *S. autumnalis*, Rich.)
Native. Pastures and downs. Not common. P. August and September. *Top. Bot.* 59.
First Record. As *Ophrys spiralis*. Dry pastures, Sapey, *Sheward, Nash, Hist. Worc., Sup.*, p. 96 (1799).
    Severn. Wyre Forest, *Jorden*. Crookbarrow Hill, *Lees*. Coppice near Bewdley.
    Malvern. Doddenham. Woodbury Hill. Malvern Link.
    Lickey. Tardebigge, *Humphreys*.
The fragrance which gives the book-name to this plant is not very powerful. The flowers are greenish-white, and their spiral arrangement at once distinguishes this and the next plant from other British orchids. The plant occurs in all neighbouring counties.

{**1538.   898. S. aestivalis**, Rich.   *Summer Lady's Tresses.*
Native. Bogs. Only known to have occurred in two localities in Britain, Wyre Forest, and the New Forest in Hampshire ; and at the former it is extinct. P. July and August. *Top. Bot.* 2.
First Record. Margin of the Great Bog (Bewdley Forest), *Jorden, Phyt. N.S.*, vol. i, p. 151 (1855).
    Severn. Wyre Forest, *Jorden*.
The plant was gathered at the beginning of August, 1854, and has now not been seen for many years. It much resembles the last species, but it is a larger, laxer plant, the flower possesses a larger lip, and there are some less conspicuous differences. It need hardly be said that it does not occur in any neighbouring county.}

EPIPOGUM, S. G. Gmel. 402.   (ἐπί, upon, πώγων, the beard, from the form of the flower.)

[**1541. E. aphyllum**, Sw.
Among decayed leaves. Most rare. P. August. *Top. Bot.* 2.
First Record. Gathered by *Mrs. Anderton Smith* by Sapey Brook, on the Herefordshire side, in 1854, *Lees' Bot. Worc.*, p. 83 (1867).
For a considerable period this was the only known locality for this plant in Britain, but it has since been seen at Ringwood Chase, near Ludlow, and once in Ireland. Sapey Brook, near which it was discovered, divides Worcestershire from Herefordshire, and Worcestershire misses by a few yards the honour of having found a home for it. It has never been seen at the spot, a glen locally called 'Paradise', since the first discovery.

Mr. Lees gives a drawing of it as the frontispiece to his *Table of Plants*. The root of the plant was forthwith transplanted to the Rectory garden at Tedstone Delamere, in which parish it was found, where of course it speedily died. One wonders that a person who was botanical enough to detect a strange orchid should have done such a very foolish thing. So it can be said that though it has not been recorded in Worcestershire, it has been found in two bordering counties, and in one very close to the county border.]

CEPHELANTHERA, Rich. 403.   (From κεφαλή, the head, and ἀνθηρά, feminine of ἀνθηρός, an anther, from the position of the anther.)

1543.   899. C. longifolia, Fritsch.   *Narrow-leaved White Helleborine.*
(*C. ensifolia*, Rich.   *Serapias ensifolia*, Linn.   *Epipactis ensifolia*, F. W. Schmidt.)
Native. Woods. Rare. P. May and June. *Top. Bot.* 40.
First Record. As *Serapias ensifolia*. In a wood on the Whitley (Witley) side of Abberley Hill, *Sheward, Nash, Hist. Worc., Sup.*, p. 96 (1799).
    Lickey. Wood near Barnt Green.
    Severn. Wyre Forest, *Jorden*. Witley Park, plantations near entrance Lodge, *Lees*. Between Mopson's Cross and the Sorb Tree. Wyre Forest, *Walcot* and *Lees*. Wilden, Stourport, *Miss Ladbury*. The New Parks, Wyre Forest.
    Malvern. Abberley, as first record. On the top of Abberley Hill. *Rev. T. Butt.*
This plant differs but little in appearance from the succeeding one, the distinctions being bracts falling short of the glabrous ovary or germen, and a yellow spot on the lip. It is recorded from every bordering county.

1544.   900. C. grandiflora, Gray.   *Large White Helleborine.*
(*C. pallens*, Rich.   *Epipactis grandiflora*, Sm.)
Native. Woods. Rare. P. May to July. *Top. Bot.* 31.
First Record. As *Serapias grandiflora*, Mr. Knight's walks at Wolverley, *Stokes's With. Bot. Arr.*, 2nd ed., p. 1000 (1787).
    Avon. Blockley.
    Severn. Wolverley, *Dr. Stokes.*
    Malvern. Sherridge, 1900 ; 'Miss B. Norbury showed the plant to me,' *Towndrow (MS.).*
This plant has been observed in Gloucester, Hereford, and Warwick.

HELLEBORINE, Hill, 404.   (ἑλλεβορίνη, a plant like Hellebore. The Latin form had no *h*, which is restored in accordance with the Greek.)

1545.   901. H. latifolia, Druce.   *Broad-leaved Helleborine.*
(*Epipactis latifolia*, All.   *Serapias Helleborine*, var. *a. latifolia*, Linn.)
Native. Woods and bushy places. Local and not common. P. August and September. *Top. Bot.* 86.
First Record. As *E. latifolia*. In a place called the Dingle at Pedmore, near Stourbridge, and in the deep shades of the Devil's Den at Clifton-on-Teme, *Lees, Illus. Nat. Hist. Worc.* (1834).

    Avon. Bredon Hill ; Broadway, *Lees*. Tiddesley Wood. The Slads. Trench Woods. Elmley Castle.
    Severn. Wyre Forest, *Jorden*. Northwick. Witley Court, *Mathews*. Northwood, Bewdley, *Mathews*. Shrawley Wood. Nunnery Wood. Ribbesford Wood. Monk Wood. Ockeridge Wood.
    Malvern. Kyre Park. Clifton-on-Teme, as first record. Pull Court, Bushley. Cowleigh Park, *Lees*. Copse near Lulsley. Bransford. Martley. Near Bransford Bridge. Crews Hill Wood. Malvern Link. Malvern Wells.
    Lickey. Wychbury Hill. Ham Dingle, as first record. Pennyfields Wood, *Mathews.*
The above records probably refer to the aggregate species. Mr. Mathews says, *Fl. Clent and Lickey Hills*, p. 44, that he has not observed in the localities he gives for *E. media* any plants which he can refer without hesitation to this form of *Epipactis*. The aggregate has been observed in all bordering counties.

Var. b. **media**, E. S. Marshall.
(*Epipactis media*, Fr.)
Native. Woods and bushy places, in drier spots than the last. Local. P. August and September. *Top. Bot.* 38 ?
First certain Record of segregate. In Uffmore Forest, near St. Kenelm's, Clent Hills, Worcestershire, *Irvine, Phyt.*, vol. ii, p. 321 (1857).
    Severn. Nunnery Wood, *Baxter*. Elmbridge Green.
    Malvern. Copse, Folly Farm, Alfrick.
    Lickey. Woods, sparingly ; Chadwick, 1900, *Humphreys*. In nearly every wood in the upper valley of the Stour, *Mathews*. Barnt Green. Chaddesley Woods.
There is some confusion in the records of this plant, with which Mr. Lees combines *E. purpurata*, Sm., now considered to be *H. violacea*, Druce, the next species. It is difficult to disentangle them, and the specimens in the Herbarium Hastings Museum all seem referable to this species. Mr. Mathews, *Mid. Nat.* xv. 68, has a note upon this. This segregate appears to have been observed only in Gloucestershire and Salop of the bordering counties.

1546.   902. H. violacea, Druce.
(*Epipactis purpurata*, Sm.   *E. violacea*, Boreau.)
Native. Open woods. Rare. P. September. *Top. Bot.* 6.
First Record. As *Epipactis purpurata*. Parasitical on the stump of a maple or hazel in a wood near the Norrest Farm at Leigh, Worcestershire, in 1807. *Rev. Dr. Abbot, Withering*, 7th ed., vol. iv, p. 41 (1828).
    Avon. Trench Woods.
    Severn. Nunnery Wood, *Baxter*. Habberley Valley, believed to be this, *Worc. Nat. Club Trans.* i. 38.
    Malvern. Malvern Link ; Mathon, *Towndrow*. Leigh. Folly Coppice, Alfrick, *Lees*. Wood below the Hornyold Arms, Malvern Wells. Copse south of Broadwas Church, in the parish of Alfrick. Newland.
    Lickey. Glover's Ride, Offmoor Wood, *Mathews.*
This segregate appears to have been observed in Herefordshire only of bordering counties.

1549.   903. H. longifolia, Rendle and Britten.   *Marsh Helleborine.*
(*Epipactis palustris*, Crantz.   *Serapias longifolia*, Linn.)
Native. Marshes and bogs. Local. P. July and August. *Top. Bot.* 66.
First Record. As *Serapias longifolia*. Swampy meadows, Robinson's Street, on the borders of Malvern Chace, *Ballard, Stokes's With. Bot. Arr.*, 2nd ed., p. 998 (1787).
    Severn. Wyre Forest, *Jorden*. Great Bog, Wyre Forest.
    Malvern. Longdon Marsh.
The chief distinction between all these *Helleborines* is the shape of the labellum, that is, the terminal segment of the lip, with slight variations in the form of the leaf. *H. longifolia* is recorded from all neighbouring counties.

ORCHIS, Linn. 405.   (From an untranslatable word, referring to the double tuberous root.)

1551.   904. O. pyramidalis, Linn.   *Pyramidal Orchis.*
Native. Dry pastures, roadsides. Local. P. June and July. *Top. Bot.* 64.
First Record. Cleeve Hill, *Purton, Midl. Flora*, p. 421 (1817).
    Avon. Near Himbleton. Bredon Hill. Near Trench Woods, *Mathews*. Craycombe. Broadway. Cleeve Hill. Roadside, Flyford Flavell. Grafton Flyford. Crowle.
    Severn. Warndon. Crookbarrow Hill, 1853, *Gissing*. Berwick's Brake, Worcester.
    Malvern. Spout Brook, Eastham. Sarn Hill, Bushley. Leigh Sinton. Limestone Banks, *Lees*. Quarry near Round Hills, Abberley. Near Half Key.
This Orchis is a plant of calcareous formations, not very often being found on sandstone. The flower has a very peculiar odour, as to the pleasing character of which opinions differ. It is recorded from all bordering counties.

1552.   905. O. ustulata, Linn.   *Dwarf dark-winged Orchis.*
Native. Calcareous soil. Rare. P. May to July. *Top. Bot.* 44.
First Record. Abberley, *Mrs. Gardner, Purton*, vol. iii, pt. ii, p. 335 (1821).
    Severn. Near Witley.
    Malvern. Abberley, as first record. West Malvern, *Mathews*. Mathon, *Towndrow.*
A distinctive mark of this plant is its dark brownish-purple colour, looking nearly as if it had been scorched in fire. Its growth is low, reaching from four to six inches in height. One writer says the faint odour of the flower is like boiled cherries. Of bordering counties, it has not been observed in Warwickshire.

1556.   906. O. morio, Linn.   *Green-winged Meadow Orchis.*
Native. Meadows, pastures, heaths, bogs. Locally plentiful. P. May to June. *Top. Bot.* 63.
First Record. Not localized, *Lees, Bot. Malv. Hills*, 1st ed., p. 47 (1843).
    Avon. Bredon Hill, *Mathews*. Pirton. Crowle. Hanbury. Fladbury. Norton, near Evesham. Pershore. Defford Common.

Severn. Monk Wood. Warndon. Crookbarrow Hill. Roadside beyond the Virgin's Tavern. Claines. Ombersley. Kempsey. Severn Stoke. Grimley. Hallow. Near Stourport. Near Bewdley. Hartlebury Common.

Malvern. With pink and white flowers, Bridges-stone, Alfrick. Spout Brook, Eastham. Near Suckley Court. Old Storridge Common. Malvern Link.

Lickey. General, *Mathews*. Dodford. Tardebigge. Chaddesley Woods.

The flowers of this Orchis are few, forming a loose spike, and it may be known by its purple sepals, which are veined with green and curved upwards to form a kind of helmet to the blossom. It is known locally by the names Butcher, Bloody Butcher, and Bloody-man's Thumbs. It is recorded from all neighbouring counties.

1557. 907. O. mascula, Linn.   *Early purple Orchis.*

Native. Open woods, bushy places, meadows, heaths. Generally distributed. P. May to July. *Top. Bot.* 108.

First Record. Not localized, *Lees, Bot. Malv. Hills*, 1st ed., p. 47 (1843).

Avon. Bredon Hill, *Mathews*. Cleeve Bank. The Slads. Trench Woods. Tiddesley Wood. Near Goosehill Wood.

Severn. With white flowers, near Holt. Near Northwick. Monk Wood. Nunnery Wood. Shrawley Wood. Ockeridge Wood. Claines. Kempsey Grove. Hadley. Chatley. Crowneast. Ombersley.

Malvern. Cowleigh Wood. Near Bransford Chapel. Near Suckley Court. Middleyards Coppice. Lord's Wood. Ravenhills Wood. Ankerdine Hill. Rosebury Rock Wood. Martley. Woodbury Hill. Abberley Hill. Madresfield. Leigh Sinton.

Lickey. In front of the Lickey Monument, *Humphreys*. Sling Pool, Belbroughton, *W. Whitwell*. Illey, Halesowen. Chaddesley Woods.

This plant is one of the commonest of our orchises. From the root of this plant and of *O. morio* was prepared the substance known to our immediate forefathers as Saloop, from which a drink and a food were made supposed to be exceedingly nutritious, which were sold about the streets of London and in special houses called Saloop houses. The name is of Indian origin, a food of the same nature being prepared in that country from another kind of Orchis. This is the only useful substance yielded by any of the English members of the tribe. In some parts of Worcestershire this plant is called Billy Butchers and Red Butchers. The roots of all these orchises consist of two tubers, one dying every year and another forming on the opposite side of the stem-bearing one, from which next year's flower-stalk will rise ; consequently an Orchis plant shifts its locality a little bit year by year, and in course of time will travel several inches. *Orchis mascula* is found in all neighbouring counties.

1559. 908. O. incarnata, Linn.

Native. Marshes and wet meadows. Local. P. June. *Top. Bot.* 71.

First Record. Detected by Mr. W. Mathews, jun., in Wyre Forest, *Lees, Bot. Worc.*, p. xlix (1867).

Severn. 'I refer to this species specimens from Harberrow Pool, Spout Mill Swamp, and Brake Mill Pool,' *Mathews, Mid. Nat.* xvi. 38. Stanklin Pool.

Mr. Mathews also notes that the allied species, *O. latifolia*, does not seem to occur in those localities he mentions. The specimens from Spout Mill in the Herb. Hast. Mus. Worc. appear to be only *O. latifolia*. It may be doubted very much whether the true *O. incarnata* grows in Worcestershire. This plant was recorded as new by Dr. Arnold Lees from 'several stations' in the *Bot. Rec. Club Rep.* for 1880, but it was not so. It differs from the allied species by its erect leaves, which approach the stem and are narrowed from a broad base and not spotted, while all the bracts usually exceed the flowers. The plant does not appear to have been observed in Gloucestershire.

1561. 909. O. latifolia, Linn.   *Marsh Orchis.*

Native. Marshes, bogs, wet meadows. Local. P. June and July. *Top. Bot.*, aggregate, 105 ; segregate as *O. maialis*, Wats. 47.

First Record. Between Battenhall and Worcester, *Stokes's With. Bot. Arr.*, 2nd ed., p. 976 (1787).

Avon. Abberton. Huddington.

Severn. Bridewell, Bewdley, *Mathews*. Blakedown Viaduct. Laughern Brook. Harberrow Swamp. Bubble Bridge. Monk Wood. Near Hampton Lovett Church. Near Doverdale Church. Wyre Forest. Stanklin.

Malvern. Longdon Marsh, *Lees*. Between Midsummer and Hollybush Hill, *Lees*. Near Bransford Bridge. Bank of Leigh Brook. Welland. Alfrick.

In this plant the leaves are lanceolate-acute and spreading, and are often only faintly spotted. The plant has been observed in all bordering counties.

× maculata.

First Record. Leigh, *Towndrow, Malv. Advert.*, Dec. 10, 1892.

Severn. Monk Wood, *Rea*.

Malvern. Leigh, *Towndrow*.

1562. 910. O. maculata, Linn.   *Spotted palmate Orchis.*

Native. Woods, thickets, meadows, marshes. Generally distributed. P. May to July. *Top. Bot.* 108 (aggregate).

First Record. Woods and meadows, *Scott, Hist. Stourb.*, p. 540 (1832).

This Orchis is so frequently met with that it needs no localization. The lilac or white flowers grow on a solid stem (the stem of the last two species is hollow), which is about a foot high, the spike lengthening as time goes by. It is one of the wild flowers that is locally called 'Cuckoo-flower'. It is recorded from all bordering counties.

OPHRYS, Linn. 407. (ὀφρύς, the eyebrow, which Pliny says this flower was used to blacken.)

1565. 911. O. apifera, Huds.   *Bee Orchis.*

(*O. insectifera*, Linn.)

Native. Fields, quarries, calcareous banks, railway embankments. Local. P. May to July. *Top. Bot.* 59.

First Record. As *O. insectifera*, var. *apifera*. In rough pastures of a clayey

soil on the south side of Great Comberton, towards Wollershill, *Nash, Hist. Worc., Int.*, p. lxxxix (1781).

Avon. Great Comberton, as first record. Craycombe. Crowle. The Slads. Hipton Hill. Bredon Hill. Bentley. Tardebigge Reservoir.

Severn. Red Hill, Worcester, *Havergal*.

Malvern. Spout Brook, Eastham. Ridge Hill, Martley, *Baxter*. Railway-side near Hayley Dingle, Leigh. Leigh Sinton Lime-works, *Lees*. Leigh Sinton, abundant, *Jeffery*. Abberley Hill. Half Key. West Malvern.

The Bee Orchis is recorded from all bordering counties.

{**1568. 912. O. muscifera,** Huds.   *Fly Orchis.*

(*O. insectifera*, Linn.)

Native. Woods, thickets. One locality only. P. May and June. *Top. Bot.* 43.

First and only Record. Eastham, *Rev. Ed. Whitehead, Purton, App.*, vol. iii (1821).

Malvern. Eastham, as first record.

This discovery is mentioned also in Sir J. E. Smith's *English Flora* (1825), where it is said the plant was gathered at Spout Brook by the Rev. Edward Whitehead, the rector there. The Spout Brook at Eastham is a very prolific locality for the rarer Orchids. The plant is recorded from Gloucester, Hereford, and Salop.}

[*Ophrys incertae speciei* is recorded from 'Wychbury Wood, Cradley Park, and fields' by *Scott, Hist. Stourb.*, p. 540.]

HABENARIA, Willd. 409. (*Habena*, a strap, which the spur sometimes resembles.)

1570. 913. H. conopsea, Benth.   *Fragrant Orchis.*

(*Orchis conopsea*, Linn. *Gymnadenia conopsea*, R. Br.)

Native. Pastures, bogs, marshes. Local. P. June and July. *Top. Bot.* 102.

First Record. As *Orchis conopsea*. Cradley Park, *Scott, Purton, Midl. Flora*, p. 422 (1817).

Avon. Saldon, near Himbleton. Stoulton, *Gissing*. White Ladies' Aston.

Severn. Bog in Wyre Forest, *Perry*.

Malvern. Western side of the Hills, *Miss Southall*. West Malvern, *Towndrow*. Near Clifton-on-Teme. Half Key.

Lickey. As first record. Frankley, *Mathews*. Webheath, *D. Mathews*.

The flower of this plant has a powerful and pleasant odour, and is especially common in the mountainous parts of Scotland. It is recorded from all bordering counties.

{**1572. 914. H. albida,** Br.   *Small White Habenaria.*

(*Satyrium albidum*, Linn.)

Doubtfully a Worcestershire plant. Hilly pastures. P. June and July. *Top. Bot.* 49.

First Record. As *Satyrium albidum*. Cradley Park, Wychbury Wood ; Hodge Hill, Blakeshall, *Scott, Hist. Stourb.*, p. 540 (1832).

Severn. Hodge Hill. Blakeshall.

Lickey. Cradley Park. Wychbury Wood.

Mr. Mathews says these records are almost certainly an error. The plant has not been seen in the localities since. It is a plant of the north of Britain, seldom occurring much below Yorkshire and North Wales. That it does not occur has not prevented it being recorded, not only in this county, but in others also, *H. viridis* probably being mistaken for it. It must not be admitted to the Worcestershire Flora except with grave suspicion. Yet it appears to have been noticed in all bordering counties except Warwickshire.}

1573. 915. H. viridis, Br.   *Green Habenaria. Frog Orchis.*

(*Satyrium viride*, Linn.)

Native. Downs, pastures, heaths. Local. P. June to September. *Top. Bot.* 101.

First Record. As *Satyrium viride*, in meadows and pastures about Great Comberton and Pershore abundantly, *Nash, Hist. Worc., Int.*, p. lxxxix.

Avon. Cookhill Priory Farm, *Mathews*. White Ladies' Aston, *Mr. Suttle*.

Severn. Wyre Forest, *Jorden*. Battenhall. Cotheridge. Trimpley.

Malvern. Near Malvern Link Station. Round Hill, Abberley. West Malvern. Cowleigh Park. Madresfield. Malvern.

Lickey. Near the Lickey Monument.

This plant is recorded from all neighbouring counties.

1574. 916. H. bifolia, Br.   *Lesser Butterfly Orchis.*

(*Orchis bifolia*, Linn.)

Native. Heathy places, wet meadows, woods. Local. P. June and July. *Top. Bot.* 94.

First Record of aggregate. As *Orchis bifolia*, in a list of Malvern plants, *Ainsworth, Edin. Phil. J.*, p. 99 (1828). First certain Record of segregate. Open pastures east and west of Malvern Hills, *Lees, Bot. Malv. Hills*, 1st ed., p. 47 (1843).

Avon. Dovedale, Blockley.

Severn. Wychbury Hill, *Mathews*. Wyre Forest.

Malvern. As first certain record, *Lees*. Woodbury Hill.

Lickey. Hagley Hill ; Offmoor ; Romsley ; Frankley, *Mathews*. Tardebigge ; Dodford, *Humphreys*.

H. bifolia has been observed in all bordering counties.

1575. 917. H. virescens, Druce.   *Large Butterfly Orchis.*

(*H. chloroleuca*, Ridley. *H. clorantha*, Bab. *Orchis bifolia*, var. γ., Linn.)

Native. Woods, thickets, bushy places. Fairly plentiful. P. June and July. *Top. Bot.* 90.

First Record. As *H. clorantha*. Woods on the Limestone, *Lees, Bot. Malv. Hills*, 1st ed., p. 47 (1843).

Avon. Craycombe Hill. Croome Perry Wood. Tiddesley Wood. Trench Woods. The Slads. White Ladies' Aston. Dovedale, Blockley.

Severn. Witley. North Wood, Bewdley, *Mathews*. Wyre Forest. Shrawley Wood. Ockeridge Wood. Monk Wood. Gardners Grove.

Malvern. Near Malvern. Woods, Clifton-on-Teme. Sarn Hill Wood. Malvern Link. Great Witley. Suckley Hill. Fries Wood. Middleyards Coppice. Lord's Wood. Cowleigh Wood. Ravenshill Wood. Ankerdine Hill. Woodbury Hill. Half Key. Bush Hill, Powick.

Between this plant and the last there is considerable general resemblance but several minute differences, and according to Darwin they require different species of moths to fertilize them. Roughly *H. bifolia* is the smaller plant, growing chiefly in grassy localities or on open heaths, and always very sweet-scented, while this plant is mostly confined to woods and grows sometimes two feet high, and the flowers are usually larger. It has been observed in all bordering counties.

80. IRIDACEAE, Lindley.

IRIS, Linn. 411. (*Ipis*, the rainbow, from the varied colours of the flowers.)

1577. 918. I. foetidissima, Linn. *Stinking Iris.*
Native. Woods, thickets. Local. P. May to July. *Top. Bot.* 49.
First Record. This grows plentifully in woods and thickets and by waysides about Great Comberton and other places in the neighbourhood of Pershore, *Nash, Hist. Worc., Int.*, p. lxxxix (1781).
Avon. Near Hanbury Church. Tiddesley Wood. Bredon Hill. Broadway. Great Comberton.
Severn. Crookbarrow Hill. Astwood.
Malvern. White House Coppice, near Powick Asylum, *Rea*. Sarn Hill, Bushley. Newland. Leigh Sinton.
This is a plant of the south of England, abundant enough in Devonshire and Dorsetshire, its leaves forming part of the ordinary hedgebank vegetation. It gets scarcer proceeding northward, and is only naturalized in Scotland. Both this and the next plant have the name Gladwine or earlier Gladdon, probably somehow derived from the Latin *gladius*, a sword. This plant was Stinking Gladdon, the next Corn Gladdon. Unbruised, some people say the odour of this plant resembles roast beef, whence a common name for it. 'Roast Beef Plant'; bruised the odour is detestable. Yet it has found its use in early medicine, and all parts were praised by the old herbalists. Gout, coughs, colds, and liver complaints were vanquished by it, though one writer naïvely observes the decoction 'somewhat hurts the stomach', and should not be taken without honey. The seeds of this Iris form a pretty object in winter, when the capsule opens and shrivels back, leaving the brilliant orange seeds fully displayed. The plant is not recorded from Staffordshire.

1579. 919. I. Pseudacorus, Linn. *Yellow Water Iris or Corn-flag.*
Native. By water and in marshy places. General throughout. P. May to July. *Top. Bot.* 112.
First Record. Not localized, *Lees, Bot. Malv. Hills*, 1st ed., p. 43 (1843).
This plant is a well-known feature of water-side vegetation, and sometimes its clumps stud thickly the marshy meadow. This, like the former, was

seized upon by the old herbalists for medicinal purposes; indeed, it was an important plant, being 'under the dominion of the sun'. The root is very astringent, and some preparation of it was used as a cosmetic, to 'clense the face of frekels', and to 'resolve the pockys and whelkys of the face'. In some parts of Worcestershire it is commonly called Sags. The plant is recorded from all bordering counties.

{**I. XIPHIUM**, Linn. *Spanish Iris.*
Recorded in *Nash, Hist. Worc., Int.*, p. lxxxix (1871). By the river's side near Fladbury and some other places in this county; first discovered by the Duchess Dowager of Portland, that great admirer of Natural History.
This plant is a native of sandy situations in South-west Europe. It has been long, and now is, extensively cultivated in gardens, and has no claim to any status among British plants.}

CROCUS, Linn. 412. (*κρόκη*, a thread, from the appearance of dried Saffron.)

{**1581. 920. C. *officinalis**, Huds. *Purple Spring Crocus.*
(*C. vernus*, All.)
Denizen. Meadows. Very local. P. March and April. Not in *Top. Bot.*
First Record. With white flowers in a low field south side of Worcester, *Lees, Mag. Nat. Hist.*, vol. iii, p. 160 (1830).
Severn. As first record.
This Crocus is a native of the meadows of Central and Southern Europe. Mr. Lees apparently observed this plant in other localities than the above, as he says it occurs 'near Worcester at times, but in proximity to occupied ground or old habitations', *Bot. Worc.*, p. xxxi. Probably the plant has been introduced everywhere where it occurs in England, even where most abundant, as near Nottingham and near Ludlow. It has been observed in Salop and Staffordshire.}

[**C. sativus**, Linn. *Saffron Crocus.*
First Record. Not localized, *Lees, Bot. Worc.*, p. xxv (1867).
Mr. Lees, *Bot. Worc.*, p. xxxi, records this plant as growing at Kyre Wyard. What Nash says, *Hist. Worc.*, ii, p. 72, is that great quantities of Wild Saffron grow about the place, but it appears nowhere to be cultivated; and he was most probably alluding to Meadow Saffron, *Colchicum autumnale*. The Saffron Crocus was formerly cultivated, especially in the east of England, for the sake of the aromatic orange-coloured stigmas which when dried constituted the Saffron so largely used in ancient cookery. Its habitat in a state of wildness is said to be from Italy eastward to Kurdestan.]

C. luteus, Lam.
In the Herbarium Hastings Mus., Worcester, is a specimen of this plant labelled The Hill, Diglis, Worcester. It has no claim to be British any more than that it is a popular garden favourite, gladdening the eyes in spring.

81. AMARYLLIDACEAE, R. Br.

NARCISSUS, Linn. 416. (From the youth Narcissus, fabled to have been changed into this plant.)

1587. 921. N. Pseudo-narcissus, Linn. *Common Daffodil.*
Native. Woods, coppices, meadows. Locally plentiful. P. February to April. *Top. Bot.* 78.
First Record. In orchards, Hanley Castle, *Ballard, Stokes's With. Bot. Arr.*, 2nd ed., p. 342 (1787).
Avon. Callion's Wood, Wadborough. Deerfold Wood. Bentley.
Severn. Kempsey. Norton by Kempsey. Kempsey, in a field opposite Clevelode. Boreley, in a field opposite Lenchford. Wyre Forest. Ockeridge Wood. Meadow near Astley Church. Ombersley.
Malvern. Mathon. Pensax. Near Whippett's Brook, North Malvern. Old Storridge.
Lickey. King's Norton. Northfield. Cradley, *Rev. J. H. Thompson*. Westminster and Brookhouse Farms, Frankley, *Mathews*. Near Yardley, *Westcombe*. Overend Wood, Cradley. Northfield.
The Daffodil is far more frequently seen in the garden than wild; there is hardly a garden in the land where it is not grown in some form or other, or some species or other. Most frequent of all is the double daffodil, which is not this plant, but *N. Telamonius*, most easy to grow in its double form, difficult to get to flower in its single and natural state. It is a darker yellow than this flower, with firmer petals. A name for this daffodil is Lent Lily, and it has received many country names, Daffadilly, Daffadowndilly, and Jonquils among them. Under cultivation of late years many varieties have been produced, and it is now a very fashionable flower. The history of the name Daffodil is curious. In early times the plant was called Affodil, a name properly belonging to, and derived from, the Asphodel, and afterwards transferred to this plant also. Then much discussion appears to have arisen among the learned of the period as to the confusion arising from two plants bearing the same name, and a compromise was arrived at whereby the Asphodel went on being called Affodil, while the Narcissus was called Daffodil; but where they got the 'd' from nobody knows, though many have conjectured. Curiously also while the word in England was gaining a letter, in France it stood in danger of losing one, *Décoction de frodilles*, being met with. This plant is recorded from all adjoining counties.

{Var. b. lobularis, Haw.
First Record. Localized in the Malvern and Teme district, *Lees, Bot. Worc., Table of Plants*, p. 28.
No one knows anything more of this in Worcestershire than is stated in the first record. It is not considered a true native of England, and scarcely differs from the Tenby Daffodil so frequently seen in gardens, which is *N. cambricus*, Haw., and frequently called also *N. obvallaris*. The var. *lobularis* has not been observed in any bordering county.}

1589. 922. N. INCOMPARABILIS, Linn. *Primrose Peerless.*
First Record. Stated by my late friend J. Roby, Esq., to be naturalized in some places about Malvern, *Lees, Bot. Malv. Hills*, 2nd ed., p. 79 (1852).

Severn. Meadow near Astley Church, *Herb. Hastings Museum*.
Malvern. Field near Bush Hill Wood, Bransford, with double flowers, *Jeffery and Rea*, 1908.
This plant has long been cultivated in English gardens. It is a native of Southern Europe, and when seen off a border may at once be branded as an Alien, Casual, and Garden escape. In such a condition it has been observed in Herefordshire.

1591. 923. N. *biflorus, Curtis. *Pale Narcissus.*
First Record. In fields near Yardley Wood Pool, Worcestershire, together with *N. pseudo-narcissus*, *Withering*, 4th ed., p. 325.
Severn. Glasshampton, Stourport, *Scott*. Near the Ketch, Worcester, *Lees*. Bagnall, near Kempsey, *Dr. Streeten*. Meadow above Kepax Ferry, Worcester, *Westcombe*. Kempsey Grove. Astley. Near Bewdley.
Malvern. Near the Hoarstone, Sapey Brook, *Lees*. Berrow, near Upton-on-Severn, *Lees*. Bransford.
Lickey. Yardley Wood Pool, *Withering*. Yardley, *Westcombe*.
Neither is this a native plant, but like the last, much grown in gardens, and not infrequently to be seen in orchards or meadows near habitations. It has so been noticed in all bordering counties except Staffordshire.

1592. 924. N. *poeticus, Linn. *Poet's Narcissus.*
First Record. Found in a dubious spot, *Lees' Cat., New Bot. Guide* (1835).
The 'dubious spot', from which it eventually vanished, was at Crowneast, on the site of a former garden. So says the finder, *Bot. Worc.*, p. xxxi. It has no claim to be a British plant, being a native of mountain meadows in Central and Southern Europe. It does not appear to have 'escaped' in Gloucestershire.

GALANTHUS, Linn. 417. (*γάλα*, milk, *ἄνθος*, a flower, from the colour of the flower.)

1593. 925. G. nivalis, Linn. *Snowdrop.*
Denizen. Orchards, woods, brook-sides. Rare. P. February and March. *Top. Bot.*, mentioned, but not numbered.
First Record. At the foot of Malvern Hills, *Nash, Hist. Worc., Int.*, lxxxix (1781). If this is in Herefordshire, then the First Record is Tutnall near Bromsgrove, *Stokes, Withering*, 2nd ed., p. 340, 1787.
Avon. Aldwick Wood, north side of Bredon Hill. Overbury Wood, abundant and as common as Bluebells are in other woods.
Severn. Astley Wood, *Hickman*. King Stephen's embankment, Henwick, *Lees*. Brickyard beyond Bevere Green, *Westcombe*, 1841, where a few plants still linger on to the present day.
Malvern. Hadleys Bottom, just below Upper Sneads Farm, Rock, *Jeffery*. Mathon.
Lickey. Coppice near Bell End. Finstall, Bromsgrove, *Stokes*. On the side of the Ridgeway, *Purton*.
Our earliest writers do not mention this as a wild flower. Yet it is a native of Belgium, Normandy, and Northern Europe, and may not impossibly extend sparsely in a native state to England, and those who have

seen it flowering in Overbury Wood would undoubtedly be impressed with its claim as a native. The early blossoming of the plant has given it one of its best-known country names, 'February Fair-maids'. It is abundant in gardens and shrubberies, and in places where gardens and shrubberies have been. It has been observed in all bordering counties.

LEUCOJUM, Linn. 418. (λευκός, white, ἴον, a violet.)

{1595. **926. L. vernum**, Linn.   *Summer Snowflake.*

Casual. River-sides, osier holts. Once recorded. P. April to June. *Top. Bot.* 2.

First Record. Localized in one place, suspicious and lost, in the Malvern district, Lees, *Bot. Worc.*, *Table of Plants*, p. 28 (1867).

There is no mention of this plant in any of Mr. Lees' books, except as above. It is frequently cultivated and frequently escapes, but its native range would not preclude its being indigenous in England. The plant has been observed in Warwickshire outside cultivation.}

82. DIOSCOREACEAE, R. Br.

TAMUS, Linn. 419. (The old Latin name used by Pliny.)

1596. **927. T. communis**, Linn.   *Black Bryony.*

Native. Hedges and thickets. Common and well distributed. P. May to July. *Top. Bot.* 69.

First Record. Hedges in the Vale of Evesham, *Pitt, Agric. Worc.*, p. 317 (1810).

This plant is so frequently seen in our hedges, that it is not necessary to enumerate localities, though except in Malvern, where he acknowledges it to be 'plentiful', Mr. Lees, in the *Table of Plants*, p. 27, will only allow it to be 'not uncommon'. The root of the plant is large, black, and ovoid, and was much valued by old physicians for stimulating plasters for several ailments. The flowers are inconspicuous, but the scarlet oval berries of the plant often festoon the hedgerows when autumn has arrived. It is recorded from all neighbouring counties.

83. LILIACEAE, Juss.

RUSCUS, Linn. 420. (Anciently *bruscus*, from the Celtic *bruskelen*, meaning Box-holly.)

1597. **928. R. aculeatus**, Linn.   *Butcher's Broom.*

Casual. Woods and thickets, on calcareous soil. Once recorded, doubtfully. P. January to March. *Top. Bot.* 29.

First Record. Hedge near Sherrard's Green, must have been planted, *Towndrow, Malv. Advert.*, December 10, 1892.

A rigid, prickly, little shrub, the solitary flower appearing in the middle of the upper surface of the leaf-like flattened upper shoots of the stem. It was formerly used by butchers to sweep out places where meat was kept, whence its name, but it had a string of other names also, Kneeholm, Knee Holly, Petigree, Knee-hulver, Wild Myrtle, and Jew's Myrtle, with others. It has been observed in Herefordshire, but in no other bordering county.

ASPARAGUS, Linn. 421. (Some derive it from the Greek ἀσφάραγος, the throat, whilst others derive it from ἀ, intensive, and σπαράσσω, to tear, in reference to the shiny prickles of some of the species.)

1598. **929. A. maritimus**, Mill.   *Asparagus.*
(*A. officinalis*, Linn.)

Denizen. Woods, garden walls. Very rare. P. June to August. *Top. Bot.* 7.

First Record. Rough walls at Great Malvern, Lees, *Bot. Worc.*, p. xxi (1867).

Avon. The Slads.

Malvern. As first record.

This plant was found growing 'in the heart of the Slads, far removed from human habitations, amidst giant bushes of the English Juniper and Hawthorn trees of great age', in 1891, *Worc. Nat. Club Trans.* i. 375. It was again noted 'in a field adjacent to the Slads, apparently wild', June 7, 1906, loc. cit., iii. 261. However the plant got into this position it is maintaining itself strongly at the present time. It is a well-known vegetable whose native home in England is the seashore of the south-west. It has a wide distribution in the northern temperate region, but is not found so far south as Greece. The history of the name of this plant is curious, as the histories of plant names often are. Very early the initial of the Greek name was lost, and in mediaeval Latin the plant was called *sparagus*, and from this form are derived the Italian and German names of the plant; and in England before 1600 by popular etymology the name was corrupted into 'Sparrow-grass'. In 1791 Walker in his Dictionary said '*Sparrow-grass* is so general that *asparagus* has an air of stiffness and pedantry'. It is only since the beginning of the last century that 'asparagus' has returned into literary and polite use, leaving 'sparrow-grass' to the uneducated and the cookery book, where even now it is to be seen. The plant has been observed outside cultivation in Gloucestershire and Warwickshire.

POLYGONATUM, Hill, 422. (πολύς, many, γόνυ, knee, from the many nodes of the stems.)

1600. **930. P. multiflorum**, All.   *Common Solomon's Seal.*

Native. Woods and thickets. Very rare. P. June and July. *Top. Bot.* 38.

First Record. Near Clifton-on-Teme, in one spot, Lees, *Bot. Worc.*, p. xxxi (1867).

Malvern. As first record.

Severn. Fenny Rough, near Chaddesley.

The latter locality was given as a first record in *Bot. Rec. Club Rep.* for 1881-2, but it was not so. The plant still flourishes there. It is a favourite garden flower, but it has lost its wondrous reputation as a healing herb. On cutting the root transversely certain marks are observable, having some resemblance to the characters of a seal; and from the ancients to mediaeval times descended the notion that they were the impress of the seal of Solomon himself. So the plant was used to seal or consolidate wounds; and the bruised roots in ale were given as a drink to the person 'of what sexe or age

A a

whatsoever' with broken bones, for it is one 'which soddeneth and gleweth together the bones in very short space and very strongly; yea, although the bones be but slenderly and unhandsomely placed and wrapped up'. The plant is recorded from all neighbouring counties.

CONVALLARIA, Linn. 424. (*Convallis*, a valley.)

1603. **931. C. majalis**, Linn.   *Lily of the Valley.*

Native. Woods and coppices. Local and rare. P. May and June. *Top. Bot.* 60.

First Record. Shrawley Wood, *Sheward, Nash, Hist. Worc., Sup.*, p. 96 (1799).

Severn. Wyre Forest, *Jorden.* Shrawley Wood, *Sheward.* Birchin Grove, *Lees.* Monk Wood, *Baxter*; there in greater quantities than ever, *Rea.* Fenny Rough. Astley Wood. Wyre Forest, Hinterhill, abundant. Stagbury Hill. Ribbesford Woods. Rainbow Hill, *Herb. Hastings Museum.*

Malvern. In the district, *Lees.*

Lickey. Brockhill Wood, Tardebigge, *D. Mathews.*

This favourite flower has not escaped the attention of the herbalists. It strengthened the memory and eased the apoplectic; and to alleviate the gout, was prepared in this odd way. 'Have filled a glass with the flowers, and being well stopped, set it for a moneth's space in an ante's hill, and after being drayned cleare, set it by for use.' A snuff made of the dried and powdered flowers was said to be good for the cure of headache. The plant is recorded from all neighbouring counties.

ALLIUM, Linn. 426. (Celtic *all*, acrid or burning, from the strong taste of the bulbs.)

[1607. **932. A. Scorodoprasum**, Linn.   *Sand Garlic.*

Casual. Sandy woods and fields. Possibly recorded in error. P. June and July. *Top. Bot.* 17.

First Record. Battenhall, near Worcester; on red marl, *Lees' Cat., New Bot. Guide* (1835).

Severn. As first record.

This is a suspicious record. The plant is not known certainly in England south of Yorkshire, and where it occurs is not by any means a common plant. Mr. Lees had made a mistake, which he afterwards recognized, as it is not mentioned in his *Bot. Worc.*, nor localized in his *Table of Plants*. It may safely be removed from the list of Worcestershire plants. It has not been observed in any neighbouring county.]

1609. **933. A. vineale**, Linn.   *Crow Garlic.*

Native. Dry banks, wall-tops, hedge-sides, cornfields. Not rare, except in the Lickey district. P. June to August. *Top. Bot.* 80.

First Record. *Allium vineale*, β., with a double head of bulbs. Near Worcester, *Stokes's, With. Bot. Arr.*, 2nd ed., p. 344.

Avon. Norton, near Evesham. Sheriff's Lench. Roadside, Fladbury. Crowle. Pershore. Stoulton. Pinvin. Stonebow Bridge, Peopleton. Upton Snodsbury.

Severn. Churchyard, Stourport. Mitton Chapel, near Hartlebury Common. Near Rushwick. Pitchcroft. Lincombe. Ombersley. Ladywood, Salwarpe. Kempsey. Hartlebury. Grimley.

Malvern. Lord's Wood. Powick. Bransford. Alfrick. Knightwick. Martley. Malvern Link. Madresfield.

This plant is recorded from all bordering counties.

Var. b. **bulbiferum**, Syme.

First Record. As below.

Malvern. Newland, *Towndrow.*

This variety has not been observed in any bordering county.

Var. c. **compactum**, Thuill.

First Record. The common form, *Towndrow, Malv. Advertiser*, December 10, 1892.

Avon. The Slads, *Towndrow.* Sheriff's Lench. Norton, near Evesham. Fladbury. Pershore. Crowle.

Severn. Ombersley. Ladywood, Salwarpe. Kempsey.

Malvern. Sherridge, *Miss B. Norbury.* Malvern Link, and many other places. Powick. Martley. Madresfield.

In this variety the umbel is without flowers, and the head bulbs have a leaf-like point. It is the more common form, and has been noticed in Salop.

1610. **934. A. oleraceum**, Linn.   *Streaked Field Garlic.*

Native. Upland meadows, borders of fields. Rare. P. July. *Top. Bot.* 52.

First Record. In the thicket below the Ketch, Lees, *A. Florence, Strang. Guide* (1828).

Avon. Craycombe and Coldknap Hills. Roadside, Pinvin. Fladbury. Stoulton. Norton, near Evesham. Badsey. Crowle Hill. Sheriff's Lench.

Severn. Wyre Forest, *Jorden.* As first record. Crookbarrow Hill, *Gissing.* With double heads at Battenhall, *Lees.* Meadows near Kempsey. Pixham. Bewdley. Ombersley. Meadows near Holt Bridge.

Malvern. Borders of fields about Malvern Wells and Hanley, rare, *Lees.* Malvern Hills. Sherridge. Meadows near Ham Castle. Martley.

Lickey. Field between Broom and Yieldingtree, probably this, 1903. *W. Whitwell.*

A. oleraceum has been observed in all bordering counties except Salop.

{1611. **935. A. *carinatum**, Linn.

Of this doubtfully native plant, the number attached to which in *Top. Bot.* is 3, there is a specimen in *Herb. Vict. Inst. Worc.*, labelled 'Severn, between Bewdley and Blackstone, T. Westcombe'. Nothing else is known of it as a Worcestershire plant, and it is not mentioned in any of the books. Mr. Westcombe was a very competent botanist. The plant is naturalized in muddy situations on the banks of the Tay, Ouse, and Esk, but is not known elsewhere in England.}

A a 2

1615. 936. A. ursinum, Linn. *Broad-leaved Garlic or Ramsons.*

Native. Damp woods and bushy places. Abundant, except the Avon district, where it is not so common, but generally distributed. P. May and June. *Top. Bot.* 109.

First Record. Near Crookbarrow Hill, *Lees, A. Florence, Strang. Guide* (1828). This is a common plant in suitable situations, and requires no special localization. The flower is showy, but the odour of the plant when bruised is very strong, and picking the flower brings it out very powerfully. The ancient name of the plant was Rams, a noun singular with a plural in -en, Ramsen. Ramsen got itself turned into Ramson, which word in time obtained another form of plural, and the plant is now Ramsons. It occurs in every bordering county.

SCILLA, Linn. 428. (Named from σκύλλω, I injure.)

1619. 937. S. non-scripta, Hoffmgg. and Link. *Blue-Bell, Wild Hyacinth.* (S. nutans, Sm. S. festalis, Salisb. Hyacinthus Non-scriptus, Linn. Endymion nutans, Dumort. Agraphis nutans, Link.)

Native. Woods and thickets, hedgebanks. Abundant and widely distributed. P. April to June. *Top. Bot.* 112.

First Record. As *Hyacinthus Non-scriptus,* Var. 2, Blossoms White. Near Worcester, *Stokes, Stokes's With. Bot. Arr.,* 2nd ed., p. 356 (1787).

This many-named plant is an abundant and beautiful ornament of our woods in spring, covering the ground with a shimmering sheet of blue. What flower was the Hyacinth of the ancients is uncertain, certainly not this, for it gets scarcer as it recedes from the west, and is absolutely unknown in Switzerland. The flower the Greeks called the Hyacinth bore some marks on the leaves resembling the word αι, 'alas!' when written in capital letters; and it is from the absence of those marks on our plant that it obtained the specific name of *Non-scriptus,* or the generic one of *Agraphis.* The whole plant is full of a slimy juice which does not appear to have been used other than as mucilage by our forefathers. Individuals with white flowers are met with sparingly in most woods, and sometimes, but rarely, with their flowers pink. The plant is abundant in all bordering counties.

ORNITHOGALUM, Linn. 429. (ὄρνις, a bird, γάλα, milk.)

1620. 938. O. nutans, Linn. *Drooping Star of Bethlehem.*

Alien. Meadows, fields. Rare. P. April. Not in *Top. Bot.*

First Record. Field at Clerkenleap, Worcester, *Lees, A. Florence, Strang. Guide* (1828).

Severn. As first record. Wyre Forest, *Jorden.* Warshill, Bewdley, *Jorden.* In an orchard and adjoining lane near Kidderminster. Danks's Orchard, Charlton. Hartlebury. Roadside at Hoarstone, near Bewdley.

Malvern. In the district, *Lees.*

Mr. Lees' Malvern locality no doubt is Bromsberrow, where the plant grows in a field next to the Rectory, an adjoining parish to Worcestershire, but in Gloucestershire. Except in Southern Europe, the plant is a weed of cultivated ground, or in England an escape from a garden. It has been observed in this state in Gloucestershire, Herefordshire, and Warwick.

1621. 939. O. *umbellatum, Linn. *Common Star of Bethlehem.*

Denizen. Meadows, pastures. Local. P. May and June. Not in *Top. Bot.*

First Record. In a meadow near Dr. Berkeley's at Cotheridge, *Mrs. Walcot, Lees, Illus. Nat. Hist. Worc.,* p. 160 (1834).

Avon. Near Badger's Bank, Fladbury.

Severn. Near Kempsey Grove. Broadheath. Pitchcroft, Worcester. Wyre Forest, *Rea.* Cotheridge. Riverside beyond the willow-bed below Diglis, *Westcombe.*

Malvern. Hedgebank, Upper Howsell.

This plant is a native of grassy places in the South of Europe, and is a weed of waste and cultivated ground and a garden escape elsewhere. It appears to show a tendency to become more freely naturalized in America than here, being abundant there in the neighbourhood of some towns and villages. In some explanation of the generic name Linnaeus imagined that the roots of this plant were the Dove's Dung which sold so dear at the siege of Samaria, 2 Kings vi. 25. They are still much used for food in the Levant. The plant has been noticed in all bordering counties except Staffordshire.

{**1622. 940. O. pyrenaicum,** Linn. *Spiked Star of Bethlehem.*

Native or denizen. Woods. Local. P. August. *Top. Bot.* 10.

First Record. About Cotheridge, *Lees, A. Florence, Strang. Guide* (1828).

Severn. Cotheridge.

'On old authority' says Mr. Lees, *Bot. Worc.,* p. xxxi. It has not been seen in the county in recent times. It is curious that the two species of Star of Bethlehem, this and the previous one, should occur in the same locality; it seems to point to some confusion, especially as the two species change places with each other in Mr. Lees' books; or else it shows they were both escapes from the same garden. The plant has been observed in Gloucestershire and Salop.}

LILIUM, Linn. 430. (Etymology unknown. Possibly from li, Celtic, white, from the colour of the principal species.)

{**1623. 941. L. *pyrenaicum,** Gouan. *Top. Bot.* 1.

First Record. Localized at one place in the Severn district, *Lees, Bot. Worc., Table of Plants,* p. 28 (1867).

Severn. One place, *Lees.* Left bank of brook falling into Dowles Brook, Wyre Forest.

This plant is a native, as its specific name implies, of the mountains of South-west Europe. As a plant of any status, except as the inhabitant of a flower border, it is unknown in Worcestershire otherwise than as recorded above. Whatever may be the derivation of the name, it has passed into nearly every European language. This plant has not been observed in any bordering county.}

FRITILLARIA, Linn. 431. (Fritillus, a dice-box, in allusion to the chequered petals.)

1625. 942. F. Meleagris, Linn. *Common Fritillary. Snake's Head.*

Denizen. Wet Meadows. Rare. P. April and May. *Top. Bot.* 21.

First Record. In a meadow near Warndon, two specimens, *Rev. F. J. Eld, Worc. Nat. Club Trans.,* i, p. 323 (1885).

Severn. Warndon.

Lickey. Alvechurch, *Professor Poynting.* Near Tardebigge, *Humphreys.*

This plant was not known as an inhabitant of Worcestershire to Mr. Lees or Mr. Mathews. It was discovered in the neighbourhood of Alvechurch in 1900, where it must have been lately introduced, as Mr. D. Mathews, who as his records showed scoured that district thoroughly, would surely have seen it. The plant is abundant in the spongy meadows about Iffley, near Oxford. Mr. Druce in his *Fl. Berks.,* p. 494, makes some interesting remarks on the distribution and status of the plant. The specific name as well as the generic one alludes to the patterned flower, being taken from the Guinea-fowl, *Numidia Meleagris.* It has been observed in all bordering counties except Herefordshire.

TULIPA, Linn. 432. (From tolibun, the Persian for 'turban', from its gay colours.)

1626. 943. T. sylvestris, Linn. *Wild Tulip.*

Denizen. Parks, plantations, orchards. Rare. P. May and June. In *Top. Bot.* mentioned, but not numbered.

First Record. On Clerkenleap Marl Cliff, *Lees, A. Florence, Strang. Guide* (1828).

Severn. Clerkenleap. Wyre Forest, *Jorden.* Near Astley Church. Pitchcroft, Worcester.

Malvern. Limestone Quarry at Mathon. Banks of Teme, Leigh, *Miss Dora Porter, Towndrow, Malv. Advert.,* Dec. 3, 1898; still there, 1908.

This is the solitary British representative of the gay tribe that makes our borders bright in spring and early summer; and even on it doubts have been cast as to its true nativity. The flower is very fragrant, bright yellow, the segments of the perianth are acute, and the plant increases by sending out long fibres, at the end of which new bulbs are formed. In a garden border it establishes itself strongly, and increases freely, often appearing in places where no runners could reach. The foliage disappears with the flowers. It has been observed in all bordering counties except Herefordshire.

GAGEA, Salisb. 433. (Named after Sir Thomas Gage, a British botanist.)

1627. 944. G. lutea, Gawler. *Yellow Star of Bethlehem, Yellow Gagea.* (G. fascicularis, Salisb. Ornithogalum luteum, Linn.)

Native. Woods. Very rare. P. March and April. *Top. Bot.* 42.

First Record. At the bottom of Purlieu Lane (Malvern), *Malv. Nat. Field Club Trans ,* Part ii, p. 19 (1858).

Malvern. Purlieu Lane, as first record. Leigh, 1900, *Towndrow.*

The plant was first discovered at Purlieu Lane by the Rev. Dr. Cradock in 1855, but in after years was lessened by the 'plunder of collectors' and finally disappeared. In 1899 it was rediscovered about a quarter of a mile away from the original habitat, but in Herefordshire. The Leigh station for the plant is some five miles away from the old one, and well within the county. The plant has not been seen in Stafford or Salop of neighbouring counties.

COLCHICUM, Linn. 435. (From Colchis, where it is said to grow abundantly.)

1629. 945. C. autumnale, Linn. *Common Meadow Saffron.*

Native. Damp meadows, thickets. Locally abundant and generally distributed. P. August and September. *Top. Bot.* 40.

First Record. 'Colchicum vulgare seu Anglicum purpureum and album, Ger. Park. Common meadow Saffron. I observed it growing most plentifully in the meadows of this county (Worcestershire).' *Camden, Britannia,* p. 528 (1695).

Avon. Bredon Hill, *Mathews.* Great Comberton, *Nash.* Fladbury. Pershore.

Severn. Meadows by Severn, *Stokes.* Dunclent. Chaddesley Corbett. Bewdley. Near Holt Mill. Northwick. Below Kempsey Grove. Upper Wick. Shatterford. Saint John's, Worcester. Boreley. Grimley. Crowneast. Opposite Clevelode. Near Dynes Green.

Malvern. Malvern Chace, *Ballard.* Bransford. Powick. Horsham. Martley. Clifton-on-Teme. Malvern Link.

Lickey. Halesowen, *Withering.* Dudley, *Booker.* Yardley, *Miss M. A. Beilby.* Clent. Finstall.

This plant is engraved in J. Carpenter's *Agriculture of Bromsgrove,* 1803, as very injurious to cattle. It was regarded in old times as an effectual cure for several maladies; in modern times its use survives for one of these, namely gout. The useful part of the plant is the bulb, and it is supposed to be in its best state when it is a year old. The plant is recorded in all bordering counties.

NARTHECIUM, Huds. 436. (From νάρθηξ, a rod; probably from the straight stem. The word is an anagram of Anthericum, Linnaeus's name.)

{**1630. 946. N. ossifragum,** Huds. *Lancashire Bog-Asphodel.* (Anthericum ossifragum, Linn.)

Native. Bogs and peaty places. Possibly extinct in the county. P. June to August. *Top. Bot.* 96.

First Record. Near Rubery Hill on the Lickey, *Purton, Midl. Fl.,* p. 172 (1817).

Malvern. Pool near Powick Ham. Leigh (Lye) Head, Rock, *Jorden.*

Lickey. Foot of Rednall Hill, up to 1854. Moseley, *Ick.* Bog north of the Old Rose and Crown Inn.

It is pretty certain that this plant exists no longer in Worcestershire. It has never been seen at Hartlebury Common, a very likely locality which is traversed week by week by some botanist or other. It has not been used for any medicinal purpose, but Gerarde called it the Water Gladiole or Grassy Rush, and said 'It is, of all others, the fairest and most beautiful to behold, and serveth very well for decking and trimming up of houses, because of the beauty and braverie thereof'. It is recorded from all neighbouring counties.}

PARIS, Linn. 438. (From *par*, equal, on account of the regularity of its leaves and flowers.)

1632. **947. P. quadrifolia,** Linn.   *Common Herb Paris.*
Native. Damp woods. Not uncommon, plentiful in the Malvern district. P. May and June. *Top. Bot.* 73.
First Record. In woods and thickets on the side of Bredon Hill, *Nash, Hist. Worc., Int.,* p. lxxxix (1781).
Avon. Croome Perry Wood. Bredon Hill. Battons Wood, Bredon Hill. Tiddesley Wood. Stoulton. The Slads. Tardebigge Reservoir.
Severn. Near Dodderhill. North Wood, Bewdley. Ribbesford Wood. Arley Castle. Shrawley Wood. Crookbarrow Hill. Ockeridge Wood. Wyre Forest.
Malvern. Stanford. Sapey Brook. Wood on Norbury Hill. Middleyards Coppice. Crewshill Wood. Rosebury Rock Wood. Ankerdine Hill. Martley. Malvern Wells.
Lickey. In most of the upland woods from Wychbury to Frankley, *Mathews.* Woods near Redditch, *D. Mathews.* Clatterbach, Clent. Woods on the Lickey, sparingly. Chaddesley Wood, fine patches, *Humphreys.* Tardebigge, *D. Mathews.* Dodford.
A curious plant, once seen, never forgotten. Although four leaves characterize the bulk of the individuals of the species, this number is by no means constant, plants with three, five, six, and seven leaves having been noticed. It is called in some parts of Warwickshire Four-leaved True Love. Herb Paris is recorded from all neighbouring counties.

84. JUNCACEAE, Juss.

JUNCUS, Linn. 439. (*Jungo*, I join, the stems being used to bind things together.)

1633. **948. J. bufonius,** Linn.   *Toad Rush.*
Native. Moist woods, pond-sides, heaths, especially where water has stood. Common and widely distributed. A. June to August. *Top. Bot.* 112.
First Record. Not localized, *Scott, Hist. Stourb.,* p. 540 (1832).
This is a common little plant, seldom six inches high, its forked panicle bearing solitary pale green flowers. Its leaves are so slender that it might be mistaken for grass, and it is sometimes called Toad-grass. It is recorded from all neighbouring counties.
b. **fasciculatus,** Bert.
First Record. As herein. Defford Common, *Herb. Hast. Mus. Worc.*
In this variety the branches are shorter and thicker, and are often fascicled.

1635. **949. J. squarrosus,** Linn.   *Heath Rush.*
Native. Heaths. Not very common. P. May to August. *Top. Bot.* 108.
First Record. Bromsgrove Lickey, *Purton, Midl. Fl.,* p. 176 (1817).
Avon. Fladbury. Near Cleeve Mill. Shell Brook.

Severn. Hartlebury Common. Pedmore Common (1854). Broadheath. Northwick. Grimley.
Malvern. Southwood. Leigh.
Lickey. Moseley Bog, *Mathews.*
The stems of this Rush are about a foot high, compressed and leafless, and it has many stout rigid radical leaves, most of which turn one way. It has been observed in all neighbouring counties.

1636. **950. J. compressus,** Jacq.   *Round-fruited Rush.*
(*Juncus bulbosus,* Linn.)
Native. Marshy meadows. Rare, except near Malvern. P. June to August. *Top. Bot.* 28.
First Record. Not localized, *Lees, Bot. Malv. Hills,* 1st ed., p. 46 (1843).
Avon. Avon at Fladbury. Defford Common. Hanbury.
Severn. Canal side, Worcester, *Baxter.* Northwick Brickyard. Canal Bank, opposite Hindlip House. Canal Bank, Droitwich.
Malvern. Quite common, *Towndrow.* Pendock. Barnard's Green. Alfrick. Powick Weir. Malvern Link. Leigh Sinton.
This Rush bears slender unbranched stems from six to twelve inches high, with one acute channelled leaf in the middle of each, and the terminal compound panicle usually falls short of the pale bract. The plant does not appear to have been observed in Gloucestershire.

1637. **951. J. Gerardi,** Loisel.   *Mud Rush.*
(*J. coenosus,* Bich.)
Native. Marshy meadows. Rare. P. May and June. *Top. Bot.* 87.
First Record. Saldon, near Himbledon, *Rev. W. Lea* (1856), *Lees, Bot. Worc.,* p. 33 (1867).
Avon. Saldon.
Severn. Banks of Droitwich Canal, *Mathews.* Droitwich Canal at Atterburn Brook.
Malvern. Rare, *Towndrow.* Upton-on-Severn.
This plant and the last are segregates of *J. bulbosus,* Linn. In this plant the terminal compound panicle usually exceeds the bract. This plant does not appear to have been observed in Herefordshire.

[*1638.* **J. tenuis,** Willd.   *Slender spreading Rush.*
This plant, which is found on sandy ground and roadsides, was recorded by Mr. Towndrow as new for Herefordshire, and very close to Worcestershire, *Bot. Rec. Club Rep.,* 1884. As yet our county cannot lay claim to it. The *Top. Bot.* number for the plant is 17, and it has not been observed in any neighbouring county except Herefordshire.]

1641. **952. J. inflexus,** Linn.   *Hard Rush.*
(*J. glaucus,* Ehrh.   *J. acutus,* Gerard.)
Native. Wet meadows, roadsides, commons. General throughout. P. June and July. *Top. Bot.* 90.
First Record. Not localized, *Lees, Bot. Malv. Hills,* 2nd ed., p. 80 (1852).
This is one of the common Rushes and needs no localization. It is some two feet high, with tough and rigid bluish green stems bearing a panicle of greenish brown flowers. It is recorded for every neighbouring county.

1642. **953. J. effusus,** Linn.   *Soft Rush.*
Native. Marshes, ditches, roadsides, woods. Common and generally distributed. P. June to August. *Top. Bot.* 112.
First Record. Not localized, *Scott, Hist. Stourb.,* p. 540 (1832).
This is a common Rush, needing no localization. The stems of this species and the next were used for making chair-bottoms and mats, and their pith, which is not interrupted as is that of the last species, was once extensively used for making the wicks of rush-lights. In earlier days rushes were strewed on the floors of all kinds of rooms, and even in churches; a practice which concealed much that our forefathers thought nothing of, as the rushes were not always even weekly removed. The great use of rushes for strewing led to country festivals called Rush Bearings, which, as was the case with many such country rejoicings, deteriorated into scenes of rowdiness and intoxication, and when rushes for rooms became displaced by carpets becoming general, they were discontinued, to the relief of the community. This Rush is recorded from all bordering counties.
×**inflexus.** *Top. Bot.* 36.
(*J. diffusus,* Hoppe.)
First Record. Newland Common, Malvern, *Towndrow, J. of B.,* vol. xvii, p. 278 (1879).
Avon. Craycombe Hill. Stoulton. Defford Common.
Severn. Wyre Forest. Northwick Brickyard. Monk Wood. Pixham. Broadheath.
Malvern. As first record. North Malvern. Old Hills. Castlemorton Common. Welland Common. Newland.
Of this Rush the stems are softer, less glaucous, and less striate than *J. inflexus,* and the pith continuous as in *J. effusus.* The hybrid has been observed in Stafford, Salop, and Warwick.

1643. **954. J. conglomeratus,** Linn.   *Common Rush.*
Native. Wet fields, ditches, heaths, marshy places. Common and generally distributed. P. May to August. *Top. Bot.* 112.
First Record. Not localized, *Scott, Hist. Stourb.,* p. 540 (1832).
This Rush is very similar to the last, but may be easily distinguished by its panicle which is condensed and globose, not diffused and branched as is usually the case with the former plant. At all events there was nothing to differentiate them in the eyes of the Rush users of early times, not even the pith, which in both Rushes is continuous. This plant is recorded from every bordering county.

1646. **955. J. bulbosus,** Linn.
(*J. supinus,* Moench.   *J. uliginosus,* Sm.)
Native. Moist places, ditches, ponds, in heathy situations. Rather uncommon. P. June to August. *Top. Bot.* 107.
First Record. As *Juncus uliginosus,* Bromsgrove Lickey, *Purton, Midl. Fl.,* p. 177 (1817).
Avon. Trench Woods. Tiddesley Wood. Abberton.
Severn. Hartlebury Common, *Lees.* Wyre Forest. Tagwell, near Droitwich. Perry Wood. Moseley Green. Wood near Castle Hill.
Malvern. Malvern Hills, *Lees.*

Lickey. Var. *subverticillatus,* Wulf., Moseley Bog, *Ick.*
The variety mentioned at the Lickey station is a floating form of the plant. Viviparous flowers have been found in Wyre Forest, at Tagwell, Perry Wood, Malvern, and Castle Hill. The type is recorded from all bordering counties.

1647. **956. J. subnodulosus,** Schrank.   *Blunt-flowered jointed Rush.*
(*J. obtusiflorus,* Ehrh.)
Native. Marshes, wet bogs. Very local. P. July to September. *Top. Bot.* 59.
First Record. As *Juncus obtusiflorus.* Welland Marshes, *Lees, Bot. Malv. Hills,* 2nd ed., p. 80 (1852).
Avon. Roadside, Craycombe Hill. Near Badsey, *Westcombe.*
Malvern. Welland, *Lees.*
This plant was recorded as new under the name *J. obtusiflorus* by Dr. Arnold Lees, 'Ditch side, marsh, west of Severn, north of Tewkesbury,' *Bot. Rec. Club Rep.,* 1883 (pub. 1884), but it was not new. The plant is recorded from all bordering counties.

1649. **957. J. articulatus,** Linn.   *Shining-fruited jointed Rush.*
(*J. lampocarpus,* Ehrh.)
Native. Wet meadows, pond-sides, ditches. General in all the districts. P. June to September. *Top. Bot.* 110.
First Record. A damp lane between Stirchley Street and King's Norton, *Ick, Analyst,* vol. vi, p. 22 (1837).
This is a common rush in suitable situations. It is recorded from all bordering counties.

1651. **958. J. sylvaticus,** Reich.   *Sharp-flowered jointed Rush.*
(*J. acutiflorus,* Ehrh.   *J. nemorosus,* Sibth.   *J. articulatus,* var. γ. Linn.)
Native. Moist woods, wet meadows, ditches. General in all the districts. P. July to September. *Top. Bot.* 111.
First Record. As *J. acutiflorus.* Not localized, *Lees, Bot. Malv. Hills,* 1st ed., p. 46 (1843).
This common Rush is recorded from all neighbouring counties.

LUZULA, DC. 440. (Altered from the Italian *lucciola,* a glowworm, because the heads of the flowers, wet with dew, sparkle in the moonlight.)

1656. **959. L. Forsteri,** DC.   *Narrow-leaved hairy Wood-rush.*
(*Juncoides Forsteri,* Kuntze.)
Native. Dry woods. Rare. P. April and May. *Top. Bot.* 29.
First Record. Cotheridge, *Rev. A. Bloxam, Lees' Cat., New Pop. Guide* (1835).
Avon. Trench Woods. The Slads. Tiddesley Wood.
Severn. Cotheridge. Perry Wood. Birchin Grove. Woods, Crowneast. Roadside, Crowneast. Shrawley Wood. Wyre Forest. Ockeridge Wood.
Malvern. The Grove, Little Malvern ; Malvern Wells, *Lees.* Ankerdine Hill. Cowleigh Wood. Dripshill Wood.
This species is characterized by its narrow leaves and erect habit when the fruit is ripe. The plant appears to have been noticed only in Gloucestershire and Herefordshire of bordering counties.

x pilosa.

(*Luzula Borreri*, Bromf.)

First Record. As *L. pilosa*, var. *Borreri*. Malvern, *Towndrow, Malv. Advert.*, December 10, 1892.

Severn. Wyre Forest. Crowneast Woods. Perry Wood. Shrawley Wood.

Malvern. As first record. West Malvern. Leigh. Mathon, *Mathews*. Cowleigh Wood. Dripshill Wood. Ankerdine Hill.

This hybrid has been observed in Herefordshire.

1657. 960. L. pilosa, Willd. *Broad-leaved hairy Wood-rush.*

(*L. vernalis*, DC.)

Native. Woods and thickets. General throughout. P. March to May. *Top. Bot.* 109.

First Record. As *Luzula pilosa*. Not localized, *Lees, Bot. Malv. Hills*, 1st ed., p. 46 (1843).

This plant may be easily distinguished from *J. Forsteri* ; in that all the capsules, especially in fruit, droop in one direction, in this they spread in all directions. It is too widely distributed to require localization, and is recorded in all neighbouring counties.

1658. 961. L. sylvatica, Gaud. *Great hairy Wood-rush.*

(*L. maxima*, DC. *Juncus pilosus*, Linn. *J. sylvaticus*, Huds.)

Native. Woods, thickets, heathy places. Local. P. April and May. *Top. Bot.* 109.

First Record. As *Juncus sylvaticus*. Witchery Hole, near Clifton-on-Teme, *Ballard, Stokes's With. Bot. Arr.*, and ed., p. 364 (1787).

Avon. Norton, near Evesham. Tiddesley Wood. Grafton Wood.

Severn. Shrawley Wood. Birchin Grove. Blackstone Rock. Redstone Rock. Bewdley. Arley Kings. Wyre Forest.

Malvern. Hollybush Wood ; Rosebury Rock, *Lees*.

The stem of this plant is from a foot to eighteen inches high, and the loose panicle of brownish flowers with yellow anthers much exceeds the bracts. It is recorded from all bordering counties.

1661. 962. L. campestris, DC. *Field Wood-rush.*

(*Juncus campestris*, Linn.)

Native. Fields, pastures, downs, heaths. Abundant everywhere. P. March to May. *Top. Bot.* 109.

First Record. As *Juncus campestris*. Meadows, *Scott, Hist. Stourb.*, p. 540 (1832).

This is a common plant among the grass in the early year, noticeable from its brown heads of flowers among the springing green. It is known locally by the names Chimneysweeps and Blackmen. It is recorded from all bordering counties.

1662. 963. L. multiflora, DC.

(*L. erecta*, Desv. *Juncus campestris*, var. γ. Linn. *J. multiflorus*, Ehrh.)

Native. Damp heaths, commons. General throughout the county. P. May and June. *Top. Bot.* 109.

First Record. As var. *congesta*. Perry Wood ; Hartlebury Common, *Lees, Illus. Nat. Hist. Worc.*, p. 160 (1834).

This plant is a segregate of *Luzula campestris*, differing chiefly in the seeds, which are nearly twice as long as broad, instead of being nearly globular. It is perhaps quite as common as the type, if not commoner. It is recorded from all bordering counties.

Var. b. congesta.

First Record. The first record of the plant above is a record of this variety of it.

This variety has been noticed in Salop, Stafford, and Warwick.

L. ALBIDA, DC.

This alien plant, native of woods and meadows in Central Europe, was observed between Kidderminster and Bewdley by Dr. Arnold Lees, June 1883.

85. TYPHACEAE, Juss.

TYPHA, Linn. 441. (τῦφος, a marsh, where these plants grow.)

1664. 964. T. latifolia, Linn. *Great Reed-mace.*

Native. Marshes, sides of ponds, sides of rivers, preferring still water. General. P. June to August. *Top. Bot.* 82.

First Record. Harborough. Various pools, *Scott, Hist. Stourb.*, p. 540 (1832).

This is a well-known and conspicuous plant needing no localization, as it is to be seen throughout the county at the tail of nearly every pool, and in still reaches of water elsewhere. The sterile and fertile portions of the spike are close together, the sterile one uppermost and yellow, the fertile one below and of rich velvety brown. It is commonly known as the Bulrush ; but that name properly and botanically belongs to *Scirpus lacustris*, and in the Bible is applied to the Papyrus. Other names for this plant are Cat's-tail, Cow-tail, and Calf's-foot. The long leaves are put to many purposes, for thatching, basket-making, and such like. The yellow pollen is exceedingly inflammable, and if ignited instantaneously produces a flash of light. The plant is recorded from every bordering county.

1665. 965. T. angustifolia, Linn. *Lesser Reed-mace.*

Native. Ponds and river-sides. Not uncommon. P. June and July. *Top. Bot.* 58.

First Record. In a small pond by the Moors, Worcester, and in the pool at Ham Castle, *Lees, Illus. Nat. Hist. Worc.*, p. 177 (1834).

Avon. Canal near Oddingley.

Severn. Blakedown and Hurcott Pools, *Mathews*. Hawford, *Lees*. Stanklin Pool. Westwood Park. Canal near Oddingley. Sharpley Pool. Grafton Pool. Glashampton Pools.

Malvern. New Pool, Malvern Chace, *Lees*. Hanley Hall Pool.

Lickey. Tardebigge, *Humphreys*. Hewell.

This is altogether a more graceful plant than the last, slenderer in all its parts, though equally tall. The great distinction, however, from it is that the sterile and fertile portions of the spike are separated from each other, there being a portion of stem between them. The plant is recorded from all bordering counties.

SPARGANIUM, Linn. 442. (σπάργανον, a little band, from its narrow leaves.)

1666. 966. S. erectum, Linn. *Branched Bur-reed.*

(*S. ramosum*, Curt.)

Native. Sides of rivers, ponds, canals. The aggregate is generally distributed. P. June and July. *Top. Bot.* 30. *Lond. Cat.*, 10th ed., aggregate, 110.

First Record. Very common in morasses, *Scott, Hist. Stourb.*, p. 540 (1832).

Avon. Defford Brook. By the Avon, Bredon. Wyre Piddle. Pirton. Cleeve Mill. Near Eckington Bridge. Near Stonebow Bridge. Upton Snodsbury. Crowle Brook.

Severn. Northwick Brickfields. Diglis. Pond by footpath from Mudwall Mill to Crowneast. Northwick. Berwick's Brake. Claines. Ombersley. Lincombe. Hill Ditch, Hartlebury. Pool near Bishop's Palace, Hartlebury. Grimley Brick Pits. Thorngrove Pool. Spetchley Park.

Malvern. Powick. Bransford. Knightwick. Martley. Longdon.

Lickey. Moat Mill Pool, Bromsgrove. About Stourbridge, *Scott*.

This plant is frequently seen by the side of water, two or three feet high, with a sturdy branched stem, and long leaves that rustle in the breeze. The seeds, at first green, become brown when ripe, and then are as large as barley-corns. The plant is recorded from all bordering counties.

Var. b. microcarpum, Neum.

First Record. As in this book.

Malvern. Malvern Link, *Towndrow, Malvern Advert.*, December 10, 1892.

This variety is smaller, the stem less branched, and the fruit is smaller with a longer beak.

1667. 967. S. neglectum, Beeby. *Branched Bur-reed.*

(*S. erectum*, var. *neglectum*, Richt.)

Native. Sides of rivers, ponds, canals. Probably nearly as common as the last. P. June and July. *Top. Bot.* 35.

First Record. Malvern Link, *Towndrow, Bot. Rec. Club Rep.* (1887).

Severn. Spennells.

Malvern. As first record.

Lickey. Hunnington.

A paper by Mr. Towndrow on the occurrence of this plant and the last in Worcestershire is in *J. of B.* xxiv. 142. The plant does not appear to have been observed either in Gloucester or Salop.

1668. 968. S. simplex, Huds. *Unbranched upright Bur-reed.*

Native. Rivers, canals, ponds, brooks. Locally common. P. July and August. *Top. Bot.* 99.

First Record. Very common in morasses, *Scott, Hist. Stourb.*, p. 540 (1832).

Avon. Stream between Throckmorton and Peopleton. Defford Brook. Pershore.

Severn. Northwick Brickfields. Grimley. Blakedown. Churchill. Hurcott.

Malvern. Malvern Chase. Malvern Link. Newland.

Lickey. Bittell Reservoir, *Humphreys*.

As its specific name indicates this plant is unbranched. It is recorded from all bordering counties.

1670. 969. S. minimum, Fr. *Floating Bur-reed.*

(*Sparganium natans*, Linn.)

Native. Ditches and ponds. Rare. July and August. *Top. Bot.* 54.

First Record. As *Sparganium natans*. Muddy pools near Cotheridge, *Mr. Walcot, Lees' Illus. Nat. Hist. Worc.*, p. 177 (1834).

When one botanist (Druce) says that two plants are the same, another (Babington) says they are distinct species, another (Hooker) that the one is a sub-species of another, and another (Bentham) that one is a variety of the other, the ordinary man can only bow his head reverently and pass on to something more definite. However affairs may stand, a plant called *S. minimum* appears to have been noticed in all bordering counties except Gloucestershire.

86. ARACEAE, Juss.

ARUM, Linn. 443. (Etymology doubtful.)

1671. 970. A. maculatum, Linn. *Cuckoo Pint, Wake Robin.*

Native. Woods, thickets, hedgebanks. Common in all the districts. P. April and May. *Top. Bot.* 84.

First Record. On ditch-banks, common, *Pitt, Agric. Worc.*, p. 317 (1810).

There is hardly a hedge-row bank throughout the county which somewhere along it is not in the spring ornamented with the glossy handsome leaves of this plant. It is well known to country children by the name of Lords and Ladies, which, though not found in books before the middle of the eighteenth century, must have been in colloquial use for a much longer time. The name is said to allude to the darker and lighter colour of the spadix, which varies from yellowish green to rich deep purple. It has also the popular names of Cows and Calves, Jack in the Pulpit, and Wake Robin. Young plants grown from seed do not appear above the ground for the first two seasons' growth, and in the third only one ovate leaf shows itself. It is not until the fourth season, and generally later, that the first sagittate leaves are found, and it rarely flowers before the seventh year. The red berries are highly poisonous, and every part of the plant is acrid. The roots are full of starch, and the well-known Arrow-root is produced from *Arum esculentum*, which is cultivated in the West Indies. Of course, being particularly acrid and nauseous to the taste, the plant was included in mediaeval pharmacy ; and the starch produced from its tubers was strong enough to give support to the immensely large ruffs of the Elizabethan period, in consequence of which Gerarde calls the plant Starch-wort. It is recorded from all bordering counties.

ACORUS, Linn. 444*. (ἀ, out, κόρη, the pupil of the eye, diseases of which it was supposed to remove.)

1673. 971. A. *Calamus, Linn. *Common Sweet-Sedge.*

Native. Sides of ponds. Several places. P. June to August. *Top. Bot.* 32.

First Record. River Avon, near Pershore, *Ballard, Stokes's With. Bot. Arr.*, 2nd ed., p. 357 (1787).

Avon. Pershore. Near Fladbury, *Mrs. G. Perrott*. Near Strensham Mill.

Severn. Hanley, *Rufford*. Grafton Manor, *Humphreys*.

Malvern. One place, *Lees*. Pond near Hanley Hall.

Lickey. Harvington Hall Moat ; Hewell Lake, *Humphreys*.

In former days this sweet-smelling plant was used to mix with the rushes with which the floors of dwelling-rooms were strewed, and the plant has probably been introduced when it is found in the moat or pond near some old habitation. The root is supposed to cure ague, and the starch from it was largely used by the makers of hair powder. It is easily recognized among other watery vegetation by the inch or so of crinkled upper edge of the leaf where it joins the stem. It is alleged that it forms no fertile seed in Europe. This plant was given as a new record, *Bot. Rec. Club Rep.*, 1880, by Dr. Fraser, from Harvington Hall. It is recorded from all bordering counties.

87. LEMNACEAE, Duby

LEMNA, Linn. 445. (λέμνα, in the Greek, said to be derived from λέμμα or λεπίς, a scale, from the shape of the plants.)

***1674. 972. L. trisulca,** Linn.   Ivy-leaved Duckweed.*

Native. Ponds and ditches, chiefly of stagnant water. Not uncommon. A. June. *Top. Bot.* 73.

First Record. Feckenham Moors ; Cookhill, *Purton, Midl. Fl.*, p. 436 (1817).

Avon. Hanbury. Dodderhill Common. Norton, near Evesham. Fladbury. Stoulton.

Severn. Pond, New Road, Worcester. Diglis Brick Pits. Stanklin. Trots Hill. Grimley Brick Pits. Northwick Brick Pits.

Malvern. Longdon Marsh. Castlemorton Common. Powick Ham. Bransford. Leigh. Martley. Newland. Sherrard's Green.

Lickey. Bittell Reservoir, *Humphreys*.

The Duckweeds are plants with which every one who in summer time has looked upon the surface of standing water must be familiar. To the ordinary gaze all four species are much alike, the present plant being more distinct in shape than the three others. They all increase very rapidly in warm weather, chiefly by their buds. In the case of this plant, the fronds are proliferous at right angles, and autumnal bulblets are produced which survive the winter, and give rise to the plants of the succeeding year. The plant is recorded from all bordering counties.

***1675. 973. L. minor,** Linn.   Lesser Duckweed.*

Native. Ponds and ditches. Abundant in all the districts. A. June and July. *Top. Bot.* 106.

First Record. Not localized, *Lees, Bot. Malv. Hills*, 1st ed., p. 43 (1843).

This is the commonest of the Duckweeds. It is recorded from all bordering counties.

***1676. 974. L. gibba,** Linn.   Gibbous Duckweed.*

Native. Ponds and ditches of stagnant water. Local. A. June to August. *Top. Bot.* 53.

First Record. Lower Bishop's Pool, Northwick ; near Worcester ; and in a Pool on the east side of Malvern Chace, *Stokes's With. Bot. Arr.*, 2nd ed., p. 1020 (1787).

Avon. One place, *Lees*. Eckington. Pirton Pool. Pond near Canal, Oddingley.

Severn. As first record. Wannerton Farm Pool. Northwick Pool. Diglis Brickpits. Trotshill. Severn Stoke. Pond near New Road, Worcester. Pond near Oldbury Villa. Laughern Brook, by the wooden bridge above Hadley's Mill.

Malvern. As first record. Longdon Marsh. Powick Ham. Bransford. Leigh. Malvern Link.

Lickey. Charford Mill Pool, Bromsgrove, *Humphreys*.

L. gibba is recorded from all neighbouring counties.

***1677. 975. L. polyrrhiza,** Linn.   Greater Duckweed.*

Native. Ponds and ditches of stagnant water. Not uncommon. P. Does not flower in Britain. *Top. Bot.* 56.

First Record. Not localized, *Lees' Cat., New Bot. Guide* (1835).

Severn. Wannerton Farm Pool. Tank at Churchill. Trotshill. Diglis Brickpits. Hartlebury Common. Near Severn Stoke. Hagley pools.

Malvern. General, *Lees*. Powick. Barnard's Green. Longdon Marsh. Sherrard's Green.

Lickey. Moat Mill Pool, Bromsgrove, *Humphreys*.

This is the largest of our four species, the fronds being half an inch long and nearly as broad, and underneath of a deep purple colour ; and its roots are many, while of the other species they are solitary. It never flowers in England, and therefore its fructification is unknown. In winter it sinks to the bottom of the pool. It is recorded from all neighbouring counties.

88. ALISMACEAE, R. Br.

ALISMA, Linn. 447. (Celtic *alis*, water.)

***1679. 976. A. Plantago-aquatica,** Linn.   Water Plantain.*

Native. Sides of water, in ponds, rivers, and wet places. General. P. May to August. *Top. Bot.* 100.

First record. As *Alisma Plantago*. Ponds, *Scott, Hist. Stourb.*, p. 540.

This plant is too common in suitable situations to need localization. The flowers are small and pink, and it grows quite in the water. It is recorded from all bordering counties.

***1680. 977. A. lanceolatum,** With.*

First Record. Upton-on-Severn, *Towndrow, Malv. Advertiser*, December 10, 1892.

Severn. Stakenbridge Pool, *Mathews*.

Malvern. As first record.

Lickey. Broome Pool, *Mathews*.

In this species the leaves are narrowed at the base, instead of being obtuse as in the type. It has been recorded in Stafford and Warwick.

B b

***1681. 978. A. ranunculoides,** Linn.   Lesser Water Plantain.*

Native. Turfy bogs. Once recorded. P. June and July. *Top. Bot.* 87.

First Record. In a ditch surrounding Feckenham Bog, *Purton, Midl. Fl.*, p. 189 (1817).

Avon. As first record. Great Comberton.

The plant was seen at Feckenham Moors by Mr. Westcombe also. Feckenham Bog has long ago been drained, and this plant, with several others that loved moisture, has disappeared. It is to be feared it is not now an inhabitant of our county ; but it has been recorded for all bordering ones.

ELISMA, Buchenau, 448. (A variation of the name of the last plant.)

***1682. 979. E. natans,** Buchen.   Floating Water Plantain.*

(*Alisma natans*, Linn.)

Native. In water. Once recorded. P. August. *Top. Bot.* 14.

First Record. As *Alisma natans*. Near Tenbury, *Mr. A. Aikin, Bot. Guide*, vol. ii, p. 656 (1805).

Severn. Diglis, *G. Reece*, sp. *Herb. Vict. Inst. Worc.*

Malvern. As first record.

This plant is not mentioned in Mr. Lees' *Bot. Worc.*, nor in his *Table of Plants*, and except Mr. Reece's specimen above mentioned, nothing more has been seen or heard of it within the county. It is to be feared that as is the case with the last plant, it is extinct. It has been noticed in Shropshire.

SAGITTARIA, Linn. 449. (From *sagitta*, an arrow, on account of the shape of its leaves.)

***1683. 980. S. sagittifolia,** Linn.   Common Arrow-head.*

Native. Shallow water. Not common. P. July and August. *Top. Bot.* 58.

First Record. Harvington (near Evesham) Mill, *Purton, Midl. Fl.*, p. 46.

Avon. Harvington. Huddington. Avon near Tewkesbury. Avon near Bredon. Eckinton Bridge. Nafford Mill. Evesham. Stonebow Bridge.

Severn. Hartlebury. Pool at Camp, Worcester. Severn, Stourport.

Lickey. Canal at California, Northfield. Canal at the Leasowes. Bittell Reservoir. Several places by the canal at Halesowen, *Mathews*.

The arrow-shaped leaves of this plant make it easy to recognize among waterside vegetation. The tubers are full of starch. It is recorded from all bordering counties.

DAMASONIUM, Hill. 450. (δαμάζω, I subdue, because it was supposed to cure the toad's poison.)

{***1684. 981. D. Alisma,** Mill.   Common Star-fruit.*

(*D. stellatum*, Pers.   *Actinocarpus Damasonium*, R. Br.)

Native. Ditches and ponds, in still water. Once recorded. P. June to August. *Top. Bot.* 13.

First and only Record. Has been found at a pool-side near Tenbury. This plant being most attached to the south-eastern counties of England scepticism might attach to a mere report by a young botanist ; but having an authentic specimen, no doubt whatever can exist on the subject, *Lees, Bot. Worc.*, p. 85 (1867).

Malvern. Tenbury, as first record.

Mr. Mathews remarks, *Mid. Nat.* xv. 46, 'This is a very curious sentence. Mr. Lees does not state that the plant was gathered by the young botanist, nor give his name, nor state in words that the plant was in his (Mr. Lees') possession. I have never seen the specimen nor received from Mr. Lees any information from which the locality could be identified. It has never been found a second time ; credit must, nevertheless, be given to the statement.' Mr. Lees, in his *Table of Plants*, localized the Star-fruit as occurring in the Severn district also, ' surely erroneous,' says Mr. Mathews. If the points mentioned by Mr. Mathews, who worked constantly with Mr. Lees, could not be elucidated by him in his day, to us who follow they must remain a mystery. The plant is not recorded from any neighbouring county except Salop, where it has been recorded from Ellesmere.}

BUTOMUS, Linn. 451. (βοῦς, an ox, τέμνω, I cut, because the sharp leaves cut the mouths of the cattle that feed on them.)

***1685. 982. B. umbellatus,** Linn.   Flowering Rush.*

Native. Marshes, wet meadows, watersides. Locally plentiful. P. June and July, *Top. Bot.* 60.

First Record. Side of the River Avon at Evesham, *Ballard, Stokes's With. Bot. Arr.*, 2nd ed., p. 420.

Avon. Evesham. Pershore. Tewkesbury. Near Throckmorton. Tardebigge Reservoir, *D. Mathews*; 'spreading,' 1900, *Humphreys*.

Severn. Crowneast. Northwick Brickfields. Canal-side, Worcester. Henwick Mill-pond. Canal near Kidderminster. Between Stourport Bridge and Leckhill, *Mrs. Gardner*. Laughern Brook, *Lees*. Below Kempsey Ford, *Dr. Streeten*. Diglis Canal Lock. Near Camp Weir.

Malvern. Castlemorton, bordering on Longdon Marsh, *Lees*. Near Powick Weir. Pendock. The Rhydd. Pool Brook. Upton-on-Severn.

Lickey. Canal-side at the Leasowes. Bittell Reservoir, *Humphreys*. Edgbaston Lane, near Averns Mill, *Ick* ; 'possibly in Worcester', *Mathews*.

The pink flowers stand two or three feet above the surface of the water, each on a partial stalk three inches or more long. The three-edged leaves are sharp enough to wound the hand of the incautious gatherer. The plant has been used in medicine, one of those considered antidotes for the bite of venomous reptiles. It is recorded from every neighbouring county.

89. NAIADACEAE, Juss.

TRIGLOCHIN, Linn. 452. (τρεῖς, three, γλωχίν, a point; from the three points of the capsules.)

1686. **983. T. palustre**, Linn.   *Marsh Arrow-grass.*
Native. Marshes, wet meadows, watersides. Locally common. P. June to August. *Top. Bot.* 110.
First Record. Boggy meadows near Sapey Brook, *Sheward, Nash, Hist. Worc., Sup.,* p. 96 (1799).
  Avon. Bredon Hill. Broadway. Salt Spring, Wadborough.
  Severn. Laughern Brook. Merriman's Hill. Droitwich Canal. Wyre Forest, *Jorden.* Bank of Worcester and Birmingham Canal near Worcester. Hawford. Dowles Brook.
  Malvern. Commons near Malvern, *Lees.* Malvern Hills, *Westcombe.* Pendock. Newland. Mathon.
  Lickey. Moseley. Bromsgrove Lickey, *Mathews.* Alvechurch, *D. Mathews.*
Triglochin palustre is recorded from all bordering counties.

POTAMOGETON, Linn. 454. (ποταμός, a river, γείτων, a neighbour. All the species grow in the water.)

1689. **984. P. natans**, Linn.   *Broad-leaved Pond-weed.*
Native. Ponds, rivers, ditches. Generally distributed. P. June and July. *Top. Bot.* 107.
First Record. Pool at Wichenford and ponds about Malvern, *Lees, Illus. Nat. Hist. Worc.,* p. 153 (1834).
It is not necessary to localize this widespread inhabitant of still or slowly-moving waters. The floating leaves are dull olive-green, and the cylindrical spikes of small flowers rise above the surface of the water. The Pond-weeds are a highly critical tribe of plants, the characters depending chiefly upon the shapes of their leaves and the forms of their nuts. Botanists appear to have been unable to make up their minds as to the gender of the name. γείτων may be masculine or feminine, but hardly neuter, as some seem to think it. The plant is recorded from all bordering counties.

1690. **985. P. polygonifolius**, Pourr.   *Oblong-leaved Pond-weed.*
(*P. oblongus*, Viv.)
Native. Ponds, ditches, pools, and swampy places. Not uncommon. P. June to August. *Top. Bot.* 108.
First Record. Ditches on Hartlebury Common bordering the bog there, *Lees, Bot. Worc.,* p. 14 (1867).
  Avon. Trench Woods. Defford Common.
  Severn. Hartlebury Common. Stanklin Pool, *Humphreys.* Wyre Forest. Boreley.
  Malvern. Mathon.
Some botanists have considered this to be a variety of the last plant, from

which it differs in the smaller size and red colour of the fruit, and in the leaves not being jointed to their stalks. It is recorded from all bordering counties.

  Var. **pseudo-fluitans**, Syme.
  First Record. As below (1900).
    Severn. Stanklin Pool, *Humphreys.*
This variety has not been observed in any bordering county. It is not given in *Lond. Cat.,* 10th ed.

1692. **986. P. coloratus**, Hornem.   *Plantain-leaved Pond-weed.*
(*P. plantagineus*, Du-Croz.)
Native. Stagnant pools. Rare. P. June to August. *Top. Bot.* 39.
First Record. As *P. plantagineus.* Bewdley Forest, *Jorden, Phyt., N. S.,* vol. i, p. 281 (1855).
  Severn. Wyre Forest.
Mr. Mathews, *Mid. Nat.,* xiii. 257, says that he supposes *P. polygonifolius* is meant. Why he should say this, seeing that this plant was already in existence and had been seen by himself at the Holy Well, Wyre Forest, September 30, 1854, is not clear; see *Worc. Nat. Club Trans.* ii. 67. It has been noticed only in Salop and Herefordshire of bordering counties.

1693. **987. P. alpinus**, Balb.   *Reddish Pond-weed.*
(*P. rufescens*, Schrad.)
Native. Streams, pools. Rare. P. July. *Top. Bot.* 75.
First Record. With other species. All I can certainly enumerate as growing in Severn Valley waters, *Lees, Bot. Worc.,* p. 56.
  Severn. As first record. Severn near Worcester, *Westcombe.* Grimley Brickpits. Holt.
  Lickey. Canal, Halesowen; Pool at Bartley Green, *Mathews.*
This plant is remarkable for its reddish-olive colour, and possibly can be better known from its aspect and hue than from any other characteristic. It has been noticed in Salop and Staffordshire.

[*1694.* **988. P. lanceolatus**, Sm.   *Lanceolate Pond-weed. Top. Bot.* 2.
This plant was recorded by Lees, *Bot. Malv. Hills,* 1st ed., p. 45 (1843). It was, however, an erroneous record, and was omitted in the 2nd and 3rd eds. of that book.]

[*1695.* **989. P. heterophyllus**, Schreb.   *Various-leaved Pond-weed.*
(*P. gramineum*, Linn.)
Native. Ponds and ditches. Rare. P. June and July. *Top. Bot.* 77.
First Record. In a pool on Welland Common, *Lees, Bot. Malv. Hills,* 1st ed., p. 45 (1843). In the 3rd ed., p. 107, he adds 'This is dubious, leaves only observed'.
  Malvern. Welland Common, as first record.
As this is the only record of the plant for the county it must stand as erroneous till further confirmation is obtained. It has been observed in all bordering counties except Warwickshire.]

1698. **990. P. lucens**, Linn.   *Shining Pond-weed.*
Native. Rivers, canals, ponds. Not common. P. June and July. *Top. Bot.* 79.
First Record. River Avon, *Purton, Midl. Fl.,* p. 105 (1817).
  Avon. In the river, as first record. At Evesham, *Lees.* Cleeve Mill. Near Eckington Bridge.
  Severn. Stanklin Pool. Severn near Worcester. Near Bevere Island. Grimley Brickpits.
  Malvern. One place, *Lees.* Severn at Upton.
  Lickey. One place, *Lees.* Canal, Tardebigge, *Humphreys.*
This is the largest of our species, and its leaves are beautifully nerved. It is a common species throughout England, and likely to be far more abundant in the county than the above records show. It is recorded from all counties that border Worcestershire.

  Var. b. **acuminatus**, Fr.
  First Record. As below.
    Avon. River Avon.
    Severn. Severn above Worcester. Stanklin.

1702. **991. P. praelongus**, Wulf.   *Long-stalked Pond-weed.*
Native. Rivers and Canals. Rare. P. May and June. *Top. Bot.* 51.
First Record. In a pool on Barnard's Green, below Devil's Oak Lane, *Lees, Bot. Malv. Hills,* 2nd ed., p. 79 (1852).
  Severn. Stanklin Pool, *Humphreys.*
  Malvern. As first record.
Mr. Mathews says of Mr. Lees' record, *Mid. Nat.* xiii. 187, 'This is a very doubtful record. Mr. Lees notices it in the 3rd ed. (*Bot. Malv. Hills*), p. 40, as one of the plants previously unrecorded in the Malvern district, but omits it from the Potamogetons on p. 107. He omits it also in the *Bot. Worc.,* 1867.' Mr. Bagnall says it 'must have been marked in error as it is not at all a likely plant for Worcestershire' (MS.). In *Top. Bot.,* p. 418, it is marked for Worcester, with a note of interrogation. Mr. Humphreys' record, however, steps into the breach, and confirms the plant for our county. It has been recorded from Salop and Staffordshire.

1703. **992. P. perfoliatus**, Linn.   *Perfoliate Pond-weed.*
Native. Ponds, rivers, ditches, canals. General. Less common in the Lickey district. P. July and August. *Top. Bot.* 99.
First Record. Not localized, *Lees, Bot. Malv. Hills,* 1st ed., p. 45 (1843).
This plant requires no localization, being generally met with, its stems long and slightly branched, and its leaves clear and of a dull olive-green when young. When dried they appear as a thin brown membrane, and are then very sensitive to moisture, curling up even if placed on the palm of the hand. It is recorded from all bordering counties.

1704. **993. P. crispus**, Linn.   *Curly Pond-weed.*
Native. Ponds, streams, ditches. General and common. P. June and July. *Top. Bot.* 100.
First Record. Not localized, *Lees, Bot. Malv. Hills,* 1st ed., p. 45 (1843).
Nor does this plant require localities to be given. The leaves are much

waved and crisped at the edges, at once a mark of distinction from other species. The plant varies and hybridizes, and the hybrids themselves vary sufficiently markedly for the forms to have received names, but none of them have yet been observed in Worcestershire. It is recorded from all bordering counties.

1705. **994. P. densus**, Linn.   *Opposite-leaved Pond-weed.*
Native. Shallow streams, ditches, ponds. Rare. P. June to August. *Top. Bot.* 61.
First Record. *P. densus,* for instance, not hitherto noticed out of the Avon division, *Lees, Bot. Worc.,* p. ix (1867).
  Avon. Defford Common; Bredon Hill; Broadway, *Lees.* Above Aldwick Wood, Bredon Hill.
This plant does not appear to have been observed in Shropshire.

1706. **995. P. zosterifolius**, Schum.   *Grasswrack-like Pond-weed.*
(*P. compressum*, Linn.)
Native. Streams, ditches, ponds. Rare. P. June. *Top. Bot.* 24.
First Record of aggregate, *P. compressum.* In the ditches at Abbot's Morton, *Purton, App.,* vol. iii, p. 16 (1821). First record of segregate. This Pond-weed has been detected in Blakedown Pool, near Kidderminster, by my friend Mr. Westcombe, and may probably occur in other pools, *Lees, Bot. Worc.,* p. 56 (1867).
  Avon. As first record. In the Avon near Evesham, *Westcombe.*
  Severn. In Harberrow, Brake Mill, Stakenbridge, Churchill, and Blakedown Pools, *Mathews.*
This plant was given as a first record in *Bot. Rec. Club Rep.,* 1881–2, but it was not new. The segregate has been observed in Salop, Stafford, and Warwick.

1708. **996. P. obtusifolius**, Mert. and Koch.   *Grassy Pond Weed.*
(*P. gramineus*, Sm.)
Native. Ponds. Local. P. July. *Top. Bot.* 68.
First Record. As *P. gramineus, Lees, Bot. Malv. Hills,* 1st ed., p. 45 (1843). First record as *P. obtusifolius.* Pool, Trimpley; pond, Malvern Link, *Thompson, Towndrow, Bot. Rec. Club Rep.,* 1883 (pub. 1884).
  Avon. Tributaries of the river, Evesham, *Lees.*
  Severn. Trimpley, as record above. Stanklin Pool, *Humphreys.* Severn, Claines. Hartlebury Common.
  Malvern. Malvern Link, as record above.
This plant is another of the segregates into which Linnaeus's *P. compressum* has been divided. There are two others, *P. acutifolius,* Link, and *P. Friesii,* Link, or *P. mucronatus,* Schrad., which have not yet been observed in Worcestershire. This plant is recorded from all bordering counties except Salop.

1711. **997. P. pusillus**, Linn.   *Small Pond-weed.*
Native. Rivers, ditches, ponds. Rare. P. June to July. *Top. Bot.* 106.
First Record. Kempsey Ford, *Lees, Illus. Nat. Hist. Worc.,* p. 153 (1834).
  Severn. Kempsey Ford. *Lees.* Porter's Mill. Shrawley. Northwick

Brickpit. Severn below Diglis. Severn, Pitchcroft. Grimley Brick-pits. Worcester and Birmingham Canal near Worcester.

Malvern. Kempsey Ford, *Lees.* Barnard's Green, *Lees.* Malvern Link.

The river at Kempsey being the boundary between the districts of Severn and Malvern, a water plant flourishing there may well be recorded under each of them. This plant is omitted in the 1st and 2nd eds. of Lees, *Bot. Malv. Hills*, but given in the 3rd as occurring in a pool at Barnard's Green. It has been observed in all neighbouring counties except Gloucestershire.

1714. 998. P. pectinatus, Linn.   *Fennel-leaved Pond-weed.*

Native. Rivers, ponds, canals. General throughout the county. P. July and August. *Top. Bot.* 90.

First Record. In a pond near the old Waterworks, Worcester, *Lees, Illus. Nat. Hist. Worc.*, p. 153 (1834). As *P. eu-pectinatus*, Bittell Reservoir, in this book.

Severn. Stanklin Pool.

Lickey. Bittell Reservoir, 1900, *Humphreys*, as above.

This plant is one of the commonest of the Pond-weeds, and the aggregate could hardly be confounded with any other species that is likely to be in our county. The leaves are bright green, and resemble Fennel, as the English name would indicate. The plant likes slowly-moving streams. It is recorded from all bordering counties.

1716. 999. P. interruptus, Kit.

(*P. flabellatus*, Bab.)

Native. Rivers and quickly-flowing streams. P. June to August. *Top. Bot.* 58.

First Record. *P. flabellatus* grow(s) in the Avon at Evesham, *Lees, Bot. Worc.*, p. 92 (1867).

Avon. As first record. Avon, Birlingham.

Severn. In the Stour below Kidderminster near the viaduct on the Kidderminster and Bewdley Railway, *Mathews.* Canal at Hawford, *Westcombe.*

Malvern. Pool Brook, Upton-on-Severn, *Towndrow.*

This plant is a segregate of the last, accorded only the position of a sub-species in *Hooker, Stud. Fl.*, p. 436. It has been observed in all neighbouring counties.

1717. 1000. P. filiformis, Nolte. *Top. Bot.* 18.   *Slender-leaved Pond-weed.*

First Record. As below.

Severn. Spetchley Pool. Severn at Worcester, sp. *Hast. Mus. Worc.*

ZANNICHELLIA, Linn. 456. (Named in honour of J. J. Zanichelli, a Venetian botanist.)

1720. 1001. Z. palustris, Linn.   *Horned Pond-weed.*

Native. Rivers, streams, pools. Not common. P. May to August. *Top. Bot.* 72.

First Record (of the aggregate). Feckenham Moors, *Purton, Midl. Fl.*, p. 434 (1817).

Avon. As first record. Trench Woods. Overbury. Rudge's Farm.

Severn. Diglis Basin, *Lees.* Upper Brake Pool, Hagley, *Mathews.* Stanklin Pool, *Humphreys.* Near Severn Stoke. Pond at Hadsor. Pool, Trotshill.

Malvern. Welland Common, *Lees.* Malvern Chace, *Westcombe.*

This plant grows entirely submerged in the water, and is not unlike *Potamogeton pusillus*, but readily distinguished from that by the small green flowers in the axils of the leaves. The two following species are segregates of the Linnaean plant. The aggregate occurs in all bordering counties.

Var. a. **brachystemon,** Gay.

Native. Ponds, streams, pools. Rare. P. May to August. Not mentioned in *Top. Bot.*

First Record. Malvern, *Towndrow, Malv. Advertiser*, Dec. 10, 1892.

Malvern. As above.

This form has subsessile fruit. It has not been observed in any bordering county.

1721. 1002. Z. pedunculata, Reichb.

Native. Ponds and streams. Rare. P. May to July. *Top. Bot.* 22.

First Record. As *Z. palustris*, var. β. Brooks and pools near Worcester, *Stokes's With. Bot. Arr.*, and ed., p. 1014 (1787).

Severn. As record above.

Malvern. Malvern Chace, *G. Reece*, sp. *Herb. Hastings Museum, Worc.*

This form, with stalked fruit, has not been observed in any bordering county.

91. CYPERACEAE, Juss.

ELEOCHARIS, Linn. 461. (ἕλος, a marsh, χαίρω, I delight, from the place of growth.)

1731. 1003. E. acicularis, Roem. and Schult.   *Least Spike-rush.*

(*Scirpus acicularis*, Linn.)

Native. Muddy bottoms or margins of water. Rather uncommon. P. July and August. *Top. Bot.* 73.

First Record. As *Scirpus acicularis*. Malvern Chace, *Stokes's With. Bot. Arr.*, and ed., p. 47 (1787).

Avon. One place, *Lees.*

Severn. Severn Stoke, *Lees.* Lake-side, Westwood.

Malvern. Bogs at the foot of Malvern Hills, *Lees.* Barnard's Green ; Welland Common ; Garret Pool, *Lees.*

Lickey. Bittell Reservoir, Cofton Reservoir ; *Mathews.*

This humble little plant, some three or four inches high, is recorded from every bordering county.

1732. 1004. E. palustris, Roem. and Schult.   *Creeping Spike-rush.*

(*Scirpus palustris*, Linn.)

Native. Shallow ponds, marshes, wet meadows. General throughout the county. P. April and May. *Top. Bot.* 111.

First Record. Banks of a pool at Kinnersley, *Dr. Streeten, Lees, Illus. Nat. Hist. Worc.*, p. 151 (1834).

It is hardly necessary to localize this widely distributed plant. It might be taken for a rush as it is destitute of leaves, and sends up rounded stems, each of which terminates in a solitary spikelet. It is recorded from all bordering counties.

1734. 1005. E. multicaulis, Sm.   *Many-stalked Spike-rush.*

(*Scirpus palustris*, β. Linn.)

Native. Marshes and wet heaths. Rare. P. July and August. *Top. Bot.* 88.

First Record. As *Scirpus palustris*, var. β. Severn Stoke, *Stokes's With. Bot. Arr.*, 2nd ed., p. 46 (1787).

Severn. As first record.

Malvern. Castlemorton Common, *Lees.* Bog, west side of Worcestershire Beacon.

This plant has been noticed in every neighbouring county except Herefordshire.

SCIRPUS, Linn. 462. (The old Latin name used by Pliny for a rush. Etymology doubtful.)

1735. 1006. S. pauciflorus, Lightf.   *Chocolate-headed Club-rush.*

(*Eleocharis pauciflora*, Link.)

Native. Marshes, heaths. Local. P. June to August. *Top. Bot.* 91.

First Record. As *Eleocharis pauciflora*. Not localized, *Bot. Malv. Hills*, 1st ed., p. 44 (1843).

Severn. Great Bog, Wyre Forest. Hartlebury Common.

Malvern. Castlemorton Common, *Lees.* Malvern Hills. Bog, west side of Worcestershire Beacon. Welland Common.

Lickey. One place, lost, *Lees.*

Scott, in his *Hist. Stourb.*, recorded this plant in the neighbourhood as occurring in ' Reservoirs, &c.', but without specifying the county, so it is doubtful if the plant was in Worcestershire. The plant is recorded from all neighbouring counties.

1736. 1007. S. caespitosus, Linn.   *Scaly-stalked Club-rush.*

Native. Heaths. Rare. P. April to August. *Top. Bot.* 104.

First Record. Bromsgrove Lickey, *Purton, Midl. Fl.*, p. 64 (1817).

Malvern. One place, *Lees.*

Lickey. As first record ; now extinct, *Humphreys.* Billesley Common, *Ick.*

This plant, mentioned in Lees' *Bot. Malv. Hills*, 1st ed., is not noticed in the 2nd and 3rd eds., and the record was possibly an error. The plant is recorded from all bordering counties.

1738. 1008. S. fluitans, Linn.   *Floating Club-rush.*

(*Isolepis fluitans*, R. Br. *Eleogiton fluitans*, Link.)

Native. Pools and ditches in heathy places. Not uncommon. P. May to July. *Top. Bot.* 91.

First Record. As *Eleogiton fluitans*. Not localized, *Lees, Bot. Malv. Hills*, 1st ed., p. 43 (1843).

Avon. Not uncommon, *Lees.*

Severn. Hartlebury Common. Near the Devil's Spittleful.

Malvern. Commons near Malvern, *Lees.* Old Hills, Powick.

Lickey. Rather uncommon, *Lees.*

This plant is recorded from all bordering counties.

1740. 1009. S. setaceus, Linn.   *Bristle-stalked Club-rush.*

(*Isolepis setacea*, R. Br.)

Native. Marshy places. Rare. P. July and August. *Top. Bot.* 108.

First Record. In a dry pool at Cookhill, *Purton, Midl. Flora*, p. 65 (1817).

Avon. Cookhill, as first record.

Severn. Wyre Forest, *Mathews.* Monk Wood, *G. Reece*, sp. *Herb. Hastings Museum, Worc.* Hartlebury Common. Moseley Green, Hallow.

Malvern. Commons near Malvern, *Lees.* Boggy places, Malvern Hills. High Wood, Alfrick.

Lickey. Rather uncommon, *Lees.*

A good imitation of a tiny rush, this plant, with very slender stems some three to six inches high, with leafy bases. It is recorded from all bordering counties.

{**1741. 1010. S. Holoschoenus,** Linn.   *Round cluster-headed Mud-rush.*

(*Isolepis Holoschoenus*, Roem. and Sch.)

Native. Marshes. Once recorded. P. September. *Top. Bot.* 2.

First Record. Scirpus romanus . . . Habitat in palustribus juxta Throgmorton in agro Worcestersensi, *Rev. D. Sheffield, Hudson's Flora Anglica*, pp. 19–20 (1762).

Avon. As first record.

This record has been discredited by Mr. H. C. Watson, *Cyb. Brit.*, vol. iii, p. 71, Mr. Mathews thinks without sufficient reason ; and Mr. Bagnall thinks the record should be omitted. It is most unlikely that Hudson would have recorded so rare a plant unless he had been convinced of its identity. The plant is mentioned in both of the lists of plants in Nash's *Hist. Worc.*, that in the *Int.* (1781) and also in the *Sup.* (1799) ; and in Smith's *Engl. Bot.*, vol. xxiii, no. 1612 (1806). Mr. Watson, *Top. Bot.*, p. 442, allows the plant to be native only in Devonshire, and says of this record that it is certainly an error. It has never been seen in the Malvern district by *Lees*, but from that part of Herefordshire which he liberally included in it. This plant is the true Bulrush, a name that country people apply to any tall-growing aquatic, especially the Reed-mace (*Typha*). Mats and rush-bottomed chairs are made of it, and it is often employed by coopers for caulking casks. It is found in every bordering county.}

1742. 1011. S. lacustris, Linn.   *Lake Club-rush, or Bulrush.*

Native. Rivers, ponds, canals. General throughout the county. P. July and August. *Top. Bot.* 101.

First Record. The margin of Avon's placid stream displays . . . labyrinths of Bullrushes, *Lees, Bot. Worc.*, p. 90 (1867).

It is hardly necessary to localize this plant, so general in the situations that it loves, though in the Malvern district it is probably absent, *Towndrow* (MS.). It was recorded twice in the books before the first record above ; once, *Scott*, as near Stourbridge but in Staffordshire ; and another time in the Malvern district by *Lees*, but from that part of Herefordshire which he liberally included in it. This plant is the true Bulrush, a name that country people apply to any tall-growing aquatic, especially the Reed-mace (*Typha*). Mats and rush-bottomed chairs are made of it, and it is often employed by coopers for caulking casks. It is found in every bordering county.

[× triqueter. *Blunt-edged Club-rush.*

(*S. carinatus,* Sm.)

This plant, of which the *Top. Bot.* number is 7, was recorded by *Scott, Hist. Stourb.* (1832), as occurring at 'Chickhill Pool, and other localities'. Cheekhill Pool is in Staffordshire, but some of the 'other localities' may be in Worcestershire. Mr. Lees, *Table of Plants,* localizes the plant in the Severn and Malvern districts, but nowhere else mentions it. Nevertheless, further confirmation is desirable before the plant can be included in the Worcestershire list.]

1743. 1012. S. Tabernaemontani, Gmel. *Glaucous Club-rush.*

Native. Rivers and ponds. Rare. P. June and July. *Top. Bot.* 57.

First Record. Pool at Northwick Brick-kilns, *Stanley's Worc. and Malv. Guide* (1853).

　Avon. Bredon. Twyning Fleet. Norton, near Evesham.

　Severn. As first record. Westwood Park. Shrawley Wood.

　Malvern. Clay Pits near the Rhydd, *Towndrow, Malv. Advert.,* Dec. 10, 1892.

This plant is very near to the foregoing, and is characterized by its glaucous colour, absence of floating leaves, scabrid glumes, and glabrous anther-tips. Mr. Mathews thinks it must often be overlooked. By some botanists it is considered only a digynous form of the preceding plant. It does not appear to have been noticed in Gloucestershire or Herefordshire.

1746. 1013. S. maritimus, Linn. *Salt-marsh Club-rush.*

Native. Marshes and stream-sides. Very local. P. July to September. *Top. Bot.* 85.

First Record. Marshes and ditches about Badsey, *Rufford, Purton, Midl. Fl.,* p. 65 (1817).

　Avon. As first record. Defford Common, *G. Reece.*

　Severn. One place, *Lees.* By side of Salwarpe, not far above Salwarpe Church. Salwarpe, near High Park.

　Malvern. Longdon Marsh.

This plant does not appear to have been met with in Herefordshire.

1747. 1014. S. sylvaticus, Linn. *Wood Club-rush.*

Native. Marshy places. Woods and river-banks. Local. P. June to August. *Top. Bot.* 77.

First Record. Near the Lodge Pool, Kidderminster, and in Wire Forest, *Perry, Mag. Nat. Hist.,* vol. iv, p. 450.

　Avon. Defford Brook. Near Evesham. Fladbury. Pershore. Trench Woods. Norton, near Evesham. Throckmorton.

　Severn. As first record. Ribbesford Wood. Lady Pool, Blakedown, *W. Whitwell.* Banks of Severn, *Lees.* Fenny Rough, *Mathews.* Shrawley Wood. Northwick Brickyard. Severn-side, Grimley. Holt. Ombersley.

　Malvern. Moist coppices, eastern base of the Hills, *Lees.* Bransford. Powick. Knightsford Bridge. Martley. Stanford Bridge. Sapey Brook.

　Lickey. Common in boggy spots, *Humphreys.* Lickey Woods.

This is a robust and handsome species, three or four feet high, with broad and flat leaves and innumerable small green spikelets, in groups of two or three. It is recorded from all bordering counties.

1748. 1015. S. compressus, Pers. *Broad-leaved Blysmus.*

(*S. Caricis,* Retz. *Blysmus compressus,* Panz. *Schoenus compressus,* Linn.)

Native. Marshy meadows. Locally abundant. P. June to August. *Top. Bot.* 53.

First Record. As *Blysmus compressus.* Abundant on the margin of springy spots on the hills, especially about the Wells, *Lees, Bot. Malv. Hills,* 1st ed., p. 44 (1843).

　Avon. Field near Broadway, *Mathews.* Bredon Hill.

　Malvern. As first record. Bog, west side of Worcestershire Beacon. Malvern Common.

This plant has been observed in all neighbouring counties.

ERIOPHORUM, Linn. 463. (ἔριον, wool, φέρω, I bear, from its white heads.)

{**1751. 1016. E. vaginatum,** Linn. *Hare-tail Cotton-grass.*

Native. Bogs and moors. Rare. Once recorded, now lost. P. May. *Top. Bot.* 90.

First Record. Not localized, *Lees' Cat., New Bot. Guide* (1835).

　Lickey. Moseley, *Miss M. A. Beilby.*

The Moseley station is no doubt the one at which Mr. Lees localizes the plant in the Lickey district in his *Table of Plants*; he gives it in *Bot. Worc.,* p. 119. The plant has disappeared from the spot before growing Birmingham, and has been seen nowhere else in the county. All bordering counties possess the plant.}

1752. 1017. E. angustifolium, Roth. *Narrow-leaved Cotton-grass.*

(*Eriophorum polystachyon,* Linn.)

Native. Bogs and marshes. Local. P. April to June. *Top. Bot.* 110.

First Record. As *Eriophorum polystachyon.* In a bog on the western side of the Hills·(Malvern), *Lees, Mag. Nat. Hist.,* vol. iii, p. 160 (1830).

　Avon. Boggy spots close to the river at Wyre Piddle.

　Severn. Falling Sands Common, *Perry.* Burnt Wood, Bewdley, *Perry.* Hartlebury Common, *Mathews.* Bog near Fenny Rough, *Perry.* Wyre Forest, *Mathews.*

　Malvern. As first record.

　Lickey. Moseley.

This plant is recorded from all bordering counties.

1753. 1018. E. latifolium, Hoppe. *Broad-leaved Cotton-grass.*

Native. Bogs and marshes. Local. P. May and June. *Top. Bot.* 109. *Lond. Cat.,* 10th ed., 59.

First Record. Bewdley Forest, *Gissing, Phytologist, N. S.,* vol. i, p. 151 (1855).

　Severn. Wyre Forest. Hartlebury Common.

　Malvern. Malvern Hills, *Lees.*

This plant was recorded by Dr. F. Arnold Lees as new from 'a bog above Shelfield Coppice, Wyre Forest,' in the *Bot. Rec. Club Report* for 1883 (pub. 1884.) So far from being new, it had been observed and recorded by nearly every botanist since Wyre Forest was first explored. The form is a segregate

---

of Linnaeus's *E. polystachyon,* and it is sometimes doubtful which segregate is meant, this or *E. angustifolium,* when that plant is mentioned; but this is the Wyre Forest plant. Perhaps the most marked distinctions between the two are that this has the branches of the cyme scaberulous and a triangular stem (*triquetrous* is the learned term, because the faces are slightly concave), whereas the former plant has the branches of the cyme smooth and a nearly round stem. In *Top. Bot.,* p. 447, Worcester is denied this plant, and Salop and Hereford of the neighbouring counties credited with it.

[**1754. 1019. E. gracile,** Roth. *Slender Cotton-grass.*

This plant, the number of which in *Top. Bot.* is 4, was recorded, not localized, in *Lees' Cat., New Bot. Guide* (1835). Mr. Mathews considers this record an error, *Mid. Nat.* xi. 280. Mr. Lees thought the specimen *E. angustifolium,* but the Rev. A. Bloxam, 'or a friend of his,' referred it to *E. gracile,* but if this be *E. gracile* of Sm., not Roth, it would equal the β. *minus* of Babington, of which there is a specimen from the Malvern Hills in Herb. Hastings Museum. Mr. Lees continues, 'It has not since been certainly observed, and even *E. angustifolium* has ceased to grow at the locality,' which was near Malvern. The plant has not been observed in any bordering county, and may safely be removed from the Worcestershire list.]

RYNCHOSPORA, Vahl, 464. (ῥύγχος, a beak, and σπορά, a seed, from the beaked fruit.)

1756. 1020. R. alba, Vahl. *White Beak-rush.*

(*Schoenus albus,* Linn.)

Native. Marshy meadows, boggy places. Local. P. June to August. *Top. Bot.* 75.

First Record. Bogs on Moseley Common, *Miss M. A. Beilby, Analyst,* vol. vi, p. 294 (1837).

　Severn. Hartlebury Common, *Lees ; Humphreys.*

　Lickey. Moseley, as first record.

At Moseley the plant has gone, like several others of the rarest plants in our county. At Hartlebury Common it still exists. The plant is not recorded for Herefordshire or Gloucestershire.

SCHOENUS, Linn. 465. (σχοῖνος, cordage, from the use of some species.)

{**1758. 1021. S. nigricans,** Linn. *Black Bog-rush.*

Native. Bogs and marshes. Very rare, extinct. P. June and July. *Top. Bot.* 76.

First Record. As *Cyperus nigricans,* Feckenham Moors, *Purton, Midl. Fl.,* p. 62 (1817).

　Avon. As first record.

　Lickey. In the district, lost, *Lees.*

Mr. Lees, in his *Table of Plants,* p. 30, localizes the plant in the Lickey district also, marking it as lost. He no doubt for the moment thought it was one of the rare plants occurring at Moseley, and so was led into error. The plant has been observed in Gloucestershire, Staffordshire, and Warwickshire.}

CLADIUM, P. Br. 466. (From κλάδος, a twig.)

{**1759. 1022. C. Mariscus,** Br., Crantz. *Prickly Twig-rush.*

(*C. jamaicense,* Crantz. *Schoenus Mariscus,* Linn.)

Native. Bogs. Once recorded. P. July. *Top. Bot.* 41.

First Record. Feckenham Bog, *Purton, Midl. Flora,* p. 61 (1817).

　Avon. As first record, now lost.

The plant has been observed in Salop, Stafford, and Warwick.}

KOBRESIA, Willd. 467. (In honour of De Kobres, a German patron of botany.)

[**1760. 1023. K. bipartita,** Dalla Torre. *Compound-headed Kobresia.*

(*K. caricina,* Willd. *Elyna caricina,* Mert. and Koch.)

This plant, whose *Top. Bot.* number is 4, and which is not found south of Yorkshire, was recorded as *Elyna caricina,* from the 'Side of New Pool near Wood Farm (Malvern),' by Mr. Lees, *Bot. Malv. Hills,* 1st ed., p. 48 (1843). The plant is not noticed in the 3rd ed., nor in his *Bot. Worc.,* and must be an error. It has not been observed, naturally, in any neighbouring county, and must not be included in the Worcestershire list.]

CAREX, Linn. 468. (From κείρω, I cut, alluding to the sharp edges of the leaves.)

1761. 1024. C. dioica, Linn. *Creeping separate-headed Carex.*

Native. Bogs. Rare. P. May and June. *Top. Bot.* 80.

First Record. Bromsgrove Lickey, *Rufford, Purton, Midl. Fl.,* p. 440 (1817).

　Severn. Pound Green Dingle, Wyre Forest, *Herb. Hastings Museum.*

　Malvern. Bog near Keysend Hill ; Declivity of the Worcestershire Beacon, *Lees.*

　Lickey. As first record ; now extinct.

This plant is neither mentioned in *Lees' Bot. Worc.* nor localized in his *Table of Plants,* p. 31. But in *Bot. Malv. Hills,* 3rd ed., p. 112, he gives the above localities for the plant in the district, and adds that specimens gathered by Mr. George Reece are in the *Herb. Worc. Nat. Hist. Soc.* Mr. Reece's specimens are not now in that Herbarium at the *Hast. Mus.* at Worcester ; and the note, which is a rather unusual one for Mr. Lees to make, must have been made with an object. The plant does not appear to have been observed in Herefordshire.

1763. 1025. C. pulicaris, Linn. *Flea Sedge.*

Native. Bogs, marshy places, wet heaths. Local. P. June and July. *Top. Bot.* 110.

First Record. Malvern Chace, *Ballard, Stokes's With. Bot. Arr.,* 2nd ed., p. 1026 (1787).

　Severn. Wyre Forest, *Jorden.* Railway cuttings, Wyre Forest.

　Malvern. Base of the Worcestershire Beacon, *Lees.* Malvern Common.

　Lickey. Harris's Wood, Frankley, *Mathews.*

This is a pretty little plant, some six or more inches high, with a very slender stem. It bears six to twelve brown shining seeds, which are distant from each other, and from the fancied resemblance they bear to a noxious insect, it takes its name. It has been recorded from every bordering county.

1769. 1026. C. disticha. Huds.   *Soft brown Sedge.*

(*Carex intermedia*, Gooden. *C. spicata*, Pollich. *C. arenaria*, Leers.)

Native. Wet meadows, marshes. Not common. P. May and June. *Top. Bot.* 83.

First Record. Boggy meadows on the side of Malvern Chace, *Ballard, Stokes's With. Bot. Arr.,* 2nd ed., p. 1028 (1787).

Avon. One place, *Lees.*

Severn. Droitwich Canal, *Mathews.* Northwick Brickfields. Bank of pond near Droitwich Canal at Hawford Bridge.

Malvern. Longdon, *Lees.* Barnard's Green and Sherrard's Green, *Towndrow.*

Lickey. Broom Mill Pool, *Mathews.* Coal Pit Coppice. Chaddesley Woods, *Humphreys.*

All through its growth the middle portion of the spike of this plant differs from the two extremities, by which peculiarity it may be distinguished from all other British species. It is recorded from every neighbouring county.

× **paniculata.**

First Record. As below.

Severn. Stanklin. Bewdley.

This hybrid is not mentioned in *Lond. Cat.,* 10th ed.

1771. 1027. C. diandra, Schrank.   *Lesser panicled Sedge.*

(*C. teretiuscula*, Good.)

Native. Boggy meadows. Very rare. P. June. *Top. Bot.* 57.

First Record. Marshes near the Heath; banks of Stour; Harborough Pool, *Scott, Hist. Stourb.,* p. 540 (1832).

Severn. Pedmore Heath; banks of Stour; Harberrow Pool, *Scott.* Stanklin Pool, *Humphreys.*

This plant differs from *C. paniculata* in the crowded inflorescence, the absence of a membranous border to the fruit, the rounded sides of the stem, and its untufted habit. Mr. Mathews here suspects an error. Mr. Lees omits, in his *Table of Plants,* p. 31, the Severn district localities of Scott, and localizes the plant in the Lickey district only with a note of interrogation. If earlier records were in error, the plant has now been secured for the county by Mr. Humphreys. It has been noticed in Salop, Stafford, and Warwick of bordering counties.

Var. **Ehrhartiana,** Hoppe.

First Record. Stanklin Pool, Kidderminster, *H. S. Thompson,* 1901, *Worc. Nat. Club Trans.* iii. 53 (1901).

The variety differs from the type in having longer looser spikelets, stems triquetrous above, and more caespitose roots. It has been observed in Warwickshire. It is not mentioned in *Lond. Cat.,* 10th ed.

1773. 1028. C. paniculata, Linn.   *Great panicled Sedge.*

Native. Sides of water, damp woods, marshy ground. Locally abundant. P. April and May. *Top. Bot.* 94.

First Record. Marshes near the Heath and near the Stour, *Scott, Hist. Stourb.,* p. 540 (1832).

Severn. As first record. About Kidderminster. Hartlebury. Stanklin Pool. Pools at Harberrow, and between Stakenbridge and Kidderminster, *Mathews.* Churchill, Kidderminster. Falling Sands Lock, near Stourport. Pool east of Hartlebury Common. Bog, Hartlebury Common. Grimley Brickpits.

Malvern. One place, *Lees.*

Lickey. Several places, *Lees.*

This plant forms great tussocks or tufts of leaves, elevated above the surface of the spongy soil in which it delights to grow. It is recorded from all bordering counties.

× **remota.**   *Boenninghausen's Sedge.*

(*C. Boenninghausiana,* Weihe.)

Native. Stream-sides in shady situations. Once recorded. Very rare. P. June. *Top. Bot.* 17.

First Record. Fenny Rough; Stone; Kidderminster, *Amphlett* and *Mathews,* June 5, 1886, *Bot. Rec. Club Rep.* (1887).

Severn. As first record.

Babington gives this plant specific rank. The spelling of Weihe's name in *Lond. Cat.,* 10th ed., and in *Index Kewensis,* would appear to be an error; the specific name is usually spelled *Boenninghauseniana.* The plant was not to be found the last time it was looked for, owing to the growth of woodland vegetation. It was originally a fine large tussock. This sedge has been recorded from Salop and Warwickshire, the latter in error, says Mr. Bagnall (MS.).

1774. 1029. C. vulpina, Linn.   *Great Sedge.*

Native. Sides of canals, rivers, ditches, in marshes and wet places. General. P. April to June. *Top. Bot.* 86.

First Record. Banks of Stour and canals, *Scott, Hist. Stourb.,* p. 540 (1832).

Avon. Bredon Hill; Trench Woods, *Mathews.* Defford Common. Crowle. Fladbury. Pershore. Norton, near Evesham.

Severn. Stanklin Pool. Monk Wood. Roadside near the Virgin Tavern, Tolladine Road. Hartlebury Common. Porter's Mill. Canalside, Worcester. Clifton, near Kempsey. Battenhall. Cotheridge.

Malvern. Common on margins of brooks, *Lees.* Longdon Marsh. Bransford. Powick. Knightsford Bridge. Martley. Abberley. Malvern.

Lickey. Frequent, *Mathews.* Chadwick. Coal Pit Coppice.

This is a coarse-looking plant with a stout rough stem, and leaves so rough at the edges that they cut the unwary hand. It is recorded from all bordering counties.

1777. 1030. C. muricata, Linn.   *Greater Prickly Sedge.*

Native. Ditches, sides of water, dry hedgebanks. Not uncommon. P. May and June. *Top. Bot.* 2.

First Record. Hill Pool, Holloway; Ismere, *Scott, Hist. Stourb.,* p. 540 (1832).

Avon. Croome Perry Wood. Fladbury. Grafton. Crowle.

Severn. Between Hagley and Kidderminster, in several places, *Mathews.* Ismere, as first record. Bubble Bridge, Worcester. Whitehall. Stourport. Moseley Green, Hallow. Ombersley Road, near *Potentilla argentea.*

Malvern. Under hedges, *Lees.* Near Powick Church. Bransford. Leigh. Alfrick. Ravenshill Wood. Martley. Shelsley Beauchamp. Malvern Link.

Lickey. Offmoor Wood. Frankley Wood. Frequent on hedgebanks in Clent, *Amphlett.* Hill Pool, Chaddesley Corbett, as first record. Rowney Green. Fairfield.

This sedge is recorded from all bordering counties. The reader's attention is called to the fact that the Census No. of this sedge in *Lond. Cat.,* 10th ed., is only 2, while that of another, *C. muricata,* auct. angl., synonym of no. 1776 in the catalogue, *C. contigua,* Hoppe, which Babington, *Man. Brit. Bot.,* 9th ed., makes equal to var. β. *virens* of the Linnaean plant, is 80, more nearly approaching the one assigned to this plant in *Lond. Cat.,* ed. 9, which was 78.

Var. **pseudo-divulsa,** Syme.

First Record. Malvern Link, *Towndrow, J. of B.,* vol. xxi, p. 246 (1883).

This variety, in which the spike is interrupted below, instead of being much interrupted, as in the following species, has been found also in West Suffolk. It is not given in the *Lond. Cat.,* 10th ed.

1778. 1031. C. divulsa, Stokes.   *Gray Sedge.*

(*Carex canescens,* Huds.)

Native. Hedgebanks, woods. General, except in the Lickey district, where it is rather uncommon. P. May and June. *Top. Bot.* 50.

First Record. Not localized, *Lees' Cat., New Bot. Guide* (1835).

Avon. Stoulton. Pinvin. Crowle.

Severn. Roadside Lane between Lower Wick and Bransford Road. Broadheath, Whittington. Eastbury, roadside. Ribbesford. Roadside, Cotheridge. Near Crowneast Woods.

Malvern. Leigh. Malvern Link.

Lickey. Coal Pit Coppice.

This is a slender species, with long narrow rough leaves, remarkable for its greyish hue. It is recorded from every bordering county.

× **vulpina.**

A hybrid between these two species, not mentioned in the books, was recorded by Mr. Towndrow, *Bot. Ech. Club Rep.* (1885).

Malvern. Newland Common, gone, *Towndrow, Malv. Advert.,* December 10, 1892.

Lickey. Coal Pit Coppice.

1779. 1032. C. echinata, Murr.   *Little prickly Sedge.*

(*Carex stellulata,* Good.)

Native. Bogs, wet heaths. Local. P. May and June. *Top. Bot.* 111.

First Record. As *C. stellulata.* Bromsgrove Lickey; Bog on the west side of Malvern Hill, *Purton, Midl. Fl.,* p. 441 (1817).

Avon. Defford Common. Near Gooseshill Wood.

Severn. Hartlebury Common; Wyre Forest, *Mathews.* Moseley Green, Hallow. Doverdale. Hadley. Hallow Bank. Witley.

Malvern. As first record. Welland Common. Malvern.

Lickey. Bromsgrove Lickey, *Purton.* Moseley Common, *Miss M. A. Beilby.* Rowney Green.

This Sedge does not appear to have been noticed in Gloucestershire.

1780. 1033. C. remota, Linn.   *Distant-spiked Sedge.*

Native. Wet woods, shady hedgebanks, damp ditches. General throughout the county. P. May and June. *Top. Bot.* 88.

First Record. In the rocky wood, Fenny Rough, near Stone, *Perry, Mag. Nat. Hist.,* vol. iv, p. 456 (1831).

Avon. Bredon Hill, *Mathews.* Pershore. Defford. Flyford Flavell. Upton Snodsbury.

Severn. North Wood, Bewdley, *Mathews.* Hartlebury Common. Stanklin. Rushwick Coppice. Monk Wood. Meadow below Bevere Green. Roadsides near Cotheridge. Hadley. Ombersley. Claines. Kempsey.

Malvern. Base of Malvern Hills. Martley. Roadside near Berrow Hill. Malvern Link.

Lickey. Copse near Harborne, *H. S. Thompson.* The Leasowes, Halesowen, *Mathews.* Randans Wood, *Mathews.* Moseley Common, *Miss M. A. Beilby.* Stony Lane, Moseley, *Ick.* Illey, near Halesowen. Coal Pit Coppice. Lickey Wood.

This plant is recorded from all neighbouring counties.

× **vulpina.**   *Axillary-clustered Sedge.*

(*C. axillaris,* Good.)

Native. Wet ditches. Local. P. May and June. *Top. Bot.* 57.

First Record. Banks of Sapey Brook; in a bog at Wyre (Piddle) near the Avon, *Lees, Illus. Nat. Hist. Worc.* (1834).

Avon. Wyre Piddle, as above.

Severn. Wyre Piddle. Norton by Kempsey, *Westcombe.* Canal-side at Kidderminster, *Dr. F. A. Lees.* Norton Lane.

Malvern. Sapey Brook. Madresfield, *Towndrow.* Bransford. Sledgemoor Brakes.

This plant was recorded as new from Madresfield by Mr. Towndrow and from canal near Kidderminster by Dr. Arnold Lees, in the *Bot. Rec. Club Rep.,* 1883 (pub. 1884). It was not a new record. Differs from *C. remota* in having a taller, stouter, more triquetrous stem, which is generally very scabrid below the spike, and the ribbed perigynia. The plant does not appear to have been seen in Staffordshire.

1781. 1034. C. elongata, Linn.   *Elongated Sedge.*

Native. Marshes, sides of ditches and ponds. Rare. P. June and July. *Top. Bot.* 18.

First Record. By a path through the fields between (Hartlebury) Common and the Severn, in 1851, but I could only perceive a single tuft of it anywhere about, *Lees, Bot. Worc.,* p. 11 (1867).

Severn. As first record. In the Lane between Hartlebury Common and Severn opposite Redstone Rock, *Herb. Hastings Museum.*

This plant is noticed again at p. 57 of Mr. Lees' *Bot. Worc.,* where there is a sentence with a treacherous full stop, which may possibly mean that the above locality is one of 'the low, wet, borders of the vale near the Severn' mentioned as places where 'lurk certain sedges' generally considered worthy; and not, as the unwary might suppose, that Northwick Brickfield is a locality for this plant. This species differs from *C. leporina* in the laxer arrangement of the spikelets, and the unwinged perignynia much exceeding the glumes. This sedge appears to have been noticed only in Salop and Warwick of our bordering counties.

1783. 1035. C. curta, Good.   *White Sedge.*

(*Carex canescens,* auct., non Linn.)

Native. Marshes, bogs, wet places on heaths. Rare. P. June and July. *Top. Bot.* 78.

First Record. Near Bewdley, *Gissing, Phytologist, O.S.,* vol. i, p. 151 (1853).

Severn. Hartlebury Common. Near Bewdley, *Gissing.* Sutton Common. Cutpursey Coppice.

Lickey. Bittell Reservoir.

This plant is recorded from all bordering counties.

1784. 1036. C. leporina, Linn.   *Oval-spiked Sedge.*

(*C. ovalis,* Good.)

Native. Moist heaths, marshes. General throughout the county. P. May and June. *Top. Bot.* 112.

First Record. Hungary Hill, *Scott, Hist. Stourb.,* p. 540 (1830).

Avon. Croome Perry Wood.

Severn. Hartlebury Common. Wyre Forest, *Mathews.* Iverley, west of Hagley Station. Bromley's Brickyard. Hagley. Roadside, Cotheridge. Roadside, Broadheath. Spetchley Common. Grimley Brickpits.

Malvern. Malvern, Malvern Link, and Castlemorton Commons, *Town-drow.* Meadow west of Powick Church. Rock Wood. Shelsley. Malvern Hills.

Lickey. Hungary Hill, Stourbridge, *Scott.*

This is a slender plant with long grassy leaves, triangular stems, and brownish-green shining spikelets without conspicuous bracts. It is recorded from every neighbouring county.

Var. **bracteata,** Syme.

First Record. On the common below Malvern Wells, *Lees, Bot. Malv. Hills,* 1st ed., p. 48 (1843).

Malvern. Seats Common, *Lees.* Castlemorton Common, *Westcombe.* Near Sledgemoor Brakes. Malvern Link Common.

At p. 69, *Bot. Worc.,* we are told that this plant was gathered on Seats Common, and was at first supposed to be *Carex argyroglochin,* or even one new to science and named *C. Malvernensis*; but that afterwards Mr. S. Gibson, of Hebden Bridge, Yorkshire, assigned it to this variety. The plant was the subject of papers in the *Phytologist,* vol. i, p. 715 and vol. ii, pp. 751, 759. This variety is characterized by the elongated bracts much exceeding the spike. It has not been observed in any bordering county. This variety is not mentioned in *Lond. Cat.,* 10th ed.

1788. 1037. C. elata, All.   *Tufted Bog Sedge.*

(*C. stricta,* Good. *C. caespitosa,* Huds. *C. gracilis,* Wirmer. *C. melanochloros,* Thuill. *C. Hudsonii,* Ar. Benn.)

Native. Marshy places, pond sides. Very rare. P. May and June. *Top. Bot.* 45.

First Record. Marshes, *Scott, Hist. Stourb.,* p. 540 (1832).

Severn. Brickyards at Northwick, *Lees, Herb. Hastings Museum.* Near the brick kilns above Worcester, *Westcombe.* Hartlebury Common.

Malvern. One place, *Lees.* Danemoor Pool, *Herb. Hastings Museum.*

Lickey. Moseley, *Ick.*

Mr. Mathews suspects Scott's record to be an error, but his reason, in *Mid. Nat.* xi. 41, is a little difficult to follow. Mr. Towndrow doubts if any one has seen a Worcestershire specimen of this plant, and we agree with him that the specimens of it in *Herb. Hast. Mus. Worc.* are certainly *C. gracilis*; their sheath edges are not filamentous. It has, however, been observed in all bordering counties.

1790. 1038. C. gracilis, Curt.   *Slender-spiked Sedge.*

(*C. acuta,* var. b. *rufa,* Linn. *C. rufa,* Richter.)

Native. Sides of water. General. P. June. *Top. Bot.* 76.

First Record. Bank of the Warwick Canal, *Miss M. A. Beilby, Analyst,* vol. vi, p. 294 (1837).

Avon. Avon, at Pershore. Pirton Pool. Fladbury. Twyning Fleet.

Severn. Brickfields at Northwick, *Lees.* Henwick Mill. Ditch below Kempsey Grove. Grimley Brickpits. Ombersley. Holt. Shrawley Brook.

Malvern. Near Doddenham Church. Powick. Bransford. Leigh. Martley. Stanford Bridge. Guarlford.

Lickey. Yardley, *Miss M. A. Beilby.*

Mr. Mathews remarks that the locality of the first record is possibly in Yardley and so in Worcestershire. It is so possible that our county may well receive the benefit of the doubt. There is an excellent note on this plant and its varieties in Mr. Druce's *Fl. Berks.,* p. 540. The plant is recorded from all bordering counties.

1796. 1039. C. Goodenowii, Gay.   *Common Sedge.*

(*C. vulgaris,* Fries. *C. acuta,* var. *nigra,* Linn. *C. caespitosa,* Good., var. *rigida,* Good., var. *Goodenovii,* Bailey.)

Native. Marshes, wet meadows, heaths. Not uncommon except in the Avon district. P. May. *Top. Bot.* 110.

First Record. As *Carex vulgaris.* Moseley Common, *Ick, Analyst,* vol. vi, p. 294 (1837).

Avon. Evesham. Norton, near Evesham. Cleeve Mill. Bredon. Pershore. Tiddesley Wood. Trench Woods. Dodderhill Common. Crowle.

Severn. Hartlebury Common. Wyre Forest. Brickyards at Northwick. Grimley Brickpits. Kempsey. Ombersley. Severn Stoke. Hadley. Salwarpe. Upton Warren. Shrawley. Holt.

Malvern. West side of Worcestershire Beacon. Powick. Bransford

Leigh. Knightsford Bridge. Abberley Hill. Woodbury Hill. Near Malvern and Malvern Link.

Lickey. Pool at Coleford Priory, *Mathews.* Copse near Harborne, *H. S. Thompson.* Harborne Reservoir, Worcester side, *Mathews.* Coal Pit Coppice, *Humphreys.*

Carex caespitosa was recorded by *Scott, Hist. Stourb.,* p. 540 (1832), a record which would forestall the first record above, if it was certain that he intended this plant. But he calls it Carnation grass, and records it from 'Uplands, near Wychbury', which precludes this being the plant he means, and Mr. Mathews suspects an error and that *C. glauca* (*C. flacca* below) was the plant. This plant occurs in all neighbouring counties.

1797. 1040. C. flacca, Schreb.   *Glaucous Heath Sedge.*

(*C. glauca,* Scop. *C. recurva,* Huds.)

Native. Pastures, roadsides, downs, heaths. General throughout. P. April and May. *Top. Bot.* 112.

First Record. As *C. recurva.* Not localized, *Lees, Bot. Malv. Hills,* 1st ed., p. 48.

Avon. Bredon Hill. Crowle. Trench Woods. Hanbury. Tiddesley Wood. Sheriff's Lench. Norton, near Evesham. North Piddle. Fladbury. Pershore. Stoulton. Defford Common. Rous Lench. Flyford Flavell.

Severn. Hartlebury Common. Monk Wood. Stanklin Pool. Wyre Forest. Perry Wood. Battenhall. Trotshill. Kempsey. Severn Stoke. Ombersley. Hallow. Claines. Grimley. Holt. Hadley. Laywood, Salwarpe. Shrawley. Bewdley. Lincombe. Crowneast.

Malvern. Powick. Madresfield. Bransford. Leigh. Leigh Sinton. Brockamin. Alfrick. Knightwick. Ankerdine. Martley. Clifton-on-Teme. Abberley Hill. Roadside, near Berrow Hill. Lord's Wood. Middleyards Coppice. Malvern Hills.

Lickey. Moseley, *Mathews.* Coal Pit Coppice. Balaams Wood.

A variety having awned female glumes is recorded by Mr. Towndrow, *Malv. Advert.,* December 10, 1892. This plant is recorded from all bordering counties.

c. **stictocarpa,** Druce.

First Record. As below.

Malvern. Malvern Hill, *Herb. Hastings Museum.*

1799. 1041. C. limosa, Linn.   *Mud Sedge.*

Native. Bogs and marshes. Rare. P. June and July. *Top. Bot.* 26.

First Record. Wychbury Uplands, *Scott, Hist. Stourb.,* p. 540 (1832).

Severn. In the district, *Bagnall* (MS.).

Lickey. As first record.

Mr. Mathews says it is difficult to imagine what can have been intended by this name, seeing that the locality indicated possesses neither bog nor marsh, and the plant is far more frequent in the north and Scotland than it is in more southern parts. It has been noticed in Salop and possibly in Staffordshire.

1801. 1042. C. digitata, Linn.   *Fingered Sedge.*

Native. Woods. Very rare. P. May. *Top. Bot.* 13.

First Record. Mr. Jorden reports *C. digitata* from North Wood, Bewdley, *Lees, Bot. Worc.,* p. 57 (1867).

Severn. North Wood, Bewdley, *Jorden.*

This is the only record for this rare plant. It has also been noticed in Gloucestershire and Herefordshire.

1804. 1043. C. montana, Linn.   *Mountain Sedge.*

(*C. collina,* Willd.)

Native. Heaths and banks. Very rare. P. April and May. *Top. Bot.* 11.

First Record. Mr. Westcombe has also gathered *C. montana* in Wyre Forest, *Lees, Bot. Worc.,* p. 57 (1867).

Severn. Wyre Forest. The New Parks, Wyre Forest. Golden Valley, Bewdley.

The above records are the only ones known as localizing this plant in Worcestershire, but it appears to have been noticed also in the adjoining counties of Gloucestershire and Herefordshire, and there are extensive areas of it in the Shropshire portion of Wyre Forest.

1805. 1044. C. pilulifera, Linn.   *Round-headed Sedge.*

(*C. montana,* Huds., not of Linn.)

Native. Heaths, dry woods, peaty ground. Local. P. April and May. *Top. Bot.* 104.

First Record. Bromsgrove Lickey, *Purton, Midl. Fl.,* p. 448 (1819).

Severn. Harberrow, *Mathews.* High part of Hartlebury Common. Wyre Forest. Warshill Wood. Brake Mill Plantation, Hagley. Habberley Valley. Crowneast Woods, *Westcombe.* Cutpursey Coppice.

Malvern. Castlemorton Common, *Lees.* Seats Common. Malvern Link Common.

Lickey. As first record. Farley Wood. Chaddesley Wood. Rowney Green, Alvechurch. Balaams Wood. Winwood Heath, *Mathews.* Hasbury Common, Halesowen, *Mathews.* Bromsgrove Lickey, *West-combe.*

This species has a slender rough stem, six or eight inches high, with several short spikelets of few flowers. It is very near the last species, being united with it by some botanists, and between the two plants there has been some confusion of nomenclature. It does not appear to have been noticed in Gloucestershire, but is frequent in Warwick and Stafford.

1807. 1045. C. caryophyllea, Latourr.   *Vernal Sedge.*

(*C. verna,* Chaix. *C. praecox,* Jacq. *C. montana,* Lightf., not of Linn. *C. saxatilis,* Huds., not of Linn.)

Native. Meadows, heaths, downs. General. P. April and May. *Top. Bot.* 96.

First Record. As *C. praecox.* Moseley Common, *Miss M. A. Beilby,* vol. vi, p. 294 (1837).

Avon. Bredon Hill. Sheriff's Lench. Hipton Hill. Rous Lench. Crowle. Huddington. Hanbury Park. Dodderhill Common. Inkberrow. Flyford Flavell. Dormstone. Craycombe. Broadway. Cleeve Banks. Norton, near Evesham. Dunstall Common.

Severn. Wyre Forest. Wassell Wood, Bewdley. Monk Wood. Near Shrawley Wood. Lane above the head of Habberley Valley. Drakelow, Kidderminster. Opposite Virgin Tavern, Worcester. Crookbarrow Hill. Kempsey. Ombersley. Holt Bank. Helbury Hill. Fenny Rough.

Malvern. In dry places all over the Hills, *Lees.* Woodbury Hill. Malvern Wells. Old Hills. Madresfield. Powick. Leigh. Alfrick. Ravenshill. Ankerdine. Woodbury Hill. Abberley Hill. Stockton. Lindridge. Madresfield.

Lickey. Hasbury Common, Halesowen.

This plant is recorded from all bordering counties.

1809. 1046. C. pallescens, Linn. *Pale Sedge.*

Native. Moist woods and shady places. Not uncommon. P. May. *Top. Bot.* 92.

First Record. Cradley Park, *Scott, Hist. Stourb.,* p. 540 (1832).

Avon. Tiddesley Wood. Trench Woods. Crowle. Rous Lench. North Piddle.

Severn. Wyre Forest. Crowneast Woods, *Westcombe.* Near the grating, lower pool, Shrawley Wood. Seckley Wood. Monk Wood. Ockeridge Wood.

Malvern. Shady places in small quantity, *Lees.* Powick. Madresfield. Bransford. Leigh. Knightsford Bridge. Near Southstone Rock.

Lickey. Offmoor Wood. Frankley Wood. Illey Mill and the Leasowes, Halesowen. Cradley Park. Twylands Wood. Coal Pit Coppice.

This slender species is well marked by its pale hue, hairy leaves and sheaths, and blunt fruit. It occurs in all bordering counties.

1810. 1047. C. panicea, Linn. *Pink-leaved Sedge.*

Native. Wet meadows, bogs, marshes. General. P. May. *Top. Bot.* 112.

First Record. Moseley Common, *Miss M. A. Beilby, Analyst,* vol. vi, p. 294 (1837).

Severn. Hartlebury Common. Wyre Forest. Great Bog, Wyre Forest.

Malvern. Borders of fields and banks, *Lees.* North-west side of Sugar Loaf Hill in a marshy copse, *Westcombe.* Malvern Hills.

Lickey. Moseley. Bromsgrove Lickey. Harborne Reservoir, Worcestershire side. Coal Pit Coppice.

The English book-name of this sedge does not, as might be supposed, refer to colour, but to the plant whose leaves it is supposed to imitate. It is fairly common in damp meadows, and occurs in every bordering county.

1814. 1048. C. pendula, Huds. *Great pendulous Sedge.*

Native. Shady woods, hedges. General throughout. P. June and July. *Top. Bot.* 75.

First Record. Witchery Hole, Ham Castle, *Stokes's With. Bot. Arr.,* 2nd ed., p. 1046 (1787).

Avon. Bredon Hill. Battons Wood, Bredon. Crowle. Near Goosehill Woods. Hanbury. Flyford Flavell. Fladbury. Tiddesley Wood. Deerfold Wood.

Severn. Westwood Park. Cutpursey Coppice. Monk Wood. Ribbesford Wood. Snuff Mill Valley (Golden Valley), Bewdley. Gladder Brook, Stourport. Battenhall. Near Crookbarrow Hill. Ombersley Bank. Wyre Forest. Hadley. Gardeners Grove. Wyre Forest.

Malvern. Damp shady places, *Lees.* Sarn Hill. Rock Wood. Shelsley. Clifton-on-Teme. Hayley Dingle. Leigh. Southwood. Leigh Sinton.

Lickey. The Valley, Bromsgrove. Offmoor Wood. Frankley Wood. Lickey Wood. Dodford.

This is one of the larger sedges, often growing four or five feet high. Its very long hanging spikelets well distinguish it. It is recorded from each neighbouring county.

1815. 1049. C. strigosa, Huds. *Loose pendulous Sedge.*

Native. Shady places and woods. Local. P. June. *Top. Bot.* 36.

First Record. Cradley Park, *Scott, Hist. Stourb.,* p. 540 (1832).

Avon. Trench Woods. Bow Wood. Tiddesley Wood. Wood west side of Defford Common. Badgers Bank.

Severn. Near Hartlebury, *Lees.* Between Worcester and Crookbarrow. Shrawley. Brickyards at Northwick. Rushwick. Crookbarrow Glen. Wood between Hawford and Ombersley. Shrawley Wood. Nunnery Wood.

Malvern. Banks of Sapey Brook, *Lees.* Cowleigh Park. Southwood. Wood base of Woodbury Hill. Malvern Link. Woodsfield.

Lickey. Cradley Park. Copse near Harborne, *H. S. Thompson.* Ham Dingle, Pedmore, *Mathews.*

This Carex is somewhat similar in appearance to *C. sylvatica,* but is easily distinguished by its blunt perigynia which are veined throughout, whereas in *C. sylvatica* they are strongly beaked and veined only at the base. This sedge apparently has not yet been observed in Warwickshire.

1817. 1050. C. sylvatica, Huds. *Wood Sedge.*

Native. Damp woods, hedges, thickets. Common in suitable places. P. May and June. *Top. Bot.* 87.

First Record. Cradley Park, *Scott, Hist. Stourb.,* p. 540 (1832).

This plant is so common that no localization is necessary. The slender drooping spikelets have long stalks, and the fruit is glabrous, and so acuminate as to terminate in a long beak. It grows a foot or eighteen inches high, and is found in all bordering counties.

1818. 1051. C. helodes, Linn. *Smooth-stalked beaked Sedge.*

(*C. laevigata,* Sm.)

Native. Shady and marshy places. Rare. P. April to June. *Top. Bot.* 62.

First Record. Moist field at Highgate, not far from the Rea, *Miss M. A. Beilby, Analyst,* vol. vi, p. 294 (1837).

Lickey. As first record. Just within the Worcestershire border, in a copse near Harborne, 1901, *H. S. Thompson,* named by Arthur Bennett.

As the first record is from one of those localities which may be either in Worcestershire or Staffordshire, Mr. Thompson's discovery comes opportunely to confirm the plant for our county. It is not noticed in Mr. Lees' *Bot. Worc.,* nor localized in his *Table of Plants.* This species is characterized by having the sheath of the leaf auricled at the base, a feature which is very uncommon with our British Carices. It has been noticed in all bordering counties except Gloucestershire.

1819. 1052. C. binervis, Sm. *Green-ribbed Sedge.*

(*Carex distans,* Light., not of Linn.)

Native. Heaths, wet roadsides. Rather uncommon. P. April to June. *Top. Bot.* 110.

First Record. Moseley Common, *Miss M. A. Beilby, Analyst,* vol. vi, p. 294 (1837).

Avon. Moory field near the railway, Oddingley, *Westcombe.*

Severn. Near Hartlebury. Brickyards at Northwick. Banks of canal, Hawford, *Westcombe.* Moseley Green, Hallow. Wyre Forest.

Malvern. Danemoor Pool, *Lees.* Newland. Malvern. Malvern Common.

Lickey. Moseley. Winwood Heath.

This is very like the succeeding plant, from which it is difficult to discriminate it, but is a much coarser plant, sometimes three feet high, with dark purple glumes with a greenish-yellow midrib, and the dark fruit has two prominent green ribs. It has been observed in all bordering counties.

1820. 1053. C. distans, Linn. *Loose Sedge.*

Native. Marshy fields. Rare. P. June and July. *Top. Bot.* 60.

First Record. Feckenham Moors, *Purton, Midl. Fl.,* p. 445 (1817).

Avon. Feckenham, as first record.

Severn. Cradley Park, *Scott.* Hartlebury, June 1898, *Rea.* Wyre Forest, *Westcombe.* Hawford Bridge. Banks of Droitwich Canal. Salwarpe Church, 1907, *Humphreys.*

This plant is not mentioned in Lees' *Bot. Worc.,* nor localized in his *Table of Plants,* p. 32. Mr. Mathews brands Scott's record as an error, and says Purton's plant was possibly *C. binervis,* Sm. He gives no reason for his disbelief, but Mr. Lees evidently agrees with him. Mr. Rea's and Mr. Humphreys' records now confirm the plant for the county. This is a much smoother plant than the preceding, never growing so tall, nor much exceeding a foot in height, and the glumes are brownish and the fruit yellowish-brown. It has been observed in all neighbouring counties.

1822. 1054. C. fulva, Host. *Tawny Sedge.*

(*C. Hornschuchiana,* Hoppe.)

Native. Marshes and bogs. Rather uncommon. P. May and June. *Top. Bot.* 85.

First Record. Cradley Park, *Scott, Hist. Stourb.,* p. 540 (1832).

Severn. Great Bog, Wyre Forest. Brickyards at Northwick.

Malvern. Near Brand Lodge, *Lees.* By west side of Worcestershire Beacon, *Westcombe.*

Lickey. Cradley Park, *Scott.* Near the Spring Farm, St. Kenelm's,

---

Clent ; Clent Hills, abounds in wet places, *Lees.* Offmoor. Coal Pit Coppice. Randans.

This plant appears not to have been observed in Shropshire.

1824. 1055. C. flava, Linn. *Yellow Sedge.*

Native. Marshes, bogs, wet heathy places. Not common. P. May and June. *Top. Bot.* 110.

First Record. Fields near Cradley, *Scott, Hist. Stourb.,* p. 540 (1832).

Severn. Hartlebury Common. Moseley Green, Hallow.

Malvern. Malvern Hills. Bogs on Malvern Hills, *Westcombe.*

Lickey. Cradley, *Scott.* Moseley Common, *Miss M. A. Beilby.* Coal Pit Coppice.

Var. b. **lepidocarpa** (Tausch).

First Record. Locally abundant in wet places, *Mathews, Fl. Clent and Lickey Hills,* p. 48 (1881).

Severn. Sp. *Herb. Hast. Mus. Worc.*

Lickey. As first record.

The type is recorded from all bordering counties.

Var. d. **oedocarpa,** And.

First Record. In this book, in note to *C. Oederi,* below.

Malvern. In the district, *Towndrow.*

1825. 1056. C. Oederi, Retz. *Top. Bot.* 38.

Native. Heaths, bogs, and marshes. Uncommon. P. June and July. *Top. Bot.* 38.

First Record. Spring by the roadside, near the turn beyond Great Malvern, *Dr. Streeten, Lees, Illus. Nat. Hist. Worc.,* p. 178 (1834).

Severn. Near the Birches, Hagley, 1904, *W. Whitwell* ; 'nearest to *Oederi',* teste *Arthur Bennett.* Wyre Forest, *Mathews.*

Malvern. *Streeten,* as first record. Herefordshire Beacon ; Bogs, Malvern Hills, *Westcombe.*

Mr. Mathews says of Dr. Streeten's plant that it is not true *Oederi,* but *C. flava,* var. *minor, Townsend,* in which he is confirmed by Mr. Towndrow, who says, ' we do not get it in the Malvern district at all. What we get is Townsend's var. *minor',* which is *C. flava,* var. *oedocarpa,* And., above. Mr. Mathews' Wyre Forest record is precisely stated, *Worc. Nat. Club Trans.* ii. 57, and if Dr. Streeten's is discredited, will stand as the First Record of the plant in Worcestershire. This sedge has been observed in Staffordshire and Warwickshire.

Var. b. **elatior,** And.

First Record. On the highest ground of Bromsgrove Lickey, *Purton, App.,* vol. iii, p. 69.

Lickey. Bromsgrove Lickey, *Purton, Bagnall.*

1827. 1057. C. hirta, Linn. *Hairy Sedge.*

Native. Meadows, ditches, bogs, roadsides, heaths. General throughout the county. P. April to June. *Top. Bot.* 98.

First Record. Banks of streams, *Scott, Hist. Stourb.,* p. 540 (1832).

This common plant needs no localization. It may easily be recognized by its broad hairy leaves and its spikelets of large downy fruit. It is not by any means confined to damp situations, and is sometimes seen on dry sandy soil. It is recorded from all bordering counties.

1828. **1058. C. Pseudo-cyperus,** Linn. *Cyperus-like Sedge.*
Native. Low marshes, sides of stagnant water, ditches. Not uncommon. P. May and June. *Top. Bot.* 48.
First Record. In the Lodge Pools, Kidderminster, *Perry, Mag. Nat. Hist.,* vol. iv, p. 450 (1831).
Avon. Pirton Pool.
Severn. Broome Mill, Harberrow, Churchill, and Hurcott Pools, *Mathews.* Brickyards at Northwick, *Lees.* Wyre Forest. Hartlebury. Stanklin Pool. Shrawley Wood. Monk Wood. Knights' Grove, Near Severn Stoke. Westwood Park.
Malvern. Road between Newland and Powick, *Lees.* Near Berrow Hill. Ditch, roadside near Madresfield.
Lickey. Copse near Harborne, *Thompson.* Upper Bittell; Hawne, Halesowen, *Mathews.* Pool near Offmoor Wood, *Lees.* Frankley.
This is a very handsome and distinct species, two or three feet high, with rough stems and leaves, and fruit so rigid as to be almost prickly. It occurs in all bordering counties.

1829. **1059. C. acutiformis,** Ehrh. *Lesser Common Sedge.*
(*C. paludosa,* Good. *C. acuta,* Curt.)
Native. Sides of water, marshes, wet places in woods. Not uncommon. P. May and June. *Top. Bot.* 78.
First Record. As *C. paludosa.* Banks of the Warwick Canal, *Miss M. A. Beilby, Analyst,* vol. vi, p. 294 (1837).
Avon. Near Fladbury. Cleeve Mill. Eckington Bridge.
Severn. Stanklin Pool. Bilford. Marshy places near the Birmingham Canal near Worcester. Between Bilford and the Canal. Belle Isle, near Stourport. Dunclent. River-side below Diglis Weir.
Malvern. Admirals Covert, Malvern Wells.
Lickey. Chaddesley Woods.
This is a tall plant, two or three feet high with blunt cylindrical fertile spikes and blunt glumes on the barren spikes. It is recorded from all neighbouring counties.

1830. **1060. C. riparia,** Curtis. *Great Common Sedge.*
(*C. crassa,* Ehrh. *C. vesicaria,* Leers, not of Linn.)
Native. Margins of water, marshes. Abundant throughout. P. May and June. *Top. Bot.* 79.
First Record. Banks of Stour, *Scott, Hist. Stourb.,* p. 540 (1832).
This is so common a plant that it needs no localities to be specified. It is taller and stouter than the last plant, sometimes as much as five feet high. The glossy brown bracts inclosing the flowers, and the golden anthers rising out of the mass of glaucous foliage, make a pleasing contrast. The plant occurs in all bordering counties.

1831. **1061. C. inflata,** Huds. *Slender-beaked Bottle Sedge.*
(*C. ampullacea,* Good. *C. obtusangula,* Ehrh. *C. rostrata,* Stokes.)
Native. Marshes and bogs. Local. P. June and July. *Top. Bot.* 106.

First Record. As *C. ampullacea.* Mill below Droitwich, *Mr. Baker, Withering,* 3rd ed., p. 110 (1796).
Severn. Hartlebury Common, *Mathews.* Near Churchill Station. Near Droitwich, *Baker.* Banks of the Salwarpe, *Lees.* Hurcott Pool, *Mathews.* Spout Mill, Hagley. Stanklin. Glasshampton Pools.
Lickey. Moseley, *Ick.* Copse near Harborne, *Thompson.*
This Sedge is recorded from all of the bordering counties.

1832. **1062. C. vesicaria,** Linn. *Short-beaked Bladder Sedge.*
Native. Sides of brooks and ditches. Not uncommon, scarcer in the Lickey district. P. May and June. *Top. Bot.* 80.
First Record. Pond on Moseley Common, *Miss M. A. Beilby, Analyst,* vol. vi, p. 294 (1837).
Avon. Tardebigge Reservoir.
Severn. Near Hartlebury. Brickyards at Northwick, *Lees.* Monk Wood. Dowles Brook. Diglis. Hartlebury Common.
Malvern. New Pool. Malvern Chace, *Lees.* Meadow west of Powick Church. Sherrard's Green.
Lickey. Moseley, as first record. Copse near Harborne, *Thompson.* Yardley.
This plant is much like the last, from which it is distinguished by the acute instead of rounded angles of its stem, and much larger inflated fruit and stout beak. It is recorded from every neighbouring county.

92. GRAMINEAE, Juss.

SETARIA, Beauv. 470*. (*Seta,* a bristle, from the bristles at the base of the flowers.)

1836. **1063. S. *viridis,*** Beauv. *Green Bristle-grass.*
Alien. Waste places and arable fields. Rare. A. July and August. *Top. Bot.* 34.
First Record. Arable fields at Henwick, &c., soon disappearing, *Lees, Bot. Worc.,* p. xxxi (1867).
Severn. Henwick, *Lees.* Dowles.
This alien plant is a native of Manchuria which has become a weed of cultivated and waste ground throughout the greater part of the north Temperate Zone. It has been observed in Gloucestershire, Herefordshire, Warwickshire, and Staffordshire.

1837. **1064. S. glauca,** Beauv. *Glaucous Bristle-grass.*
Casual. Waste places. Rare. A. July. Not in *Top. Bot.*
First Record. *Setaria glauca . . .* appears for a season or two, only to die away, *Lees, Bot. Worc.,* p. 58 (1867).
Severn. As first record. Dunghill near Toll Bar, Bath Road, Worcester. Near Kidderminster Station. Eastbury, near Worcester.
Malvern. Garden at Mathon, 1897, *Towndrow, Malv. Advert.,* Nov. 20, 1897.
A visitor from the grassy hills of Southern China. It has been observed as a casual in Gloucestershire, Salop, and Warwick.

1838. **1065. S. verticillata,** Beauv. *Rough Bristle-grass.*
Casual. Waste ground. Rare. A. July and August. Not in *Top. Bot.*
First Record. Not localized, *Scott, Hist. Stourb.,* p. 540 (1832).
This plant is a frequent weed of waste and cultivated ground in Central and South Europe. It has not been noticed in any bordering county.

PHALARIS, Linn. 473. (φαλός, shining, from the nature of the seed, or φαλαρός, patched with white.)

1843. **1066. P. canariensis,** Linn. *Cultivated Canary Grass.*
Casual. Waste ground near houses. Introduced. A. June to August. Not in *Top. Bot.*
First Record. Near Gregory's Mill, Worcester, *Lees, Illus. Nat. Hist. Worc.,* p. 151 (1834).
Avon. Evesham. Pershore.
Severn. Bromwich, near Worcester. On the soil thrown out for cutting Camp Lock, 1845. Wilden, Stourport. Pitchcroft, Worcester.
Malvern. As first record. Near Malvern, *Westcombe.* Mamble.
Always owing its origin to the cage of the captive bird, this grass never establishes itself in England. It is a weed of cultivated ground all over Central and Southern Europe. Mr. Lees, in his *Table of Plants,* p. 32, marks it as occasional in all the districts. Records would show that birds are kept in cages in every neighbouring county also.

1846. **1067. P. arundinacea,** Linn. *Reed Canary Grass.*
(*Arundo colorata,* Sobol.)
Native. Sides of water. Common in all the valleys. P. June and July. *Top. Bot.* 110.
First Record. As *Arundo colorata.* Near (Bel)broughton Village, *Scott, Hist. Stourb.,* p. 540 (1832).
This grass grows by most of our streams, to the height of four or five feet. The well-known Ribbon Grass of our gardens is a form of this plant, sometimes obtaining the less elegant name of 'Gardener's Garters'. Still, it has some compensation in the fact that it is also called 'Ladies' Laces'. It occurs in every bordering county.

ANTHOXANTHUM, Linn. 474. (ἄνθος, a flower, ξανθός, yellow, from the yellowish hue of the spikes in old age.)

1847. **1068. A. odoratum,** Linn. *Sweet-scented Vernal-grass.*
Native. Meadows, pastures, heaths, open woods. Abundant and widely distributed. P. April and June. *Top. Bot.* 112.
First Record. Grassfields, *Scott, Hist. Stourb.,* p. 540 (1832).
This grass is common in all our hayfields, one of the earliest to blossom, and it gives to new-mown hay its delicious odour, though other grasses contribute to this. It is as common in bordering counties as it is in our own.

1848. **1069. A. *aristatum,*** Boiss.
(*A. Puelii,* Lecoq and Lamotte.)
Casual. Cultivated fields. Rare and not permanent. A. June to August. *Top. Bot.* 13.

First Record. *Thompson* and *Mathews,* Hazley (Hagley), *Bot. Rec. Club Rep.* (1877).
Severn. Hagley, as above.
Malvern. New Road, West Malvern, 1897, *Towndrow, Malv. Advert.,* Nov. 20, 1897.
The locality indicated in the first record was a field near the Birches, which at the date 1877 was full of the plant. A native of Southern Europe it appears to be spreading throughout England, a fact which is not to be wondered at since its seed is largely used to adulterate that of the preceding plant. It does not appear, however, to have been recognized as yet in Salop.

ALOPECURUS, Linn. 476. (ἀλώπηξ, a fox, and οὐρά, a tail, from the shape of the spikes.)

1850. **1070. A. myosuroides,** Huds. *Slender Foxtail Grass.*
(*A. agrestis,* Linn.)
Native. Cultivated fields, roadsides. Plentiful, except in the Lickey district where it is not so common. P. April to October. *Top. Bot.* 68.
First Record. As *Alopecurus agrestis.* Cornfields at Brook End, near Kempsey, *Lees, Illus. Nat. Hist. Worc.,* p. 151 (1834).
This is sometimes a troublesome weed with the farmer, coming up early in the spring with wheat and clover. The glumes are of a delicate sea-green colour, often tipped with purple, and the leaves are often of a purplish-green hue. It occurs in all bordering counties.

1851. **1071. A. aequalis,** Sm. *Orange-spiked Foxtail Grass.*
(*A. paludosus,* Beauv. *A. fulvus,* Sm.)
Native. Margins of ponds. Local. P. June to August. *Top. Bot.* 28.
First Record. On the borders of New Pool (Malvern), *Lees, Bot. Malv. Hills,* 2nd ed., p. 77 (1852).
Severn. One place, *Lees.*
Malvern. As first record.
Lickey. Harborne Reservoir, *Mathews.* Edgbaston.
This grass is closely allied to the following, and very much commoner, plant. It has been noticed in Herefordshire, Stafford, and Warwick.

1852. **1072. A. geniculatus,** Linn. *Floating Foxtail Grass.*
Native. Wet places, margins of pools, sometimes on drier land. General throughout the county. P. May to September. *Top. Bot.* 112.
First Record. Severn Meadows, Kempsey, *Lees, Illus. Nat. Hist. Worc.,* p. 151 (1834).
This is quite a common grass in the situations it loves, though it is not infrequently found on drier land, though not in such cases growing with its ordinary luxuriance. The stem is always kneed, a fact giving rise to the specific name, and a ready mark whereby to recognize the plant. It occurs in every bordering county.

1854. **1073. A. pratensis,** Linn. *Meadow Foxtail Grass.*
Native. Meadows, pastures, roadsides. Abundant everywhere. P. June to October. *Top. Bot.* 109.

First Record. Not localized, *Lees, Bot. Malv. Hills*, 1st ed., p. 44 (1843).

This is an early grass to flower, and sometimes constituting the greater portion of the grass of the hay-field. Village children strip the glumes off the young spikes, and secretly twist the stem into the loose hair of a companion, causing him an unexpected twinge. The plant is common in all bordering counties.

MILIUM, Linn. 477. (Latin, it was supposed, for Millet.)

1856. 1074. M. effusum, Linn. *Spreading Millet-grass.*
Native. Woods and shady places. Common. P. May to July. *Top. Bot.* 89.
First Record. Wood near Powick Ham, &c., *Lees, A. Florence, Strang. Guide* (1828).
Avon. Bredon Hill, *Mathews.* Trench Woods. Tiddesley Wood. Croome Perry Wood. Deerfold Wood. Norton, near Evesham. Stoulton. The Slads. Rous Lench Woods. Bow Wood.
Severn. Fenny Rough. Crookbarrow Hill. Lark Hill, Worcester. Shrawley Wood. Whittington, near Worcester. Kempsey. Perry Wood. Nunnery Wood. Foot of Helbury Hill. Rough, near Kempsey. Monk Wood. Ockeridge Wood. Wyre Forest. Lincombe.
Malvern. Brockhill Wood, *Lees.* Ankerdine Hill. Lord's Wood. Middleyards Coppice. Dripshill Wood. Crews Hill Wood. Woodbury. Abberley Hill.
Lickey. Beoley; Harris's Wood, Halesowen, *Mathews.*

This is a tall slender grass, conspicuous in moist shady woods, where it is often abundant, its stem rising to a height of three feet or more, bearing a panicle of numerous small green spikelets on long stems. It is recorded from every neighbouring county.

PHLEUM, Linn. 478. (From φλέως, a name formerly applied to the Reed-mace, which this grass distantly resembles.)

1859. 1075. P. pratense, Linn. *Common Cat's-tail Grass,* or *Timothy Grass.*
Native. Meadows, pastures, roadsides, downs. Plentiful in all the districts. P. June to October. *Top. Bot.* 111.
First Record. Not localized, *Lees, Bot. Malv. Hills*, 1st ed., p. 45 (1843).
This is one of the commonest of our meadow grasses, growing well on dry poor soils. It owes one of its book-names to Mr. Timothy Hanson, who cultivated it extensively in America. It is recorded from all bordering counties.
Var. b. **nodosum,** Linn.
First Record. Not localized, *Scott, Hist. Stourb.,* p. 540 (1832).
Malvern. Powick, *Towndrow.*
Scott's record leaves it indefinite whether the plant was in Staffordshire or Worcestershire; or, if in the latter county, whether in the Severn or the Lickey district. Mr. Towndrow's record brings it certainly into our county. The variety does not seem to have been observed in Staffordshire of our bordering counties.

AGROSTIS, Linn. 480. (From ἀγρός, a field, being a name given by the Greeks to grasses.)

1864. 1076. A. canina, Linn. *Brown Bent-grass.*
(*Agrostis vinealis,* With.)
Native. Heaths, dry sandy fields, rough pastures, heathy woods. General throughout the county, not so common in the Lickey district. P. July and August. *Top. Bot.* 101.
First Record. Not localized, *Lees, Bot. Malv. Hills*, 1st ed., p. 45 (1843).
This is an abundant grass on meadows, with lower leaves setaceous, often gathered for its delicate beauty, the airy clusters being formed of numerous small spikelets on threadlike branches varying through every tint from green to purple. The word 'bent', from the earliest appearance of Northern literature, had the meaning of a grassy field or surface, but was not applied to what grew upon them till the fifteenth century; and this is the meaning of 'bent' in the name of this grass; it has nothing to do with the verb 'to bend'. The word has been applied to numbers of plants of various genera, and now has come to mean in common talk any withered grass stalks, or even stalks of other herbage, which, brown and dry, cover the fields in winter. This grass is recorded from every bordering county.

1865. 1077. A. alba, Linn. *Marsh Bent-grass.*
(*A. palustris,* Huds.)
Native. Moist meadows, heaths, woods. General in Avon and Severn, less common in Malvern and Lickey. P. June and July. *Top. Bot.* 112.
First Record. As *Agrostis alba.* Not localized, *Scott, Hist. Stourb.,* p. 540 (1832).
This is a stouter, taller grass than the preceding, having no setaceous lower leaves, and is equally abundant in suitable situations. The panicles also vary in colour as do those of *A. canina.* It is recorded from all bordering counties.
Var. b. **stolonifera,** Linn.
First Record. As *Agrostis stolonifera.* Not localized, *Scott, Hist. Stourb.,* p. 540 (1832).
This record suffers from similar disabilities as that of Scott for *Phleum pratense,* var. *nodosum* above. Let us give Worcestershire the benefit of the doubt, and say nothing about the district. The plant has been observed in Salop, Staffordshire, and Warwickshire.

1866. 1078. A. tenuis, Sibth. *Fine Bent-grass.*
(*A. vulgaris,* With.)
Native. Dry pastures, roadsides, heaths. Common and generally distributed. P. June to August. *Top. Bot.* 112.
First Record. As *A. vulgaris.* Not localized, *Lees, Bot. Malv. Hills*, 1st ed., p. 45 (1843).
The short truncate ligule of the leaves easily separates this from the other species of *Agrostis.* It is a very common grass, which occurs in every bordering county.

D d

Var. **pumila,** Linn.
First Record. As *Agrostis pumila.* Astwood Common, *Purton, Midl. Fl.,* p. 71 (1817).
Malvern. In the district, *Towndrow (MS.).* Malvern Hills, *Herb. Hastings Museum.*
Lickey. As first record.
This variety is based on a diseased condition of *Agrostis tenuis* caused by the smut *Tilletia decipiens* (Pers.) Körn. It appears to have been observed only in Warwickshire. The variety is not given in *Lond. Cat.,* 10th ed.

1867. 1079. A. nigra, With.
Native. Dry pastures and heaths. P. July and August. *Top. Bot.* 20.
First Record. Not localized, *Scott, Hist. Stourb.,* p. 540 (1832).
Avon. Bredon Hill. Craycombe Hill. Cleeve Banks.
Severn. In the district, *Bagnall (MS.).* West of Hagley station. Stagbury Hill. Woodbury Hill.
Malvern. Madresfield. Malvern Link.
Lickey. In the district, *Bagnall (MS.).*
This species differs from *A. tenuis* in having the sheaths of the leaves rough, the ligule long, and the pedicels of the flowers toothed. The plant was recorded as new to the county by Mr. Bagnall, *Bot. Rec. Club Rep.,* 1881–2, but it was not so. It has been observed in all bordering counties except Gloucestershire.

POLYPOGON, Desf. 481. (From πολύς, much, πώγων, a beard, from the appearance of the panicle.)

1869. 1080. P. monspeliensis, Desf. *Annual Beard-grass.*
Casual. Usually in moist pastures near the sea. Once recorded. A. July and August. *Top. Bot.* 7.
First Record. On a rubbish heap at Hoo Mill, Kidderminster, 1875. Introduced with wool, *Mathews, Fl. Clent and Lickey Hills,* p. 50 (1881).
This plant is a native of wet sandy ground from the Mediterranean region along the west coast of Europe as far as our Island, but is a weed of wider distribution over waste and cultivated ground. It has been observed in Gloucestershire.

CALAMAGROSTIS, Adans. 482. (Named from κάλαμος, a reed, and ἀγρωστις, a genus of grasses.)

1870. 1081. C. epigeios, Roth. *Wood Small-reed.*
(*Arundo epigeios,* Linn.)
Native. Damp woods and thickets. Local. P. June and July. *Top. Bot.* 60.
First Record. As *Arundo epigeios.* Not localized, but said to be 'not rare', *Purton, Midl. Fl.,* p. 78 (1817).
Avon. Saldons, Droitwich; Trench Woods, *Mathews.* Evesham, roadside about three miles from Spetchley.
Severn. Monk Wood; Grimley, *Lees.* Perry Wood, *Baxter.* Base of Helbury Hill. Spetchley Pool.

Malvern. A wet field at Berrow, *Lees.* Old Hills, Powick.
Lickey. Butler's Hill Wood, Alvechurch.
A handsome plant with a round erect stem, sometimes five feet high, and harsh leaves. It is recorded from all neighbouring counties.

1871. 1082. C. canescens, Druce. *Purple-flowered Small-reed.*
(*C. lanceolata,* Roth.)
Native. Wet places. Rare. P. July. *Top. Bot.* 40.
First Record. As below.
Severn. In the district, *Bagnall (MS.).*
This plant is recorded by Scott, *Hist. Stourb.,* p. 540 (1832), from Pensnett Reservoir near that town, but the locality is unfortunately in Staffordshire. It has been observed in every bordering county except Herefordshire.

GASTRIDIUM, Beauv. 484. (γαστρίδιον, a little swelling, as seen at the base of the spikelet.)

1874. 1083. G. lendigerum, Gaud. *Awned Nit-grass.*
(*Milium lendigerum,* Linn. *G. australe,* Beauv.)
Denizen. Usually in damp places near the sea. Rare. P. August. *Top. Bot.* 24.
First Record. Sarnhill, near Longdon, *Lees, Bot. Malv. Hills,* 1st ed., p. 45 (1843).
Severn. Two places, *Lees.*
Malvern. Sarn Hill, as first record.
In his *Table of Plants,* p. 33, Mr. Lees omits the Sarn Hill locality and places the plant as above in the Severn district, as to which there is no mention in his *Bot. Worc.* Of neighbouring counties the plant has been observed in Gloucestershire, Herefordshire, and Warwickshire.

APERA, Adans. 485. (ἀ, not, πηρός, maimed, alluding to the entire floral glume.)

1875. 1084. A. Spica-venti, Beauv. *Spreading silky Bent-grass.*
(*Agrostis Spica-venti,* Linn. *Agrostis Anemagrostis,* Syme.)
Denizen. Sandy cornfields. Once recorded. A. July to September. *Top. Bot.* 17.
First Record. As localized in the Lickey district by *Lees, Bot. Worc., Table of Plants,* p. 33 (1867).
Lickey. As first record.
This plant is probably only native in Central and South-east Europe. North-west of this the plant is confined to cultivated ground and roadsides. It has been observed in Shropshire and Warwickshire.

AMMOPHILA, Host. 486. (ἄμμος, sand, φιλέω, I love.)

[1877. 1085. A. arenaria, Link. *Common Sea-grass, Marram,* or *Mat-weed.*
(*A. arundinacea,* Host.)
Casual, if at all. Sandy sea-shores, binding the sand. P. July. *Top. Bot.* 64.
First Record. Not localized, *Lees' Cat., New Bot. Guide* (1835).

D d 2

As Mr. Mathews says, this must be an error, though some parts of Hartlebury Common might appear to be suitable soil for it. It is not mentioned in Lees, *Bot. Worc.*, nor is it recorded for any neighbouring county.]

AIRA, Linn. 488.   (αἴρω, I destroy, first applied to *Lolium temulentum*, and afterwards transferred to this genus.)

1880. 1086. A. caryophyllea, Linn.   *Silvery Hair-grass.*
Native. Dry heathy places. Rather uncommon. A. May and June. *Top. Bot.* 110.
First Record. Not localized, but near Malvern, *Lees, Mag. Nat. Hist.*, vol. iii, p. 160 (1830).
    Severn. Blakedown; Bissell; Harberrow, *Mathews*. Near Bewdley. Pedmore Common. Sutton Common. Sandy fields east of Kidderminster Water Works. Near Holt Bridge.
    Malvern. Ankerdine Hill. Malvern Hills.
This grass is recorded from all bordering counties.

Var. c. **aggregata** (Tim.).
First Record. Malvern Wells and Malvern Link, *Towndrow, Malv. Advert.*, 1892.
    Severn. Sandy fields near Kidderminster.
    Malvern. Near the Wyche. Malvern Link. Malvern Wells.
This variety does not appear to have been observed in any bordering county.

1881. 1087. A. praecox, Linn.   *Early Hair-grass.*
Native. Dry heathy places, commons, dry banks. Not common. A. April to June. *Top. Bot.* 111.
First Record. Commons, *Scott, Hist. Stourb.*, p. 540 (1832).
    Severn. Blakedown; Bissell, *Mathews*. Hartlebury Common, *Lees*. Bewdley. Hartlebury.
    Malvern. Ankerdine Hills. Malvern Hills. West Malvern.
    Lickey. Winwood Heath; Clent Hills, *Mathews*.
A little grass hardly three inches high, with a few-flowered panicle of pale silvery-green, withering early in dry seasons. Among this grass on the Clent Hills and neighbouring eminences is usually found *Moenchia erecta* where it occurs, one plant somewhat imitating the other not unsuccessfully. *A. praecox* is recorded from all bordering counties.

DESCHAMPSIA, Beauv. 490.   (Named in honour of H. Deschamps, a French chemist.)

1883. 1088. D. caespitosa, Beauv.   *Tufted Hair-grass.*
(*Aira caespitosa*, Linn.)
Native. Meadows, pastures, open places in woods. Very common. P. June to August. *Top. Bot.* 112.
First Record. As *Aira caespitosa*. Not localized, *Lees, Bot. Malv. Hills*, 1st ed., p. 45 (1843).
This grass is known to country people as Hassock or Tussock Grass,

---

forming matted tufts of leaves. Horses refuse it, and cattle only eat it as a last resource. It is recorded from all bordering counties.

1886. 1089. D. flexuosa, Trin.   *Waved Hair-grass.*
(*Aira flexuosa*, Linn.)
Native. Heaths and fields on sandy soil. General. P. June and July. *Top. Bot.* 110.
First Record. As *Aira flexuosa*. Helbury Hill, and Wood on the Broadheath road, *Lees, A. Florence, Strang. Guide* (1828).
The spikelets of this grass are larger than those of the preceding species, on wavy branches hardly thicker than a thread. It occurs in all bordering counties.

HOLCUS, Linn. 491.   (From ἕλκω, I extract, because it was supposed to draw thorns out of the flesh.)

1887. 1090. H. mollis, Linn.   *Creeping Soft-grass.*
Native. Woods, heaths, fields. Rather local. P. June and July. *Top. Bot.* 110.
First Record. Meadows on the western side of Severn, *Lees, A. Florence, Strang. Guide* (1828).
This grass is one of the pests of the farmer of light sandy lands which in our county is called, with several others, Scutch, the Couch or Quitch of other districts. It is very similar to the next plant, but is easily distinguished from it by its pointed glumes and the exserted awn of the flowering glumes. It is not so frequently met with as *H. lanatus*, seldom growing in meadows, the true home of that. It is recorded from every bordering county.

[*H. bulbosus* is recorded in *Scott, Hist. Stourb.*, p. 540 (1832), a plant which Mr. Mathews supposed, *Mid. Nat.* xi. 42, to be a bulbous variety of the above. The *Index Kewensis* makes *H. bulbosa*, Schrad., equal to *Arrhenatherum avenaceum*, Beauv.]

1888. 1091. H. lanatus, Linn.   *Meadow Soft-grass.*
Native. Meadows, cultivated and fallow fields, waste places. Abundant. P. May to July. *Top. Bot.* 112.
First Record. Not localized, *Scott, Hist. Stourb.*, p. 540 (1832).
The presence of this plant in the meadow usually indicates damp, poor soil. The small spikelets, crowded together, are tinged with pink, and with the leaves covered with soft down. Cattle leave it for other grasses more to their taste. It is sometimes called Yorkshire Fog. It is recorded from all bordering counties.

TRISETUM, Pers. 492.   (*Tres*, three, *seta*, a bristle, from the three-awned flowering glume.)

1889. 1092. T. flavescens, Beauv.   *Yellow Oat-grass.*
(*T. pratense*, Pers. *Avena flavescens*, Linn.)
Native. Meadows, pastures, roadsides. General throughout the county. P. June and July. *Top. Bot.* 95.
First Record. As *Avena flavescens*. Not localized, *Scott, Hist. Stourb.*, p. 540 (1832).
This plant occurs in every neighbouring county.

AVENA, Linn. 493.   (The Latin name from *aveo*, I long for, because the animals seek out this grass.)

1890. 1093. A. pubescens, Linn.   *Downy Oat-grass.*
Native. Dry pastures and downs. Local. P. May and June. *Top. Bot.* 94.
First Record. Meadow near Bubble Bridge, *Stanley's Worc. and Malc. Guide* (1853).
    Avon. Bredon Hill, *Mathews*. Rudge's Farm. Near Oddingley to the eastward, *Westcombe*. Norton, near Evesham. Pershore.
    Severn. Park Hall, near Kidderminster; Hurcott, *Mathews*. Battenhall. Warndon. Ombersley. Kempsey. Claines. Elmley Lovett. Crowneast. Hallow.
This species differs from *A. pratensis* in having hairy leaves and sheaths, and the lower outer glume is the only one blind. This grass is to be found in all neighbouring counties.

1891. 1094. A. pratensis, Linn.   *Narrow-leaved perennial Oat-grass.*
Native. Dry pastures and downs, preferring calcareous soil. Local. P. May to July. *Top. Bot.* 76.
First Record. Broadway Hills, *Rufford, Purton, App.*, vol. iii, p. 13 (1821).
    Avon. Tredington, *Bagnall*. Bredon Hill. Craycombe. Broadway. Crowle. Dormstone. Inkberrow.
    Severn. Spetchley Common. Warndon.
    Malvern. Calcareous eminences, *Lees*. Old Hills, Powick.
This grass is recorded from all neighbouring counties.

1893. 1095. A. *fatua, Linn.   *Wild Oat-grass.*
Colonist. Cultivated ground, roadsides, and waste places. Often seen, except in the Lickey district. A. June to August. *Top. Bot.* 77.
First Record. In hard-tilled cornfields, *Pitt, Agric. Worc.*, p. 317 (1810).
The native home of this grass is unknown, but in temperate Europe, Asia, and Africa it is found, as it is in England, a weed of waste and cultivated ground. It is a handsome plant with bright green leaves, marked with fine lines, and the chaffy glumes are also striped. It is so like the cultivated Oat, *Avena sativa*, that it has been thought to be a variety of that plant, and farmers have said that it is unsafe to cultivate oats, as they leave behind them a troublesome crop of wild ones. The plant is recorded from all bordering counties.

Var. a. **pilosissima,** Gray.
First Record. As below.
    Malvern. Near Malvern, *Towndrow, Malvern Advert.*, January 20, 1894.
In this variety the lower pales are hairy, becoming dark brown. It has been observed in Herefordshire and Warwickshire.

Var. b. **intermedia,** Lindgr.
First Record. As below.
    Malvern. Powick, *Rev. E. F. Linton, Towndrow, Malv. Advert.*, January 20, 1894.
In this variety the lower pales are almost glabrous, becoming pale yellowish-olive. It has been observed in Herefordshire and Warwickshire.

ARRHENATHERUM, Beauv. 494.   (ἄρρην, male, ἀθήρ, a beard of corn, from the long awned glume of the male flower.)

1894. 1096. A. elatius, Mert. and Koch.   *Common oat-like Grass.*
(*A. avenaceum*, Beauv. *Avena elatior*, Linn.)
Native. Roadsides, hedges, pastures, thickets. Everywhere abundant. P. May to July. *Top. Bot.* 112.
First Record. As *Avena elatior*. Near Dudley Castle, *Booker, Dud. Cast.*, p. 107 (1825).
A tall conspicuous plant, with panicles often a foot or more long. It is found abundantly in every bordering county.

Var. b. **bulbosum,** Presl.
First Record. *Lees*, localized in *Bot. Worc., Table of Plants*, p. 33 (1867) in all the districts.
The creeping roots of this variety, with their strings of little bulb-like knobs into which the lower joints of the stem are swollen, are exceedingly troublesome to the farmer of light land or poor soils, and known as Onion Scutch. It is most difficult to eradicate, as every little bit left on the ground will grow with great rapidity. It does not appear to have been noticed in Gloucestershire or Herefordshire.

SIEGLINGIA, Bernh. 496.

1896. 1097. S. decumbens, Bernh.   *Decumbent Heath-grass.*
(*Festuca decumbens*, Linn. *Triodia decumbens*, Beauv. *Poa decumbens*, E. B.)
Native. Heaths, bogs, heathy pastures. Local. P. April to July. *Top. Bot.* 110.
First Record. As *Poa decumbens*. Not localized, *Scott, Hist. Stourb.*, p. 540 (1832).
    Avon. Bredon Hill. Broadway. Near White Ladies' Aston.
    Severn. Hartlebury Common. Habberley Valley. Crookbarrow Hill. Broadheath. Wyre Forest.
    Malvern. Worcestershire Beacon; Malvern Chase, *Lees*. Near Ham Castle. Malvern Link Common. Newland Common.
    Lickey. Bromsgrove Lickey, *Mathews*.
This is one of the few grasses in which the ligule is replaced by a circle of hairs, and it is further characterized by the three-toothed flowering glume. It is recorded from all bordering counties.

PHRAGMITES, Adans. 497.   (φραγμίτης, an enclosure, this plant being used for making such.)

1897. 1098. P. communis, Trin.   *Common Reed.*
(*Arundo Phragmites*, Linn.)
Native. Sides of water and wet hedgebanks. Plentiful in the Avon, Severn, and Malvern districts, rare in the Lickey district. P. June to August. *Top. Bot.* 104.
First Record. As *Arundo Phragmites*. South of Worcester, *Scott, Hist. Stourb.*, p. 540 (1832).
    Avon. Plentiful in the district. Cleeve Mill. Norton, near Evesham. Evesham. Fladbury. Pershore. Eckington. Bredon. Feckenham Bog.

Severn. Longdon, *Lees.* Stanklin Pool. Shrawley Wood. Captains Pool, south of Kidderminster. Spetchley Pool. Birmingham Canal, near Worcester. Droitwich Canal. Banks of the Salwarpe. Ombersley. Laughern Brook. Grimley. Holt. Stourport.
Malvern. Longdon Marsh. Bransford. Powick. Martley. Abberley. Madresfield. Newland Common. Southwood.
Lickey. Hewell Park, *D. Mathews.*

In the east of England the Reed is very plentiful, and used for a number of domestic purposes, while from the long stems are formed embankments against the water. The plant entered largely into the economy of our Anglo-Saxon forefathers, and the plant must have been far more abundant in earlier times than it is at present. Drainage has diminished it. A common country appellation for the plant was 'Windle-straw'; and its name has passed on to numberless significations in our language, notably in the domain of music. The Reed is recorded from all bordering counties.

CYNOSURUS, Linn. 499. (κύων, a dog, and οὐρά, a tail, from the shape of the spike.)

1899. 1099. C. cristatus, Linn.　　*Crested Dog's-tail Grass.*
Native. Dry fields, roadsides, downs. Common and generally distributed. P. May to August. *Top. Bot.* 112.
First Record. Roads, fields, &c., *Scott, Hist. Stourb.,* p. 540 (1832).
This common grass is recorded from all neighbouring counties.

KOELERIA, Pers. 500. (Named after G. L. Koeler, a German writer on grasses.)

1903. 1100. K. gracilis, Pers.　　*Crested Koeleria.*
(*K. cristata,* Pers. *Aira cristata,* Linn. *Airochloa cristata,* Link.)
Native. Dry pastures and downs. Local. P. June and July. *Top. Bot.* 90.
First Record. As *Poa cristata.* On the edge of a Marle Rock, Clarkton Leap (Clarkenleap), near Worcester, *Stokes's With. Bot. Arr.,* and ed., p. 91 (1787).
Avon. Bredon Hill. Broadway. Craycombe. Near Feckenham.
Severn. As first record. Spetchley Common.
Malvern. One place, *Lees.* Near the Wyche, Malvern, *Westcombe.* Malvern Common. Old Hills, Powick.
Lickey. One place, *Lees.*
This is a grass most frequent in the north. It is recorded from all bordering counties.

MOLINIA, Schrank, 501. (Named in honour of Don G. I. Molina, who wrote a Natural History of Chili, published in 1782.)

1904. 1101. M. caerulea, Moench.　　*Purple Melic-grass.*
(*M. varia,* Schrank. *Aira caerulea,* Linn.)
Native. Bogs, marshes, wet heathy places. More common in the Severn district than elsewhere, but local throughout. P. July to September. *Top. Bot.* 110.

First Record. Moseley Common, *Miss M. A. Beilby, Analyst,* vol. vi, p. 294 (1837).
Severn. Wyre Forest, *Jorden.* Broadheath. Crowneast Wood, *Lees.* Hartlebury Common. Near Bewdley. Great Bog, Wyre Forest. Near Bewdley. Doverdale.
Malvern. Pendock. Longdon.
Lickey. Bromsgrove Lickey, *Mathews.* Pool, near the Independent College, Moseley, 1858, *Mathews.*

In the case of this grass all the leaves spring from the base or from a single joint immediately above it, and they are destitute of a ligule or the ligule is represented by a tuft of hairs. The panicle is bluish-purple, deeper in hue than any other native grass. It is recorded from all bordering counties.

CATABROSA, Beauv. 502. (κατάβρωσις, a gnawing, from the appearance of the tops of the glumes.)

1905. 1102. C. aquatica, Beauv.　　*Water Whorl-grass.*
(*Aira aquatica,* Linn.)
Native. Muddy margins of ditches and ponds. Rather local, except in the Avon and Severn districts, where it is plentiful. P. May to July. *Top. Bot.* 94.
First Record. As *Aira aquatica.* About Stourbridge, *Purton, Midl. Fl.,* p. 74 (1817).
Severn. Churchill and Blakedown Pools. Hurcott Pool. Lane between Cotheridge and Broadheath. Near Hartlebury Church. Pond, Hartlebury. Droitwich Canal.
Malvern. In wet places on the western side of the hills, rare, *Lees.* Below West Malvern Church. Malvern Wells. Malvern Link. Mathon.
Lickey. Farley Wood, Romsley, *Mathews.*
This is quite an aquatic grass, its flat, blunt leaves floating on the water to a considerable length, or sprawling over the wet mud of the waterside ; but never except in the water itself attaining any luxuriance. It is recorded from every bordering county.

MELICA, Linn. 503. (The Italian name, *Melica* or *Melliga,* for *Sorghum vulgare,* Durra, or Indian Millet, which was applied by Linnaeus to this somewhat allied genus.)

1906. 1103. M. montana, Huds.　　*Mountain Melic-grass.*
(*M. nutans,* auct.)
Native. Hedgebanks, shady woods. Rare. P. May to July. *Top. Bot.* 49.
First Record. As *M. nutans.* Not localized, *Scott, Hist. Stourb.,* p. 540 (1832).
Severn. Wyre Forest, *Jorden.* Holy Well, Wyre Forest, *Mathews.* Snuff Mill Valley, Bewdley, 1904. Shrawley Wood. Ribbesford Woods.

Lickey. Near Stourbridge, *Scott,* gone.
Mr. Mathews, who knew the district intimately, discredited the Stourbridge record. ' It is surprising ', he says, *Mid. Nat.* xii. 42, ' that this rare grass should be inserted without locality. It does not now grow in the Stourbridge neighbourhood.' Possibly if the record is true, it was not in Worcestershire ; but luckily the other localities allow us still to claim the plant for the county, and it flourishes in some abundance in the Worcestershire portion of Wyre Forest. It does not appear to have been observed in Warwickshire.

1907. 1104. M. nutans, Linn.　　*Wood Melic-grass.*
(*M. uniflora,* Retz.)
Native. Shady Woods. General. P. May and June. *Top. Bot.* 96.
First Record. Blackstone Rock, near Bewdley, *Perry, Mag. Nat. Hist.,* vol. iv, p. 450 (1831).
Avon. General, *Lees.* Tiddesley Wood. Elmley Castle. Trench Woods. Gooshill Wood. Crowle. Rous Lench Woods. Bow Wood.
Severn. As first record. Wyre Forest. Snuff Mill Valley, Bewdley, 1904. Perry Wood. Between Habberley Valley and Kidderminster. Roadside, Astley. Near Worcester. Crookbarrow Hill. Wood opposite the Ketch. Shrawley Wood. Ockeridge Wood. Nunnery Wood.
Malvern. Woody places, abundant, *Lees.* The Leap, near Martley. Old Storridge. Abberley. Ankerdine Hill. Clifton-on-Teme. Weymans Wood. Near Southstone Rock. Middleyards Coppice.
Lickey. Moseley, *Mathews.* Tardebigge. Bournheath.
This grass is recorded from all neighbouring counties.

DACTYLIS, Linn. 504. (δάκτυλος, a finger, from the shape of the spikes.)

1908. 1105. D. glomerata, Linn.　　*Rough Cock's-foot Grass.*
Native. Pastures, waste places, roadsides. Abundant and generally distributed. P. May and July. *Top. Bot.* 112.
First Record. Commons and fields, *Scott, Hist. Stourb.,* p. 540 (1832).
This is one of the common unmistakable grasses met with in every meadow and pasture. It prefers shade and therefore loves the orchard. Nutritious and plentiful it makes good fodder, but farmers dislike it for its coarseness and roughness. It grows in every neighbouring county.

BRIZA, Linn. 505. (βρίζα was the Greek name for some kind of corn ; perhaps from βρίζω, I nod.)

1910. 1106. B. media, Linn.　　*Common Quaking-grass.*
Native. Pastures, roadsides, boggy places, downs. Common and widely distributed. P. May to July. *Top. Bot.* 111.
First Record. Not localized, *Lees, Bot. Malv. Hills,* 1st ed., p. 45 (1843).
This grass is a favourite with all who love the country side, but it is a sign that the soil on which it grows is poor. It is also called Doddering Grass, and sometimes Maiden's Hair, while the shape of the spikelets,

something like a heart, have caused pretty fancies to be woven round it by country people. It occurs in every neighbouring county.

[**1911. 1107. B. minor,** Linn.　　*Small Quaking-grass.*
Native. Pastures. Rare. P. June. *Top. Bot.* 7.
First Record. Grass fields, *Scott, Hist. Stourb.,* p. 540 (1832).
Malvern. *Lees, Bot. Malv. Hills,* 1st ed., p. 45 (1843).
Lickey. As first record.
Mr. Mathews, *Mid. Nat.* xi. 42, discredits Scott's record. It must be an error for *B. media,* he says. Mr. Lees' record is also no doubt erroneous. In the 2nd and 3rd eds. of his book he terms the plant *B. media,* var. *abortiva,* a variety of his own invention, with the locality ' Bog at the base of the Worcestershire Beacon ', omitting *B. minor* altogether. Nor is it in Lees' *Bot. Worc.* The plant is nearly confined to the extreme south-west of the kingdom. It has been noticed only in Gloucestershire of our bordering counties.]

POA, Linn. 506. (πόα, grass.)

1912. 1108. P. annua, Linn.　　*Annual Meadow-grass.*
Native. Meadows, pastures, waste places, roadsides. Abundant everywhere. A. January to December. *Top. Bot.* 112.
First Record. Not localized, *Lees, Bot. Malv. Hills,* 1st ed., p. 45 (1843).
This is probably the commonest grass in the county, and perhaps throughout the whole world, disputing with *Capsella Bursa-pastoris* the honour of being the commonest plant on earth. It is the plant to which we might refer when we say ' green as grass ', for it is always bright in hue, never turning yellow, never tinged with purple. It is found in our neighbouring counties as well as everywhere else.

1919. 1109. P. nemoralis, Linn.　　*Wood Meadow-grass.*
Native. Woods and shady hedgebanks. Not uncommon. P. April to June. *Top. Bot.* 31.
First Record. Not localized, *William Addison, Trans. Prov. Med. Surg. Ass.,* p. 82 (1836).
Avon. Tiddesley Wood. Trench Woods. Crowle. Huddington. Gooshill Wood. Badgers Bank. Craycombe. Bredon Hill. Deerfold Wood. Croome Perry Wood. Dodderhill Common. Rous Lench. Harvington. Hipton Hill.
Severn. Crookbarrow Hill. Whittington, near Worcester. Vallombrosa. Severn-side, Claines. Northwick Brickpits. Perry Wood. Fenny Rough. Coppice below Diglis. Nunnery Wood. Holt. Ombersley. Shrawley Wood. Wyre Forest. Monk Wood. Crowneast.
Malvern. Middleyards Coppice. Crews Hill Wood. Ankerdine Hill. Berrow Hill, Martley. Weyman's Wood. Bayton. Near Corn Brook. Abberley Hill. Stanford Bridge. Leigh. Dripshill Wood. Cowleigh Park. Malvern Wells, Malvern.
This is a variable plant easily known by the short ligule and the upper leaf being longer than its sheath. Many forms have been given divers names by divers botanists. The type occurs in every bordering county.

Var. *β*. angustifolia, Parn.
First Record. As in this book.
Avon. Bredon Hill.
Severn. Ketch.
In this variety the stem and panicles are very slender, the leaves long and narrow, the uppermost knot near the panicle, and the spikelets one to two-flowered. This variety is not given in *Lond. Cat.*, 10th ed.

1920. 1110. P. compressa, Linn.   *Flat-stemmed Meadow-grass.*
Native. Dry fields, wall-tops. Not uncommon. P. May to August. *Top. Bot.* 70.
First Record. Not localized, *Lees, Bot. Malv. Hills*, 1st ed., p. 45 (1843).
This grass may be distinguished by its flattened stem ; it grows about a foot high in favourable situations, and in others, where it is not very conspicuous, is liable to be overlooked. It is recorded from all bordering counties.

1922. 1111. P. pratensis, Linn.   *Smooth-stalked Meadow-grass.*
Native. Meadows, pastures, waysides, woods, wall-tops, heaths. Abundant in all the districts. P. May to August. *Top. Bot.* 112.
First Record. Not localized, *Scott, Hist. Stourb.*, p. 540 (1832).
This is a very common grass in meadows and rich pastures, and like *P. nemoralis* it has a short ligule, but the upper leaf is shorter than its sheath. It is recorded from all bordering counties.

Var. b. subcaerulea, Sm.
First Record. As herein.
Severn. Habberley Valley.
This variety is a small glaucous form.

Var. c. angustifolia, Linn.
First Record. As herein.
Avon. Bredon Hill.
This is a form with slender leaves.

1924. 1112. P. trivialis, Linn.   *Roughish Meadow-grass.*
Native. Meadows, pastures, borders of fields, woods, roadsides. Common in all the districts. P. May to July. *Top. Bot.* 110.
First Record. Not localized, *Scott, Hist. Stourb.*, p. 540 (1832).
Easily distinguished by its rough stem and sheaths, and the oblong acute ligule. This grass, which is very common in meadows and pastures, grows in every adjoining county.

GLYCERIA, R. Br. 507. (γλυκερός = γλυκύς, sweet.)

1925. 1113. G. fluitans, R. Br.   *Floating Meadow-grass.*
(*Festuca fluitans*, Linn. *Poa fluitans*, Scop.)
Native. Margins of water, often floating. Frequent. P. June to August. *Top. Bot.* 112.
First Record. Lusbridge Brook ; (Bel)broughton Brook, near the village, *Scott, Hist. Stourb.*, p. 540 (1832).

This is a common grass. It grows abundantly on the continent, and in Holland and some parts of Germany and Poland its seeds are used as food. From it is prepared the farinaceous food called *Mannacroup*, sometimes finding its way into England and made into milk puddings. The last syllable of the word is the Russian word for groats. The grass grows in all bordering counties.

× plicata.
Var. *pedicellata*, Townsend.
First Record. Localized as occurring in one place in the Avon district, *Lees, Bot. Worc., Table of Plants*, p. 34 (1867).
Severn. Near Camp Weir.
Malvern. Commoner than the type, *Towndrow, Malv. Advert.*, Dec. 10, 1892.
This hybrid has been observed in every bordering county except Gloucestershire. It is a plant that never fruits.

1926. 1114. G. plicata, Fr.   *Floating Meadow-grass.*
(*Poa fluitans*, Scop. *Glyceria fluitans*, Sm.)
Native. Ditches, margins of ponds and slow streams. General throughout the county. P. June to October. *Top. Bot.* 75.
First Record. Wet low spots exhibit . . . *Glyceria plicata, Lees, Bot. Worc.*, p. 58 (1867).
Avon. Dunhampstead.
Severn. Broadwas. Side Basin, Diglis. Berwicks Pool, near Worcester.
Malvern. In water, not frequent, *Towndrow, Malv. Advert.*, Dec. 10, 1892.
This grass is a segregate of the preceding one, the chief distinction being the rough furrowed sheath, and the three-toothed lower pale, which in this is twice as long as broad, while in the type it is three times as long as broad, is sub-entire, and the sheaths are nearly smooth and striate. It is found in all bordering counties.

1928. 1115. G. aquatica, Wahlenberg.   *Reed Meadow-grass.*
(*Poa aquatica*, Linn.)
Native. Sides of rivers, canals, and brooks. Many places. P. July and August. *Top. Bot.* 79.
First Record. Rivers and morasses, *Scott, Hist. Stourb.*, p. 540 (1832).
Avon. Abundant by the Avon, *Lees.* Shell Brook. Stonebow Bridge. Flyford Flavell. Norton, near Evesham. Fladbury. Pershore.
Severn. Canal-side near Kidderminster, *Lees.* Banks of Severn. Pool near Severn Stoke. Canal near Worcester. Grimley Brickpits. Diglis. Holt. Near Redstone Rock. Shrawley Wood.
Malvern. Severn End. Ditches towards the Severn, *Lees.* Near the Rhydd. Leigh Sinton. Southwood. Powick Ham. Brockamin.
Lickey. Canal-side at Haleswoen, *Mathews.*
This tall grass, in localities where it is abundant, gets mingled with the stems of other large grasses and sedges and put to the purposes they are. It is particularly plentiful in the Fen district, where under favourable conditions it grows something like six feet high. It is recorded from all bordering counties.

1929. 1116. G. maritima, Mert. and Koch.   *Creeping Sea Meadow-grass.*
(*Sclerochloa maritima*, Lindl. *Poa maritima*, Huds.)
Denizen. Usually in damp places near the sea. Once recorded. P. June and July. *Top. Bot.* 68.
First Record. As *Poa maritima.* Near the canal from Droitwich to the Severn, *Mr. Baker, Withering*, 3rd ed., p. 147 (1796).
Not only is this the only record of this plant for Worcestershire, but it is also interesting as the first record of a maritime plant in the saline waters of Droitwich. It is recorded only in Gloucestershire of our bordering counties.

1932. 1117. G. distans, Wahlenb.   *Reflexed Meadow-grass.*
(*Poa distans*, Linn. *Sclerochloa distans*, Bab.)
Denizen. Usually near the sea. Rare. P. July to September. *Top. Bot.* 56.
First Record. As *Poa distans.* Not localized, *Scott, Hist. Stourb.*, p. 540 (1832).
Avon. Saldons, near Droitwich, *Mathews.* Salt spring at Wadborough.
Severn. By the Droitwich Canal, plentiful. Droitwich Canal, half a mile east of Hawford Bridge. Porters Mill. River Salwarpe, west of Droitwich. Tagwell, south of Droitwich.
Lickey. Near Stourbridge, *Scott.*
Mr. Mathews regards Scott's record as one of his many errors, and at any rate it is possibly not in Worcestershire. In the detached part of Worcestershire surrounding Dudley, an island in Staffordshire, is a wood named the Saltwells, from a decidedly briny spring which rises there. Before the Black Country was so black it is not improbable that maritime plants grew there as well as in other parts of the drainage area of the Severn. But in Staffordshire alone of all the bordering counties, this plant has not been noticed.

FESTUCA, Linn. 508. (A Latin name of uncertain origin meaning hay or straw.)

1935. 1118. F. rigida, Kunth.   *Hard Meadow-grass.*
(*Poa rigida*, Linn. *Sclerochloa rigida*, Griseb. *Glyceria rigida*, Sm.)
Native. Dry ground, wall-tops, railway ballast. Local. A. May to July. *Top. Bot.* 91.
First Record. As *Glyceria rigida.* Near the Ketch on the Bath Road, between Kempsey and Worcester, *Dr. Streeten, Lees, Illus. Nat. Hist. Worc.*, p. 152.
Avon. Crowle. South Littleton. Lenchwick Bottom. Bredons Norton. Bredon. Ripple. Elmley Castle. Pinvin. Inkberrow. Hanbury.
Severn. Habberley Valley, *Mathews.* Near Coningree Wood. Rainbow Hill. London Road. Roadside, Redhill, Worcester. Kempsey. Severn Stoke. Blackstone Hill. Holt. Grimley. Ombersley.
Malvern. Malvern Hills. Martley Hill. Ankerdine. Leigh Sinton. Powick. Abberley. Knighton-on-Teme. Eastham. Malvern Link.
This is a wiry little grass some four or five inches high, and rooting very superficially. It is found in all bordering counties.

1936. 1119. F. rottboellioides, Kunth.   *Meadow Fescue-grass.*
(*F. elatior*, Koch. *F. arundinacea*, Schreb. *F. pratensis*, Huds. *F. loliacea*, Huds.)
Native. Wet pastures, sides of rivers, marshes, cultivated fields. General throughout the county. P. June to August. *Top. Bot.* 41.
First Record. As *F. loliacea.* Badsey Fields, *Rufford, Purton, Midl. Flor.*, p. 83 (1817). As *F. pratensis.* Fields, *Scott, Hist. Stourb.*, p. 540 (1832). (First records as *F. elatior* and *F. arundinacea*, below.)
These *Festucas* are difficult plants. No doubt the synonyms given above are mixed up ; but it is not easy to do anything else but so to use them. The grasses represented by the four synonyms are treated of by Mr. Watson in *Top. Bot.*, 2nd ed., p. 497. Speaking of the first pair, he says, 'I much doubt whether any botanist could separate those named as above into two quasi-specific groups to be satisfactory to all other botanists'. Of the last pair he says, 'Every grade of intermediate form may be found in various places, and one form of inflorescence can be converted into the other on the self-same root under experimental culture'. Of all four he says : 'Unfortunately the binary yoking of the four names fails to remove all the difficulties in the nomenclature of the grasses themselves. The two names *pratensis* and *elatior* (one from each pair) are often used interchangeably ; and I have seen specimens so intermediate as to prevent me from putting either of these two names to them with any confidence'. He also says, 'To give the names of counties with personal authorities in detail would less show the distribution of the four grasses themselves than the use of four names by individual botanists, too often only the doubting use of them '. See *F. elatior*, below. This particular form, *F. rottboellioides*, whichever of these grasses it may correspond to, does not appear to have been noticed by that name in any adjoining county, though of course grasses have been named according to the synonyms in all of them.

1937. 1120. F. membranacea, Druce.   *Single-glumed Fescue-grass.*
(*F. uniglumis*, Soland.)
Casual. Usually on sandy sea-shores. Once recorded. A. June. *Top. Bot.* 20.
First Record. I have gathered *Festuca uniglumis*, by no means common with us, on a wall capped with earth, at Eckington, *Lees, Bot. Worc.*, p. 97 (1867).
Avon. Eckington, as first record.
Here, as in several cases, Gloucestershire, by reason of its possessing in the west a shore which is practically that of the sea, is the only bordering county in which a maritime plant occurs.

1940. 1121. F. Myuros, Linn.   *Wall Fescue-grass.*
(*Festuca pseudo-myurus*, Soy.-Will.)
Native. Wall-tops, dry sandy soil, usually near houses. Not common. A. June to August. *Top. Bot.* 52.
First Record. Commons and roadsides, *Scott, Hist. Stourb.*, p. 540 (1832).
Avon. Bredon. Crowle. South Littleton. Inkberrow.
Severn. Plentiful near Old Sutton Toll-house, Kidderminster ;

between Blakedown and Kidderminster, *Mathews.* Near Bewdley. Red Hill, Worcester. Stourbridge.

Malvern. Cottage roof, Ankerdine Hill. Churchyard wall, Lower Sapey. Malvern Common. Malvern Hills.

Lickey. As first record, *Scott.*

This plant appears to have been noticed in every bordering county except Gloucestershire.

1941. **1122. F. bromoides,** Linn. *Barren Fescue-grass.*

(*F. sciuroides,* Roth.)

Native. Sandy pastures, heaths, walls, dry banks. Locally common. A. May and June. *Top. Bot.* 108.

First Record. Astwood, *Purton, Midl. Fl.,* p. 82 (1817).

Avon. Badsey. Pershore. Pinvin. Huddington.

Severn. Not infrequent about Hartlebury, *Lees.* Between Blakedown and Kidderminster, *Mathews.* Near Bewdley. Oldbury Lane, Worcester. Farleys Lane, Worcester. Bewdley Railway Station. Witley.

Malvern. Ankerdine Hill. Leigh Sinton. Ankerdine. Martley. Powick.

The preceding grass, *F. Myuros,* is sometimes considered a form of this one, differing in the shape of the panicle, which in both species turns one way, in this erect and spreading and separate from the sheath of the highest leaf, in that drooping at the end and spike-like, and the panicle included in the sheath of the highest leaf. This grass occurs in all neighbouring counties.

1942. **1123. F. ovina,** Linn. *Sheep's Fescue-grass.*

Native. Dry pastures, heaths, commons. Locally abundant and not uncommon. P. May and June. *Top. Bot.* 112.

First Record. Malvern Hill and Chase, *Ballard, Stokes's With. Bot. Arr.,* 2nd ed., p. 97 (1787).

A variable grass, puzzling to botanists, young and old. It forms the chief part of the turf of the Clent Hills and of many other eminences in the county. The leaves are all setaceous, more or less curved, usually turning more or less yellow in autumn. With the following plant it is usually the grass sown for lawns, as their herbage is fine and they seldom grow many inches high. This grass is recorded from all bordering counties.

1943. **1124. F. rubra,** Linn.

(*Festuca duriuscula,* Syme.)

Native. Dry pastures, roadsides, heaths, wall-tops, dry woods. Locally common and widely distributed. P. June to August. *Top. Bot.* 103.

First Record. As *Festuca duriuscula.* Walls of Dudley Castle, *Withering,* 3rd ed., p. 153 (1796). As *Festuca rubra.* Commons, *Scott, Hist. Stourb.,* p. 540 (1832).

This is also a very variable plant, differing in appearance from the preceding one, but deficient in characters by which it may be distinguished from it, the chief distinction being an underground one, as its lowest sheaths are hairy and send out long underground shoots, ending in suckers. This plant is taller than *F. ovina,* with a more spreading panicle and pale red

spikelets. Some botanists have united this and the preceding plant. This grass is not assigned any locality in *Lees' Table of Plants.* It occurs in every neighbouring county.

{*1945.* **1125. F. sylvatica,** Vill. *Reed Fescue-grass.*

(*Festuca calamaria,* Sm.)

Native. Woods. One locality. P. July. *Top. Bot.* 30.

First Record. Worcestershire, *Stokes's With. Bot. Arr.,* 2nd ed., p. 102 (1787).

Severn. Shrawley Wood.

Sir James Smith, in *E.B.,* vol. xiv, under no. 1005 (1802), writes : ' Mr. Moseley, of Glasshampton, favoured us lately with living plants from the ledges of a lofty red sandstone rock in Shrawley Wood near his residence.' The plant has been diligently searched for in the locality for very many years, but has not been seen there since May 29, 1865, when specimens were gathered at a meeting of the Worc. Nat. Club. It was searched for at another meeting, August 9, 1870, but was not found then, nor has it been seen since. It differs from *F. arundinacea* in the long ligule to the leaf and the hairy top of the ovary. The plant has been observed in Gloucestershire and Herefordshire and possibly in Staffordshire.}

1946. **1126. F. pratensis,** Huds. *Meadow Fescue-grass.*

Native. Meadows and cultivated fields. Frequent. P. June to August. *Top. Bot.* 102.

First Record. Not localized, *Lees, Bot. Malv. Hills,* 1st ed., p. 48 (1843). Coupled with *F. loliacea,* and marked ' general' in all the districts, *Lees, Table of Plants,* p. 34 (1867).

The *Index Kewensis* makes this plant equal to the succeeding one, *F. elatior,* Linn. A Fescue-grass purporting to be it has been recorded in every neighbouring county except Gloucestershire.

x **Lolium perenne.**

(*F. loliacea,* Curt.)

First Record. As in this book.

Severn. Near Worcester. Meadows near Cotheridge. Near Tibberton. In this hybrid the spikelets are sessile, solitary, and awnless.

1947. **1127. F. elatior,** Linn. *Tall Fescue-grass.*

Native. Wet pastures, cultivated fields, damp hedgerows. Not uncommon. P. June to August. *Top. Bot.* 103.

First Record. Not localized, *Lees, Bot. Malv. Hills,* 1st ed., p. 45 (1843).

With regard to this plant, see Mr. Watson's remarks under *F. rottboellioides* above, p. 415. ' No one botanist,' he says, ' can separate (this plant and its variety *arundinacea*) to the satisfaction of all other botanists.' Practically the chief difference between these two plants is that in *F. elatior* the panicle is close, simple, and the spikelets linear, whereas in the var. *arundinacea* the panicle is spreading, drooping, much branched, and the spikelets ovate-lanceolate. This plant has been observed in Gloucestershire, Stafford, and Warwick. Mr. Watson does not allow it in Worcestershire, nor any bordering county except Gloucestershire.

E e

Var. b. **arundinacea,** Schreb.

Native. Sides of rivers and wet ditches. Rare. June and July. *Top. Bot.* 84.

First Record. As in this book.

Avon. Tiddesley Wood. Crowle. Fladbury. Pershore. Defford. Shell Brook. Pinvin.

Severn. In the district, *Bagnall.* Damp pastures, occasional, *Mathews.* Banks of Severn above Kepax. Meadows near Worcester. Ombersley Road. Droitwich Road. Banks of Droitwich Canal. Holt. Ombersley. Kempsey. Grimley. Stourport. Severn Stoke.

Malvern. Powick. Leigh. Southwood. Stanford Bridge.

Lickey. In the district, *Bagnall.* Damp pastures, occasional, *Mathews.*

Mr. Mathews's district, treated of in his *Fl. Clent and Lickey Hills,* extended over that part of the Severn district north of Kidderminster and east of the Stour ; so that when he generalizes a plant, as in the present case, it may with all probability be taken to occur, in that part of the Severn district included in his area as well as in the Lickey district proper. This plant has been recorded in Staffordshire and Warwickshire, but with regard to it, and with regard also to the above records, Mr. Watson's remarks quoted under preceding plants apply.

BROMUS, Linn. 509. (βρόμος, a kind of oat, from the same root as βρῶμα, food.)

1948. **1128. B. giganteus,** Linn. *Tall bearded Fescue-grass.*

(*Festuca gigantea,* Vill.)

Native. Shady hedgebanks, woods. Rather local. P. July and August. *Top. Bot.* 105.

First Record. As *Festuca gigantea.* On the limestone rock north of Malvern, *Lees, Mag. Nat. Hist.,* vol. iii. p. 160 (1830).

Avon. Overbury. Near Tiddesley Wood. Near Goosehill Wood.

Severn. Hurcott Wood, *Mathews.* Near Birchin Grove. Roadside near Spetchley. Banks of Severn. Ombersley Bank. Lane near Thorngrove.

Malvern. Malvern. Valley west of North Hill. Cowleigh Park. Woodbury Hill. Southwood. Newland.

Lickey. Twyland and Uffmoor Woods ; Ham Dingle, Pedmore ; The Leasowes and Pools, Halesowen, *Mathews.* The Valley, near Bromsgrove. River Stour, Lyde Green, Cradley.

This is a tall grass, in suitable localities reaching a height of five or six feet, of which the panicle is large and loose, and the awn of the glumes is double their length. It is recorded from all bordering counties.

Var. b. **triflorus,** Syme.

First Record. As *Festuca gigantea,* var. *triflorus.* Madresfield, *Towndrow, Malv. Advertiser,* December 10, 1892.

Avon. Overbury, *Herb. Hastings Museum.*

Malvern. As first record.

The chief distinction of this variety is that the panicle is smaller and more erect, while the spikelets consist of about three flowers. It has not been observed in any bordering county.

1949. **1129. B. ramosus,** Huds. *Hairy wood Brome-grass.*

(*B. asper,* Murr. *B. serotinus,* Benck. *B. hirsutus,* Curt.)

Native. Woods, hedges, shady places. General. P. May to August. *Top. Bot.* 103.

First Record. As *Bromus asper.* Perry Wood, *Lees, Illus. Nat. Hist. Worc.,* p. 152 (1834).

This is a common grass, its gracefully pendant flowering stem and open panicle adorning many an old hedgebank and coppice border. The root leaves are rough to the touch, and the sheaths have hairs pointing downwards. Cattle rarely touch it. It is recorded from all bordering counties.

1950. **1130. B. erectus,** Huds. *Upright Brome-grass.*

Native. Calcareous pastures, railway-banks. Local. P. June to August. *Top. Bot.* 50.

First Record. Not localized, *Lees' Cat., New Bot. Guide* (1835).

Avon. Bredon Hill, *Mathews.* Broadway.

Severn. Sandy fields east of Kidderminster Water Works. South of Droitwich. Spetchley.

Malvern. Ridge Hill, Martley ; Silurian eminences, *Lees.* Crews Hill Wood. Martley.

Lickey. Railway-bank, Hawne Colliery, Halesowen ; The same between Hagley and Churchill Stations, *Mathews.*

This plant is characterized by its erect panicle, and sheaths with hairs pointing upwards, while the root-leaves are broader than those of the stem. It is a harsh plant, rejected by cattle. It is recorded from every bordering county except Staffordshire.

1951. **1131. B. madritensis,** Linn. *Upright Brome-grass.*

(*B. ciliatus,* Huds. *B. diandrus,* Curt.)

Casual. Sandy waste places. Once recorded. A. June and July. *Top. Bot.* 11.

First Record. Severn Stoke, *Stokes's With. Bot. Arr.,* 2nd ed., p. 107 (1787).

Severn. Severn Stoke, as first record.

This grass is a native of the Mediterranean region, and is also a weed there of waste and cultivated ground, but it is absent in Northern France, Belgium, and Holland. Although admitted into the books, it is doubtless no native of England. It has been observed in Gloucestershire of our bordering counties.

1954. **1132. B. sterilis,** Linn. *Barren Brome-grass.*

Native. Waysides, wall-tops, waste and cultivated ground. Common. A. June to August. *Top. Bot.* 108.

First Record. Woods, &c., *Scott, Hist. Stourb.,* p. 540 (1832).

The stem of this grass is between one and two feet high, and its cluster of large, drooping, long-awned spikelets, sometimes when young delicately tinged with purple, is often to be seen on roadside waste or bordering hedgebank. During all their existence the awns are conspicuous. The plant is recorded from all neighbouring counties.

1955. 1133. B. secalinus, Linn. *Smooth Rye Brome-grass.*
(*Serrafalcus secalinus,* Bab.)
Colonist. Cornfields, pastures, waste places. Very local. A. May to July. *Top. Bot.* 80.
First Record. Var. 3, *hordaceus.* Near Kempsey and Ridd (Rhydd) Green, *Stokes's With. Bot. Arr.,* 2nd ed., p. 104 (1787).
Avon. Craycombe, *Lees.*
Severn. Shrawley. Ombersley. Claines. Kempsey.
Malvern. Powick. Martley. Rhydd Green.
This grass is characterized by the lax arrangement of the glumes in the spikelets, which are longer than the awns. Its seeds are said to impart a bitterness to flour if they are mingled with the wheat. It is mainly a grass of tillage fields, and though widely spread in England, as well as in the rest of Northern Europe and North America, is only a weed of cultivation, whose true home is in Mediterranean meadows. It has been recorded from every bordering county.

Var. b. **velutinus,** Schrad.
First Record. Helbury Hill, *Lees, A. Florence, Strang. Guide* (1828).
Severn. As first record. Near Worcester, 1892, *Towndrow.* Bromsgrove Station. White's Nursery, Lower Wick. Hallow. Broadheath.
This variety possesses a nearly simple panicle, and larger and downy flowers. It has been observed in Herefordshire and Warwickshire.

1956. 1134. B. racemosus, Linn. *Smooth Brome-grass.*
(*Serrafalcus racemosus,* Parl. *B. arvensis,* Knapp, not of Linn.)
Native. Meadows and pastures. Locally common. B. June and July. *Top. Bot.* 24.
First Record. Severn meadows, Kempsey, *Lees, Illus. Nat. Hist. Worc.* p. 152 (1834).
Avon. Between Trench Wood and Himbleton, *Mathews.* Dunhampstead. Pershore. Defford.
Severn. Kempsey, *Lees.* Pitchcroft. Porter's Mill. Lower Wick. Meadows near Cotheridge. Severn meadows below Diglis.
Malvern. Powick Ham. Powick. Bransford. Leigh.
The differences of these Brome-grasses depend upon somewhat minute characteristics. This plant is very similar to *B. hordaceus,* but the panicle is narrow, not effuse, as in *B. secalinus,* the close imbricate glumes equal the slender awn, and the lower empty glume is lanceolate. It is recorded from all bordering counties.

1957. 1135. B. commutatus, Schrad. *Tumid Field Brome-grass.*
(*Serrafalcus commutatus,* Bab. *B. racemosus,* var. *commutatus,* Doell.)
Native. Meadows, field-borders, cultivated ground. Locally common. B. May to September. *Top. Bot.* 97.
First Record. As *Serrafalcus commutatus.* Not localized, *Lees, Bot. Malv. Hills,* 2nd ed., p. 78 (1852).
Avon. Defford Common.
Severn. Warndon. Near Monk Wood. Towing-path, side of Severn, above Worcester. Spetchley. Hallow.
Mr. Watson, quoted in Anne Pratt, *Grasses* (no date, no ed. given), p. 50,

says, 'This species is known by its glossy grey-green spikelets acquiring a brownish tinge in sunny spots, its longer and harsher peduncles than *B. mollis* and *B. racemosus,* and glumellas larger and more inflated than in *B. secalinus* and *B. arvensis.*' In Mr. Druce's *Fl. Berks.,* p. 592 et seq., are some interesting notes on this and other allied Brome-grasses. They are clearly matters for the specialist to worry over ! This grass is recorded from all bordering counties.

1958. 1136. B. hordeaceus, Linn. *Soft Brome-grass.*
(*Serrafalcus mollis,* Parl. *B. mollis,* Linn.)
Native. Meadows, waysides, waste ground, cultivated fields. An abundant species. A. or B. May to August. *Top. Bot.* 112.
First Record. Var. 2, *nanus.* Barren soil near Stourbridge, *Stokes's With. Bot. Arr.,* 2nd ed., p. 106 (1787). First Record of type. As *Bromus mollis. Lees, Bot. Malv. Hills,* 1st ed., p. 45 (1843).
This is one of the commonest grasses, and is easily known amongst its allies by the long awn much exceeding the length of the glumes. It is recorded from all bordering counties.

Var. b. **glabratus,** Doell.
First Record.
Severn. Hallow, *Herb. Hastings Museum.*
Malvern. Malvern Wells, *Towndrow, Malv. Advert.,* Nov. 20, 1897.
This variety possesses glabrous spikelets. It has not been observed in any bordering county.

1960. 1137. B. ARVENSIS, Linn. *Field Brome-grass.*
Colonist. Rare. P. July and August. Not in *Top. Bot.*
First Record. As below.
Severn. Churchill and Hartlebury Common, *Herb. Hastings Museum.*
Characterized by the long branches bearing the spikelets and its outspreading growth when in fruit.

BRACHYPODIUM, Beauv. 510. (βραχύς, short, πόδιον, a foot, from the sessile or nearly sessile spikelet.)

1961. 1138. B. sylvaticum, Roem. and Schult. *Slender false Brome-grass.*
(*B. gracile,* Beauv. *B. gracilis,* Weigel. *Triticum sylvaticum,* Parnell. *Festuca sylvatica,* Huds.)
Native. Shady places, hedgebanks, woods, rough pastures. Not uncommon ; general in the Avon, Severn, and Malvern districts. P. June to August. *Top. Bot.* 111.
First Record. As *Festuca sylvatica.* Cleeve Hill, *Purton, Midl. Fl.,* p. 83 (1817).
The differences between the two British species of *Brachypodium* are as follows : *B. sylvaticum* has the awns longer than the glumes, spike drooping, and leaves broad and hairy ; and *B. pinnatum* has the awns shorter than the glumes, spike erect, and leaves involute and almost glabrous. This plant is recorded from all bordering counties.

1962. 1139. B. pinnatum, Beauv. *Heath false Brome-grass.*
(*Bromus pinnatus,* Linn. *Festuca pinnata,* Huds.)
Native. Dry pastures, roadsides. Local ; fond of calcareous soil. P. July and August. *Top. Bot.* 37.
First Record. As *Bromus pinnatus.* Abundant in almost every rough pasture of a clayey soil in the neighbourhood of Great Comberton and Pershore, *Nash, Hist. Worc., Int.* lxxxix (1781).
Avon. As first record. Craycombe. Shemington Hill, Alderminster, *Cheshire.* Bredon Hill. Broadway. Badsey, *Purton.* Pirton. Crowle Hill. Tiddesley Wood. Copse near Defford Common. Goosehill Wood. Trench Woods. Fladbury. Badsey.
Severn. Many places, *Lees.* Shrawley Wood. Wyre Forest. Boreley Banks. Holt. Severn Stoke. Ombersley Banks. Lincombe. Monk Wood. Nunnery Wood.
Malvern. Many places, *Lees.* Ankerdine Hill Woods. Weymans Wood. Abberley Hill. Berrow Hill. Sarn Hill. Martley. Little Malvern.
Lickey. Rather uncommon, *Lees.* Mr. Mathews does not mention it in his district.
This is an elegant grass and indicates poor soil, disappearing as land is improved. It has been noticed in Herefordshire, Staffordshire, and Warwickshire.

LOLIUM, Linn. 511. (' Quasi *dolium,* δόλιον, quod *dolosum* sit vel *adulterinum.* Fit enim e corruptis Tritici ac Hordei seminibus.' This is Hooker and Arnott's derivation ; but others derive it from *loloa,* Celtic for Rye-grass, or ὄλλυμι, I kill, from its effect on animals.)

1963. 1140. L. perenne, Linn. *Perennial or beardless Rye-grass.*
Native. Pastures, roadsides, cultivated ground. Abundant. P. June to August. *Top. Bot.* 112.
First Record. Grass fields, *Scott, Hist. Stourb.,* p. 540 (1832).
The grasses of this genus are easily distinguished by the single outer glume and wavy rachis of the main stem. In this species the outer glumes are smaller than the spikelet, whereas in *L. temulentum* they are longer and awned. *Lolium perenne* is a common grass in all the neighbouring counties.

Var. c. MULTIFLORUM, Lam.
First Record. In cultivated fields, introduced, *Lees, Bot. Malv. Hills,* 2nd ed., p. 187 (1852).
Severn. Lower Wick. Near Worcester. Near Smith's Nursery.
Malvern. As first record. Near Powick, *Westcombe.*
Lickey. Twyland Wood, Frankley, *Mathews.*
A South European grass, largely introduced for agricultural purposes. It is not known in a wild state. It has been observed as introduced in Warwickshire.

Var. italicum, Braun. *Italian Rye-grass.*
First Record. On Bisshill (Bissell), near Kidderminster, but introduced, *Lees, Bot. Worc.,* p. 14 (1867).
Severn. Claines, *Towndrow.* Bissell, near Kidderminster, *Lees.* Rail-

way, Worcester. Hartlebury Common. Stourport. Churchill, Kidderminster.
This variety, kept separate in 9th ed., is equated with the foregoing in the *Lond. Cat.,* 10th ed. The grass is localized in all the districts as introduced throughout, by Mr. Lees, in his *Table of Plants,* p. 35. It has been observed in all the bordering counties.

1965. 1141. L. *temulentum, Linn. *Darnel.*
Colonist. Cornfields and waste places. Rare. A. June to August. *Top. Bot.* 64.
First Record. Not localized, *Lees' Cat., New Bot. Guide* (1835).
Avon. Sheriff's Lench. Crowle. Near Pershore.
Severn. Ombersley. Spetchley.
Malvern. Callow End. Powick.
The stem of this grass is two or three feet high. Our forefathers believed that in wet summers the wheat degenerated into Darnel. The plant is suspected to be the only poisonous British grass ; bad symptoms have undoubtedly ensued when the seeds have been eaten, but it is more than probable that these have arisen from their being infected with ergot, *Claviceps purpurea,* or bunt, *Tilletia lolii,* rather than from any deleterious property in the seeds themselves. There is no doubt that the plant is the ' infelix lolium ' of Vergil, and it is supposed to be the plant translated ' tares ' in the Bible. It does not appear to have been seen in Staffordshire.

Var. b. *arvense, With.
First Record. Fields about Badsey, rare, *Purton, Midl. Fl.,* p. 87 (1817).
Avon. As first record.
Malvern. In the district, *Towndrow* (MS.). Near Doddenham Church.
This variety has the awn short or wanting, and has been seen in Salop, Staffordshire, and Warwick.

AGROPYRON, J. Gaertn. 512. (ἀγρός, a field, πυρός, wheat ; wild wheat.)

1966. 1142. A. caninum, Beauv. *Fibrous-rooted Wheat-grass.*
(*Triticum caninum,* Linn.)
Native. Hedges, borders of woods. General, except in the Lickey district where it is less common. P. July. *Top. Bot.* 91.
First Record. As *Triticum caninum.* Woodlands, *Scott, Hist. Stourb.,* p. 540 (1832).
A common grass in woods and hedgerows, distinguished from the next grass by its roots, which consist of a mass of downy fibres. It is recorded from all bordering counties.

1968. 1143. A. repens, Beauv. *Creeping Wheat-grass, or Couch-grass.*
(*Triticum repens,* Linn.)
Native. Cultivated fields, hedges. Abundant, especially on badly farmed arable land. P. June to August. *Top. Bot.* 112.
First Record. As *Triticum repens.* Not localized, *Scott, Hist. Stourb.,* p. 540 (1832).
This is one of the grasses with creeping roots, and perhaps the most common of them, which are known in Worcestershire by the name of 'Scutch'.

Other names are Twitch, Quickens, Quitch, Squitch, and in some places Stroil. The point of the root is covered with hard scales, which enable it to overcome all obstructions. The plant varies considerably, and it would appear that there is no accordance among specialists as to the limits of the forms of this plant and their nomenclature. The plant occurs in all bordering counties.

Var. **barbatum**, Duval-Jouve.
First Record. As below.
Malvern. In the district, *Towndrow* (*MS.*).
The varieties chiefly depend upon the shapes of the glumes and the pales, which vary considerably, the extreme forms being named. This variety has the 'glumes very attenuate, subulate, or awned, pales long-awned'. It is not given in *Lond. Cat.*, 10th ed. It has been observed in Salop.

1970. 1144. A. junceum, Beauv.
Denizen, if anything. Sandy seashores. Once recorded. P. July and August. *Top. Bot.* 50.
First Record. Fields between Pedmore and Hagley, *Scott, Hist. Stourb.*, p. 540.
'A very doubtful record,' says Mr. Mathews. So doubtful that the plant might well be removed from the Worcestershire list if it were not for Mr. Lees' note in the case of the succeeding plant, though if Professor Buckman's identification be accepted, that must be sacrificed to retain this. This plant, as in the case of the next, has not been observed in any bordering county.

x **repens**.
(*Triticum laxum*, Fries. *A. acutum*, auct. Angl.)
Denizen. Usually on sandy seashores. Recorded from the Droitwich Canal. P. July and August. *Top. Bot.* 20.
First Record. As *T. laxum*. Droitwich Canal, *Lees, Bot. Worc.*, p. 38 (1867).
Severn. Droitwich Canal, *Lees*.
There is a footnote in *Bot. Worc.* at the first record saying, 'This *Triticum* is in Professor Buckman's list referred to *T. junceum*.' The plant is sometimes treated as a sub-species of *A. junceum*. It has not been observed in any bordering county.

NARDUS, Linn. 514. (νάρδος was the name of an odoriferous substance, how applicable is not known.)

1972. 1145. N. stricta, Linn.   *Mat-grass.*
Native. Heaths and commons. Not uncommon except in the Avon district, where it has not been noticed. P. May to August. *Top. Bot.* 107.
First Record. Malvern Chace, *Ballard, Stokes's With. Bot. Arr.*, 2nd ed., p. 54 (1787).
Severn. Hartlebury Common. Broadheath. Harberrow Hill. Moseley Green, Hallow. Pedmore Common, 1854, *Rev. J. H. Thompson*.
Malvern. Barnard's Green, *Lees*. West side of Worcestershire Beacon. Malvern Common.
Lickey. Bromsgrove Lickey; near St. Kenelms, *Mathews*. Clent Hills. Moseley, *Mathews*.

It is refreshing to come to a grass which is always itself, and easily recognizable. This is a rigid grass, growing in tufts, five or six inches high, with the sessile spikelets forming a unilateral spike, sometimes bronzed or purplish. The foliage is so harsh that even sheep pass it by. It is recorded from every neighbouring county.

HORDEUM, Linn. 515. (The old Latin name for Barley, of unknown origin, probably connected with *horridus*, bristly, because of the awned glumes.)

1973. 1146. H. europaeum, All.   *Lyme-grass, or Wood Barley.*
(*H. sylvaticum*, Huds.   *Elymus europaeus*, Linn.)
Native. Woods and thickets on calcareous soil. Rare. P. July and August. *Top. Bot.* 29.
First Record. As *Elymus europaeus*. Wood Lyme Grass. Malvern Hills near the Wych, *Lees, A. Florence, Strang. Guide* (1828).
Malvern. As first record. In several of the Western Woods, rare, *Lees*.
This grass appears to have been noticed only in Gloucestershire, Hereford, and Salop of bordering counties.

1974. 1147. H. nodosum, Linn.   *Meadow Barley.*
(*H. pratense*, Huds.   *H. secalinum*, Schreb.)
Native. In rich meadows and pastures. Plentiful in the Severn district; not uncommon in Avon and Malvern, scarcer in the Lickey district. P. June and July. *Top. Bot.* 62.
First Record. As *Hordeum pratense*. Not localized. Common. *Purton, Midl. Fl.*, p. 88 (1817).
This plant is easily recognized by its hairy sheaths, and is recorded from all neighbouring counties.

1975. 1148. H. murinum, Linn.   *Wall Barley.*
Native. Waste places, roadsides, walls. Common and widely distributed. B. June to November. *Top. Bot.* 78.
First Record. Dry banks on roadsides, *Pitt, Agric. Worc.*, p. 317 (1810).
This grass is hardly ever seen away from villages or buildings, but in these neighbourhoods is common enough. The awns of the Barleys are thickly set with a double row of minute spines, which not only take away its value as a fodder plant, from the irritation they cause to the tongue and throat of the animal eating it, but lead to a common sport of village children, who put an ear of Barley grass into the sleeve, and allow it to work its way up to the shoulder. The plant is recorded from all bordering counties.

1976. 1149. H. marinum, Huds.   *Sea-side Barley.*
(*H. maritimum*, With.)
Denizen. Usually in waste places by the sea. B. June. *Top. Bot.* 27.
First Record. As below.
Severn. Banks of the Droitwich Canal, *Humphreys*.
This plant has been observed in Gloucestershire and Herefordshire.

ACOTYLEDONES, Jussieu

93. FILICES, Linn.

PTERIS, Linn. 520. (From πτερόν, a plume or feather.)

1982. 1150. P. aquilina, Linn.   *Common Brake Fern.*
Native. Bushy places, commons, open woods, parks. Plentiful. P. July and August. *Top. Bot.* 112.
First Record. In a record of *Botrychium Lunaria*, which is said to occur 'On the north side of Breedon Hill in several places, particularly above Woollershill in rough ground amongst the Pteris aquilina or Common Brakes', *Nash, Hist. Worc., Int.*, lxxxix (1781).
This sometimes grows to a considerable height. Some is recorded, *W. N. C. Trans.*, I, p. 38, twelve feet high in Habberley Valley in 1856. In other places, on parts of the Clent Hills for instance, the fronds hardly reach twelve inches. Years gone by there was a large local industry in the Clent and Lickey district in burning the bracken for the sake of the alkali in the ashes, out of which were formed round balls the size of a cricket ball, called 'Ess-balls', which found a ready sale in the neighbouring Black Country to be used in the way of soap. Modern chemistry has disestablished this use, as well as others connected with the fern, including several in the way of medicine. The young fronds when properly cooked, are said to afford an excellent vegetable. The portion of the stem just below the surface of the earth, when cut across, discloses a pattern of dark tubes and thread, which are held in various places to have varying similitudes; sometimes J. C., in reference to our Saviour; sometimes the figure of an oak, and the flight of King Charles is mentioned; sometimes an eagle, whence its specific name. The plant seldom varies into tasselled or other forms as several ferns abundantly do, and are frequently in fern gardens; but it is sometimes found in a bifid condition. It grows in all bordering counties.

CRYPTOGRAMME, R. Br. 521. (κρυπτός, concealed, γραμμή, a line, from the concealed line of sori.)

{**1983. 1151. C. crispa**, R. Br.   *Curled Rock-brake, Parsley Fern.*
Native. Among loose stones on mountains and on stone walls. On Malvern Hill, now lost. P. July. *Top. Bot.* 59.
First Record. On the Malvern Hills, *Phyt. O. S.*, vol. i, p. 46 (July 12, 1841).
Malvern. As first record.
In Mr. Lees' *Bot. Malv. Hills*, 1st ed., p. 50, he gives the exact locality. 'In a fissure of crumbling rock on one of the eastern buttresses of the Herefordshire Beacon, above the Priory Farm, Little Malvern.' This well-known fern of Limestone mountains and the Lake District can no longer be claimed as an inhabitant of our county. It has long disappeared from the Malvern Hills.}

BLECHNUM, Linn. 522. (βλῆχνον, a Greek name for a fern.)

1984. 1152. B. Spicant, With.   *Northern Hard-fern.*
(*B. boreale*, Swartz. *Osmunda Spicant*, Linn. *Lomaria Spicant*. Desv.)
Native. Woods and heaths. Many places, except in the Avon district. P. July and August. *Top. Bot.* 112.
First Record. As *Blechnum boreale*. In the lanes about Bromsgrove Lickey, *Purton, Midl. Fl.*, p. 517 (1817).
Severn. Hartlebury Common. Shrawley. Harberrow Pool-side. Wannerton Downs. Shrawley Wood. Wyre Forest. Seckley Wood.
Malvern. Boggy ground about the Hills, *Lees*. West side of Malvern Hills.
Lickey. The Lower Lickey. The Valley, Bromsgrove. Ell Wood. Farley Wood. Ham Dingle. In abundance near Alvechurch, *D. Mathews*.
This fern is recorded from all bordering counties.

ASPLENIUM, Linn. 523. (ά, not, σπλήν, the spleen; the plant having been supposed to be useful in affections of that organ.)

1987. 1153. A. Adiantum-nigrum, Linn.   *Black-stalked Spleenwort.*
Native. Walls and dry banks. Local, except in the Avon and Malvern districts, where it is sometimes abundant. P. June and July. *Top. Bot.* 109.
First Record. On the rocks of the Malvern Hill, abundantly, though small; and in other localities, *Lees, Illus. Nat. Hist. Worc.*, p. 180 (1834).
Avon. Evesham. Hanbury Church.
Severn. West of Hagley Village. Farmyard wall at the Spout Farm, Hagley, 1906, *W. Whitwell*. Lane leading from Sinton Green to Monk Wood. Roadside near The Castle, Hartlebury. Habberley Valley. Claines. Spetchley. Bevere. Chatley.
Malvern. Common on the rocks, *Lees*. Ham Castle. Stanford Bridge. Lane near Powick. Tower of Abberley Church. North Hill. Malvern Wells. Newland.
Lickey. Cottage walls in the Lickey Monument grounds, 1900, *Humphreys*. Clent, gone. Finstall.
This fern is late in unfolding its fronds, but these are sub-evergreen and persist through the winter, being tough and leathery. It is recorded from all neighbouring counties.

{**1989. 1154. A. viride**, Huds.   *Green Spleenwort.*
Native. Rocks in mountainous districts. On Ham Bridge, gone. P. June to September. *Top. Bot.* 47.
First Record. Found by the late Mr. T. B. Stretch growing on Ham Bridge, Clifton. It is, however, eradicated, for some improver has plastered a coat of whitewash over every part of the bridge, *Lees, Illus. Nat. Hist. Worc.*, p. 179 (1834).
Malvern. Ham Bridge.

The plant was not eradicated, as Mr. Lees supposed, at that time. He gives the subsequent history of the fern, *Bot. Worc.*, p. 87. It was in 1827, after seeing a specimen in Mr. Stretch's Herbarium, that he first visited the bridge, only to find it whitewashed, and no trace of the plant remaining. In 1835 it had reappeared on both sides of the wall, and so it remained till 1853, when the wall on which it grew was rebuilt and the fern finally exterminated. But while the destruction was going on Mr. J. S. Haywood, a nurseryman at Worcester, happened to pass by, and seeing the fern among the bricks in the road, he carried it home and for a long time it flourished in his fernery at Wick, and finally died away there a few years ago. During its resuscitation it was an object of pilgrimage to botanists. It formed the subject of papers in the *Phytologist*, vol. i, p. 46 (1841), and vol. iv, p. 947 (1853). Edward Newman, in his *History of British Ferns* (1840), gives an account of his visit to the spot. It has not been seen in the locality since 1853. It has been noticed in Herefordshire and Staffordshire.}

1990. 1155. A. Trichomanes, Linn.   *Common Wall Spleenwort. Maiden-hair Spleenwort.*

Native. Old walls, hedgebanks. General in the Severn and Malvern districts, less common elsewhere. P. May to September. *Top. Bot.* 109.

First Record. Badsey, *Purton, Midl. Fl.*, p. 513 (1817).

Avon. Badsey, *Purton.*

Severn. Wall near Witley Court. Wall near the Cathedral, 1907. Blackstone Rock.

Malvern. On the shady rocks on the Hills, *Lees*. Eardiston. Eastham. Rosebury Rock Wood. North Hill. Ham Castle. Malvern Link. Malvern. Mathon.

Lickey. Walls at Clent, *Amphlett*. Garden Wall, Pedmore.

This fern is recorded from all bordering counties.

1991. 1156. A. Ruta-muraria, Linn.   *Wall-rue Spleenwort.*

Native. Walls. General and locally abundant. P. May to September. *Top. Bot.* 111.

First Record. Walls at Bewdley, *Withering*, 3rd ed., p. 769 (1796).

In some places this fern grows in great quantities, on the walls round Lea Castle at Wolverley, and the Infirmary wall, Worcester, for instance ; and generally is found far more frequently in such places than in any more natural habitat. It is recorded from all bordering counties.

ATHYRIUM, Roth. 524.   (ἀ, not, θύρα, a door, without a door, because the membrane which covers the sori is very little developed.)

1994. 1157. A. Filix-foemina, Roth.   *Lady Fern.*

(*Polypodium Filix-foemina*, Linn. *Aspidium Filix-foemina*, Swartz. *Asplenium Filix-foemina*, Bernh.)

Native. Woods and lane sides, on damp soil. Generally distributed. P. June to August. *Top. Bot.* 110.

First Record. As *Aspidium Filix-foemina*. Burcot and the wet lanes near Bromsgrove Lickey. Very common, *Purton, App.*, vol. iii (1821).

This is a well-known fern, very variable, assuming hundreds of branched, tasselled, or depauperated forms that are seen in the fern garden, but few of which have any botanical value. From an extremely tasselled form, *multifidum*, to the variety called by the garden name of *Victoriae*, where the pinnae are reduced to mere little leaves arranged crosswise up the stem like the backbone of a fish, every intermediate form is met with. It occurs in every bordering county.

Var. b. erectum, Syme.

(Var. *rhaeticum*, Roth.)

First Record. As below.

Malvern. Near Malvern, *Towndrow*. North Malvern. Malvern Hills.

A few varieties are given in Hooker's *Stud. Handbook* and Babington's *Manual*, but this does not appear to be among them. Nevertheless it has been noticed in Herefordshire, Staffordshire, and Warwickshire.

CETERACH, Willd. 525.   (Supposed to be the Chetherak of the Arabians.)

1997. 1158. C. officinarum, Willd.   *Common Ceterach.*

(*Asplenium Ceterach*, Linn. *Grammitis Ceterach*, Sw. *Scolopendrium officinarum*, E. B.)

Native. Old walls. Widely distributed, but thinly. P. April to October. *Top. Bot.* 68.

First Record. As *Scolopendrium officinarum*. On walls at Badsey, *Rufford, Purton, Midl. Fl.*, p. 516 (1817).

Avon. Badsey, *Rufford*. Wall near Badsey Church.

Severn. Garden Wall, Pedmore. Birch's Farm, Churchill. Dowles Bridge. Hanbury.

Malvern. Powick. Malvern Abbey Church, *Newman*. Old Storrage, *Haywood*. Near the Firs, Great Malvern, *Lees*. Warner's Farm, Leigh. Millham Farm, Alfrick. Abundant on the wall in the stable yard at the Westminster Arms, West Malvern. Malvern Link.

Lickey. The Leasowes, Halesowen, 1850. Drayton House, Chaddesley. Clent House and Clent Cottage, Clent.

In 1865 this fern was supposed to be lost to Worcestershire. Dowles Bridge had been washed away ; the wall at the Leasowes had been repaired, and the plant probably destroyed ; and its condition in other known habitats was despaired of. So in the autumn of that year there was rejoicing when the Powick locality was discovered. Really it is a widely distributed fern, though never growing in much quantity, especially in the north of the county ; but it hides itself from observation. It grows on many an old fold yard or garden wall on private premises, unobserved of passers-by, and unnoted until one who knows happens to visit the spot. Usually in the Lickey district it does not grow more than eight or ten feet above the ground, and nearly invariably it prefers the north side of the wall it grows upon. The plant appears to be the true Spleenwort of the ancients to which they attributed so great an effect upon disorders of the spleen. Cretan swine, when feeding upon it, were said to lose that organ altogether ! It is a widely distributed plant, ranging from the latitude of Belgium eastward

---

to the Himalayas, and southward into North Africa, and in warmer climes than ours it grows to a much larger size. The plant is recorded from all bordering counties.

PHYLLITIS, Hill. 526.   (φύλλον, a leaf, from the leaf-like appearance of the fronds. The specific name is given because the lines of fructification resemble the legs of a σκολόπενδρα, or centipede.)

1998. 1159. P. Scolopendrium, Newm.   *Hart's-tongue Fern,*

(*Scolopendrium vulgare*, Symons.)

Native. Woods, hedgebanks, brickwork of wells, brick walls. General throughout, but scarce in the Lickey district. P. July to September. *Top. Bot.* 101.

First Record. Rocks near Bell's Mill ; Chaddesley ; Hill Pool ; very rare, *Scott, Hist. Stourb.*, p. 540 (1832).

In the Lickey district this plant is usually seen inside wells, and in a stunted condition on old walls ; seldom in its natural situation in the hedgebank or by the streamside. It is not nearly so often seen in the county as in some parts of England ; it is not one of the common hedgebank plants. The plant varies very much in cultivation, and in natural conditions is not unfrequently seen with bifid fronds, but in the fern garden a multitude of varieties are cultivated, presenting wonderful departures from the normal state. The Hart's-tongue occurs in every bordering county.

CYSTOPTERIS, Bernh. 528.   (κύστις, a bladder, πτερίς, a fern, from the bladder-like involucre.)

{**2001. 1160. C. fragilis**, Bernh.   *Brittle Bladder-fern.*

(*Polypodium fragile*, Linn.)

Native. Rocks and walls. Rare, perhaps extinct. P. July and August. *Top. Bot.* 82.

First Record. As *Polypodium fragile*. Road from Bourne Heath to Worm's Ash near Bromsgrove, *Miss Read, Withering*, 3rd ed., p. 779 (1796).

Avon. Bredon Hill, north escarpment ; Bredon, fissure of the precipice below the Prospect House, *Lees*.

Malvern. Teme Valley, *Lees*.

Lickey. As first record, near Catshill.

There was a second station for this fern besides the one mentioned in the first record, on a sandstone rock in the valley above Offad's Well, between the well and Bourneheath, where it existed about 1860, but for certainly forty years it has not been seen there. Nor in recent years has it been observed anywhere else. Several well-marked varieties are in garden cultivation. The plant has been observed in all bordering counties.}

POLYSTICHUM, Roth. 529.   (πολύς, many, στίχος, a row, from the arrangement of the sori.)

2005. 1161. P. aculeatum, Syme.   *Soft prickly Shield-fern.*

(*Polypodium aculeatum*, Linn. *Aspidium aculeatum*, Willd.)   *Top. Bot.* 104.

First Record. As *Polystichum aculeatum*. In a ditch in a meadow in the Valley, near Bromsgrove, *Miss Read, Withering*, 3rd ed., p. 777 (1796).

Avon. Tiddesley Wood. Coppice by the Avon, Norton, near Evesham.

Severn. Boggy Glen near Crookbarrow. Claines. Astley. Whittington, near Worcester. Linacre, Claines. Shrawley Wood.

Malvern. Near Malvern, *Towndrow*. Rosebury Rock Wood. Woodbury. Abberley.

Lickey. As first record. The Lickey, *Humphreys*. Occasional, *Mathews*. Yardley, *Miss M. A. Beilby*. Tardebigge. Finstall. Shut Mill.

There is considerable confusion about this plant and its varieties. Some botanists treat the following *P. angulare* as a variety of this fern. Others give only two main species, making them *P. aculeatum* and *P. angulare*, as here, making *lobatum* a variety of the first, with less divided pinnae ; while a variety *lonchitidoides* has been formed with the pinnae nearly entire, imitating *P. Lonchitis*. To the unspecialising botanist, however, there are two well-marked forms met with, *P. aculeatum* and *P. angulare*. The rest is a question of degree. Var. *lobatum* is characterized by its narrower rigid lanceolate fronds and its decurrent sessile pinules, whereas in the type the rigid broader fronds are either decurrent or are tapered to a wedge-shaped base and attached to the rachis by the point of the wedge. In *P. angulare* the pinnules are distinctly stalked and have awned teeth, and the shaggy reddish chaffy scales of the stem are continued throughout the upper part of the lax frond. This fern is recorded from all bordering counties.

Var. b. lobatum, Presl.   *Close-leaved prickly Shield-fern.*

(*Polypodium lobatum*, Huds. *Aspidium lobatum*, Swartz.)

Native. Woods and shady hedges. Many places. P. July and August. *Top Bot.* (aggregate), 104.

First Record. Growing magnificently in the shady dingles by the Spout Brook at Eastham, *Lees, Illus. Nat. Hist. Worc.*, p. 179 (1834).

Var. genuinum, Syme.

First Record. As herein.

Malvern. Near Malvern, *Towndrow (MS.)*. Not given in *Lond. Cat.*, 10th ed.

2006. 1162. P. angulare, Presl.   *Angular-leaved Shield-fern.*

Native. Dry woods, shady banks. Rather local. P. July and August. *Top. Bot.* 65.

First Record. In the woods at Suckley and Leigh Sinton, *Lees, Illus. Nat. Hist.*, p. 179 (1834).

Avon. Bredon Hill, *Mathews*.

Severn. Shrawley Wood. Oldbury Lane. Astley. Between Rushwick and Crowneast. Hallow.

Malvern. Leigh Sinton ; Suckley, *Lees*. The Gullet Pass, Abberley Hill. Southstone Rock. Alfrick. Old Storridge Common. Near Powick. Clifton-on-Teme. Rosebury Rock. Mathon.

Lickey. Ham Dingle, Pedmore. Beoley. Wychbury Wood. Clent Hills. Beacon Hill. Warren Wood. Dodford. Shut Mill.

This fern is recorded in all bordering counties.

LASTRAEA, Presl 530. (Named after Monsieur de Lastre of Chatillon-haut.)

2008. 1163. L. montana, T. Moore.   *Heath Shield-fern, or Mountain-fern.*
(*L.* Oreopteris, Presl. *Aspidium Oreopteris*, Swartz. *A. montanum*, Asch. *Nephrodium Oreopteris*, Desv. *N. montanum*, Baker. *Polypodium montanum*, Vogler. *P.* Oreopteris, Ehrh. *Dryopteris montana*, Kuntze.)
Native. Heathy places, woods, thickets. Local. P. July and August. *Top. Bot.* 102.
First Record. As *Polypodium Oreopteris*. In a wood at Old Foot's (Offad's) Well, near Bromsgrove, *Miss Read, Withering*, 3rd ed., p. 775 (1796).
Severn. Dripping Well, Shrawley, possibly extinct, *Rea*. Wyre Forest. Ribbesford Wood. Habberley Valley. Hartlebury Common. Stanklin Pool, first time, 1907, *Humphreys*.
Malvern. Pudford Hill, Martley. Abberley Hill. West side of Worcestershire Beacon, *Westcombe*. Malvern Hill, *Rev. F. K. Clarke*.
Lickey. Little Farley Wood, 1901, *Humphreys*. Near Sling Common, gone ; Great Farley, *Amphlett*. Moseley, 1858, *Mathews*. Warren Wood.
The locality for this fern near the Sling Common in Belbroughton, where it grew finely among bracken on the roadside waste over some little space, was absolutely cleared of the fern one spring by the peripatetic fern gatherer, and it no doubt was taken to be sold in Birmingham Market Hall, and to wither in suburban gardens. It still grows finely, three or four feet high, at the Great Farley station. The fern is recorded from all bordering counties.

2009. 1164. L. Filix-mas, Presl.   *Male Fern.*
(*Dryopteris Filix-mas*, Schott. *Aspidium Filix-mas*, Swartz. *Polypodium Filix-mas*, Linn. *Filix-mas vulgare*, Park. *Nephrodium Filix-mas*, Richard.)
Native. Woods, thickets, hedges, shady places. Generally distributed. P. June and July. *Top. Bot.*, 9th ed., 112 ; 10th ed., no number.
First Record. Not localized, *Lees, Bot. Malv. Hills*, 1st ed., p. 51 (1843).
This is the commonest of our Worcestershire ferns, as it is in most parts of England ; and is distributed all over the north temperate regions, extending to India and Africa, while it also is found in the Andes in South America. It held great renown with our ancestors, not only for its supposed medicinal virtues, but in connexion with superstitious practices also. Here also ' serpents' came in, not by way of cure but of prevention, for, being burned, it drove them away, together with the lesser evils of 'gnats and other noisome creatures'. A great number of varieties of this fern are to be seen in the fern garden, some of them beautifully crested and tasselled. It is recorded from all bordering counties.

Var. **Borreri**, Newm.
First Record. As below.
Severn. Near Bewdley.
This variety is characterized by the very scaly rachis, bright golden yellow fronds, and very obtuse, almost truncate pinnules. It is a very handsome plant and is a well-marked form. The variety is not given in *Lond. Cat.*, 10th ed., but is recorded for Herefordshire.

{**2012. 1165. L. cristata**, Presl.   *Crested Shield-fern.*
(*Aspidium cristatum*, Sw.)
Native. Bogs and boggy heaths. Very rare. P. July. *Top. Bot.* 10.
First Record. As *Aspidium cristatum*, Moseley Common, *Miss M. A. Beilby, Analyst*, vol. vi, p. 294 (1837). Recorded also, by *Lees*, in Crow's Nest (Crowneast) Wood, between Worcester and Cotheridge, according to the Herbarium of the late Mr. T. Stretch, of Worcester, *Bot. Malv. Hills*, 1st ed., p. 51 (1843).
Miss Beilby's record Mr. Mathews suspects to be an error for *L. spinulosa*, and by the older botanists *L. aristata* was known by this name, while Mr. Lees' record was dropped in the 2nd and 3rd eds. of *Bot. Malv. Hills*, nor does he localize the plant in his *Table of Plants*. Therefore it will be well to await further confirmation before this fern is definitely added to the Worcestershire list. It has been observed in Staffordshire.}

2013. 1166. L. spinulosa, Presl.   *Prickly-toothed Shield-fern.*
(*Dryopteris spinulosa*, Kuntze. *Dryopteris intermedia*, Asa Gray. *Aspidium spinulosum*, Sm. *Nephrodium spinulosum*, Strempel. *Polypodium spinulosum*, Muell.)
Native. Damp woods, thickets, heaths. Local. P. June to August. *Top. Bot.* 84.
First Record. As *Aspidium spinulosum*. In a cave on the right-hand side of the road from Kidderminster to Bewdley, *Perry, Mag. Nat. Hist.*, vol. iv, p. 450 (1831).
Avon. Trench Woods.
Severn. Shrawley Wood. Wyre Forest. Ockeridge Wood. Crowneast. About Kidderminster. Near Bewdley. Lady Pool, Blakedown. Stanklin. Chaddesley Woods.
Malvern. In the dry parts of the western woods, *Lees*. Also in eastern woods.
Lickey. Offmoor Wood. Hagley Wood. The Valley, Bromsgrove. Fouchers Pool, Stourbridge. Lower Lickey. Warren Wood.
This and the next species, *L. aristata*, are inextricably mixed up. The next plant is frequently considered a sub-species of this ; and a fern that is not found in Worcestershire, *L. aemula*, Brack., or *L. Foenisecii*, Bab., intrudes itself into the fray. One writer has said (*Hook and Arn., Brit. Fl.*, 6th ed., p. 570) : ' No one has studied (these) Ferns with a candid and unbiassed mind, but must be satisfied that uniformity of opinion as regards the due limitation of their species is not to be looked for among botanists. The conclusions to be drawn from a careful investigation of *Aspidium spinulosum* and its allies would be as various as the individuals who examine them.' The main differences between these two ferns is the following. *L. spinulosa* has erect pale-green oblong lanceolate fronds eglandular on the under-side, the stem is sparingly furnished with semi-transparent bluntly ovate scales of a uniform brown colour, and the involucre is not gland-ciliate. *L. aristata* has drooping darker and brighter green ovate lance-shaped fronds glandular on the underside, the stem is much thickened at the base and densely clothed with entire lance-shaped pointed scales, which are dark-brown in the centre but nearly transparent at the margin, and the involucre is gland-ciliate. Great

F f

scope for error among local botanists here ! And this must be borne in mind when dealing with these ferns, either in records, the field, or in Herbaria. At all events, this form is recorded from all bordering counties.

2014. 1167. L. aristata, Rendle and Britten.   *Broad Shield-fern.*
(*L. dilatata*, Presl.)
Native. Woods, thickets, heaths. General, but less common in the Avon district. P. July to September. *Top. Bot.* 111.
First Record. As *Aspidium dilatatum*. Blackstone Rock, Bewdley, *Perry, Mag. Nat. Hist.*, vol. iv, p. 450 (1831).
Avon. Tiddesley Wood. Rous Lench Beeches. Deerfold Wood.
Severn. Shrawley Wood. Ockeridge Wood. Blackstone Rock, Hartlebury. Hartlebury Common. Nunnery Wood. Holt. Areley Kings. Wyre Forest. Ribbesford Woods. Astley Wood. Stanklin.
Malvern. Below West Malvern Church. Malvern Wells. Shelsley. Crews Hill Wood. Ankerdine Hill. Martley. Abberley Hill. Malvern.
Lickey. Frankley. Rubery. Great Farley Wood. Moseley. The Valley, Bromsgrove. Woods in Clent. Warren Wood.
This form occurs in all bordering counties.

Var. **c. dumetorum**, Moore.
First Record. As here.
Malvern. Glen between the North and End Hills, *Westcombe*.
This variety is comparatively small, with oblong-ovate or ovate triangular fronds, covered with glands, and the stem clothed with narrow-pointed pale-coloured scales.

POLYPODIUM, Linn. 531.   (πολύς, many, πούς, a foot, from the numerous roots.)
2016. 1168. P. vulgare, Linn.   *Common Polypody.*
Native. Walls, banks, roofs, pollard willows, tree trunks. Generally distributed. P. August to October. *Top. Bot.* 112.
First Record. β. Wings doubly serrated, Worcestershire, *Stokes's With.*, 2nd ed., vol. iii, p. 55 (1792).
This well-known fern requires no localization. The fern is often seen with the lobes more or less serrated, the variety b. *serratum* of the *London Catalogue*. Two well-marked varieties occur, var. *cambricum*, in which the lobes are irregularly toothed, a form which is always sterile ; and var. *hibernicum*, where the frond is more regularly twice or thrice pinnate, and is fertile. The fern is recorded from all bordering counties.

PHEGOPTERIS, Presl. 532.   (φηγός, a kind of oak, and πτερίς, a fern.)
2017. 1169. P. Dryopteris, Fée.   *Tender three-branched Polypody, Oak Fern.*
(*Polypodium Dryopteris*, Linn. *Aspidium Dryopteris*, Baumg.)
Native. Shady woods. Very rare. P. July to September. *Top. Bot.* 76.
First Record. As *Polypodium Dryopteris*. In considerable plenty among the loose stones occupying the glen between the North and End Hills, Malvern. Pointed out by Mr. Salisbury, *Lees, Illus. Nat. Hist. Worc.*, p. 179 (1834).

Severn. Wyre Forest, 'a plentiful supply,' *W. N. C. Trans.* i. 88 (1864). Rooted up in Shrawley Wood before May, 1865. Above the Quarry, Shrawley, *Westcombe*. By the spring in Shrawley Wood, *Westcombe* ; *Rea*.
Malvern. As first record. Western declivity of the Worcestershire Beacon, *Lees*.
Lickey. Dales Wood, Romsley ? *Mathews* (*Lees, Bot. Worc.*, p. 123). The Valley, near Bromsgrove. Offad's Well, gone. Pepper Wood, Belbroughton, one plant, 1900, *Humphreys*.
This delicate little fern, universally known as the Oak Fern, has to a great extent disappeared from the county, owing, it is to be feared, to the ravages of the fern-hunter rather than to any inherent cause in the plant itself. Forty years ago it was fairly plentiful in the valley above Offad's Well, and the records tell of its eradication at Shrawley Wood, but we have grounds to believe that it is not exterminated there. It has certainly been eradicated at the Malvern station. It has been noticed in all bordering counties.

[**2018. 1170. P. Robertiana**, Braun.   *Rigid three-branched Polypody.*
(*P. calcarea*, Fée. *Polypodium Robertianum*, Hoffm. *Polypodium calcareum*, Sm. *Aspidium Robertianum*, Luerssen.)
Native. On broken limestone ground. Very rare. P. May to August, *Top. Bot.* 24.
First Record. Among the loose stones occupying the glen between the North and End Hills, Malvern, Mr. Salisbury, *Lees, Illus. Nat. Hist. Worc.* p. 179 (1834).
Mr. Lees' note upon this plant in *Bot. Malv. Hills*, 3rd ed., p. 118, is this : ' I have gathered growing in the above spot with *P. Dryopteris* a fern that has been called *P. calcareum* by several botanists, but which appears to me to be really *intermediate* between the very rigid fern of *P. calcareum* of other localities and the tender, less luxuriant *P. Dryopteris*.' Mr. Mathews will not accept this plant, saying, 'the true *P. calcareum* has not yet been found.' Still, curiously enough, Mr. Mathews accepts as 'first true record' a notice in *Bot. Worc.*, p. 104, that the plant occurs at Snowshill in Gloucestershire, an adjoining parish to Broadway, saying one locality was possibly in Worcester, and it is known that it occurs in dense masses in several portions of the Cotswold Hills. But there is no mention of such locality in Mr. Lees' book to which Mr. Mathews was referring, and where he got this statement from is unknown. True *calcarea* has yet to be found in this county. It is recorded from every bordering county except Warwickshire.]

[**2019. 1171. P. polypodioides**, Fée.   *Pale Mountain Polypody, Beech Fern.*
(*Phegopteris Phegopteris*, Underw. *Polypodium Phegopteris*, Linn. *Aspidium Phegopteris*, Baumg.)
Native. Shady woods. Very rare. P. July to September. *Top. Bot.* 76.
First Record. ' *Polypodium Phegopteris* I have myself never seen within Worcestershire, but it has been *reported* in the Teme valley near Shelsley,' *Lees, Bot. Worc.*, p. 88.
This fern, commonly called the Beech Fern, must wait outside the Worcestershire list until something more than a report entitles the plant to a place in it. It is recorded in *Top. Bot.*, p. 509, from every bordering county, but Worcestershire is not mentioned.]

F f 2

OSMUNDA, Linn. 534. (Probably given in honour of some person. Osmund, in Saxon, means *domestic peace*; from *os*, a house, and *mund*, peace. Others derive it from Osmunder, another name of the Scandinavian god Thor; and others from *os*, mouth, *munda*, clean, from its supposed medicinal value.)

{**2021.** **1172. O. regalis**, Linn.   *Common Osmund-Royal*, or *Flowering Fern*.
Native. Boggy spots, shady woods. Lost to the county. P. June to September. *Top. Bot.* 89.
First Record. This plant, though before not to be found for many miles around Birmingham, lately appeared on a butt on Moseley Common, artificially made with mud from a deep pit, in which the seeds had probably lain for a great length of time. It continued to flourish so long as the butt was permitted to remain, but has probably now again disappeared, *Withering*, 4th ed., p. 747 (1801).
Severn. Near Lower Broad Water Forges, Kidderminster. Comberton, 1853.
Lickey. Moseley Common, *Withering*.
The plant existed at Moseley for forty years after the time of Withering's record. Miss Beilby recorded it in 1837. Edward Newman, *Phytologist*, vol. i, p. 508 (1843), gives the names of four botanists besides Mr. Lees who had noticed it. Then the plant disappeared on the enclosure of Moseley Common about the year 1840. For something like a dozen years after this the plant was unknown in the county; but in *Phytologist*, vol. iv, part 2 (1852), p. 715, is an account of a meeting of the Phytologist Club, at which the President, Mr. Edward Newman, announced that the plant had been gathered the previous week in Lower Broadwater Forges, near Kidderminster, but the name of the discoverer was not mentioned. Mr. Lees, *Bot. Worc.*, p. 9, says the finder was Mr. Bennett Williams, and gives the locality as Cookley. The locality, however, was probably not near either of these places, but at Comberton, near Kidderminster, where the plant was gathered by Mr. Mathews, July 29, 1853, a few yards south of the Bromsgrove turnpike road, in a ditch leading towards the Hoo Mill, *Mid. Nat.* xiii. 202. In his *Fl. Clent and Lickey Hills*, p. 54, however, Mr. Mathews seems to consider Broadwaters a locality for the fern distinct from the Comberton one. They are a couple of miles apart. At the present time the fern is not to be found at either of them. The plant has been noticed in all bordering counties.}

OPHIOGLOSSUM, Linn. 535. (ὄφις, a serpent, and γλῶσσα, a tongue, from the shape of the spike of fructification.)

2022. **1173. O. vulgatum**, Linn.   *Adder's Tongue*.
Native. Meadows, pastures, grassy rides in woods. Locally common. P. May to July. *Top. Bot.* 88.
First Record. Broadmoor, near Birmingham, *Withering*, 2nd ed., vol. iii, p. 45 (1792).
Avon. Borrow Hill, Feckenham, *Humphreys*. Cookhill Priory Farm, *Mathews*. Kite Wood, Dormstone. Near Stoulton Station.
Severn. Grimley. Hadsor. Tibberton. Near Drakelow, Wolverley. Stoke Prior. Northwick Brickyard. Near Bewdley. Spetchley

Common. Meadows bordering the Severn at Claines, Ombersley, and Kempsey. Hallow. Astley meadows.
Malvern. Hollybush Hill. Longdon Marsh. Malvern Link. Purlieu Lane. Near Alfrick Church. Folley Copse, Alfrick. Near Ham Castle.
Lickey. Redditch. Romsley. Halesowen. In the plantation at Frankley Beeches, *Mathews*. Upland meadows at Clent. Bordesley Park. In front of the Lickey Monument, 1900, *Humphreys*. Dodford.
This plant is far more common in meadows and pastures than is supposed; it requires some searching before its pale-green blade is distinguished among the growing grass; sheep are very fond of its sweet leaves when the grass is grazed; and when in mowing grass it is too inconspicuous to be easily seen. It was much prized by herbalists of old, its special virtue being the healing of 'greene wounds', and the juice of the leaves mingled with distilled Horse-tail water was a 'singular remedy' for internal wounds. In all European countries its name has reference to snakes; yet, curiously enough, we hear nothing of any use of the plant against the injuries inflicted by the ever-present 'serpent'. It is a widely distributed plant ranging over the northern and southern temperate regions, except in the East of Asia. It occurs in every bordering county.

BOTRYCHIUM, Sw. 536. (Βότρυς, a bunch of grapes, from the appearance of the cluster of sori.)

2024. **1174. B. Lunaria**, Sw.   *Common Moonwort*.
(*Osmunda Lunaria*, Linn.)
Native. Dry grassy pastures, heaths, woods among bracken. Very uncommon. P. May to July. *Top. Bot.* 103.
First Record. On the north side of Breedon Hill, in several places, particularly above Woollershill, in rough ground amongst the *Pteris aquilina*, or Common Brakes, *Nash, Hist. Worc., Int.*, lxxxix (1781).
Avon. As first record.
Severn. Warshill. Bewdley. Habberley Valley. Hartlebury Common. Astley, below the corner of Abberley Hill.
Malvern. Cockshoot Hill, Shelsley. Rowburrow Wood, Mathon.
Lickey. Bordesley. Near the Lickey Monument, 1900, *Humphreys*; still flourishing, 1907.
At the Lickey Monument, it is said, where the plant existed in some abundance before 1870, it was wantonly destroyed by the agent of the owner of the soil. Of late years it has reappeared both in the original locality and in adjoining grounds. In an adjoining plantation in Herefordshire near to the British Camp an immense number of specimens of this fern were found amongst dead leaves, certainly over one hundred, but although specimens of these fruited and flourished when placed in a Wardian case at home, no plants came up at the locality in succeeding years. It was regarded as a plant of wondrous potency by our forefathers. Moonwort put in keyholes would open fastened locks; it would draw the shoes from horses' feet that trod upon it. It entered into the concoctions of the alchemists, and by them was called Martagon, which was not the Martagon lily or Turk's Cap (which Martagon means) Lily, but this plant. The word is met with in 1477, while

the Lily was not introduced to this country before 1596. The plant has a very wide distribution, being found on both sides of the Tropics, in the north and south temperate regions, and in Europe extends far up towards the north. It is recorded from all bordering counties.

94. EQUISETACEAE, Deland

EQUISETUM, Linn. 537. (*Equus*, a horse, *seta*, a bristle, whence the English name 'Horsetail', from its resemblance to the same.)

2027. **1175. E. maximum**, Lam.   *Great Water Horsetail*.
(*Equisetum Telmateia*, Ehrh. *E. fluviatile*, Sm.)
Native. Wet shady places, swampy woods. Not uncommon. P. March to April. *Top. Bot.* 84.
First Record. As *E. fluviatile*. Boggy glen near Crookbarrow. Also at Malvern, Alfrick, &c., *Lees, Illus. Nat. Hist. Worc.*, p. 180 (1834).
Avon. Base of Bredon Hill. Tiddesley Wood. Shell Brook. Wyre Piddle.
Severn. Witley. Fenny Rough. Wyre Forest. Battenhall. Crookbarrow. Whittington, Worcester. Ribbesford Wood.
Malvern. Near Malvern. Alfrick. Abberley Hill. Newland.
Lickey. Frankley Woods. St. Kenelm's. The Leasowes. Ham Dingle. Coombs Wood, Halesowen. Warren Woods.
This is a fine plant when growing luxuriantly, as it does in Fenny Rough, giving the idea of a grove of fir-trees in miniature. It is recorded from all bordering counties.

2028. **1176. E. arvense**, Linn.   *Corn Horsetail*.
Native. Cultivated and waste ground. Railway-banks and railway ballast. General and common. P. Fruiting stems appear in March. *Top. Bot.* 111.
First Record. In corn-fields, *Pitt, Agric. Worc.*, p. 317 (1810).
This is a common field plant, a pestilent weed in light sandy land, vying with Scutch in difficulty of eradication. The harshness of the stem and branches made it useful for cleaning and polishing, and it was used under the name of 'Shave-grass' for cleaning wooden spoons and platters and other kinds of ware, and in the north by dairymaids for cleaning wooden milk-pails. The plant is recorded from all bordering counties.

2030. **1177. E. sylvaticum**, Linn.   *Branched Wood Horsetail*.
Native. Shady woods, damp heathy places. Local. P. Fertile spike appears in April. *Top. Bot.* 97.
First Record. Moseley Bog, *Ick, Midl. Counties Herald*, August 5 (1838).
Severn. Wyre Forest. Between Bewdley and Kidderminster, *Jorden*. Wribbenhall. Moseley Common. Ribbesford Hall. Shrawley Wood.
Malvern. Clifton-on-Teme.
Lickey. Frankley. Romsley. Woods near Alvechurch and Redditch, *D. Mathews*. The Lickey. Clent Hills. Moseley. Between Romsley and Walton Hill Coppice at Rubery. Doctors Close, Hunnington. Path to Oldenhall, Cradley Park.
This plant is recorded from all bordering counties.

2031. **1178. E. palustre**, Linn.   *Marsh Horsetail*.
Native. Marshy and boggy places, river banks. Several places throughout the county. P. June to August. *Top. Bot.* 106.
First Record. In bogs, *Pitt, Agric. Worc.*, p. 317.
Avon. Broadway.
Severn. Hartlebury Common. Rudges Farm. Northwick Brickpits. Near Porters Mill. Wyre Forest. Bubble Bridge, Worcester. Pedmore Common. Banks of Severn below Diglis.
Malvern. By the Teme, and near brooks, *Lees*. New Pool.
Lickey. Marshy places, common, *Mathews*. Campion's Pool, Wildmoor; Reservoir, Bilberry Hill, 1900, *Humphreys*. Broom.
This Equisetum is easily distinguished by its slightly rough stem being deeply grooved with from five to twelve furrows, whereas in *E. limosum* the smooth stem is only faintly marked with ten to thirty striae. This Horsetail is recorded from every neighbouring county.
Var. b. **polystachyum**, Weigel.
First Record. Newland, *Towndrow, Malv. Advertiser*, Dec. 10, 1892.
Malvern. Newland. Near Leigh Mill.
This variety differs from the type in its branch-bearing spikes. It has been observed in Warwickshire.

2032. **1179. E. limosum**, Sm.   *Smooth naked Horsetail*.
Native. Ditches, ponds, slow streams, preferring still water. Plentiful; less so in the Avon district. P. June to August. *Top. Bot.* 107.
First Record. Near Kidderminster, *Purton, Midl. Fl.*, p. 501 (1817).
Avon. Bank of Avon, Pershore. Nafford Mill. Dodderhill Common.
Severn. Bromleys Brickyard. Kempsey. Henwick Mill Pond. Northwick Brickpits. Marshy spot west bank of Severn below Diglis Weir. Near Broadwas. Saint Georges Lane. Merrymans Pond.
Malvern. New Pool.
Lickey. Cofton Hackett Reservoir. Halesowen.
This is the next largest of our Horsetails to *E. maximum*, but it has fewer angles and teeth and fewer branches in a whorl, while the fruit-spikes are on stems that are similar to the barren ones. It is recorded from all bordering counties.
Var. b. **fluviatile**, Linn.
First Record. Plentiful in boggy woods near Worcester, Great and Little Malvern; indeed, generally, *E. Lees*; near Worcester, *T. Westcombe, Phyt., O. S.*, vol. i, p. 508 (1843).
Avon. Dodderhill Common.
Severn. As first record. Boggy Glen near Crookbarrow Hill.
Malvern. As first record. Mathon, *Towndrow*. North-west of Malvern Hills.
In this variety the stems are scaberulous above, the tapering branches longer than the internodes, and the cones peduncled. This variety has been observed in Salop, Stafford, and Warwick.

2033. **1180. E. hyemale**, Linn.   *Rough Horsetail*.
Native. Damp banks and woods. Rare. P. July and August. *Top. Bot.* 41.

First Record. Moseley Bog, *Ick, Midland Counties Herald*, August 5 (1838).
Severn. Wyre Forest, *Jorden*. North Wood, near Bewdley.
Malvern. Sapey Brook, *Westcombe*.
Lickey. Moseley Bog. By the side of a pond on the Upper Hill Farm, Frankley, close to the Frog Mill Dingle ; Dayhouse Farm, Romsley, *Mathews*.

This plant possesses a simple stem without the whorls of branches that characterize other species, though sometimes one is produced from the base of one of the sheaths. The sheaths are white with black tip and base, and the cones are acute. It is the roughest of the tribe and used formerly to be imported for cleaning purposes under the name of 'Dutch Rushes'. It grows in great abundance on the banks by the sides of canals, where its long matted roots are of use in binding the soil together and counteracting the wearing effects of water. This plant appears to have been recorded in all bordering counties except Gloucestershire.

95. LYCOPODIACEAE, Sw.

LYCOPODIUM, Linn. 538. (λύκος, a wolf, πόδιον, a little foot, from the supposed appearance of some species.)

{**2038. 1181. L. Selago,** Linn.   *Fir Club-moss.*
Native. Heathy places. Very rare. P. July and August. *Top. Bot.* 89.
First Record. Bog on Moseley Common. *Miss M. A. Beilby, Analyst*, vol. vi, p. 294.
Severn. Near Bewdley, gone, *J. B. Duncan.*
Lickey. As first record.
The claim of Worcestershire to this plant rests on these two records alone. With other rare plants that once existed at Moseley it has now disappeared, and the last time it was looked for at the Bewdley locality it could not be seen. It has been noticed in every bordering county except Gloucestershire.}

2039. 1182. L. inundatum, Linn.   *Marsh Club-moss.*
Native. Marshy spots on peaty heaths. Very rare. P. August and September. *Top. Bot.* 57.
First Record. In the large bog on Hartlebury Common, *Lees, Illus. Nat. Hist. Worc.*, p. 180 (1834).
Severn. Hartlebury Common.
This inconspicuous little plant still exists in some quantity in the spot where it is first recorded, lying close to the ground and only rewarding careful search. It has been noticed in Staffordshire and Warwickshire.

2041. 1183. L. clavatum.   *Common Club-moss.*
Native. Heathy places, commons, woods. Very rare. P. July to August. *Top. Bot.* 95.
First Record. On a sandstone cliff by the Severn, at Winterdyne, near Bewdley, *T. Robinson*, from whom I have a specimen, *E. Lees*. Bog on Hartlebury Common, *R. J. N. Streeten*. Moseley Common, *W. Southall, junr., Phytologist, O. S.*, vol. i, p. 508.
Severn. As first record, Bewdley and Hartlebury Common. Gladder Brook, *J. B. Duncan.*

Lickey. Moseley, as first record. Walton Hill, Clent, *Mrs. John Amphlett*, 1881. Randans Woods, 1885, *Humphreys*. Upper Lickey, 1885, *Rev. Henry Boyden.*

This plant is familiar enough to the tourist in mountain districts under the name of Stag's-horn Moss, but in the midland counties it is very far from being so frequently met with. Of the stations mentioned above for the plant, it has disappeared from Moseley ; and from Walton Hill also, where there were two plants, of which one was utterly destroyed by the hoof of a horse, while the turf on which the other grew was taken up by a rustic and applied face downwards to mending a neighbouring hedgebank. It has not been seen at the Randans for many years ; and on Hartlebury Common, where until a few years ago there was a considerable patch of it, it has lately been destroyed by fire, lighted no doubt by one of the gipsies that infest that place. Still, it may reappear in some of these or other localities. On Walton Hill nothing was seen of it for at least ten years before its rediscovery in 1881. The plant has been noticed in all bordering counties except Gloucestershire.

2042. 1184. L. alpinum, Linn.   *Savin-leaved Club-moss.*
(*L. complanatum*, Linn. *L. complanatum*, var. *fallax*, Celak.)
Native. Usually on stony moors. Twice recorded. P. August. *Top. Bot.* 58.
First Record. Once gathered at Hartlebury Common in 1836 by the *Rev. Churchill Babington and Mrs. Waller* of Stourport, *Lees, Bot. Worc.*, p. 10 (1867).
Severn. Hartlebury Common.
Malvern. Worcestershire Beacon, August 1893, *Freeman Roper, Malvern Advert.*, January 20, 1894.

A specimen of the Hartlebury plant is in Professor Cardale Babington's Herbarium, but it has been said not to be true *L. alpinum.* No sign of any such plant has been seen at the common by any other observer, and, whatever it was, it has been extinct there for very many years. Both *L. alpinum* and *L. complanatum* are mentioned for Worcestershire in *Top. Bot.*, p. 527, but both as requiring further investigation.

SALVINIACEAE, Bartling

AZOLLA, Willd. (ἄζω, I dry up, ὅλλυμι, I kill, because a dry condition causes the plants to perish.)

1185. A. caroliniana, Willd.
First Record. Mr. Carleton Rea reported that Miss Maud Bates had discovered a large quantity of *Azolla caroliniana* growing in Perdiswell Pool, *Worc. Nat. Club Trans.* iii, p. 268 (1906).
Severn. Perdiswell. Barbourne Brook.
This stranger is a native of the United States from New York southward and of South America to Buenos Ayres. It is only recently that it has been seen in England, but it appears to have a tendency to spread. Possibly the plant is an escape from an aquarium whenever it appears in this country, if not sometimes an attempt at naturalization, so injudiciously advocated by some of the popular botanists (!) of to-day.

98. CHARACEAE, L. C. Rich.

CHARA, Linn. 542. (χαρά, delight, from the graceful form of the plants when growing in the water.)

2048. 1186. C. fragilis, Desv.
(*C. globularis*, Thuill.)
Native. Ditches, brooks, ponds, canals, rivers. Rare. June to August. *Top. Bot.* 94.
First Record. Localized in the Malvern District, *Lees, Bot. Worc., Table of Plants*, p. 36 (1867).
Severn. Near Foxberry (Fockbury), *Worc. Nat. Club Trans.* i. 5. Great Bog, Wyre Forest. Pool near Norton Barracks. Perdiswell Pool. Westwood Pool.
Malvern. Welland, *Towndrow*. Berrington's Pool, Little Malvern. Mathon.
This plant has been noticed in every bordering county.

2051. 1187. C. aspera, Willd.
Native. Pools. Very rare. July and August. *Top. Bot.* 42.
First Record. As herein.
Severn. Stanklin Pool, Aug. 25, 1900, *Humphreys.*
This plant has not been seen in any neighbouring county.

{**2055. 1188. C. tomentosa,** Linn.
Native. Slow and still water. Very rare ; extinct. *Top. Bot.*, only in Ireland.
First Record. Feckenham Bog, *Purton, Midl. Fl.*, p. 435 (1817).
Avon. Feckenham Bog.
This plant in *Babington's Manual*, 9th ed., p. 542, is said to occur only in Ireland. As this book is edited by Messrs. H. and J. Groves, who have made a special study of the *Characeae*, it must be taken as a last word upon the plant, and one which will throw doubt upon the accuracy of the above record. At all events it has gone the way of all the rarities at Feckenham Bog, even if it once existed there. It has not been observed in any bordering county.}

{**2056. 1189. C. hispida,** Linn.   *Prickly Chara.*
Native. Ponds, artificial water. Very rare. July to September. *Top. Bot.* 52.
First Record. Ditches of Longdon Marsh, *Lees, Bot. Malv. Hills*, 1st ed., p. 49 (1843).
Severn. Longdon Marsh.
The above is the only record of the occurrence of this plant in this county, and no doubt it has disappeared with the drainage of Longdon Marsh. The plant has been observed only in Staffordshire of our bordering counties.}

2058. 1190. C. vulgaris, Linn.   *Common Chara.*
(*C. foetida*, A. Br.)
Native. Ponds, ditches, marshes, canals. Rare. July to September. *Top. Bot.* 84.

First Record. Pools and bog-holes, *Lees, Bot. Malv. Hills*, 1st ed., p. 49 (1843).
Severn. Several places, *Lees.*
Malvern. Several places ; Commons near Malvern, *Lees*. Berrington's Pool, Little Malvern.
This plant has been observed in Herefordshire, Staffordshire, and Warwickshire.

Var. b. **longibracteata**, Kuetz.
First Record. Malvern, *J. of B.*, vol. xxiii (1885).
Malvern. Leigh Sinton, *Towndrow.*
This variety has been observed also in Herefordshire and Warwickshire.

NITELLA, Agardh. 546. (*Niteo*, I shine.)

2073. 1191. N. flexilis, Agardh.   *Flaccid Chara.*
(*Chara flexilis*, Linn.)
Native. Pools, ditches. Rare. July and August. *Top. Bot.* 35.
First Record. As *Chara flexilis*, in a stew at Cookhill, *Purton, Midl. Fl.*, p. 435 (1817).
Avon. As first record. Pit near the Droitwich Canal at Hanbury, *Worc. Nat. Club Trans.* i. 60.
Severn. One locality, *Lees.*
Malvern. One locality, *Lees*. Ditches near Malvern.
The Droitwich Canal at Hanbury forms the boundary between the Avon and Severn districts, the Avon district being on the east of the canal, on which side the party of botanists appears to have been investigating. The plant has been noticed in every bordering county except Gloucestershire.

2074. 1192. N. opaca, Agardh.
Native. Ponds and ditches. Very rare. May to July. *Top. Bot.* 82.
First Record. As herein. Newland, determined by James Groves, *Towndrow.*
Malvern. Newland, *Towndrow.*
This plant has been observed in Staffordshire and Warwickshire of our bordering counties.

MUSCI

I. SPHAGNACEAE

SPHAGNUM, Dill. (σφάγνος, a kind of tree-moss, probably a *lichen*.)

1. **S. acutifolium**, Russ. and Warnst.
Wyre Forest. Jockey Brook, Pensax. Lickey Hill.
Var. **viride**, Warnst.
Foxholes, Habberley.

2. **S. quinquefarium**, Warnst.
Wyre Forest.

3. **S. subnitens**, Warnst.
Rock Coppice, Bewdley. Wyre Forest.
Var. **griseum**, Warnst.
Wyre Forest.

4. **S. squarrosum**, Pers.
Feckenham, *Purton*. Malvern, *Lees*. Lickey Hill. Moseley Bog.
Var. **spectabile**, Russ. and Warnst.
Rock Coppice, Bewdley.

5. **S. cuspidatum**, Russ. and Warnst.
Lickey Hill.

6. **S. recurvum**, Russ. and Warnst.
Lickey Hill.

7. **S. molluscum**, Bruch.
Hartlebury Common.

8. **S. subsecundum**, Limpr.
Hartlebury Common. Wyre Forest. Seckley Wood.

9. **S. inundatum**, Warnst.
Wyre Forest. Hartlebury Common. Wythall Heath.

10. **S. Gravetii**, Warnst.
Wyre Forest. Seckley Wood. Arley Wood. Lickey Hill. Wythall Heath.

11. **S. rufescens**, Warnst.
Seckley Wood. Arley Wood. Lickey Hill. Wythall Heath.

12. **S. crassicladum**, Warnst.
Wyre Forest. Seckley Wood. Cofton Hackett. Lickey Hill. Pepper Wood. Bell End.

13. **S. cymbifolium**, Warnst.
Lickey Hill.

14. **S. papillosum**, Lindb.
Rock Coppice, Bewdley. Wyre Forest.

II. ANDREAEACEAE

ANDREAEA, Ehrh. (After Andreä, an apothecary of Hanover.)

15. **A. Rothii**, Web. and Mohr.
Habberley Valley, *J. B. Duncan*.
Rocks. Very rare.

III. TETRAPHIDACEAE

TETRAPHIS, Hedw. (τέτρα, for τέσσαρες, four ; φύω, I produce ; the peristome has four teeth.)

16. **T. pellucida**, Hedw.
Turfy banks and rotting stumps in woods, and on sandstone rocks. Frequent, barren.

IV. POLYTRICHACEAE

CATHARINEA, Ehrh. (After the Empress Catharine II of Russia.)

17. **C. undulata**, W. and Mohr.
On sandy soil or clay in woods, on heaths, &c. Very common. The variety is rare.
Var. **Haussknechtii**, Dixon.
Chatley Green. Seckley Wood. Offmoor Wood. Northfield.

POLYTRICHUM, Dill. (πολύς, many, θρίξ, hair, from the hairy calyptra.)

18. **P. nanum**, Neck.
Rhydd Covert, Bewdley. Wyre Forest. Cookley. Habberley Valley. Old Storridge Common. Bromsgrove. Lickey Hill. Barnt Green.
Sandy heathy ground, usually in woods. Not uncommon, but the variety is rare.
Var. **longisetum**, Hampe.
Rhyd Covert, Bewdley. Wyre Forest. Cookley. Old Storridge Common. Barnt Green.

19. **P. aloides**, Hedw.
Habberley Valley. Seckley Wood. Wyre Forest. Old Storridge Common. Lickey Hill. Rubery.
Sandy or clayey banks, disused quarries. Frequent.

20. **P. urnigerum**, Linn.
Wyre Forest. Over railway tunnel, Bewdley. Lickey Hill.
Dry and stony places. Rare and always barren.

21. **P. piliferum**, Schreb.
Church Lench. Kempsey Common. Shrawley Wood. Wyre Forest. Cowleigh Park, *Towndrow*. Abberley Hill. Lickey Hill.
Dry heaths. Common.

22. **P. juniperinum**, Willd.
Shrawley Wood. Habberley Valley. Wyre Forest. Pedmore Common. Old Storridge Common. Pensax. Lickey Hill.
Dry heaths and waste places. Common.

23. **P. gracile**, Dicks.
Wyre Forest.
Peaty turf in woods. Very rare.

24. **P. formosum**, Hedw.
Trench Woods. Hadley. Farm Wood, Ombersley. Shrawley Wood. Habberley Valley. Wyre Forest. Seckley Wood. Old Storridge Common ; Cowleigh Park, *Towndrow*. Randans. Rubery.
Dry woods. Common.

25. **P. commune**, Linn.
Shrawley Wood. Seckley Wood. Wyre Forest. Old Storridge Common. Abberley Hill. Acock's Green. Lickey Hill.
Damp peaty places. Not common.

V. BUXBAUMIACEAE

BUXBAUMIA, Haller. (Named in honour of Buxbaum.)

26. **B. aphylla**, Linn.
Wyre Forest, *J. B. Duncan*. On the ground amongst long heather at ' The Devil's Spadeful ', Bewdley, January 1909, *J. B. Duncan*. On the ground or on rotting wood ; not reappearing often in the same locality. Very rare. In the Wyre Forest locality, after appearing for three or four seasons, the plant has now disappeared.

VI. DICRANACEAE

ARCHIDIUM, Brid. (ἀρχή, the type, εἶδος, form.)

27. **A. alternifolium**, Schimp.
Wyre Forest. Ribbesford Wood. Malvern, *Lees*.
Damp, bare patches on sandy or clayey ground, usually in woods. Rare.

PLEURIDIUM, Brid. (πλευρόν, a rib, εἶδος, form, alluding to the capsule being apparently lateral.)

28. **P. axillare**, Lindb.
Pepper Wood. Twylands and Cocks Wood, Frankley. Illey. Truman's Heath. Stirchley Street. Moseley. Broadheath, near Tenbury. Wyre Forest.
Damp clayey fields and margins of ditches. Not uncommon.

29. **P. subulatum**, Rab.
North Wood, Bewdley. Wyre Forest. Seckley Wood. Old Storridge Common. Frankley. Moseley. Acock's Green.
Bare sandy banks, fields, quarries, and stony places. Common.

30. **P. alternifolium**, Rab.
Wyre Forest. Folly Point, Bewdley. Cocks Wood, Frankley, 1870. Offmoor Wood, 1870.
Damp sandy or clayey ground. Rare.

DITRICHUM, Timm. (διτριχιάω in Greek means ' I have a double row of hairs ', alluding to the divided filiform peristome teeth.)

31. **D. homomallum**, Hampe.
Ravens Hill, Alfrick.
Very rare. On damp sandy or gravelly banks.

32. **D. flexicaule**, Hampe.
Cleeve Prior. Rous Lench. Broadway. Old quarries, Ravenshill, Suckley, and Martley. Eastham. Worcestershire Beacon, *Lees*.
Limestone rocks and dry calcareous banks. Frequent on calcareous ground.

SELIGERIA, B. and S. (Named after Seliger, a Silesian bryologist.)

33. **S. pusilla**, B. and S.
Bredon Hill, *H. H. Knight*.
Shady rocks. Very rare.

CERATODON, Brid. (κέρας, a horn, ὀδούς, a tooth.)

34. **C. purpureus**, Brid.
Very common everywhere.

35. **C. conicus**, Lindb.
Croome Park wall. Wych and North Hill, Malvern. Highter Heath, Kings Norton.
Bare places and mud-capped walls. Rare.

CYNODONTIUM, B. and S. (κύων, a dog, ὀδούς, a tooth.)

36. **C. Bruntoni**, B. and S.
Bredon Hill. Wribbenhall. Seckley Wood. Wyre Forest. Blakeshall Common. Blackstone Rock, Bewdley, *Rev. J. H. Thompson*. North Hill.
Shady rocks. Rare. Frequent on sandstone rocks near Bewdley, Wolverley, and Kidderminster.

DICHODONTIUM, Schimp. (διχῆ, in two, ὀδούς, a tooth.)

37. **D. pellucidum**, Schimp.
Wyre Forest. Seckley Wood. Arley Wood. North Wood. Fenny Rough. Pensax Dingle. Menith Wood, Pensax. Sapey Brook. Rubery. Frankley.
Wet rocks or sandy detritus by streams. Frequent in the north-west of the county.

DICRANELLA, Schimp. (A Latin diminutive of the Greek δίκρανος, two-pointed.)

38. D. heteromalla, Schimp.
Sandy banks in woods, &c. Very common. The two varieties are rare.
Var. stricta, B. and S.
Wythall Heath. Weoley Quarries.
Var. interrupta, B. and S.
Alvechurch. Wood near Barnt Green. Wythall Heath.

39. D. cerviculata, Schimp.
Hartlebury Common. Malvern, *Lees*. Canal siding near Alvechurch.
Peaty ground. Rare. The variety is very rare.
Var. pusilla, Schimp.
Hartlebury Common.

40. [D. secunda, Lindb.
Malvern Hill, *Dr. Griffiths*.
Stony ground on mountains. Very rare. This is an old record, and requires confirmation.]

41. D. rutescens, Schimp.
Seckley Wood. Gladder Brook, Bewdley. Malvern Hill. Lickey Hill.
Wet clayey ground. Rare.

42. D. varia, Schimp.
Broadway Hill. Armscote. Seckley Wood. Clows Top. Warshill. Park Attwood. Eastham. Eardiston. Ravenshill, Alfrick. Old Storridge Common. Pensax. Lickey Hill. Rubery. Offmoor. Halesowen.
Heavy clay soil in woods and by streams. Common.

43. D. Schreberi, Schimp.
Porchbrook Rock ; Hanley Dingle, *J. B. Duncan*.
Damp clay banks and sides of ditches. Rare.
Var. elata, Schimp.
Seckley Wood.
In damper situations. Very rare.

44. D. squarrosa, Schimp.
Wyre Forest, October, 1908, *J. B. Duncan*.
Sandy margins of streams. Very rare.

DICRANOWEISIA, Lindb. (The Greek δίκρανος, two-pointed, followed by a word derived from a surname.)

45. D. cirrata, Lindb.
On trees, rocks, walls, and fences ; a very common species.

[D. crispula, Lindb.
Malvern Hill, *Lees*.
Mr. Lees records *Weisia crispula*, but a plant gathered at Malvern in company with Mr. Lees by the Rev. J. H. Thompson proved on examination to be only the common species, *Weisia viridula*.]

CAMPYLOPUS, Brid. (καμπύλος, crooked, πούς, a foot, referring to the cygneous seta.)

46. C. flexuosus, Brid.
Near Bewdley. Habberley Valley. Wyre Forest. Park Attwood. Shakenhurst. Lickey Hill.
Shady rocks and boulders. Rare.

47. C. pyriformis, Brid.
Habberley Valley. Wyre Forest. Seckley Wood. Blakeshall Common. Hartlebury Common. Canal-side, Alvechurch. Lickey Hill.
Heathy ground. Not common.

48. C. fragilis, B. and S.
Winterdyne, Bewdley. Habberley Valley.
Shady rocks. Rare.

DICRANUM, Hedw. (δίκρανος, two-pointed, referring to the peristome teeth.)

49. D. Bonjeani, De Not.
Defford Common. Ribbesford. Habberley Valley. Wyre Forest. Old Storridge Common. Lickey Hill. Beoley.
Boggy places and heaths. Not common ; the three varieties recorded are rare.
Var. juniperifolium, Braithw.
Kempsey Common. Warshill Wood.
Var. calcareum, Braithw.
Limestone rocks, Old Storridge Common.
Var. rugifolium, Bosw.
Wyre Forest. Malvern Hill.

50. D. scoparium, Hedw.
Ribbesford Wood. Carpenter's Hill Wood. Small form of type in Seckley and Arley Woods, on tree-roots. Malvern Hill. Old Storridge Hill. Abberley Hill. Near Alvechurch. Randans. Pitcher Oak Hill.
Woods, heaths, and tree trunks. Very common. The variety *orthophyllum*, however, is not common, and the two others are rare.
Var. paludosum, Schimp.
Kempsey Common. Blakeshall Common.
Var. orthophyllum, Brid.
Kempsey Common. Arley Wood. Seckley Wood. Malvern Hills.
Var. spadiceum, Boul.
Wyre Forest.

51. D. majus, Turn.
Shrawley Wood. North Wood. Ribbesford Wood. Wyre Forest. Seckley Wood. Habberley Valley. Malvern Hills. Old Storridge. Cofton Hackett.
In woods, on shady banks, and near streams. Not very common.

G g

52. D. fuscescens, Turn.
Trench Woods.
Rocks ; frequent in mountain districts. Very rare in other places.

53. [D. Scottianum, Turn.
Malvern, *Lees*.
Rocks in sub-alpine regions. Doubtful ; an old record and requiring confirmation.]

54. D. strictum, Schleich.
Rotting tree-trunks near Hampton Lovett.
Tree-trunks and old railings. Very rare.

55. D. flagellare, Hedw.
Near Jackman's Hill, Droitwich.
Rotten tree-trunks. Very rare.

56. D. montanum, Hedw.
Trench Woods. Ockeridge Wood. Carpenter's Hill Wood. Warshill Wood. Ribbesford Wood. Seckley Wood.
Near the base of tree-trunks and on rotting stumps in woods. Rare.

LEUCOBRYUM, Hampe. (λευκός, white, βρύον, moss.)

57. L. glaucum, Schimp.
Shrawley Wood. Rhydd Covert, Bewdley. Ribbesford Wood. Wyre Forest. Welland Common, *Lees*. Lickey Hill.
Heaths and woods. Very local.

VII. FISSIDENTACEAE

FISSIDENS, Hedw. (*Fissus*, cleft, *dens*, a tooth.)

58. F. exilis, Hedw.
Gladder Brook. North Wood. Upper Arley. Pirton Pool. Dick Brook, Pensax. Barnt Green. Kings Norton. Grosty Hill. Offmoor Wood. Halesowen.
Clay banks in woods. Not common.

59. F. viridulus, Wahl.
Clows Top. Wyre Forest. Coldridge Wood. Eastham. Pensax Dingle. Near Hewell Grange. Near St. Kenelms. Halesowen.
Clay banks and shady rocks near streams. Frequent. The variety is very rare.
Var. Lylei, Wils.
Pensax Wood.

60. F. pusillus, Wils.
Seckley Wood. Gladder Brook. Upper Arley. Near Stanford-on-Teme. Dick Brook, Pensax. Southstone Rock.
Shady sandstone rocks. Not uncommon, but the variety is very rare, and occurs on dripping rocks.
Var. madidus, Spruce.
Spout Farm, Eardiston.

61. F. incurvus, Starke.
Blockley. Whittington. Norton, near Worcester. Malvern. Martley. Blackwell. Alvechurch. Near Kings Norton. Offmoor. Clay banks. Not common.
Var. tamarindifolius, Braithw.
Ockeridge Wood. Ombersley. Bewdley. Ravenshill, Alfrick. Wythall. St. Kenelms.

62. F. bryoides, Hedw.
Clay banks and woods. Frequent.

63. F. crassipes, Wils.
Boulders in the Severn by Seckley Wood. Bewdley Bridge. Boulders in the Stour, Halesowen, 1870.
On stones in the larger streams. Rare.

64. F. adiantoides, Hedw.
Defford. Wyre Forest. Seckley Wood. Ribbesford Wood. St. Kenelms.
Bogs and wet rocks. Not common.

65. F. decipiens, De Not.
Limestone Quarries, Broadway. Bredon Hill. Cleeve Prior. Church Lench. Gladder Brook. Ribbesford. Wyre Forest. Seckley Wood. Beoley.
Damp rocks. Not common.

66. F. taxifolius, Hedw.
Woods and banks on clayey soil. Frequent.

OCTODICERAS, Brid. (ὀκτώ, eight, and δίκερας, a double horn.)

67. O. Julianum, Brid.
River Severn at Bewdley, *J. B. Duncan* (1901), first record for Britain.
Aquatic. Rare. On rocks, stones, and tree-roots at many points in the bed of the river Severn. Sometimes attached to old barges and floating stages. This moss has been recorded from the Severn at Shrewsbury and also at Gloucester ; quite recently it has been detected in the river Thames.

VIII. GRIMMIACEAE

GRIMMIA, Ehrh. (Named in honour of Grimm.)

68. G. apocarpa, Hedw.
Wood Norton. Bredon Hill. Croome D'Abitot. Seckley. Shatterford. Longdon Marsh. Ravenshill, Alfrick. Knightwick. Barnt Green. Offmoor.
Stones and walls. Common. The varieties occur on rocks near streams, and are rare.
Var. rivularis, W. and M.
Bed of the river Severn near Upper Arley, Seckley, and Bridewell. Dowles Brook.
Var. gracilis, W. and M.
Severn Banks, Upper Arley. Leigh Sinton. Pensax Dingle.

G g 2

69. G. pulvinata, Smith.
   Rocks, walls, and roofs. Very common. The variety is rare.
   Var. obtusa, Hübn.
   Walls near Blakeshall. Walls near Offmoor.

70. G. trichophylla, Grev.
   Bewdley. Wyre Forest. Cowleigh Park, *Lees.* Worcestershire
   Beacon. Old Storridge Common.
   Walls. Rare.

71. G. subsquarrosa, Wils.
   North Hill, *Ley.* Worcestershire Beacon.
   Siliceous rocks. Very rare.

72. [G. ovata, Schwaeg.
   Malvern, *Lees.*
   Siliceous rocks on mountains. Rare. An old record which
   requires confirmation.]

[G. leucophaea, Grev.
   Malvern Hill, *Thompson.*
   A specimen of Mr. Thompson's which is so named, in the *Herb.
   Hast. Mus. Worc.,* proves to be a poor form of *Rhacomitrium hetero-
   stichum,* Brid. The record is probably an error.]

RHACOMITRIUM, Brid. (ῥάκος, a rag, μιτρίον, a small belt, from the
   fringed calyptra.)

73. R. aciculare, Brid.
   Wyre Forest. Malvern Hill, *Purton, Midl. Fl.,* iii. 89. Walls, New
   Brook, Frankley, 1870.
   Rocks and stones in and near streams. Rare.

74. R. fasciculare, Brid.
   Sandstone rocks near Blakeshall. Walls, New Brook, Frankley.
   Rocks and walls. Rare.

75. R. heterostichum, Brid.
   Cleeve Hill. Railway, Bewdley. Wyre Forest. Blakeshall.
   Cowleigh Park, *Ley.* New Brook, Frankley.
   Rocks and walls. Not common. The variety is rare.
   Var. alopecurum, Hübn.
   Railway-cutting, Bewdley. Malvern Hills. New Brook, Frankley.

76. R. lanuginosum, Brid.
   Bredon Hill. Railway cutting, Bewdley. Wyre Forest. Malvern
   Hill. New Brook, Frankley.
   Stony heaths, rocks and walls. Not common.

77. R. canescens, Brid.
   Hartlebury Common. Wyre Forest. Habberley Valley. Malvern
   Hills, *Purton*; *Bagnall.* New Brook, Frankley. Rubery. Cali-
   fornia.
   Barren heaths. Not common.

PTYCHOMITRIUM, B. and S. (πτύξ, a fold, μιτρίον, a little belt,
   referring to the plicate calyptra.)

78. P. polyphyllum, Fürn.
   Blockley. Bredon Hill. Railway cutting, Bewdley. Wyre
   Forest. Near Seckley. New Brook, Frankley.
   Rocks and walls. Rare.

HEDWIGIA, Ehrh. (Named after Johänn Hedwig.)

79. H. ciliata, Ehrh.
   Malvern Hills, *Bagnall.* Cowleigh Park, *Ley.* Hollybush Hill.
   North Hill.
   Siliceous rocks. Rare.

IX. TORTULACEAE

ACAULON, C. M. (ἄκαυλος, without stalk, the plants are very minute
   and bulbiform.)

80. A. muticum, C. M.
   Kings Norton, *Webb* ; *Bagnall.* Higher Heath, near Kings Norton.
   Bare sandy ground. Rare.

PHASCUM, Schreb. (φάσκον, a kind of *lichen* on trees.)

81. P. cuspidatum, Schreb.
   Common in fallow fields and on waysides.

82. P. curvicolle, Ehrh.
   Not uncommon on Bredon Hill, *H. H. Knight.* Hanley Child.
   Mud-capped walls and bare calcareous ground. Rare.

POTTIA, Ehrh. (After J. F. Pott, a German botanist.)

83. P. truncatula, Lindb.
   Frequent in fallow fields and on waysides.

84. P. intermedia, Fürnr.
   Blockley Fields. Near Suckley. Martley. Highter Heath, near
   Kings Norton. Winnall Green. Selly Oak. Rubery. Grosty
   Hill. Halesowen. Bartley Green.
   In similar situations to the preceding, but less common.

85. [P. Wilsoni, B. and S.
   Crews Hill, Alfrick. Moseley, *Badger.*
   Sandy ground, usually near the sea. Rare. Doubtful records
   requiring confirmation.]

86. P. minutula, Fürn.
   Newbould-on-Stour. Blockley. Preston-on-Stour. Trench Woods.
   Ombersley. Westwood Park. Arley Wood. Near Suckley.
   Malvern Hill, *Thompson.* Illey. Dudley Castle.
   Banks and fallow fields. Not common.

87. [P. Starkeana, C. M.
   Malvern, *Lees.*
   Bare ground. Rare. An old record which requires confirmation.]

88. P. lanceolata, C. M.
   Shatterford. Defford Common. Littleton. Trench Woods. Broad-
   way. Overbury. Sapey, *F. S. Lea.* Eastham. Hanley Child.
   Suckley. Old Storridge Common. Martley. Knightwick.
   Grosty Hill.
   Frequent on limestone soil.

TORTULA, Hedw. (emend. Lindb.). (Diminutive of *tortus,* twisted,
   from the form of the peristome.)

89. T. pusilla, Mitt.
   Blockley. Paxford. Broadway. Newbould-on-Stour. North and
   South Littleton. Hanley Child. Malvern, *Lees.*
   Frequent on mud-capped walls in the south-east of the county.
   Var. incana, Nees and Hornsch.
   Overbury. Bredon. Newbould-on-Stour. North and South
   Littleton.

90. T. lamellata, Lindb.
   Paxford. Broadway. North Littleton. Wooferlow, near Hanley
   Child.
   Mud-capped walls and banks in calcareous districts.

91. T. rigida, Schrad.
   Paxford. Newbould-on-Stour. Cleeve Prior. North Littleton.
   In similar situations to the last.

92. T. ambigua, Ångstr.
   Blockley. Newbould-on-Stour. Church Lench. Littleton. Near
   Droitwich. Wyre Forest. Shatterford. Suckley. Knight-
   wick. Rubery.
   In similar situations to the preceding.

93. T. aloides, De Not.
   Blockley. Littleton. Wyre Forest. Trimpley. Blackstone Rock.
   Seckley. Ravenshill. Crews Hill. Alfrick. Knightwick.
   Hanley Child. Martley. Pensax. Rubery. Frankley. Clent.
   Calcareous rocks and banks and on old walls. Not uncommon.

94. T. cuneifolia, Roth.
   Pickersleigh, *Towndrow.* Malvern Wells. Knightwick, *Binstead.*
   Grosty Hill.
   Rocks and walls. Rare.

95. T. marginata, Spruce.
   Whittington. Shrawley. Seckley Wood. Blakedown. Churchill
   (Kidderminster). Lincombe. Cooper's Hill. Barnt Green.
   Woodcote Hill. Northfield. Offmoor. Clent.
   Shady sandstone rocks and walls. Not common.

96. T. muralis, Hedw.
   Rocks, walls and stones. Very common.
   Var. rupestris, Schultz.
   Broadway. Bredon Hill. Defford. Littleton. Malvern. Han-
   ley Child. Kings Norton.
   Walls and rocks in limestone districts.
   Var. aestiva, Brid.
   Bredon Hill. Wribbenhall. Seckley Wood. Hartlebury. Bell
   End. Clent. Halesowen.
   Shady sandstone and calcareous rocks. Rare.

97. T. subulata, Hedw.
   Broadway. Ribbesford. Rhydd Covert. Blackstone. Lye Head,
   Bewdley. Holt Fleet. Hanley Child. Worcestershire
   Beacon. Lower Wych. North Hill. Stanford Bridge. Dod-
   ford. Woodcote Green.
   Sandy banks. Common.

98. T. angustata, Wils.
   Hedgebank, Pickersleigh, *Towndrow* ; *Bagnall.*
   Hedgebanks. Very rare.

99. T. mutica, Lindb.
   Armscote. Cleeve Prior. Harvington. Peopleton. Pershore.
   Feckenham. Bewdley. Ombersley. Seckley. Pickersleigh,
   Towndrow ; *Bagnall.* Stanford-on-Teme. Halesowen.
   Tree-roots and rocks by streams, where liable to be submerged
   during floods. Not uncommon.

100. T. laevipila, Schwaeg.
   Alderminster. Harvington. Pershore. Wyre Forest. Crundalls,
   Bewdley. Blakeshall. Longdon Marsh. Pensax. South-
   stone Rock. Dick Brook.
   Tree-trunks in the open. Common.

101. T. intermedia, Berk.
   Broadway. Bredon. Alderminster. Whittington. Worcester,
   Towndrow. Shatterford. Wyre Forest. Alfrick. Old Stor-
   ridge Common. Great Cowleigh, *Ley.* Stanford Bridge.
   Calcareous rocks and soil. Frequent.

102. T. ruralis, Ehrh.
   Bredon. Defford. Aston Magna. Alderminster. Preston-on-
   Stour. Blakeshall. Habberley Valley. Shatterford. Hanley
   Child. Malvern Hills. Southstone Rock. Shelsley.
   Stony ground. Walls and roofs of farm buildings. Frequent.

103. T. papillosa, Wils.
   Broadway. Alderminster. Cleeve Prior. Longdon Marsh. Near
   Tewkesbury.
   Trunks of trees. Rare.

BARBULA, Hedw. (*Barbula*, a little beard, alluding to the peristome.)

104. **B. lurida**, Lindb.
     Wyre Forest. Wribbenhall, on walls. Lincombe Weir. By the Rea at Shakenhurst. Frankley Hill.
     Rocks by streams, rarely on walls. Rare. Fruiting in several stations.

105. **B. rubella**, Mitt.
     Shaded rocks and walls. Frequent.

     [Var. ruberrima, Braithw.
     Several records are to be found for this as a Worcestershire plant, but they are certainly, in one way or another, all errors.]

106. **B. tophacea**, Mitt.
     Alderminster. Wyre Forest. Railway-banks near Bewdley. Hanley Child. Sapey Brook. Alvechurch. Barnt Green. Hopwood. Northfield.
     Wet calcareous rocks and walls. Not uncommon.

107. **B. fallax**, Hedw.
     Clay banks, waste places, walls, &c. Very frequent.

     Var. brevifolia, Schultz.
     Dick Brook. Wyre Forest. Hanley Dingle. Offmoor Wood. Offmoor. Rare.

108. **B. spadicea**, Mitt.
     Bredon Hill. Harvington. Gladder Brook. Eymore Wood. Old Storridge Common. Knighton - on - Teme. Barnt Green. Frankley Hill.
     Rocks and boulders in beds of streams. Rare.

109. **B. rigidula**, Mitt.
     Bredon Hill. Peopleton. Tardebigge. Dick Brook. Rubery. Dudley Castle.
     Walls and rocks. Rare.

110. **B. cylindrica**, Schimp.
     Blockley. Broadway. Bewdley. Holt Fleet. Wyre Forest. High Habberley. Cowleigh, *Towndrow*. Ravenshill. Malvern. Hagley.
     Walls and rocks near streams. Frequent.

111. **B. vinealis**, Brid.
     Broadway. Wyre Forest. Frequent on walls near Bewdley. Suckley. Malvern Hill. Pensax. Knightwick. Frankley Hill. Alvechurch. Lower Hagley.
     Sandstone rocks and walls. Frequent.

112. **B. sinuosa**, Braithw.
     Alderminster. Bow Bridge. Rous Lench. Dick Brook. Croome Park. Eymore Wood. Pensax. Hanley Dingle.
     Rocks and walls. Rare.

113. **B. Hornschuchiana**, Schultz.
     Goose Hill, near Droitwich. Worcester, *Towndrow*. Croome Park. Shatterford. Alfrick. Frankley Hill.
     On bare ground, old quarries, and on walls. Rare.

114. **B. revoluta**, Brid.
     Blockley. Bewdley. Habberley Valley. Longdon. Martley. Hopwood. Frankley Hill.
     Limestone rocks and walls and on mortar. Abundant in calcareous districts.

115. **B. convoluta**, Hedw.
     Blockley. Habberley Valley. Hartlebury Common. Wyre Forest. Wolverley. Ravenshill. Pensax. Offmoor Wood.
     On the ground and on wall tops. Frequent.

116. **B. unguiculata**, Hedw.
     Banks, walls, and bare ground. Very common.

     Var. cuspidata, B. and S.
     Defford. Shrawley. Ravenshill. Alfrick. North Malvern.

LEPTODONTIUM, Hampe. (λεπτός, slender, and ὀδούς, a tooth.)

117. **L. flexifolium**, Hampe.
     Wyre Forest. Habberley Valley. Hartlebury Common.
     On peaty soil. Rare.

WEISIA, Hedw. (From a personal name.)

118. **W. crispa**, Mitt.
     Bredon Hill, *H. H. Knight*. Malvern, *Lees*.
     Bare ground. Rare or overlooked.

119. **W. squarrosa**, C. M.
     Canal-bank near Droitwich.
     Fallow fields and bare places. Rare or overlooked.

120. **W. microstoma**, C. M.
     Stagbury. Westwood Park, Droitwich. Bewdley. Near Arley Wood. Harts Green. Malvern Hill, *Griffiths*. Hagley. Moseley.
     Banks and bare ground. Not common.

121. **W. tortilis**, C. M.
     Bredon Hill.
     Calcareous ground. Rare.

122. **W. viridula**, Hedw.
     Hedgebanks and sandy ground. Common. The variety is rare.

     Var. amblyodon, B. and S.
     Seckley Wood. Bell End.

123. **W. mucronata**, B. and S.
     Dick Brook. Gladder Brook. Ribbesford Wood. Wyre Forest. Seven Hills, near Bewdley. Pensax. Moseley.
     Clayey ground. Rare.

124. **W. tenuis**, C. M.
     Near Lincombe Lock. Hanley Dingle. Canal-bank, Alvechurch. Quarry near Halesowen. Rubery.
     On faces of sandstone or calcareous rocks. Rare.

125. **W. verticillata**, Brid.
     Upper Arley. Wyre Forest. Lincombe. Eymore Wood. Sapey Bridge, *Ley*. Hanley Dingle. Southstone. Martley. Hanley Child.
     Damp calcareous rocks. Not uncommon. Fruiting in several localities.

TRICHOSTOMUM, B. and S. emend. (θρίξ, hair, στόμα, the mouth, alluding to the peristome.)

126. **T. crispulum**, Bruch.
     Rous Lench. Church Lench. Wyre Forest. Arley Wood. Ravenshill and Crews Hill, Alfrick.
     Calcareous rocks. Rare.

127. **T. mutabile**, Bruch.
     Wyre Forest. Upper Arley. Frankley, 1870.
     Shady rocks. The plant and the variety are very rare.

     Var. littorale, Dixon.
     Severn. Wyre Forest.

128. **T. tortuosum**, Dixon.
     Wyre Forest. Malvern Hills, *Lees*. Worcestershire Beacon.
     Shady rocks. Very rare. The variety is also very rare, in more exposed situations.

     Var. fragilifolium, Dixon.
     Wyre Forest.

129. **T. tenuirostre**, Lindb.
     Seckley Wood, *J. B. Duncan*.
     Shady rocks near streams. Very rare.

CINCLIDOTUS, P. Beauv. (κιγκλίς, a latticed gate, from the appearance of the base of the peristome.)

130. **C. Brebissoni**, Husn.
     Armscote. Cleeve Prior. Pershore. Peopleton. Feckenham. Hampstall Feruy. Westwood Park. Upper Arley. Stour near Cookley. Near Stockton-on-Teme, *Towndrow*. Leigh Sinton. Stanford-on-Teme. Knightwick.
     On rocks and tree-roots by streams. Not common.

131. **C. fontinaloides**, P. Beauv.
     Cleeve Prior. Hamstall Bridge. Brockenhurst. Stourport. Upper Arley. Lincombe. Bewdley. Stockton-on-Teme, *Towndrow*. River Rea near Shakenhurst. River Rea, *Moseley*.
     Rocks in larger streams, where frequently submerged. Not common.

X. ENCALYPTACEAE

ENCALYPTA, Schreb. (ἐν, within, καλύπτω, I cover, referring to the large calyptra.)

132. **E. vulgaris**, Hedw.
     Blockley. Broadway. Bredon Hill. Habberley Valley. Drakelow. Ribbesford. Hanley Child. Near Clent.
     Calcareous rocks, banks and occasionally on walls. Not common.

133. **E. streptocarpa**, Hedw.
     Blockley. Broadway. Bredon Hill. Arley Wood. Bewdley. Dowles. Hanley Child. Shakenhurst. Menith Hill. Abberley Hill. Hollybush Hill. Lickey Hill.
     Limestone rocks and banks, and on mortar of walls. Frequent.

XI. ORTHOTRICHACEAE

ZYGODON, Hook and Tayl. (ζυγόν, a join, ὀδούς, a tooth.)

134. **Z. Mougeotii**, B. and S.
     Wyre Forest. Worcestershire Beacon, *Lees*. Alfrick.
     Damp rocks. Very rare.

135. **Z. viridissimus**, R. Brown.
     Aston Magna, with fruit. Wyre Forest. High Habberley. Foxholes, near Kidderminster. Hanley Child. Knightwick. Alvechurch, *Russell*.
     Trees, sometimes on rocks. Not uncommon. Fruit very rare.

     Var. rupestris, Hartm.
     Wyre Forest. Ravenshill, Alfrick.
     On rocks. Very rare.

136. **Z. Stirtoni**, Schp.
     Wyre Forest ; Park Attwood, *J. B. Duncan*.
     Rocks. Very rare.

137. **Z. Forsteri**, Mitt.
     Near Harvington, *H. H. Knight*, 1905.
     Tree-trunks. Exceedingly rare.

ULOTA, Mohr. (οὐλότης, curliness, from the curled leaves when dry.)

138. **U. Bruchii**, Hornsch.
     On trees, Dick Brook. Wyre Forest. Cowleigh Park, *Lees*. Eastham.
     On branches of trees near streams. Very rare.

139. **U. crispa**, Brid.
     North Wood, Bewdley. Shelsley Walsh. Near Blackwell. Offmoor Wood.
     Both the plant and the variety occur on trees, and are rare.

Var. intermedia, Braithw.
Dick Brook. North Wood, Bewdley. Wyre Forest. Upper Arley. Pensax. Blackwell.

ORTHOTRICHUM, Hedw. (ὀρθός, straight, θρίξ, hair, from the erect hairs on the calyptra.)

140. O. rupestre, Schleich.
Rocks. Very rare. This record is from a list compiled from the British Museum Herbarium ; the locality is not stated.

141. O. anomalum, Hedw.
Alderminster. Middle Hill, *Lees.*
Rocks. Very rare. Mr. Lees records this from Malvern Hill, but the record is untrustworthy.

Var. saxatile, Milde.
Broadway, *Lees.* Blockley. Aston Magna. Bredon Hill. New-bould-on-Stour. Malvern, *Lees.* Leigh Sinton. Martley.
Calcareous rocks and walls. Not uncommon.

142. O. cupulatum, Hoffm.
Old walls, Newbould-on-Stour. Bredon Hill. Hartlebury Common. Malvern Hill, *Lees.*
Calcareous rocks and walls. Both type and variety are rare, the latter occurring on rocks.

Var. nudum, Braithw.
By the Stour, Alderminster. Dowles Brook. Wyre Forest. Habberley Valley.

143. O. leiocarpum, B. and S.
Dick Brook. Spout Farm, Eardiston. Stoke Bliss.
Tree-trunks. Rare.

144. O. Lyellii, H. and T.
On trees. Frequent, always barren.

145. O. affine, Schrad.
On trees. Frequent in south Worcestershire ; local in the north· The variety, which occurs on trees by streams, is rare.

Var. rivale, Wils.
Near Alderminster. Dick Brook.

146. O. rivulare, Turn.
Alderminster. Goosehill Green. Westwood Park, Droitwich. River Severn, Seckley. Near Stockton-on-Teme, *Towndrow.* Shakenhurst. Lickey Hills, *Lees.* River Rea, near Moseley.
On rocks and tree-roots by water. Rare.

147. O. Sprucei, Mont.
Tree-roots by Severn at Upper Arley.
Trees by water. Very rare.

148. O. stramineum, Hornsch.
Trees by Piddle Brook.
Trees. Very rare.

149. O. tenellum, Bruch.
On ash trees by Tredington.
Trees. Very rare.

150. O. diaphanum, Schrad.
Broadway. Alderminster. Cleeve Prior. Harvington. Pershore. Wyre Forest. Blackstone. Hartlebury. Longdon Marsh. Pensax. Stanford Bridge. Moseley.
Tree-trunks, wooden fences, rarely on walls. Frequent.

151. O. obtusifolium, Schrad.
On ash trees near Shipston-on-Stour.
Trees. Very rare.

XIV. FUNARIACEAE

DISCELIUM, Brid. (δι- two, and σκέλος, a leg.)

152. D. nudum, Brid.
Near Lincombe Weir, *J. B. Duncan.*
Clay banks. Very rare.

EPHEMERUM, Hampe. (ἐφήμερος, living but a day, short-lived.)

153. E. serratum, Hampe.
Stagbury Hill. Wyre Forest. Warshill. Chatley Green. Mount Segg, near Kidderminster. Barnt Green. The Lickey. Northfield. Kings Norton. Acocks Green.
Bare ground and fallow fields. Not common.

PHYSCOMITRELLA, B. and S. (Latinized diminutive of the name of the next genus.)

154. P. patens, B. and S.
Blackstone. Bewdley. Moseley, *Webb.* Near Offmoor Wood. Kings Norton.
Margins of pools and clay soil. Very rare.

PHYSCOMITRIUM, Brid. (φύσκη, the stomach, μιτρίον, a cap.)

155. P. pyriforme, Brid.
Paxford. Cleeve Prior. Stagbury Hill. Hoo Brook. Lincombe. Lickey Hill. Higher Heath and Wythall Heath, near Kings Norton. Bell End. Moseley. Clent.
Clay banks and sides of ditches. Not common.

FUNARIA, Schreb. (*Funarius*, a rope-maker ; from the seta being twisted when dry.)

156. F. fascicularis, Schimp.
Near Warshill Wood. Malvern, *Lees.* Higher Heath, near Kings Norton.
Fallow fields and bare ground. Rare.

157. F. hygrometrica, Sibth.
Waste ground, walls, heaths, &c., especially where the ground has been burnt. Very common. The variety is rare.

Var. calvescens, B. and S.
Rous Lench. Dunley. Near Lickey Hill.

XV. MEESIACEAE

AMBLYODON, P. Beauv. (ἀμβλύς, blunt, ὀδούς, a tooth.)

158. A. dealbatus, P. Beauv.
Canal-side near Hopwood Dingle and near Alvechurch.
Boggy places and springs. Very rare. Mr. Lees mentions this moss in his Botany of Malvern, but gives no locality.

AULACOMNIUM, Schwaeg. (αὐλαξ, a furrow, μνίον, moss, from the striate capsule.)

159. A. palustre, Schwaeg.
Kempsey Common. Hartlebury Common. Pedmore. Seckley. Spinney, near Arley. Wyre Forest. Malvern, *Bloxam.* Lickey Hill. Acocks Green.
Bogs. Not uncommon.

160. A. androgynum, Schwaeg.
Heathy and sandy ground and on sandstone rocks. Frequent in the sandy districts in the north of the county, probably rare or absent in parts of it.

XVII. BARTRAMIACEAE

BARTRAMIA, Hedw. (From a personal name.)

161. B. pomiformis, Hedw.
Near Cookley Wood, *Lees.* Stourport. Kidderminster. Churchill. Blakeshall Common. Wyre Forest. Malvern Hills. Ankerdine. Bell End. Moseley.
Sandy banks and rocks. Not common.

PHILONOTIS, Brid. (φίλος, loving, νοτίς, dampness.)

162. P. fontana, Brid.
Bredon Hill. Wyre Forest. Seckley. Malvern Hills. Old Storridge Common. Randans. Lickey Hill. Wythall Heath. Rubery.
Bogs and sides of rills. Not common.

163. P. caespitosa, Wils.
Near Newbould-on-Stour. Malvern Hill. Butlers Hill, near Alvechurch.
Bogs and wet heaths. Rare.

164. P. calcarea, Schimp.
Dick Brook. Menith Wood.
Bogs and springs on calcareous ground. Rare.

165. P. capillaris, Lindb.
Wyre Forest. Rhydd Covert, near Blakeshall. On limestone rocks, Old Storridge Common.
Rocks and sandy ground. Rare.

XVIII. BRYACEAE

LEPTOBRYUM, Wils. (λεπτός, slender, and *Bryum.*)

166. L. pyriforme, Wils.
Shrawley. Blackstone Rock. Bewdley. Spout Farm, Eardiston. Orleton. Canal-side, Alvechurch. Barnt Green. Coles Green. Bell End. Oldswinford.
Sandstone rocks. Often appears in greenhouses. Rare.

WEBERA, Hedw. (From a personal name.)

167. [W. cruda, Schwaeg.
Malvern, *Lees.*
Clefts of rocks on mountains. Old record, and requiring confirmation.]

168. W. nutans, Hedw.
Peaty and sandy soil. Frequent. The variety occurs on wet heaths, and is rare.

Var. longiseta, B. and S.
Hartlebury Common. Rubery.

169. W. annotina, Schwaeg.
Wyre Forest. Ombersley. Rhydd Covert, Bewdley. Near Tardebigge. Barnt Green. Wythall Heath.

Var. erecta, Correns.
Wyre Forest. Railway-banks, Bewdley, *J. B. Duncan.*

170. W. proligera, Bryhn.
Foxholes near Kidderminster. Railway cuttings near Bewdley and Hartlebury, *J. B. Duncan.*
Sandy ground and sandstone rocks. Not common.

171. W. carnea, Schimp.
Broadway. Feckenham. Defford Common. Churchill. Bewdley. Tanners Brook. Heightington. Longdon Marsh. Ravenshill. Alvechurch. Offmoor.
Damp clay banks. Not common.

172. W. albicans, Schimp.
Alderminster. Whittington. Wyre Forest. Seckley Wood. Railway-banks, Bewdley. Pensax. Knightwick. Alvechurch. Lickey Hill. Rubery.
Clay banks, ditches, and margins of streams. Common.

BRYUM, Dill. (*βρύον*, a kind of mossy sea-weed.)

B. PTYCHOSTOMUM

173. **B. pendulum**, Schimp.
Church Lench. Hartlebury. Oldswinford. Lincombe. North Hill. Barnt Green. Halesowen. Kings Norton.
Sandy ground, sandstone rocks and walls. Not common.

C. CLADODIUM

174. **B. lacustre**, Brid.
Broadway Hill. Bewdley. Harborne.
Sandy ground. Rare.

175. **B. inclinatum**, Bland.
Bredon Hill. Habberley Valley. North Hill. Barnt Green. Rubery. Frankley. Northfield.
Dry heaths, walls and banks. Not uncommon. A small variety is abundant in drains at Frankley.

176. **B. uliginosum**, B. and S.
Frankley Hill.
Damp sandy ground. Very rare.

D. LEUCODONTIUM

177. **B. pallens**, Sw.
Broadway. Warshill Wood. Wyre Forest. Churchill. Malvern. Stanford Bridge. Southstone Rock. Canal-side, Alvechurch. Frankley Hill.
Bogs, springs, and banks of streams. Not common.

178. [**B. turbinatum**, Schwaeg.
Bromsgrove Lickey, *Purton*, iii. 82.
An old record, which requires confirmation.]

E. EU-BRYUM

179. **B. pseudo-triquetrum**, Schwaeg.
Teddington. Defford Common. Wyre Forest. Stockton Pool. Hopwood Dingle.
Bogs and wet places. Not common.

180. **B. bimum**, Schreb.
Teddington. Defford Common. Jockey Brook, Bewdley. Wyre Forest. Stockton Pool.
Bogs and pools. Not common.

181. **B. affine**, Lindb.
Near Churchill. North Hill.
Damp walls and rocks. Rare.

182. **B. intermedium**, Brid.
Oldswinford. Railway cutting near Bewdley. Barnt Green. Frankley. Northfield.
Damp ground and walls. Rare.

183. **B. caespiticium**, Linn.
Walls, rocks, and dry banks. Frequent.

184. **B. capillare**, Linn.
Walls, banks, rocks, and tree-trunks. Very common.
Var. **macrocarpum**, Hübn.
Ravenshill, Alfrick.
Var. **flaccidum**, B. and S.
Broadway. Bransford Bridge. Near Alfrick.

185. **B. Donianum**, Grev.
Knightwick, *Rev. C. H. Binstead*.
Sandstone rocks. Very rare.

186. **B. erythrocarpum**, Schwaeg.
Hartlebury Common. Wyre Forest.
Sandy ground and heaths. Rare.

187. **B. atropurpureum**, W. and M.
Broadway. Littleton. North Wood, Bewdley. Ribbesford. Stanford-on-Teme. Eastham. Bewdley. Hagley Wood.
Banks, roadsides, and waste ground. Not uncommon.
Var. **gracilentum**, Tayl.
Footways, Defford. Near Lincombe Lock. Wribbenhall. New Brook, Frankley. Halesowen Railway Station.

188. **B. murale**, Wils.
Midsummer Hill, *Ley*. Suckley Hill. Near Tundridge. Leigh Sinton. Worcestershire Beacon. Hanley Child. Walls, Kidderminster. Canal Bridge, Acocks Green.
Limestone rocks and on the mortar of walls. Not common.

[**B. Mildeanum**, Jur.
Wribbenhall, on walls.
This moss has been recorded in error. It is not known in Worcestershire.]

189. **B. argenteum**, Linn.
Footpaths, waste ground, wall-tops, &c. Frequent.
Var. **majus**, B. and S.
Worcester, *Towndrow*. Canal-side near Hewell Grange. Near Wythall.
Var. **lanatum**, B. and S.
Old walls, Alderminster. Seckley Wood. Offmoor. Halesowen Railway.

F. RHODOBRYUM

190. **B. roseum**, Schreb.
Shrawley Wood. Golden Valley. Wyre Forest. Eymore Wood. Ribbesford Wood. Sandbourne. Old Storridge Common. Fenny Rough.
Wet and shady bushy ground, wooded banks, and sandy hedge-banks.

H h

MNIUM, Linn. (emend. B. and S.). (*μνίον*, moss.)

191. **M. affine**, Bland.
Shrawley Wood. Warshill Wood. Canal-side near Alvechurch. Frankley Hill.
Damp woods and bogs. Not uncommon, but the variety is rare.
Var. **elatum**, B. and S.
Stanklin Pool.

192. **M. cuspidatum**, Hedw.
Gladder Brook. Areley Kings. Seckley Wood. Upper Arley. Worcestershire Beacon. Hanley Court.
On the ground in woods and near streams. Rare or overlooked.

193. **M. rostratum**, Schrad.
Shrawley Wood. Bewdley. Warshill Wood. Churchill. Stanklin Pool. Near Arley. Wyre Forest. Southstone Rock. Martley. Newnham Bridge. Hagley. Frankley Hill.
Shady banks and damp rocks. Frequent.

194. **M. undulatum**, Linn.
Bewdley. In fruit near Nash Elm, Upper Arley. Wyre Forest. Martley, in fruit. Menith Wood. Malvern, in fruit, *Lees*. Hopwood Dingle. Lickey Hill.
Shady banks and woods. Frequent. Very rare in fruit.

195. **M. hornum**, Linn.
Banks, rocks, tree-roots, &c., in woods and shady places. Very common.

196. **M. serratum**, Schrad.
By the Severn at Seckley, Blackstone, and Lincombe. Park Attwood. *J. B. Duncan*.
Rocks near streams. Rare.

197. **M. stellare**, Reich.
Shrawley Wood. Churchill. Seckley. Wyre Forest. In fruit, Gladder Brook, *J. B. Duncan*. Old Storridge Common, *Towndrow*. Tundridge, near Suckley. Pensax. Southstone Rock. Woodcote Green. Frankley Hill. Clent.
Shady banks, woods, and on rocks near streams. Not uncommon. Fruit very rare.

198. **M. punctatum**, Linn.
Shrawley Wood. Wyre Forest. Near Upper Arley. Stanklin Pool. Warshill Wood. Old Storridge Common. Pensax. Near Alvechurch. Hopwood Dingle. Frankley Hill.
Damp rocks and shady banks, and in bogs. Frequent.

199. **M. subglobosum**, B. and S.
Rock Coppice, Bewdley. Wyre Forest. Railway - bank near Bewdley. Old Storridge Common. Mathon, now in Herefordshire. Frankley Hill. Acocks Green.
Bogs and springs. Rare.

XIX. FONTINALACEAE

FONTINALIS, Dill. (A word formed from *fons*, a fountain.)

200. **F. antipyretica**, Linn.
Cleeve Prior. Goose Hill, Hanbury. River Severn, Bewdley. Wyre Forest. Canal at Wolverley. Near Malvern Link. Stockton-on-Teme. Barnt Green. River Rea, Moseley.
Attached to stones or wood in streams and pools. Not uncommon. The variety is rare.
Var. **gracilis**, Schimp.
River Severn, Upper Arley.

201. **F. squamosa**, Linn.
River Severn, between Bewdley and Arley.
Streams. Rare.

XX. CRYPHAEACEAE

CRYPHAEA, Mohr. (*κρυφαῖος*, hidden; the capsule being hidden by the involucral leaves.)

202. **C. heteromalla**, Mohr.
Quarries at Broadway Hill. Shipston-on-Stour. Near Upper Arley. On a wall at Shatterford. Cowleigh Park, *Griffiths*. Hollybush Hill, *Lees*.
Trunks of trees. Rare.

XXI. NECKERACEAE

NECKERA, Hedw. (Named after N. J. Necker, a German botanist.)

203. **N. crispa**, Hedw.
Bredon Hill, *Thompson*. Broadway. Wyre Forest. Seckley Wood. Shelsley Walsh. St. Kenelm's.
Calcareous hills and shady rocks. Not common.
Var. **falcata**, Boul.
Copse, Bredon Hill.

N. pumila, Hedw.
For this moss, which is given in *J. of B.*, December 1903, for Worcestershire, Mr. Lees mentions in his Malvern botany that the only station is Brock Hill Wood, Colwall, which locality is in Herefordshire.]

204. **N. complanata**, Hübn.
Cleeve Prior. Harvington. Peopleton. Wyre Forest. North Wood. Gladder Brook. Park Attwood. Shrawley Wood. Suckley. Knightwick. Martley. Rubery. St. Kenelm's.
Rocks, banks, and trunks of trees. Common.

HOMALIA, Brid. (*ὁμαλός*, smooth-surfaced, from the complanate leaves.)

205. **H. trichomanoides**, B. and S.
Tree-trunks and rocks in shady situations. Frequent.

H h 2

XXII. HOOKERIACEAE

PTERYGOPHYLLUM, Brid. (πτέρυξ, a wing, φύλλον, a leaf; from the pinnate arrangement of the leaves.)

206. P. lucens, Brid.
Shrawley Wood. Golden Valley, Bewdley. Wyre Forest. Seckley Wood. North Wood. Ribbesford Wood. Hanley Dingle. Moseley, *Westcott.*
Moist sandy ground in shady woods and dingles. Not uncommon in the north-west of the county.

XXIII. LEUCODONTACEAE

LEUCODON, Schwaeg. (λευκός, white, ὀδούς, a tooth; from the pale peristome.)

207. L. sciuroides, Schwaeg.
Frequent in the Avon district, rare or local in North Worcestershire. Wyre Forest. Dick Brook. Crundalls, Bewdley.
Trunks of trees, occasionally on walls.

PTEROGONIUM, Swartz. (πτέρυξ, a wing, γόνυ, a knee.)

208. P. gracile, Swartz.
Ragged Stone Hill, *Fraser.*
Rocks. Very rare.

ANTITRICHIA, Brid. (ἀντί, opposite, and τρίχια, filaments.)

209. A. curtipendula, Brid.
Malvern, *Lees.*
Rocks and tree-trunks. Very rare.

POROTRICHUM, Brid. (πόρος, a pore, or perforation, and θρίξ, hair; from the perforated processes of the inner peristome.)

210. P. alopecurum, Mitt.
North Wood, Bewdley. Wyre Forest. Seckley Wood. Dunley. Sapey Brook, *Lea.* Leigh Sinton. Hanley Dingle. Knightwick. Frankley. Clent, fruits occasionally.
Shady woods and rocks by waterfalls. Common.

XXIV. LESKEACEAE

LESKEA, Hedw. (Named after N. C. Leske.)

211. L. polycarpa, Ehrh.
Frequent on tree-roots by streams and pools.

Var. paludosa, Schimp.
Pershore. Peopleton. Westwood Park, Droitwich. Stanklin Pool. Seckley Wood. Near Kidderminster.
The variety is a larger lax form, not differing in any essential particular from the type.

ANOMODON, Hook. and Tayl. (ἄνομος, lawless, ὀδούς, a tooth; erroneously supposed to have an abnormal peristome.)

212. A. viticulosus, Hook. and Tayl.
Wolverton, near Worcester. Bredon Hill. Armscote. Aston Magna. Wyre Forest. Westwood Park, Droitwich. Gladder Brook. Lincombe. Habberley Valley. Longdon Marsh. Suckley. Alfrick. Eastham. Stanford-on-Teme. Pensax. Sapey Brook. Clifton-on-Teme.
Rocks and tree-trunks. Fruits occasionally. Not uncommon.

HETEROCLADIUM, B. and S. (ἕτερος, the other, κλάδος, a young shoot, from the dimorphous leaves.)

213. H. heteropterum, B. and S.
Wyre Forest, *Fraser.* Seckley Wood.
Damp shady rocks. Very rare. The variety is a very slender form.

Var. fallax, Milde.
Wyre Forest. North Wood, Bewdley. Seckley Wood. Near Upper Arley.
Less rare than the type.

THUIDIUM, B. and S. (θυία, a juniper or cypress, εἶδος, form.)

214. T. tamariscinum, B. and S.
Shady woods. Common.

215. T. recognitum, Lindb.
Wolverton, near Worcester. Bredon Hill, *H. H. Knight.* Wyre Forest, *J. B. Duncan.*
Banks and shady rocks. Very rare.

XXV. HYPNACEAE

CLIMACIUM, Web. and Mohr. (κλῖμαξ, a staircase, alluding to the structure of the inner peristome.)

216. C. dendroides, Web. and Mohr.
Stanklin Pool, near Kidderminster. Churchill. Blakedown. Malvern Hills. Barnt Green. Rubery.
Damp marshy ground. Not common.

CYLINDROTHECIUM, B. and S. (κύλινδρος, a cylinder, and θηκίον, a little vessel, from the form of the capsule.)

217. C. concinnum, Schp.
Bredon Hill, *H. H. Knight* (January 1909).
Among grass on calcareous soil. Very rare.

CAMPTOTHECIUM, B. and S. (καμπτός, bent, and θηκίον, a little vessel, from the curved capsule.)

218. C. sericeum, Kindb.
Rocks, walls, and tree-trunks. Frequent

219. C. lutescens, B. and S.
Armscote. Broadway Hill. Wyre Forest. Baches Hill, near Kidderminster. Lincombe. Suckley. Martley. Knightwick. Menith Wood. Eastham.
Dry banks and quarries. Frequent on calcareous soil.

220. C. nitens, Schimp.
Near Frankley. *E. Cleminshaw.*
Bogs and marshes. Exceedingly rare. At Frankley this moss occurs on the swampy sides of a railway cutting, and is perhaps now extinct owing to the drainage of the locality.

BRACHYTHECIUM, B. and S. (βραχύς, short, θηκίον, a little vessel, from the short capsule.)

221. B. glareosum, B. and S.
Whittington, near Worcester. Tardebigge. Churchill. Wyre Forest. Dunley. Suckley. Malvern Hills. Pensax. Abberley Hills. Menith Wood. Eardiston. California, near Harborne.
Calcareous banks, quarries, and sometimes on stiff clay. Not common.

222. B. albicans, B. and S.
Habberley Valley, fruiting.
Dry sandy ground. Local throughout the county. Very rare in fruit.

223. B. salebrosum, B. and S.
Croome Park. Habberley Valley. Malvern, *Lees.* Suckley. Knightwick. Rubery. Twylands Wood, Frankley.
Stones, sandy banks, and tree-roots. Rare. The variety occurs in damp clayey fields, and is rare.

Var. palustre, Schimp.
Bredon. Kempsey Green. North Wood, Bewdley. Wyre Forest. Seckley Wood. Malvern Link, *Towndrow.* Near Abberley Hall.

224. [B. campestre, B. and S.
Seckley Wood, near Upper Arley.
This is a doubtful species, which only differs from the preceding in the slightly rough seta.]

225. B. rutabulum, B. and S.
On earth, walls, rocks, or trees. Very common.

Var. robustum, B. and S.
Defford Common. Kempsey Green. Pensax. Alvechurch. Frankley.

226. B. rivulare, B. and S.
Blockley. Wolverton. Upper Arley. Dick Brook. Wyre Forest. Gladder Brook. North Wood. Pensax. Southstone Rock. Woodcote Green. Offmoor. Fenny Rough.
Rocks and stones by the margins of streams. Not common. The variety occurs on marshy ground, and is very rare.

Var. chrysophyllum, Bagnall.
Lane to Upper Arley. Wyre Forest.

227. B. velutinum, B. and S.
Tree-trunks, rocks, walls, and banks. Frequent.

228. B. populeum, B. and S.
Wyre Forest. Habberley Valley. Seckley Wood. Park Attwood. Shatterford. Cowleigh Park, *Lees.* Old Storridge Common. Malvern Hills. Menith Wood, Pensax. Beoley. Frankley Hill. Selly Oak. Dudley Castle.
Stones, rocks, and walls. Not common.

229. B. plumosum, B. and S.
Gladder Brook. Wyre Forest. Dick Brook. North Wood. Cowleigh, *Towndrow.* Old Storridge Common. Pensax. Offmoor Wood. Frankley.
Rocks in and near streams. Not common. A var. *homomallum*, which is a form only, not usually considered worthy of mention, has been recorded from Old Storridge Common.

230. B. caespitosum, Dixon.
Feckenham. Peopleton. Kempsey Common. Near Shrawley. Wyre Forest. Westwood Park. Hadley, near Droitwich. River Severn, Bewdley to Arley. Stanford-on-Teme.
Rocks near streams, where liable to be submerged in floods. Not common.

231. B. illecebrum, De Not.
Harvington. Cleeve Hill. Blackstone Rock. Habberley Valley. Blakeshall Common. Lincombe. Wolverley. Malvern Wells. Malvern Link. Pensax.
On the ground and on rocks. Rare.

232. B. purum, Dixon.
Broadway, in fruit. Stagbury and Wyre Forest, in fruit. Martley, in fruit. Near Redditch. Frankley. Clent Hills.
Banks and grassy places. Very common. Fruit rare.

HYOCOMIUM, B. and S. (ὑοκόμος, a lover of moisture, alluding to the habitat of *H. flagellare.*)

233. H. flagellare, B. and S.
Cleeve Mill, near Cleeve Prior.
Wet rocks by streams. Very rare.

EURHYNCHIUM, B. and S. (emend. Milde). (εὖ, well, ῥυγχίον, a little beak. The capsule lid has a long beak.)

234. E. piliferum, B. and S.
Damp shady banks. Frequent in woods, &c.

235. E. crassinervium, B. and S.
Shrawley Wood. Seckley Wood. Gladder Brook. Dick Brook. Hartlebury. Near St. Anne's Well.
Boulders and rocks near streams. Sometimes on walls in shady situations. Rare.

236. E. speciosum, Schimp.
Canal-side near Alvechurch.
On stones near water. Very rare.

237. **E. praelongum**, Hobk.
　　Woods, banks, and hedgerows.　Very common.　The variety occurs in woods, and is rare.
　　Var. **Stokesii**, Brid.
　　　Wyre Forest.　Seckley Wood.　Arley Wood.　Cowleigh Park, *Towndrow.*　Pensax.　Frankley Hill.　World's End, near Harborne.

238. **E. Swartzii**, Hobk.
　　In similar situations to the last.　Common.

239. **E. abbreviatum**, Schimp.
　　Old Storridge Common.
　　Shady woods.　Rare.

240. **E. pumilum**, Schimp.
　　Peopleton.　Rous Lench.　Coles Green, near Worcester.　Wyre Forest.　Lincombe.　Bewdley.　Southstone Rock.　Leigh Brook.　Menith Wood, Pensax.　Barnt Green.
　　Damp shady rocks.　Rather rare.

241. **E. curvisetum**, Husn.
　　Rous Lench.　Ribbesford Wood.
　　Rocks and stones near streams.　Very rare.

242. **E. Teesdalii**, Schimp.
　　Gladder Brook.　Dick Brook.　Upper Arley.　Seckley Wood, *Fraser.*　North Wood.　Eastham.　Hanley Dingle.　Shelsley Walsh.
　　Rocks and stones near streams.　Rare.

243. **E. tenellum**, Milde.
　　Bredon Hill.　Wyre Forest.　Trimpley Green.　Seckley Wood.　Park Attwood.　Lincombe.　Western slopes of Malvern, *Lees.*　Martley Quarries.　Southstone Rock.　Eastham.
　　Rocks and stones, chiefly in calcareous places.　Not common.

244. **E. myosuroides**, Schimp.
　　Seckley Wood.　Wyre Forest, Tundridge.　Alfrick.　Offmoor.　Halesowen.
　　Rocks and tree-roots.　Frequent.　A variety *rivulare* was recorded, *J. of B.,* December 1903, from Ockeridge Wood ; Wyre Forest, on wet rocks ; and Seckley Wood, on stones in streams ; but certainly in error.　Its usual habitat is by mountain streams.

245. **E. myurum**, Dixon.
　　Tree-roots, rocks and banks in shady situations.　Frequent.

246. **E. striatum**, B. and S.
　　Frequent.　In woods and on hedgebanks.

247. **E. rusciforme**, Milde.
　　Rocks and stones in and near streams.　Frequent.　The variety is rare.
　　Var. **inundatum**, Brid.
　　Stream on Frankley Hill.

248. **E. murale**, Milde.
　　Blockley.　Broadway.　Bredon.　Aston Magna.　Dick Brook.　Bewdley.　Hartlebury.　Tundridge.　Alfrick.　Eastham.
　　Rocks and walls in shady situations.　Not common.
　　Var. **julaceum**, B. and S.
　　Walls, Bockleton near Tenbury.

249. **E. confertum**, Milde.
　　Besford.　Bewdley.　Shatterford.　Holt Fleet.　High Habberley.　Malvern Link, *Towndrow.*　Woodcote Green.　Alvechurch.
　　Stones, walls, tree-trunks, and banks.　Not uncommon.

250. **E. megapolitanum**, Milde.
　　Dowles Brook.　Upper Arley.　Frankley Hill.
　　Stony and sandy ground.　Rare.

PLAGIOTHECIUM, B. and S.　(πλάγιος, slanting, θηκίον, a little box, from the inclined capsule.)

251. **P. depressum**, Dixon.
　　Stonework of a drain near Shatterford.　North Wood.　Wyre Forest.
　　On stones and rocks in shady situations.　Rare.

252. **P. elegans**, Sull.
　　Shrawley Wood.　North Wood.　Bewdley.　Wyre Forest.　Arley Wood.　Habberley Valley.　Hanley Dingle, Stanford-on-Teme.　Malvern.　Lickey Hill.　Frankley.
　　On sandstone rocks and peaty soil in woods.　Not common.

253. **P. denticulatum**, B. and S.
　　Woods, frequent.
　　Var. **aptychus**, Spruce.
　　Bewdley.　Seckley Wood.　Cowleigh.　Pensax.　Beoley.　Frankley.
　　[Var. **majus**, Boul.
　　Trench Woods.　Highter Heath, near Kings Norton.　Quarry at Halesowen.
　　These records are doubtful.]

254. **P. silvaticum**, B. and S.
　　North Wood, Bewdley.　Seckley Wood.　Shatterford.　Eymore Wood.　Wyre Forest.　Pensax Wood.　Shortwood Coppice, near Alvechurch.
　　Woods and shaded rocks and banks.　Rather rare.

255. **P. undulatum**, B. and S.
　　Near Ombersley.　Seckley Wood.　Arley Wood.　Wyre Forest.　Foxholes and Park Attwood, near Kidderminster.　Lickey Hill.　Romsley Hill.
　　Damp woodland banks.　Not common.

256. **P. latebricola**, B. and S.
　　Near Ombersley.　Hurcott Pool, near Kidderminster.　Wyre Forest.　Seckley Wood.　Warshill Wood.　Near Franche.　Twylands Wood, Frankley.　Offmoor Wood.　Clent.

　　At the base of trees in woods and at roots of sedges by pools.　Rare.　The form *gemmascens*, Ryan and Hagen, with numerous cylindrical green jointed gemmae from the tips of the leaves or axillary, occurs in the county.

AMBLYSTEGIUM, B. and S.　(ἀμβλύς, blunt, στέγη, a roof, from the form of the capsule lid.)

257. **A. compactum**, Aust.
　　Blackstone, near Bewdley.　Park Attwood, *J. B. Duncan.*
　　Shady recesses of calcareous rocks.　Very rare.

258. **A. serpens**, B. and S.
　　On the ground, stones, tree-stumps, &c.　Common.

259. **A. Juratzkanum**, Schimp.
　　Coles Green, near Worcester.　Alfrick.　Barnt Green.
　　Damp walls, tree-roots, and rotting wood.　Rare.

260. **A. varium**, Lindb.
　　Stonebow Bridge, Peopleton.　Stanklin Pool, near Kidderminster.　Near Blackstone, Bewdley.　Near Alvechurch.　Near Bromsgrove.　Wythall Heath and West Heath, near Kings Norton.
　　On the ground, tree-roots and stumps in moist places.　Rare.

261. **A. irriguum**, B. and S.
　　Peopleton.　Coles Green, near Worcester.　Dick Brook.　Gladder Brook.　Kidderminster.　Leigh Sinton.　Southstone.　Pedmore.　Alvechurch.　Kings Norton.　Twylands Wood, Frankley.
　　Stones and rocks in streams.　Rare.

262. **A. fluviatile**, B. and S.
　　River Severn near Bewdley and Seckley Wood.
　　Stones and rocks in River Severn.　Sometimes attached to floating stages.　Rare.

263. **A. filicinum**, De Not.
　　Bewdley.　Wyre Forest.　Gladder Brook.　Shatterford.　Old Storridge, *Towndrow.*　Longdon Marsh.　Castlemorton.　Southstone.　Tardebigge Canal.　Lickey Hill.　Hopwood Dingle.　Frankley Hill.
　　Swampy ground, stream-sides, damp rocks and stones.　Common.
　　Var. **Vallisclausae**, Dixon.
　　Wyre Forest.　Southstone Rock.　Great Cowleigh.　Canal-side, Hopwood.　Near Rubery.
　　In calcareous springs.　Rare.
　　Var. **trichodes**, Brid.
　　Wyre Forest.　Frankley Hill.

HYPNUM, Linn. (emend. B. and S.).　(ὕπνον, a kind of moss.)

A. CAMPYLIUM

264. **H. riparium**, Linn.
　　Pershore.　Wyre Forest.　Churchill.　Stanklin Pool.　Stanford Bridge.　River Severn.　River Rea near Shakenhurst.　Tardebigge.　Frankley Hill.

　　Marshes, margins of pools on tree - roots and stones.　Not common.
　　Var. **longifolium**, Schimp.
　　Gorse Hill Green, Hanbury.　Holt Fleet.　Stanford Bridge.
　　Var. **splendens**, De Not.
　　River Severn, near Upper Arley.　Knightwick.　Pitts Wood, near Harborne.

265. **H. polygamum**, Schimp.
　　Stanklin Pool, near Kidderminster.
　　Marshes.　Very rare.

266. **H. stellatum**, Schreb.
　　Defford.　Wyre Forest.　Near Arley.　Castlemorton Common.　Old Storridge Common.　Frankley Hill.　Offmoor Wood.
　　Bogs and marshes.　Not common.
　　Var. **protensum**, Röhl.
　　Trench Woods.　Wyre Forest.　Old Storridge Common.　Suckley.　Frankley.　Offmoor Wood.
　　Damp calcareous rocks.

267. **H. chrysophyllum**, Brid.
　　Bredon Hill.　Ockeridge Wood.　Wyre Forest.　Dick Brook.　Dunley.　Ravenshill.　Alfrick.　Eastham.　Menith Wood, Pensax.　Martley.　Frankley Hill.
　　On calcareous ground.　Not common.

268. **H. hispidulum**, Brid.
　　Var. **Sommerfeltii**, Myr.
　　Limestone rocks, Broadway, *E. Cleminshaw.*
　　Very rare.

B. HARPIDIUM

269. **H. aduncum**, Hedw. non Linn.
　　Wyre Forest.　Billesley Common, Moseley.
　　Marshes and pools.　Rare.
　　Var. **Kneiffii**, Schimp.
　　Kyre Common, near Tenbury.
　　Var. **paternum**, Sanio.
　　Clay-pit, Armscote.

270. **H. Sendtneri**, Schimp.
　　St. Kenelm's.
　　Bogs and marshes.　Rare.

271. **H. fluitans**, Linn.
　　Wyre Forest.　Knightwick.　Bittell Reservoir.　Alvechurch.
　　Bogs and pools.　Rare.
　　Var. **Jeanbernati**, Ren.
　　Dowles Brook.　Wyre Forest.
　　Forma **tenella**.
　　Dowles Brook.　Wyre Forest.

272. **H. exannulatum**, Gümb.
    Wyre Forest. Stream by Twylands Wood, Frankley.
    Bogs. Rare.
  Var. **pinnatum**, Boul.
    Near Shelsley Beauchamp.
  Var. **falcifolium**, Ren.
    Rock Coppice, Bewdley. Stream by Twylands Wood, Frankley.

273. **H. uncinatum**, Hedw.
    Near Dowles Brook. Wyre Forest, 1890. Malvern, *Lees*. Moseley, *Webb*.
    Shady rocks. Rare.

274. [**H. vernicosum**, Lindb.
    Dowles Brook. Wyre Forest.
    Bogs. Very rare. A doubtful record.]

275. **H. revolvens**, Swartz.
    Seckley Wood, near Bewdley. Wyre Forest.
    Bogs. Rare. The variety, *Cossoni*, Ren., is mentioned, *J. of B.*, December 1903, from Beaucastle, Wyre Forest.

276. **H. intermedium**, Lindb.
    Wyre Forest. Hartlebury Common.
    Bogs. Very rare.

277. **H. commutatum**, Hedw.
    Blockley. Wyre Forest. Seckley Wood. North Wood. Dick Brook. Nash Elm, Arley. Old Storridge Common. Martley. Stanford-on-Teme. Knightwick. Clifton-on-Teme. Hopwood Dingle. Frankley Hill.
    Calcareous bogs and springs. Not uncommon.

278. **H. falcatum**, Brid.
    Wyre Forest. Castlemorton Common. Hopwood Dingle.
    In similar situations to the preceding. Rare.
  [Var. **gracilescens**, Schimp.
    Wyre Forest. Stanford-on-Teme.
    This record is too doubtful to give this moss a status in Worcestershire.]
  Var. **virescens**, Schimp.
    Blockley. Stream in Wyre Forest. Southstone Rock.
    The Worcestershire plants are not well marked, and might be more properly described as 'approaching var. *virescens*'.

C. DREPANIUM

279. **H. cupressiforme**, Linn.
    Stones, rocks, walls, tree-trunks, and on grassy or heathy banks. Very common.
  Var. **resupinatum**, Schimp.
    Defford. Walls, Croome Park. Ribbesford. Seckley Wood. Wyre Forest.

  Var. **filiforme**, Brid.
    Bredon Hill. Croome Park. Whittington. Wyre Forest. Seckley Wood. Pensax.
  Var. **minus**, Wils.
    Shrawley Wood.
  Var. **ericetorum**, B. and S.
    Kempsey Common. Ribbesford Wood. North Wood, Bewdley. Wyre Forest. Hartlebury Common. Ravenshill. Pensax Wood. Lickey Hill. Randans Wood. Pepper Wood.
    This variety is common on heathy ground.
  Var. **tectorum**, Brid.
    Bredon Hill. Trench Woods. Kempsey Common. Old Storridge Common. Abberley Hill. Offmoor Wood. Pitcher Oak.
  Var. **elatum**, B. and S.
    Kempsey Common. Pensax Wood.

280. **H. Patientiae**, Lindb.
    Near Pershore. Trench Woods. Shrawley Wood. Ribbesford Wood. North Wood. Seckley. Habberley Valley. Wyre Forest. Alfrick. Abberley Hill. Kyrebatch, Tenbury. Alvechurch. Hagley Wood. Frankley.
    Clayey ground in woods. Not common.

281. **H. molluscum**, Hedw.
    Frequent on limestone soil and on rocks near streams.
  Var. **erectum**, *Bagnall*, is given in the list in *J. of B.*, December 1903, from Crews Hill, Alfrick.
  Var. **condensatum**, Schp.
    Wyre Forest, *J. B. Duncan*.
    Rocks by streams. Very rare.

D. LIMNOBIUM

282. **H. palustre**, Huds.
    Wyre Forest. Seckley Wood. North Wood. Stream at Pensax. Twylands Wood, Frankley. Offmoor Wood. Bell End. Illey Mill, near Halesowen.
    Stones and rocks in or near streams. Not common.
  Var. **hamulosum**, B. and S.
    Dowles Brook, Wyre Forest. Stream on Old Storridge Common. Stream at Pensax. Twylands Wood. Frankley.
  Var. **subsphaericarpon**, B. and S.
    North Wood, Bewdley.

283. **H. scorpioides**, Linn.
    Feckenham Bog, *Purton*. Moseley Bog, *Webb*.
    Bogs. Very rare. It is probably extinct.

E. CALLIERGON

284. **H. stramineum**, Dicks.
    Bilberry Hill, 1870.
    Bogs. Very rare.

285. **H. cordifolium**, Hedw.
    Stourport. Stanklin Pool, near Kidderminster. California.
    Marshes and pools. Rare.

286. **H. giganteum**, Schimp.
    Stanklin Pool. Railway cutting, Acocks Green, 1871.
    Marshes and pools. Very rare.

287. **H. cuspidatum**, Linn.
    Marshes and wet ground. Very common.

288. **H. Schreberi**, Willd.
    Wyre Forest, in fruit, *Webb*.
    Heaths and woods. Frequent.

HYLOCOMIUM, B. and S. (ὑλόκομος, an inhabitant of woods.)

289. **H. splendens**, B. and S.
    Heaths and woods. Frequent.

290. **H. brevirostre**, B. and S.
    Trench Woods. Wyre Forest. Seckley Wood. Malvern Hill, *Lees*.
    Among stones and boulders in shady woods. Rare.

291. **H. loreum**, B. and S.
    Monk Wood, *Thompson*. Ribbesford Wood. Gladder Brook. Wyre Forest. Eymore Wood. Coldridge Wood. Pensax Dingle. Worcestershire Beacon. Cowleigh Park, *Lees*.
    Woods. Not common.

292. **H. squarrosum**, B. and S.
    Grassy banks. Very common.
  Var. **calvescens**, Hobk.
    Ockeridge Wood. Wribbenhall. Shatterford. Old Storridge Common. Offmoor Wood.

293. **H. triquetrum**, B. and S.
    Ankerdine, in fruit. Sapey Brook, in fruit. Wyre Forest, in fruit. Gladder Brook, in fruit. Clent, in fruit.
    Woods and banks. Very frequent. Fruit not common.

HEPATICAE

I. MARCHANTIALES

RICCIACEAE

RICCIA Linn. (Named after Ricci, an Italian Botanist.)

1. **R. glauca** Linn.
    Shrawley. Near Churchill. Trimpley. Park Attwood. Lickey Hill. Clent. Dick Brook.
    Fallow fields on clayey soil. Locally abundant.

2. **R. sorocarpa** Bisch.
    Fields at Hoarstone, Bewdley, and near Hartlebury, *J. B. Duncan*. Fallow fields. Rare.

3. **R. fluitans** Linn.
    Longdon, *H. H. Knight*.
    Floating in ponds amongst Lemna. Very rare.

MARCHANTIACEAE

TARGIONIA Linn. (Named after an Italian botanist, Targioni-Tozetti.)

4. **T. hypophylla** Linn.
    Moist sandstone rocks, Habberley, *Lees*. Habberley Valley. Near Blakeshall. Malvern Hill.
    Dry sandstone rocks. Rare.

REBOULIA Raddi. (Named after Reboule, a French naturalist.)

5. **R. hemisphaerica** (Linn.) Raddi.
    Cookley Wood, *Lees*. Shrawley Wood. Near Habberley Valley. Malvern, *Lees*. Near Hagley.
    Sandstone rocks and sandy banks. Rare.

CONOCEPHALUM Neck. (κῶνος, a cone, κεφαλή, the head.)

6. **C. conicum** (Linn.) Dum.
    Frequent on wet rocks and damp shady banks.

LUNULARIA Adans. (From *lunula*, a little moon, in allusion to the crescent-shaped patches of green gemmae on the frond.)

7. **L. cruciata** (Linn.) Dum.
    Frequent, and always barren. On rocks by streams, damp shady paths in gardens, neglected flower-beds, and often very abundant in greenhouses. It is a Mediterranean species, and may not be native in Britain, and possibly has been introduced with garden plants, as it has been into other North European countries. The distribution of this plant was treated of by Mr. S. M. Macvicar, *J. of B.*, December 1908.

MARCHANTIA Linn.   (Named in honour of Marchant.)

8. **M. polymorpha** Linn.
  Wyre Forest.   Seckley Wood.   Bewdley.   Malvern, *Lees*.   Old
  Storridge Common.
  Damp earth and rocks.   Not unfrequent.

II. JUNGERMANNIALES

JUNGERMANNIACEAE ANACROGYNAE

ANEURA Dum.   (ἀ, without, νεῦρον, a nerve.)

9. **A. pinguis** (Linn.) Dum.
  Seckley Wood.   Wyre Forest.   Wet rocks, railway cutting near
  Bewdley.   Valley of the Teme, *Lees*.   Malvern Hill.   Lickey
  Hill.   Gannow Green.   Frankley Hill.   Hagley Wood.
  Boggy places, wet rocks and margins of rills.   Not common.

10. **A. multifida** (Linn.) Dum.
  Chatley Green.   Valley of the Teme, *Lees*.   Malvern Hill.   Twy-
  lands Wood.   Offmoor.
  Wet loamy soil in woods.   Rare.

11. **A. sinuata** (Dicks) Limpr.
  Seckley Wood.
  Wet rocks near streams.   Very rare.

METZGERIA Raddi.   (Named after Johann Metzger, a German botanist.)

12. **M. furcata** (Linn.) Lindb.
  Wyre Forest.   North Wood.   Seckley Wood.   Blakeshall.   Malvern
  Hills.   Shelsley Walsh.   Offmoor Wood.
  Trunks of trees and on rocks.   Not uncommon.

13. **M. conjugata** Lindb.
  Wyre Forest, *J. B. Duncan*.
  Shady rocks.   Very rare.

[**M. hamata** Lindb.
  This very rare plant, found on rocks near mountain streams, has
  been recorded from Hanley Dingle, but probably in error.]

PELLIA Raddi.   (πηλός, clay, from the usual habitat.)

14. **P. endiviaefolia** (Dicks) Dum.
  Cleeve Prior.   Shrawley Wood.   Wyre Forest.   Seckley Wood.
  Near Wolverley and Blakeshall.   Offmoor Wood.   Frankley
  Hill.   Chaddesley Wood.
  Wet banks and rocks.   Not common.

15. **P. epiphylla** (Linn.) Dum.
  Shrawley Wood, *Lees*.   Wyre Forest.   Habberley Valley.   North
  Wood.   Seckley Wood.   Teme Valley, *Lees*.   Old Storridge
  Common.
  Wet shady rocks and banks.   Common.

BLASIA Linn.   (A genus dedicated to Blasius.)

16. **B. pusilla** Linn.
  Seckley Wood.   Near railway between Bewdley and Kidder-
  minster.
  Wet clayey banks.   Rare.

FOSSOMBRONIA Raddi.   (In honour of Fossombroni.)

17. **F. Wondraczekii** (Corda) Dum.
  Trench Woods.   Ockeridge Wood.   Wyre Forest.
  On the ground in woods.   Very rare.

18. **F. pusilla** (Linn.) Dum.
  Wyre Forest.   Shrawley Wood.   Seckley.   Blakeshall Common.
  Hoarstone.   Hartlebury.
  Fallow fields and bare ground in woods.   Not uncommon.

19. **F. caespitiformis** De Not.
  Wyre Forest.   Offmoor Wood.
  Bare earth in woods.   Very rare.

JUNGERMANNIACEAE ACROGYNAE

MARSUPELLA Dum.   (Diminutive from *marsupium*, a pouch.)

20. **M. emarginata** (Ehrh.) Dum.
  Wyre Forest.   Seckley Wood, *J. B. Duncan*.
  Wet rocks by streams ; sometimes on clayey soil.   Very rare.

NARDIA S. F. Gray.

21. **N. scalaris** (Schrad.) Gray.
  Butlers Hill, Newbold.   Wyre Forest.   Habberley Valley.   Hartle-
  bury Common.   Old Storridge Common.
  Damp banks, heaths, and rocks.   Not uncommon.

22. **N. hyalina** (Lyell) Carr.
  Wyre Forest.   Seckley Wood, *J. B. Duncan*.
  Damp shady banks, near streams.   Very rare.

APLOZIA Dum.   (ἁπλόος, simple, ὄζος, a bough.)

23. **A. autumnalis** (DC.) Schffn.
  Skeys Wood ; Seckley Wood, *J. B. Duncan*.
  Dry rocks in shady situations, sometimes at base of tree-trunks.

24. **A. crenulata** (Sm.) Dum.
  Newbold.   Trench Woods.   Ockeridge Wood.   Wyre Forest.   Arley
  Wood.   Blakeshall Common.   Lickey Hill.   Offmoor Wood.
  Wythall Heath.   Hagley Wood.
  Damp sandy soil in woods and on heaths.   Not uncommon.
  Var. gracillima Sm.
  Wyre Forest.   Seckley Wood.   Arley Wood.   Shrawley Wood.
  Hartlebury Common.   Old Storridge Common.   Lickey Hill.
  Hagley Wood.   Wythall Heath.

25. **A. sphaerocarpa** (Hook.) Dum.
  Newbold.   Shrawley Wood.   Wyre Forest.   Seckley Wood.   Arley
  Wood.   The Heath, Bewdley.   Old Storridge Common.   Lickey
  Hill.   Offmoor Wood.   Bell End.   Hagley.
  Banks of streams and damp soil in woods.   Rather rare.

26. [**A. cordifolia** (Hook.) Dum.
  River Severn near Seckley.
  Mountain springs and rills.   A doubtful record.]

27. **A. riparia** (Tayl.) Dum.
  Wyre Forest.   River Severn near Seckley.   Dowles Brook.   Seckley
  Wood.   Twylands Wood.   Stream near Hagley.
  Rocks and stones in and near streams.   Not common.

28. **A. pumila** (With.) Dum.
  Wyre Forest.   Ribbesford Wood.   Arley Wood.
  Shady rocks near streams.   Very rare.

LOPHOZIA Dum.   (λόφος, a plume, ὄζος, a bough.)

29. **L. inflata** (Huds.) Howe.
  Wyre Forest.   Hartlebury Common.   Upper Arley.   Blakeshall
  Common.   Railway cuttings, Bewdley.   Old Storridge Common.
  Lickey Hills.   Wythall Heath.   Hagley Wood.
  Wet peaty soil, sometimes on heathy ground and dry sandstone
  rocks.

30. **L. turbinata** (Raddi) Steph.
  Cleeve Prior.   Trench Woods.   Wyre Forest.   With perianths,
  Upper Arley.   Stockton-on-Teme.   Pensax.   Twylands Wood,
  Frankley.
  Damp calcareous rocks.   Not common.

31. **L. ventricosa** (Dicks.) Dum.
  Wyre Forest.   Seckley Wood.   Habberley Valley.   Blakeshall
  Common.   Malvern Hills.   Old Storridge Common.   Lickey
  Hills.   Wythall Heath.
  Rocks and banks in woods and on heathy ground.   Not uncom-
  mon.

32. **L. bicrenata** (Schmid.) Dum.
  Blakeshall Common.   Warshill.   Wyre Forest.   Hartlebury Com-
  mon.   Spring Grove, Bewdley.
  Sandy heaths.   Rare.

33. **L. excisa** (Dicks.) Dum.
  Wyre Forest.   Sandbourne, Bewdley.   Blakeshall Common.   Ravens-
  hill Wood, Suckley.
  Sandy ground in woods and on heaths.   Rare.

34. **L. incisa** (Schrad.) Dum.
  Wyre Forest.
  Amongst mosses on shady rocks.   Very rare.

35. **L. quinquedentata** (Huds.) Cogn.
  Wyre Forest ; Seckley Wood, *J. B. Duncan*.
  Amongst mosses on shady rocks.   Very rare.

36. **L. barbata** (Schmid.) Dum.
  Rocks near Churchill (Kidderminster).
  Amongst mosses on rocks or heathy banks.   Very rare.

37. **L. gracilis** (Schleich.) Steph.
  Habberley Valley.   Blakeshall Common.   Spring Grove and Black-
  stone, Bewdley.
  Sandstone rocks.   Not uncommon in the Kidderminster district.

PLAGIOCHILA Dum.   (πλάγιος, slanting, χεῖλος, a lip.)

38. **P. spinulosa** (Dicks.) Dum.
  Wyre Forest.   Gladder Brook.   Sandstone rocks, Upper Arley.
  Shady rocks, amongst mosses.   Very rare.

39. **P. asplenioides** (Linn.) Dum.
  Wyre Forest.   Seckley Wood.   Shrawley Wood.   Arley Wood.   Old
  Storridge Common.   Lickey Hill.
  Rocks and shady banks.   Common.
  Var. major Nees.
  Seckley Wood.   Wyre Forest.   North Wood.
  Damp shady banks in woods.   Not common.

LOPHOCOLEA Dum.   (λόφος, a plume, κολεός, a sheath.)

40. **L. bidentata** (Linn.) Dum.
  Trench Woods.   Common about Evesham.   Wyre Forest.   Shrawley
  Wood.   Seckley Wood.   Lickey Hills.   Twylands Wood.
  Shady banks and woods.   Very common.

41. **L. cuspidata** Limpr.
  Common about Evesham.   Wyre Forest.   Seckley Wood.   Shrawley
  Wood.   Coles Green, near Worcester.
  Shady banks and rocks.   Common.

42. **L. heterophylla** (Schrad.) Dum.
  Common about Evesham.   Wyre Forest.   Warshill.   Seckley Wood.
  Stockton-on-Teme.   Wythall Heath.
  Tree-trunks, rotting wood, damp rocks and banks.

CHILOSCYPHUS Corda.   (χίλιοι, a thousand, σκύφος, a cup.)

43. **C. polyanthos** (Linn.) Corda.
  Wyre Forest.   Seckley Wood.   North Wood.   Old Storridge Com-
  mon.   Lickey Hills.   Frankley Hill.   Wythall Heath.
  Rocks in and near streams.   Not common.
  Var. rivularis Nees.
  Stream, Frankley Hill.
  In streams.   Very rare.

SACCOGYNA Dum. (σάκκος, a bag, γυνή, a woman.)

44. **S. viticulosa** (Sm.) Dum.
Wyre Forest. Seckley Wood. Gladder Brook. North Wood. Twylands Wood, near Frankley.
Shady mossy banks near streams. Rare.

CEPHALOZIA Dum. (κεφαλή, the head, ὄζος, a bough.)

45. **C. bicuspidata** (Linn.) Dum.
Wyre Forest. Habberley Valley. Park Attwood. Hartlebury Common.
Rocks, stones, and peaty soil in woods and on heaths. Frequent.

46. **C. connivens** (Dicks.) Spruce.
Malvern, *Lees*. Lickey Hill. Wythall Heath.
Wet heaths amongst mosses. Very rare.

47. **C. lunulaefolia** Dum.
Seckley Wood. Arley Wood. Foxholes, near Kidderminster. Lickey Hill. Wythall Heath.
Damp rocks and heaths. Very rare.

CEPHALOZIELLA Schffn. (Diminutive of the name of the foregoing genus.)

48. **C. byssacea** (Roth.) Warnst.
Trench Woods. Wyre Forest. North Wood. Gladder Brook. Habberley Valley. Old Storridge Common. Ravenshill. Pensax. Offmoor Wood. Bell Heath.
Dry rocks and heathy ground. Frequent.

49. **C. Limprichtii** Warnst.
Bredon Hill, February 1908, *H. H. Knight*; confirmed by *Mr. S. M. Macvicar*, January 1909.
Calcareous soil. Very rare.

ODONTOSCHISMA Dum. (ὀδούς, a tooth, σχίσμα, a cleft.)

50. **O. sphagni** (Dicks.) Dum.
Hartlebury Common, *J. B. Duncan*.
Wet bogs amongst *Sphagnum*. Very rare.

KANTIA S. F. Gray.

51. **K. trichomanis** (Linn.) Gray.
Wyre Forest. Hartlebury Common. Foxholes, near Kidderminster. Shelsley Walsh. Pensax. Lickey Hill.
Damp rocks and woodland banks. Not uncommon.

52. **K. Sprengelii** (Mart.) Pears.
Wyre Forest. Foxholes, near Kidderminster, *J. B. Duncan*.
Wet rocks. Rare.

53. **K. arguta** (Nees and Mont.) Lindb.
Wyre Forest. Seckley Wood. Habberley Valley. Arley Wood. Twylands Wood. Offmoor Wood. Hagley Wood.
Damp clayey soil in woods. Rather rare.

BAZZANIA S. F. Gray.

54. **B. trilobata** (Linn.) Gray.
Wyre Forest. Seckley Wood. Gladder Brook.
Damp mossy banks in woods. Very rare.

LEPIDOZIA Dum. (λεπίς, a scale, ὄζος, a bough.)

55. **L. reptans** (Linn.) Dum.
Trench Woods. Wyre Forest. Warshill. Shrawley Wood. Dunley. Seckley Wood. Arley Wood.
Shady rocks and banks. Not uncommon.

BLEPHAROSTOMA Dum. (βλέφαρον, the eyelids, στόμα, the mouth.)

56. **B. trichophyllum** (Linn.) Dum.
Wyre Forest. Seckley Wood.
Shaded mossy rocks. Very rare.

PTILIDIUM Nees. (πτίλον, a feather.)

57. **P. ciliare** (Linn.) Hampe.
Habberley Valley. Hartlebury Common. Blakeshall. Malvern Hills. Walton Hill. Heathy footways near Kings Heath.
Heaths. Rare.

TRICHOCOLEA Dum. (θρίξ, the hair, κολεός, a sheath.)

58. **T. tomentella** (Ehrh.) Dum.
Wyre Forest. Seckley Wood. High Grove Hill and the Gullet, Malvern Hills, *Lees*.
Woodland bogs. Rare.

DIPLOPHYLLUM Dum. (διπλόος, double, φύλλον, a leaf.)

59. **D. albicans** (Linn.) Dum.
Wyre Forest. Habberley Valley. Hartlebury Common. Blakeshall Common. Shrawley Wood. North Wood. Lickey Hills. Bell Heath.
Rocks and banks in woods and on heaths. Very common on siliceous formations, but rare on calcareous ground.

60. **D. obtusifolium** (Hook.) Dum.
Habberley Valley, *J. B. Duncan*.
Sandstone rocks. Very rare.

SCAPANIA Dum. (σκαπάνη, a hoe.)

61. **S. compacta** (Roth.) Dum.
Seckley Wood. Wassall Wood. Sandstone rocks near Bewdley and Kidderminster. Old Storridge Common.
Sandstone rocks and on boulders in woods. Rare.

62. [**S. aspera** Bernet.
Wyre Forest. Shrawley Wood. Blakeshall Common.
Dry calcareous rocks. Very rare. A doubtful plant.]

63. **S. gracilis** (Lindb.) Kaal.
Wyre Forest. Seckley Wood. Hartlebury Common. Blakeshall Common. Offmoor Wood. Cofton Hackett.
Rocks and boulders in woods, occasionally amongst heather on heaths. Not common.

64. **S. nemorosa** (Linn.) Dum.
Trench Woods. Wyre Forest. Seckley Wood. Shrawley Wood. Arley Wood. Blakeshall Common. Old Storridge Common. Offmoor Wood. Hagley Wood. Lickey Hills.
Shaded banks and rocks. Not common.

65. **S. undulata** (Linn.) Dum.
Wyre Forest. Seckley Wood. West Heath. Lickey Hills.
Wet rocks. Very rare.

66. **S. irrigua** (Nees) Dum.
Wyre Forest. Seckley Wood. Arley Wood. Spring Grove, Bewdley. Old Storridge Common. Lickey Hills. Wythall Heath. Hagley Wood.
Wet sandy ground, and on heathy soil in woods. Not common.

67. **S. curta** (Mart.) Dum.
Trench Woods. Wyre Forest. North Wood. Seckley Wood. Rhydd Covert, Bewdley. Old Storridge Common. Bell Heath.
Sandy soil and borders of paths in woods. Rare.

68. **S. umbrosa** (Schrad.) Dum.
Habberley Valley and Foxholes, near Kidderminster, *J. B. Duncan*. Cofton Hackett.
Shady sandstone rocks. Very rare.

RADULA Dum. (*Radula*, a razor; requiring a razor to shave it off the bark of trees.)

69. **R. complanata** (Linn.) Dum.
Piddle Brook. Not uncommon near Evesham. Wyre Forest. Seckley. Shrawley Wood. Hartlebury Common. Eardiston. Wythall Heath.
Trunks of trees. Not common.

MADOTHECA Dum. (μαδός, loose, θήκη, a box.)

70. **M. laevigata** (Schrad.) Dum.
Skeys Wood, Wyre Forest, *J. B. Duncan*.
Shady rocks and boulders amongst mosses. Very rare.

71. **M. platyphylla** (Linn.) Dum.
Common in Evesham district. Wyre Forest. Seckley Wood. Park Attwood. Habberley Valley. Southstone Rock. Hagley.
Dry rocks and on tree-trunks. Not uncommon.

LEJEUNIA Lib. (Named after Lejeune, a Belgian botanist.)

72. **L. cavifolia** (Ehrh.) Lindb.
Wyre Forest. Seckley Wood. North Wood. Gladder Brook. Hanley Dingle. Shelsley Walsh.
Rocks near streams in very shady situations. Rare.

73. [**L. patens** Lindb.
Hanley Dingle.
Rocks and stones in damp shady situations. Very rare. A doubtful record]

FRULLANIA Raddi.

74. **F. tamarisci** (Linn.) Dum.
Bredon Hill, *H. H. Knight*. Wyre Forest. Seckley Wood. Worcestershire Beacon. St. Kenelm's.
Rocks and boulders. Rare.

75. **F. dilatata** (Linn.) Dum.
Common about Evesham. Wyre Forest. Seckley Wood. Hartlebury. Dunley. Hanley Dingle. Pensax. Wythall Heath. Fenny Rough.
Trunks of trees. Common.

III. ANTHOCEROTALES

ANTHOCEROTACEAE

ANTHOCEROS Linn. (ἄνθος, a flower, κέρας, a horn.)

76. **A. punctatus** Linn.
Trimpley. Hoarstone, near Bewdley. Churchill. Wolverley. Near Wannerton Downs. Malvern.
Damp fallow fields. Rare.

LICHENES

Fam. I. COLLEMACEI

II. COLLEMEI

COLLEMA (Ach.) Nyl. (*κόλλημα*, that which is glued together.)
1. C. cheileum Ach.
    Malvern, *Dr. Holl.*
2. C. pulposum (Bernh.).
    Herefordshire Beacon, *Rev. J. H. Thompson.* Claines; Hawford Bridge, *Dr. Holl.*
3. C. crispum Ach.
    Lane above Great Malvern, *Rev. J. H. Thompson.*
4. C. microphyllum Ach.
    On an old tree, Hindlip, near Worcester, *Dr. Holl.* On Beech, Bredon Hill, *Rev. J. H. Thompson.*
5. C. nigrescens Linn.
    Malvern, *Dr. Holl.* On Willow, Henwick, *Rev. J. H. Thompson.*

LEPTOGIUM Ach.

6. L. muscicola (Sw.).
    Eymore Wood; Malvern, *Rev. J. H. Thompson.*
7. L. turgidum (Ach.).
    Near Little Malvern, *Dr. Holl.*
8. L. fragrans (Sm.).
    Laughern Brook, *Rev. J. H. Thompson.*
9. L. subtile (Schrad.).
    Near Worcester, *Dr. Holl.*
10. L. lacerum (Ach.).
    Var. lophaeum Ach.
    Castlemorton Common, *Dr. Holl.*
11. L. sinuatum (Huds.).
    Malvern Hills, *Rev. J. H. Thompson.*
12. L. tenuissimum (Dicks.).
    St. Ann's Well, *Rev. J. H. Thompson*
13. L. Schraderi (Bernh.).
    Holly Bush Hill, *Dr. Holl.*

Fam. III. LICHENACEI

I. CALICIEI

SPHINCTRINA (Fr.) D. N. (*σφιγκτήρ*, that which binds tightly.)
14. S. turbinata Pers.
    Powick, *Lees.*

CALICIUM (Ach.) Nyl.

15. C. chrysocephalum Ach.
    Norton juxta Kempsey, *Dr. Holl.*
16. C. phaeocephalum Borr.
    Cookhill Wood, Worcestershire, *Purton.*
17. C. trichiale Ach.
    Var. ferrugineum (Borr.).
    Little Malvern, *Lees.*
18. C. hyperellum Ach.
    Eymore Wood, *Rev. J. H. Thompson.*
19. C. trachelinum Ach.
    Hollow tree, Laughern Brook, *Rev. J. H. Thompson.*
20. C. curtum Borr.
    Hallow, *Lees.* Hatfield, near Worcester, *Dr. Holl.*

CONIOCYBE (Ach.) Nyl. (*κόνις*, dust, *κύβη*, the head.)

21. C. furfuracea Ach.
    Near Stourbridge, Worcestershire, *Scott.* Near Winds Point, Malvern, *Dr. Holl.* Eymore Wood, Powick, *Rev. J. H. Thompson.*

TRACHYLIA (Fr.), Nyl.

22. T. tympanella Fr.
    Cowleigh Park. Oak Gate post, Lovington, near Hallow, *Rev. J. H. Thompson.*

II. SPHAEROPHOREI

SPHAEROPHORON Pers. (*σφαῖρα*, a ball, *φέρω*, I bear.)
23. S. compressum Ach.
    Malvern Hills, *Purton, Rev. J. H. Thompson.*
24. S. coralloides Pers.
    Malvern Hills, *Purton, Rev. J. H. Thompson, Dr. Holl.*

III. BAEOMYCEI

BAEOMYCES Pers. (*βαιός*, little, *μύκης*, fungus.)
25. B. rufus DC.
    Near St. Ann's Well, *Rev. J. H. Thompson.*

IV. CLADONIEI

CLADONIA (Hffm.) Nyl. (*κλαδών*, a branch.)
26. C. (Cladonia) endiviaefolia Fr.
    Cookley. Malvern Hills. Hartlebury Common, *Rev. J. H. Thompson.*
27. C. (Cladonia) cervicornis Schaer.
    Wyre Forest, *Rev. J. H. Thompson.*
28. C. (Cladonia) alcicornis Flk.
    Whippets Brook, Malvern, *Rev. J. H. Thompson.* Malvern Hills, *Dr. Holl.*
29. C. (Cladonia) pyxidata Fr.   *Cup Moss.*
    Common everywhere.
    Var. chlorophaea Flk.
    Coppice near Crowneast, *Dr. Holl.*
30. C. (Cladonia) gracilis Hffm.
    Malvern Hills; Hartlebury Common, *Rev. J. H. Thompson.*
    Var. chordalis (Ach.).
    Malvern Hills, *Purton.*
31. C. (Cladonia) furcata Hffm.
    Malvern Hills; Wyre Forest; Canal-bank, Salwarpe; Areley Kings; Hartlebury Common, *Rev. J. H. Thompson.*
32. C. (Cladonia) cornucopioides Fr.   *Red Cup Moss.*
    Bromsgrove Lickey; Ridgway; Malvern Hills, *Purton.* Hartlebury Common, *Rev. J. H. Thompson.*
33. C. (Cladonia) deformis Hffm.
    Hartlebury Common, *Rev. J. H. Thompson.*
34. C. (Cladonia) digitata Hffm.
    Crowneast Wood; Rosebury Rock; North Hill; Pirton; Wyre Forest, *Rev. J. H. Thompson.*
    Var. macilenta Hffm.
    Hawksbatch Dingle, *Rev. J. H. Thompson.*
35. C. (Cladina) rangiferina Hffm.   *Reindeer Moss.*
    Crowneast Wood; Hartlebury Common; Lower Lickey, *Rev. J. H. Thompson.* Malvern Hills, *Dr. Holl.*
36. C. (Cladina) uncialis Hffm.
    Above St. Ann's Well; North Hill; Hartlebury Common, *Rev. J. H. Thompson.* Bromsgrove Lickey, *Purton.*

V. STEREOCAULEI

STEREOCAULON Schreb. (*στερεός*, solid, *καυλός*, a stalk.)
37. S. nanum Ach.
    Blackstone; Rosebury Rock, *Rev. J. H. Thompson.* Malvern, *Dr. Holl.*

VIII. USNEEI

USNEA Hffm.
38. U. barbata Fr.   *Tree-beard Moss.*
    General.

IX. RAMALINEI

ALECTORIA (Ach.) Nyl. (*ἀλέκτωρ*, unmarried, from the uncertainty respecting the male elements.)
39. A. jubata Ach.   *Rock Hair.*
    Malvern Hills, *Purton.* Worcestershire Beacon, *Rev. J. H. Thompson.* The rock on the Worcestershire Beacon on which this plant used to grow was used as the base of a bonfire to celebrate the marriage of Earl Beauchamp in February 1868, and the plant was destroyed, *Lees.*

EVERNIA (Ach.) Nyl. (*εὐερνής*, sprouting well.)
40. E. prunastri (Linn.).
    Malvern, *Dr. Holl.* Laughern Brook, near Worcester; Malvern Hills; Monk Wood, *Rev. J. H. Thompson.*

RAMALINA (Ach.) Fr. (*Ramus*, a branch.)
41. R. calicaris Fr.
    North Hill, *Rev. J. H. Thompson.*
    Var. farinacea (Linn.).
    Bush Hill, Powick; South of Cabbage Lane, Powick, *Rev. J. H. Thompson.*
42. R. fraxinea Fr.
    Near the Asylum, Powick, *Rev. J. H. Thompson.*
43. R. fastigiata Fr.
    Powick; Malvern Hills; Hallow; Laughern Brook, near Worcester, *Rev. J. H. Thompson.*
44. R. pollinaria Ach.
    Bush Hill, Powick; Eastbury; Hallow, *Rev. J. H. Thompson.*

X. CETRARIEI

CETRARIA (Ach.) Nyl. (*Cetra*, a short Spanish shield.)
45. C. aculeata Fr.
    Malvern Hills, *Purton.* Hartlebury Heath, *Dr. Holl.*
    f. 2. muricata Ach.
    Malvern Hills, *Purton.*

PLATYSMA (Hffm.) Nyl. (*πλάτυσμα*, a flat piece, or plate.)
46. P. triste (Web.).
    Malvern Hills, *Purton.* Hartlebury Common, *Rev. J. H. Thompson.*
47. P. glaucum (Linn.).
    Malvern Hills, *Purton.* North Hill, *Rev. J. H. Thompson.*
    f. fallax (Web.).
    North Hill, *Rev. J. H. Thompson.*

XI. PELTIGEREI

NEPHROMIUM Nyl. (νεφρός, a kidney.)

48. **N. laevigatum** Ach.
Malvern Hills, *Rev. J. H. Thompson.*

PELTIGERA (Hffm.) Ach. (*Pelta*, a shield, *gero*, I carry.)

49. **P. canina** (Linn.).
Canal-banks, Hindlip, *Dr. Holl.* Lower Lickey ; Ribbesford Wood,
Rev. J. H. Thompson.

50. **P. rufescens** (Hffm.).
Hindlip, *Dr. Holl.* Drakelow, near Kidderminster ; Laughern
Brook, near Worcester ; Cowleigh Park, *Rev. J. H. Thompson.*

51. **P. polydactyla** (Hffm.).
Hindlip, *Dr. Holl.* Shrawley Wood ; Drakelow, near Kidderminster ;
Rosebury Rock, *Rev. J. H. Thompson.*

52. **P. horizontalis** (Linn.).
Seckley Wood ; North Malvern ; Malvern, *Rev. J. H. Thompson.*

XII. PARMELIEI

STICTINA Nyl.

53. **S. fuliginosa** (Dicks.).
Worcestershire Beacon, *Rev. J. H. Thompson.*

STICTA (Ach.). (στικτός, spotted.)

54. **S. pulmonacea** Ach.
Rosebury Rock ; On an old Pear-tree south of Cabbage Lane,
Powick ; On Lime, Hawford, *Rev. J. H. Thompson.*

PARMELIA (Ach.) Nyl. (πάρμη, a small shield.)

55. **P. caperata** (Linn.).
Powick ; Pirton ; North Hill, *Rev. J. H. Thompson.*

56. **P. olivacea** (Linn.).
North Hill ; Henwick Mill, *Rev. J. H. Thompson.*

Var. **exasperata** (Ach.).
Near Worcester, *Dr. Holl.*

57. **P. stygia** (Linn.).
North Hill ; Malvern Hills, *Rev. J. H. Thompson.*

58. **P. physodes** (Linn.).
On rails, Bilberry Hill ; Malvern Hills ; Midsummer Hill ; Swin-
yards Hill ; Pirton ; Seckley Wood ; Romsley Hill, *Rev. J. H.
Thompson.* Trees, Malvern ; Old rails, Lickey Hills, *Dr. Holl.*

Var. **labrosa** Ach.
Malvern Hills, *Sir J. E. Smith.*

59. **P. perlata** (Linn.).
Holly Bush Hill ; Malvern Hills, *Rev. J. H. Thompson.*

60. **P. aleurites** Ach.
Malvern Hills, *Rev. J. H. Thompson.*

61. **P. Borreri** (Turn.).
Birchin Grove, near Worcester, *Dr. Holl.* On an old tree, Bush
Hill, Powick ; Blackstone Rock, *Rev. J. H. Thompson.*

62. **P. conspersa** (Ehrh.).
Rocks, Herefordshire Beacon, *Dr. Holl.* On palings, Shrawley Wood ;
Rocks, North Hill, *Rev. J. H. Thompson.*

63. **P. acetabula** (Neck.).
Near Worcester, *Dr. Holl.*

64. **P. saxatilis** (Linn.).
Temple Laughern ; Hallow ; Powick ; Malvern Hills ; St. Ann's
Well, *Rev. J. H. Thompson* ; Malvern Hills, *Dr. Holl.*

Var. **omphalodes** (Linn.).
Malvern Hills, *Rev. J. H. Thompson, Rev. W. S. Rufford.*

PHYSCIA Nyl. (φύσκη, a blister.)

65. **P. flavicans** (Siv.).
North Hill, *Rev. J. H. Thompson.*

66. **P. parietina** (Linn.).
Hartlebury Common ; Laughern Brook, near Worcester ; New Pool,
Malvern, *Rev. J. H. Thompson.*

Var. **aureola** Ach.
Near Worcester, *Dr. Holl.*

Var. **lychnea** Ach.
Malvern Hills, *Dr. Holl.*

Var. **polycarpa** (Ehrh.).
Bradley Green, *Rev. W. S. Rufford.*

67. **P. ciliaris** (Linn.).
Laughern Brook ; Near the Asylum, Powick ; Monk Wood, *Rev. J. H.
Thompson.*

f. **verrucosa** Ach.
Littleton and Badsey, *Purton.*

68. **P. pulverulenta** (Schreb.).
Cowleigh Park ; Powick ; Pirton ; Laughern Brook, near Worcester,
Rev. J. H. Thompson. Pixham, *Dr. Holl.*

f. **muscigena** (Whlnb.).
Malvern Hills, *Dr. Holl.*

69. **P. obscura** (Ehrh.).
Norton juxta Kempsey, *Dr. Holl.*

Var. **virella** Ach.
Norton juxta Kempsey, *Dr. Holl.*

70. **P. stellaris** (Linn.).
Laughern Brook, near Worcester ; Hartlebury Common, *Rev. J. H.
Thompson.*

Var. **tenella** (Scop.).
Eastbury ; Laughern Brook, near Worcester ; Hartlebury Common,
Rev. J. H. Thompson.

Var. **caesia** (Hffm.).
Malvern Hills, *Dr. Holl.*

XIII. GYROPHOREI

UMBILICARIA Hffm.

71. **U. pustulata** Hffm.
Rocks, North Hill, *Dr. Holl.* Malvern Hills ; North Hill, *Rev. J. H.
Thompson.*

XIV. LECANOREI

PSOROMA (Fr.) Nyl. (ψώρα, a cutaneous disease.)

72. **P. hypnorum** (Vahl).
Malvern, *Dr. Holl.*

PANNARIA Del. (*Pannus*, a cloth.)

73. **P. pezizoides** (Web.).

Var. **coronata** (Ach.).
Malvern, *Dr. Holl.*

74. **P. nigra** (Huds.).
Edgiock, *Purton.* Bridge near Upton-on-Severn, *Lees.* Malvern,
Dr. Holl.

AMPHILOMA (Fr.) Nyl. (ἀμφί, about, λῶμα, a fringe.)

75. **A. lanuginosum** (Ach.).
Malvern Hills, *Dr. Holl.*

SQUAMARIA DC. (*Squama*, a scale.)

76. **S. saxicola** (Poll.).
Malvern Hills ; North Hill, *Dr. Holl.* Hartlebury Common, *Rev.
J. H. Thompson.*

PLACODIUM (DC.) Nyl. ·(πλακώδης = πλακόεις, flat.)

77. **P. murorum** (Hffm.).
Gravestone, churchyard, Evesham ; Hartlebury Common, *Rev. J. H.
Thompson.*

78. **P. citrinum** (Ach.).
Worcester, *Dr. Holl.*

79. **P. candicans** (Dicks.).
Malvern ; Worcester, *Varenne.*

LECANORA (Ach.) Nyl. (λεκάνη, a dish.)

80. **L. vitellina** Ach.
Malvern, *Rev. J. H. Thompson.*

81. **L. candelaria** Ach.
On Elm, Bredon Hill ; Powick ; Malvern, *Rev. J. H. Thompson.*

82. **L. glaucocarpa** (Whlnb.).

f. **pruinosa** (Sm.).
Near Malvern and Worcester, *Dr. Holl.*

83. **L. squamulosa** (Schrad.).
Malvern Hills, *Dr. Holl.*

f. **smaragdula** Whlnb.
Malvern Hills, *Dr. Holl.*

84. **L. sarcopis** (Whlnb.).
Old palings, Crowle, *Dr. Holl.*

85. **L. tartarea** (Linn.).   *Cudbear Lichen.*
Malvern Hills, *Rev. J. H. Thompson, Purton.*

86. **L. parella** (Linn.).
St. Ann's Well ; Horsley, near Kidderminster, *Rev. J. H. Thompson.*
Malvern Hills, *Purton.*

87. **L. varia** (Ehrh.).
Battenhall, Worcester, *Dr. Holl.*

88. **L. atra** (Huds.).
Malvern Hills ; Hartlebury Common, *Rev. J. H. Thompson.*

89. **L. sulphurea** (Hffm.).
Malvern Hills, *Dr. Holl.*

90. **L. subfusca** (Linn.).
On Trees about Worcester, *Dr. Holl.* Malvern Hills ; Porter's Mill ;
Perry Wood ; Bredon Hill, *Rev. J. H. Thompson.*

f. **angulosa** (Ach.).
Kempsey, *Dr. Holl.*

f. **albella** (Pers.).
Kempsey, *Dr. Holl.*

Var. **crenulata** Dicks.
Alfrick, *Dr. Holl.*

91. **L. galactina** (Ach.).
Walls, Great Malvern, *Dr. Holl.*

92. **L. calcarea** (Linn.).
St. Ann's Well, *Rev. J. H. Thompson.* Alfrick, *Dr. Holl.*

f. **Hoffmanni** Ach.
Malvern Hills, *Lees.*

93. **L. gibbosa** (Ach.).
Herefordshire Beacon ; Malvern Hills, *Dr. Holl.* Horsley Bank,
near Kidderminster, *Rev. J. H. Thompson.*

94. **L. badia** Ach.
Malvern Hills, *Dr. Holl.*

95. **L. glaucoma** (Hffm.).
Malvern Hills, *Dr. Holl.*

96. **L. ferruginea** (Huds.).
Malvern Hills ; North Hill ; Hartlebury Common, *Rev. J. H.
Thompson.* Raggedstone Hill, *Dr. Holl.* Malvern Hills, *Purton.*

97. **L. cerina** (Ehrh.).
    Kempsey, *Dr. Holl.*

98. **L. sophodes** (Ach.).
    Powick, *Rev. J. H. Thompson.*
    f. **exigua** (Ach.).
    On Ash near Worcester, *Dr. Holl.*

99. **L. haematomma** (Ehrh.).
    On Rocks, Malvern, *Rev. J. H. Thompson.*

URCEOLARIA (Ach.) Nyl. (*Urceolus*, a little pitcher.)

100. **U. scruposa** (Linn.).
    Wall at Hartlebury Common ; Horsley, near Kidderminster, *Rev. J. H. Thompson.* Malvern Hills, *Dr. Holl.*
    f. **bryophila** Ach.
    Malvern, *Dr. Holl.*

PERTUSARIA DC. (*Pertusus*, perforated.)

101. **P. dealbata** (Ach.).
    North Hill, *Rev. J. H. Thompson.*

102. **P. communis** DC.
    Common on trees.

103. **P. fallax** (Pers.).
    Battenhall, *Dr. Holl.*

104. **P. faginea** (Linn.).
    Hartlebury Common. On Beech, Overbury ; Seckley Wood ; Pirton ; Malvern Hills ; Laughern Brook ; Arley, *Rev. J. H. Thompson.*

105. **P. globulifera** (Turn.).
    Herefordshire Beacon, *Dr. Holl.* Powick ; Vallombrosa ; Monk Wood ; Spetchley ; Ankerdine Hill ; Laughern Brook ; Henwick, *Rev. J. H. Thompson.*

PHLYCTIS Wallr.

106. **P. agelœa** (Ach.).
    Norton juxta Kempsey, *Dr. Holl.*

107. **P. argena** (Ach.).
    Battons Wood, Bredon Hill ; On Alder, Ronks Wood ; On Willow, river Salwarpe, near Droitwich, *Rev. J. H. Thompson.*

THELOTREMA Ach. (θηλή, a teat, τρῆμα, a perforation.)

108. **T. lepadinum** Ach.
    Seckley Wood ; Holly Bush Hill, *Rev. J. H. Thompson.*

XV. LECIDEINEI

LECIDEA (Ach.) Nyl. (λεκίς, a little dish.)

109. **L. ostreata** (Hffm.).
    Norton juxta Kempsey, *Dr. Holl.*

110. **L. dispansa** Nyl.
    Near Bewdley, *Dr. Holl.*

111. **L. lucida** Ach.
    Near Knightsford Bridge, *Dr. Holl.*

112. **L. flexuosa** (Fr.).
    f. **aeruginosa** (Borr.).
    Bromsgrove Lickey, *Leighton.*

113. **L. decolorans** Flk.
    Malvern Wells ; The Lickey Hills, *Dr. Holl.* Malvern Hills ; St. Ann's Well, *Rev. J. H. Thompson.*

114. **L. prasina** Fr.
    Coppice at Crowneast, *Dr. Holl.*

115. **L. querna** (Dicks.).
    Old gatepost, Lovington, near Hallow ; Sinton, *Rev. J. H. Thompson.* Near Worcester and Malvern ; South end of the Malvern Hills, *Dr. Holl.*

116. **L. viridescens** (Schrad.).
    Beyond Franche, Kidderminster, *Rev. J. H. Thompson.*

117. **L. parasema** (Ach.).
    Lovington, near Hallow ; Laughern Brook, near Worcester, *Rev. J. H. Thompson.*

118. **L. uliginosa** (Schrad.).
    Malvern Hills ; North Hill, *Dr. Holl.* Midsummer Hill, *Rev. J. H. Thompson.*
    f. **fuliginea** (Ach.).
    Bevere, *Dr. Holl.*

119. **L. coarctata** (Sm.).
    f. **elacista** (Ach.).
    Raggedstone Hill, *Dr. Holl.*
    f. **ornata** (Smmrf.).
    Malvern Hills ; North Hill, *Dr. Holl.*

120. **L. panaeola** Ach.
    North Hill, *Rev. J. H. Thompson.*

121. **L. tenebrosa** (Flot.).
    Malvern Hills, *Dr. Holl.*

122. **L. contigua** Fr.
    Malvern Hills, *Dr. Holl.*

123. **L. confluens** (Webr.).
    Hartlebury Common, *Rev. J. H. Thompson.*

124. **L. calcivora** (Ehrh.).
    Stourport, *Rev. J. H. Thompson.*

125. **L. canescens** (Dicks.).
    Laughern Brook, near Worcester ; Northwick ; Ankerdine Hill ; Pirton, *Rev. J. H. Thompson.* Whittington, near Worcester, *Dr. Holl.*

126. **L. atroalba** Ach.
    Near Oakley Wood, Droitwich, *Rev. J. H. Thompson.*

127. **L. myriocarpa** (DC.).
    Malvern ; Battenhall, *Dr. Holl.*

128. **L. grossa** Pers.
    Near Malvern, *Dr. Holl.*

129. **L. tricolor** (With.).
    Horsley Bank, near Kidderminster ; Malvern Hills, *Rev. J. H. Thompson.* Malvern, *Dr. Holl.*

130. **L. pulverea** Borr.
    Rosebury Rock, *Rev. J. H. Thompson.*

131. **L. Caradocensis** Leight.
    Ombersley ; Whippets near North Malvern, *Dr. Holl.*

132. **L. incompta** Borr.
    Claines ; Kempsey, *Dr. Holl.*

133. **L. alboatra** (Hffm.).
    On Oak, canal-side, Droitwich, *Rev. J. H. Thompson.* Whittington, near Worcester, *Dr. Holl.*
    f. **epipolia** (Ach.).
    Malvern Hills ; Old wall, Spetchley, *Dr. Holl.*

134. **L. aromatica** (Sm.).
    Whittington, *Dr. Holl.*

135. **L. sphaeroides** Smrf.
    Bevere Green, *Dr. Holl.*

136. **L. pachycarpa** (Duf.).
    Laughern Brook, near Worcester ; Palings, Henwick ; St. Ann's Well, *Rev. J. H. Thompson.*

137. **L. premnea** Ach.
    Hatfield, near Worcester ; Dynes Green, *Dr. Holl.*

138. **L. milliaria** Fr.
    Claines ; Battenhall, *Dr. Holl.*

139. **L. endoleuca** Nyl.
    Churchill, near Worcester, *Dr. Holl.*

140. **L. rosella** (Pers.).
    Worcester ; Malvern, *Dr. Holl.*

141. **L. rubella** (Ehrh.).
    Old Storridge, *Rev. J. H. Thompson.* Near Malvern ; Kempsey, *Dr. Holl.*

142. **L. muscorum** (Sw.).
    Malvern Hills, *Dr. Holl.*

143. **L. effusa** (Sm.).
    Var. **caesio-pruinosa** Mudd.
    Pirton, *Dr. Holl.*

144. **L. geographica** (Linn.).
    Malvern Hills, *Purton.*
    Var. **urceolatum** Schrad. Malvern Hills, *Dr. Holl.*
    Var. **atrovirens** (Linn.). Malvern Hills, *Dr. Holl.*

145. **L. petraea** (Wulf.).
    Railway Bridge, Arley, *Rev. J. H. Thompson.* Near Old Storridge, *Dr. Holl.*

146. **L. truncigena** (Ach.).
    Castlemorton Common, *Dr. Holl.*

147. **L. Parmeliarum** Smrf.
    North Hill, *Dr. Holl.*

XVI. GRAPHIDEI

GRAPHIS (Ach.) Nyl. (γραφίς, drawing, from the resemblance of the apothecia to forms of certain Oriental alphabets.)

148. **G. elegans** (Sm.).
    On Holly beyond the Herefordshire Beacon ; Cowleigh Park ; Holly Bush Hill, *Rev. J. H. Thompson.* Berrow Wood, *Lees.* Midsummer Hill, *Dr. Holl.*

149. **G. scripta** Ach.
    Mathon Lodge ; Ronk's Wood ; Porters Mill ; Hagley Wood, *Rev. J. H. Thompson.*
    f. **varia** Leight.
    Near Little Malvern, *Lees.*
    f. **horizontalis** Leight.
    Holly Bush Hill, *Lees.*
    f. **serpentina** Ach.
    Near Little Malvern, *Lees.*
    f. **eutypa** Ach.
    Berrow, Malvern Hills, *Lees.*

150. **G. inusta** Ach.
    Malvern Hills, *Lees.*

151. **G. sophistica** Nyl.
    Var. **pulverulenta** (Sm.).
    On Beech, Overbury Wood, *Rev. J. H. Thompson.* Near the Berrow, Malvern Hills, *Lees.*
    Var. **flexuosa** Leight.
    On Beech, Overbury Wood. *Rev. J. H. Thompson.*

OPEGRAPHA (Ach.) Nyl. (ὀπή, a hole, γραφή, drawing, from the shape of the apothecia.)

152. **O. herpetica** Ach.
    f. **rubella** Pers.
    Berrow, Malvern Hills, *Lees.*

153. **O. atra** Pers.
    Franche ; Rosebury Rock ; Holly Bush Hill ; Fearnall Heath ; on Holly, Powick and Sinton, *Rev. J. H. Thompson.*
    f. **hapalea** Ach.
    Holly Bush Hill, *Lees.*

154. **O. Turneri** Leight.
    Near Worcester, *Lees.*

155. **O. varia** Pers.
    Powick; Holly Bush Hill; Lovington, near Hallow, *Rev. J. H. Thompson.*

    f. **tigrina** Ach.
        Norton juxta Kempsey, *Dr. Holl.*

156. **O. vulgata** Ach.
    On Maple, Cabbage Lane, Powick, *Rev. J. H. Thompson.* Malvern Hills, *Lees.*

157. **O. lyncea** (Sm.).
    On great Lime tree, Hawford; on Oak, Norton juxta Kempsey, *Rev. J. H. Thompson.* Great Malvern, *Lees.*

STIGMATIDIUM Mey. (στίγμα, a spot, from the punctiform apothecia.)
158. **S. crassum** Dub.
    Norton juxta Kempsey; near Malvern, *Dr. Holl.*

159. **S. venosum** (Ach.).
    Bredon Hill; Porters Mill; Holly Bush Hill, *Rev. J. H. Thompson.*

ARTHONIA, Ach.

160. **A. epipasta** (Ach.).
    Monk Wood, *Rev. J. H. Thompson.*

161. **A. cinnabarina** (Wallr.).
    Bredon Hill, *Rev. J. H. Thompson.*

XVIII. PYRENOCARPEI

ENDOCARPON (Hedw.) Nyl. (ἔνδον, within, καρπός, fruit, from the apothecia being immersed in the thallus.)
162. **E. miniatum** (Linn.).
    Var. **complicatum** (Sw.).
        Rocks, North Malvern, *Rev. J. H. Thompson.*

VERRUCARIA (Pers.) Nyl. (*Verrucaria*, driving away warts.)
163. **V. mutabilis** Borr.
    Near Norton juxta Kempsey.

164. **V. nigrescens** (Pers.).
    Bromsgrove Lickey, *Leighton.*

165. **V. fuscella** Turn.
    Malvern, *Lees.*

166. **V. viridula** (Schrad.).
    Norton; Bevere, *Dr. Holl.*

167. **V. rupestris** Schrad.
    Malvern, *Lees.* Bromsgrove Lickey, *Leighton.*

    Var. **muralis** Ach.
        Malvern, *Lees.*

168. **V. gemmata** Ach.
    On Willow, Lovington, near Hallow, *Rev. J. H. Thompson.* Malvern, *Lees.*

169. **V. biformis** Borr.
    Malvern, *Lees.* Near Worcester, *Dr. Holl.*

170. **V. Salweii** Leight.
    Malvern, *Lees.*

171. **V. nitida** (Weig.).
    Holly Bush Hill, *Dr. Holl.*

UNIDENTIFIED RECORDS

In the Herbarium at the Hastings Museum, Worcester, are specimens named as under, which cannot be recognized, nor arranged with regard to modern nomenclature and method.

Callopisma vitellinellum Mudd.
    Elm trees, Fearnall Heath; Malvern Hills, *Dr. Holl.*

Callopisma luteoalbum Turn.
    Elm trees near Malvern, *Dr. Holl.*

Buellia microcarpa DC.
    Near the Pale Farm, Upper Howsell, Malvern; near Worcester, *Dr. Holl.*

Buellia atrogrisea Delisc.
    White Ladies' Aston, *Dr. Holl.*

Buellia luteola var. **fuscella** Fr.
    Hindlip, *Dr. Holl.*

Acolium stigonellum Ach.
    Hindlip, *Dr. Holl.*

Lecanora fumosa Hffm.
    Malvern Hills, *Dr. Holl.*

Lecothecium nigrum Huds.
    Cowleigh Park, *Dr. Holl.*

Cyphelinum chlorellum Wahlb.
    Kempsey, *Dr. Holl.*

Baeomyces byssoides Linn.
    The Hills, Malvern, *Dr. Holl.*

Opegrapha media.
    Laughern Brook, *Rev. J. H. Thompson.*

Stereocaulon botryosporum Hooker.
    Malvern Hills, *Rev. J. H. Thompson.*

FUNGI

Division I. EUMYCETAE

A. TELEOMYCETAE

BASIDIOMYCETAE

HYMENOMYCETAE

AGARICACEAE

AMANITA Pers. (ἀμανῖται were certain fungi found on Mount Amanus in Cilicia.)
1. **A. phalloides** (Vaill.) Fr.
    Shrawley Wood. Monk Wood. Great Farley Wood. Dripshill Wood. Perry Wood. Nunnery Wood.
    Common. In woods. Very poisonous. The active principle of the poison is a toxic albumen called phalline. It is characterized by the free volva splitting into deltoid segments.
    Var. **verna** Lam.
        The Slads.
        Rather rare. It is a white and early form. Very poisonous.

2. **A. mappa** (Batsch) Fr.
    Shrawley Wood. Wyre Forest. Nunnery Wood. Trench Woods. Little Malvern. Chaddesley Corbett Woods. Monk Wood. Ockeridge Wood.
    Common. In woods. Very poisonous. The toxic principle is known as mappine. It is distinguished from *A. phalloides* by its close-cut volva.

3. **A. porphyria** (A. and S.) Fr.
    Wyre Forest. Shrawley Wood.
    Rather uncommon. In woods. Poisonous. It is much smaller than the two former and has a brownish ring.

4. **A. recutita** Fr.
    Alfrick Woods.
    Very rare. Chiefly in pine woods. Poisonous. The volva is very distinctly cut round close to the base of the stem.

5. **A. muscaria** (Linn.) Pers.   *Fly Agaric.*
    Shrawley Wood. Wyre Forest. Great Farley Wood. Trench Woods. Bevere Green. Dripshill Wood. Wood above Little Malvern. Crews Hill Wood.

Very common. It grows in greatest abundance under birches and Scotch pines. Very poisonous. The extract of the poisonous principle is known as muscarine. This noble ornament of our woods in autumn must have attracted the attention of all observers by the glory of its crimson pileus covered with patches of the white volva on its top. In former days British housewives used to make fly-papers from this fungus, hence its English name The Fly Agaric.

    Var. **aureola** Kalchbr.
        First Record for Britain, Habberley Valley, Kidderminster, *Rea, Trans. Worc. Nat. Club,* vol. i, p. 419.
        Wyre Forest. Shrawley.
        This is an extreme form of *A. muscaria* with a free volva instead of the usual friable one.

6. **A. pantherina** (DC.) Fr.
    Wyre Forest. Shrawley Wood. Crews Hill Wood.
    Rather uncommon. In woods. Very poisonous.

7. **A. rubescens** Pers.
    Churchill Wood. Bevere Green. Shrawley Wood. Trench Woods. Rous Lench Beeches. Ockeridge Wood. Monk Wood. Nunnery Wood. Dripshill Wood. Middleyards Coppice.
    Common. In woods and on lawns. Edible, but its flavour is somewhat insipid.

8. **A. spissa** Fr.
    Shrawley Wood. Wyre Forest. Ribbesford Wood.
    Uncommon. In woods. Poisonous.

9. **A. nitida** Fr.
    Nunnery Wood. Sheriff's Lench. Wyre Forest.
    Uncommon. In woods and pastures.

AMANITOPSIS Roze. (*Amanita*, ὄψις like, but destitute of a ring.)
10. **A. vaginata** (Bull.) Roze.
    Common. In woods. Generally distributed. Esculent and one of the most delicious of fungi.
    Var. **nivalis** Grev.
        Nunnery Wood. Monk Wood. Wyre Forest.
    Var. **plumbea** Schaeff.
        Wyre Forest. Shrawley Wood. Tiddesley Wood. Trench Woods. Perry Wood.
    Var. **fulva** Schaeff.
        Wyre Forest. Shrawley Wood. Dripshill Wood. Nunnery Wood. Ockeridge Wood. Habberley Valley.

11. **A. strangulata** (Fr.) Mass.
    Tiddesley Wood. Wyre Forest. Shrawley Wood. Trench Woods.
    Not common. In woods. Esculent and delicious.

12. **A. adnata** (W. G. Sm.) Mass.
    Badgers Bank. Near Crews Hill Wood.
    Rare. In woods and oaks.

LEPIOTA Fr. (λεπίς, a scale, οὖς, the ear, from the scaly epidermis of many of the species.)

13. **L. procera** (Scop.) Pers. *Parasol Mushroom.*
Common. In woods and on commons amongst heather and bracken. Edible, excellent. Characterized by the transversely scaly stem and white flesh.

14. **L. rachodes** (Vitt.) Fr.
Sharpway Gate. Wood above Little Malvern. Fries Wood. Shrawley Wood. Temple Laughern. Near Tiddesley Wood. Near Evesham.
Not uncommon. In woods and under trees in pastures. Edible, flavour very nice. It differs from *L. procera* in having a smooth stem and the flesh turning reddish when broken.

Var. **puellaris** Fr.
Hampton Lovett. Cowleigh Park. Wyre Forest.
This variety is smaller and is entirely white.

15. **L. prominens** (Viv.) Fr.
Fries Wood. Wyre Forest.
Rather uncommon. In woods and on heaths. Edible. The stem is scaly with distinct scales, but not caused to be transversely scaly by the later expansion of the stem as in *L. procera.*

16. **L. permixta** Barla.
Malvern Wells. Hartlebury Common.
Rare. Chiefly on commons and open downs. Edible. It has the transversely scaly stem of *L. procera*, but the reddish flesh of *L. rachodes.*

17. **L. excoriata** (Schaeff.) Pers.
Sharpway Gate. Wyre Forest. Shrawley Wood. Trench Woods. Cowleigh Park. Hartlebury Common. Chaddesley Corbett Woods.
Not uncommon. In woods and on heaths. Edible. The cuticle of the pileus does not grow as rapidly as the rest of the pileus, so it has the appearance of being torn away from the margin of the pileus.

18. **L. gracilenta** (Kromb.) Fr.
Wyre Forest. Malvern Wells. Defford Common. Trench Woods.
Not common. In woods, but chiefly on heaths. Edible. It is characterized by its prominent umbo and scaly stem. The Parasol Mushroom is the common English name for all the species enumerated under numbers 13 to 18 inclusive, and they are all very delicious esculents.

19. **L. acutesquamosa** (Weinm.) Fr.
Wyre Forest. Oldington Wood. Shrawley Wood.
Uncommon. In oak woods.

20. **L. Friesii** (Lasch.) Fr.
Shrawley Wood. The Manor House Farm, Sheriff's Lench.
Rare. On bare ground.

21. **L. Badhami** Berk.
Northwick.
Rare. Under a hawthorn hedge, not a usual habitat. Characterized by turning blood-red when bruised or cut.

22. **L. clypeolaria** (Bull.) Fr.
Shrawley Wood. Claines.
Rather rare. In woods and arable fields. Edible.

23. **L. pratensis** (Bull.) Fr.
Wyre Forest. Malvern Wells. Defford Common.
Not uncommon. Amongst short grass on hills and commons. Edible. This is treated by most authors as a variety of *L. clypeolaria*, but it differs from that species in its floccose pileus and stem.

24. **L. metulaespora** B. and Br.
Wyre Forest. Tiddesley Wood.
Uncommon. In woods.

25. **L. submarasmioides** Speg.
First Record. *Rea, Trans. Worc. Nat. Club*, vol. iii, p. 49.
Valley of the Whiteleaved Oak.
Very rare. Amongst short grass. The spores of this species are deltoid in shape and of a light straw colour when viewed in mass on a spore map deposited on black paper.

26. **L. felina** Pers.
Malvern. Wyre Forest. Shrawley Wood. Trench Woods. Tiddesley Wood.
Not uncommon. In woods.

27. **L. cristata** (A. and S.) Fr.
Bevere Green. Great Farley Wood. Crews Hill Wood. Shrawley Wood. Ockeridge Wood. Malvern Wells. Little Malvern. Trench Woods.
Common. Amongst short grass and in woods.

28. **L. erminea** Fr.
Malvern Wells. Claines. Near Ham Castle.
Rather uncommon. Amongst short grass.

29. **L. holosericea** Fr.
Lower Wick, near Worcester.
Very uncommon. On bare ground and in arable fields. Edible.

30. **L. leucothites** (Vitt.) Fr.
First Record. *Trans. Brit. Myc. Soc.*, vol. i, p. 20.
Near Holly Bush pass, Raggedstone Hill. Wyre Forest. The Slads.
Rare. Amongst short grass. Edible. It has probably been elsewhere erroneously recorded as *Agaricus cretaceus* Fr.

31. **L. cepaestipes** (Sow.) Fr.
Woodside, Worcester. The Manor House, Sheriff's Lench. White's Nursery, Lower Wick.
Common. In gardens and hot-houses.

Var. **cretacea** (Bull.) Fr.
Oakden, Kidderminster. South Bank, Worcester.
Not uncommon. In gardens and on rubbish heaps.

32. **L. licmophora** B. and Br.
Oakden, Kidderminster.
Rather uncommon. The plants recorded grew on cocoa-nut fibre in a greenhouse.

33. **L. carcharias** Pers.
Wyre Forest. Shrawley Wood. Tiddesley Wood. Trench Woods. Ockeridge Wood.
Not uncommon. In woods especially under Conifers.

34. **L. granulosa** (Batsch) Pers.
Shrawley Wood. Wyre Forest. Great Farley Wood. Trench Woods.
Common. In woods and on commons. Edible.

35. **L. amianthina** (Scop.) Fr.
Shrawley Wood. The Randans Wood. Malvern Wells. Perry Wood. Wyre Forest. Trench Woods.
Very common. In woods and mossy pastures. Edible.

36. **L. haematosperma** (Bull.) Boud.
Shrawley Wood.
Rare. In woods and gardens.

37. **L. seminuda** (Lasch.) Fr.
Wyre Forest. Shrawley Wood. Monk Wood.
Not common. In woods.

38. **L. Bucknalli** B. and Br.
Canal-side between Cookley and Wolverley.
Rare. Amongst grass. It has a strong smell, like gas tar.

39. **L. illinita** Fr.
Northwick Hall, Claines.
Rare. In a hedgerow.

ARMILLARIA Fr. (*Armilla*, a ring, from the ring on the stem.)

40. **A. robusta** (A. and S.) Fr. var. minor Fr.
Shrawley Wood. Record somewhat doubtful.
Rare. In woods under Conifers. Edible.

41. **A. mellea** (Vahl.) Fr.
Common everywhere. In woods and on old stumps in hedgerows. Edible but not very palatable. This is a destructive parasite and damages the value of timber. The disease is known as White Rot.

42. **A. mucida** (Schrad.) Fr.
Hagley. Rous Lench.
Not common in this county. On beeches. Edible.

TRICHOLOMA Fr. (θρίξ, a hair, λῶμα, fringe, because the margin of the pileus generally bears traces of the universal veil.)

43. **T. sejunctum** (Sow.) Fr.
Wyre Forest. Shrawley Wood. Nunnery Wood. Monk Wood. Crews Hill Wood. Trench Woods. Tiddesley Wood.
Not uncommon. In woods. Edible.

44. **T. portentosum** Fr.
Wyre Forest.
Rare. In woods. Edible.

45. **T. spermaticum** Fr.
Shrawley Wood. Wyre Forest.
Rather uncommon. In woods. Edible.

46. **T. resplendens** Fr.
Dripshill Wood. Wyre Forest. Crews Hill Wood.
Rather uncommon. In woods.

47. **T. acerbum** (Bull.) Fr.
Middleyards Coppice. Wyre Forest. Chaddesley Corbett Woods. Shrawley Wood.
Not common. In woods. Edible.

48. **T. flavobrunneum** Fr.
Wyre Forest. Shrawley Wood. Great Farley Wood. Monk Wood. Nunnery Wood.
Common. In woods. Poisonous.

49. **T. albobrunneum** (Pers.) Fr.
Shrawley Wood. Crews Hill Wood. Tiddesley Wood. Trench Woods. Wyre Forest.
Common. In woods. Edible.

50. **T. ustale** Fr.
Wyre Forest.
Rare. In woods.

51. **T. rutilans** (Schaeff.) Fr.
Chatley. Shrawley Wood. Wyre Forest. Bishampton Wood. Bevere Green. Great Farley Wood. Ockeridge Wood.
Common. In pine woods.

52. **T. columbetta** Fr.
Wood above Little Malvern. Wyre Forest. Shrawley Wood. Cowleigh Park. Randans Wood.
Uncommon. In woods. Edible.

53. **T. murinaceum** (Bull.) Fr.
Wyre Forest.
Rare. In woods. Edible.

54. **T. terreum** (Schaeff.) Fr.
Ockeridge Wood. Shrawley Wood. Wyre Forest. Crews Hill Wood. Sheriff's Lench.
Common. In woods, especially of fir. Edible.

Var. **atrosquamosum** Chev.
Wyre Forest. Shrawley Wood.

Var. **argyraceum** Bull.
Wyre Forest.

Var. **chrysites** Jungh.
Near Mrs. Lea's, Hallow. Shrawley.

55. **T. squarrulosum** Bres.
First Record. *Rea, Trans. Brit. Myc. Soc.*, vol. ii, p. 62, pl. 4.
Black House Hill Wood.
Rare. In woods under oaks and nut-trees.

56. **T. saponaceum** Fr.
Wyre Forest. Ockeridge Wood. Shrawley Wood. Crews Hill
Wood. Trench Woods. Monk Wood. Nunnery Wood. Drips-
hill Wood.
Common. In woods.

57. **T. cartilagineum** (Bull.) Fr.
Wyre Forest.
Rare. In dry and grassy woods. Edible.

58. **T. cuneifolium** Fr.
Wyre Forest. 34 Foregate Street, Worcester. Great Farley Wood.
Crews Hill Wood. Shrawley Wood. Tiddesley Wood.
Not uncommon. In woods and pastures. Edible.
Var. **cinereo-rimosum** (Batsch) Fr.
Shrawley Wood. Wyre Forest.
Not uncommon. In woods and pastures.

59. **T. sulphureum** Fr.
Wood above Little Malvern. Shrawley Wood. Perry Wood. Wyre
Forest. Nunnery Wood.
Not uncommon. In woods. Poisonous.

60. **T. lascivum** Fr.
Wyre Forest. Perry Wood.
Uncommon. In mixed woods.

61. **T. carneum** (Bull.) Fr.
Keys End. Malvern Wells. Sheriff's Lench. Wyre Forest. Kempsey
Common. Hartlebury Common.
Not uncommon. In woods and pastures.

62. **T. gambosum** Fr.   *Saint George's Mushroom.*
Whittington. Malvern Wells. Oldbury Road. Dodderhill Common.
Cowleigh Park. Stanford Park. Near Tiddesley Wood. Bredon
Hill.
Not uncommon. In pastures. Edible and delicious.

63. **T. boreale** Fr.
Wyre Forest.
Rather rare. In woods amongst grass.

64. **T. patulum** Fr.
Pitchcroft, Worcester.
Very rare. Amongst grass.

65. **T. album** (Schaeff.) Fr.
Wyre Forest. Shrawley Wood. Ockeridge Wood. Tiddesley Wood.
Trench Woods.
Not uncommon. In woods. Poisonous.

66. **T. leucocephalum** Fr.
Wyre Forest. Garden, 34 Foregate Street, Worcester.
Uncommon. In woods and on lawns.

67. **T. personatum** Fr.   *Blue Legs.*
Sheriff's Lench. Ladywood, Salwarpe. Chaddesley Corbett. Alfrick.
Shrawley Wood. Near Tiddesley Wood. Old Hills.
Common. In pastures. Edible, but rather insipid in flavour.

68. **T. glaucocanum** Bres.
Wyre Forest. Trench Woods. Perry Wood. Tiddesley Wood.
Common. In woods. Edible and excellent. This species is
exactly intermediate between *Tricholoma personatum* and *T. nudum*,
combining the characters of both.

69. **T. nudum** (Bull.) Fr.   *Blewits, Blue Hats.*
Wyre Forest. Trench Woods. Randans Wood. Crews Hill Wood.
Shrawley Wood. Trench Woods. Fries Wood. Perry Wood.
Nunnery Wood. Tiddesley Wood. Dripshill Wood.
Common. In woods, occasionally in pastures. Esculent and
delicious.

70. **T. panaeolum** Fr.
Chatley. Hanbury Park. Near Randans Wood. Spetchley Park.
Cowleigh Park.
Locally common. In pastures. Edible, but rather soft in texture.

71. **T. melaleucum** (Pers.) Fr.
The Randans Wood. Tiddesley Wood. Shrawley Wood. Wyre
Forest. Near Monk Wood. Near Perry Wood.
Fairly common. In pastures and amongst short grass in woods.
Edible.
Var. **polioleucum** Fr.
Ladywood. Shrawley Wood. Wyre Forest. Trench Woods. Tid-
desley Wood. Nunnery Wood.
The common form.

72. **T. grammopodium** (Bull.) Fr.
Battenhall. Ham Wood. Shrawley Wood. Wyre Forest. Lady-
wood, Salwarpe.
Not uncommon. In woods and pastures. Edible.

73. **T. brevipes** (Bull.) Fr.
Malvern Link. Near Nunnery Wood.
Uncommon. On cinder paths and bare ground. Edible.

74. **T. humile** (Pers.) Fr.
Comer Gardens, Worcester.
Uncommon. In gardens and on bare ground. Edible.

75. **T. subpulverulentum** (Pers.) Fr.
Porter's Mill, Claines.
Uncommon. In pastures. Edible.

76. **T. sordidum** (Schum.) Fr.
Chatley. Trench Woods. Claines. Sheriff's Lench. Defford.
Hartlebury. Kempsey.

Common. In pastures and in gardens often amongst manure.
Edible.
Var. **lilacea** (Quél.).
Near Hallow Ford, Claines. Ladywood, Salwarpe. Near Tiddesley
Wood.
Not uncommon. In pastures amongst short grass. Edible.

CLITOCYBE Fr. (κλιτύς, a hill-side or declivity, κύβη, head, from
the decurrent gills.)

77. **C. nebularis** (Batsch) Fr.   *Clouded Agaric.*
Shrawley Wood. Perry Wood. Nunnery Wood. Wyre Forest.
Ockeridge Wood. Fries Wood.
Common. In woods. Edible and delicious.

78. **C. clavipes** (Pers.) Fr.
Shrawley Wood. Wyre Forest. Chaddesley Corbett Woods. Great
Farley Wood. Malvern Wells. Tiddesley Wood. Trench
Woods.
Common. In woods, especially pine.

79. **C. hirneola** Fr.
Chaddesley Corbett Woods. Ockeridge Wood.
Rare. Amongst moss in woods and pastures. Edible.

80. **C. odora** (Bull.) Fr.
Wood above Little Malvern. Shrawley Wood. Fries Wood. Perry
Wood. Nunnery Wood. Trench Woods. Wyre Forest.
Not uncommon. In woods. Edible and delicious.

81. **C. rivulosa** (Pers.) Fr.
Madresfield Park. Chatley. Ladywood, Salwarpe. Claines. Malvern
Wells. Defford Common. Wyre Forest.
Not uncommon. Chiefly in pastures and by roadsides.

82. **C. cerussata** Fr.
Shrawley Wood. Wyre Forest. Trench Woods. Perry Wood.
Crews Hill Wood. Nunnery Wood.
Not uncommon. In woods.

83. **C. phyllophila** Fr.
Wyre Forest. Shrawley Wood. Trench Woods. Monk Wood.
Tiddesley Wood.
Uncommon. In woods. Poisonous.

84. **C. pithyophila** Fr.
Dripshill Wood. Oldington Wood.
Rare. In pine woods.

85. **C. tornata** Fr.
Ockeridge Wood. Shrawley Wood. Wyre Forest.
Not common. In woods.

86. **C. candicans** (Pers.) Fr.
Shrawley Wood. Trench Woods. Wyre Forest. Perry Wood. Nun-
nery Wood. Monk Wood. Tiddesley Wood.
Common. In woods. Easily known by its cartilaginous stem.

87. **C. dealbata** (Sow.) Fr.
Ladywood, Salwarpe. Shrawley Wood. Wyre Forest. Near Nunnery
Wood. Fields between Ockeridge Wood and Monk Wood.
Common. Amongst short grass in woods, pastures and lawns. Edible.
Var. **minor** Cke.
Wyre Forest. Hartlebury Common. Dodderhill Common. Chatley.
Common. Amongst short grass. Edible.

88. **C. gallinacea** (Scop.) Fr.
Chatley. Dripshill Wood. Old Hills. Nunnery Wood.
Rather uncommon. Amongst moss and short grass.

89. **C. decastes** Fr.
Ladywood, Salwarpe. Wyre Forest.
Rare. In woods and open grassy places. Edible.

90. **C. aggregata** (Schaeff.) Fr.
The Manor House, Sheriff's Lench.
Uncommon. In woods and gardens.

91. **C. fumosa** (Pers.) Fr.
Wyre Forest. Crews Hill Wood. Trench Woods. Tiddesley Wood.
Monk Wood. Shrawley Wood.
Not uncommon. In woods.

92. **C. maxima** (Gärtn. and Mey.) Fr.
Hadley Bowling Green. Shrawley Wood. Storridge Common.
Cowleigh Park. Spetchley Park.
Not uncommon. In pastures, seldom in woods. Edible and
delicious.

93. **C. infundibuliformis** (Schaeff.) Fr.
Bridgestone. Trench Woods. Wyre Forest. Crews Hill Wood.
Wyre Forest. Perry Wood. Nunnery Wood. Shrawley Wood.
Dripshill Wood. Old Hills. Monk Wood. Malvern Hills.
Hartlebury Common. Kyre Park.
Common. In woods and fields. Edible and delicious.

94. **C. incilis** Fr.
Malvern Wells.
Rather local. In woods.

95. **C. sinopica** Fr.
Malvern Wells. Swineyards Hill. Eymore Wood.
Uncommon. In woods and open places.

96. **C. geotropa** (Bull.) Fr.
Ladywood, Salwarpe. Trench Woods. Wyre Forest. Chaddesley
Corbett Woods. Crews Hill Wood. Shrawley Wood. Fries
Wood. Perry Wood. Nunnery Wood. Westwood Park.
Spetchley Park.
Common. In woods and pastures. Edible and delicious.

97. **C. inversa** (Scop.) Fr.
Shrawley Wood. Old Hills. Great Farley Wood. Ockeridge
Wood. Wyre Forest. Shrawley.
Not uncommon. In woods and under furze bushes on commons.

98. C. flaccida (Sow.) Fr.
Old Hills. Shrawley Wood. Wyre Forest. Chaddesley Corbett Woods. Ockeridge Wood.
Not uncommon. In woods.

99. C. cyathiformis (Bull.) Fr.
Ockeridge Wood. Shrawley Wood. Wyre Forest. Arley. The Randans. Ockeridge Wood. Perry Wood. Nunnery Wood.
Common. In woods and pastures. Edible.

100. C. brumalis Fr.
Wyre Forest. Ockeridge Wood. Nunnery Wood. Tiddesley Wood. Trench Woods. Perry Wood.
Not uncommon. In woods. Edible.

101. C. metachroa Fr.
Trench Woods. Wyre Forest. Great Farley Wood. Shrawley Wood. Oldington Wood.
Not uncommon. In pine woods.

102. C. ditopa Fr.
Chaddesley Corbett Woods. Ockeridge Wood. Wyre Forest. Shrawley Wood. Dripshill Wood.
Not uncommon. In pine woods.

103. C. fragrans (Sow.) Fr.
Chatley. Trench Woods. Wyre Forest. Shrawley Wood. Perry Wood. Nunnery Wood. Great Farley Wood. The Randans. Monk Wood. Old Hills. Westwood Park.
Common. In woods and pastures. Edible and delicious.

LACCARIA Berk. (From a resinous substance which characterizes most of the species, similar to *Lac*, an Eastern dye.)

104. L. laccata (Scop.) B. and Br.
Common everywhere. In woods and heaths. Edible.

Var. amethystina (Vaill.) B. and Br.
Bevere Green. Wyre Forest. Shrawley Wood. Trench Woods. Dripshill Wood. Great Farley Wood.

Var. tortilis (Bolt.) B. and Br.
Wyre Forest. Bevere Green. Nunnery Wood. Shrawley Wood.

COLLYBIA Fr. (κόλλυβος, a small coin, from the shape of these plants.)

105. C. radicata (Relh.) Fr.
Perry Wood. Shrawley Wood. Trench Woods. Nunnery Wood. Ockeridge Wood. Monk Wood. Dripshill Wood. Wyre Forest.
Common. In woods.

106. C. longipes (Bull.) Fr.
Ladywood, Salwarpe. Between Knightwick station and Ravenshill Wood. Wyre Forest. Shrawley Wood. Nunnery Wood.
Uncommon. On heaths, in hedgerows, and in woods. Edible.

107. C. eriocephala Rea.
First Record. As *C. veluticeps Rea, Trans. Brit. Myc. Soc.*, vol. i, p. 157.
Claines, Lane leading to Hallow Ford, *W. H. Edwards.*
Rare. On old stumps of elm.

108. C. platyphylla Fr.
Nunnery Wood. Wyre Forest. Shrawley Wood. Monk Wood. Ockeridge Wood. Dripshill Wood. Trench Woods. Perry Wood.
Common. In woods, on rotten wood or stumps.

109. C. semitalis Fr.
Trench Woods.
Rare. In woods.

110. C. fusipes (Bull.) Fr.
Wyre Forest. Shrawley Wood. Tiddesley Wood. Great Farley Wood. Ockeridge Wood.
Common. In woods and under oaks in fields. Edible and palatable.

111. C. maculata (A. and S.) Fr.
Malvern Wells. Habberley Valley. Great Farley Wood. Shrawley Wood. Ockeridge Wood. Wyre Forest.
Common. In pine and mixed woods. Edible, but very bitter and unpalatable.

112. C. prolixa (Fl. Dan.) Fr.
Shrawley Wood. Wyre Forest.
Uncommon. Amongst dead leaves.

113. C. butyracea (Bull.) Fr.
Common everywhere. In woods.

114. C. velutipes (Curt.) Fr.
Ladywood, Salwarpe. Chaddesley Corbett Woods. Kempsey Grove. Sheriff's Lench. Worcester. Malvern. Madresfield.
Common. On trunks, stumps, and palings. Edible.

115. C. vertirugis Cke.
Shrawley Wood. Wyre Forest.
Uncommon. Growing on bracken stems cut off the previous year.

116. C. stipitaria Fr.
Perry Wood. Ockeridge Wood. Roadside at top of Red Hill, Worcester. Wyre Forest. Sheriff's Lench.
Common. On dead grass and twigs.

117. C. confluens (Pers.) Fr.
Westwood. Shrawley Wood. Wyre Forest. Perry Wood. Nunnery Wood. Tiddesley Wood. Trench Woods.
Common. In woods amongst the dead leaves.

118. C. conigena (Pers.) Fr.
Hadley. Wyre Forest. Oldington Wood. Shrawley Wood. Dripshill Wood. Trench Woods.
Not uncommon. Growing on old buried cones.

119. C. cirrhata (Pers.) Fr.
Shrawley Wood. Trench Woods. Nunnery Wood. Ockeridge Wood. Wyre Forest.
Common. On the bare ground.

120. C. tuberosa (Bull.) Fr.
Shrawley Wood. Wyre Forest. Perry Wood. Dripshill Wood. Trench Woods. Nunnery Wood. Monk Wood. Tiddesley Wood.
Common. On dead agarics. It springs from a sclerotium, whereas *C. cirrhata* does not.

121. C. xanthopoda Fr.
Shrawley Wood.
Rare. In pine woods.

122. C. esculenta (Wulf.) Fr.
Ockeridge Wood. Shrawley Wood. Oldington Wood. Wyre Forest.
Not uncommon. On decaying cones. Esculent, but as this species is rarely more than half an inch across the pileus it is difficult to conceive of it as a valuable addition to our food supply.

123. C. acervata Fr.
Wyre Forest. Fries Wood.
Uncommon. In pine woods, &c. Edible.

124. C. dryophila (Bull.) Fr.
Common everywhere. In woods and pastures under oaks. Edible.

Var. funicularis Fr.
This variety is characterized by sulphur-coloured gills; it was found in Broughton Wood.

125. C. aquosa Fr.
Nunnery Wood. Shrawley Wood. Wyre Forest. Old Hills. Trench Woods. Monk Wood.
Not uncommon. In woods and under oaks. This seems rather a variety of *C. dryophila* than a distinct species, as the hygrophanous pileus and striate margin seem the only constant characters by which it is generally separated from it. Edible.

126. C. extuberans Fr.
Malvern Wells. Sharpway Gate. The Slads. Witley. Old Hills. Wyre Forest.
Not uncommon. In woods and pastures. Edible.

127. C. ocellata Fr.
Middleyards Coppice. Ockeridge Wood.
Uncommon. In clearings and open places in woods.

128. C. rancida Fr.
Shrawley Wood. Trench Woods. Perry Wood. Wyre Forest. Nunnery Wood. Crews Hill Wood. Middleyards Coppice. Dripshill Wood.
Common. In woods.

129. C. atrata Fr.
Perry Wood. Tiddesley Wood. Shrawley Wood. Old Hills. Wyre Forest.
Not uncommon. On burnt ash heaps and charcoal heaps. In this species the pileus is umbilicate, the flesh is brown, and the margin of the pileus involute.

130. C. ambusta Fr.
Tiddesley Wood. Wyre Forest. Shrawley Wood. Trench Woods. Dripshill Wood. The Slads.
Not uncommon. On charcoal heaps and burnt ground. The pileus in this species is umbonate, the margin of the pileus straight, and the stem pruinose.

131. C. clusilis Fr.
Malvern Hills. Wyre Forest. Hartlebury Common.
Uncommon. In woods and downs. Known by its broad gills and umbilicate pileus.

MYCENA Fr. (μύκης, a fungus.)

132. M. pelianthina Fr.
Great Farley Wood. Ockeridge Wood.
Not uncommon. In woods, especially beech.

133. M. olivaceo-marginata Mass.
Near Northwick Hall, Claines.
Not common. In pastures.

134. M. rubro-marginata Fr.
Shrawley Wood.
Uncommon. On rotten wood, pine stumps, and dead pine needles.

135. M. strobilina Pers. var. coccinea (Sow.) Fr.
Shrawley Wood.
Uncommon. On fir cones and twigs.

136. M. pura (Pers.) Fr.
Ockeridge Wood. Shrawley Wood. Wyre Forest. Crews Hill Wood. Malvern Wells. Perry Wood. Nunnery Wood. Tiddesley Wood. Monk Wood. Trench Woods.
Common. In woods.

137. M. pseudopura Cke.
Shrawley Wood. Perry Wood. Wyre Forest.
Common. In woods. This is really only a form of *M. pura*.

138. M. lineata (Bull.) Fr.
Wyre Forest. Malvern Wells. Hartlebury Common.
Uncommon. Amongst short grass and moss.

139. M. flavo-alba Fr.
Rimells Farm, Crowneast. Westwood Park. Malvern Wells Common. Ladywood, Salwarpe.
Not uncommon. In mossy pastures.

140. **M. luteo-alba** (Bolton) Pers.
Malvern. Oldington Wood. Spetchley Park. Trench Woods.
Not common. Amongst moss, chiefly under pines.

141. **M. lactea** Pers.
Trench Woods. Great Farley Wood. Shrawley Wood. Hadley Bowling Green.
Not uncommon. On conifer needles.

142. **M. parabolica** Fr.
Malvern Hills above the Wells. Perry Wood. Trench Woods. Shrawley Wood.
Uncommon. On stumps, chiefly of conifers.

143. **M. polygramma** (Bull.) Pers.
Ockeridge Wood. Shrawley Wood. Trench Woods. Wyre Forest. Great Farley Wood. Crews Hill Wood. Perry Wood. Nunnery Wood. Monk Wood.
Common. On trunks and stumps.

144. **M. galericulata** (Scop.) Pers.   *Old Woman's Bonnet.*
Common everywhere. On stumps and trunks.

var. **calopa** Fr.
Trench Woods. Wyre Forest. Shrawley Wood.
Not uncommon. On stumps.

145. **M. sudora** Fr.
Perry Wood. Wyre Forest. Nunnery Wood.
Rare. On trunks and stumps, especially birch.

146. **M. rugosa** Fr.
Shrawley Wood. Great Farley Wood. Ockeridge Wood. Wyre Forest. Perry Wood. Nunnery Wood.
Very common. On and near trunks and stumps.

147. **M. prolifera** (Sow.) Fr.
Perry Wood.
Very rare. On a rotten oak stump.

148. **M. tenuis** (Bolton) Fr.
Chatley.
Uncommon. On a damp hedgebank.

149. **M. umbellifera** (Schaeff.) Quél. (=aetites Fr.).
Near Bevere Green. Malvern Wells. Hartlebury Common.
Uncommon. Amongst grass and damp moss.

150. **M. peltata** Fr.
Spetchley Park.
Uncommon. Amongst grass.

151. **M. cinerea** Mass and Crossl.
Near Randans.
Uncommon. Amongst short grass.

152. **M. ammoniaca** Fr.
Dripshill. Wyre Forest. Trench Woods. Malvern Wells.
Abundant in places where it occurs. On the ground, especially under conifers.

153. **M. alcalina** Fr.
Shrawley Wood. Ockeridge Wood. Great Farley Wood. Wyre Forest. Chaddesley Corbett Woods. Crews Hill Woods. Perry Wood. Monk Wood. Trench Woods.
Common. On trunks and stumps and amongst dead leaves.

154. **M. plicosa** Fr.
Great Malvern.
Very uncommon. Amongst short grass.

155. **M. filopes** Bull.
Shrawley Wood. Trench Woods. Perry Wood. Ockeridge Wood. Nunnery Woods. Monk Wood. Great Farley Wood. Chaddesley Corbett Woods. Wyre Forest.
Not uncommon. In woods amongst leaves.

156. **M. Iris** Berk.
Shrawley Wood. Wyre Forest.
Rather uncommon. On fir stumps.

157. **M. amicta** Fr.
Shrawley Wood. Wyre Forest. Trench Woods. Nunnery Wood.
Uncommon. Amongst moss on the ground.

158. **M. debilis** Fr.
Perry Wood. Nunnery Wood. Monk Wood. Shrawley Wood. Wyre Forest.
Not uncommon. Amongst moss on the ground.

159. **M. vitilis** Fr.
Nunnery Wood. Shrawley Wood. Wyre Forest.
Not uncommon. In woods in damp places.

160. **M. acicula** (Schaeff.) Fr.
Hawford. Middleyards Coppice. Wyre Forest.
Uncommon. On fallen twigs and rotten wood.

161. **M. leucogala** Cke.
Wyre Forest. Great Farley Wood. Shrawley Wood. Perry Wood. Nunnery Wood. Trench Woods. Monk Wood.
Not uncommon. In woods on the ground and on stumps.

162. **M. galopa** Pers.
Shrawley Wood. Perry Wood. Trench Woods. Wyre Forest. Great Farley Wood. Crews Hill Wood. Ockeridge Wood. Nunnery Wood.
Common. In woods. The plant varies in all shades of colour to a pure white.

163. **M. sanguinolenta** (A. and S.) Fr.
Wood above Little Malvern. Shrawley Wood. Wyre Forest. Perry Wood. Nunnery Wood. Monk Wood. Ockeridge Wood.
Common. In woods. It is the smallest of the Mycenae which have red juiced stems, and the margin of the gill is dark red.

164. **M. haematopa** Pers.
Shrawley Wood. Trench Woods. Wyre Forest.
Not uncommon. In woods and heaths, chiefly on rotting birch.

165. **M. rorida** Fr.
Ockeridge Wood. Shrawley Wood. Nunnery Wood. Trench Woods. Wyre Forest.
Not uncommon. In woods on dead twigs.

166. **M. vulgaris** Pers.
Wyre Forest. Trench Woods. Oldington Wood. Hadley Bowlin Green. Shrawley Wood.
Abundant locally. Under pines in woods and on lawns.

167. **M. pelliculosa** Fr.
Wyre Forest. Hartlebury Common. Castlemorton Common.
Uncommon. In damp places amongst heather.

168. **M. clavicularis** Fr.
Malvern Wells. Wyre Forest. Trench Woods. Nunnery Wood.
Uncommon. In woods.

169. **M. epipterygia** (Scop.) Pers.
Wyre Forest. Ockeridge Wood. Shrawley Wood. Great Farley Wood. Crews Hill Wood. Wyre Forest. Perry Wood. Nunnery Wood. Monk Wood. Trench Woods. Tiddesley Wood.
Common. In woods and pastures.

170. **M. pterigena** Fr.
Wyre Forest. Shrawley Wood.
Rare. On dead bracken.

171. **M. discopa** (Lév.) Fr.
Nunnery Wood. Perry Wood. Shrawley Wood. Wyre Forest. Monk Wood. Dripshill Wood.
Common. In woods on twigs.

172. **M. tenerrima** Berk.
Camp. Claines. Shrawley Wood. Wyre Forest. Trench Woods. Tiddesley Woods. The Slads.
Not uncommon. On pollarded willow heads, fir cones, sticks, and elm-trunks.

173. **M. stylobates** Pers.
Perry Wood. Grimley. Nunnery Wood. Wyre Forest. Shrawley Wood. Tiddesley Wood.
Not uncommon. On dead twigs and sticks. Known by its radiately striate disk at the base of the stem.

174. **M. capillaris** Fr.
Wyre Forest. Knightwick. Shrawley Wood. Wyre Forest. Rous Lench Beeches.
Locally common. On beech leaves.

175. **M. hiemalis** (Osbeck.) Fr.
Barnard's Green, Malvern. Ladywood, Salwarpe. Wyre Forest. Shrawley Wood.
Not uncommon. On bark of trees, chiefly elm.

176. **M. corticola** Pers.
Wyre Forest. Sharpway Gate. Malvern Wells. Tiddesley Wood.
Uncommon. On bark of trees.

OMPHALIA Fr. (ὀμφαλός, the navel, from the usual shape of the pileus.)

177. **O. hydrogramma** Fr.
Malvern Hills.
Uncommon. Amongst dead leaves.

178. **O. Postii** Fr.
Wyre Forest.
Uncommon. Amongst grass in swampy places.

179. **O. pyxidata** (Bull.) Fr.
Wyre Forest.
Not common. Amongst short grass and heather.

180. **O. rustica** Fr.
Wyre Forest. Trench Woods. Hartlebury Common.
Not uncommon. On the bare soil.

181. **O. hepatica** (Batsch) Fr.
Shrawley Wood. Wyre Forest.
Not common. Amongst short grass.

182. **O. muralis** (Sow.) Fr.
Salwarpe Mill. Fernhill Heath. Bridge over Dick Brook. St. John's, Worcester.
Not uncommon. Amongst moss on walls.

183. **O. umbellifera** (Linn.) Fr.
Wyre Forest.
Not common. Amongst grass in swamps and on decaying thatched roofs.

184. **O. pseudoandrosacea** (Bull.) Fr.
Barbourne, Worcester.
Uncommon. On the soil in a pot of ferns.

185. **O. griseo-pallida** Desmaz.
Fernhill Heath.
Uncommon. Amongst moss on walls.

186. **O. camptophylla** Berk.
Wyre Forest.
Uncommon. On dead twigs.

187. **O. grisea** Fr.
Ockeridge Wood. Shrawley Wood. Wyre Forest. Trench Woods.
Common. On the ground and amongst short grass

188. **O. umbratilis** Fr.
Northwick.
Uncommon. Amongst short grass.

189. **O. fibula** (Bull.) Fr.
Wyre Forest. Hadley Bowling Green. Chaddesley Corbett Woods. Shrawley Wood. Trench Woods.
Common. Amongst short grass and on charcoal heaps.

Var. **Swartzii** Fr.
Old Hills. Shrawley Wood. Chaddesley Corbett Woods.
Not uncommon. Amongst short grass.

190. O. integrella (Pers.) Fr.
Nunnery Wood. Shrawley Wood. Dripshill Wood.
Not uncommon. On the ground and on twigs.

PLEUROTUS Fr. (πλευρόν, a side, οὖς, an ear, from the shape of the pileus and position of the stem when present.)

191. P. corticatus Fr.
Bourne's Field, Claines. Shrawley Wood.
Uncommon. On oak-stumps. Edible.

192. P. dryinus (Pers.) Fr.
Lock, Ladywood, Salwarpe. Kepax Ferry, Worcester. Hanbury Park.
Uncommon. On walnut and oak trees. Edible.

193. P. ulmarius (Bull.) Fr.
Spetchley Park. Bewdley. Ripple. Westwood Park.
Not uncommon. On dying elms. Edible.

194. P. subpalmatus Fr.
Ladywood, Salwarpe. Sharpway Gate, Stoke. Bevere.
Not uncommon. On elm posts and stumps.

195. P. ostreatus (Jacq.) Fr.   *Oyster Mushroom.*
Upper Wick. Porter's Mill. Saint Oswald's, Worcester. Cowleigh Park. Near Birtsmorton Court. Elmley Castle. Way up to Saint Ann's, Malvern.
Common. On trunks. Edible and delicious.

Var. glandulosus (Bull.) Fr.
Shrawley Wood.
Uncommon. On rotten stumps.

Var. euosmus (Berk.) Cke.
River Salwarpe near Mildenham Mill.
Uncommon. On a prostrate elm. The spores are lilac-coloured when deposited in mass.

196. P. tremulus (Schaeff.) Fr.
Wyre Forest.
Uncommon. On mosses and on bare soil in ruts.

197. P. acerosus Fr.
Nunnery Wood. Tiddesley Wood. Monk Wood.
Not common. On mosses and amongst dead leaves.

198. P. septicus Fr.
Spetchley Park. Perry Wood. Ladywood, Salwarpe.
Uncommon. On dead rotten wood.

199. P. cyphellaeformis Berk.
Shrawley Wood.
Uncommon. On dead sticks and nettle stems.

200. P. applicatus (Batsch) Fr.
Shrawley Wood. Nunnery Wood.
Not uncommon. On decaying wood and twigs.

201. P. chioneus (Pers.) Fr.
Shrawley Wood. Trench Woods. Nunnery Wood.
Not uncommon. On fallen branches and leaves.

VOLVARIA Fr. (*Volva*, a wrapper, from the universal veil which remains distinct at the base of the stem.)

202. V. bombycina (Schaeff.) Fr.
Astwood, Claines.
Rare. On an elm stump.

203. V. Taylori Berk.
Ladywood. Upton-on-Severn.
Uncommon. On the bare soil. Poisonous.

204. V. speciosa Fr.
Ombersley.
Rather rare. On a manure heap. Poisonous.

205. V. parvula (Weinm.) Fr.
Merryman's Hill, Worcester.
Uncommon. Amongst short grass.

206. V. media (Schum.) Fr.
Sharpway Gate, Stoke.
Not common. Amongst short grass.

PLUTEUS Fr. (*Pluteus*, a pent-house, from the conical pileus.)

207. P. cervinus (Schaeff.) Fr.
Shrawley Wood. Perry Wood. Monk Wood. Nunnery Wood. Trench Woods. Wyre Forest. Tiddesley Wood.
Common. On stumps and buried wood.

Var. Bullii Berk.
Nunnery Wood.
Uncommon. On stumps.

208. P. ephebeus Fr.
Lane leading to Porters Mill.
Rare. On stumps.

209. P. salicinus (Pers.) Fr.
Shrawley Wood. Wyre Forest.
Uncommon. On willow and alder trunks and sticks.

210. P. chrysophaeus (Schaeff.) Fr.
Wyre Forest.
Rare. On fallen sticks.

ENTOLOMA Fr. (ἐντός, within, λῶμα, a fringe, probably referring to the innate character of the partial veil.)

211. E. sinuatum Fr.
Wyre Forest. Crews Hill Wood. Trench Woods. Shrawley Wood. Tiddesley Wood.
Common. In deciduous woods. Poisonous.

212. E. lividum (Bull.) Fr.
Crews Hill Wood. Shrawley Wood. Wyre Forest.
Not uncommon. In dry woods. Poisonous.

213. E. prunuloides Fr.
Wyre Forest.
Uncommon. Amongst grass and moss.

214. E. porphyrophaeum Fr.
Shrawley Wood. Near Dripshill Wood. Little Malvern.
Not uncommon. In woods and pastures.

215. E. Bloxami B. and Br.
Near Crews Hill Wood.
Uncommon. In pastures.

216. E. ardosiacum (Bull.) Fr.
Great Bog, Wyre Forest.
Rare. In swampy places amongst grass.

217. E. Saundersii Fr.
Timber-yard, near Stoke Works Station.
Rare. On sawdust heap. The gills are almost free.

218. E. jubatum Fr.
Bevere Green. Wyre Forest. Malvern Wells. Near Tiddesley Wood. Crowneast. Old Hills.
Not uncommon. In pastures and in clearings in woods.

219. E. griseo-cyaneum Fr.
Wyre Forest. Near Ockeridge Wood. Near Dripshill Wood. Old Hills.
Not uncommon. In pastures and in woods.

220. E. clypeatum (Linn.) Fr.
Rubery. Shrawley Wood. Castlemorton Common.
Not uncommon. Amongst grass, often caespitose and occurring in spring and early summer.

221. E. rhodopolium Fr.
Seckley Wood, Wyre Forest. Shrawley Wood. Nunnery Wood.
Not uncommon. In woods.

222. E. costatum Fr.
Wyre Forest. Woodbury Hill. Ockeridge Wood. Malvern Wells.
Not uncommon. In pastures and open places in woods.

223. E. sericeum (Bull.) Fr.
Randans Wood. Wyre Forest. Woodbury Hill. Crews Hill Wood. Trench Woods. Chatley. Kempsey. Malvern Wells. Defford Common. Ladywood, Salwarpe.
Common. In grassy places in woods and pastures.

224. E. nidorosum Fr.
Wyre Forest. Perry Wood. Shrawley Wood. Monk Wood. Ockeridge Wood. Nunnery Wood. Trench Woods.
Common. In woods.

225. E. speculum Fr.
Monk Wood.
Uncommon. Amongst grass and twigs.

CLITOPILUS Fr. (κλίτος, a declivity, πῖλος, a cap, from the decurrent gills.)

226. C. prunulus (Scop.) Fr. (including orcella (Bull.) Fr.).   *Sweetbread.*
Common everywhere thoughout the county in woods and pastures. Edible and delicious.

227. C. popinalis Fr.
Cowleigh Park. Bredon Hill.
Uncommon. On downs and amongst short grass.

228. C. cancrinus Fr.
Between Perry and Nunnery Woods. New Parks, Wyre Forest.
Rather uncommon. In pastures and open glades.

229. C. stilbocephalus B. and Br.
Hampton Lovett.
Rare. Amongst grass in pastures, forming rings.

LEPTONIA Fr. (λεπτός, slender, because most of the species are of medium size.)

230. L. lampropa Fr.
Wyre Forest. Shrawley Wood. Malvern Wells. Ladywood, Salwarpe. Hartlebury Common.
Not uncommon. In pastures and grassy glades in woods.

231. L. aethiops Fr.
Malvern Wells. Dodderhill Common. Bevere Green. Wyre Forest.
Not common. In pastures and on commons.

232. L. solstitialis Fr.
Shrawley Wood. Between Perry and Nunnery Woods. Dodderhill Common.
Uncommon. Amongst short grass.

233. L. serrulata Fr.
Wyre Forest. Shrawley Wood. Little Malvern. Near Dripshill.
Not uncommon. Amongst short grass in woods and pastures and on lawns.

234. L. euchroa (Pers.) Fr.
Perry Wood. Trench Woods. Nunnery Wood. Shrawley Wood. Monk Wood. Ockeridge Wood.
Not uncommon. On stumps and branches of alder, hazel, and birch.

235. L. lazulina Fr.
Wyre Forest.
Uncommon. Amongst short grass.

236. L. incana Fr.
Wyre Forest. Crews Hill Wood. Trench Woods. Monk Wood. Ockeridge Wood.
Not uncommon. Amongst short grass. The smell is exactly that of a mousy cupboard.

237. **L. sericella** (Fr.) Quél.
  Wyre Forest. Nunnery Wood. Perry Wood. Trench Woods. Shrawley Wood. Tiddesley Wood.
  Common. Amongst short grass.

NOLANEA Fr. (*Nola*, a little bell, from the shape of the pileus in most of the species.)

238. **N. pascua** (Pers.) Fr.
  Common everywhere. In woods and pastures.

239. **N. mammosa** (Linn.) Fr.
  Wyre Forest. Shrawley Wood. Nunnery Wood. Trench Woods.
  Not uncommon. In woods and pastures.

240. **N. pisciodora** (Cesati) Fr.
  Wyre Forest. Shrawley Wood. Randans Wood. Trench Woods. Perry Wood. Nunnery Wood.
  Not uncommon. In woods amongst dead leaves.

241. **N. icterina** Fr.
  Wyre Forest. Monk Wood.
  Uncommon. Amongst short grass.

242. **N. picea** Kalchbr.
  Shrawley Wood. Trench Woods.
  Uncommon. In woods amongst short grass. Distinguished from *N. pisciodora* by the dark smooth pileus and stem.

ECCILIA Fr. (*ἔγκοιλος*, hollowed out, because most of the species have an umbilicate pileus.)

243. **E. carneo-grisea** B. and Br.
  Wyre Forest. Shrawley Wood. Trench Woods.
  Uncommon. In woods under pines.

244. **E. griseo-rubella** (Lasch.) Fr.
  Hadley Bowling Green. Malvern Wells.
  Uncommon. Under Scotch pines.

245. **E. atropuncta** (Pers.) Fr.
  Shrawley Wood.
  Uncommon. In woods amongst *Mercurialis perennis*.

CLAUDOPUS W. G. Smith. (*Claudus*, lame, *πούς*, a foot, from the crooked or absent stem.)

246. **C. variabilis** (Pers.) W. G. Sm.
  Common everywhere. On dead leaves, decaying twigs and branches.

PHOLIOTA (Fr.). (*φολίς*, a scale, because the pileus or stem in many species is scaly.)

247. **P. terrigena** Fr.
  Common Hill, Northwick, Worcester. Malvern Wells. Wyre Forest. Nunnery Wood. Middleyards Coppice.
  Not uncommon. Amongst grass in pastures and woods.

248. **P. erebia** Fr.
  Crews Hill Wood.
  Uncommon. In woods and grassy places.

249. **P. togularis** (Bull.) Fr.
  Knightwick. Shrawley Wood. Wyre Forest. Perry Wood. Nunnery Wood.
  Not uncommon. In woods and amongst short grass.

250. **P. dura** (Bolton) Fr.
  Near Worrells Mill. Malvern Wells. Defford Common. Claines. Hindlip.
  Not uncommon. In fields and gardens.

251. **P. praecox** (Pers.) Fr.
  Hallow Ford. Wyre Forest. Kempsey. Helbury Hill. Monk Wood. Tiddesley Wood. Malvern Wells.
  Common. In woods, fields, and gardens. Not confined to spring as the specific name suggests.

252. **P. radicosa** (Bull.) Fr.
  Shrawley Wood. Wyre Forest. Nunnery Wood. Monk Wood. Ockeridge Wood. Trench Woods.
  Uncommon. In woods, generally near a stump.

253. **P. aegerita** Fr.
  Perdiswell Lane. Great Farley Wood. Kempsey Grove. Spetchley Park.
  Uncommon. On elm, ash, and poplar trunks.

254. **P. squarrosa** (Müll.) Fr.
  Acacia House, Foregate Street, Worcester. Chatley. Ladywood, Salwarpe. Shrawley Wood. Crews Hill Wood.
  Common. Caespitose at the base of trees, especially apple.

255. **P. grandis** Rea.
  First Record. *Rea, Trans. British Myc. Soc.*, vol. ii, p. 37.
  Saint John's, Worcester. Shrawley Wood. Near Clerkenleap. Pershore.
  Not uncommon. Caespitose at the base of trees, especially ash.

256. **P. spectabilis** Fr.
  Bevere Green. Chatley. Clevelode. Alfrick Pound. Sheriff's Lench. Ombersley
  Common. Subcaespitose at the base of trees and stumps.

257. **P. adiposa** Fr.
  Ladywood, Salwarpe. Hanbury Park. Malvern Wells.
  Uncommon. Caespitose on trunks of trees, especially ash.

258. **P. mutabilis** (Schaeff.) Fr.
  Shrawley Wood. Crews Hill Wood. Dripshill Wood. Trench Woods. Tiddesley Wood.
  Common. Caespitose on stumps, especially lime. Edible.

259. **P. marginata** (Batsch) Fr.
  Shrawley Wood. Wyre Forest. Trench Woods. Crews Hill Wood. Ockeridge Wood.
  Not uncommon. In woods, chiefly under conifers.

INOCYBE Fr. (*ἴς*, fibre, *κύβη*, head, from the fibrous pileus.)

260. **I. hystrix** Fr.
  Shrawley Wood. Trench Woods. Ockeridge Wood. Wyre Forest.
  Uncommon. In woods.

261. **I. lanuginosa** (Bull.) Fr.
  Shrawley Wood. Crews Hill Wood. Wyre Forest.
  Not uncommon. On the ground in woods and shady places.

262. **I. cincinnata** Fr.
  Monk Wood. Crews Hill Wood. Shrawley Wood. Wyre Forest.
  Not uncommon. In woods and shady places.

263. **I. pyriodora** (Pers.) Fr.
  Wyre Forest. Tiddesley Wood. Elmley Castle. Shrawley Wood. Nunnery Wood.
  Not uncommon. In woods.

264. **I. corydalina** Quél.
  First record for Britain, Crews Hill Wood, *Rea, Trans. Brit. Myc. Soc.*, vol. ii, p. 63, pl. iv.
  Rare. In woods, chiefly under birch.

265. **I. incarnata** Bres.
  Shrawley Wood. Crews Hill Wood.
  Uncommon. In woods.

266. **I. rhodiola** Bres.
  Wood Norton.
  Rare. In woods and shady places.

267. **I. scabra** (Müll.) Fr.
  Crews Hill Wood.
  Uncommon. In woods.

268. **I. maritima** Fr.
  Oldington Wood. Wyre Forest.
  Uncommon. In woods.

269. **I. flocculosa** Berk.
  Dripshill Wood. Monk Wood. Nunnery Wood.
  Not uncommon. In woods and shady places on the naked soil.

270. **I. mutica** Fr.
  Wyre Forest. Trench Woods.
  Uncommon. In woods.

271. **I. cervicolor** (Pers.) Quél.
  Shrawley Wood. Wyre Forest. Nunnery Wood. Perry Wood. Trench Woods. Tiddesley Wood.
  Not uncommon. On the bare soil in woods and shady places.

272. **I. fastigiata** (Schaeff.) Fr.
  Shrawley Wood. Wyre Forest. Nunnery Wood.
  Uncommon. In woods.

273. **I. Godeyi** Gillet.
  Wood Norton.
  Rare. In woods.

274. **I. rimosa** (Bull.) Fr.
  Common everywhere. In woods and shady places.

275. **I. brunnea** Quél.
  Trench Woods.
  Rare. In woods.

276. **I. duriuscula** Rea.
  First record for Britain, Monk Wood, *Rea, Trans. Brit. Myc. Soc.*, vol. iii, p. 44, pl. III.
  Trench Woods.
  Uncommon. In woods.

277. **I. asterospora** Quél.
  Shrawley Wood. Perry Wood. Crews Hill Wood. Wyre Forest. Nunnery Wood. Monk Wood. Trench Woods.
  Not uncommon. In woods.

278. **I. eutheles** B. and Br.
  Wyre Forest. Ockeridge Wood.
  Uncommon. In woods.

279. **I. calospora** Quél.
  Nunnery Wood.
  Rare. In woods.

280. **I. geophylla** (Sow.) Fr.
  Wyre Forest. Shrawley Wood. Trench Woods. Ockeridge Wood. Great Farley Wood. Chaddesley Corbett Woods.
  Common. In woods and shady places. Both the white and blue forms are equally common.

281. **I. scabella** (Fr.) Bres.
  Bevere Green. Perry Wood. Nunnery Wood. Shrawley Wood.
  Not uncommon. In woods and shady places.

282. **I. fulvella** Bres.
  Wyre Forest. Nunnery Wood.
  Uncommon. In woods.

283. **I. petiginosa** (Fr.) Quél.
  Shrawley Wood. Wyre Forest. Trench Woods. Nunnery Wood.
  Not uncommon. In woods and shady places.

HEBELOMA Fr. (*ἥβη*, youth, *λῶμα*, a fringe, from the veil.)

284. **H. musivum** Fr.
  Shrawley Wood.
  Rare. In woods.

285. **H. fastibile** Fr.
　　Ockeridge Wood. Shrawley Wood. Perry Wood. Wyre Forest. Randans Wood. Great Farley Wood.
　　Common. In woods. Poisonous.

286. **H. glutinosum** (Pers.) Fr.
　　Wyre Forest. Shrawley Wood. Nunnery Wood. Trench Woods.
　　Not common. In woods amongst dead leaves.

287. **H. versipelle** Fr.
　　Trench Woods. Shrawley Wood.
　　Uncommon. In woods.

288. **H. mesophaeum** Fr.
　　Shrawley Wood. Wyre Forest. Trench Woods. Tiddesley Wood. Nunnery Wood.
　　Not uncommon. In woods on the ground and on charcoal heaps.

289. **H. sinapizans** Fr.
　　Shrawley Wood.
　　Rare. In woods.

290. **H. crustuliniforme** (Bull.) Fr.
　　Perry Wood. Shrawley Wood. Wyre Forest. Crews Hill Wood. Chaddesley Corbett Woods. Nunnery Wood.
　　Common. In woods and pastures.

　　Var. minor Cke.
　　Ockeridge Wood. Shrawley Wood. Wyre Forest.

291. **H. elatum** Fr.
　　Bevere Green. Trench Woods.
　　Uncommon. In woods, chiefly under conifers.

292. **H. longicaudum** (Pers.) Fr.
　　Shrawley Wood. Wyre Forest. Tiddesley Wood. Monk Wood.
　　Not uncommon. In woods. Edible.

293. **H. nauseosum** Cke.
　　Wyre Forest. Shrawley Wood.
　　Not common. In woods.

FLAMMULA Fr.　(*Flamma*, a flame.)

294. **F. gummosa** (Lasch.) Fr.
　　Shrawley Wood. Ockeridge Wood. Dripshill Wood.
　　Not uncommon. In woods and about old stumps.

295. **F. carbonaria** Fr.
　　Ockeridge Wood. Wyre Forest. Shrawley Wood. Hanley Dingle. Trench Woods. Monk Wood.
　　Common. On charcoal heaps and where wood has been burned.

296. **F. fusa** (Batsch) Fr.
　　Wyre Forest.
　　Uncommon.

297. **F. rubicundula** Rea.
　　First Record. Wyre Forest, *Grevillea*, vol. xxii, p. 40.
　　Rare. In woods under scrub oaks.

298. **F. alnicola** Fr.
　　Teme-side, Lower Wick. Shrawley Wood.
　　Not common. On stumps of alder and willow, &c.

299. **F. flavida** (Schaeff.) Fr.
　　Middleyards Coppice. Shrawley Wood. Chaddesley Corbett Woods. Nunnery Wood.
　　Not uncommon. On buried pieces of wood, generally in cart-tracks in woods.

300. **F. inopa** Fr.
　　Bevere Green. Wyre Forest. Great Farley Wood. Shrawley Wood. Ockeridge Wood.
　　Not uncommon. On stumps. Said to be only a form of *Hypholoma fasciculare.*

301. **F. sapinea** Fr.
　　Wyre Forest. Shrawley Wood. Trench Woods.
　　Not common. On stumps and fallen branches of conifers.

302. **F. ochrochlora** Fr.
　　Perry Wood. Shrawley Wood. Dripshill Wood.
　　Uncommon. In woods and near stumps.

303. **F. tricholoma** (A. and S.) Karst.
　　Shrawley Wood.
　　Uncommon. In woods.

NAUCORIA Fr.　(*Naucum*, a trifle, from the almost obsolete veil.)

304. **N. melinoides** (Bull.) Fr.
　　Trench Woods. Bevere Green. Sheriff's Lench. Fernhill Heath. Ladywood, Salwarpe. Hartlebury Common.
　　Common. Amongst short grass in woods, pastures, and lawns.

305. **N. badipes** Fr.
　　Fries Wood. Tiddesley Wood. Middleyards Coppice. Willow bed, Claines.
　　Not uncommon. In damp places in woods and swampy ground.

306. **N. striaepes** Cke.
　　Ladywood, Salwarpe. Defford Common.
　　Uncommon. Amongst grass.

307. **N. pediades** Fr.
　　Ladywood, Salwarpe. Kempsey.
　　Not common. In pastures.

308. **N. semiorbicularis** (Bull.) Fr.
　　Westwood Park. Northwick Lane. Hanbury Park. Chatley. Kempsey. Ladywood, Salwarpe.
　　Common. In pastures.

M m

309. **N. tabacina** (DC.) Fr.
　　Hadley Bowling Green. Wyre Forest. Shrawley Wood. Grimley Brickyard.
　　Not uncommon. In damp places and roadsides.

310. **N. tenax** Fr.
　　Hartlebury Common.
　　Uncommon. On dead stems of *Potentilla palustris* growing in the bog.

311. **N. myosotis** Fr.
　　Hartlebury Common.
　　Uncommon. Amongst *Sphagnum* and *Potentilla palustris* in the bog.

312. **N. temulenta** Fr.
　　Nunnery Wood. Middleyards Coppice. Dripshill Wood. Tiddesley Wood.
　　Not uncommon. Amongst twigs and chips in damp places in woods.

313. **N. escharoides** Fr.
　　Shrawley Wood. Wyre Forest. Cowleigh Park.
　　Not uncommon. In moist places under alders.

PLUTEOLUS Fr.　(*Pluteolus*, diminutive of *pluteus*, a pent-house or shed.)

314. **P. aleuriatus** Fr.
　　Shrawley Wood.
　　Uncommon. On rotten sticks.

GALERA Fr.　(*Galerus*, a cap, from the shape of the pileus in most species.)

315. **G. tenera** (Schaeff.) Fr.
　　Common everywhere. In pastures and woods.

316. **G. campanulata** Mass.
　　Malvern Wells.
　　Uncommon. By roadsides.

317. **G. ovalis** Fr.
　　Ladywood, Salwarpe. Sheriff's Lench. Hartlebury Common. Defford Common.
　　Not uncommon. On dung and amongst grass.

318. **G. spicula** (Lasch.) Fr.
　　Kidderminster, *Goodwin.*
　　Rare. On cocoa-nut fibre.

319. **G. spartea** Fr.
　　Perry Wood. Bevere Green. Chatley. Ladywood, Salwarpe. Hartlebury Common. Westwood Park.
　　Not uncommon. In pastures and in woods.

320. **G. hypnorum** (Schrank) Fr.
　　Common everywhere. Amongst moss in fields and woods.

　　Var. sphagnorum (Pers.) Fr.
　　Shrawley Wood. Wyre Forest. Hartlebury Common.

TUBARIA W. G. Smith.　(*Tuba*, a trumpet.)

321. **T. furfuracea** (Pers.) W. G. Sm.
　　Common everywhere. On the ground, twigs and chips, throughout the year.

　　Var. trigonophylla Fr.
　　Almost as common as the type.

322. **T. paludosa** Fr.
　　Hartlebury Common. Wyre Forest.
　　Uncommon. In swamps and bogs amongst *Sphagnum.*

323. **T. stagnina** Fr.
　　Hartlebury Common.
　　Uncommon. In swamps amongst *Sphagnum.*

324. **T. autochthona** (B. and Br.) W. G. Sm.
　　Trench Woods.
　　Uncommon. On naked soil.

325. **T. crobula** Fr.
　　Ockeridge Wood. Nunnery Wood. Shrawley Wood. Trench Woods. Dripshill. Monk Wood.
　　Not uncommon. In moist places and in woods.

326. **T. inquilina** (Fr.) W. G. Sm.
　　Nunnery Wood. Middleyards Coppice. Wyre Forest.
　　Uncommon. On twigs in damp places, chiefly in woods.

CREPIDOTUS Fr.　(*κρηπίς*, a sandal.)

327. **C. mollis** (Schaeff.) Fr.
　　Claines. Chatley. Hanbury. Tiddesley Wood. Shrawley Wood. Monk Wood. Nunnery Wood.
　　Common. On rotten stumps, trunks, and twigs.

328. **C. calolepis** Fr.
　　Wyre Forest. Shrawley Wood.
　　Uncommon. On rotten trunks, stumps, and twigs.

329. **C. epigaeus** (Pers.) Fr.
　　Rous Lench Beeches.
　　Rare. On the ground.

AGARICUS (Linn.) Mass.　(*ἀγαρικόν*, a region of Sarmatia.)

330. **A. augustus** Fr.
　　Worcester. Bishampton. Croome Park.
　　Uncommon. In fields and woods. Edible.

331. **A. campestris** Linn.　*Mushroom.*
　　Common everywhere. In fields. Edible.

　　Var. silvicola Vitt.
　　Shrawley Wood. Dripshill Wood. Nunnery Wood. Wyre Forest. Tiddesley Wood.
　　Common. In woods. Edible.

Var. hortensis Cke.
Sheriff's Lench. Woolashall. Hindlip. Madresfield. Witley.
This is the common cultivated form. Its flavour is not so good as the two previous ones.

Var. rufescens Berk.
Lucas's garden, Britannia Square, Worcester.
Uncommon. In pastures. Edible.

332. **A. perrarus** Schulz.
First Record for Britain. Claines, Worcestershire, August 26, 1901, Rea, *Trans. Brit. Myc. Soc.*, vol. i, p. 200.
Claines. Pasture near Crews Hill Wood.
Uncommon. Under oaks in pastures. The scales on the pileus are as strongly marked as those of a *Lepiota*.

333. **A. arvensis** Schaeff. *Horse Mushroom. Abrahams.*
Common everywhere. In pastures under trees, often running to a very large size. Edible, and considered by most mycophagists as superior in flavour to *A. campestris*.

Var. purpurascens Cke.
Shrawley Wood.
Very uncommon. This is a very beautiful variety and almost worthy of specific rank.

334. **A. xanthoderma** Genév. *Yellow-stained Horse Mushroom.*
Common everywhere. In pastures and woods. This species differs from *A. arvensis* in having lactiferous vessels. Edible, but causing indisposition with some people, just as there are some persons who cannot digest pork, shell crustaceans, or fish.

335. **A. silvaticus** Schaeff.
Hindlip. Claines. Dripshill Wood. Shrawley Wood. Nunnery Wood.
Uncommon. In woods and pastures. Edible.

336. **A. pratensis** Schaeff.
Shrawley Wood.
Uncommon. Amongst grass in lane leading into the wood from the inn. Edible.

337. **A. haemorrhoidarius** Schulz.
The Noke Hill, Martley. Fries Wood. Near Crews Hill Wood. Sheriff's Lench. Chatley. Ladywood, Salwarpe.
Not uncommon. Chiefly in pastures, sometimes in woods and there especially under conifers. Known by its brown pileus and the flesh turning to a blood colour. Edible and the finest in flavour of all our mushrooms.

338. **A. comtulus** Fr.
Claines. Ladywood, Salwarpe. Between Perry Wood and Nunnery Wood. Near Trench Woods. Ombersley. Kempsey.
Not uncommon. In pastures. A small species with a pleasant aromatic scent. Edible and delicious.

STROPHARIA Fr.   (στρόφος, a sword-belt, referring to the ring.)

339. **S. aeruginosa** (Curt.) Fr.
Common everywhere. In woods and pastures. Generally admired on account of its beautiful colour. Said to be poisonous.

340. **S. squamulosa** Mass.
Malvern Wells.
Uncommon. In woods and pastures. It differs from *S. aeruginosa* in being dry, not viscid, and the gills at maturity are brown, not purplish.

341. **S. albo-cyanea** (Desm.) Fr.
Perry Wood. Shrawley Wood. Randans Wood. Wyre Forest. Tiddesley Wood. Nunnery Wood.
Common. In woods.

342. **S. inuncta** Fr.
Wyre Forest. Tiddesley Wood. Shrawley Wood. Trench Woods. Crews Hill Wood. Nunnery Wood.
Not uncommon. In woods chiefly, and occasionally in pastures and hedgerows.

343. **S. coronilla** (Bull.) Fr.
Grimley. Shrawley Wood. Malvern Wells. Hanbury Park. Sheriff's Lench. Ladywood, Salwarpe.
Not uncommon. In pastures and woods.

344. **S. melasperma** (Bull.) Fr.
Wyre Forest. Claines. Chatley. Sheriff's Lench. Clerkenleap. Ladywood, Salwarpe.
Not uncommon. In pastures. Poisonous.

345. **S. squamosa** (Pers.) Fr.
Shrawley Wood. Wyre Forest. Nunnery Wood. Tiddesley Wood.
Not uncommon. In woods.

Var. thrausta Fr.
Shrawley Wood. Nunnery Wood.
Not uncommon. In woods.

Var. aurantiaca Cke.
Shrawley Wood.
Uncommon. In woods.

346. **S. merdaria** Fr.
Shrawley Wood. Nunnery Wood.
Uncommon. On dry dung in drives in woods.

347. **S. stercoraria** Fr.
Ladywood, Salwarpe. Sheriff's Lench. Chatley. Hadley. Hanbury Park. Hallow.
Common. On dung in pastures.

348. **S. semiglobata** (Batsch) Fr.
Common everywhere. On horse-dung throughout the year.

349. **S. scobinacea** Fr.
Kings Norton.
Rare. At base of an ash-tree. Caespitose.

HYPHOLOMA Fr.   (ὑφή, a web, λῶμα, a fringe, because the partial veil is apparent at the margin of the pileus in most of the species.)

350. **H. sublateritium** Fr.
Common everywhere. On and about old stumps in woods and hedgerows. Poisonous.

351. **H. capnoides** Fr.
Dripshill Wood. Wyre Forest. Great Farley Wood. Woodbury Hill.
Not uncommon. In pine woods and on pine-trunks.

352. **H. epixanthum** (Paul.) Fr.
Hanbury Park. Shrawley Wood. Wyre Forest. Ockeridge Wood.
Not uncommon. In woods.

353. **H. fasciculare** (Huds.) Fr.
Common everywhere. On old stumps throughout the year. Poisonous.

354. **H. dispersum** Fr.
Ockeridge Wood. Shrawley Wood.
Uncommon. In woods on pine twigs. Characterized by its scattered habit, and dark stem covered with whitish flocci.

355. **H. lacrymabundum** Fr. (= Stropharia cotonea Quél.)
Shrawley Wood.
Uncommon. In woods under conifers. The base of the stem is yellowish.

356. **H. velutinum** (Pers.) Fr. (= lacrymabundum (Bull.) Quél.).
Martley. Wyre Forest. Monk Wood. Middleyards Coppice. Shrawley Wood. Nunnery Wood.
Common. In fields, woods, and roadsides.

Var. leiocephalum B. and Br.
Shrawley Wood.
Uncommon. On the ground and about old stumps.

357. **H. pyrotrichum** (Holmsk.) Fr.
Bransford Court. Tiddesley Wood. Nunnery Wood. Perry Wood. Trench Woods. Monk Wood.
Common. In woods and pastures.

358. **H. cascum** Fr.
Between Perry and Nunnery Woods. Monk Wood. Near Middleyards Coppice.
Uncommon. In woods and pastures.

359. **H. appendiculatum** (Bull.) Fr.
Holt. Lincombe. Wyre Forest. Tiddesley Wood. Middleyards Coppice. Monk Wood. Randans Wood. Ribbesford Woods. Nunnery Wood.
Common. In woods and shady places. Edible and delicious.

360. **H. pilulaeforme** (Bull.) Fr.
Dripshill Wood.
Rare. In woods.

361. **H. hydrophilum** (Bull.) Quél.
Common everywhere. On stumps in woods and shady places.

PSILOCYBE Fr.   (ψιλός, naked, κύβη, head, because no veil is apparent on the pileus.)

362. **P. sarcocephala** Fr.
Bevere Green. Shrawley Wood. Hanbury Park. Spetchley Park. Malvern Wells.
Not uncommon. About the base of trees and stumps. Edible.

363. **P. subericaea** Fr.
Castlemorton Common. Hartlebury Common. The Marsh Common.
Not uncommon. On damp wastes and marshy fields.

364. **P. uda** (Pers.) Fr.
Hartlebury Common.
Uncommon. On boggy ground.

365. **P. coprophila** (Bull.) Fr.
Chatley. Ladywood, Salwarpe.
Uncommon. On manure, especially cow-dung.

366. **P. bullacea** (Bull.) Fr.
Hollybush Hill. Habberley Valley.
Uncommon. On manure.

367. **P. physaloides** (Bull.) Fr.
Malvern Wells. Old Hills. Dodderhill Common.
Uncommon. Amongst short grass.

368. **P. semilanceata** Fr.   *Liberty-cap.*
Common everywhere. In pastures and woods. Poisonous.

Var. caerulescens Cke.
The Slads.
Uncommon. Amongst grass.

369. **P. spadicea** Fr.

Var. hygrophila Fr.
Wyre Forest.
Uncommon. In woods about stumps.

370. **P. foenisecii** (Pers.) Fr.
Common everywhere. In pastures and on lawns. Edible.

PSATHYRA Fr.   (ψαθυρός, friable.)

371. **P. conopilea** Fr.
Deadmans Coppice, Ladywood, Salwarpe. Kempsey.
Not common. In ditches and shady places.

372. P. corrugis (Pers.) Fr.
    Wyre Forest. Ockeridge Wood. Monk Wood. Shrawley Wood. Nunnery Wood. Trench Woods. Tiddesley Wood.
    Common. In woods and pastures on the ground and amongst sticks.

373. P. bifrons Berk.
    Shrawley Wood. Crowneast. Wyre Forest.
    Not uncommon. On the ground and on twigs and chips.
    Var. semitincta Phill.
    Nunnery Wood. Wyre Forest. Tiddesley Wood.

374. P. fibrillosa (Pers.) Fr.
    Wyre Forest. Shrawley Wood. Nunnery Wood. Trench Woods. Monk Wood. Ockeridge Wood.
    Common. In woods on the ground or on twigs.

375. P. semivestita B. and Br.
    Wyre Forest. Middleyards Coppice.
    Not common. In woods and shady places.

376. P. pennata Fr.
    Perry Wood. Nunnery Wood. Trench Woods.
    Not uncommon. In woods on the bare soil.

377. P. gossypina (Bull.) Fr.
    Shrawley Wood. Dripshill Wood. Crowneast Wood.
    Uncommon. In woods and shady places.

ANELLARIA Karst. (*Anellus*, a little ring, from the ring on the stem.)

378. A. separata (Linn.) Karst.
    Common everywhere. In pastures and woods, chiefly on horse-dung.

379. A. fimiputris (Bull.) Karst.
    Common everywhere. In pastures, roadsides, and occasionally in woods on dung.

PANAEOLUS Fr. (πᾶν, all, αἰόλος, variegated, from the appearance of the gills.)

380. P. phalaenarum (Bull.) Fr.
    Chatley, Westwood Park. Ladywood, Salwarpe. Sheriff's Lench. Bransford. Croome Park.
    Not uncommon. In pastures on dung.

381. P. retirugis Fr.
    Diglis Lock. Hanbury Park. Spetchley Park. Ladywood, Salwarpe.
    Not uncommon. In pastures amongst grass and on dung.

382. P. sphinctrinus Fr.
    Nunnery Wood. Tything, Worcester. Powick Ham. Ankerdine Hill. Defford Common.
    Common. In pastures and on dung.

383. P. papilionaceus (Bull.) Fr.
    Chatley. Ladywood, Salwarpe. Hartlebury Common. Sheriff's Lench. Grimley. Crowle.
    Common. In pastures, gardens, and on dung.

384. P. campanulatus (Linn.) Fr.
    Common everywhere. On dung in pastures and woods.

PSATHYRELLA Fr. (ψαθυρός, friable.)

385. P. gracilis (Pers.) Fr.
    Common everywhere. Roadsides, pastures, gardens, and cartways in woods.

386. P. atomata Fr.
    Common everywhere. Roadsides, hedges, pastures, manure and waste heaps, and in woods.

387. P. disseminata (Pers.) Fr.
    Common everywhere. Amongst mosses on old stumps, ditch-sides, and damp places. Gregarious.

COPRINUS Pers. (κόπρος, dung, the habitat of most of the species.)

388. C. comatus (Fl. Dan.) Pers.   *Maned Agaric.*
    Pitchcroft. Sheriff's Lench. Barbourne. Chaddesley Corbett Woods. Ladywood, Salwarpe. Malvern.
    Common. Amongst grass in fields, gardens, woods, and especially abundant near refuse tips. Edible and delicious.

389. C. atramentarius (Bull.) Fr.
    Wyre Forest. Monk Wood. Shrawley Wood. Kempsey Grove. Nunnery Wood. Trench Woods. Westwood Park.
    Common. In fields and woods. Caespitose. Edible and delicious.

390. C. squamosus Morgan.
    First Record. Hanbury Park, *Rea, Trans. Brit. Myc. Soc.*, vol. i, p. 158.
    Uncommon. At the base of elms.

391. C. fuscescens (Schaeff.) Fr.
    Wyre Forest. Shrawley Wood. Nunnery Wood. Monk Wood. Tiddesley Wood. Trench Woods.
    Not uncommon. In woods on and near stumps. Edible.

392. C. picaceus (Bull.) Fr.   *Magpie Coprinus.*
    Nunnery Wood. Chaddesley Corbett Woods. Shrawley Wood.
    Not uncommon. In woods. Poisonous.

393. C. fimetarius (Linn.) Fr.
    Droitwich. Chatley. Ladywood, Salwarpe.
    Uncommon. In gardens.

394. C. cinereus (Schaeff.) Fr.
    Nunnery Wood. Perry Wood. Shrawley Wood. Monk Wood. Wyre Forest. Trench Woods. Tiddesley Wood. Hartlebury Common.
    Common. On the ground in woods and on dung. This is

generally considered in the textbooks as only a variety of *C. fimetarius*, but it is very different from that species and is never found associated with it.

395. C. niveus (Pers.) Fr.
    Common everywhere. On horse-dung in fields and woods appearing through the greater part of the year.

396. C. roseotinctus Rea.
    First Record. Temple Laughern, *Rea, Trans. Brit. Myc. Soc.*, vol. i, p. 23.
    Uncommon. On the ground.

397. C. micaceus (Bull.) Fr.
    Common everywhere. On and about stumps in fields and woods.

398. C. papillatus (Batsch) Fr.
    Perry Wood. Spetchley Park. Inkberrow. Pershore.
    Uncommon. On dung.

399. C. deliquescens (Bull.) Fr.
    Wyre Forest. Shrawley Wood. Chatley. Hadley. Habberley Valley. Nunnery Wood.
    Not uncommon. In woods and hedgerows.

400. C. Hendersonii Berk.
    White's Nursery, Lower Wick.
    Rare. In gardens.

401. C. lagopus Fr.
    Shrawley Wood. Wyre Forest. Nunnery Wood. Ockeridge Wood. Monk Wood. Trench Woods. Tiddesley Wood.
    Not uncommon. On the ground and on dung.

402. C. radiatus (Bolton) Fr.
    Worcester. Monk Wood. Wyre Forest. Chaddesley Corbett Woods. Nunnery Wood. Trench Woods.
    Common. On dung and manure heaps.

403. C. domesticus (Pers.) Fr.
    The Tything, Worcester.
    Uncommon. Damp walls.

404. C. stercorarius (Bull.) Fr.
    Trench Woods.
    Uncommon. On dung.

405. C. ephemerus (Bull.) Fr.
    Wyre Forest. Lower Wick. Crowle. Sheriff's Lench. Sharp Gate. Stourport. Kempsey.
    Not uncommon. On manure heaps.

406. C. plicatilis (Curt.) Fr.
    Common everywhere. In pastures and woods.

407. C. filiformis B. and Br.
    Powick Ham. Malvern Wells.
    Uncommon. On manure.

BOLBITIUS Fr. (βόλβιτον, cow-dung, because the habitat of the species is on dung, but rarely that of the cow.)

408. B. flavidus (Bolton) Mass.
    Knightwick. Rock.
    Uncommon. In fields and on dunghills. Deliquescing.

409. B. Boltoni Fr.
    Northwick Lane. Nunnery Wood. Near Tiddesley Wood. Wyre Forest. Monk Wood.
    Not uncommon. In woods and fields. Becoming dry and papery at maturity.

410. B. fragilis (Linn.) Fr.
    Common everywhere. In pastures, woods and near manure heaps. Known by its glabrous stem and rust-coloured spores.

411. B. titubans (Bull.) Fr.
    Near Nunnery Wood. Monk Wood. Trench Woods. Sharpway Gate. Malvern Wells.
    Not uncommon. In pastures, woods, and near old manure heaps. Characterized by its white stem and fragile tottering pileus.

412. B. tener Berk.
    Common everywhere. In woods and pastures. Characterized by its whitish cylindrical pileus.

CORTINARIUS Fr. (*Cortina*, a veil, because these plants are characterized by an arachnoid veil.)

PHLEGMACIUM Fr. (φλέγμα, clammy moisture, from the viscidity of the pileus in this subgenus.)

413. C. (Phlegmacium) triumphans Fr.
    Shrawley Wood. Wyre Forest.
    Uncommon. In woods chiefly under birches.

414. C. (Phlegmacium) balteatus Fr.
    Trench Woods. Tiddesley Wood.
    Uncommon. In woods. Edible.

415. C. (Phlegmacium) varius (Schaeff.) Fr.
    Perry Wood. Shrawley Wood. Wyre Forest. Chatley.
    Not uncommon. In woods and pastures. This species has to be carefully distinguished from *C. (Phlegmacium) largus*, but its salient features seem to be the squat habit of the stem, white flesh, and ochraceous or rust-coloured gills with violet margin.

416. C. (Phlegmacium) cyanopus (Secr.) Fr.
    Trench Woods. Shrawley Wood.
    Uncommon. In woods.

417. C. (Phlegmacium) largus (Buxbaum) Fr.
    Trench Woods. Wyre Forest. Shrawley Wood. Nunnery Wood.
    Not uncommon. In woods. Subcaespitose. Edible.

418. C. (Phlegmacium) infractus (Pers.) Fr.
Wyre Forest. Nunnery Wood. Monk Wood.
Not uncommon. In woods.

419. C. (Phlegmacium) anfractus Fr.
Shrawley Wood. Crews Hill Wood. Wyre Forest. Tiddesley Wood.
Not uncommon. In woods. Quélet unites this with *C. infractus*, which Fries says differs from this in the margin of the pileus being surrounded by a brown zone and broad crowded gills which in *C. anfractus* are distant and crisped.

420. C. (Phlegmacium) multiformis Fr.
Crews Hill Wood. Shrawley Wood. Seckley Wood.
Not common. In woods.

421. C. (Phlegmacium) talus Fr.
Perry Wood. Shrawley Wood.
Not common. In woods.

422. C. (Phlegmacium) glaucopus (Schaeff.) Fr.
Shrawley Wood. Nunnery Wood.
Not common. In woods.

423. C. (Phlegmacium) calochrous (Pers.) Fr.
Crews Hill Wood.
Rare. In beech woods.

424. C. (Phlegmacium) caerulescens (Schaeff.) Fr.
Wyre Forest. Trench Woods. Nunnery Wood.
Not uncommon. In woods. Edible.

425. C. (Phlegmacium) purpurascens Fr.
Bevere Green. Wyre Forest. Ockeridge Wood. Shrawley Wood. Trench Woods. Tiddesley Wood. Perry Wood.
Common. In woods.

Var. subpurpurascens (Batsch) Fr.
Wyre Forest. Nunnery Wood. Shrawley Wood.
Not common. In woods.

426. C. (Phlegmacium) fulgens (A. and S.) Fr.
Crews Hill Wood.
Uncommon. In woods.

427. C. (Phlegmacium) fulmineus Fr.
Trench Woods.
Uncommon. In woods.

428. C. (Phlegmacium) testaceus Cke.
Crews Hill Wood. Black House Hill Wood.
Uncommon. In woods.

429. C. (Phlegmacium) prasinus (Schaeff.) Fr.
Crews Hill Wood. Black House Hill Wood.
Uncommon. In woods.

430. C. (Phlegmacium) scaurus Fr.
Wyre Forest. Trench Woods. Shrawley
Not common. In woods.

431. C. (Phlegmacium) emollitus Fr.
Perry Wood. Nunnery Wood. Trench Woods. Wyre Forest.
Not uncommon. In woods.

432. C. (Phlegmacium) cristallinus Fr.
Wyre Forest. Ockeridge Wood.
Uncommon. In woods, generally under beeches.

433. C. (Phlegmacium) decolorans (Pers.) Fr.
Wyre Forest. Shrawley Wood. Trench Woods.
Uncommon. In woods under conifers.

434. C. (Phlegmacium) porphyropus (A. and S.) Fr.
Perry Wood.
Uncommon. In woods.

MYXACIUM Fr. (μύξα, slime, from the glutinous stem and pileus of this subgenus.)

435. C. (Myxacium) collinitus (Pers.) Fr.
Ockeridge Wood. Nunnery Wood. Crews Hill Wood.
Not uncommon. In woods. Edible.

436. C. (Myxacium) mucifluus Fr.
Trench Woods. Ockeridge Wood. Crews Hill Wood. Shrawley Wood. Little Malvern. Wyre Forest.
Common. In woods.

437. C. (Myxacium) elatior Fr.
Ockeridge Wood. Shrawley Wood. Perry Wood. Nunnery Wood. Monk Wood. Trench Woods. Tiddesley Wood. Wyre Forest. Crews Hill Wood.
Common. In woods and under trees in pastures.

438. C. (Myxacium) livido-ochraceus Berk.
Perry Wood.
Uncommon. In woods.

439. C. (Myxacium) pluvius Fr.
Wyre Forest. Perry Wood. Nunnery Wood. Monk Wood. Shrawley Wood.
Not uncommon. In woods.

INOLOMA Fr. (ἴς, a fibre, λῶμα, a fringe, from the rough surface of the pileus in this subgenus.)

440. C. (Inoloma) argentatus (Pers.) Fr.
Shrawley Wood. Wyre Forest.
Uncommon. In woods.

441. C. (Inoloma) violaceus (Linn.) Fr.
Shrawley Wood. Wyre Forest. Ockeridge Wood.
Uncommon. In woods.

442. C. (Inoloma) alboviolaceus (Pers.) Fr.
Perry Wood. Wyre Forest. Shrawley Wood. Trench Woods. Tiddesley Wood.
Not uncommon. In woods.

443. C. (Inoloma) callisteus Fr.
Wyre Forest.
Rare. In woods.

444. C. (Inoloma) Bulliardi (Pers.) Fr.
Crews Hill Wood.
Uncommon. In woods. It is one of the few Cortinarii that have blood-red mycelium.

445. C. (Inoloma) pholideus Fr.
Nunnery Wood. Trench Woods. Wyre Forest. Tiddesley Wood. Shrawley Wood.
Not uncommon. In woods, generally under birches.

DERMOCYBE Fr. (δέρμα, skin, κύβη, head, from the smooth surface of the pileus of this subgenus.)

446. C. (Dermocybe) ochroleucus (Schaeff.) Fr.
Wyre Forest. Ockeridge Wood. Perry Wood. Monk Wood. Nunnery Wood.
Not uncommon. In woods.

447. C. (Dermocybe) tabularis (Bull.) Fr.
Trench Woods. Ockeridge Wood. Perry Wood. Nunnery Wood. Wyre Forest.
Not uncommon. In woods.

448. C. (Dermocybe) camurus Fr.
Nunnery Wood. Wyre Forest.
Uncommon. In woods.

449. C. (Dermocybe) caninus Fr.
Perry Wood. Ockeridge Wood. Shrawley Wood. Wyre Forest. Crews Hill Wood. Nunnery Wood. Monk Wood.
Common. In woods. Edible.

450. C. (Dermocybe) anomalus Fr.
Bevere Green. Chatley. Wyre Forest. Perry Wood. Monk Wood. Shrawley.
Not uncommon. In woods and pastures under oaks.

451. C. (Dermocybe) lepidopus Cke.
Shrawley Wood. Trench Woods. Ockeridge Wood. Perry Wood. Wyre Forest. Tiddesley Wood. Monk Wood.
Common. In woods and under oaks in pastures.

452. C. (Dermocybe) miltinus Fr.
Shrawley Wood. Wyre Forest. Tiddesley Wood.
Uncommon. In woods, chiefly under birches.

453. C. (Dermocybe) cinnabarinus Fr.
Wyre Forest. Shrawley Wood. Monk Wood.
Uncommon. In woods. Distinguished from *C. sanguineus* by the crenulate margin of the gills, stuffed stem, and scent of radishes.

454. C. (Dermocybe) sanguineus (Wulf.) Fr.
Wyre Forest. Shrawley Wood. Nunnery Wood. Tiddesley Wood. Trench Woods.
Not uncommon. In woods.

455. C. (Dermocybe) anthracinus Fr.
Wyre Forest. Shrawley Wood.
Uncommon. In woods.

456. C. (Dermocybe) cinnamomeus (Linn.) Fr.
Shrawley Wood. Ockeridge Wood. Wyre Forest. Perry Wood. Nunnery Wood. Trench Woods. Tiddesley Wood. Monk Wood.
Common. In woods.

Var. croceus Fr.
Nunnery Wood. Tiddesley Wood.
Not common. In woods.

Var. semisanguineus Fr.
Wyre Forest. Shrawley Wood. Ockeridge Wood. Monk Wood. Trench Woods.
Not uncommon. In woods.

457. C. (Dermocybe) uliginosus Berk.
West Malvern. Wyre Forest.
Uncommon. In woods and swamps amongst Sphagna.

458. C. (Dermocybe) cotoneus Fr.
Trench Woods. Perry Wood. Shrawley Wood.
Uncommon. In woods.

459. C. (Dermocybe) raphanoides (Pers.) Fr.
Shrawley Wood. Perry Wood. Wyre Forest.
Uncommon. In woods.

TELAMONIA Fr. (τελαμών, a bandage, from the ring which is generally apparent on the stem in most species.)

460. C. (Telamonia) torvus Fr.
Shrawley Wood. Crews Hill Wood. Perry Wood. Wyre Forest. Nunnery Wood. Trench Woods. Monk Wood.
Common. In woods.

461. C. (Telamonia) evernius Fr.
Black House Hill Wood.
Uncommon. In woods.

462. C. (Telamonia) quadricolor (Scop.) Fr.
Trench Woods. Shrawley Wood. Wyre Forest.
Uncommon. In woods.

463. C. (Telamonia) haematochelis (Bull.) Fr.
Trench Woods.
Uncommon. In woods.

464. C. (Telamonia) limonius Fr.
Shrawley Wood.
Rare. In woods.

465. C. (Telamonia) hinnuleus (Sow.) Fr.
     Ockeridge Wood. Shrawley Wood. Weymans Wood. Wyre Forest.
     Crews Hill Wood. Perry Wood. Nunnery Wood.
     Common. In woods and pastures.

466. C. (Telamonia) bovinus Fr.
     Crews Hill Wood.
     Uncommon. In woods.

467. C. (Telamonia) brunneus (Pers.) Fr.
     Wyre Forest. Shrawley Wood. Nunnery Wood. Trench Woods.
     Ockeridge Wood. Crews Hill Wood.
     Not uncommon. In woods chiefly under conifers.

468. C. (Telamonia) brunneofulvus Fr.
     Monk Wood. Nunnery Wood.
     Uncommon. In woods.

469. C. (Telamonia) injucundus Weinm.
     Trench Woods.
     Uncommon. In woods.

470. C. (Telamonia) flexipes (Pers.) Fr.
     Perry Wood. Wyre Forest. Shrawley Wood.
     Not uncommon. In woods.

471. C. (Telamonia) psammocephalus (Bull.) Fr.
     Wyre Forest. Ockeridge Wood.
     Not uncommon. In woods, especially on old charcoal heaps.

472. C. (Telamonia) incisus (Pers.) Fr.
     Ockeridge Wood. Perry Wood. Wyre Forest. Monk Wood. Shraw-
     ley Wood. Hartlebury Common.
     Not uncommon. In woods and swampy places.

473. C. (Telamonia) hemitrichus (Pers.) Fr.
     Wyre Forest. Ockeridge Wood. Perry Wood. Nunnery Wood.
     Monk Wood. Tiddesley Wood. Trench Woods.
     Common. In woods and under birches.

474. C. (Telamonia) rigidus (Scop.) Fr.
     Bevere Green. Wyre Forest. Shrawley Wood. Cowleigh Park.
     Not uncommon. In woods and swampy places.

475. C. (Telamonia) paleaceus Fr.
     Shrawley Wood. Perry Wood. Wyre Forest. Monk Wood. Nun-
     nery Wood. Trench Woods.
     Not uncommon. In woods.

HYDROCYBE Fr. (ὕδωρ, water, κύβη, head, because of the hygrophanous
          appearance of most of the plants in this subgenus.)

476. C. (Hydrocybe) subferrugineus (Batsch) Fr.
     Trench Woods. Crews Hill Wood.
     Uncommon. In woods.

477. C. (Hydrocybe) armeniacus (Schaeff.) Fr.
     Wyre Forest. Ockeridge Wood. Trench Woods.
     Uncommon. In woods, chiefly under conifers.

478. C. (Hydrocybe) duracinus Fr.
     Black House Hill Wood. Wyre Forest. Shrawley Wood.
     Not common. In woods.

479. C. (Hydrocybe) castaneus (Bull.) Fr.
     Shrawley Wood. Trench Woods. Bevere Green.
     Not uncommon. In woods and pastures. Edible.

480. C. (Hydrocybe) bicolor Cke.
     Shrawley Wood. Wyre Forest.
     Uncommon. In woods.

481. C. (Hydrocybe) dolabratus Fr.
     Ockeridge Wood. Tiddesley Wood.
     Uncommon. In woods.

482. C. (Hydrocybe) leucopus (Pers.) Fr.
     Shrawley Wood. Crews Hill Wood. Wyre Forest. Perry Wood.
     Nunnery Wood. Trench Woods.
     Uncommon. In woods.

483. C. (Hydrocybe) scandens Fr.
     Nunnery Wood.
     Uncommon. In woods.

484. C. (Hydrocybe) erythrinus Fr.
     Nunnery Wood. Wyre Forest. Ockeridge Wood.
     Not uncommon. In woods.

485. C. (Hydrocybe) decipiens (Pers.) Fr.
     Trench Woods. Ockeridge Wood. Great Farley Wood. Perry Wood.
     Nunnery Wood. Wyre Forest.
     Common. In woods.

486. C. (Hydrocybe) obtusus Fr.
     Wyre Forest. Shrawley Wood. Trench Woods.
     Uncommon. In woods.

487. C. (Hydrocybe) acutus (Pers.) Fr.
     Shrawley Wood. Trench Woods. Wyre Forest. Nunnery Wood.
     Tiddesley Wood. Monk Wood.
     Common. In woods.

GOMPHIDIUS Fr. (γόμφος, a large bolt or nail, from the shape
          of the pileus.)

488. G. glutinosus (Schaeff.) Fr.
     Dripshill Wood.
     Uncommon. In woods under conifers.

489. G. viscidus (Linn.) Fr.
     Witley. Hadley. Shrawley Wood. Wyre Forest. Oldington Wood.
     Ockeridge Wood.
     Common. In woods and on lawns under conifers. Edible.

N n

490. G. roseus Fr.
     Dripshill Wood. Trench Woods. Shrawley Wood.
     Not uncommon. In woods and on lawns under conifers.

491. G. gracilis Berk.
     Between Martley and Witley. Rous Lench. Little Malvern.
     Not uncommon. In woods and heaths under conifers.

PAXILLUS Fr. (Paxillus, a small stake, from the shape of the pileus.)

LEPISTA. (Lepista, a drinking-vessel, a goblet.)

492. P. (Lepista) giganteus (Sow.) Fr.
     The Old Hills. Hanbury Park. Churchyard, Rock.
     Not common. Amongst grass. Edible, excellent.

493. P. (Lepista) lepista Fr.
     Trench Woods. Shrawley Wood.
     Uncommon. Amongst grass in woods.

494. P. (Lepista) lividus Cke.
     Shrawley Wood.
     Uncommon. In woods.

TAPINIA. (ταπεινός, low, short.)

495. P. (Tapinia) involutus (Batsch) Fr.
     Common everywhere in woods and on commons. Edible, but rather
     mucilaginous in texture.
     Var. excentricus Fr.
     Nunnery Wood. Trench Woods. Wyre Forest.
     Not uncommon. In woods, chiefly on mossy stumps.

496. P. (Tapinia) leptopus Fr.
     Shrawley Wood. Wyre Forest. Monk Wood. Nunnery Wood.
     Trench Woods.
     Not uncommon. In woods on stumps and mossy trunks. Edible.

497. P. (Tapinia) atrotomentosus (Batsch) Fr.
     Oldington Wood.
     Uncommon. In woods, chiefly on pine-stumps.

498. P. (Tapinia) panuoides Fr.
     Sawpit in timber-yard near Stoke Works Station.
     Not common. On sawdust and rotten rails.

HYGROPHORUS Fr. (ὑγρός, moist, φέρω, I bear, because most of the
          species are hygrophanous.)

LIMACIUM. (Limax, a slug, slimy.)

499. H. (Limacium) chrysodon (Batsch) Fr.
     Crews Hill Wood. Shrawley Wood. Wyre Forest. Trench Woods.
     Not common. In woods. Edible.

500. H. (Limacium) eburneus (Bull.) Fr.
     Shrawley Wood. Wyre Forest. Tiddesley Wood. Trench Woods.
     Nunnery Wood.
     Not uncommon. In woods. Edible.

501. H. (Limacium) cossus (Sow.) Fr.
     Shrawley Wood. Crews Hill Wood. Wyre Forest. Trench Woods.
     Perry Wood. Monk Wood.
     Common. In woods.

502. H. (Limacium) erubescens Fr.
     Crews Hill Wood.
     Rare. In woods. Edible, according to Quélet, but when cooked it
     tastes very bitter and astringent, and is certainly not palatable.

[H. (Limacium) penarius (Sow.) Fr. was recorded in error in the list of
     fungi for the Victoria County History, C. Rea.]

503. H. (Limacium) glutinifer Fr.
     Brickhill Coppice, Crowle.
     Uncommon. In woods.

504. H. (Limacium) arbustivus Fr.
     Middleyards Coppice. Wyre Forest. Shrawley Wood.
     Not common. In woods. Edible.

505. H. (Limacium) discoideus (Pers.) Fr.
     Wyre Forest. Ockeridge Wood. Shrawley Wood. Trench Woods.
     Tiddesley Wood.
     Not uncommon. In woods.

506. H. (Limacium) olivaceo-albus Fr.
     Wyre Forest. Crews Hill Wood. Nunnery Wood.
     Not uncommon. In woods.

507. H. (Limacium) hypothejus Fr.
     Shrawley Wood. Wyre Forest. Trench Woods. Ockeridge Wood.
     Great Farley Wood.
     Not uncommon. In woods and open places under conifers.
     Edible.

CAMAROPHYLLUS. (καμάρα, a vault, φύλλον, a leaf, from the
          shape of the gills.)

508. H. (Camarophyllus) nemoreus (Lasch) Fr.
     Shrawley Wood. Wyre Forest.
     Uncommon. In woods and pastures. Edible.

509. H. (Camarophyllus) pratensis (Pers.) Fr.
     Common everywhere. In pastures, woods, and on heaths. Edible
     and delicious.

510. H. (Camarophyllus) virgineus (Wulf.) Fr.
     Common everywhere. In pastures, woods, and on lawns. Edible
     and delicious.
     Var. roseipes Mass.
     Wyre Forest. Shrawley Wood. Between Perry Wood and Nunnery
     Wood. Holt Bank.

N n 2

511. H. (Camarophyllus) niveus (Scop.) Fr.
Common everywhere. On lawns, in pastures, and in woods. Edible and delicious. *H. niveus* is much smaller than *H. virgineus*, and differs in the hollow stem and umbilicate pileus.

512. H. (Camarophyllus) russo-coriaceus B. and Br.
Wyre Forest. Rimells Farm, Crowneast. Near Chaddesley Corbett Woods. Westwood Park. Spetchley Park.
Not uncommon. In pastures and heathy places. Edible and delicious. It is characterized by its strong smell of Russian leather.

513. H. (Camarophyllus) fornicatus Fr.
Near Bubble Brook, Crowneast. Field near Perry Wood. Shrawley Wood. Wyre Forest. Dripshill Wood.
Uncommon. In fields and woods.

514. H. (Camarophyllus) ovinus (Bull.) Fr.
Cowleigh Park. Wyre Forest. Field between Monk Wood and Ockeridge Wood. Dodderhill Common.
Not uncommon. In pastures and in woods.

515. H. (Camarophyllus) subradiatus (Schum.) Fr.
Var. lacmus Fr.
Wyre Forest. Shrawley Wood. Woodbury Hill.
Uncommon. In woods and on downs. It is somewhat curious that the type is very scarce in England, although the variety is fairly often recorded.

HYGROCYBE. (ὑγρός, moist, κύβη, head, from the hygrophanous pileus.)

516. H. (Hygrocybe) laetus (Pers.) Fr.
Hewell Park. Malvern Wells. Wyre Forest. Shrawley Wood. Stagbury Hill.
Not uncommon. On commons, in pastures and grassy rides in woods.

517. H. (Hygrocybe) vitellinus Fr.
Wyre Forest. Middleyards Coppice. Trench Woods.
Uncommon. In woods and on heaths.

518. H. (Hygrocybe) ceraceus (Wulf.) Fr.
Bevere Green. Fields near Ravenshill Woods. Chatley. Ladywood, Salwarpe. Claines. Ombersley. Old Hills.
Common. In pastures. Edible and delicious.

519. H. (Hygrocybe) coccineus (Schaeff.) Fr.
Common everywhere. In pastures and woods. Edible and delicious.

520. H. (Hygrocybe) miniatus Fr.
Bevere Green. Temple Laughern. Wyre Forest. Trench Woods. Shrawley Wood. Tiddesley Wood. Old Hills.
Common. In pastures, on commons, and in woods. Edible and delicious.

521. H. (Hygrocybe) turundus Fr.
Hartlebury Common. Wyre Forest. Ribbesford Wood.
Uncommon. In moist places in the open and in woods.
Var. mollis B. and Br.
Hartlebury Common. Wyre Forest.

522. H. (Hygrocybe) puniceus Fr.
Bevere Green. Field near Shrawley Brook. Shrawley Wood. Trench Woods. Tiddesley Wood. Holt Bank.
Common. In pastures and in woods.

523. H. (Hygrocybe) obrusseus Fr.
Malvern Wells. Crews Hill Wood. Shrawley Wood. Wyre Forest.
Uncommon. In grassy drives in woods and on commons.

524. H. (Hygrocybe) conicus (Scop.) Fr.
Common everywhere. In pastures and woods. The form with crimson pileus is the common one in Britain and this county, and the form with yellowish pileus somewhat scarce. Both forms are characterized by every part turning black when bruised or broken.

525. H. (Hygrocybe) calyptraeformis Berk.
Bevere Green. Shrawley Wood. Wyre Forest. Ockeridge Wood. Old Hills.
Uncommon. In open drives in woods and on commons.
Var. niveus Cke.
Bevere Green.
Very brittle and hard to bring home perfect. It is a very distinct species, and certainly not connected with *H. conicus* as Quélet and Massee seem to think.

526. H. (Hygrocybe) chlorophanus Fr.
Common everywhere. In pastures and woods. Edible and delicious.

527. H. (Hygrocybe) psittacinus (Schaeff.) Fr.
Common everywhere. In pastures, woods, and on lawns and commons. Edible and delicious. A remarkable variety with light blue pileus and stem was gathered near Mawley Hall, Salop.

528. H. (Hygrocybe) unguinosus Fr.
Bevere Green. Shrawley Wood. Wyre Forest.
Not uncommon. On grassy drives in woods and on open commons.

529. H. (Hygrocybe) nitratus (Pers.) Fr.
Field near Perry Wood.
Uncommon. In pastures.

LACTARIUS Fr. (*Lac*, milk, because the gills in this genus give forth milk when wounded or cut.)

PIPERITES. (*Piperitis*, pepper-wort, from the taste of the milk.)

530. L. (Piperites) scrobiculatus (Scop.) Fr.
Wyre Forest. Ockeridge Wood.
Not common. In woods and on commons under birches. **Poisonous.**

531. L. (Piperites) intermedius Krombh.
Cowleigh Wood.
Uncommon. In woods.

532. L. (Piperites) torminosus (Schaeff.) Fr.
Trench Woods. Shrawley Wood. Monk Wood. Wyre Forest. Ockeridge Wood. Tiddesley Wood. Crews Hill Wood.
Common. In woods and on commons.

533. L. (Piperites) turpis (Weinm.) Fr.
Wyre Forest. Shrawley Wood. Great Farley Wood. Randans Wood. Nunnery Wood. Trench Woods.
Common. In woods and on commons, chiefly under birches. Edible.

534. L. (Piperites) controversus Pers.
Worcestershire side of Dowles Brook near to Bewdley. Pasture near Crews Hill Wood.
Uncommon. In pastures and woods, chiefly under Poplars.

535. L. (Piperites) pubescens Fr.
Shrawley Wood. Wyre Forest. Randans Wood. Crews Hill Wood. Trench Woods.
Common. In woods and on commons, chiefly under birches.

536. L. (Piperites) insulsus Fr.
Wyre Forest. Crews Hill Wood. The Slads. Shrawley Wood. Chatley. Spetchley Park.
Not uncommon. In woods and pasture. **Poisonous.**

537. L. (Piperites) blennius Fr.
Shrawley Wood. Wyre Forest. Nunnery Wood. Trench Woods. Dripshill Wood. Monk Wood. Fries Wood.
Common. In woods.

538. L. (Piperites) hysginus Fr.
Wyre Forest.
Rare. In woods.

539. L. (Piperites) trivialis Fr.
Wyre Forest. Shrawley Wood.
Uncommon. In woods.

540. L. (Piperites) circellatus (Battara) Fr.
Wyre Forest. Holy Well, Malvern Wells. Shrawley Wood.
Not uncommon. In woods and under beeches.

541. L. (Piperites) uvidus Fr.
Shrawley Wood. Wyre Forest. Perry Wood. Monk Wood. Trench Woods. Tiddesley Wood.
Not uncommon. In woods. **Poisonous.**

542. L. (Piperites) pyrogalus (Bull.) Fr.
Perry Wood. Monk Wood. Nunnery Wood. Wyre Forest. Shrawley Wood. Crews Hill Wood. Trench Woods.
Common. In woods and pastures. **Poisonous.**

543. L. (Piperites) chrysorheus Fr.
Wyre Forest. Ockeridge Wood. Shrawley Wood. Perry Wood. Nunnery Wood. Monk Wood. Trench Woods.
Common. In woods, chiefly under oaks.

544. L. (Piperites) pargamenus (Swartz) Fr.
Wyre Forest. Shrawley Wood. Trench Woods.
Uncommon. In woods. Edible.

545. L. (Piperites) piperatus (Scop.) Fr.
Nunnery Wood. Shrawley Wood. Wyre Forest. Perry Wood. Ockeridge Wood.
Not uncommon in some seasons. In woods. Edible.

546. L. (Piperites) vellereus Fr.
Common everywhere. In woods.

DAPETES. (*Daps*, food.)

547. L. (Dapetes) deliciosus Fr.
Hartlebury Common. Hadley Bowling Green. Wyre Forest. Shrawley Wood. Ockeridge Wood. Dripshill Wood.
Common. In woods and under Scotch pines. Edible and delicious.

RUSSULARES. (*Russus*, red, from the colour of the pileus in many species of this subgenus.)

548. L. (Russularia) pallidus (Pers.) Fr.
Ockeridge Wood. Shrawley Wood. Perry Wood. Nunnery Wood. Monk Wood. Wyre Forest. Malvern Wells.
Common. In woods, chiefly under beeches.

549. L. (Russularia) quietus Fr.
Common everywhere. In woods and open ground under trees.

550. L. (Russularia) aurantiacus (Fl. Dan.) Fr.
Wyre Forest. Shrawley Wood. Monk Wood.
Uncommon. In woods.

551. L. (Russularia) theiogalus (Bull.) Fr.
Wyre Forest. Shrawley Wood. Nunnery Wood. Tiddesley Wood. Trench Woods.
Not uncommon. In woods. The milk in this species very slowly turns to a yellow colour, whereas in *L. chrysorheus* it flows out on the slightest touch and becomes very quickly of a rich sulphur colour; the pileus in *L. chrysorheus* is zoned with spots and is lighter in colour than the present species, which is not zoned. The French mycologists incorrectly call *L. chrysorheus, L. theiogalus.*

552. L. (Russularia) vietus Fr.
Perry Wood. Shrawley Wood. Wyre Forest.
Uncommon. In woods.

553. L. (Russularia) rufus (Scop.) Fr.
Shrawley Wood. Wyre Forest. Oldington Wood. Chaddesley Corbett Woods. Trench Woods.
Common. In woods, especially under Scotch pines. **Poisonous.**

554. L. (Russularia) helvus Fr.
Wyre Forest.
Uncommon. In woods.

555. L. (Russularia) glyciosmus Fr.
Shrawley Wood. Ockeridge Wood. Monk Wood. Wyre Forest.
Tiddesley Wood. Trench Woods. Chaddesley Corbett Woods.
Common. In woods. Edible.

556. L. (Russularia) fuliginosus Fr.
Trench Woods. Hewell Park. Wyre Forest. Crews Hill Wood.
Perry Wood. Tiddesley Wood. Monk Wood.
Common. In woods.

557. L. (Russularia) volemus Fr.
Wyre Forest. Shrawley Wood. Trench Woods. Ockeridge Wood.
Monk Wood. Nunnery Wood.
Not uncommon. In woods. Edible and delicious.

558. L. (Russularia) serifluus (DC.) Fr.
Ockeridge Wood. Wyre Forest. Perry Wood. Shrawley Wood.
Dripshill Wood. Trench Woods. Monk Wood.
Common. In woods and on commons. Characterized by its watery
milk.

559. L. (Russularia) mitissimus Fr.
Ockeridge Wood. Shrawley Wood. Perry Wood. Wyre Forest.
Trench Woods. Little Malvern. Nunnery Wood.
Common. In woods. Edible.

560. L. (Russularia) subdulcis (Bull.) Fr.
Perry Wood. Ockeridge Wood. Shrawley Wood. Wyre Forest.
Great Farley Wood. Chaddesley Corbett Woods. Crews Hill
Wood.
Common. In woods and pastures. Edible.

561. L. (Russularia) camphoratus (Bull.) Fr.
Shrawley Wood. Wyre Forest.
Not common. In woods.

562. L. (Russularia) cimicarius (Batsch) Fr.
Oldington Wood. Nunnery Wood. Spetchley Wood.
Not uncommon. In woods and pastures.

563. L. (Russularia) minimus W. G. Sm.
Shrawley Wood. Wyre Forest. Nunnery Wood. Ockeridge Wood.
Not uncommon. In woods.

RUSSULA Pers. (*Russulus*, reddish, from the reddish appearance
of many of the species.)

COMPACTAE. (*Compingo*, I put together, compact.)

564. R. (Compactae) nigricans (Bull.) Fr.
Shrawley Wood. Perry Wood. Nunnery Wood. Monk Wood. Wyre
Forest. Trench Woods. Dripshill Wood. Goosehill Wood.
Common. In woods.

565. R. (Compactae) adusta (Pers.) Fr.
Nunnery Wood. Shrawley Wood. Crews Hill Wood. Wyre Forest.
Tiddesley Wood.
Not uncommon. In woods.

Var. albo-nigra (Krombh.) Fr.
Shrawley Wood.

566. R. (Compactae) densifola Secr.
Shrawley Wood. Wyre Forest.
Uncommon. In woods.

567. R. (Compactae) delica (Vaill.) Fr.
Wyre Forest. Nunnery Wood. Shrawley Wood.
Uncommon. In woods.

568. R. (Compactae) chloroides (Krombh.) Bres.
Holt Bank Wood. Perry Wood. Nunnery Wood. Wyre Forest.
Trench Woods. Shrawley Wood. Dripshill Wood.
Common. In woods. Edible.

FURCATAE. (*Furca*, a fork, from the forked gills.)

569. R. (Furcatae) olivascens Fr.
Nunnery Wood.
Uncommon. In woods.

570. R. (Furcatae) furcata (Lam.) Pers.
Shrawley Wood. Lords Wood. Ockeridge Wood. Wyre Forest. Cow-
leigh Park. Tiddesley Wood. Stagbury Hill.
Common. In woods and pastures. Poisonous.

571. R. (Furcatae) sanguinea (Bull.) Fr.
Monk Wood. Wyre Forest.
Uncommon. In woods.

572. R. (Furcatae) rosacea (Bull.) Fr.
Nunnery Wood. Wyre Forest. Monk Wood. Ockeridge Wood.
Uncommon. In woods and pastures.

573. R. (Furcatae) sardonia Fr.
Westwood Park. Monk Wood. Nunnery Wood. Shrawley Wood.
Tiddesley Wood.
Not uncommon. In woods and pastures. Easily known by the gills
and stem becoming spotted with yellow when bruised.

574. R. (Furcatae) depallens (Pers.) Fr.
Shrawley Wood. Middleyards Coppice. Wyre Forest. Nunnery
Wood. Trench Woods. Spetchley Park.
Common. In woods and pastures. Edible.

575. R. (Furcatae) caerulea (Pers.) Fr.
Wyre Forest.
Uncommon. In woods. Edible.

576. R. (Furcatae) drimeia Cke. (= expallens Gillet).
Nunnery Wood. Shrawley Wood. Dripshill Wood. Wyre Forest.
Not uncommon. In woods.

RIGIDAE. (*Rigidus*, rigid.)

577. R. (Rigidae) lactea (Pers.) Fr.
Tiddesley Wood. Shrawley Wood. Ockeridge Wood.
Uncommon. In woods. Edible.

578. R. (Rigidae) incarnata Quél.
Shrawley Wood.
Uncommon. In woods.

579. R. (Rigidae) virescens (Schaeff.) Fr.
Wyre Forest. Tiddesley Wood. Great Farley Wood.
Not common. In woods. Edible.

580. R. (Rigidae) lepida Fr.
Shrawley Wood. Ockeridge Wood. Wyre Forest. Chaddesley
Corbett Woods. Crews Hill Wood.
Not uncommon. In woods. Edible and delicious even when raw.

581. R. (Rigidae) rubra (DC.) Fr.
Shrawley Wood. The Slads. Trench Woods.
Not common. In woods, chiefly under conifers.

582. R. (Rigidae) rubra (Krombh.) Bres.
Monk Wood. Nunnery Wood. Tiddesley Wood. Shrawley Wood.
Not uncommon. In woods.

583. R. (Rigidae) xerampelina (Schaeff.) Fr.
Nunnery Wood. Monk Wood. Tiddesley Wood. Wyre Forest.
Chatley.
Not uncommon. In woods and pastures under trees. Edible.

HETEROPHYLLAE. (ἕτερος, different, φύλλον, a leaf, because the gills
consist of many shorter ones mixed with longer ones.)

584. R. (Heterophyllae) vesca Fr.
Ockeridge Wood. Nunnery Wood. Wyre Forest. Shrawley Wood.
Crews Hill Wood.
Common. In woods. Edible and delicious.

585. R. (Heterophyllae) azurea Bres.
Wyre Forest. Chaddesley Corbett Woods. Great Farley Wood.
Uncommon. In woods.

586. R. (Heterophyllae) cyanoxantha (Schaeff.) Fr.
Trench Woods. Wyre Forest. Cowleigh Park. Ribbesford Wood.
Crews Hill Wood. Tiddesley Wood.
Common. In woods and pastures under trees. Edible and delicious ;
it tastes like a nut when raw.

587. R. (Heterophyllae) heterophylla Fr.
Monk Wood. Nunnery Wood. Ockeridge Wood.
Uncommon. In woods. Edible.

588. R. (Heterophyllae) galochroa (Bull.) Fr.
Wyre Forest. Shrawley Wood. Great Farley Wood. Trench Woods.
Not uncommon. In woods. Edible.

589. R. (Heterophyllae) consobrina Fr.
Shrawley Wood. Roadside, Claines. Dodderhill Common.
Not uncommon. In woods and under conifers.

Var. sororia Fr.
Westwood Park.

Var. intermedia Cke.
The Slads.

590. R. (Heterophyllae) foetens Pers.
Wyre Forest. Shrawley Wood. Little Malvern Wood. Monk
Wood. Trench Woods. Crews Hill Wood. Lickey Monument.
Common. In woods and pastures under trees.

591. R. (Heterophyllae) fellea Fr.
Shrawley Wood. Great Farley Wood. Wyre Forest. Hewell Park.
Dodderhill Common. Churchill Wood.
Common. In woods and pastures, especially under beeches. Known
by the straw colour of all its parts. Poisonous.

FRAGILES. (*Fragilis*, fragile or brittle.)

592. R. (Fragiles) emetica Fr.
Shrawley Wood. Tiddesley Wood. Malvern Wells. Hanbury Park.
Dodderhill Common. Great Farley Wood.
Common. In woods and pastures, chiefly under beeches. Poisonous.

593. R. (Fragiles) orchroleuca Pers.
Shrawley Wood. Ockeridge Wood. Wyre Forest. Monk Wood.
Great Farley Wood. Randans Wood.
Common. In woods.

594. R. (Fragiles) granuiosa Cke.
Trench Woods. Crews Hill Wood. Wyre Forest. Monk Wood.
Shrawley Wood. Dripshill Wood.
Common. In woods. It is very doubtful if this be not a form of
the preceding species altered in outward facies by being exposed to sun
and wind and other agents.

595. R. (Fragiles) citrina Gillet.
Nunnery Wood. Shrawley Wood.
Uncommon. In woods.

596. R. (Fragiles) fragilis (Pers.) Fr.
Common everywhere. In woods, pastures, and on lawns. Poisonous.

Var. nivea Pers.
Ockeridge Wood. Perry Wood. Nunnery Wood. Monk Wood.
Shrawley Wood. Wyre Forest.

Var. violacea Quél.
Ockeridge Wood. Shrawley Wood. Wyre Forest. Perry Wood.
Nunnery Wood. Monk Wood. Trench Woods.

Var. fallax Cke.
Ockeridge Wood. Nunnery Wood. Wyre Forest. Shrawley Wood.

597. R. (Fragiles) veternosa Fr.
Wyre Forest.
Uncommon. In woods.

598. R. (Fragiles) decolorans Fr.
Rous Lench Beeches. Tiddesley Wood.
Uncommon. In woods.

599. R. (Fragiles) integra (Linn.) Fr.
Wyre Forest. Perry Wood. Monk Wood. Trench Woods. Tiddesley Wood. Crews Hill Wood,
Common. In woods and pastures under trees.

600. R. (Fragiles) aurata (With.) Fr.
Wyre Forest. Perry Wood. Ribbesford Wood. Tiddesley Wood. Cleeve Banks. Dripshill Wood.
Not uncommon. In woods.

601. R. (Fragiles) nitida (Pers.) Fr.
Wyre Forest. Randans Wood. Near Holt Mill. Monk Wood.
Not uncommon. In woods.
Var. cuprea Cke.
Nunnery Wood.
Uncommon. In woods.
Var. pulchralis Britz.
Wyre Forest. Monk Wood. Nunnery Wood.
Rather uncommon. In woods.

602. R. (Fragiles) alutacea Pers.
Shrawley Wood. Hewell Park. Wyre Forest. Nunnery Wood. Monk Wood.
Not uncommon. In woods. Edible

603. R. (Fragiles) armeniaca Cke.
Perry Wood. Wyre Forest. Shrawley Wood. Ockeridge Wood. Perry Wood. Monk Wood. Trench Woods.
Common. In woods.

604. R. (Fragiles) puellaris Fr.
Wyre Forest. Great Farley Wood. Warshill Wood. Monk Wood. Nunnery Wood. Ockeridge Wood.
Not uncommon. In woods. It much resembles R. fragilis var. violacea, but the taste is mild and the stem foxes when touched.

605. R. (Fragiles) lutea (Huds.) Fr.
Shrawley Wood. Wyre Forest. Ockeridge Wood. Crews Hill Wood. Monk Wood. Perry Wood. Nunnery Wood. Trench Woods.
Common. In woods. Edible.

606. R. (Fragiles) chamaeleontina Fr.
Sharpway Gate. Kempsey Common.
Very uncommon. In woods, pastures, and commons.

CANTHARELLUS, Adans. (κάνθαρος, a sort of drinking-cup, which some species resemble.)

607. C. cibarius Fr. *Chantarelle.*
Common everywhere in woods. Edible and delicious.

608. C. aurantiacus (Wulf) Fr.
Hadley. Wyre Forest. Shrawley Wood. Hartlebury Common. Great Farley Wood. Malvern Wells.
Common. On commons and in woods under conifers.

609. C. carbonarius (A. and S.) Fr.
Wyre Forest.
Not common. On charcoal heaps.

610. C. tubaeformis (Bull.) Fr.
Wyre Forest. Shrawley Wood. Ockeridge Wood.

611. C. infundibuliformis (Scop.) Fr.
Wyre Forest. Trench Woods. Monk Wood.
Not uncommon. In woods.

612. C. cinereus (Pers.) Fr.
Shrawley Wood.
Uncommon. In woods.

613. C. Houghtoni Phil.
Trench Woods.
Rare. In woods.

614. C. muscigenus (Bull.) Fr.
Wyre Forest.
Uncommon. On moss.

NYCTALIS Fr. (νύξ, night, from inhabiting dark places.)

615. N. parasitica (Bull.) Fr.
Wyre Forest. Perry Wood. Shrawley Wood. Nunnery Wood. Monk Wood. Trench Woods.
Common. On dead Russulae and Lactarii in woods.

616. N. asterophora Fr.
Wyre Forest. Perry Wood. Trench Woods.
Not uncommon. On dead Russulae in woods.

MARASMIUS Fr. (μαραίνω, I wither or shrivel, because the species of this genus are not putrescent.)

617. M. urens (Bull.) Fr.
Oldington Wood.
Uncommon. In woods.

618. M. peronatus (Bolt.) Fr.
Common everywhere. In woods.

619. M. porreus Fr.
Wyre Forest. Nunnery Wood.
Not uncommon. Amongst dead oak and birch leaves in woods.

620. M. oreades (Bolt.) Fr. *Fairy-ring Champignon.*
Common everywhere. In pastures and woods. Edible and delicious.

621. M. prasiosmus Fr.
Shrawley Wood.
Uncommon. Amongst dead beech leaves. Edible.

622. M. erythropus (Pers.) Fr.
Trench Woods. Shrawley Wood. Great Farley Wood. Tiddesley Wood. Monk Wood.
Not uncommon. In woods. Edible.

623. M. archyropus (Pers.) Fr.
Shrawley Wood.
Uncommon. Amongst dead leaves.

624. M. calopus (Pers.) Fr.
Trench Woods. Middleyards Coppice. Monk Wood.
Not uncommon. On twigs and dead wood.

625. M. Vaillantii (Pers.) Fr.
West Malvern.
Uncommon. On twigs and leaves in woods.

626. M. foetidus (Sow.) Fr.
Little Malvern Wood.
Uncommon. On dead twigs and leaves in woods.

627. M. ramealis (Bull.) Fr.
Shrawley Wood. Ockeridge Wood. Perry Wood. Trench Woods. Wyre Forest. Tiddesley Wood. The Randans.
Common. On twigs, leaves, and stems of bramble.

628. M. candidus (Bolt.) Fr.
Wyre Forest. Shrawley Wood.
Not uncommon. On twigs and leaves.

629. M. alliaceus (Jacq.) Fr.
Shrawley Wood.
Uncommon. On rotten wood and amongst dead leaves in woods.

630. M. rotula (Scop.) Fr.
Common everywhere. On fallen twigs in woods and hedgerows.

631. M. graminum (Libert) Berk.
Nunnery Wood. Monk Wood.
Not uncommon. On leaves and stems of grasses.

632. M. androsaceus (Linn.) Fr.
Common everywhere. On fallen leaves and twigs in woods and hedgerows.

633. M. epiphyllus Fr.
Common everywhere. On dead fallen leaves and twigs in woods and hedgerows.

LENTINUS Fr. (*Lentus*, tough or pliant from the consistency of the species of this genus.)

634. L. lepideus Fr.
Hadley Firs. Wyre Forest.
Not uncommon. On pine and fir stumps and on railway sleepers.

635. L. cochleatus (Pers.) Fr.
Shrawley Wood. Perry Wood. Nunnery Wood. Wyre Forest. Trench Woods.
Common. In woods on stumps. Edible and delicious.

PANUS Fr. (A name given to an arboreal fungus by Pliny.)

636. P. torulosus (Pers.) Fr.
Holt. Wyre Forest. Shrawley Wood.
Rather uncommon. On willow and beech stumps. Edible.

637. P. rudis Fr.
Shrawley.
Uncommon. On beech-stumps.

638. P. stipticus (Bull.) Fr.
Ockeridge Wood. Shrawley Wood. Wyre Forest. Monk Wood. Nunnery Wood. Trench Woods.
Common. On stumps and twigs.

SCHIZOPHYLLUM Fr. (σχίζω, I split, φύλλον, a leaf, from the divided edge of the gill.)

639. S. commune Fr.
Shrawley Wood.
Uncommon. On felled oak. This species has generally previously been reported only on imported timber or near timber yards.

LENZITES Fr. (After Lenz, a German botanist.)

640. L. betulina (Linn.) Fr.
Common everywhere. On trunks, stumps, and palings.

641. L. flaccida (Bull.) Fr.
Shrawley Wood. Wyre Forest. Trench Woods.
Not uncommon. On stumps and trunks. This is a thinner species, paler and more strigose on the pileus than *L. betulina*, but it is very doubtful if it is a distinct species from it.

642. L. saepiaria (Wulf) Fr.
Merrymans Hill, Worcester. Cowleigh Park.
Uncommon. On palings.

POLYPORACEAE

BOLETUS Dill. (βῶλος, a clod, from the round form of the pileus.)

643. B. luteus Linn.
Wyre Forest. Pines near Wyche Pass, Malvern. Ockeridge Wood. Shrawley Wood.
Common. In woods and under pines on lawns. Edible, but rather soft and mushy.

644. **B. elegans** Schum.
Shrawley Wood. Wood, Little Malvern. Wyre Forest. Great Farley Wood. Trench Woods. Hewell Park. Ockeridge Wood.
Common. In woods and pastures under conifers. Edible, but rather mucilaginous.

645. **B. granulatus** Linn.
Hadley. Wyre Forest. Trench Woods. Rous Lench.
Not uncommon. In woods, on commons, and in gardens under pines. Edible and excellent.

646. **B. tenuipes** Cke.
Wyre Forest. Hadley.
Uncommon. In woods and pastures.

647. **B. bovinus** Linn.
Wyre Forest.
Not uncommon. In woods and on commons under pines. Edible, but very mucilaginous in texture.

648. **B. badius** Fr.
Shrawley Wood. Wyre Forest. Great Farley Wood. Trench Woods. Ockeridge Wood. Tiddesley Wood.
Common. In woods, chiefly under conifers. Edible.

649. **B. piperatus** Bull.
Bevere Green. Shrawley Wood. Ockeridge Wood. Wyre Forest. Nunnery Wood. Trench Woods.
Not uncommon. In woods, chiefly under pines.

650. **B. variegatus** Swartz.
Tiddesley Wood.
Uncommon. In woods and on commons, chiefly under pines.

651. **B. chrysenteron** Bull.
Common everywhere. In woods and pastures. Edible, but rather mucilaginous in texture.
Var. **nanus** Mass.
Chatley.
Uncommon. In pastures under trees.

652. **B. versicolor** Rostk.
Pasture near Crowneast Wood.
Uncommon. In woods and pastures. Edible.

653. **B. subtomentosus** Linn.
Common everywhere. In woods and pastures. Edible, but rather mucilaginous in texture.

654. **B. radicans** Pers.
Bransford. Trench Woods. Tiddesley Wood.
Uncommon. In woods and pastures.

655. **B. pachypus** Fr.
Wyre Forest.
Uncommon. In woods.

656. **B. edulis** Bull.
Wyre Forest. Shrawley Wood. Perry Wood. Tiddesley Wood. Rous Lench Beeches. Cowleigh Park.
Common. In woods. Edible and very delicious.
Var. **laevipes** Mass.
Wyre Forest. Trench Woods. Shrawley Wood.
Common. In woods.
Var. **crassus** Mass.
Wyre Forest. Shrawley Wood. Cowleigh Park.
Common. In woods.

657. **B. impolitus** Fr.
Wyre Forest. Tiddesley Wood. Shrawley Wood. Cowleigh Park.
Not uncommon. In woods. Edible and delicious.

658. **B. aestivalis** Fr.
Rous Lench Beeches. Wyre Forest.
Not uncommon. In woods and pastures. Edible and delicious.

659. **B. satanas** Lenz.
Shrawley Wood. Tiddesley Wood. Rous Lench Beeches.
Uncommon. In woods. Poisonous.

660. **B. luridus** Schaeff.
Common everywhere. In woods and pastures.
Var. **erythropus** (Pers.) Fr.
Wyre Forest. Trench Woods. Cowleigh Wood. Shrawley Wood. Ockeridge Wood.
Common. In woods and pastures.

661. **B. purpureus** Fr.
Tiddesley Wood.
Uncommon. In woods and pastures. Poisonous.

662. **B. laricinus** Berk.
Sheriff's Lench. Shrawley Wood. Malvern Wells.
Common. In woods under larches. Edible, but very mucilaginous.

663. **B. duriusculus** Schulz.
Wyre Forest. Trench Woods. Shrawley Wood.
Rather uncommon. In woods. Edible.

664. **B. versipellis** Fr.
Wyre Forest. Tiddesley Wood. Shrawley Wood. Crews Hill Wood. Hartlebury Common.
Common. In woods and on commons. Edible and delicious.

665. **B. scaber** Bull.
Common everywhere. In woods and on heaths. Edible, but very mucilaginous in texture.

666. **B. rugosus** Fr.
Wyre Forest. Black House Hill Wood. Oldington Wood.
Rather uncommon. In woods. Edible. This differs from *B. scaber* in the dry pileus and costate rugose stem stippled, as it were, with black punctations.

667. **B. felleus** Bull.
Shrawley Wood. Wyre Forest.
Rather uncommon In woods. Poisonous.

668. **B. alutarius** Fr.
Shrawley Wood.
Uncommon. In woods.

669. **B. castaneus** Bull.
Little Malvern Wood. Nunnery Wood.
Uncommon. In woods. Edible.

STROBILOMYCES Berk. (στρύβιλος, a pine-cone, μύκης, a fungus, from the scaly pileus.)

670. **S. strobilaceus** (Scop.) Berk.
Shrawley Wood.
Rather rare. In woods.

FISTULINA Bull. (*Fistula*, a pipe, from the pipe-like character of the tubes.)

671. **F. hepatica** (Huds.) Fr. *Beefsteak Fungus.*
Common everywhere. On and inside old oaks. Edible, but with scarcely any flavour of its own. When it is cooked too young, it is too astringent to be pleasant from the tannic acid which it contains.

POLYPORUS (Micheli) Fr. (πολύς, many, πόρος, a pore, from the shape of the hymenium.)

672. **P. rufescens** (Pers.) Fr.
Ladywood, Salwarpe. Nunnery Wood. Shrawley Wood. Chatley. Pershore. Evesham.
Not uncommon. In woods and orchards on cut-down apple stumps and other stubs.

673. **P. squamosus** (Huds.) Fr.
Common everywhere. On many different trees. Alleged by Mrs. Hussey to be edible and as good as saddle-flaps.

674. **P. picipes** Fr.
Severn-side near Shrawley. Holt. Claines. Powick. Defford Brook, near Defford. Evesham.
Common. On pollarded willows and sometimes on alders.

675. **P. varius** (Pers.) Fr.
Wyre Forest. Shrawley Wood. Kempsey. Abberton. Crowle.
Not uncommon. On willows and ash chiefly.

676. **P. elegans** (Bull.) Fr.
Holt. Wyre Forest.
Uncommon. On stumps and fallen branches, especially birch.
Var. **nummularius** (Bull.) Fr.
Dripshill Wood. Wyre Forest.
Not uncommon. On willow and beech.

677. **P. intybaceus** Fr.
Hanbury Park. Hewell Park.
Not common. On beech. Edible and delicious.

678. **P. giganteus** (Pers.) Fr.
Arley Castle. Worcester.
Not uncommon. At the base of trees, chiefly beeches.

679. **P. sulphureus** (Bull.) Fr.
Common everywhere. On many different trees. A destructive parasite.

680. **P. spongia** Fr.
Wyre Forest.
Rather rare. On dead pine-stumps.

681. **P. dryadeus** (Pers.) Fr.
Sharpway Gate. Shrawley Wood. Alfrick. Kyre Park. Hanbury Park. Westwood Park.
Common. On oaks. A destructive parasite, destroying the timber.

682. **P. hispidus** (Bull.) Fr.
Worcester. Trench Woods. Warndon. Malvern Wells. Ombersley. Kempsey. Claines.
Common. On ash and apple trees. A parasite on wounds.

683. **P. cuticularis** (Bull.) Fr.
Shrawley Wood. Wyre Forest.
Not common. On oaks and beeches.

684. **P. quercinus** (Schrad.) Fr.
Between Broadheath and Berrow Green.
Rare. On pollarded oak.

685. **P. crispus** (Pers.) Fr.
Shrawley Wood. Wyre Forest. Nunnery Wood.
Not uncommon. On oak, beech, and birch stumps.

686. **P. nidulans** Fr.
Shrawley Wood. Slashes Wood.
Not common. On fallen branches.

687. **P. mollis** (Pers.) Fr.
Shrawley Wood. Wyre Forest. Hadley.
Not uncommon. On pine-stumps.

688. **P. rutilans** (Pers.) Fr.
Wyre Forest. Tiddesley Wood. Trench Woods.
Not uncommon. On dead branches of poplar and birch.

689. **P. destructor** (Schrad.) Fr.
Arley. Dripshill.
Uncommon. On rotting wood.

690. **P. betulinus** (Bull.) Fr.
Common everywhere. On dead and dying birches.

691. **P. borealis** (Wahlb.) Fr.
Great Farley Wood. Trench Woods.
Not uncommon. On larch and pine stumps. This destructive parasite causes the White Rot disease of the spruce.

692. **P. fumosus** (Pers.) Fr.
Wyre Forest. Trench Woods. Shrawley Wood. Tiddesley Wood. Middleyards Coppice.
Not uncommon. On old stumps.

693. **P. adustus** (Willd.) Fr.
Common everywhere. On old stumps and palings.

694. **P. chioneus** Fr.
Hipton Hill. Wyre Forest. Ockeridge Wood. Monk Wood. Trench Woods.
Not uncommon. On birch-stumps.

695. **P. caesius** (Schrad.) Fr.
Wyre Forest. The Slads. Shrawley Wood. Dripshill Wood. Nunnery Wood.
Common. On old stumps.

696. **P. spumeus** (Sow.) Fr.
White's Nursery, Lower Wick, Worcester.
Uncommon. On a rotten felled pear-trunk.

697. **P. lacteus** Fr.
Castle Hill Wood. Perry Wood. Monk Wood. Wyre Forest. Shrawley Wood.
Not uncommon. On stumps and fallen branches.

698. **P. fragilis** Fr.
Wyre Forest. Shrawley Wood. Ockeridge Wood. Trench Woods.
Not uncommon. On stumps and fallen branches of conifers.

FOMES Fr. (*Fomes*, tinder, because many of the species were so employed.)

699. **F. lucidus** (Leyss.) Fr.
Brickkiln Covert. Chatley. Hanbury Park. Alfrick. Lower Wick. Worcester.
Not uncommon. On and at the base of many kinds of trees.

700. **F. ulmarius** Fr.
Old Malvern Road, Worcester. Kempsey. Ribbesford. Sinton Green. Hindlip. Madresfield.
Not uncommon. On and at the base of old elms and elm-stumps.

701. **F. populinus** Fr.
Acacia House, Foregate Street, Worcester.
Uncommon. On *Robinia pseudacacia*.

702. **F. connatus** Fr.
Hadley. Wyre Forest. Cowleigh Park. Old Hills.
Not uncommon. On poplar, apple, and other kinds of trees.

703. **F. fomentarius** (Linn.) Fr.
Shrawley Wood. Hagley Park. Northwick Park. Hanbury Park.
Common. On old trees, especially beeches.

704. **F. igniarius** (Linn.) Fr.
Common everywhere. On different sorts of trees.

Var. **pomaceus** Pers.
Common everywhere. On plums and other fruit trees.

705. **F. nigricans** Fr.
Worcester. Wyre Forest. Shrawley Wood. Defford. Hartlebury.
Not uncommon. On birches.

706. **F. annosus** Fr.
Little Malvern Wood. Hadley. Wyre Forest. Great Farley Wood.
Common. At the base of conifers and on their stumps. It causes the Red Rot disease of conifers.

707. **F. applanatus** (Pers.) Wallr.
Wyre Forest. Nunnery Wood. Salwarpe. Sheriff's Lench.
Not uncommon. On trunks of oak and ash.

708. **F. ferruginosus** Fr.
Shrawley Wood. Wyre Forest. Trench Woods. Tiddesley Wood. Monk Wood.
Not uncommon. On stumps and fallen branches.

POLYSTICTUS Fr. (πολύς, many, στικτός, pricked, from the appearance of the hymenium.)

709. **P. perennis** (Linn.) Fr.
Shrawley Wood. Wyre Forest. Tiddesley Wood. Trench Woods. Ockeridge Wood.
Common. On charcoal heaps and burnt places.

710. **P. versicolor** (Linn.) Fr.
Common everywhere. On stumps and fallen branches. Very protean in colour.

711. **P. radiatus** (Sow.) Fr.
Laughern Brook, St. Johns, Worcester. Claines. Defford. Powick. Ladywood, Salwarpe.
Common. On alder, hazel, and birch trunks and stumps.

712. **P. hirsutus** (Schrad.) Fr.
Whittington, Wyre Forest. Hallow. Clerkenleap. Hartlebury. Nunnery Wood.
Not uncommon. On stumps, fallen branches, and sticks.

713. **P. velutinus** (Pers.) Fr.
Perry Wood. Nunnery Wood. Shrawley Wood. Wyre Forest. Tiddesley Wood.
Not uncommon. On stumps and twigs, especially of birch.

714. **P. abietinus** (Dicks.) Fr.
Wyre Forest. Dripshill Wood.
Common. On dead wood of conifers and beeches. Quélet says this plant is only a form of No. 755, *Irpex fusco-violaceus*.

PORIA Pers. (πόρος, a passage, because the plant is mainly made up of tubes.)

715. **P. vaporaria** Pers.
Common everywhere. On fallen twigs and branches.

Var. **secernibilis** B. and Br.
Shrawley Wood. Wyre Forest, Trench Woods. Malvern Wells.
Not uncommon. On fallen branches and stumps.

716. **P. mollusca** (Pers.) Fr.
Wyre Forest. Shrawley Wood. Trench Woods. Oldington Wood.
Not uncommon. On stumps and twigs of conifers.

717. **P. vulgaris** Fr.
Nunnery Wood. Wyre Forest. Shrawley Wood. Ockeridge Wood.
Not uncommon. On fallen branches.

718. **P. medulla-panis** (Pers.) Fr.
Shrawley Wood. Trench Woods. Monk Wood. Wyre Forest.
Not uncommon. On dead stumps and fallen branches.

719. **P. vitrea** Pers.
Stoke Works.
Uncommon. On rotten wood.

720. **P. hibernica** B. and Br.
Wyre Forest. Nunnery Wood. Trench Woods.
Rather uncommon. On dead wood.

721. **P. blepharistoma** B. and Br.
Wyre Forest. Nunnery Wood. Monk Wood.
Not uncommon. On dead wood and branches.

722. **P. callosa** Fr.
Bredon Hill.
Uncommon. On rotten oak branches.

723. **P. obducens** Pers.
Worcester. Wyre Forest. Spetchley. Shrawley Wood.
Not uncommon. On rotten wood.

724. **P. terrestris** (DC.) Fr.
Wyre Forest. Nunnery Wood. Trench Woods.
Uncommon. On the ground and on rotten wood.

725. **P. sanguinolenta** A. and S.
Landing-stage near Shrawley Wood.
Uncommon. On wood of landing-stage.

726. **P. micans** (Ehbg.) Fr.
Shrawley Wood.
Uncommon. On rotten wood.

727. **P. armeniaca** Berk.
Wyre Forest. Nunnery Wood. Trench Woods. Monk Wood.
Not uncommon. On rotten stumps and branches

728. **P. bombycina** Fr.
Nunnery Wood. Wyre Forest.
Uncommon. On rotten wood.

TRAMETES Fr. (*Trama*, the weft, because the generic distinction depends on the trama, the pores being sunk into it and not forming a distinct layer.)

729. **T. gibbosa** (Pers.) Fr.
Bredon Hill. Tiddesley Wood.
Not common. On stumps and posts.

730. **T. Bulliardi** Fr.
Wyre Forest. Laughern Brook.
Uncommon. On willows and alders.

731. **T. suaveolens** (Linn.) Fr.
Ladywood, Salwarpe. Whittington. Berwick's Brake. Claines. Grimley. Holt. Crowle.
Common. On pollarded willows.

732. **T. serpens** Fr.
Shrawley Wood. Wyre Forest. Tiddesley Wood. Nunnery Wood.
Not uncommon. On fallen oak branches.

733. **T. mollis** (Sommerf.) Fr.
Trench Woods. Shrawley Wood. Monk Wood.
Rather uncommon. On fallen branches.

DAEDALEA Pers. (δαίδαλος, curiously wrought, from the elaborate hymenium.)

734. **D. quercina** (Linn.) Pers.
Wyre Forest. Shrawley Wood. Trench Woods. Nunnery Wood. Tiddesley Wood. Holt.
Common. On dead oak-stumps and trunks.

735. **D. cinerea** Fr.
Shrawley Wood.
Uncommon. On dead oak and beech trunks.

736. **D. unicolor** (Bull.) Fr.
Shrawley Wood. Wyre Forest.
Uncommon. On rotten stumps, especially birch.

MERULIUS Fr. (*Merula*, a blackbird, because of the colour of some species.)

737. **M. lacrymans** (Wulf.) Fr. *Dry-rot.*
Worcester. Hallow. Sheriff's Lench. Wyre Forest. Droitwich. Keybridge. Malvern.
Common. On trunks and worked wood. A dangerous saprophyte.

738. **M. Guillemoti** Boud.
First Record for Britain. *Rea, Trans. Brit. Myc. Soc.*, vol. ii, p. 38.
Worcestershire Cricket Club Ground, Worcester.
Rare. On the wooden uprights to the stands.

739. **M. tremellosus** Schrad.
　　Wyre Forest. Shrawley Wood. Monk Wood. Ockeridge Wood. Nunnery Wood.
　　Common. On stumps and fallen branches.

740. **M. corium** (Pers.) Fr.
　　Shrawley Wood. Wyre Forest. Nunnery Wood. Trench Woods. The Slads.
　　Common. On stumps and fallen branches.

HYDNACEAE

HYDNUM Linn. (ὕδνον, truffle.)

741. **H. repandum** Linn.
　　Common everywhere. In woods. Edible and delicious.

742. **H. rufescens** Pers.
　　Shrawley Wood. Wyre Forest. Perry Wood. Nunnery Wood. Trench Woods.
　　Common. In woods. Edible and delicious.

743. **H. auriscalpium** Linn.
　　Sheriff's Lench. Trench Woods. Hadley. Wyre Forest. Shrawley Wood.
　　Common. On fallen pine-cones.

744. **H. ochraceum** Gmel.
　　Shrawley Wood. Wyre Forest. Trench Woods. Oldington Wood.
　　Not uncommon. On fallen branches.

745. **H. alutaceum** Fr.
　　Wyre Forest. Trench Woods. Shrawley Wood.
　　Uncommon. On dead pine-stumps and fallen branches.

746. **H. viride** (A. and S.) Fr.
　　Shrawley Wood.
　　Uncommon. On rotten wood.

747. **H. melleum** B. and Br.
　　Nunnery Wood. Monk Wood.
　　Uncommon. On rotten wood.

748. **H. udum** Fr.
　　Shrawley Wood. Wyre Forest. Middleyards Coppice.
　　Not uncommon. On fallen branches.

749. **H. niveum** Pers.
　　Shrawley Wood. Ockeridge Wood. Middleyards Coppice. Dripshill Wood. Nunnery Wood.
　　Common. On rotten wood and fallen branches.

750. **H. farinaceum** Pers.
　　Wyre Forest. Shrawley Wood. Trench Woods. Dripshill Wood.
　　Not uncommon. On rotten wood, especially pine.

751. **H. argutum** Fr.
　　Nunnery Wood. Monk Wood. Wyre Forest.
　　Not uncommon. On rotten wood and fallen branches, especially birch.

CALDESIELLA Sacc.

752. **C. ferruginosa** (Fr.) Sacc.
　　Shrawley Wood. Wyre Forest. Nunnery Wood. Dripshill Wood. Trench Woods.
　　Not uncommon. On fallen branches, especially under the bark.

IRPEX Fr. (*Irpex*, a harrow, from the arrangement of the teeth on the hymenium.)

753. **I. spathulatus** (Schrad.) Fr.
　　Wyre Forest. Shrawley Wood. Nunnery Wood. Trench Woods.
　　Not uncommon. On fallen branches.

754. **I. obliquus** (Schrad.) Fr.
　　Common everywhere. On stumps and fallen branches.

755. **I. fusco-violaceus** (Schrad.) Fr.
　　Oldington Wood. Great Farley Wood.
　　Uncommon. On pine-trunks and fallen branches. Quélet says this plant is the perfect condition of No. 714, *Polystictus abietinus*.

RADULUM Fr. (*Radula*, a root, from the root-like appearance of the processes of the hymenium.)

756. **R. orbiculare** Fr.
　　Holt. Aislehurst Coppice. Wyre Forest. Shrawley Wood. Tiddesley Wood.
　　Common. On fallen branches of alder, birch, and firs.

757. **R. quercinum** (Pers.) Fr.
　　Wyre Forest. Nunnery Wood. Shrawley Wood. Monk Wood. Dripshill Wood.
　　Common. On fallen oak branches.

758. **R. tomentosum** Fr.
　　Shrawley Wood.
　　Uncommon. On dead apple branches.

PHLEBIA Fr. (φλέψ, a vein, from the veiny appearance of the hymenium.)

759. **P. merismoides** Fr.
　　Malvern. Westwood Park. Wyre Forest.
　　Not uncommon. On stumps and fallen branches.

760. **P. radiata** Fr.
　　Shrawley Wood. Wyre Forest. Nunnery Wood. Monk Wood.
　　Not uncommon. On fallen branches of birch and alder.

761. **P. contorta** Fr.
　　Shrawley Wood. Wyre Forest.
　　Uncommon. On fallen birch branches.

762. **P. vaga** Fr.
　　Wyre Forest. Monk Wood. Ockeridge Wood.
　　Not uncommon. On rotten wood.

GRANDINIA Fr. (*Grando*, hail, from the granular character of the hymenium.)

763. **G. granulosa** (Pers.) Fr.
　　Common everywhere. On dead wood and fallen branches.

764. **G. ocellata** Fr.
　　Ockeridge Wood.
　　Uncommon. On rotten wood.

765. **G. crustosa** (Pers.) Fr.
　　Wyre Forest. Shrawley Wood. Ockeridge Wood. Dripshill Wood.
　　Not uncommon. On fallen branches.

766. **G. mucida** Fr.
　　Trench Woods. Wyre Forest. Shrawley Wood.
　　Not uncommon. On rotten wood.

THELEPHORACEAE

CRATERELLUS Fr. (*Crater*, a bowl, from the shape of some of the species.)

767. **C. cornucopioides** (Linn.) Pers.
　　Common everywhere. In woods. Edible and very delicate in flavour.

768. **C. sinuosus** Fr.
　　Shrawley Wood. Wyre Forest.
　　Uncommon. In woods. This is incorrectly recorded as *C. clavatus* Fr. in *Trans. Worc. Nat. Club*, vol. ii, p. 27.

THELEPHORA Ehrh. (θηλή, a teat, φέρω, I bear, from the surface of the hymenium.)

769. **T. caryophyllea** (Schaeff.) Pers.
　　Shrawley Wood. Tiddesley Wood.
　　Rather uncommon. On the ground in deciduous woods.

770. **T. palmata** (Scop.) Fr.
　　Wyre Forest. Trench Woods.
　　Uncommon. On the ground in woods under pines.

771. **T. laciniata** Pers.
　　Nunnery Wood. Great Farley Wood. Wyre Forest. Shrawley Wood. The Slads.
　　Common. On twigs, stumps, and running up heather stems.

772. **T. terrestris** Ehrh.
　　Shrawley Wood.
　　Uncommon. On the ground under conifers.

SOPPITTIELLA Mass. (After the late Mr. H. T. Soppitt, a Yorkshire botanist.)

773. **S. sebacea** (Pers.) Mass.
　　Shrawley Wood. Wyre Forest.
　　Not uncommon. On stumps, twigs, and grasses in woods.

774. **S. caesia** (Pers.) Mass.
　　Wyre Forest. Dripshill Wood.
　　Uncommon. On the ground and running over mosses and twigs in woods.

775. **S. fastidiosa** (Pers.) Mass.
　　Wyre Forest. Nunnery Wood. Bredon Hill. Ockeridge Wood.
　　Not uncommon. On the ground and running over mosses and dead leaves.

776. **S. cristata** (Pers.) Mass.
　　Wyre Forest. Shrawley Wood. West Malvern. Hanbury Park.
　　Not uncommon. On the ground and running over dead leaves and grasses.

777. **S. crustacea** (Schum.) Mass.
　　Shrawley Wood. Trench Woods.
　　Uncommon. On the ground and running over mosses and dead leaves.

STEREUM Pers. (στερεός, hard, from the nature of the plant.)

778. **S. Sowerbei** (Berk.) Mass.
　　Shrawley Wood. Wyre Forest.
　　Uncommon. On the ground in woods.

779. **S. hirsutum** (Wild.) Pers.
　　Common everywhere. In woods and hedgerows on stumps and fallen branches.

780. **S. ochroleucum** Fr.
　　Worcester. Shrawley Wood. Wyre Forest. Nunnery Wood. Trench Woods. The Slads.
　　Common. On dead trunks, stumps, and fallen branches.

781. **S. purpureum** Pers.
　　Common everywhere. On stumps and fallen branches, especially of birch.

782. **S. sanguinolentum** (A. and S.) Fr.
　　Ladywood, Salwarpe. Great Farley Wood. Trench Woods. Dripshill Wood.
　　Not uncommon. On stumps and fallen branches of conifers.

783. **S. rugosum** Pers.　　*Oak Leather.*
　　Common everywhere. On stumps, rotten wood, and fallen branches.

784. **S. spadiceum** (Pers.) Fr.
　　Wyre Forest. Nunnery Wood. Shrawley Wood. Middleyards Coppice. Cowleigh Wood.
　　Common. On trunks, stumps, and fallen branches.

785. **S. disciforme** (DC.) Fr.
Nunnery Wood.
Probably not uncommon. On fallen oak branches.

786. **S. vorticosum** Fr.
Roadside between Spetchley Road and Nunnery Wood. Worcester.
Uncommon. On dead wood, stumps, and twigs.

CONIOPHORA (DC.) Pers. (κόνις, dust, φέρω, I bear, from the dust-like character of the hymenium.)

787. **C. arida** (Fr.) Karst.
Shrawley Wood. Oldington Wood. Trench Woods. Dripshill Wood.
Not uncommon. On rotten pine wood.

788. **C. sulphurea** (Pers.) Mass.
Wyre Forest. Spetchley Park. Nunnery Wood. Shrawley Wood. Middleyards Coppice.
Common. On rotten wood, fallen branches, and worked timber.

789. **C. laxa** (Fr.) Quél.
First Record for Britain. As in this book.
Near Perry Wood, *Rea*.
Uncommon. On wooden steps.

790. **C. puteana** (Schum.) Quél.
Common everywhere. On rotten and worked wood, and running over leaves.

PENIOPHORA Cke. (πηνίον, the quill on which the thread is wound, φέρω, I bear, from the prominent cystidia on the hymenium.)

791. **P. quercina** (Pers.) Cke.
Common everywhere. On fallen branches, especially of oak.

792. **P. gigantea** (Fr.) Mass.
Common everywhere. On pine-stumps and running over pine-needles.

793. **P. rosea** (Pers.) Mass.
Trench Woods. Wyre Forest.
Uncommon. On stumps and fallen branches.

794. **P. incarnata** (Pers.) Mass.
Old Hills. Perry Wood. Wyre Forest. Dripshill Wood.
Common. On fallen branches, logs, and rails.

795. **P. ochracea** (Fr.) Mass.
Wyre Forest. Shrawley Wood. Dripshill Wood, Worcester. Spetchley Park. Stoke.
Common. On rotten wood.

796. **P. cinerea** (Pers.) Cke.
Wyre Forest. Nunnery Wood. Shrawley Wood. Monk Wood. Ockeridge Wood. Trench Woods.
Common. On fallen branches and logs.

797. **P. velutina** (DC.) Cke.
Wyre Forest. Tiddesley Wood. Monk Wood.
Not uncommon. On stumps, trunks, and fallen branches.

HYMENOCHAETE Lév. (ὑμήν, a membrane, χαίτη, long flowing hair, from the coloured cystidia on the hymenium.)

798. **H. rubiginosa** (Dicks.) Lév.
Stoulton. Dripshill Wood. Bransford. The Dene Wood. Wyre Forest. Shrawley Wood.
Common. On stumps and fallen branches, especially oak.

799. **H. tabacina** (Sow.) Lév.
Wyre Forest. Nunnery Wood. Holt Bank. Ockeridge Wood. Clerkenleap. Claines.
Not uncommon. On trunks and stumps.

CORTICIUM (*Cortex*, bark, from the usual habitat of these species.)

800. **C. porosum** B. and Curt.
Monk Wood. Nunnery Wood.
Rather uncommon. On fallen branches.

801. **C. calceum** (Pers.) Fr.
Near Ockeridge Wood. Nunnery Wood. Hindlip. Kempsey. Madresfield. Hallow.
Common. On stumps and fallen timber.

802. **C. lacteum** Fr.
Nunnery Wood. Perry Wood. Wyre Forest. Shrawley Wood. Tiddesley Wood. Ockeridge Wood.
Common. On fallen branches and leaves.

803. **C. laeve** Pers.
Trench Woods. Monk Wood. Nunnery Wood. Wyre Forest.
Not uncommon. On rotten wood, especially birch.

804. **C. nudum** Fr.
Hindlip. Bevere. Kempsey. Bewdley. Claines. Spetchley Park.
Not uncommon. On stumps and fallen branches.

805. **C. arachnoideum** Berk.
Nunnery Wood. Wyre Forest. Shrawley Wood.
Not uncommon. On fallen branches and leaves.

806. **C. sambuci** Fr.
Holt. Diglis. Shrawley Wood. The Slads. Hallow. Crowle. Ombersley.
Common. On elders.

807. **C. sanguineum** Fr.
Ockeridge Wood. Wyre Forest. Trench Woods. Shrawley Wood.
Common. On rotten wood and fallen branches.

808. **C. caeruleum** (Schrad.) Fr.
Dripshill Wood. Crowneast. Perry Wood. Ankerdine Hill. The Slads.
Not uncommon. On fallen branches and twigs.

809. **C. comedens** (Nees.) Fr.
Common everywhere. Bursting the bark off dead and fallen oak, hazel, and plum branches.

CYPHELLA Fr. (κυφός, a goblet, because the plants are cup-shaped.)

810. **C. capula** (Holmsk.) Fr.
Nunnery Wood. Tiddesley Wood. Aislehurst Coppice. Knightsford Bridge.
Not uncommon. On dying and dead herbaceous stems.

811. **C. muscigena** (Pers.) Fr.
Near Fernhill Heath. Hadley Mill. Hartlebury Common.
Not uncommon. On various mosses.

812. **C. villosa** (Pers.) Karst.
Wyre Forest. Shrawley Wood.
Not uncommon. On fallen branches.

813. **C. muscicola** Fr.
Perry Wood.
Uncommon. On various mosses.

EXOBASIDIUM Woronin. (*Ex*, outside, *basidium*, basidium, because the basidia only are exposed on the surface of the host.)

814. **E. vaccinii** Woronin.
Wyre Forest.
Uncommon. On living leaves of *Vaccinium Myrtillus*.

SOLENIA Hffm. (σωλήν, a pipe, because the plants are tubular in shape.)

815. **S. fasciculata** Pers.
Wyre Forest. Tiddesley Wood. Trench Woods.
Not uncommon. On rotten wood.

816. **S. anomala** (Pers.) Fckl.
Shrawley Wood. Nunnery Wood. Perry Wood. Wyre Forest. Dripshill Wood. Monk Wood.
Common. On fallen branches.

CLAVARIACEAE

CLAVARIA Linn. (*Clava*, a club, because many of the species have this shape.)

817. **C. amethystina** (Holmsk.) Pers
Shrawley Wood. Spetchley Park. Hanbury Park.
Not common. In woods and pastures. Edible and delicious.

818. **C. fastigiata** Linn.
Common everywhere. In pastures amongst short grass.

819. **C. muscoides** Linn.
Common everywhere. In pastures amongst short grass.

820. **C. coralloides** Linn.
Bevere Green. Wyre Forest.
Uncommon. In woods on the ground. Edible.

821. **C. cinerea** Bull.
Common everywhere. In woods. Edible and delicious.

822. **C. cristata** (Holmsk.) Pers.
Common everywhere. In woods. Edible and delicious.

823. **C. rugosa** Bull.
Common everywhere. In woods and pastures under trees. Edible and delicious.

824. **C. subtilis** Pers.
Wyre Forest near Seckley rapid. Ockeridge Wood. Trench Woods. Crews Hill Wood.
Uncommon. In woods on the bare ground.

825. **C. pyxidata** Pers.
Shrawley Wood.
Rare. On rotten wood and on the ground.

826. **C. aurea** Schaeff.
Crews Hill Wood.
Uncommon. In woods. Edible.

827. **C. formosa** Pers.
Wyre Forest. Crews Hill Wood.
Uncommon. On the ground in woods. Edible and delicious.

828. **C. abietina** Pers.
Wyre Forest. Eymore Wood. Oldington Wood.
Not uncommon. In woods under conifers. Edible.

829. **C. flaccida** Fr.
Dripshill Wood. Trench Woods.
Rather uncommon. In woods under conifers. Edible.

830. **C. fusiformis** Sow.
Shrawley Wood. Wyre Forest. Hartlebury Common. Cowleigh Park. The Slads. Holt Bank.
Common. In woods and pastures. Edible.

831. **C. inaequalis** Fl. Dan.
Bevere Green. Shrawley Wood. Wyre Forest. Witley. Holt. Chaddesley Corbett. Alfrick.
Common. In grassy places in woods and pastures. Edible and delicious.

832. **C. dissipabilis** Britz.
Wyre Forest. Shrawley Wood. Ockeridge Wood. Monk Wood. Nunnery Wood. Hartlebury Common.
Common. In grassy places in woods and pastures. This differs from *C. inaequalis* in having echinulate spores.

833. **C. luteo-alba** Rea.
First Record for Worcestershire as below.
Wyre Forest. Hartlebury Common. Cowleigh Park. Defford Common.
Not uncommon. In woods and pastures amongst short grass. See *Trans. Brit. Myc. Soc.*, vol. ii, p. 66, plate 3; and vol. iii, p. 30. 'It was

my observation of this plant in Worcestershire that induced me after close study to decide that it had hitherto been undescribed though very markedly distinct.' *C. Rea.*

834. C. argillacea Fr.
Hartlebury Common. Castlemorton Common. Hanbury Park.
Not common. On commons and amongst short grass.

835. C. vermicularis Scop.
Common everywhere. In pastures, and occasionally in woods.
Edible, and when cooked tastes exactly like cheese-straws. Easily recognized by its extreme fragility.

836. C. fragilis Holmsk.
Crowle. Trench Woods. Alfrick. Kyre Park. Wyre Forest. Old Hills. Kempsey.
Common. In pastures and in woods amongst short grass.

837. C. fumosa Pers.
Black House Hill Wood.
Uncommon. Under oaks and hazel.

838. C. pistillaris Linn.
Middleyards Coppice. Coppice by the side of the road between Leigh Sinton and the New Inn. Crews Hill Wood.
Uncommon. On the ground in woods. Edible and delicious.

839. C. ligula Schaeff.
Wyre Forest. Crews Hill Wood.
Uncommon.

TYPHULA Fr. (*Typha*, the reed-mace, which it somewhat resembles in miniature.)

840. T. erythropus (Pers.) Fr.
Rosebury Rock Wood. Leigh Sinton. Perry Wood. Wyre Forest. Monk Wood.
Common. On twigs and fallen leaves in woods.

841. T. phacorrhiza (Reichard) Fr.
Wyre Forest. Shrawley Wood. Middleyards Coppice.
Not uncommon. On *Sclerotium scutellatum* on dead leaves in woods.

842. T. Grevillei Fr.
Wyre Forest. Nunnery Wood. Boughton Park. Eymore Wood
Not uncommon. On dead leaves and herbaceous stems.

843. T. muscicola (Pers.) Fr.
Hawford Bank.
Uncommon. On various mosses.

PISTILLARIA Fr. (*Pistillum*, a pestle, from the shape of the species.)

844. P. tenuipes (B. and Br.) Mass.
Wyre Forest. Shrawley Wood. Trench Woods. Dripshill Wood.
Not uncommon. On charcoal heaps and bare heathy ground.

845. P. quisquiliaris Fr.
Shrawley Wood. Little Malvern Wood. Rosebury Rock Wood.
Not uncommon. On dead fern-stems.

846. P. puberula Berk.
Wyre Forest. Elmley Castle.
Uncommon. On dead bracken-stems.

TREMELLACEAE

AURICULARIA Bull. (*Auricula*, the ear, from the form of the species.)

847. A. mesenterica (Dicks.) Pers.
Common everywhere. On stumps, especially of elm.

848. A. lobata Sommerf.
Holt. Bransford. Ombersley. Hampton Lovett.
Not uncommon. On stumps.

HIRNEOLA Fr. (*Hirneolus*, a small jug, from the shape.)

849. H. auricula-Judae (Linn.) Berk. *Jew's Ear.*
Near Dog and Duck Ferry, Worcester. Diglis. Holt Bank. Shrawley Wood. Malvern. Perry Wood. Saltwells Wood.
Common. Chiefly on living elders, occasionally on beech and elm. Edible, but with no pronounced flavour.

EXIDIA Fr. (ἐξιδίω, I exude ; from the nature of the receptacle.)

850. E. glandulosa (Bull.) Fr. *Witches' Butter.*
Shrawley Wood. Wyre Forest. Ockeridge Wood. Tiddesley Wood. Dripshill Wood.
Common. On dead branches of oak, lime, &c.

851. E. recisa (Ditmar) Fr.
Middleyards Coppice. Wyre Forest.
Uncommon. On dead branches of willow, sloe, &c.

852. E. albida (Huds.) Bref.
Shrawley Wood. Wyre Forest. Helbury Hill. Nunnery Wood. Trench Woods. Tiddesley Wood.
Common. On fallen branches.

ULOCOLLA Bref. (οὖλος, all, κόλλα, glue, from the consistency of the plants.)

853. U. foliacea (Pers.) Bref.
Wyre Forest. Shrawley Wood. Cowleigh Park.
Not uncommon. On stumps and fallen branches, especially of conifers.

TREMELLA (Dill.) Fr. (*Tremo*, I tremble, because of its gelatinous consistency.)

854. T. frondosa Fr.
Wyre Forest. Shrawley Wood. Nunnery Wood. Trench Woods.
Not uncommon. On stumps and fallen branches of oak.

P p

855. T. lutescens Pers.
Shrawley Wood. Dodderhill Common. Wyre Forest. Nunnery Wood. Spetchley Park.
Common. On fallen branches and fallen timber.

856. T. mesenterica Retz.
Shrawley Wood. Trench Woods. Hartlebury Common. Monk Wood. Wyre Forest.
Common. On fallen branches and dead heather and gorse stems. Edible.

857. T. intumescens Eng. Bot.
Crowneast. Wyre Forest.
Uncommon. On stumps and fallen branches.

858. T. indecorata Sommerf.
Shrawley Wood.
Uncommon. On fallen willow and poplar branches.

859. T. tubercularia Berk.
Wyre Forest. Perry Wood. Shrawley Wood. Nunnery Wood. Trench Woods. Holt Bank.
Common. On fallen branches, especially oak.

860. T. sarcoides Sm.
Common everywhere. On fallen branches. Probably the conidial stage of *Coryne sarcoides*.

NAEMATELIA Fr. (ναιμά, a word coined by Fries from ναίω, and applied to the gelatinous substance which surrounds the nucleus ; εἰλέω, to roll or wrap round.)

861. N. nucleata Fr.
Wyre Forest. Shrawley Wood.
Uncommon. On rotten wood.

862. N. virescens (Schum.) Cda.
Wyre Forest.
Uncommon. On rotten wood.

DACRYOMYCES Nees. (δάκρυ, a tear, μύκης, a fungus, from the appearance of the species.)

863. D. deliquescens (Bull.) Duby.
Common everywhere. On dead and worked wood.

864. D. stillatus Nees.
Common everywhere. On dead and squared pine wood.

865. D. chrysocomus (Bull.) Fr.
Wyre Forest.
Uncommon. On rotten wood.

866. D. torta (Berk.) Mass.
Monk Wood.
Uncommon. On dead oak branches.

CALOCERA Fr. (καλός, beautiful, κέρας, a horn, from the shape of some of the species.)

867. C. viscosa (Pers.) Fr.
Malvern Wells. Wyre Forest. Great Farley Wood.
Common. On stumps of conifers.

868. C. cornea (Batsch) Fr.
Shrawley Wood. Wyre Forest. Ockeridge Wood. Trench Woods. Tiddesley Wood. Nunnery Wood.
Common. On trunks and fallen branches.

869. C. stricta Fr.
Wyre Forest. Shrawley Wood. Nunnery Wood. Spetchley Park. Monk Wood. Ockeridge Wood.
Common. On stumps, fallen branches, and felled timber.

GASTEROMYCETAE

NIDULARIACEAE, *Bird's Nest Fungi*

CYATHUS Hall. (κύαθος, a cup, from the shape of the species.)

870. C. striatus (Huds.) Hffm.
Wyre Forest. Shrawley Wood. Trench Woods. Dripshill Wood. The Slads.
Not uncommon. On stumps, twigs, and rotten wood.

871. C. vernicosus (Bull.) DC.
Ladywood, Salwarpe. Worcester. Wyre Forest. Shrawley Wood. Monk Wood. Ockeridge Wood.
Not uncommon. On the ground, on stubble, and on rotten wood.

CRUCIBULUM Tul. (*Crucibulum*, a melting-pot, from the shape of the plants.)

872. C. vulgare Tul.
Claines. Shrawley Wood. Hallow Ford. Great Farley Wood. Tiddesley Wood.
Not uncommon. On dead wood and twigs.

SPHAEROBOLUS Tode. (σφαῖρα, a ball, βολή, throw, because the endoperidium is ejected at maturity.)

873. S. stellatus Tode.
Wyre Forest. Great Farley Wood. Shrawley Wood. Nunnery Wood. Trench Woods. Dripshill Wood.
Not uncommon. On rotten wood and twigs.

THELEBOLUS Tode. (θηλή, a nipple, βολή, throw, from the way in which the endoperidium protrudes through the outer.)

874. T. terrestris (A. and S.) Tode.
Wyre Forest. Shrawley Wood.
Uncommon. On wood and on the ground.

P p 2

LYCOPERDACEAE

MYRIOSTOMA. (μυρίος, numberless, στόμα, mouth, because the endo-peridium dehisces by numerous mouths.)

875. **M. coliformis** (Dicks.) Desv.
Hanley Castle, *Messrs. Ballard and Rufford, Withering,* 2nd ed., vol. iv, p. 460.
Very rare. On the ground.

GEASTER Mich. (γῆ, earth, ἀστήρ, star, because the outer peridium splits in a star-like manner.) *Earth-stars.*

876. **G. Bryantii** Berk.
Leap Gate, Hartlebury. Lane near Baldwin's Works, Wilden.
Uncommon. On the ground and amongst leaves on the top of an oak-stump.

877. **G. fornicatus** (Huds.) Fr.
Stoke Heath. White's Nursery, Oldbury Road, Worcester. Hartlebury.
Uncommon. On the ground amongst trees.

878. **G. mammosus** Chev.
Overbury Park.
Rare. Under Beeches.

879. **G. fimbriatus** Fr.
Stoke. The Foxholes, Kidderminster.
Not uncommon. Amongst leaves.

LYCOPERDON Tournf. (λύκος, a wolf, πέρδομαι, I break wind; because some of the older writers believed that these fungi developed from the dung of the wolf.) *Puff-balls.*

880. **L. echinatum** Pers.
Near Crews Hill Wood.
Not uncommon. Amongst leaves in woods and plantations.

881. **L. Hoylei** Berk.
Cowleigh Park. Shrawley Wood.
Uncommon. Amongst leaves in woods.

882. **L. atropurpureum** Vitt.
Crews Hill Wood.
Uncommon. In woods amongst leaves.

883. **L. excipuliforme** Scop.
Wyre Forest. Alfrick. Randans Wood. Shrawley Wood. Westwood Park.
Not uncommon. In woods and pastures. Edible and delicious.
Var. **flavescens** Quél.
Near Monk Wood. Hartlebury Common. Hagley.

884. **L. saccatum** Vahl.
Shrawley Wood. Wyre Forest. Tiddesley Wood. Ockeridge Wood.
Common. In woods. Edible and delicious.

885. **L. gemmatum** Batsch.
Bevere Green. Great Farley Wood. Wyre Forest. Shrawley Wood.
Not uncommon. In woods and on commons.

886. **L. pyriforme** Schaeff.
Common everywhere. In woods and pastures on stumps and in hedgerows. Edible and delicious.
Var. **tessellatum** Pers.
Ladywood. Hadley. Wyre Forest. Tiddesley Wood. Crowneast Wood.
Var. **excipuliforme** (Desm.).
Shrawley Wood. Wyre Forest. Trench Woods.

887. **L. perlatum** Pers.
Trench Woods. Wyre Forest. Shrawley Wood. Randans Wood. Cowleigh Park. Tiddesley Wood.
Common. In woods, and occasionally in pastures. Edible and delicious.

888. **L. umbrinum** Pers.
Shrawley Wood. Wyre Forest. Perry Wood. Holt. Monk Wood. Nunnery Wood. Tiddesley Wood.
Common. In woods.

889. **L. hyemale** (Bull.) Vitt. (= **depressum** Bon.).
First Record for Britain. Kempsey Common, *Rea, Trans. Brit. Myc. Soc.,* vol. ii, p. 99.
Kempsey Common. Shrawley Wood. Near Crews Hill Wood. Hartlebury Common. Wyre Forest.
Common. In pastures, on commons, and in grassy places in woods.

890. **L. caelatum** Bull.
Sheriff's Lench. Hagley Hall. Near Monk Wood. Claines. Clerkenleap. Old Hills.
Not uncommon. In pastures, rarely in woods. Edible and delicious.

891. **L. bovista** Linn.
Ladywood, Salwarpe. Worcester. The Old Hills. Kyre. Claines. Castlemorton.
Not uncommon. In pastures. Edible and very delicious. This is one of the largest of our British Fungi and two Worcestershire specimens have weighed respectively eight pounds two ounces and seven pounds, whilst the girth of the former was forty-two inches by thirty-eight inches and of the latter forty-seven inches by thirty-seven inches.

892. **L. pusillum** Batsch.
Dodderhill Common. Hartlebury Common. Kempsey Common. Randans Wood.
Not uncommon. On commons amongst short grass.

BOVISTA Dill. (*Bos,* a cow, because the species grow in pastures.)

893. **B. plumbea** Pers.
Common everywhere. In pastures.

894. **B. nigrescens** Pers.
Bevere Green. Cowleigh Park. Dunhampstead. Ladywood, Salwarpe. Claines. Ombersley.
Common. In pastures.

SCLERODERMATACEAE

SCLERODERMA Pers. (σκληρός, hard, δέρμα, skin, from the firm peridium.)

895. **S. vulgare** Fl. Dan.
Shrawley Wood. Trench Woods. Ockeridge Wood. Bevere Green. Hewell Park. Randans Wood.
Common. In woods and pastures.

896. **S. verrucosum** (Bull.) Pers.
Trench Woods. Shrawley Wood. Bevere Green. Wyre Forest. Cowleigh Wood. Dripshill Wood.
Common. In woods and pastures.

897. **S. geaster** Fr.
Shrawley Wood.
Uncommon. In woods and pastures.

PHALLOIDALES

PHALLACEAE

ITHYPHALLUS Fisch. (ἰθύς, straight, φαλλός, the penis, from the shape of the mature plant.)

898. **I. impudicus** (Linn.) Fisch. *Stink-horn.*
Shrawley Wood. Wyre Forest. Malvern Wells. Great Farley Wood. Habberley Valley. Hewell Park.
Common. In woods and under trees, especially pines. Edible in the egg state.

MUTINUS Fisch. (*Mutinus,* penis, from the shape of the plants.)

899. **M. caninus** (Huds.) Fr.
Shrawley Wood. Perry Wood. Trench Wood. Tiddesley Wood. Nunnery Wood.
Not uncommon. In woods.

LYSURUS Fr. (λύσις, setting free, οὐρά, a tail, because the arms of these plants are free at the ends.)

900. **L. australiensis** Cke. and Mass.
First Record. *D. P. Goodwin,* Kidderminster, *Trans. Brit. Myc. Soc.,* vol. ii, p. 57, plate 3.
Kidderminster.
Very rare. The first record of this plant for Europe.

UREDINALES

MELAMPSORACEAE

MELAMPSORA Cast. (μέλας, black, σωρός, heap, from the dark colour of the teleutospore beds.)

901. **M. helioscopiae** (Pers.) Wint.
Claines. Ombersley. Powick. Grimley. Hallow. Knightsford Bridge.
Not uncommon. On *Euphorbia Peplus, exigua,* and *Helioscopia.*

902. **M. lini** (Pers.) Tul.
The Slads. Shrawley Wood. Old Hills. Holt Bank. Abberley Hill.
Not uncommon. On *Linum catharticum.*

903. **M. farinacea** (Pers.) Schröt.
Monk Wood, near Wyre Piddle Lock. The Rhydd.
Not uncommon. On *Salix cinerea* and *caprea.*

904. **M. euonymi-caprearum** Kleb.
Monk Wood. Bransford. Tiddesley Wood.
Not uncommon. On *Euonymus europaeus.*

905. **M. tremulae** Tul.
Monk Wood. Deerfold Wood. Trench Woods. Nunnery Wood.
Common. On *Populus tremula.*

906. **M. aecidioides** (DC.) Schröt.
Monk Wood. Trench Woods.
Uncommon. On *Populus alba.*

907. **M. Rostrupii** Wagner.
Abberley Hills. Cowleigh Wood. Trench Woods. Astley.
Not uncommon. On *Mercurialis perennis* and *Salix cinerea.*

908. **M. populina** (Jacq.) Cast.
Mildenham Mill. Near Deerfold Wood.
Not uncommon. On *Populus nigra.*

MELAMPSORIDIUM Kleb.

909. **M. betulinum** (Pers.) Kleb.
Common everywhere. On *Betula alba.*

COLEOSPORIACEAE

COLEOSPORIUM Lév. (κολεός, a sheath, σπόρος, seed, because the teleutospores are enclosed in a membrane.)

910. **C. senecionis** (Pers.) Lév.
Common everywhere. On *Senecio vulgaris, sylvatica,* and *Jacobaea,* and *Pinus sylvestris.*

911. **C. tussilaginis** (Pers.) Lév.
Crowle. Cotheridge. Powick. Leigh Sinton. Bransford.
Common. On *Tussilago Farfara.*

912. **C. petasitidis** de Bary.
Knightsford Bridge. Grimley. Hallow.
Not uncommon. On *Petasites ovatus.*

913. **C. sonchi-arvensis** (Pers.) Wint.
Common everywhere. On *Sonchus oleraceus* and *arcensis.*

914. **C. campanulae** (Pers.) Lév.
Near Worcester. Shrawley Wood. Alfrick. Sheriff's Lench. Witley.
Not uncommon. On *Campanula rotundifolia* and *Trachelium.*

915. **C. euphrasiae** (Schum.) Wint.
Alfrick. Knightwick.
Not uncommon. On *Euphrasia officinalis.*

PUCCINIACEAE

UROMYCES Link. (οὐρά, a tail, μύκης, fungus, from the form of the teleutospores.)

916. **U. fabae** (Pers.) Cke.
Sheriff's Lench. Ombersley. Claines. Kempsey. Welland.
Not uncommon. On *Faba vulgaris,* and *Vicia sepium* and *sativa.*

917. **U. polygoni** (Pers.) Fckl.
Claines. Sheriff's Lench. Pershore.
Not common. On *Polygonum aviculare.*

918. **U. geranii** (DC.) Cke.
Hallow Ford. Ham Bridge. Holt. Stanford Bridge.
Not uncommon. On *Geranium pratense* and *molle.*

919. **U. betae** (Pers.) Kühn.
Near New Inn, Claines.
Not common. On *Beta vulgaris.*

920. **U. valerianae** (Schum.) Wint.
Hurcott Wood. Grimley Brickpits.
Not uncommon. On *Valeriana sambucifolia.*

921. **U. dactylidis** Otth.
Salwarpe. Astley. Leigh. Ombersley.
Not uncommon. On *Ranunculus bulbosus* and *Dactylis glomerata.*

922. **U. poae** Rabh.
Salwarpe. Ladywood, Salwarpe. Grimley.
Not uncommon. On *Ranunculus Ficaria* and *repens,* and *Poa pratensis* and *annua.*

923. **U. ficariae** (Schum.) Lév.
Common everywhere. On *Ranunculus Ficaria.*

924. **U. scillarum** (Grev.) Wint.
Wyre Forest. Monk Wood. Shrawley Wood. Perry Wood. Tiddesley Wood.
Common. On *Scilla non-scripta.*

925. **U. behenis** (DC.) Ung.
Blackstone. Wyre Forest Station.
Uncommon. On *Silene latifolia.*

926. **U. ornithogali** (Wallr.) Lév.
Aisleshurst Plantation.
Not common. On *Gagea lutea.*

GYMNOSPORANGIUM Hedw. (γυμνός, naked, σπόρος, seed, ἄγγος, a vessel, from the shape of the teleutospore receptacle.)

927. **G. sabinae** (Dicks.) Wint.
Worcester. Ombersley. Ladywood.
Uncommon. Only the aecidiosporous condition has been observed on *Pyrus communis.*

928. **G. clavariaeforme** (Jacq.) Rees.
Sheriff's Lench. The Slads. Holt. Wyre Forest.
Uncommon. On *Crataegus monogyna* and *Juniperus communis.*

929. **G. juniperinum** (Linn.) Wint.
Malvern.
Uncommon. Only the aecidiosporous condition has been observed on *Pyrus Aucuparia.*

PUCCINIA Pers. (Puccini, an Italian botanist.) *Rusts.*

930. **P. asparagi** DC.
Sheriff's Lench.
Uncommon. On *Asparagus maritimus.*

931. **P. calthae** Link.
Bubble Brook near Mudwall Mill.
Not uncommon. On *Caltha palustris.*

932. **P. lapsanae** (Schultz.) Fckl.
Bilford Lane, Claines. Kempsey. Powick. Holt.
Not uncommon. On *Lapsana communis.*

933. **P. variabilis** Grev.
Grimley Brickpits. Between Spetchley and Churchill.
Not uncommon. On *Taraxacum officinale.*

934. **P. pulverulenta** Grev.
Grimley Brickpits. Spetchley. Seckley Wood.
Not uncommon. On *Epilobium hirsutum* and *montanum.*

935. **P. violae** (Schum.) DC.
Between Martley and Witley. Abberley Hill. Middleyards Coppice. Lords Wood. Cowleigh Park.
Common. On *Viola odorata, sylvestris* and *Riviniana.*

936. **P. albescens** (Grev.) Plow.
Holt Mill. Laughern Brook above Henwick Mill. Bog near Doverdale Church.
Common. On *Adoxa Moschatellina.*

937. **P. pimpinellae** (Strauss) Link.
Salwarpe. Cowleigh Wood. Claines. Old Hills.
Not uncommon. On *Heracleum Sphondylium* and *Anthriscus sylvestris.*

938. **P. menthae** Pers.
Leigh Mill. Brockamin. Fladbury. Worcester. Diglis. Holt.
Common. On *Mentha spicata, hirsuta,* and *arvensis.*

939. **P. primulae** (DC.) Grev.
Trench Woods. Vallombrosa. Holt Bank. Monk Wood. Middleyards Coppice.
Common. On *Primula vulgaris.*

940. **P. saniculae** Grev.
Middleyards Coppice. Perry Wood. Eymore Wood.
Uncommon. On *Sanicula europaea.*

941. **P. graminis** Pers.
Common everywhere. On our cereals and grasses. The aecidium grows on *Berberis vulgaris.*

942. **P. coronata** Cda.
Wyre Forest. Trench Woods. Holt. Ombersley.
Common. On *Holcus mollis* and *Agropyron repens.* The aecidium grows on *Rhamnus Frangula.*

943. **P. coronifera** Kleb.
The Slads. Dunhampstead. Wyre Forest. Sheriff's Lench. Oddingley.
Common. On *Avena sativa* and *pratensis,* and *Lolium perenne.* The aecidium grows on *Rhamnus catharticus.*

944. **P. dispersa** E. and H.
Ombersley. Hartlebury. Sheriff's Lench.
Not uncommon. On *Secale cereale* and *Triticum vulgare.* The aecidium grows on *Lycopsis arvensis.*

945. **P. phalaridis** Plow.
Purlieu Lane. Whittington. Spetchley Park. Severn below Diglis.
Not uncommon. On *Phalaris arundinacea.*

946. **P. poarum** Nielsen.
Norton juxta Kempsey. Kempsey Grove. Powick.
Not uncommon. On *Tussilago Farfara,* and *Poa trivialis* and *pratensis.*

947. **P. caricis** (Schum.) Rebent.
Laughern Brook. Canal-side, Oldington. Claines Brickpits.
Not uncommon. On *Carex riparia.*

948. **P. obscura** Schröt.
Between Monk Wood and Ockeridge Wood. Helbury Hill. Bransford. Kempsey.
Not uncommon. On *Luzula campestris.*

949. **P. phragmitis** (Schum.) Körn.
Northwick Brickpits. Grimley Brickpits. Birmingham Canal, Worcester.
Uncommon. On *Phragmites communis.*

950. **P. persistens** Plow.
Severn-side opposite Kepax Ferry. Grimley Brickpits.
Uncommon. On *Agropyron repens.*

951. **P. suaveolens** (Pers.) Wint.
Common everywhere. On *Cnicus arvensis* and *Centaurea Cyanus.*

952. **P. hieracii** (Schum.) Mart.
Middleyards Coppice. Wyre Forest.
Not uncommon. On *Hieracium vulgatum* and *boreale.*

953. **P. taraxaci** Plow.
Helbury Hill. Kempsey Grove.
Not uncommon. On *Taraxacum officinale.*

954. **P. glumarum** (Schum.) E. and H.
Common everywhere. On *Triticum vulgare, Secale cereale,* and *Hordeum vulgare.*

955. **P. simplex** Kornicke.
Common everywhere. On *Hordeum vulgare.*

956. **P. lychnidearum** Link.
Hartlebury. Shrawley Wood. Wyre Forest. The Slads. Holt Bank.
Not uncommon. On *Lychnis dioica.*

957. **P. chrysanthemi** Roze.
Worcester. Kidderminster.
Not uncommon. On *Chrysanthemum sinense.*

958. **P. tragopogi** (Pers.) Wint.
Grimley.
Uncommon. On *Tragopogon minus.*

959. **P. betonicae** (A. and S.) DC.
Monk Wood. Trench Woods. Shrawley Wood. Tiddesley Wood.
Common. On *Stachys officinalis.*

960. **P. aegopodii** (Schum.) Link.
Near Lincombe Lock.
Uncommon. On *Aegopodium Podagraria.*

961. **P. umbilici** Guép.
Knightwick.
Uncommon. On *Cotyledon Umbilicus-Veneris.*

962. **P. fusca** (Relhan) Wint.
Middleyards Coppice. Slashes Wood.
Not uncommon. On *Anemone nemorosa.*

963. **P. adoxae** DC.
Doverdale Marsh.
Uncommon. On *Adoxa Moschatellina.*

964. **P. malvacearum** Mont.
Common everywhere. On *Malva moschata, sylvestris,* and *rotundifolia,* and *Althaea rosea.*

965. **P. buxi** DC.
Worcester.
Not uncommon. On *Buxus sempervirens.*

PHRAGMIDIUM Link. (φραγμός, a fence, from the partitions between the numerous cells in each spore.)

966. **P. fragariastri** DC.
Clifton-on-Teme. Abberley Hills. Gullet Pass. Middleyards Coppice. Wyre Forest.
Common. On *Potentilla sterilis.*

967. **P. violaceum** (Schultz.) Wint.
Little Malvern Wood. Withy-bed, Eastbury.
Not uncommon. On *Rubus fruticosus.*

968. **P. rubi** (Pers.) Wint.
Common everywhere.   On *Rubus fruticosus.*

969. **P. sanguisorbae** (DC.) Schröt.
Abberley Hill.   Cotheridge.   Spetchley.
Not uncommon.   On *Poterium Sanguisorba.*

970. **P. subcorticatum** (Schrank) Wint.
Common everywhere.   On *Rosa canina.*

XENODOCHUS Schlecht. (ξενοδόχος, a host.)

971. **X. carbonarius** Schlecht.
Between Perry Wood and Nunnery Wood.
Uncommon.   On *Poterium officinale.*

TRIPHRAGMIUM Link. (τρί, three, φραγμός, a fence, because the spores are three-celled.)

972. **T. ulmariae** (Schum.) Link.
Grimley.   Claines.   Bransford.   Glasshampton.   Shrawley Wood.
Lincombe.
Common.   On *Spiraea Ulmaria.*

ENDOPHYLLUM Lév. (ἔνδον, within, φύλλον, a leaf, because these parasites are at first developed under the epidermis.)

973. **E. euphorbiae** (DC.) Lév.
Trench Woods.   Middleyards Coppice.   Deadman's Coppice, Lady-
Wood.   Wyre Forest.
Common.   On *Euphorbia amygdaloides.*

974. **E. leucospermum** (DC.) Sopp.
Middleyards Coppice.   Perry Wood.   Wyre Forest.
Not uncommon.   On *Anemone nemorosa.*

UREDO Pers. (*Uredo*, a blight.)   *Rusts.*

975. **U. symphyti** DC.
Severn-side Claines.   Grimley.   Leigh.   Fladbury.
Common.   On *Symphytum officinale.*

976. **U. mülleri** Schröt.
Wyre Forest.   Shrawley Wood.   Cowleigh Park.   Tiddesley Wood.
Common.   On *Rubus fruticosus.*

MILESIA B. White.   (Named after the Rev. Miles Joseph Berkeley.)

977. **M. scolopendri** Fckl.
Wyre Forest.
Uncommon.   On *Blechnum Spicant.*

USTILAGINALES
USTILAGINACEAE

USTILAGO Pers. (*Ustus*, burnt, from the scorched appearance of the organs of the hosts in which the spores are developed.)   *Smuts.*

978. **U. longissima** (Sow.) Tul.
Canal-side, Selly Oak.   Twyning Fleet.   Northwick Brickpits.   Kemp-
sey.   Powick Ham.
Common.   On *Glyceria aquatica* and *fluitans.*

979. **U. hypodytes** (Schlecht.) Fr.
Ombersley.   Hadley.   Grimley.   Claines.   Kempsey.
Not uncommon.   On *Agropyron repens.*

980. **U. tritici** (Pers.) Jensen.
Common everywhere.   On *Triticum vulgare.*

981. **U. hordei** (Pers.) Jensen.
Common everywhere.   On *Hordeum vulgare.*

982. **U. nuda** Jensen.
Common everywhere.   On *Hordeum vulgare.*

983. **U. avenae** (Pers.) Jensen.
Common everywhere.   On *Avena sativa.*

984. **U. scabiosae** (Sow.) Wint.
Near Shrawley Wood.   Knightwick.
Not uncommon.   In the anthers of *Scabiosa arvensis.*

985. **U. flosculorum** (DC.) Wint.
Wyre Forest.   Chatley.   Martley.
Not uncommon.   In the anthers of *Scabiosa Succisa.*

986. **U. utriculosa** (Nees) Tul.
Trench Woods.   Tiddesley Wood.   Powick.   Hallow.   Kempsey.
Ombersley.
Not uncommon.   On *Polygonum Persicaria* and *Hydropiper.*

987. **U. violacea** (Pers.) Tul.
Conderton Camp.   Wyre Forest.   Blackstone Hill.
Not uncommon.   In the anthers of *Silene latifolia.*

988. **U. tragopogi** (Pers.) Wint.
Bransford Court.   Old Powick Bridge.
Not uncommon.   On *Tragopogon minus.*

SPHACELOTHECA de Bary. (σφάκελος, mortification, θήκη, a box, from the destruction that it causes to the ovaries of the hosts.)

989. **S. hydropiperis** (Schum.) Schröt.
Trench Woods.   The Slads.
Locally very common.   On *Polygonum Hydropiper.*

SOROSPORIUM Rud. (σωρός, a heap, σπορά, seed, from the mass of spores.)

990. **S. scabies** (Berk.) F. v. Wald.
Worcester.
Not uncommon.   On tubers of *Solanum tuberosum.*

TILLETIACEAE

TILLETIA Tul.   (Named after Mathieu Tillet, a French botanist of the eighteenth century.)

991. **T. tritici** (Bjerk.) Wint.   *Bunt.*
Not very common, but present everywhere.   On *Triticum vulgare.*

992. **T. decipiens** (Pers.) Körn.
Malvern.   Wannerton Down.   Bishampton Banks.
Not uncommon.   On *Agrostis tenuis,* causing the variety *pumila* of phanerogamic botanists.

ENTYLOMA De Bary. (ἐντός, within, λῶμα, the hem, because the parasite produces its spores within the epidermis of the host.)

993. **E. ranunculi** (Bon.) Schröt.
Northwick.   Lower Wick.
Not uncommon.   On *Ranunculus sceleratus* and *Ficaria.*

UROCYSTIS Rab. (οὖρον, urine, κύστις, a vessel, from the vesicular spores.)

994. **U. anemones** (Pers.) Wint.
Monk Wood.   Wyre Forest.
Not uncommon.   On *Anemone nemorosa* and *Ranunculus repens.*

995. **U. violae** (Sow.) F. v. Wald.
Powers Mill, Ombersley.   Shrawley Wood.   Bransford.
Not uncommon.   On *Viola odorata, sylvestris,* and *Riviniana.*

ASCOMYCETAE
GYMNOASCALES
EXOASCACEAE

EXOASCUS Fckl.   (ἔξω, on the outside, ἀσκός, a bag, because these plants consist of asci only, which are developed on the outside of the host.)

996. **E. pruni** (Tul.) Fckl.   *Pocket Plums.*
Worcester.   Evesham.   Pershore.   Ombersley.
Not uncommon.   On fruit of *Prunus domestica.*

997. **E. deformans** (Berk.) Fckl.   *Leaf curl.*
Worcester.   Grimley.   Malvern.
Not uncommon.   On leaves of *Prunus domestica* and *Persica.*

998. **E. turgidus** Sadeb.   *Witches' brooms.*
Common everywhere.   On *Betula alba.*

GYMNOASCACEAE

GYMNOASCUS Baran. (γυμνός, naked, ἀσκός, a bag, because the plants have no stroma.)

999. **G. Reesii** Baran.
Claines.   Worcester.   Ladywood.
Not uncommon.   On horse-dung.

PYRENOMYCETAE
ERYSIPHACEAE

SPHAEROTHECA Lév. (σφαῖρα, a ball, θήκη, a case, from the round shape of these plants.)

1000. **S. pannosa** (Wallr.) Lév.   *Rose Mildew.*
Common everywhere.   On *Rosa canina.*

1001. **S. castagnei** Lév.   *Hop Mildew.*
Common everywhere.   On *Humulus Lupulus.*

1002. **S. mors-uvae** B. and C.   *American Gooseberry Mildew.*
Evesham.   Fladbury.   Pershore.   Harvington.   Worcester.   Om-
bersley.
Common.   On *Ribes Grossularia* and its var. b. *Uva-crispa.*

PODOSPHAERA Kunze. (πούς, a foot, σφαῖρα, a ball, because the perithecia are round, and furnished with appendages.)

1003. **P. oxyacanthae** (DC.) de Bary.
Claines.
Not uncommon.   On *Crataegus monogyna.*

ERYSIPHE (Hedw.) DC. (ἐρυθρός, red, σίφων, a tube.)

1004. **E. graminis** DC.
Common everywhere.   On various grasses.

1005. **E. Martii** Lév.
Near Nunnery Wood.   Holt Bank.
Not uncommon.   On *Hypericum.*

1006. **E. communis** (Wallr.) Fr.
Common everywhere.   On *Circaea lutetiana* and *Polygonum aviculare.*

1007. **E. cichoracearum** DC.
Hawford.   Brockamin.
Not uncommon.   On *Arctium majus* and *minus.*

MICROSPHAERA Lév. (μικρός, small, σφαῖρα, a ball, from the shape of the receptacle.)

1008. **M. grossulariae** (Wallr.) Lév.
Swinesherd.   Worcester.
Not uncommon.   On *Ribes Grossularia.*

UNCINULA Lév. (*Uncus*, hooked, from the curved ends of the appendages.)

1009. **U. salicis** (DC.) Wint.
Trench Woods. Hawford Brook.
Not uncommon. On *Populus* and *Salix*.

1010. **U. aceris** (DC.) Sacc.
Common everywhere. On *Acer campestre* and *Pseudo-platanus*.

PHYLLACTINIA Lév. (φύλλον, a leaf, ἀκτίς, a ray, from the single appendages.)

1011. **P. suffulta** (Rebent.) Sacc.
Common everywhere. On *Crataegus*, *Betula*, *Quercus*, *Alnus*, and *Fraxinus*.

PERISPORIACEAE

EUROTIUM Link. (εὐρώς, mould.

1012. **E. herbariorum** (Wigg.) Link.
Common everywhere. On plants in herbaria, rotten fruit, and mouldy bread. The conidial forms of *Eurotium* and *Penicillium* cause the well-known Blue Mould on jams, cheese, and fruit.

1013. **E. repens** de Bary.
Worcester.
Not uncommon. On *Campanula patula* in my herbarium, *Rea*.

PENICILLIUM Link. (*Penicillum*, a painter's brush, from the shape of the conidial form.)

1014. **P. crustaceum** (Linn.) Fr.
Common everywhere. On decaying substances.

LASIOBOTRYS Kunze. (λάσιος, shaggy with hair, βότρυς, a bunch of grapes, because the perithecia are seated on the margin of a hairy stroma.)

1015. **L. lonicerae** Kunze.
Sheriff's Lench.
Uncommon. On *Lonicera Periclymenum*.

CAPNODIUM Mont. (καπνός, smoke, from the colour of these plants.)

1016. **C. salicinum** (A. and S.) Mont.
Shrawley Wood. Worcester.
Not uncommon. On various trees and shrubs.

HYPOCREACEAE

ELEUTHEROMYCES Fckl. (ἐλεύθερος, free, μύκης, a fungus.)

1017. **E. subulatus** (Tode) Fckl.
Perry Wood.
Uncommon. On dead agarics.

GIBBERELLA Sacc. (*Gibber*, crook-backed, from the shape of the ascospores.)

1018. **G. pulicaris** (Fr.) Sacc.
Diglis. Shrawley Wood.
Not uncommon. On dead branches of *Sambucus nigra*.

NECTRIA Fr.

1019. **N. cinnabarina** (Tode) Fr.
Common everywhere. On dead branches and twigs.

1020. **N. coccinea** (Pers.) Fr.
Oldbury Road. Wyre Forest.
Not uncommon. On fallen branches and dead twigs.

1021. **N. ditissima** Tul. *Apple-tree Canker.*
Common everywhere. On various trees, especially *Pyrus Malus* and *Fagus sylvatica*.

1022. **N. aquifolii** (Fr.) Berk.
Raggedstone Hill.
Not uncommon. On *Ilex Aquifolium*.

1023. **N. sanguinea** (Sibth.) Fr.
Wyre Forest.
Not uncommon. On fallen branches.

1024. **N. episphaeria** (Tode) Fr.
Wyre Forest. Shrawley Wood. Eymore Wood.
Not uncommon. On *Diatrype stigma* and *Ustulina vulgaris*.

1025. **N. peziza** (Tode) Fr.
Wyre Forest. Shrawley Wood.
Uncommon. On fallen branches and dead wood.

HYPOMYCES Fr. (ὑπό, under, μύκης, fungus, because the general habitat of these plants is to grow upon another fungus.)

1026. **H. rosellus** (A. and S.) Tul.
Shrawley Wood. Monk Wood. Wyre Forest.
Common. On *Stereum hirsutum*, *Russulae*, *Lactarii*, etc.

1027. **H. chrysospermus** Tul.
Common everywhere. On *Boleti*.

1028. **H. asterophorus** Tul.
Common everywhere. On *Nyctalis parasitica*.

1029. **H. aurantius** (Pers.) Tul.
Wyre Forest. Spetchley Park. Shrawley Wood. Trench Woods. Nunnery Wood.
Common. On *Stereum hirsutum*, *Polyporus squamosus* and *adustus*, etc.

1030. **H. torminosus** (Mont.) Tul.
Trench Woods. Shrawley Wood. Monk Wood. Ockeridge Wood. Dripshill Wood.
Common. Always on *Lactarius pubescens*.

HYPOCREA Fr. (ὑπό, under, κρέας, flesh, because the perithecia are sunk in the stroma.)

1031. **H. rufa** (Pers.) Fr.
Common everywhere. On fallen branches and dead wood.

1032. **H. fungicola** Karst.
Perry Wood. Wyre Forest.
Uncommon. On decayed *Polyporus*.

1033. **H. alutacea** (Pers.) Tul.
Wyre Forest.
Uncommon. On *Clavaria ligula*.

POLYSTIGMA DC. (πολύς, many, στίγμα, a prick, because the perithecia are plentifully developed on the fleshy stroma.)

1034. **P. rubrum** (Pers.) DC.
Ombersley. Kempsey. Norton, near Evesham. Sheriff's Lench. Pershore.
Not uncommon. On *Prunus spinosa* and *domestica*.

EPICHLOE Fr. (ἐπί, upon, χλόη, a shoot, from the position of the plants upon the stems of various grasses.)

1035. **E. typhina** (Pers.) Tul.
Common everywhere. On stems of various grasses.

CLAVICEPS Tul. (*Clavus*, a nail, *caput*, the head, from the shape of the plant.)

1036. **C. purpurea** (Fr.) Tul. *Ergot.*
Common everywhere. On cereals and wild grasses.

1037. **C. microcephala** (Wallr.) Tul.
Monk Wood. Broad Heath. Birmingham Canal near Worcester.
Not uncommon. On *Phragmites communis* and *Molinia caerulea*.

CORDYCEPS Fr. (κορδύλη, a club, *caput*, head, from the shape of the plants.)

1038. **C. entomorrhiza** (Dicks.) Link.
Sarn Hill Woods.
Uncommon. On *pupae*.

1039. **C. militaris** (Linn.) Link.
Common everywhere. On *pupae*.

1040. **C. ophioglossoides** (Ehrh.) Link.
Shrawley Wood.
Not uncommon. On *Elaphomyces cervinus*.

1041. **C. capitata** (Holmsk.) Link.
Wyre Forest.
Uncommon. On *Elaphomyces variegatus*.

SPHAERIACEAE

CHAETOMIACEAE

CHAETOMIUM Kunze. (χαίτωμα, hair, because of the hairs on the perithecia.)

1042. **C. elatum** Kze.
Worcester.
Uncommon. On a chip box that had remained in 880 ammonia for over a month.

SORDARIACEAE

SORDARIA Ces. and de Not. (*Sordes*, dirt, because the species live on manure.)

1043. **S. fimicola** (Rob.) Ces. and de Not.
Claines.
Not uncommon. On horse-dung.

PODOSPORA Ces. (πούς, a foot, σπορά, seed, because the spores have a stalked appendage.)

1044. **P. coprophila** (Fr.) Ces. and de Not.
Kempsey. Pitchcroft.
Not uncommon. On cow- and horse-dung.

HYPOCOPRA Fr. (ὑπό, under, κόπρος, dung, because these plants are developed under a stroma on dung.)

1045. **H. merdaria** Fr.
Shrawley Wood.
Not uncommon. On rabbit dung.

TRICHOSPHAERIEAE

LASIOSPHAERIA Ces. and de Not. (λάσιος, shaggy, σφαῖρα, a ball, because the perithecia are hairy.)

1046. **L. flavescens** (Fr.) Sacc.
Eymore Wood.
Uncommon. On fallen branches.

LEPTOSPORA (Fckl.). (λεπτός, thin, σπορά, seed, from the slender spores.)

1047. **L. spermoides** (Hoffm.) Fckl.
Shrawley Wood. Wyre Forest.
Not uncommon. On old stumps and fallen branches.

1048. **L. ovina** (Pers.) Fckl.
Wyre Forest. Shrawley Wood. Eymore Wood.
Common. On fallen branches and dead twigs.

1049. **L. canescens** (Pers.) Wint.
Wyre Forest. Shrawley Wood.
Not uncommon. On rotten branches and old stumps.

CHAETOSPHAERIA Tul. (χαίτη, long, flowing hair, σφαῖρα, a ball, because the perithecia are hairy.)

1050. **C. phaeostroma** (Dur. and Mont.) Fckl.
Wyre Forest.
Common. On fallen branches and rotten stumps.

MELANOMMEAE

ROSELLINIA Ces. and de Not.

1051. **R. aquila** (Fr.) de Not.
Wyre Forest. Shrawley Wood.
Common. On fallen branches and rotten wood.

1052. **R. mammiformis** (Pers.) Ces. and de Not.
Ockeridge Wood. Wyre Forest.
Not uncommon. On stumps and fallen branches.

1053. **R. pulveracea** (Ehrh.) Fckl.
Shrawley Wood. Rous Lench Beeches.
Not uncommon. On fallen branches, especially of *Fagus sylvatica*.

1054. **R. clavariae** (Tul.) Wint.
Nunnery Wood. Trench Woods. Wyre Forest. Crews Hill Wood.
Common. On *Clavariae*.

BOMBARDIA Fr.

1055. **B. fasciculata** Fr.
Shrawley Wood. Wyre Forest.
Not uncommon. On rotten stumps and fallen branches.

BERTIA de Not.

1056. **B. moriformis** (Tode) de Not.
Eymore Wood.
Not uncommon. On fallen branches and dead wood.

MELANOPSAMMA Niessl. (μέλας, black, ψάμμος, sand, from the macroscopic appearance of the species.)

1057. **M. pomiformis** (Pers.) Sacc.
Uncommon. On stumps and fallen branches.

MELANOMMA Fckl. (μέλας, black, ὄμμα, the eye, from the form of the perithecia.)

1058. **M. pulvis-pyrius** (Pers.) Fckl.
Wyre Forest. Shrawley Wood. Tiddesley Wood.
Common. On stumps and rotten wood.

CERATOSTOMEAE

CERATOSTOMELLA Sacc. (κέρας, a horn, στόμα, a mouth, from the long orifice of the perithecia.)

1059. **C. rostrata** (Fr.) Sacc.
Wyre Forest. Shrawley Wood.
Uncommon. On dead branches, stumps, and rotten wood.

AMPHISPHAERIEAE

AMPHISPHAERIA Ces. and de Not. (ἀμφί, on both sides, σφαῖρα, a ball, because of the roundness of the perithecia.)

1060. **A. applanata** (Fr.) Ces. and de Not.
Worcester.
Uncommon. On *Rubus idaeus*.

TREMATOSPHAERIA Fckl. (τρῆμα, pierced, σφαῖρα, a ball, from the form of the orifice of the perithecia.)

1061. **T. pertusa** (Pers.) Fckl.
Wyre Forest.
Not uncommon. On a decorticated log.

1062. **T. mastoidea** (Fr.) Wint.
Shrawley Wood.
Not uncommon. On *Fraxinus excelsior*.

CUCURBITARIEAE

CUCURBITARIA Gray. (*Cucurbita*, a gourd.)

1063. **C. laburni** (Pers.) Ces. and de Not.
Worcester.
Not uncommon. On *Cytisus Laburnum*.

SPHAERELLOIDEAE

STIGMATEA Fr. (στίγμα, a prick, from the appearance of the perithecia.)

1064. **S. robertiani** Fr.
Ombersley.
Uncommon. On *Geranium Robertianum*.

SPHAERELLA Ces. and de Not. (σφαῖρα, a ball, from the shape of the perithecia.)

1065. **S. fragariae** (Tul.) Sacc.
Common everywhere. On *Fragaria vesca*.

1066. **S. punctiformis** (Pers.) Sacc.
Shrawley Wood.
Not uncommon. On *Quercus*.

1067. **S. grossulariae** (Fr.) Auersw.
Common everywhere. On fallen leaves of *Ribes Grossularia*.

PLEOSPOREAE

LEPTOSPHAERIA Ces. and de Not. (λεπτός, slender, σφαῖρα, a ball, from the small size of the perithecia.)

1068. **L. arundinacea** (Sow.) Sacc.
Wyre Forest.
Uncommon. On *Phragmites communis*.

1069. **L. maculans** (Sow.) Karst.
Avon, opposite Twyning Fleet.
Uncommon. On *Scirpus lacustris*.

1070. **L. doliolum** (Pers.) Ces. and de Not.
Osier-bed beyond Boughton Park.
Not uncommon. On dead stems of *Urtica dioica*.

1071. **L. vagabunda** Sacc.
Worcester. Sheriff's Lench. Pershore. Fladbury.
Common. On *Ribes Grossularia*.

1072. **L. acuta** (Moug. and Nestl) Karst.
Willow-bed beyond Boughton Park.
Not uncommon. On dead stems of *Urtica dioica*.

PLEOSPORA Rabh. (πλέω, more, σπορά, seed.)

1073. **P. herbarum** (Pers.) Rabh.
Shrawley Wood. Wyre Forest.
Common. On dead stems.

OPHIOBOLUS Riess. (ὄφις, a snake, βάλλω, I throw, because the asci eject long snake-like spores.)

1074. **O. porphyrogonus** (Tode) Sacc.
Wyre Forest.
Uncommon. On dead haulm of *Solanum tuberosum*.

CLYPEOSPHAERIEAE

HYPOSPILA Fr. (ὑπό, under, σπίλος, a spot, because the perithecia are developed under a thickened spot.)

1075. **H. pustula** (Pers.) Karst.
Astley.
Not uncommon on dead leaves of *Quercus Robur*.

VALSACEAE

VALSA Fr.

1076. **V. (Eutypa) lata** (Pers.) Nitschke.
Common everywhere. On wood and fallen branches.

1077. **V. (Cryptosphaeria) populina** (Pers.) Wint.
Shrawley Wood.
Not uncommon. On fallen branches of *Populus*.

1078. **V. (Euvalsa) salicina** (Pers.) Fr.
Cowleigh Park.
Uncommon. On fallen branches of *Salix*.

1079. **V. (Leucostoma) nivea** (Pers.) Fr.
Tiddesley Wood.
Not uncommon. On fallen branches of *Populus tremulus*.

RHYNCHOSTOMA Karst. (ῥύγχος, a snout, στόμα, a mouth, from the blunt orifice of the perithecia.)

1080. **R. anserina** (Pers.) Wint.
Shrawley Wood. Wyre Forest.
Not uncommon. On dead wood and fallen branches.

DIATRYPEAE

QUATERNARIA Tul. (*Quater*, four, from the appearance of the mouths of the perithecia.)

1081. **Q. Persoonii** Tul.
Shrawley Wood. Rous Lench.
Not uncommon. On fallen trunks of *Fagus sylvatica*.

DIATRIPELLA Ces. and de Not. (διατριβή, a wearing away, because the species develop under the bark and force it off.)

1082. **D. quercina** (Pers.) Nitschke.
Common everywhere. On fallen oak branches.

1083. **D. verrucaeformis** (Ehrh.) Nitschke.
Perry Wood. Shrawley Wood. Wyre Forest.
Common. On fallen branches, especially of *Corylus Avellana*.

DIATRYPE Fr. (διατριβή, a wearing away, because the species develop under the bark and force it off.)

1084. **D. stigma** (Hoffm.) de Not.
Common everywhere. On fallen branches.

1085. **D. disciformis** (Hoffm.) Fr.
Seckley Wood. West Malvern.
Not uncommon. On fallen branches of *Fagus sylvatica*.

XYLARIACEAE

HYPOXYLON Nitschke. (ὑπόξυλος, wooden underneath, because all the species grow on wood.)

1086. **H. (Endoxylon) semiimmersum** Nitschke.
Shrawley Wood.
Not uncommon. On *Fagus sylvatica*.

1087. **H. (Epixylon) multiforme** Fr.
Common everywhere. On stumps and fallen branches.

1088. **H. (Euhypoxylon) rubiginosum** (Pers.) Fr.
Shrawley Wood.
Not uncommon. On stumps and fallen branches.

1089. **H. (Euhypoxylon) fuscum** (Pers.) Fr.
  Common everywhere. On fallen branches and dead twigs, especially
  of *Corylus Avellana* and *Crataegus monogyna*.

1090. **H. (Euhypoxylon) coccineum** Bull.
  Common everywhere. On fallen branches of *Fagus sylvatica*.

1091. **H. (Daldinia) concentricum** (Bolton) Grev.
  Ladywood, Salwarpe. Shrawley Wood. The Slads.
  Not uncommon. On trunks and stumps, especially of *Fraxinus
  excelsior*.

USTULINA Tul. (*Ustulo*, I scorch, from the burnt appearance of the
        stroma.)

1092. **U. vulgaris** Tul.
  Common everywhere. On stumps and logs.

PORONIA Willd.

1093. **P. punctata** (Linn.) Fr.
  Defford Common.
  Uncommon. On horse-dung.

XYLARIA Hill. (ξύλον, wood, because of the hardness of the stroma.)

1094. **X. hypoxylon** (Linn.) Grev.
  Common everywhere. On stumps, fallen branches, posts and rails.

1095. **X. carpophila** (Pers.) Fr.
  Shrawley Wood.
  Uncommon. On old decaying nuts of *Fagus sylvatica*.

1096. **X. digitata** (Linn.) Grev.
  Sheriff's Lench. Bransford.
  Not uncommon. On posts and rails.

1097. **X. polymorpha** (Pers.) Grev.
  Broadway. Wyre Forest. Shrawley Wood.
  Not uncommon. On stumps and pollarded trees.

DOTHIDEACEAE

PHYLLACHORA Nitschke. (φύλλον, a leaf, χώρα, a place, because the
        species grow on leaves.)

1098. **P. graminis** (Pers.) Fckl.
  Common everywhere. On leaves of various grasses.

1099. **P. junci** (Fr.) Fckl.
  Hartlebury Common.
  Not common. On *Juncus*.

1100. **P. trifolii** (Pers.) Fckl.
  Near Evesham.
  Not uncommon. On *Trifolium*.

DOTHIDELLA Speg.

1101. **D. betulina** (Fr.) Sacc.
  Common everywhere. On fallen leaves of *Betula alba*.

1102. **D. ulmi** (Duv.) Wint.
  Common everywhere. On fallen leaves of *Ulmus campestris*.

DOTHIDEA Fr.

1103. **D. ribesia** (Pers.) Fr.
  Malvern. Worcester. Kempsey.
  Not uncommon. On dead branches of *Ribes*.

RHOPOGRAPHUS Nke. (ῥώψ, a low shrub, γραφή, writing, from the
    dark marks of the stroma on the stems of *Pteris aquilina*.)

1104. **R. Pteridis** (Sow.) Wint.
  Common everywhere. On *Pteris aquilina*.

TUBERALES

ONYGENACEAE

ONYGENA Pers. (ὄνυξ, a hoof, γένος, race, from the habitat of
      these species.)

1105. **O. equina** (Willd.) Pers.
  Shrawley Wood. Ombersley.
  Not common. On owl pellets and decayed horse-hoof.

ELAPHOMYCETACEAE

ELAPHOMYCES Nees. (ἔλαφος, a deer, μύκης, a fungus, from the
      colour of the peridium.)

1106. **E. cervinus** (Pers.) Schröt. (=**granulatus** Fr.).
  Shrawley Wood. Wyre Forest.
  Not uncommon. Underground. Its presence is often indicated by
  Cordyceps ophioglossoides, a parasitic fungus that appears above ground.

1107. **E. variegatus** Vitt.
  Wyre Forest.
  Not uncommon. Underground. Like the last species its presence
  is often disclosed by the parasitic fungus, *Cordyceps capitata*.

TUBERACEAE

HYDNOTRYA B. and Br.

1108. **H. Tulasnei** B. and Br.
  Ham Dingle, Stourbridge.
  Rare. Underground.

HYSTERIALES

HYSTERIACEAE

HYSTERIUM Tode. (ὕστερος, later.)

1109. **H. pulicare** Pers.
  Nunnery Wood.
  Not uncommon. On fallen branches.

1110. **H. angustatum** A. and S.
  Monk Wood.
  Uncommon. On bark of *Quercus Robur*.

HYSTEROGRAPHIUM Cda. (ὑστέρα, the womb, γραφή, writing.)

1111. **H. fraxini** (Pers.) de Not.
  Wyre Forest. Aileshurst Coppice.
  Common. On dead branches of *Fraxinus excelsior*.

DICHAENACEAE

DICHAENA Fr. (διχαίνω, I yawn, because the ascophores dehisce by
      an elongated slit.)

1112. **D. quercina** (Pers.) Fr.
  Common everywhere. On living trunks and branches of *Quercus
  Robur*.

1113. **D. faginea** Fr.
  Shrawley Wood. Dodderhill Common.
  Not uncommon. On living trunks and branches of *Fagus sylvatica*.

DISCOMYCETAE

HELVELLACEAE

MORCHELLA Dill. (*Morchel*, German for Morel.) *Morel.*

1114. **M. crassipes** (Venten.) Pers.
  Worcester.
  Not common. On the ground. Edible and delicious.

1115. **M. Smithiana** Cke.
  Lower Wick, Worcester. Ombersley.
  Not uncommon. On the ground. Edible and delicious.

1116. **M. esculenta** (Linn.) Pers.
  Worcester.
  Not uncommon. On the ground. Edible and delicious.

MITROPHORA Lév. (μίτρα, head-dress, φέρω, I bear, from the
      shape of the pileus.)

1117. **M. hybrida** (Sow.) Lév.
  Near Holt Mill.
  Not common. On the ground. Edible and delicious.

HELVELLA Linn. (*Helvella*, used by Cicero to denote some kind
      of fungus.)

1118. **H. crispa** (Scop.) Fr.
  Wyre Forest. Shrawley Wood. Trench Woods. Chaddesley Corbett
  Woods.
  Common. On the ground. Edible and delicious.

1119. **H. lacunosa** Afzl.
  Shrawley Wood. Crews Hill Wood.
  Uncommon. On the ground. Edible.

1120. **H. atra** König.
  Wyre Forest. Shrawley Wood.
  Uncommon. In woods.

1121. **H. elastica** Bull.
  Wyre Forest. Shrawley Wood.
  Not uncommon. In woods. Edible and delicious.

1122. **H. macropus** (Pers.) Karst.
  Perry Wood. Shrawley Wood. Dripshill Wood. Wyre Forest.
  Trench Woods.
  Common. In woods.

GEOGLOSSUM Pers. (γῆ, the earth, γλῶσσα, tongue, from a fancied
    resemblance of the ascophores.) *Earth-tongues.*

1123. **G. ophioglossoides** (Linn.) Sacc. (=**glabrum** Pers.).
  Clent Cottage, Clent.

SPATHULARIA Pers. (σπάθη, a broad blade, from the shape of
      the ascophore.)

1124. **S. clavata** (Schaeff.) Sacc.
  Shrawley Wood. Dripshill Wood. Ockeridge Wood.
  Common. On dead fir needles, especially those of *Larix europaea*.

MITRULA (Fr.). (μίτρα, a head-dress, from the shape of the ascophore.)

1125. **M. cucullata** (Batsch) Fr.
  Shrawley Wood.
  Uncommon. On decaying needles of *Pinus sylvestris*.

1126. **M. serpentina** (O. F. Muell.) Mass. *Green Earth-tongue.*
  Wyre Forest.
  Uncommon. Amongst short grass and dead leaves in woods.

LEOTIA Hill. (λειότης, smoothness.)

1127. **L. gelatinosa** Hill (=**lubrica** Pers.).
  Wyre Forest. Shrawley Wood. Nunnery Wood. Tiddesley Wood.
  Trench Woods.
  Common. On the ground in woods.

1128. **L. acicularis** Pers.
  Wyre Forest. Perry Wood. Trench Woods. Monk Wood. Shrawley
  Wood.
  Common. On stumps and rotten branches.

PEZIZACEAE

ACETABULA Fr. (*Acetabulum*, a vinegar cup, from the shape of the ascophore.)

1129. **A. vulgaris** Fckl.
Weymans Wood. Hanley Dingle. Shrawley Wood. Wyre Forest. Abberley.
Not uncommon. On the ground. Edible.

GEOPYXIS Pers. (γῆ, earth, πυξίς, a box.)

1130. **G. coccinea** (Jacq.) Mass.
Severn Stoke Hill. Cookley. Witley. Eymore Wood.
Not uncommon. On rotten branches in hedges and in woods.

1131. **G. cupularis** (Linn.) Sacc.
Shrawley Wood. Dick Brook.
Not uncommon. On the ground in damp places amongst moss.

PEZIZA Dill. (πέζα, the foot, bottom, base, that which rests on its base sessile, a word used by Pliny to denote various kinds of puff-balls.)

1132. **P. vesiculosa** (Bolt.) Bull.
Malvern Wells. Great Farley. Alfrick.
Not uncommon. On manure heaps and amongst rotten leaves. Edible and delicious.

1133. **P. cerea** Sow.
Worcester. Shrawley Wood.
Not uncommon. On manured ground and amongst rotten leaves.

1134. **P. sepiatra** Cke.
Wyre Forest. Worcester.
Not uncommon. On charcoal heaps and in shady places.

1135. **P. venosa** Pers.
Madams Hill Wood. Weymans Wood. Aisleshurst Coppice.
Not uncommon. On the ground. Edible and delicious.

1136. **P. ampliata** Pers.
Tutnall. Worcester. Severn Stoke.
Not uncommon. On rotten wood.

Var. **tectoria** (Cke.) Mass.
Worcester.
Not uncommon. On damp walls.

1137. **P. mellea** Cke. and Plow.
Shrawley Wood.
Uncommon. On rotten wood.

1138. **P. badia** Pers.
Malvern Wells. Nunnery Wood. Worcester. Sheriff's Lench. Ombersley. Crews Hill Wood.
Common. On the ground. Edible and delicious.

1139. **P. succosa** Berk.
Shrawley Wood. Wyre Forest.
Not uncommon. On the ground in woods.

OTIDEA Pers. (ὠτίον, a little ear, from the shape of the ascophores.)

1140. **O. leporina** (Batsch) Fckl.
Shrawley Wood. Wyre Forest. Tiddesley Wood. Trench Woods.
Not uncommon. Amongst dead leaves in woods.

1141. **O. cochleata** (Linn.) Fckl.
Rosebury Rock Wood. Lower Wick. Trench Woods. Worcester.
Common. On the ground. Edible and delicious.

1142. **O. umbrina** (Pers.) Boud.
Rosebury Rock Wood.
Not uncommon. On the ground.

1143. **O. onotica** (Pers.) Fckl.
Tiddesley Wood. Shrawley Wood.
Not uncommon. Amongst dead leaves in woods.

1144. **O. aurantia** (Müll.) Mass.
Leigh Sinton. Hollybush Hill. Chatley. Shrawley Wood. Dripshill Wood.
Common. On stumps and on the ground. Edible and delicious.

BARLAEA Sacc. (In honour of J. B. Barla, a botanist of Nice.)

1145. **B. constellatio** (B. and Br.) Sacc.
Wyre Forest. Tiddesley Wood.
Not uncommon. On the ground amongst moss and on charcoal heaps.

1146. **B. Crouani** (Cke.) Mass.
Hadley Mill. Fernhill Heath.
Not uncommon. Amongst moss on walls.

HUMARIA Fr. (*Humus*, earth, from the usual habitat.)

1147. **H. humosa** (Fr.) Cke.
Wyre Forest.
Uncommon. On damp ground.

1148. **H. rutilans** (Fr.) Sacc.
Wyre Forest. Shrawley Wood.
Not uncommon. Amongst moss.

1149. **H. carbonigena** (Berk.) Sacc.
Trench Woods. Wyre Forest.
Uncommon. On charcoal heaps.

1150. **H. fusispora** (Berk.) Sacc.
Wyre Forest.
Not uncommon. On charcoal heaps.

1151. **H. omphalodes** (Bull.) Mass.
Wyre Forest. Great Farley Wood. Ribbesford Wood.
Common. On charcoal heaps, staining the heap a deep blood-red.

1152. **H. granulata** (Bull.) Quél.
Common everywhere. On dry cow- and horse-dung.

1153. **H. violacea** (Pers.) Sacc.
First Record. Wyre Forest, *Rea, Massee's Brit. Fung. Flora*, vol. iv, p. 417.
Wyre Forest.
Uncommon. On charcoal heaps.

SEPULTARIA Cke. (*Sepultus*, buried, from being immersed.)

1154. **S. semiimmersa** Karst.
Walton Hill.
Uncommon. On the ground and on horse-dung.

1155. **S. coronaria** (Jacq.) Mass.
Hanbury Park.
Uncommon. Amongst dead leaves of *Fagus sylvatica*. Edible and delicious.

LACHNEA Fr. (λάχνη, down, from the villous or hairy exterior of the ascophore.)

1156. **L. stercorea** (Pers.) Gill.
Chatley. Ladywood, Salwarpe. Ombersley. Claines. Malvern. Powick. Bransford.
Common. On cow-dung.

1157. **L. scutellata** (Linn.) Gill.
Wyre Forest. Tardebigge Reservoir. Claines. Shrawley Wood. Fladbury.
Common. On stumps and rotten logs.

1158. **L. hemisphaerica** (Wigg.) Gill.
Shrawley Wood. Wyre Forest.
Not uncommon. On the ground.

NEOTTIELLA Cke. (νεοττιά, a nest.)

1159. **N. polytrichi** (Schum.) Mass.
Wyre Forest.
Uncommon. Amongst mosses.

1160. **N. nivea** (Romell) Sacc.
Trench Woods.
Uncommon. On the ground amongst dead leaves.

PITYA Fckl. (πίτυς, a pine or fir tree, because the ascophores grow on conifers.)

1161. **P. cupressi** (Batsch) Rehm.
Lower Wick, Worcester.
Uncommon. On living twigs of conifers.

SPHAEROSPORA Sacc. (σφαῖρα, a ball, σπορά, a seed, from the shape of the ascospores.)

1162. **S. trechispora** (B. and Br.) Sacc.
Wyre Forest. Shrawley Wood. Malvern Hills.
Not uncommon. On the ground.

DASYSCYPHA Fr. (δασύς, hairy, σκύφος, cup, because the ascophores are hairy on the outside.)

1163. **D. virginea** (Batsch) Fckl.
Common everywhere. On rotten wood, stumps, and dead twigs.

1164. **D. scintillans** Mass.
Eymore Wood.
Uncommon. On dead twig of *Quercus Robur*.

1165. **D. nivea** (Hedw.) Sacc.
Common everywhere. On rotten wood, fallen branches, and twigs.

1166. **D. acutipila** (Karst.) Sacc.
Birmingham Canal, Worcester.
Uncommon. On dead *Phragmites communis*.

1167. **D. bicolor** (Bull.) Fckl.
Middleyards Coppice. Shrawley Wood. Wyre Forest. Trench Woods.
Not uncommon. On fallen branches.

1168. **D. leuconica** (Cke.) Mass.
Shrawley Wood.
Uncommon. On dead wood.

1169. **D. ciliaris** (Schrad.) Sacc.
Wyre Forest.
Uncommon. On dead leaves of *Quercus Robur*.

1170. **D. acuum** (A. and S.) Sacc.
Trench Woods.
Uncommon. On fallen needles of *Pinus sylvestris*.

1171. **D. aspidiicola** (B. and Br.) Sacc.
Shrawley Wood.
Not uncommon. On decaying stems of *Lastraea Filix-mas*.

1172. **D. hyalina** (Pers.) Mass.
Common everywhere. On decaying wood, fallen branches, and inside bark.

1173. **D. calycina** (Schum.) Fckl.
Near Camp Weir. Dripshill Wood. Shrawley Wood. Madresfield. Great Farley Wood.
Common. On *Pinus sylvestris* and *Larix europaea*.

1174. **D. sulfurea** (Pers.) Mass.
Wyre Forest. Ockeridge Wood.
Not uncommon. On dead herbaceous plants.

1175. **D. puberula** (Lasch) Quél.
Kings Norton.
Uncommon. On dead leaves of *Quercus Robur*.

ERINELLA Sacc. (ἠρινός, in the spring, from the time when the ascophores appear.)

1176. **E. Nylanderi** Rehm.
Aisleshurst Coppice.
Uncommon. On dead stems.

PLECTANIA Fckl. (πλεκτάνη, anything twined or plaited, from the convolutions of the hyphae at the base of the ascophores.)

1177. **P. melastoma** (Sow.) Fckl.
Trench Woods.
Uncommon. On dead twigs.

TAPESIA Pers. (τάπης, a carpet, because the mycelium forms a dense subiculum.)

1178. **T. fusca** (Pers.) Fckl.
Common everywhere. On dead wood, fallen branches, and twigs.

1179. **T. caesia** (Pers.) Fckl.
Shrawley Wood. Wyre Forest.
Not uncommon. On chips and logs of *Quercus Robur*.

CHLOROSPLENIUM Fr. (χλωρός, green, σπλήν, the spleen, from the colour of the ascophores.)

1180. **C. aeruginosum** (Oed.) de Not.
Shrawley Wood. Wyre Forest. Stanford Woods. Saltwell Wood.
Not uncommon. On fallen branches. The mycelium stains the wood a deep verdigris-green, and in this condition is very common throughout Worcestershire. Wood so stained was formerly used in the production of Tunbridge Wells ware, but the effect is now chemically produced.

SCLEROTINIA Fckl. (σκληρότης, hardness, because the mycelium forms hard compact bodies which are the resting condition of the mycelium.)

1181. **S. tuberosa** (Hedw.) Fckl.
Clent Cottage, Clent. Rosebury Rock Woods. Aisleshurst Coppice.
Not uncommon. On the rhizomes of *Anemone nemorosa*.

1182. **S. sclerotiorum** (Lib.) Mass.
Ladywood, Salwarpe. Wyre Forest.
Not uncommon. On sclerotia inside the stems of *Solanum tuberosum*.

1183. **S. Curreyana** (Berk.) Karst.
Hartlebury Common.
Uncommon. On sclerotia on *Juncus lamprocarpus*.

CIBORIA Fckl. (κιβώριον, a drinking-cup, from the shape of the ascophores.)

1184. **C. ochroleuca** (Bolt.) Mass.
Wyre Forest. Shrawley Wood. Trench Woods. Perry Wood. Nunnery Wood.
Common. On fallen branches of *Quercus Robur*.

1185. **C. echinophila** (Bull.) Sacc.
Shrawley Wood.
Not uncommon. On decaying involucres of *Castanea vesca*.

1186. **C. amentacea** (Balb.) Fckl.
Wyre Forest. Shrawley Wood.
Uncommon. On fallen catkins of *Alnus glutinosa*.

CYATHICULA de Not. (κύαθος, a cup, from the shape of the ascophores.)

1187. **C. coronata** (Bull.) de Not.
Osier-bed beyond Boughton Park. Wyre Forest.
Not uncommon. On dead herbaceous stems.

HELOTIUM Fr. (ἧλος, a nail, from the shape of the ascophores.)

1188. **H. claro-flavum** (Grev.) Berk.
Shrawley Wood. Cowleigh Wood. Wyre Forest.
Not uncommon. On stumps and fallen branches.

1189. **H. laburni** B. and Br.
Worcester.
Uncommon. On dead branches of *Cytisus Laburnum*.

1190. **H. lenticulare** (Bull.) Fr.
Dodderhill Common.
Uncommon. On fallen branches of *Fagus sylvatica*.

1191. **H. citrinum** (Hedw.) Fr.
Wyre Forest. Trench Woods. Tiddesley Wood. Dripshill Wood. Shrawley Wood.
Common. On stumps and fallen branches.

1192. **H. luteolum** Currey.
First Record. 'On branches buried among moss near Worcester, Mr. Carleton Rea, September 1894.' *Massee's Brit. Fung. Fl.*, vol. iv, p. 240.
Uncommon.

1193. **H. serotinum** (Pers.) Fr.
Shrawley Wood. Hanbury Park.
Not uncommon. On fallen branches of *Fagus sylvatica*.

1194. **H. salicellum** Fr.
Willow-plantation beyond Northwick Brick-pits.
Uncommon. On dead and fallen branches of *Salix*.

1195. **H. virgultorum** (Vahl) Karst.
Wyre Forest. Shrawley Wood. Trench Woods. Dripshill Wood. Perry Wood.
Common. On dead twigs and fallen branches.
Var. **fructigenum** (Bull.) Karst.
Shrawley Wood. Nunnery Wood. Wyre Forest. Cowleigh Park.
Common. On fallen acorns and beechmast.

1196. **H. calyculus** (Sow.) Berk.
Holt. Wyre Forest.
Not uncommon. On wood, bark, and fallen branches.

1197. **H. moniliferum** (Fckl.) Rehm.
Shrawley Wood.
Uncommon. On a stump among *Bispora monilioides*.

R r

1198. **H. cyathoideum** (Bull.) Karst.
Osier-bed beyond Boughton Park. Grimley Brick-pits.
Common. On dead herbaceous stems.

1199. **H. herbarum** (Pers.) Fr.
Trench Woods. Shrawley Wood. Wyre Forest.
Common. On dead herbaceous stems.

1200. **H. epiphyllum** (Pers.) Fr.
Perry Wood. Wyre Forest. Shrawley Wood.
Common. On dead leaves of *Quercus Robur*.

1201. **H. renisporum** Ellis.
Shrawley Wood.
Uncommon. On dead leaves of *Quercus Robur*.

1202. **H. fagineum** (Pers.) Fr.
Shrawley Wood Dodderhill Common.
Not uncommon. On fallen nuts of *Fagus sylvatica*.

BELONIDIUM Mont. and Dur. (βελονίς, a needle, from the shape of the ascospores.)

1203. **B. pruinosum** (Jerdon) Mass.
Eymore Wood.
Uncommon. On *Diatrype stigma*.

MOLLISIA Fr. (*Mollis*, soft, from the consistency of the ascophores.)

1204. **M. melaleuca** (Fr.) Sacc.
Eymore Wood. Perry Wood.
Not uncommon. On rotten stumps, dead wood, and fallen branches.

1205. **M. cinerea** (Batsch) Karst.
Common everywhere. On rotten stumps, dead wood, and fallen branches.

1206. **M. lignicola** Phil.
Perry Wood. Crown East.
Not uncommon. On old posts and rails.

1207. **M. atrocinerea** (Cke.) Phil.
Grimley Brick-pits. Little Hadley Mill.
Not uncommon. On dead herbaceous stems.

1208. **M. atrata** (Pers.) Karst.
Hallow, near Hallow Ford.
Uncommon. On dead stems of *Spiraea Ulmaria*.

1209. **M. filicum** Phil.
Shrawley Wood. Wyre Forest. Tiddesley Wood.
Not uncommon. On dead stems of *Lastraea Filix-mas*.

PSEUDOPEZIZA Fckl. (ψευδός, false, *Peziza*, because the ascophores resemble *Phacidium*.)

1210. **P. trifolii** (Bernh.) Fckl.
Claines. Ladywood, Salwarpe. Sheriff's Lench.
Not uncommon. On living leaves of *Trifolium*.

1211. **P. medicaginis** (Lib.) Sacc.
Between Offenham and Cleeve Mill.
Not common. On living leaves of *Medicago sativa*.

1212. **P. radians** (Rob.) Karst.
Near Worcester.
Uncommon. On *Campanula patula*.

ASCOBOLACEAE

ASCOBOLUS Pers. (ἀσκός, a leathern bag, βόλος, a throw, from the projecting asci.)

1213. **A. vinosus** Berk.
Wyre Forest.
Uncommon. On rabbit dung.

1214. **A. glaber** Pers.
Selly Oak.
Uncommon. On dung.

1215. **A. stercorarius** (Bull.) Schröt. (=furfuraceus Pers.).
Common everywhere. On cow-dung.

BULGARIACEAE

OMBROPHILA Fr. (ὄμβρος, rain, φίλος, loving.)

1216. **O. clavus** (A. and S.) Cke.
Common everywhere. On leaves, twigs, and dead herbaceous stems.

1217. **O. brunnea** Phil.
Shrawley Wood.
Uncommon. On dead herbaceous stems.

ORBILIA Fr.

1218. **O. coccinella** (Sommerf.) Karst.
Trench Woods.
Uncommon. On dead wood.

1219. **O. leucostigma** Fr.
Common everywhere. On dead wood and bark, especially *Betula alba*.
Var. **xanthostigma** Fr.
Equally common. On dead wood and bark.

1220. **O. inflatula** Karst.
Wyre Forest. Shrawley Wood. Nunnery Wood. Trench Woods.
Not uncommon. On rotten stumps, dead wood, and bark.

CALLORIA Fr. (κάλλος, beauty, because the ascophores are beautifully coloured.)

1221. **C. fusarioides** (Berk.) Fr.
Common everywhere. On dead stems of *Urtica dioica*.

R r 2

CORYNE Tul. (κορύνη, a club, from the shape of the ascophores.)

1222. **C. urnalis** (Nyl.) Sacc.
Wyre Forest. Shrawley Wood. Trench Woods. Tiddesley Wood. Monk Wood.
Common. On stumps and dead wood.

1223. **C. sarcoides** (Jacq.) Tul.
Common everywhere. On stumps and felled timber.

1224. **C. atrovirens** (Pers.) Sacc.
Shrawley Wood. Wyre Forest.
Not uncommon. On rotten stumps, fallen branches, and dead twigs.

BULGARIA (Fr.). (*Bulga*, a leathern bag, from the consistency of the ascophores.)

1225. **B. polymorpha** (Oeder) Wetts.
Crowle. Spetchley Park. Shrawley Wood. Wyre Forest. Ockeridge Wood.
Common. On dead trunks of trees.

DERMATACEAE

CENANGIUM Fr. (κενός, empty, ἄγγος, a vessel, from the hollow ascophore.)

1226. **C. furfuraceum** (Roth) de Not.
Boughton Park. Near Camp Weir. Wyre Forest. Hawford Hallow.
Common. On dying and dead branches of *Alnus glutinosa*.

1227. **C. populneum** (Pers.) Rehm.
Grimley.
Uncommon. On dead branches of *Populus*.

1228. **C. dryinum** (Cke.) Mass.
Nunnery Wood.
Uncommon. On bark of living *Quercus Robur*.

TYMPANIS Tode. (τύμπανον, a kettle-drum, from the general appearance of the ascophores.)

1229. **T. conspersa** Fr.
Tiddesley Wood. Wyre Forest.
Uncommon. On living bark of *Betula alba*.

1230. **T. alnea** (Pers.) Fr.
Wyre Forest.
Uncommon. On *Alnus glutinosa*.

SCLERODERRIS Fr. (σκληρός, hard, δέρρις, a leathern covering, from the consistency of the ascophores.)

1231. **S. ribesia** (Pers.) Karst.
Evesham.
Not uncommon. On *Ribes rubrum*.

PATELLARIACEAE

HETEROSPHAERIA Grev. (ἕτερος, different, σφαῖρα, a ball, different from the normal *Sphaeriae*.)

1232. **H. patella** (Tode) Grev.
Middleyards Coppice.
Uncommon. On dead stems of *Angelica sylvestris*.

STICTIDACEAE

PROPOLIS Fr. (*Propolis*, bee-glue, which the hymenium is like.)

1233. **P. faginea** (Schrad.) Karst.
Shrawley Wood. Wyre Forest. Tiddesley Wood. The Slads. Cowleigh Park.
Common. On wood, fallen branches, and chips.

PHACIDIACEAE

COCCOMYCES de Not. (κόκκος, a berry, μύκης, a fungus, from the shape of the ascophores.)

1234. **C. coronatus** (Schum.) de Not.
Shrawley Wood.
Uncommon. On dead leaves of *Quercus Robur*.

1235. **C. dentatus** (Kze. and Schmidt) Sacc.
Wyre Forest.
Not uncommon. On dead leaves of *Quercus Robur*.

PHACIDIUM Fr. (φακός, a lentil, εἶδος, like, from the shape of the ascophores.)

1236. **P. multivalve** (DC.) Kze. and Schmidt.
Common everywhere. On dead leaves of *Ilex Aquifolium*.

TROCHILA Fr. (τροχύς, anything round or circular, from the form of the ascophores.)

1237. **T. craterium** (DC.) Fr.
Common everywhere. On dead leaves of *Hedera Helix*.

1238. **T. laurocerasi** (Desm.) Fr.
White's Nursery, Lower Wick, Worcester.
Not uncommon. On fallen leaves of *Prunus Laurocerasus*.

1239. **T. ilicis** (Chev.) Crouan.
Common everywhere. On fallen leaves of *Ilex Aquifolium*.

RHYTISMA Fr. (ῥύτισμα, a patch, from the appearance of the host leaves.)

1240. **R. acerinum** (Pers.) Fr.
Common everywhere. On leaves of *Acer Pseudoplatanus*.

1241. **R. punctatum** (Pers.) Fr.
Bevere Island. Malvern. Witley Court.
Not uncommon. On leaves of *Acer Pseudoplatanus* and *campestre*.

1242. **R. salicinum** (Pers.) Fr.
Northwick Brickpits. Osier-bed beyond Boughton Park.
Common. On dead leaves of various species of *Salix*.

COLPOMA Wallr. (κόλπος, a hollow, from the slit in the ascophore at maturity.)

1243. **C. quercinum** (Pers.) Wallr.
Nunnery Wood. Shrawley Wood. Wyre Forest.
Not uncommon. On fallen branches and twigs of *Quercus Robur*.

PHYCOMYCETAE

ZYGOMYCETAE

MUCORACEAE

PILOBOLUS Tode. (*Pila*, a ball, βόλος, a throw, because the sporangia are ejected at maturity.)

1244. **P. crystallinus** (Wiggers) Tode.
Common everywhere. On cow- and horse-dung.

1245. **P. roridus** (Bolt.) Pers.
Pitchcroft, Worcester.
Uncommon. On horse-dung.

PILAIRA Van Tiegh.

1246. **P. anomala** (Ces.) Schröt.
Wyre Forest. Old Hills.
Not uncommon. On rabbit dung.

MUCOR Mich. (*Mucor*, mould.)

1247. **M. mucedo** Linn.
Common everywhere. On rotten fruit, jam, and dung.
Var. **caninus** Pers.
Worcester.
Not uncommon. On dog dung.

SPINELLUS Van Tiegh. (*Spina*, a thorn, because some of the branches are furnished with spiny outgrowths.)

1248. **S. fusiger** (Lk.) Van Tiegh.
Shrawley Wood. Wyre Forest. Middleyards Coppice. Perry Wood.
Not uncommon. On decaying *Agarics*.

SPORODINIA Link. (σπορά, scattered.)

1249. **S. aspergillus** (Scop.) Schröt.
Common everywhere. On decaying fungi.

RHIZOPUS Ehr. (ῥίζα, a root, πούς, a foot, because the hyphae give off numerous rhizoids.)

1250. **R. nigricans** Ehr.
Shrawley Wood.
Uncommon. Amongst decaying leaves.

ENTOMOPHTHORACEAE

EMPUSA Cohn. (ἔμπουσα, a hobgoblin.)

1251. **E. muscae** Cohn.
Common everywhere. On dead flies.

OOMYCETAE

PERONOSPORACEAE

CYSTOPUS Lév. (κύστις, a bladder, πούς, a foot.)

1252. **C. candidus** (Pers.) Lév.
Common everywhere. On various *Cruciferae*, especially *Capsella Bursa-pastoris*.

1253. **C. tragopogonis** (Pers.) Schröt.
Old Powick Bridge. Camp. Bewdley.
Not uncommon. On *Tragopogon minus*.

1254. **C. lepigoni** De Bary.
Hartlebury Common.
Not uncommon. On *Spergularia rubra*.

PHYTOPHTHORA De Bary. (φυτόν, a plant, φθορά, destruction, from the damage caused by this parasite.)

1255. **P. infestans** (Mont.) De Bary.   *Potato disease.*
Common everywhere. On leaves of *Solanum tuberosum*.

PLASMOPARA Schröt.

1256. **P. pygmaea** (Unger) Schröt.
Middleyards Coppice. Aisleshurst Plantation.
Not uncommon. On leaves of *Anemone nemorosa*.

1257. **P. nivea** (Unger) Schröt.
Berwick's Brake, Worcester. Claines.
Not uncommon. On *Aegopodium Podagraria* and *Anthriscus sylvestris*.

PERONOSPORA Cda. (περόνη, pointed, σπόρος, seed, from the shape of the conidia.)

1258. **P. parasitica** (Pers.) Tul.
Claines.
Uncommon. On *Nasturtium*.

1259. **P. effusa** (Grev.) Rabh.
Near Bewdley.
Uncommon.   On *Atriplex patula*.

1260. **P. urticae** (Lib.) De Bary.
Near the Ketch, Worcester.
Uncommon.   On *Urtica dioica*.

SAPROLEGNIACEAE

LEPTOMITUS Agardh.   (λεπτός, thin, μίτος, a thread.)

1261. **L. lacteus** Ag.
Hawford.
Uncommon.   On *Potamogeton*.

SAPROLEGNIA Nees.   (σαπρός, rotten.)

1262. **S. ferax** Nees.
Common everywhere.   On dead flies and fish.

PYTHIUM Pringsh.   (πύθω, I rot.)

1263. **P. De-Baryanum** Hesse.
Worcester.
Not uncommon.   On seedlings of *Lepidium sativum*.

CHYTRIDIACEAE

SYNCHYTRIUM De Bary and Woron.   (σύν, together, χυτρίον, a little earthen pot, from the appearance of the spots.)

1264. **S. anemones** De Bary and Woron.
Perry Wood.
Uncommon.   On *Anemone nemorosa*.

1265. **S. mercurialis** Fckl.
Aisleshurst Coppice.
Uncommon.   On *Mercurialis perennis*.

1266. **S. taraxaci** De Bary and Woron.
Claines.   Hanbury.   Fladbury.
Not uncommon.   On leaves of *Taraxacum officinale*.

PROTOMYCETACEAE

PROTOMYCES Unger.   (πρῶτος, first, μύκης, a fungus, from its low development.)

1267. **P. macrosporus** Unger.
Near Berwick's Brake.
Uncommon.   On *Aegopodium Podagraria*.

B.   DEUTEROMYCETAE, or *Fungi imperfecti*

SPHAEROPSIDALES

SPHAERIOIDACEAE

PHYLLOSTICTA Pers.   (φύλλον, a leaf, στικτός, spotted, from the appearance of the leaves of the host plant.)

1268. **P. sambuci** Desm.
Holt Bank.
Uncommon.   On leaves of *Sambucus nigra*.

1269. **P. primulaecola** Desm.
Monk Wood.   Middleyards Coppice.
Not uncommon.   On leaves of *Primula acaulis*.

PHOMA Fr.   (φωίς, a blister, from the shape of the plants.)

1270. **P. samararum** Desm.
Malvern Wells.   Claines.
Not uncommon.   On Samaras of *Fraxinus excelsior*.

1271. **P. nebulosum** Berk.
Worcester.
Uncommon.   On dead stems of *Impatiens Roylei*.

1272. **P. longissima** Berk.
Claines.   Holt.   Monk Wood.
Common.   On dead stems of *Umbelliferae*.

SPHAERONEMA Fr.   (σφαῖρα, a ball, νῆμα, a thread, because the spores ooze out in a globule.)

1273. **S. subulatum** Tode.
Perry Wood.
Uncommon.   On dead, dried *Agarics*.

CONIOTHYRIUM Cda.   (κόνις, dust, θύριον, a little door.)

1274. **C. vagabundum** Sacc.
First record for Britain, *Rea*, as *C. ribicolum*, P. Brun., Sheriff's Lench, Worcestershire, in *Trans. Brit. Myc. Soc. II*, p. 168; and the perfect ascosporous condition, *Leptosphaeria vagabunda* Sacc., was recorded by *Rea* as occurring at Worcester, in *Trans. Brit. Myc. Soc. III*, p. 42.
Sheriff's Lench.   Worcester.   Castle Hill.   Claines.   Ombersley.   Pinvin.
Common.   On *Ribes Grossularia*.   This is a dangerous pest and probably more harmful to the Gooseberry bushes than the American Gooseberry Mildew (*Sphaerotheca mors-uvae*).   The best remedy for this disease is to uproot the plants and burn them.

1275. **C. hederae** Desm., Sacc.
Trench Woods.
Uncommon.   On *Hedera Helix*.

SPHAEROPSIS Lév.   (σφαῖρα, a ball, ὄψις, appearance, because the perithecia look like a *Sphaeria*.)

1276. **S. malorum** Berk.
Worcester.
Uncommon.   On fruit of *Pyrus Malus*.

DIPLODIA Fr.   (διπλόος, double, from the two-celled spores.)

1277. **D. vulgaris** Lév.
Wyre Forest.
Not uncommon.   On twigs.

1278. **D. herbarum** (Cda.) Lév.
Hartlebury.
Not uncommon.   On dead stems of *Urtica dioica*.

ASCOCHYTA Lib.   (ἀσκός, a leathern bag, χυτός, poured out, because the spores collect at the mouth of the perithecia.)

1279. **A. pisi** Lib.
Worcester.
Uncommon.   On pods of *Pisum sativum*.

1280. **A. scabiosae** Rabh.
Ombersley.
Uncommon.   On leaves of *Scabiosa arvensis*.

SEPTORIA Fr.   (*Septum*, a wall, because the spores are many-celled.)

1281. **S. ulmi** Kze.
Claines.   Kempsey.
Not uncommon.   On leaves of *Ulmus campestris*.

1282. **S. hippocastani** B. and Br.
Worcester.
Not uncommon.   On leaves of *Aesculus Hippocastanum*.

1283. **S. fraxini** Desm.
Claines.
Uncommon.   On leaves of *Fraxinus excelsior*.

1284. **S. ribis** Desm.
Claines.
Uncommon.   On leaves of *Ribes nigrum*.

CEUTHOSPORA Fr.   (κεύθω, I hide, σπορά, seed, from the position of the spores.)

1285. **C. lauri** Grev.
Madresfield.   Arley Castle.
Not uncommon.   On dead leaves of *Prunus Lauro-cerasus*.

LEPTOSTROMATACEAE

LEPTOSTROMA Fr.   (λεπτός, thin, στρῶμα, layer.)

1286. **L. spiraeae** Fr.
Grimley Brick-pits.
Not uncommon.   On dead stems of *Spiraea Ulmaria*.

MELANCONIALES

MELANCONIACEAE

GLOEOSPORIUM Desm.   (γλοιός, sticky, σπορά, a seed, because the spores stick together in tendrils.)

1287. **G. fructigenum** Berk.   *Apple-rot*.
Common everywhere.   On fruit of *Pyrus Malus*.

COLLETOTRICHUM Cda.   (κολλητός, glued together, θρίξ, hair, because the conidia are surrounded by setae.)

1288. **C. lycopersici** Chester.
Near Worcester.
Uncommon.   On haulms of *Solanum tuberosum*.

1289. **C. Lindemuthianum** Sacc. and Magnus.
Claines.   Pershore.
Not uncommon.   On pods of *Phaseolus vulgaris*.

MYXOSPORIUM Pers.   (μύξα, mucus, σπορά, seed, because the spores ooze out in large tendrils.)

1290. **M. croceum** (Pers.) Link.
Shrawley Wood.   Rous Lench.
Common.   On trunks of *Fagus sylvatica*.

HYPHOMYCETAE

MUCEDINACEAE

FUSIDIUM Link.   (*Fusus*, a spindle, from the shape of the conidia.)

1291. **F. griseum** Link.
Perry Wood.   Shrawley Wood.   Wyre Forest.
Common.   On dead leaves of *Quercus Robur*.

MONILIA Pers.   (*Monile*, a necklace, from the chains of conidia.)

1292. **M. aurea** (Link) Gmel.
Wyre Forest.
Uncommon.   On bark.

1293. **M. fructigena** Pers.
Worcester.
Common.   On fruit of *Pyrus Malus*.

CYLINDRIUM Bon.   (κύλινδρος, a cylinder, from the shape of the conidia.)

1294. **C. cylindricum** (Cda.) Lindau.
Perry Wood.   Shrawley Wood.
Not uncommon.   On dead leaves of *Quercus Robur*.

1295. **C. aeruginosum** (Link) Lindau.
Knightwick.   Nunnery Wood.
Not uncommon.   On dead leaves of *Quercus Robur*.

OIDIUM (Link) Sacc.   (Diminutive of ᾠόν, an egg, from the shape of the conidia.)

1296. **O. erysiphoides** Fr.
Holt Mill.
Not uncommon.   On *Trifolium*.

1297. **O. Tuckeri** Berk.   *The Vine Mildew.*
Worcester.
Not uncommon.   On the leaves and fruit of *Vitis vinifera*.

1298. **O. farinosum** Cke.
Worcester.
Not uncommon.   On twigs and leaves of *Pyrus Malus*.

1299. **O. balsamii** Mont.
Worcester.
Uncommon.   On leaves of *Verbascum Thapsus*.

BOTRYOSPORIUM Cda.   (βότρυς, a bunch of grapes, σπορά, seed, from the appearance of the conidia.)

1300. **B. diffusum** (Grev.) Cda.
Wyre Forest.
Uncommon.   On rotten twigs.

ASPERGILLUS Micheli.   (*Aspergo*, I sprinkle, because of the numerous conidia.)

1301. **A. candidus** (Pers.) Link.
Worcester.
Uncommon.   On Cayenne pepper.

STERIGMATOCYSTIS Cram.   (στήριγμα, a support, κύστις, a bag, from the inflated tips carrying the conidia on sterigmata.)

1302. **S. dubia** Sacc.
Perry Wood.   Wyre Forest.
Not uncommon.   On rotten fungi and dung.

PENICILLIUM Link.   (*Penicillum*, a painter's brush, from the arrangement of the conidia.)

1303. **P. crustaceum** (Linn.) Fr. (= glaucum Link.)
Common everywhere.   On fruit, leaves, and decaying organic matter.
Var. **coremium** Sacc.
Worcester.   Hartlebury Common.
Common.   On jam and rotten apples.

1304. **P. candidum** Link.
Shrawley Wood.
Not uncommon.   On dead leaves.

ACREMONIUM Link.   (ἀκρέμων, a small branch, from the form of the fungus.)

1305. **A. verticillatum** Link.
Worcester.
Uncommon.   On dead wood put in a damp chamber.

RHINOTRICHUM Cda.   (ῥίς, a nose, θρίξ, hair, because the conidia arise from minute spinules.)

1306. **R. repens** Preuss.
Wyre Forest.
Uncommon.   On dead rotten wood.

1307. **R. Thwaitesii** B. and Br.
Wyre Forest.
Uncommon.   On rotten wood.

SPOROTRICHUM (Link).   (σπορά, a seed, θρίξ, hair, because the conidia are scattered about on the hyphae.)

1308. **S. sulphureum** Grev.
Worcester.
Uncommon.   On rotten cork in wine cellar.

MONOSPORIUM Bon.   (μόνος, single, σπορά, seed, because the conidia are not in chains.)

1309. **M. olivaceum** Cke. and Mass.
Nunnery Wood.
Uncommon.   On *Corticium*.

BOTRYTIS Mich.   (βότρυς, a bunch of grapes, from the clusters of conidia.)

1310. **B. vulgaris** Fr.
Common everywhere.   On rotten fruit, fungi, dead stems, and leaves.

1311. **B. cinerea** Pers.
Wyre Forest.   Claines.   Nunnery Wood.   Trench Woods.
Common.   On rotten sticks, putrid fungi, and decaying leaves.
Var. **sclerotiophila** Sacc.
Wyre Forest.   Ombersley.   Ladywood.   Claines.
Common.   On dead haulms of *Solanum tuberosum* and inside stems of various *Umbelliferae*.

1312. **B. fascicularis** Sacc.
Shrawley Wood.
Uncommon.   On dead herbaceous stems.

OVULARIA Sacc.   (*Ovum*, an egg, from the shape of the conidia.)

1313. **O. veronicae** (Fckl.) Sacc.
Claines.   Kempsey.
Not uncommon.   On *Veronica hederifolia*.

1314. **O. obliqua** (Cke.) Oudem.
Northwick Brick-pits.
Uncommon.   On *Rumex crispus*.

VERTICILLIUM Nees.   (*Verticillus*, the whirl of a spindle, because the plants are branched in a verticillate manner.)

1315. **V. agaricinum** (Link) Cda.
Wyre Forest.   Shrawley Wood.
Not uncommon.   On decayed *Agarics*.

1316. **V. epimyces** B. and Br.
Wyre Forest.
Uncommon.   On rotten *Stereum*.

1317. **V. Marquandii** Massee.
Rimells Farm.   Claines.   Holt.
Not uncommon.   On *Hygrophorus virgineus* and *niveus*.

ACROSTALAGMUS Cda.   (ἄκρος, at the top, σταλαγμός, a drop, because the conidia are involved in mucus at the tips of the branches.)

1318. **A. cinnabarina** Cda.
Wyre Forest.
Uncommon.   On rotten stems.

DIPLOCLADIUM Bon.   (διπλόος, twofold, κλαδίον, a young branch, because the plant is branched.)

1319. **D. melleum** (B. and Br.) Sacc.
Shrawley Wood.
Uncommon.   On rotten *Stereum*.

TRICHOTHECIUM Link.   (θρίξ, hair, θήκη, a box, from the solitary conidia at the apex of the fertile hyphae.)

1320. **T. roseum** Link.
Wyre Forest.   Shrawley Wood.
Not uncommon.   On rotten leaves and bark.

MYCOGONE Link.   (μύκης, a fungus, γονή, seed.)

1321. **M. rosea** Link.
Wyre Forest.
Uncommon.   On decaying *Lactarius*.

1322. **M. cervina** Ditm.
Shrawley Wood.
Uncommon.   On *Helvella macropus*.

DACTYLIUM Nees.   (δάκτυλος, a finger, from the verticillate branching of the hyphae.)

1323. **D. dendroides** (Bull.) Fr.
Shrawley Wood.   Wyre Forest.   Perry Wood.
Common.   On decaying *Agarics*.

RAMULARIA Unger.   (*Ramulus*, a little branch, because the hyphae are sparingly branched.)

1324. **R. hellebori** Fckl.
Southstone Rock.
Uncommon.   On *Helleborus foetidus*.

1325. **R. calcea** (Desm.) Ces.
Common everywhere.   On *Nepeta Glechoma*.

DERMATIACEAE

TORULA Pers.   (*Torulus*, a little elevation, from the shape of the chains of conidia.)

1326. **T. monilioides** Cda.
Chatley.   Shrawley Wood.   Monk Wood.
Common.   On rotten wood and fallen branches.

1327. **T. pulveracea** Cda.
Shrawley Wood.
Uncommon.   On fallen branches.

1328. **T. herbarum** Link.
Shrawley Wood.
Not uncommon.   On dead stems of *Pteris aquilina*.

ACROSPEIRA B. and Br.   (ἄκρος, at the top, σπεῖρα, a coil, from the somewhat spirally coiled tips of the fertile hyphae.)

1329. **A. mirabilis** B. and Br.
Common.   On fruit of *Castanea vesca*.

ZYGODESMUS Cda.   (ζυγόν, a yoke, δεσμός, a band.)

1330. **Z. fuscus** Cda.
Shrawley Wood.   Wyre Forest.   Monk Wood.   Ockeridge Wood.
Common.   On rotten wood and fallen branches.

BISPORA Cda.   (*Bis*, twice, σπορά, seed, because the conidia are two-celled.)

1331. **B. monilioides** Cda.
Common everywhere.   On stumps and felled trees.

FUSICLADIUM Bon.   (*Fusus*, spread out, κλάδος, a young branch.)

1332. **F. dendriticum** (Wallr.) Fckl.
Common everywhere.   On leaves and fruit of *Pyrus Malus* and *communis*.

1333. **F. pyrinum** (Lib.) Fckl.
Worcester.   Worcestershire (Miss A. Lorrain Smith).
Common.   On fruit and leaves of *Pyrus communis*.

POLYTHRINCIUM Kze. and Schm.   (πολύν, many, θριγκίον, a little coping.)

1334. **P. trifolii** Kze. and Schm.
Holt.   Offenham.   Ladywood, Salwarpe.   Kempsey.
Common.   On living leaves of *Trifolium*.

CLADOSPORIUM Link.   (κλάδος, a young branch, σπορά, seed.)

1335. **C. fulvum** Cke.
Boughton Park.   Worcester.   Pershore.   Sheriff's Lench.
Not uncommon.   On living leaves of *Solanum lycopersicum*.

1336. C. herbarum (Pers.) Link.
Common everywhere. On decaying herbaceous plants, wood, fungi, &c.

HELMINTHOSPORIUM Link. (ἔλμινς, a worm, σπορά, seed, from the shape of the conidia.)

1337. H. velutinum Link.
Shrawley Wood. Wyre Forest. Eymore Wood.
Not uncommon. On rotten wood.

1338. H. exasperatum B. and Br.
Wyre Forest. Nunnery Wood.
Not uncommon. On dead stems.

1339. H. Rousselianum Mont.
Shrawley Wood.
Uncommon. On dead wood.

1340. H. fusiforme Cda.
Shrawley Wood. Wyre Forest.
Not uncommon. On fallen branches.

HETEROSPORIUM Klotzsch. (ἕτερος, different, σπορά, seed, from the different forms of the conidia.)

1341. H. epimyces Cke. and Mass.
Shrawley Wood.
Not common. On Gomphidius viscidus.

SPOROSCHISMA B. and Br. (σπορά, seed, σχίσμα, division, because the conidia are produced by division within the hyphae.)

1342. S. mirabile B. and Br.
Shrawley Wood.
Uncommon. On rotten wood.

MACROSPORIUM Fr. (μακρός, long, σπορά, seed, from the shape of the conidia.)

1343. M. commune Rabh.
Common everywhere. On decaying stems and leaves.

1344. M. tomato Cke.
Worcester. Claines. Malvern.
Not uncommon. On fruit of Solanum lycopersicum.

FUMAGO Pers. (Fumus, smoke, ago, I make, from the appearance of the leaves of the host plant.)

1345. F. vagans Pers.
Common everywhere. On living leaves of various trees.

TRIPOSPORIUM Cda. (τρίπους, with three feet, σπορά, seed, from the shape of the conidia.)

1346. T. elegans Cda.
Monk Wood.
Uncommon. On fallen branches of Quercus Robur.

STILBACEAE

STILBUM Tode. (στίλβω, I shine.)

1347. S. tomentosum Schr.
Wyre Forest. Ockeridge Wood.
Not uncommon. On Trichia.

1348. S. erythrocephalum Ditm.
Shrawley Wood. Wyre Forest. Coleridge Wood.
Not uncommon. On rabbit dung.

1349. S. fimetarium.
Shrawley Wood. Wyre Forest.
Not uncommon. On rabbit dung.

ISARIA Pers. (ἴσος, equal, from the likeness of all its organs.)

1350. I. arachophila Ditm.
Holt Bank.
Uncommon. On spiders.

1351. I. citrina Pers.
Shrawley Wood.
Uncommon. On trunk of a fallen tree.

GRAPHIUM Cda. (γραφεῖον, a writing style, from the shape of the stroma.)

1352. G. subulatum (Nees) Sacc.
Wyre Forest.
Uncommon. On an old boot.

STYSANUS Cda.

1353. S. stemonitis (Pers.) Cda.
Perry Wood.
Uncommon. On fallen branches.

TUBERCULARIACEAE

TUBERCULARIA Tode. (Tubercularia, a small swelling.)

1354. T. brassicae Lib.
Claines.
Uncommon. On decaying stalks of Brassica oleracea.

AEGERITA Pers.

1355. A. candida Pers.
Wyre Forest. Shrawley Wood.
Not uncommon. On rotting wood and bark, especially of Sambucus nigra.

FUSARIUM Link. (Fusus, a spindle, from the shape of the conidia.)

1356. F. sarcochroum (Desm.) Sacc.
Shrawley Wood.
Uncommon. On fallen twigs.

s s

1357. F. solani (Mart.) Sacc.
Wyre Forest.
Uncommon. On dead haulms of Solanum tuberosum.

EPICOCCUM Link. (ἐπί, upon, κόκκος, a kernel, from the shape of the stroma.)

1358. E. micropus Cda.
Perry Wood.
Uncommon. On Lactarius.

Division II. MYXOMYCETAE Wallr.
= MYCETOZOA De Bary

EXOSPOREAE

CERATIOMYXACEAE

CERATIOMYXA Schröter. (κέρας, a horn, μύξα, mucus, from the shape and consistency of the sporophores.)

1359. C. mucida (Pers.) Schröt.
Chatley. Monk Wood.
Common. On rotten wood.

ENDOSPOREAE

PHYSARACEAE

BADHAMIA Berk. (After C. D. Badham, M.D.)

1360. B. utricularis (Bull.) Berk.
Trench Woods. Fernhill Heath.
Not uncommon. On bark of fallen trees and on Stereum hirsutum.

1361. B. macrocarpa (Ces.) Rost.
Wyre Forest. Nunnery Wood.
Uncommon. On dead wood.

PHYSARUM Pers. (φυσάριον, a little air bubble, from the shape of some of the species.)

1362. P. nutans Pers.
Common everywhere. On stumps, dead wood, leaves, &c.

FULIGO Haller. (Fuligo, soot.)

1363. F. septica (Linn.) Gmel. *Flowers of tan.*
Worcester. Claines. Shrawley Wood. Kidderminster.
Common. Amongst dead leaves, spent tan, and rotten wood.

CRATERIUM Trent. (κρατήριον, a little wine bowl, from the shape of the sporangia.)

1364. C. pedunculatum Trent.
Wyre Forest. Shrawley Wood.
Uncommon. On dead leaves, sticks, &c.

1365. C. leucocephalum (Pers.) Ditm.
Shrawley Wood.
Uncommon. On wood and amongst dead leaves.

LEOCARPUS Link. (λεῖος, smooth, καρπός, fruit, from the smooth wall of the sporangium.)

1366. L. vernicosus (Pers.) Link.
Wyre Forest. West Malvern.
Not uncommon. On dead leaves, grass stems, &c.

CHONDRIODERMA Rost. (χόνδρος, a corn, δέρμα, the skin, from the round fine granules of the outer sporangium wall.)

1367. C. spumarioides (Fr.) Rost.
Shrawley Wood.
Common. On dead leaves and twigs.

1368. C. Michelii (Lib.) Rost.
Ham Dingle, near Stourbridge. Shrawley Wood.
Common. On dead leaves and twigs.

1369. C. floriforme (Bull.) Rost.
Wyre Forest.
Uncommon. On rotten branches.

DIDYMIACEAE

DIDYMIUM Schrad. (δίδυμος, double, from the two layers of the sporangium wall.)

1370. D. difforme (Pers.) Duby.
Bevere Lane. Deadmans Coppice, Ladywood, Salwarpe.
Common. On dead leaves and herbaceous stems.

1371. D. clavus (A. and S.) Rost.
Trench Woods.
Uncommon. On dead leaves.

1372. D. farinaceum Schrad.
Valley of the Rocks, Abberley Hall.
Not uncommon. On dead leaves.

1373. D. effusum Link.
Perry Wood. Valley of the Rocks, Abberley Hall.
Common. On dead leaves and wood.

SPUMARIA Persoon. (Spuma, foam, from the appearance of the aethalia.)

1374. S. alba (Bull.) DC.
Bevere Green. Croome. Perry Wood. Wichenford.
Common. On grass, dead leaves, herbaceous stems, and twigs.

s s 2

STEMONITACEAE

STEMONITIS Gleditsch. (στήμων, the warp, in allusion to the capillitium.)

1375. **S. fusca** Roth.
        Shrawley Wood. Worcester.
        Common. On dead leaves, old stumps, and fallen branches.

1376. **S. herbatica** Peck.
        Shrawley Wood.
        Uncommon. On rotten wood.

COMATRICHA Preuss. (κόμη, hair, θρίξ, hair, from the gyrose capillitium.)

1377. **C. obtusata** (Fr.) Preuss.
        Shrawley Wood. Monk Wood. Wyre Forest. Trench Woods.
        Common. On dead wood and fallen branches.

1378. **C. typhoides** (Bull.) Rost.
        Shrawley Wood. Eymore Wood.
        Not uncommon. On dead wood and fallen branches.

LAMPRODERMA Rost. (λαμπρός, brilliant, δέρμα, skin, from the sporangium wall shining with iridescent colours.)

1379. **L. irideum** (Cke.) Mass.
        Shrawley Wood.
        Common. On dead leaves, especially those of *Fagus sylvatica*.

AMAUROCHAETACEAE

BREFELDIA Rost. (After Professor O. Brefeld, an eminent German mycologist.)

1380. **B. maxima** (Fr.) Rost.
        Roadside Common Hill, Worcester.
        Not uncommon. Swarming over grass at the base of *Ulmus campestris*.

HETERODERMACEAE

CRIBRARIA Pers. (*Cribrum*, a sieve, from the appearance of the upper portion of the sporangium.)

1381. **C. argillacea** Pers.
        Wyre Forest. Shrawley Wood.
        Uncommon. On dead wood.

RETICULARIACEAE

DICTYDIAETHALIUM Rost. (δίκτυον, a network, διά, through, αἴθαλος, soot.)

1382. **D. plumbeum** (Schum.) Rost.
        Wyre Forest.
        Uncommon. On stumps of *Picea excelsa*.

ENTERIDIUM Ehrenb. (ἐντερίδια, the little intestines, from the form of the walls of the sporangia in the aethalium.)

1383. **E. olivaceum** Ehrenb.
        Nunnery Wood.
        Not uncommon. On dead wood.

RETICULARIA Bull. (*Reticulum*, a little net, from the way in which the walls of the sporangia appear in the aethalium.)

1384. **R. Lycoperdon** Bull.
        Sharpway Gate. Perdiswell. Worcester. Boreley.
        Common. On posts, rails, dead wood, and trunks of trees.

LYCOGALACEAE

LYCOGALA Micheli. (λύκος, a wolf, γάλα, milk.)

1385. **L. miniatum** Pers.
        Shrawley Wood. Stanford Woods. Hallow. Claines. Wyre Forest. Holt.
        Common. On dead wood.

TRICHIACEAE

TRICHIA Haller. (θρίξ, hair, because the sporangia are filled with elaters.)

1386. **T. affinis** De Bary.
        Wyre Forest.
        Not uncommon. On dead wood.

1387. **T. persimilis** Karst.
        Wyre Forest. Ockeridge Wood.
        Not uncommon. On dead wood, leaves, &c.

1388. **T. scabra** Rost.
        Wyre Forest.
        Uncommon. On dead wood.

1389. **T. varia** Pers.
        Perry Wood. Wyre Forest. Shrawley Wood.
        Common. On dead wood.

1390. **T. fallax** Pers.
        Shrawley Wood. Wyre Forest.
        Not uncommon. On dead wood.

1391. **T. Botrytis** Pers.
        Wyre Forest. Ockeridge Wood.
        Not uncommon. On wood.

HEMITRICHIA Rost. (ἡμι-, half, θρίξ, hair, because the elaters are only partially free.)

1392. **H. rubiformis** (Pers.) List.
        Eymore Wood.
        Not uncommon. On dead wood.

1393. **H. clavata** (Pers.) Rost.
        Dudley.
        Uncommon. On dead wood.

ARCYRIACEAE

ARCYRIA Hill. (ἄρκυς, a net, from the profuse capillitium.)

1394. **A. albida** Pers.
        Shrawley Wood. Wyre Forest.
        Not uncommon. On dead wood, leaves, &c.

1395. **A. punicea** Pers.
        Perry Wood. Wyre Forest. Shrawley Wood.
        Common. On dead wood.

1396. **A. incarnata** Pers.
        Wyre Forest. Shrawley Wood.
        Not uncommon. On dead wood, sticks, and bark.

1397. **A. flava** Pers.
        Shrawley Wood. Battenhall.
        Not uncommon. On dead wood.

PERICHAENA Fr. (περιχαίνω, I open the mouth wide, because some of the species dehisce by a well-defined lid.)

1398. **P. populina** Fr.
        Shrawley Wood.
        Not uncommon. On dead wood and bark.

1399. **P. variabilis** Rost.
        Wyre Forest.
        Uncommon. On and amongst dead leaves and decaying stems.

FRESH-WATER ALGAE

Class I. *CHLOROPHYLLOPHYCEAE*

Order I. COCCOPHYCEAE

Family I. PALMELLACEAE

PALMELLA Lyngb. (παλμός, vibration, from the loosely gelatinous nature of these plants.)

1. **P. aestivalis** Lees.   *Pond Palmella.*
    Snead's Green, Mathon, 1854. This alga is not mentioned in Cooke.

PORPHYRIDIUM Näg. (πορφύρεος, purple.)

2. **P. cruentum** Näg.   *Gory Dew.*
    Not uncommon at the base of damp rocks and walls. It forms patches of deep red or purple colour, as if blood or wine had been poured upon the ground. Superstitions have been attached to it in all ages. It has had many names.

BOTRYDINA Breb. (βοτρυδόν, like a bunch of grapes.)

3. **B. vulgaris** Breb.
    Mentioned as a Worcestershire alga, *Vict. Hist. Worc.*, vol. i, p. 69.

TETRASPORA Link. (τέτρα, four, σπορά, seed, in allusion to the arrangement of the cells.)

4. **T. lubrica** Roth.
    In little stagnant pools and ditches, *Lees.*

5. **T. explanata** Agardh.
    Lower Wick, in Mr. J. H. White's nurseries, *Rea, Trans. W. N. C.* vol. ii, p. 3. This was the first record of the plant in Britain.

COCCOCHLORIS Spreng. (κόκκος, a berry, χλωρός, light green.)

6. **C. protuberans** Spreng.
    On the ground among mosses, not uncommon in moist weather, *Lees.* *Coccochloris* is not given as a genus in Cooke, but is placed among the *Palmellaceae* by Mr. Lees.

7. **C. botryoides** Lyngb.
    On the ground upon dead mosses, *Lees.*

8. **C. muscicola** Menegh.
    On fallen leaves in the thick wood on Hollybush Hill in early spring, *Lees.*

Family II. PROTOCOCCACEAE

CHLOROCOCCUM Fries. (χλωρός, light green, κόκκος, a berry.)

9. **C. vulgare** Grev.
   General, *Lees*. Not under this name in Cooke.

10. **C. murorum** Grev.
   Frequent, *Lees*.

JOLITHUS Linn. (ἴον, violet, λίθος, stone.)

11. **J. lichenoideus** *Lees*.   *Lichen Jolithus.*
   On very old tombstones in Eldersfield Churchyard, *Lees*. *Jolithus* is not given as a genus in Cooke.

12. **J. odoratus** Ag.
   On branches of trees between Malvern and Worcester, *Lees*.

Family III. VOLVOCINEAE

VOLVOX Linn. (*Volvere*, to roll or turn about.)

13. **V. globator** Linn.
   Plentiful at Malvern Wells and Upton-on-Severn, *W. J. Farthing*. In clear pools and ponds. The life-history of this alga has been subjected to close scrutiny. Its powers of spontaneous movement caused it in the past to be placed in the animal kingdom.

PANDORINA, Ehrb.

14. **P. morum** Ehr.
   In standing water. Plentiful at Malvern Wells and Upton-on-Severn, *W. J. Farthing*.

Order II. ZYGOPHYCEAE

Family I. DESMIDIEAE

CLOSTERIAE

SPIROTAENIA Breb. (σπεῖρα, a coil, ταινία, a band.)

15. **S. condensata** Breb.
   Fairly common. Barnard's Green ; Malvern Wells ; Little Malvern and Castlemorton ; *W. J. Farthing*.

PENIUM Breb. (πηνίον, the thread of the woof.)

16. **P. interruptum** Breb.
   Scarce. One specimen only at Twyning Fleet ; *W. J. Farthing*.

CLOSTERIUM Nitsche. (κλωστήρ, thread, in allusion to the attenuated form of the frond.)

17. **C. costatum** Cda.
   Generally distributed over the south of the county ; *W. J. Farthing*.

18. **C. lunula** Ehr.
   Scarce. New Pool, Malvern ; Dunstall Common ; *W. J. Farthing*.

19. **C. setaceum** Ehr.
   Scarce. One specimen only at Dunstall Common ; *W. J. Farthing*.

COSMARIAE

MICRASTERIAS Agardh. (μικρός, small, ἄστρον, star.)

20. **M. denticulata** Breb.
   Fairly common. Little Malvern ; Kidderminster ; *W. J. Farthing*.

EUASTRUM Ehr. (εὖ, well, ἄστρον, star.)

21. **E. oblongum** Grev.
   Common. Blackmore Park ; Birtsmorton ; Malvern Wells ; Hall Green, Yardley ; *W. J. Farthing*.

22. **E. didelta** Ralfs.
   Scarce. Two specimens at Little Malvern ; *W. J. Farthing*.

DESMIDIAE

HYALOTHECA Ehr. (ὕαλος, glass, θήκη, box, from its transparency.)

23. **H. dissiliens** (Smith) Ralfs.
   Fairly common. Malvern Wells ; Little Malvern ; Marsh Common, Defford ; *W. J. Farthing*.

Family II. ZYGNEMACEAE

ZYGNEMA Kutz. (ζυγός, a yoke, νῆμα, a thread, because the threads, separate at first, are afterwards yoked together.)

24. **Z. nitidum** Ag.
   In ditches ; *Lees*. Not given under this name in Cooke.

ZYGOGONIUM Kutz. (ζυγός, a yoke, γωνία, an angle.)

25. **Z. (Conferva) ericetorum** Kutz.
   On heaths and hilly places. Castlemorton Common ; *Lees*.

26. **Z. (Conferva) rivularis** Linn.   *River Conferva.*
   Common in running water, *Lees*. Not given in Cooke.

Order III. SIPHOPHYCEAE

Family I. BOTRYDIACEAE

BOTRYDIUM (Wallr.). (βοτρυδόν, like a bunch of grapes.)

27. **B. granulatum** Linn.
   Mostly on the bottom of dried-up pools during the summer. On the bottom of a pool by the Link ; Blackpole Green, Old Hills, Powick ; *Lees*.

Family II. VAUCHERIACEAE

VAUCHERIA DC. (After Professor Jean Pierre Étienne Vaucher of Geneva, an algologist.)

28. **V. dichotoma** Lyngb.
   In ponds and ditches, *Lees*.

29. **V. Dillwynii** Ag.
   On bark of trees in shady places at Little Malvern, *Lees*.

30. **V. sessilis** Vauch.
   Not uncommon in ditches or on the ground, *Lees*.

31. **V. geminata** Vauch.
   In stagnant water, *Lees*.

32. **V. terrestris** Lyngb.
   On the ground in moist places, *Lees*. It prefers damp clay soil.

33. **V. ovoidea** Vauch.
   In ditches and in shallow, stagnant, or running water, *Lees*. The plant is not mentioned under this name in Cooke.

Order IV. NEMATOPHYCEAE

Family I. PRASIOLEAE

PRASIOLA Ag. (πράσον, a leek.)

34. **P. calophylla** Spreng.
   As *Ulva calophylla*, Old Hills ; Holy Well, Henwick, Worcester, *Lees*. Mentioned in *Vict. Hist. Worc.*, vol. i, p. 69. Occurs on damp stones, rocks, &c.

ULVA Linn. (*Ul*, Celtic for water.)

35. **U. bullosa** Roth.
   Pools on Welland Common, *Lees*. Not given in Cooke.

ENTEROMORPHA Link. (ἔντερον, intestine, μορφή, form.)

36. **E. intestinalis** Linn.
   Very general, *Lees*. Floating in ponds in summer, with long green inflated fronds, often considerably distended. It also occurs in ditches, and is not averse to brackish water.

CLADOPHORA Kutz. (κλάδος, a branch, φορέω, I bear.)

37. **C. fracta** Dillw.
   Mentioned as a Worcestershire plant in *Vict. Hist. Worc.*, vol. i, p. 69. Mentioned as common in ponds by *Lees* under the name of *C. crispata*, Sm., a name given in Cooke, with the authority (Roth.).

38. **C. glomerata** Linn.
   In running water, attached to stones and walls and spreading out, *Lees*. Tewkesbury ; near Ripple ; *W. J. Farthing*.

Family VII. CHROOLEPIDAE

CHROOLEPUS Ag.

39. **C. aureus** Linn.
   Byssus aurea, E. B. *Callithamnium aureum*, *Vict. Hist. Worc.*
   Golden red or orange, on walls, chips, bark, &c. Mr. Lees says, *Bot. Malv. Hills*, 3rd ed., 158, 'The most unobservant eye cannot but notice this golden growth of old Time's beard, fringing many of the rocks in a beautiful manner.'

40. **C. ebenea** Ag.   *Black Byssus.*
   On damp rock, plentiful, especially at the base of the Holly-bush Hill, *Lees*. Not mentioned in Cooke under this name.

41. **C. barbatus** Sm.
   Byssus barbatus, E. B. ; *Ozonium auricomum*, Pers. ; *Callithamnium barbatus*, *Vict. Hist. Worc.*
   Mr. Lees marks this with a (?) and says it seems confined to timber, while *C. aureus* affects rocks and stone. He had specimens taken from a beam in an outhouse at Wheatfields, Powick, gathered by Abraham Edwards of Worcester. It is probably a fungus, he continues, an immature state of *Merulius lacrymans*. It is not mentioned under any of the above names in Cooke. It is now known to be a condition of the mycelium of several *Coprini*.

PROTONEMA Ag. (πρῶτος, first, νῆμα, thread.)

42. **P. muscicola** Ag.
   In shady places among mosses, very abundant, and crowding about the bases of their stems, *Lees*. This name is not given in Cooke, and is placed by Mr. Lees immediately under the above plants.

Family VIII. CHAETOPHORACEAE

DRAPARNALDIA Ag. (After Draparnald.)

43. **D. glomerata** Ag.
   In pools and ditches, *Lees*.

CHAETOPHORA Roth. (χαίτη, mane, φορέω, I bear.)

44. **C. elegans** Roth.
   On sticks in pools, *Lees*. Malvern Wells ; near Ripple ; *W. J. Farthing*.

45. **C. endivaefolia** Ag.
   In little streams, covering stones in the water in the form of little green protuberances irregularly lobed at the extremities, *Lees*.

46. **C. rivularis.**
   Mentioned in *Vict. Hist. Worc.*, vol. i, p. 69. Not mentioned in Cooke.

47. **C. capillaris.**
   Mentioned in *Vict. Hist. Worc.*, vol. i, p. 69. Not given in Cooke.

Order II. NEMATOGENAE

Family I. NOSTOCEAE

NOSTOC Vauch. (A name used by Paracelsus.)

48. **N.** muscorum Ag.
Mentioned in *Vict. Hist. Worc.*, vol. i, p. 69.

49. **N.** commune Vauch.
On gravelly walks and roads after rain, *Lees.* He says, 'Its appearance is so sudden, and its growth so rapid, that it is often supposed to have fallen from the sky.'

Family II. LYNGBYAE

OSCILLARIA Bosc. (*Oscillum*, a swing.

50. **O.** limosa Ag.
Mentioned in *Vict. Hist. Worc.*, vol. i, p. 69.

MICROCOLEUS Desm. (μικρός, small, κολεός, a sheath.)

51. **M.** repens Harv.
On damp places by roadsides, forming a dull green decumbent stratum in moist weather, *Lees.* Not given in Cooke.

LYNGBYA Ag. em Thuret. (After Hans Christian Lyngbye, a Danish botanist.)

52. **L.** muralis.
Not uncommon in interstices of the rocks to the summit of the hills; determined by the Rev. M. J. Berkeley, *Lees.* This species is not given in Cooke.

53. **L.** zonata Hass.
Growing in running water and not uncommon, *Lees.* Not given in Cooke.

Family III. SCYTONEMEAE

STIGONOMA Ag. (στίγων, dotted, νῆμα, thread.)

54. **S.** atrovirens Ag.
On the face of a wet rock on the north side of the Worcestershire Beacon ; local and rare ; *Lees.* This plant, placed by Mr. Lees in the above family, is neither generically nor specifically mentioned in Cooke. It is given in *Vict. Hist. Worc.*, vol. i, p. 69.

Class III. *RHODOPHYCEAE*

Family IV. BATRACHOSPERMEAE

BATRACHOSPERMUM Roth. (βάτραχος, a frog, σπέρμα, spawn.)

55. **B.** vagum Roth.
In rivulets near Purlieu Lane and about Mathon, *Lees.*

56. **B.** atrum Harv.
Mentioned in *Vict. Hist. Worc.*, vol. i, p. 69.

Family V. LEMANEACEAE

LEMANEA Bory. (After M. Leman, a French algologist.)

57. **L.** fluviatilis Ag.
Mentioned in *Vict. Hist. Worc.*, vol. i, p. 69.

58. **L.** torulosa Ag.   *Beaded Lemanea.*
On the weir in the Teme above the old bridge at Powick ; at Bransford Weir, *Lees.*

DIATOMACEAE

EUNOTIA Ehr.

59. **E.** tetraodon Ehr.
Fairly common near Defford and Strensham, *W. J. Farthing.*

AMPHIPLEURA Kütz. (ἀμφί, on both sides, πλευρά, a rib, from the markings on the valves.)

60. **A.** pellucida Breb.
Near Defford and Strensham, *W. J. Farthing.*

COCCONEIS Ehr.

61. **C.** pediculus Sm.
Near Defford and Strensham, fairly common, *W. J. Farthing.*

COCCONEMA Ehr. (κόκκος, grain, νῆμα, a thread.)

62. **C.** lanceolatum Sm.
Near Ripple, scarce, *W. J. Farthing.*

PINNULARIA Ehr. (*Pinnula*, a little plume.)

63. **P.** major Ralfs.
Fairly numerous, Marsh Common, Defford, *W. J. Farthing.*

64. **P.** gigas.
Fairly numerous, Marsh Common, Defford, *W. J. Farthing.*

65. **P.** vividis W. Sm.
Ripple, fairly common, *W. J. Farthing.*

NAVICULA Bory. (*Navicula*, a small vessel.)
Several species plentiful over the south end of the county, *W. J. Farthing.*

PLEUROSIGMA Sm. (πλευρά, a rib, σίγμα, the letter S.)
Sundry species plentiful at the south end of the county, *W. J. Farthing.*

COLLETONEMA Sm. (κολλητός, glued together, νῆμα, a thread, from the nature of the plants.)
A few supposed specimens were taken at Malvern Wells, but were not satisfactorily determined, *W. J. Farthing.*

INDEX

This Index contains the Latin names of the Genera, and English names, both Book-names and Folk-names.